D1215316

Extracellular Matrix Biochemistry

Extracellular Matrix Biochemistry

Editors

Karl A. Piez
Collagen Corporation, Palo Alto, California

A. H. Reddi
National Institutes of Health, Bethesda, Maryland

Elsevier
New York • Amsterdam • Oxford

Elsevier Science Publishing Co., Inc.
52 Vanderbilt Avenue, New York, New York 10017

Distributors outside the United States and Canada:

Elsevier Science Publishers B.V.
P.O. Box 211, 1000 AE, Amsterdam, The Netherlands

Library of Congress Cataloging in Publication Data

Main entry under title:

Extracellular matrix biochemistry.
Includes bibliographical references and index.
1. Connective tissues. 2. Collagen. 3. Collagen diseases.
I. Piez, Karl A. II. Reddi, A. H., 1942–
QP88.23.E953 1984 599'.01'852 83-20618
ISBN 0-444-00799-7

Manufactured in the United States of America

Contents

Preface

The outline of this book is based largely on a course taught at the Foundation for Advanced Education in the Sciences (FAES) Graduate School at the National Institutes of Health (NIH) by Dr. Vincent C. Hascall and one of the editors (KAP). The other editor (AHR) has been a frequent guest lecturer in the course. We are much indebted to Dr. Hascall for his contributions to the outline and general philosophy of the course, and thus to the book.

The book is addressed to the laboratory and clinical researcher who has a good background in biochemistry. That person may be a graduate student wishing to broaden her or his horizons, a postdoctoral fellow who needs specific information in the field, or someone already experienced in one aspect of extracellular matrix biochemistry who wants knowledge of another. The book can be both a text and a reference for individual reading. To our knowledge, no other book in the field is directed to the same end.

With this view in mind, authors of individual chapters were not asked to write a critical review of recent developments with the usual comprehensive bibliography. Rather, each chapter is meant to be inclusive in breadth but selective in detail, presenting the background didactically with sufficient recent research to show current trends and pose important questions. The degree to which this has been accomplished is, of course, our responsibility since we have instructed the authors and edited their manuscripts for both content and style. At the same time, as will be evident, the individuality of the authors has not been suppressed. References have generally been limited to reviews, important early articles, and enough recent papers to gain access to the literature.

The reason for this book is that extracellular matrix biochemistry has come into its own with the realization that cell biology extends beyond the cell. The plasma membrane is increasingly less of a boundary and more of a pause along a continuum of biochemical events. Indeed, multicellular animals have evolved and can function only because single cells learned to synthesize, maintain, and utilize connections among them; in so doing, they have become dependent upon them. These connections serve both a mechanical and a biologically interactive role in development, steady-state existence, and disease. We need to know the biochemical bases for these functions. This book recounts the background and current status of our knowledge.

As usually defined, extracellular matrix is all material outside of cells and inside the epithelial and endothelial boundaries of multicellular animals. This includes proteins and other macromolecules associated specifically or nonspecifically with cell surfaces but not embedded in the plasma membrane. Thus, the obvious connective tissues are dermis, tendon, cartilage, and bone. But also composed of collagen, elastin, proteoglycan, polysaccharide, and glycoprotein are a myriad of smaller structures, including, for example, elastic ligaments; the cornea and other structures in the eye; basement membranes and other pericellular structures throughout the body; reticular networks within organs; blood vessel and intestinal walls; sheaths associated with muscle, nerve, and tendon; adventitia supporting various organs; and membranes through which pass gases in the lung, waste products in the kidney, and nutrients in the placenta and gastrointestinal tract.

Our acknowledgments to individuals are too many to list. However, the stimulus provided by students, and their demand for knowledge and rigorousness of thought, should not go unrecognized. The NIH and the FAES must also be credited for their support in recognizing the need to respond to these demands.

Karl A. Piez
A. H. Reddi

Contributors

William G. Carter, Ph.D.
Assistant Member and Assistant Professor, Fred Hutchinson Cancer Research Center, and University of Washington, Seattle, Washington

Minoru Fukuda, Ph.D.
Research Scientist, Fred Hutchinson Cancer Research Center, and University of Washington, Seattle, Washington

John M. Gosline, Ph.D.
Associate Professor, Department of Zoology, University of British Columbia, Vancouver, B.C., Canada

Sen-Itiroh Hakomori, M.D., Ph.D.
Professor and Program Head, Biochemical Oncology, Fred Hutchinson Cancer Research Center, Departments of Pathobiology, Microbiology, and Immunology, University of Washington, Seattle, Washington

Dick Heinegård, M.D., Ph.D.
Professor, Department of Physiological Chemistry, University of Lund, Lund, Sweden

Kari I. Kivirikko, M.D., Ph.D.
Professor, Department of Medical Biochemistry, University of Oulu, Oulu, Finland

Stephen M. Krane, M.D.
Professor of Medicine, Harvard Medical School, and Physician and Chief, Arthritis Unit, Massachusetts General Hospital, Boston, Massachusetts

Edward J. Miller, Ph.D.
Professor of Biochemistry, Senior Scientist, Institute of Dental Research, University of Alabama in Birmingham, Birmingham, Alabama

Raili Myllylä, Ph.D.
Docent, Department of Medical Biochemistry, University of Oulu, Oulu, Finland

Mats Paulsson, Dr. Med. Sci.
Department of Physiological Chemistry, University of Lund, Lund, Sweden

Karl A. Piez, Ph.D.
Director, Connective Tissue Research Laboratories, Collagen Corporation, Palo Alto, California

Hari Reddi, Ph.D.
Chief, Bone Cell Biology Section, National Institute of Dental Research, National Institutes of Health, Bethesda, Maryland

Joel Rosenbloom, M.D., Ph.D.
Professor of Biochemistry, University of Pennsylvania Dental School, Philadelphia, Pennsylvania

Kiyotoshi Sekiguchi, Ph.D.
Associate, Biochemical Oncology, Fred Hutchinson Cancer Research Center, and University of Washington, Seattle, Washington

Rupert Timpl, Ph.D.
Max-Plank-Institut für Biochemie, Abt. Bindegewebsforschung, Martinsreid, München, West Germany

Arthur Veis, Ph.D.
Department of Biochemistry and Molecular Biology, Chairman, Department of Oral Biology, Northwestern University Dental and Medical Schools, Chicago, Illinois

David E. Woolley, Ph.D.
Professor, University Department of Medicine, University Hospital of South Manchester, Manchester, England

Abbreviations

These abbreviations are used throughout the book. Abbreviations used in a single chapter are not included but are defined on initial use.

Amino Acids

Ala	Alanine
Arg	Arginine
Asn	Asparagine
Asp	Aspartic acid
Cys	Cysteine
Gln	Glutamine
Glu	Glutamic acid
Gly	Glycine
His	Histidine
Hyl	Hydroxylysine
Hyp	Hydroxyproline
Ile	Isoleucine
Leu	Leucine
Phe	Phenylalanine
Pro	Proline
PSer	Phosphoserine
pGlu	Pyroglutamic acid
Ser	Serine
Tyr	Tyrosine
Trp	Tryptophan
Val	Valine
X, Y, or Xxx	Unknown or undefined

Carbohydrates

Fuc	Fucose
Gal	Galactose
Glc	Glucose
GlcNac	N-Acetylglucosamine
GalNac	N-Acetylgalactosamine
GlcN	Glucosamine
GalN	Galactosamine
GlcUa	Glucuronic acid
Man	Mannose
NeuAc, NANA	N-Acetylneuraminic acid
Xyl	Xylose
IdUa	Iduronic acid

Extracellular Matrix Biochemistry

1

Molecular and Aggregate Structures of the Collagens

Karl A. Piez

1.1. Introduction

The prototype for the collagens is type I collagen, since it is the best understood collagen and is present in the largest amount in most vertebrate tissues. Indeed, until about 1970 (Chapter 2), nothing was known about the other collagens; there was not even direct evidence that there was more than one among vertebrate species. Therefore, most of what is presented in this chapter relates to type I collagen. Data on other collagens are given when available. In general, types II and III are similar to type I, while type IV (Section 1.2.6) is quite different. Very little is known about type V structure.

What the collagens have in common is that they all contain a characteristic triple-chain helix. They also contain nonhelical regions in varying amounts. It is convenient, therefore, to define a protein as a collagen if it consists largely of triple-chain helix. Other proteins that contain some triple-chain helix, but would not be considered collagens, will be mentioned later.

This definition of a collagen is based on a particular structure and requires good criteria for its identification. The best criterion is the x-ray diffraction pattern given by the collagen helix. Nearly as good is the characteristic and necessary repeating triplet amino acid sequence of the chains, Gly-X-Y, where X and Y can be any amino acid and are often proline and hydroxyproline, respectively. For historical reasons, these chains are referred to as α chains. Empirically, even the amino acid composition of a purified protein is a good criterion since collagens are one-third glycine, and hydroxyproline and hydroxylysine are nearly unique to these proteins. However, criteria based on appearance with the electron microscope must be approached cautiously. These points will become clear later, but it is important to note now that collagens are specific molecules that can be described by atomic coordinates corresponding, at least, to an average

From the National Institute of Dental Research, National Institutes of Health, Bethesda, Maryland.

or typical structure. It is at that level of structure that we seek to understand them and the interactions that determine their biological roles.

Since it will be necessary to use nomenclature and refer to information developed in other chapters, the reader may wish to read ahead to gain some understanding of collagen types, procollagens, and aspects of collagen synthesis. The three-letter abbreviations for amino acids used in this chapter can be found at the front of the book. A review by Bornstein and Traub (1979) gives detailed background information for much of what is discussed here.

1.2. Molecular Structure

1.2.1. The Collagen Triple-Chain Helix

Type I collagen was one of the earliest biological materials to be examined by x-ray diffraction, as it was readily available from tendon as oriented fibers, and was recognized as helical (*See* Bear 1952). However, that the helix had three chains, each being a left-handed helix itself with three residues per turn (about 1.0 nm), was not recognized until Ramachandran and Kartha (1954) deduced these facts from the x-ray diffraction pattern and from the observations that type I collagen had one-third glycine and large amounts of proline and hydroxyproline. How these factors enter into the structure will be described in Section 1.2.2. It was then recognized (Ramachandran and Kartha 1955; Rich and Crick 1955; Cowan et al. 1955) that the chains are supercoiled with a pitch in the range of 30–40 residues (about 10 nm).

The high angle diffraction pattern shown in Figure 1.1 reflects these features. The strong arc on the meridian at 0.286 nm is the residue spacing. The strong layer lines at about 0.4 to 1.0 nm arise from helical repeats. These features are

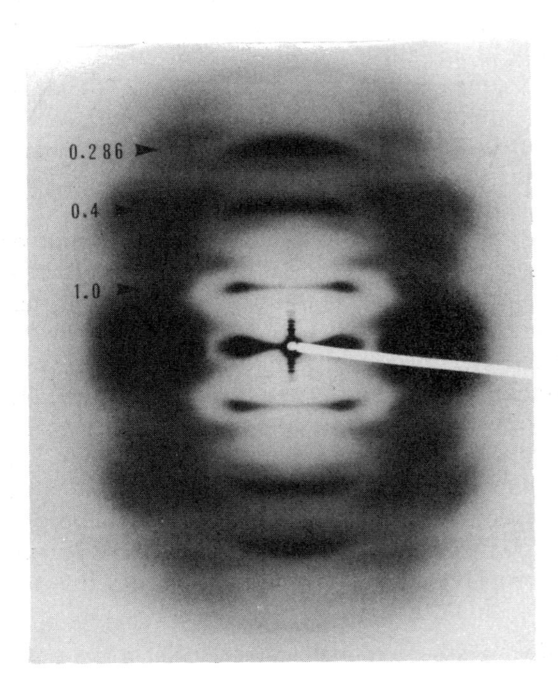

Figure 1.1. A high angle x-ray diffraction pattern of oriented fibers of rat tail tendon collagen. The layer line at 0.286 nm corresponds to the amino acid residue spacing along the fiber axis. The layer lines at 0.4 and 1.0 nm arise from helical repeats in the molecule. The central region can be seen expanded in a medium angle pattern (Figure 1.13). (Courtesy of R. D. B. Fraser.)

diagnostic of the triple helix. A more detailed discussion of collagen diffraction patterns will be given later (Section 1.3.1).

Details of how the presently accepted coordinates were worked out have been discussed by Ramachandran (1967), Traub and Piez (1971), and Fraser and MacRae (1973). The coordinates usually used were determined from poly(Gly-Pro-Pro) by Yonath and Traub (1969). More recently, Fraser et al. (1979) have refined the coordinates for tendon collagen stretched 10%. These quite similar sets of coordinates describe the main features of the collagen helix, but it is also clear that there must be local variations in structure imposed by the local sequence, molecular packing constraints, internal motion, and functional groups such as carbohydrate.

Short segments of the collagen helix are shown by three different kinds of models in Figures 1.2, 1.3, and 1.4 to illustrate the main features. These can be summarized as follows.

1. Three α chains with a repeating Gly-X-Y sequence staggered by one residue relative to one another are required to make a collagen molecule. An individual α chain is not stable as a helix. The pitch of the triple-chain supercoil

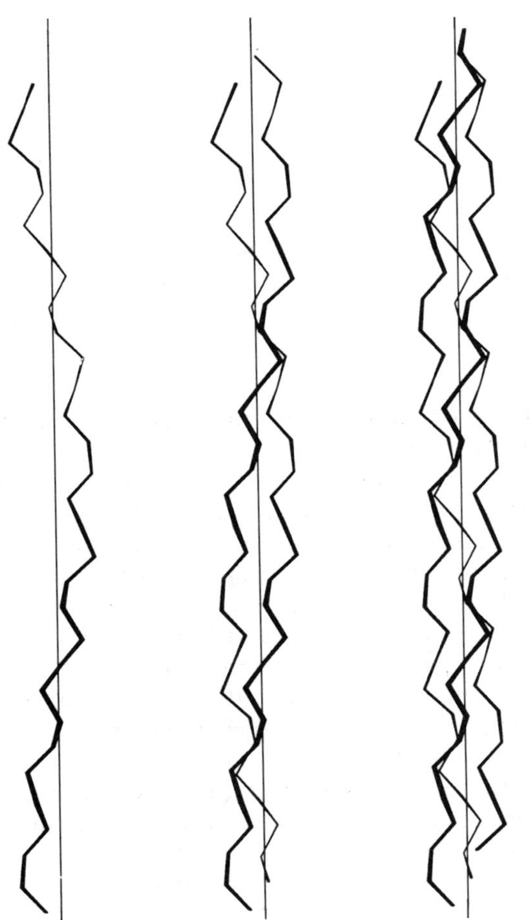

Figure 1.2. The backbone of the collagen triple helix showing one, two, and three chains *(left to right)*. The lines connect C_α atoms in successive residues. The thin vertical lines represent the molecular axes. The individual chains are left-handed helices and are wrapped around each other in a right-handed superhelix. (Courtesy of B. L. Trus.)

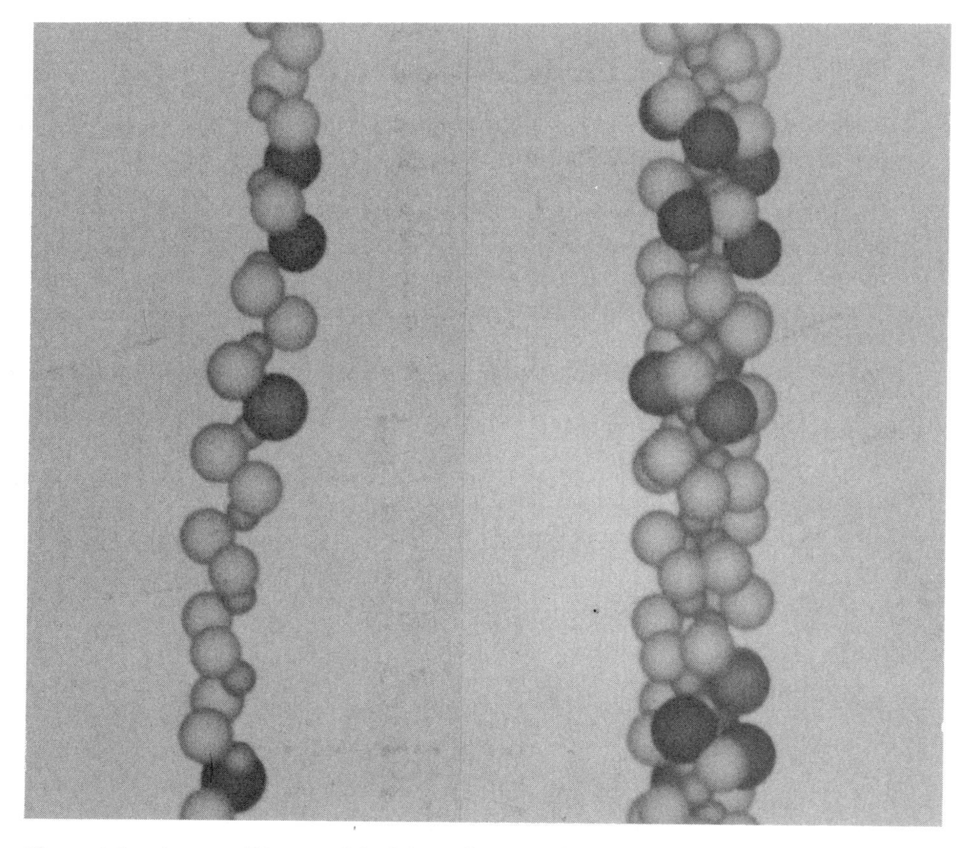

Figure 1.3. A space-filling model of the collagen triple helix showing one chain *(left)* and three chains *(right)*. Each sphere represents an amino acid residue. The small spheres up the center are glycine residues. The larger spheres on the outside can be any amino acid. (Courtesy of B. L. Trus.)

is right handed but its value is not known. Sequence analysis (Piez and Trus 1978) suggests 39 residues (11.2 nm).

2. The glycine residues, contributed alternately by the three chains, form a shallow helix up the center of the three-chain structure where there is no room for a larger residue. There is an absolute requirement for glycine in every third position, since glycine is the only amino acid with no side chain. Any other amino acid would disrupt the helix.
3. The side chains of the amino acids in the X and Y positions are directed outward where they can participate in various interactions, both intra- and intermolecular. They form a helical ridge (Figure 1.4) which may be important in packing of molecules under tension when ridges would fit into grooves.
4. Insofar as can be told by examining the structure, any amino acid can precede or follow glycine in the X and Y positions. In practice, some preferences have been observed which apply to types I, II, and III collagen. Glutamic acid, histidine, leucine, and phenylalanine are usually in the X position, while threonine, lysine, and arginine are usually in the Y position. Proline would

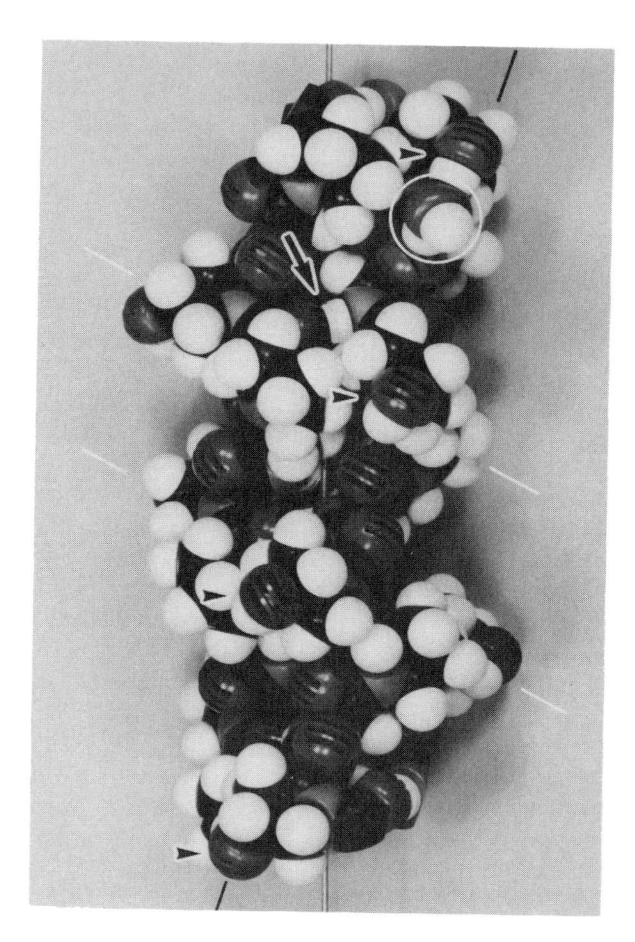

Figure 1.4. A CPK space-filling model of the collagen triple-helix, showing all the atoms in a ten-residue segment of the repeating triplet sequence (Gly-Pro-Hyp)$_n$. The arrow shows an interchain hydrogen bond. The arrow heads identify the hydroxy groups of hydroxyproline in one chain. The circle shows a hydrogen-bonded water molecule. The short white lines identify the ridge of amino acid side chains responsible for the 1.0-nm layer line in the x-ray diffraction pattern (Figure 1.1). The short black lines indicate the supercoil of one chain.

be equally distributed but when it precedes glycine it is usually hydroxylated to 4-hydroxyproline after translation (Chapter 3).

5. Water molecules can be fitted into the structure in various ways. Figure 1.4 illustrates one hydrogen-bonded water bridge from the hydroxy group of hydroxyproline to a backbone carbonyl oxygen, one residue back in the same chain.

Amino acid sequencing has shown that the α chains of types I, II, and III collagen have just over 1000 amino acids in an unbroken triplet array with short nontriplet regions at each end (Chapter 2). These molecules would then be expected to be long rods with calculated molecular weights of about 290,000. Type IV collagen contains noncollagenous domains and will be discussed separately (Section 1.2.6). Type V collagen is not yet well characterized.

Consistent with this model of type I collagen are a variety of earlier physicochemical studies, beginning with the first definitive study done by Boedtker and Doty (1956) (see Piez 1967; von Hippel 1967) and direct visualization of individual collagen molecules with the electron microscope, first done by Hall (1958). Figure 1.5 shows rotary-shadowed type I collagen molecules using a

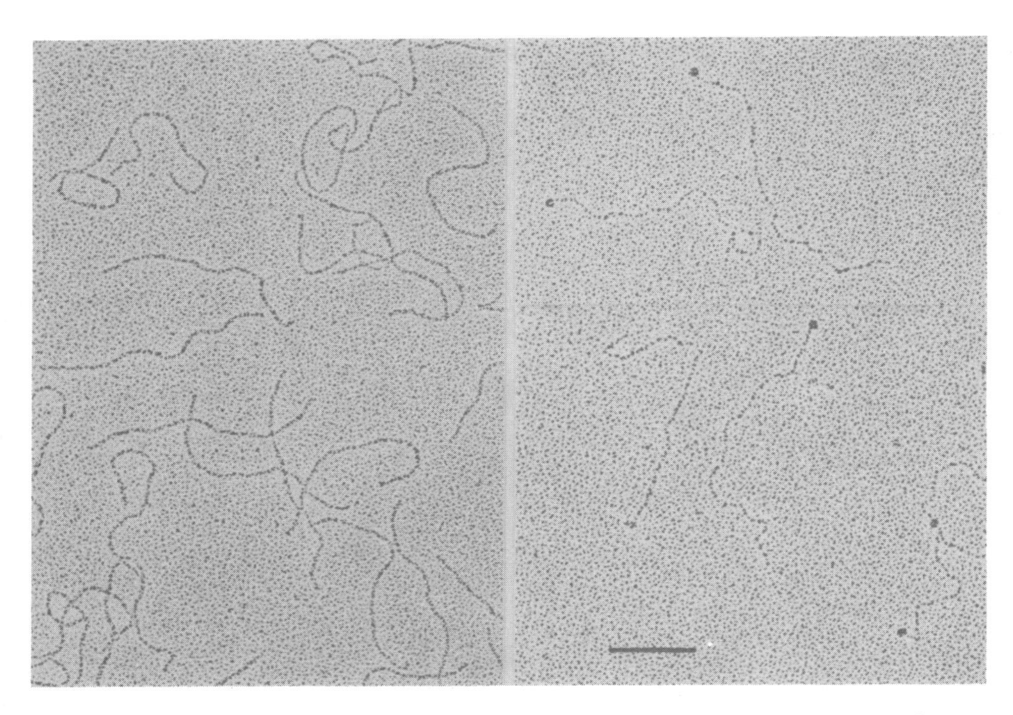

Figure 1.5. Electron micrographs of collagen molecules sprayed in solution onto mica and rotary shadowed with platinum. *Left:* type I collagen, 300 nm long, from calf skin. *Right:* intact type IV collagen, 400 nm long, from teratocarcinoma parietal yolk sac cells. The bar is 100 nm long. (Courtesy of H. Wiedemann, I. Oberbäumer, R. Timpl, and K. Kühn.)

recently improved procedure (Kühn et al. 1981). The conclusion drawn from such micrographs and from chemical and physical chemical studies is that the type I collagen molecule is a somewhat flexible rod 300 nm long and about 1.5 nm in diameter. The diameter of an equivalent solid cylinder would be 1.2 nm, but amino acid side chains make the effective diameter larger.

Type I collagen has nonidentical chains referred to as α1(I) and α2(I). There are two α1(I) chains and one α2(I) chain in each molecule. They are homologous but different in important ways (*See* Chapter 2). Since the three chains are staggered by one residue, they are not equivalent in the type I molecule. That is, the distribution of side chains on the surface will be different depending on which chain is α2(I). This could be important in molecular packing (Section 1.3.2), but which chain is α2(I) is not known. Types II and III collagens each have three identical chains but are otherwise very similar to type I collagen in molecular structure.

1.2.2. Helix Stabilization

Figure 1.6 shows the chemical formula of a triplet sequence common in the collagens, Gly-Pro-Hyp. This sequence is useful to illustrate the interactions and constraints that stabilize the collagen helix. They may be summarized as follows.

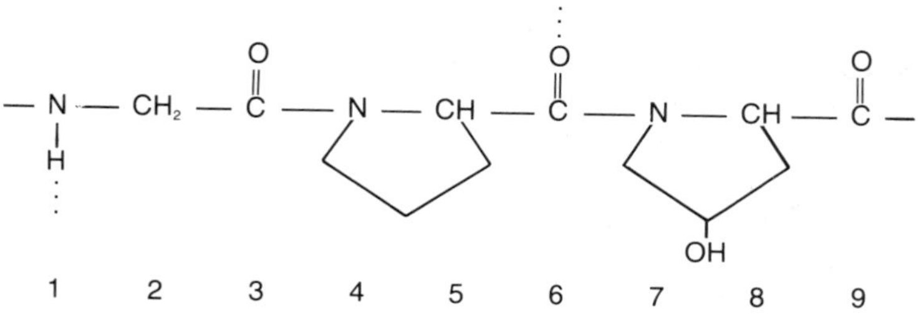

O O O
— N — CH_2 — C — N — CH — C — N — CH — C —
H
OH
1 2 3 4 5 6 7 8 9

Figure 1.6. The triplet sequence Gly-Pro-Hyp illustrating elements of collagen triple-helix stabilization. The numbers identify peptide backbone atoms. The conformation is determined by *trans* peptide bonds (3–4, 6–7, and 9–1); fixed rotation angle of bond in proline ring (4–5); limited rotation of proline past the C = O group (bond 5–6); interchain hydrogen bonds *(dots)* involving the NH hydrogen at position 1 and the C = O at position 6 in adjacent chains; and the hydroxy group of hydroxyproline, possibly through water-bridged hydrogen bonds (Figure 1.4).

1. Peptide bonds. As is true for all native proteins, the peptide bonds in the collagen helix are all *trans*. There have been suggestions that some *cis* peptide bonds occur in collagen (*See* Ramachandran 1976), but this is unlikely and could not be confirmed by nuclear magnetic resonance studies (Di Blasi and Verdini 1979). The *trans* peptide bond is energetically most stable in a near planar configuration and so fixes the conformation of every third bond in any polypeptide chain. (However, random coil α chains contain *cis* peptide bonds.)
2. Hydrogen bonds. The number of interchain hydrogen bonds was an early point of contention. It is now generally agreed (*See* Ramachandran 1976) that there is one per triplet between the peptide NH of glycine and the C = O oxygen of the second residue in the triplet of a neighboring chain. These two positions are indicated in Figure 1.6. Local differences are not ruled out.
3. Proline. The ring structure prevents rotation about the N—C_α bond in a polypeptide chain. In addition, model studies and energy calculations show that rotation about the C_α — C = O bond is restricted. Proline, therefore, is a major element in stabilizing the conformation. In fact, synthetic polyproline forms a very stable single-chain helix in aqueous solution (polyproline II) that is very similar to the helix of an α chain in a collagen (*See* Carver and Blout 1967).
4. Hydroxyproline. It might be expected that hydroxyproline would have the same conformational-directing characteristics as proline. However, this does not seem to be the case in the collagen helix where it precedes glycine. It is apparently the position that is important, since poly(Gly-Pro-Ala) will make a stable helix whereas poly(Gly-Ala-Pro) will not (*See* Traub and Piez 1971). Similarly, poly(Ala-Hyp-Gly) makes a stable triple helix while poly(Ala-Pro-Gly) does not (Rao and Adams 1979). Also, collagen that is underhydroxylated is markedly less stable than normal collagen (Berg and Prockop 1973; Rosenbloom et al. 1973); the melting temperature is 20°C lower. Therefore, hydroxyproline contributes to stability in some way that is different from proline. It is not known how, but one suggestion is that the hydroxy group is involved

in a water-bridged hydrogen bond, as illustrated in Figures 1.4 and 1.6. There must also be some difference whose significance is not understood between the conformations of the X and Y positions in the triplet sequence. In any case, it follows that Gly-Pro-Hyp is the most stable triplet of all.

5. Water. At 50% relative humidity, collagen binds about 20% water and under vacuum the last few percent are held tenaciously (*See* Gustavson 1956). One site involving hydroxyproline where water may be bound has already been mentioned. Ramachandran (1976) has suggested several others. There is evidence that these sites are occupied at least part of the time (Grigera and Berendsen 1979). Tritium exchange studies show two slowly exchanging hydrogens per triplet (*See* Traub and Piez 1971). If one is the direct interchain hydrogen bond, the other could represent an average of one water-bridged bond per triplet. The contribution to stability could, therefore, be considerable. Another view is that water does not bind to specific sites but has a more subtle effect through intrinsic environmental factors (Engel et al. 1977).
6. Glycine. Although glycine is entropically destabilizing because there are many rotational degrees of freedom about its NH — C_α and C_α — C = O bonds, the absence of a side chain permits close packing of the three chains and is thus critical to structure.
7. Side-chain interactions. Model building using known α chain sequences shows that it is frequently possible to make hydrogen bonds, charged-pair interactions, and hydrophobic interactions between side chains on the surface of the type I collagen molecule. There is no direct evidence that these provide a significant degree of stability to molecular structure. However, charged amino acids occur as groups and opposite pairs much more frequently than would be expected by chance (Hulmes et al. 1973) which is suggestive of intra- and intermolecular charge clustering (Section 1.3.3).
8. Disulfide bonds. Type III collagen is a special case in that its α chains have two adjacent cysteine residues at the C-terminal end of its triplet region. These form interchain disulfide bonds in the native molecule and undoubtedly stabilize that end of the helix. Type IV must also have disulfide bonds since reduction is required to dissociate its chains.

1.2.3. Synthetic Collagen-Like Polymers

A wide variety of polymers with glycine in every third position have been made and studied. Some of these, and the information gained from them, have already been mentioned. The major conclusion is that the requirements for a collagen helix are only two: a repeating Gly-X-Y sequence and sufficient proline in the X position or hydroxyproline in the Y position for stability. A detailed discussion of both structural and biochemical studies has been presented by Bhatnagar and Rapaka (1976).

The crystal structure of $(\text{Pro-Pro-Gly})_{10}$ has recently been determined to high resolution by x-ray diffraction (Okuyama et al. 1981). The polymer makes a triple helix very similar to the collagen helix discussed here (Section 1.2.1). It has nearly the same residue translation (0.287 nm) and interchain hydrogen bond pattern (Figure 1.6). However, the chain supercoil is 21 residues (60 nm), a value outside the range of 30–40 residues estimated for collagen. Thus, the two structures may be described as allomorphic.

1.2.4. Denaturation and Renaturation

When a solution of native type I collagen is heated at a pH below approximately 5, it denatures. Melting also occurs above pH 5 if the concentration is low enough to prevent the aggregation that would normally occur (Section 1.3.4) or if some substance is used to inhibit aggregation (Hayashi and Nagai 1973). Melting at neutral pH is several degrees higher than at acid pH and 1–2°C above body temperature. Unlike most proteins, the products are soluble. This is a fortunate characteristic which has aided in the chemical characterization of the chains (Chapter 2). Types II and III collagen behave in a similar manner.

Because of the length of the helix and the periodic nature of its structure, melting can be treated as a phase transition and followed by a variety of physicochemical methods (*See* Traub and Piez 1971). Optical rotation (or its related property, circular dichroism) and viscosity have been most commonly used. Figure 1.7 shows the circular dichroic spectra of native and denatured type I collagen. By selecting an appropriate wavelength, denaturation can be followed as a function of temperature. A typical melting curve is shown in Figure 1.8. Since denaturation is not instantaneous, such curves will be time dependent. If melting is carried out by heating at a constant rate of less than about 6°C/h (either continuously or in small steps), this factor can be minimized. Since the curves are dependent on the pH and ionic environment, detailed comparisons must be made under the same conditions. The curves are characterized by the midpoint of the melting transition (T_m), the temperature required to go from

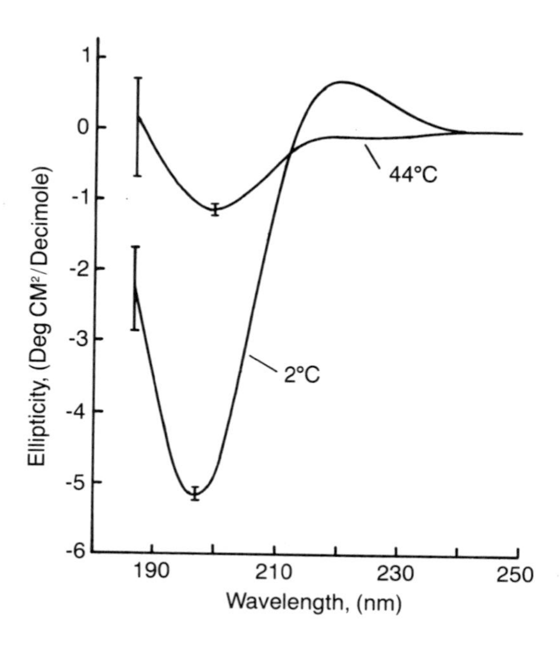

Figure 1.7. Circular dichroic spectra of native (2°C) and denatured (44°C) type I collagen. The transition on heating is sharp (Figure 1.8). *(Source: Redrawn, Piez and Sherman (1970) Biochemistry 9:4129–4133, with permission.)*

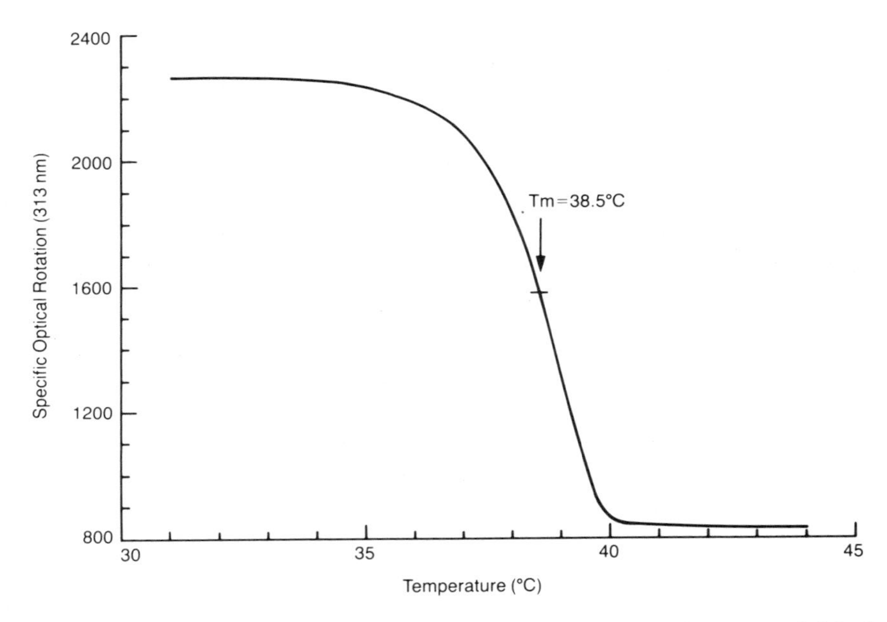

Figure 1.8. Melting transition of type I collagen from rat tail tendon, 0.8 mg/ml in 5 mM acetic acid. The temperature was increased continuously at 4°C per hour. T_m is the midpoint of the transition. (Courtesy of S. L. Lee.)

one-quarter to three-quarters of the total transition, and the total change in the property being measured. These values are good criteria of the degree of native helicity.

Treatment of melting as a phase transition implies that a given molecule is either all helix or all random coil and denaturation is statistical. This may be convenient for many purposes, but in actual fact it is known that some regions of the helix are less stable than others. In general, these regions would correspond to regions with low content of proline and hydroxyproline. Biologically, helix stability may be an important feature of a recognition site; the animal collagenase site may be an example (Chapter 4).

A given T_m is proportional to the highest environmental temperature at which the animal normally lives. Of course, this will be body temperature for mammals. The correlation is illustrated in Figure 1.9. For example, type I collagen from codfish skin melts at about 15°C while collagen from rat tail tendon melts at about 38°C. Collagen from chick skin (not shown) melts about 4°C higher than collagen from rat skin; the body temperatures of these two species differ by about the same amount, showing the close correlation and suggesting an important biological reason (Miller et al. 1967). Presumably, collagen must be stable enough to remain native at the usual temperatures encountered but not so stable that it cannot be turned over. In general, T_m is regulated by the proline plus hydroxyproline content. Other factors may enter into stability, particularly for certain unusual collagens from the invertebrates (Rigby 1968).

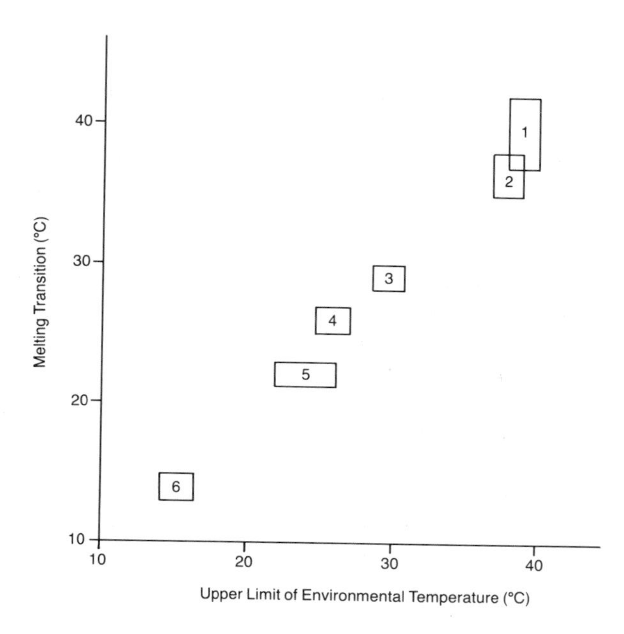

Figure 1.9. Relationship between the midpoint of the melting transition (T_m) of several collagens and the environmental temperature of the animal or tissue. 1: hog intestine and its parasite *Ascaris*. 2: skin (rat, human, bovine). 3: *Helix* aspersa. 4: tuna skin. 5: earthworm and shellfish. 6: cod skin. *(Source: Redrawn from Rigby (1968) Symposium on Fibrous Proteins, Australia: Butterworths, with permission.)*

In tissues, the collagens exist in fibrillar or some other aggregated state. However, a property related to T_m can be measured for collagenous tissues. If a length of wet tendon or strip of skin, for example, is suspended in solvent and heated, it will abruptly shrink to a fraction of its length as the collagen helix melts at some temperature designated the shrinkage temperature (T_s). The T_s is usually about 20°C higher than T_m, reflecting the added stability contributed by molecular packing and tissue structure. Although measurements of T_s are important historically, they are rarely made today since it is too complex a function to be readily interpreted.

Renaturation, defined as helix formation from melted collagen chains in vitro, has been extensively studied. (*See* Traub and Piez 1971) The process is inefficient if the starting point is cooled random coil α chains. In general, short lengths of triple-helix joined by unassociated chains are obtained. This structure in three dimensions traps water forming the common cold gelatin. At low concentrations and with slow cooling, some molecules with a helix content similar to that of native type I collagen can be obtained, but perfect alignment of chains may not be achieved.

If three chains are covalently crosslinked to maintain chain register, renaturation can be rapid and efficient. The time is proportional to length and the rate is limited by *cis–trans* isomerization of the peptide bond (Bächinger et al. 1980). Molecules crosslinked in this way are found in some soluble type I collagen preparations (*See* Piez 1967) or may be made artificially and are referred to as γ-components. Type III collagen with its disulfide bonds at one end of the helix is another example. In vivo, the difficult to achieve, specific association of three α chains is avoided since chains are made as larger pro α chains and alignment is determined by the pro domains (Section 1.2.5 and Chapter 3).

1.2.5. Procollagens

The α chains of types I, II, and III collagens are synthesized as pro α chains which are longer at the N-terminal end by about 150 amino acids and at the C-terminal end by about 250 amino acids. These regions are referred to as the pN and pC peptides or domains. Sequence data and schematic representations can be seen in Chapter 2, Figures 2.1 and 2.2. After assembly in the cell of three pro α chains to form a procollagen and secretion, the pN and pC domains are removed to give a collagen (Chapter 3). The procollagens can thus be considered intracellular transport forms with properties suitable for that role. For example, the procollagens are soluble under physiological conditions while the collagens are not. Biological control is exercised is this way.

The pN peptide of pro α1(I) has a globular region with intrachain disulfide bonds, then a region of 16 triplets of the same Gly-X-Y form found in the body of the molecule, and finally a few amino acids not in triplet form where the protease cleaves in the later conversion of the procollagen to collagen. Normally, the triplet regions in the pN peptide of two pro α1(I) chains associate with a similar region in the pN peptide of α2(I) during helix formation.

This association of the pN peptides might be thought to confer specificity on the process of molecular formation. However, this does not seem to be the case. Chain association apparently begins in the pC domain (Chapter 3). Furthermore, in the absence of pro α2(I), three pro α1(I) chains will form what is referred to as type I trimer. This apparent lack of complete specificity is curious and presently unexplained. The reason for the short helical region within the pN peptides is also not clear. It is very stable owing to a high proline and hydroxyproline content and thus may stabilize that end of the molecule. Disulfide bonds are also present in the pN peptides but are intrachain except in type III collagen, where both inter- and intrachain disulfides occur.

The pC peptides are entirely globular but of unknown structure. These domains contain inter- and intrachain disulfide bonds which stabilize the association. Carbohydrate typical of the amide-linked, high-mannose glycoproteins (Chapter 7) is also present.

1.2.6. Type IV Collagen

Basement membranes form a filtration and structural barrier between epithelial cells and the underlying connective tissue. Similar or identical membranes underlie endothelial cells and surround other cell types such as fat and muscle cells. Basement membranes are usually 20–30 nm thick but in some cases where they have a special structural role, such as the parietal yolk sac and the lens capsule, they may be as much as 5 μm thick. They are complex structures containing a glycoprotein called laminin (Chapter 7), a heparan sulfate proteoglycan (Chapter 8), and a collagen referred to as type IV. Details concerning morphology and function can be found in recent reviews (Kefalides et al. 1979; Timpl and Martin 1981; Grant et al. 1982).

Type IV collagen contains at least two similar chains, α1(IV) and α2(IV). They apparently form separate molecules but are not yet well characterized. Their chemistry is discussed in Chapter 2. There has been considerable controversy about molecular structure because degradative procedures have been used to make soluble preparations and it was not possible to relate the various fragments to one another. It was evident, however, that the collagen contained large non-collagenous domains. The availability of improved sources of intact, or nearly intact, type IV collagen and electron microscopy of rotary shadowed molecules and fragments have answered many of the problems (Timpl et al. 1981). Figure 1.5 shows an electron micrograph of an intact molecule. Overall, the molecule is about 400 nm long with a large globular head much like the pC domain of type I collagen. The remainder appears to be largely triple-helical, but a bend about 60 nm from the end opposite to the globular end and sequence studies (Chapter 2) show that the helix is interrupted by nontriplet regions. The location and size of these nonhelical domains is not known.

1.2.7. Other Proteins Containing Collagen Helix

The best characterized protein that would not be considered a collagen but contains collagen helix is subcomponent C1q of the first component of complement. Each molecule contains six identical subunits which consist of bent rods 1.5 × 25 nm with a globular head. Electron microscopy has shown that the rods are close-packed at one end but the heads are spread out much like a bunch of flowers. The heads interact with antigen–antibody complexes and the rod ends with the C1r-C1s complex in the complement system (Reid and Porter 1981). Each subunit has three nonidentical polypeptide chains about 200 residues long. About 80 residues near one end of each chain are Gly-X-Y triplets and have been sequenced (Reid 1979).

Like the collagens, these sequences contain 4-hydroxyproline and hydroxylysine. All of the original lysine residues preceding glycine are hydroxylated and many appear to be glycosylated. On the other hand, many of the prolines preceding glycine are not hydroxylated. This is opposite to types I, II, and III collagens, in which such prolines rarely escape hydroxylation while lysine often does (Chapters 2 and 3). The unusual feature of the sequence is the break about halfway through the triplet sequence of the A and C chains. The latter has an Ala where a Gly would normally be, and the former has an extra residue between triplets, presumably associated with the bend in the rods. The structure in this region is unknown. Similar breaks apparently occur in the helix of type IV collagen (Section 1.2.6).

Another protein shown to have collagen helix is acetylcholinesterase (Bon et al. 1979; Anglister and Silman 1978; Mays and Rosenberry 1981). In this case, up to 12 catalytic subunits are attached in groups of four by disulfide bonds to a collagen tail about 50 nm long, corresponding to about 150 amino acid residues. Hydroxyproline and hydroxylysine are present and the structure is susceptible to bacterial collagenase. It is speculated that the tail anchors the enzyme to basement membrane.

These are the only proven examples of collagen helix in noncollagenous proteins. Others may exist, but at present it seems that the helix is not often found except in the collagens.

1.3. The Native Collagen Fibril

1.3.1. Axial Structure

Because of their axial order, tendon collagen fibrils (type I) were among the first biological materials to be studied by x-ray diffraction and electron microscopy. Both procedures show a periodic band pattern with a polarized repeat (referred to as D) of about 64 nm in dry samples (*See* Bear 1952). The accepted value today, obtained by x-ray diffraction of wet samples stretched just enough to remove any slack, is close to 67 nm (Miller 1976).

An electron micrograph of positively stained fibrils is shown in Figure 1.10. In positive staining, heavy metal ions bind to charged groups. The bands, therefore, locate clustered charged groups along collagen molecules; protein is not seen. The pattern arises from staggered molecules (Schmitt et al. 1955; Hodge and Schmitt 1960; Kühn and Zimmer 1961) spanning about four periods, leading to what has been called the quarter-stagger model. As refined by Hodge and Petruska (1963), a collagen molecule in a stained and dried fibril is 4.4 D long, giving rise to an 0.4 D overlap and an 0.6 D gap region with a predicted mass ratio of 5:4 (Figure 1.11). Before the molecular stagger was understood, Tomlin and Worthington (1956) derived a step function for the electron density in a D period from the low angle x-ray diffraction pattern, as discussed below. The portion corresponding to the overlap is 0.47 D in wet fibrils.

An electron micrograph of negatively stained collagen fibrils is also shown in Figure 1.10. In negative staining, a heavy metal salt such as a tungstate or uranyl compound fills voids in and around the object, thus outlining it and its internal structure to the extent that stain can penetrate. The alternating light and dark bands, therefore, represent regions of different protein mass, the darker regions

Figure 1.10. Electron micrographs of reconstituted type I collagen fibrils. The arrows, pointing in the C-terminal direction, represent collagen molecules 300 nm long in their approximate relationship to the band pattern (see Figure 1.12). *Left:* Positively stained with uranyl acetate. *Right:* Negatively stained with phosphotungstic acid at neutral pH. (Courtesy of J. A. Chapman.)

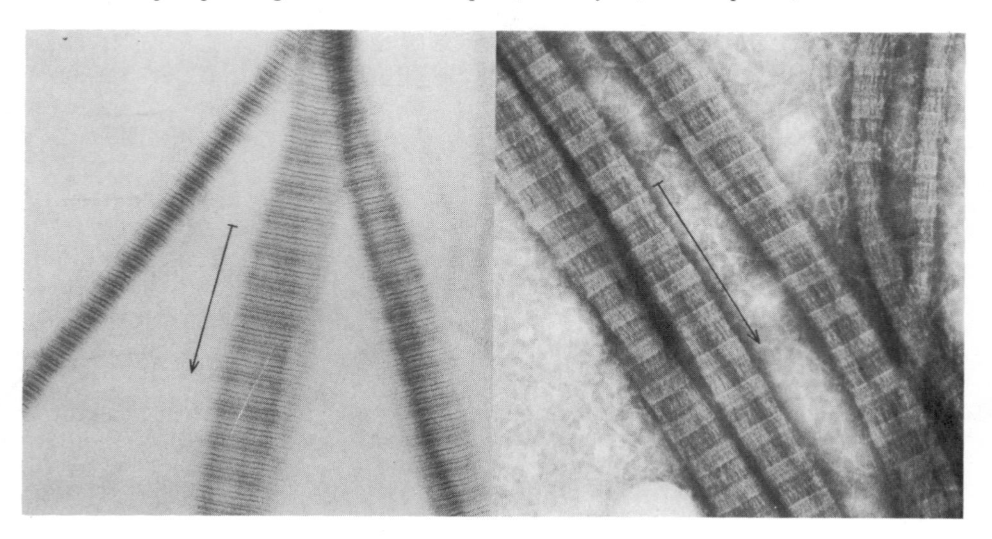

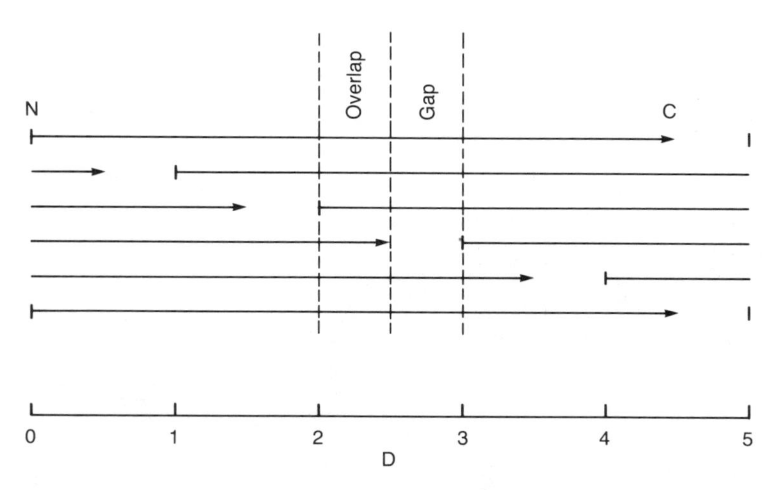

Figure 1.11. Schematic view of the axial stagger relationship of type I collagen molecules *(arrows)* in the native fibril. The molecules are about 4.5 D long, where D = 67 nm. The nonintegral length results in overlap and gap regions with different protein densities. It is believed that the 1 D and 4 D staggers, as shown, are important nearest-neighbor relationships (Figure 1.18), but they may not be the only ones in the fibril.

holding more stain because of less protein and a more open structure. It should be noted that the stain density is the reverse of the protein density. The appearance of fibrils varies with the stain used and the amount held around the fibrils. Since negatively stained fibrils generally show superimposed positively stained bands, the two types of preparations can be aligned, as in Figure 1.12.

Types I, II, and III collagen can all form similar fibrils as shown by reconstitution from soluble preparations (Section 1.3.4). Their band patterns are visually identical (Kühn and von der Mark 1978), although a careful analysis should show differences since their sequences differ somewhat. They also cannot be distinguished in tissue sections by the band pattern, although identity can often be inferred from other features, or from a knowledge of the identity of the tissue and its chemistry. Type IV collagen does not form fibrils but rather a mat-like network through associations between molecular ends (Timpl et al. 1981).

Since the amino acid sequences of the α1(I) and α2(I) chains are known, and it can be assumed that the triple helix is sufficiently regular to space residues equally in axial projection, it is possible to compare the observed band pattern of a positively stained type I collagen fibril with the pattern predicted from the sequence. This was done by Piez and Miller (1974) and by Doyle et al. (1974b) using the α1(I) sequence, and in greater detail using both the α1(I) and α2(I) sequences by Meek et al. (1979). The observed and predicted patterns agree remarkably well (Figure 1.12). The best agreement is obtained when the assumed stagger between molecules is 234–235 residues. Thus, every amino acid in the sequence can be accurately assigned a position along the fibril.

An independent measure in residues of the stagger between molecules composed of three α1(I) chains was obtained by Hulmes et al. (1973) by calculating charged and hydrophobic interactions between collagen molecules, assuming

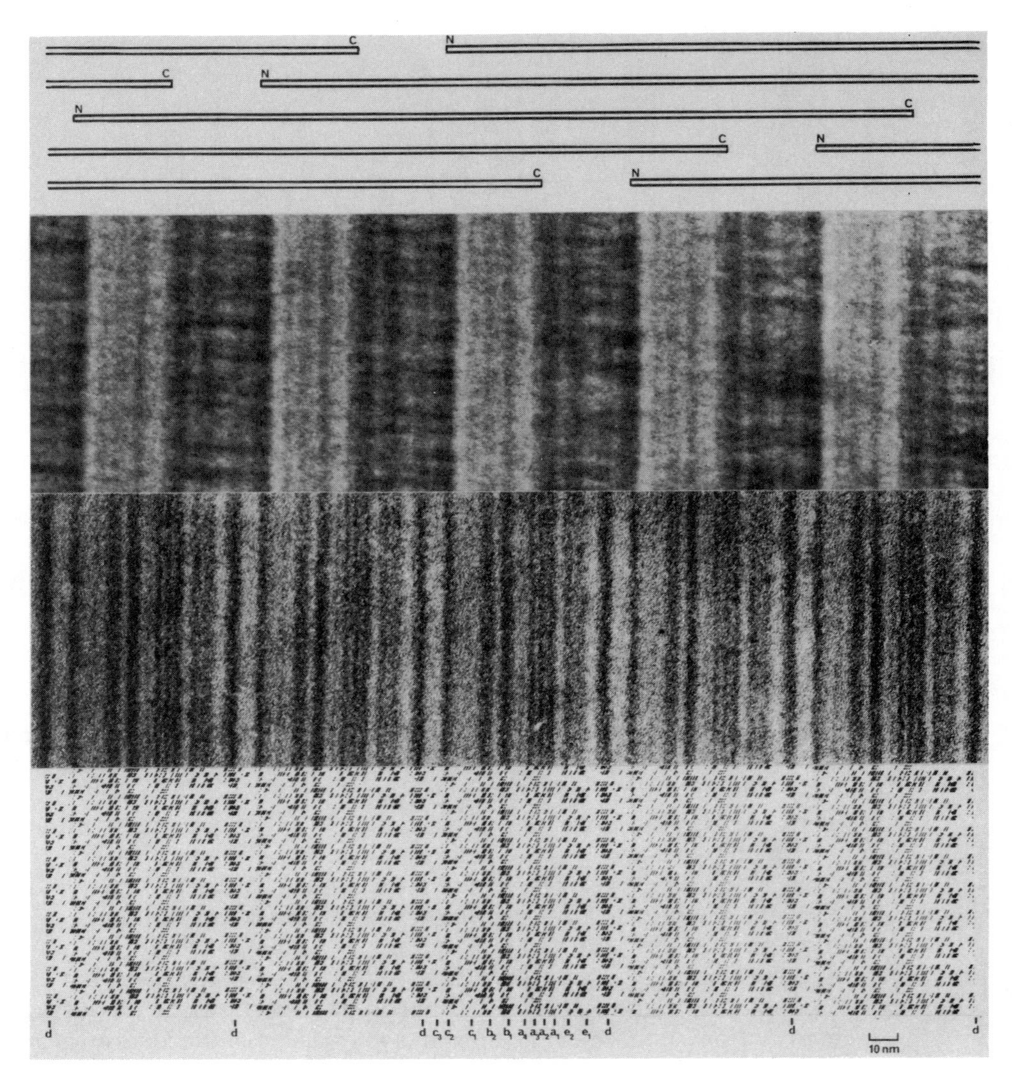

Figure 1.12. Type I collagen fibril structure. *From top to bottom:* Schematic D-stagger pattern (see Figure 1.11); electron micrograph of a negatively stained fibril (phosphotungstic acid); electron micrograph of a positively stained fibril (phosphotungstic acid and uranyl acetate); a computer reconstruction of the distribution of charged groups in the α1(I) and α2(I) chains in molecules staggered by 234 residues. The patterns are aligned showing the axial relationships of amino acid sequence to fibril structure. The band nomenclature appears at the bottom. (Courtesy of J. A. Chapman.)

again a regular spacing of residues. A more detailed analysis including the α2(I) chain and the type III molecule was done by Hofmann et al. (1980). The results show that interactions are maximal at staggers of multiples of 234 residues. Zero stagger is also favored, which can be explained by the fact that many of the positive and negative charges cluster together along the molecule in pairs or groups and thus neutralize the charge repulsion that would normally occur between aligned molecules. In the confined environment inside a fibril, charge clustering may contribute positively to structure (Piez and Torchia 1975).

X-ray diffraction presently offers the best information about molecular packing. Since the technique is not widely understood, it is worth summarizing some of the major characteristics of collagen fiber diffraction patterns. A more complete discussion can be found in the book by Fraser and MacRae (1973).

1. Spots or reflections represent constructive interference of scattered x-rays arising from periodicities in the structure. Electrons are responsible for the scattering as the x-ray beam passes through the fiber or fiber bundle. A fiber, by definition, is a bundle of fibrils. The scattered electrons are captured on film to form a characteristic pattern. The shorter the periodicity, the farther out on the pattern the reflection falls. Thus, one speaks of reciprocal or Fourier space in making measurements on diffraction patterns; it is simply the inverse of real space. Since long range order (>1.0 nm) gives rise to reflections near the origin, they are best seen in low (or medium) angle patterns where the fiber to film distance is long. On the other hand, short range order falls far out in the pattern and is best seen at high angles using a short fiber to film distance. The high angle pattern in Figure 1.1 has already been mentioned as showing short range molecular periodicities related to the triple-chain helix (Section 1.2.1).
2. The meridian (displayed vertically) corresponds to the axis of the fiber or fiber bundle. Thus, reflections on or near the meridian arise from axial order, which is good if fibers are aligned. Equatorial reflections (the horizontal direction) arise from lateral order (perpendicular to the axis, or nearly so). Order in this direction, as will be seen, is generally poor. The meridian and the equator are mirror lines and thus the four quadrants contain identical information, except that intensities may differ owing to tilting of the sample relative to the beam.
3. The pattern contains two sets of structural information—one pertaining to the molecule and the other pertaining to the lattice on which molecules are located. These two patterns (or transforms) are said to be convoluted, or more simply, multiplied, so that one cannot be seen where the other is zero. The result is that the positions of reflections are determined by the lattice, as defined by the unit cell or repeating unit, while the intensities of the reflections correspond to features of the molecular structure and its arrangement in the unit cell. The sharper the spot, the greater the crystallinity.
4. The pattern contains a great deal of diffuse x-ray scatter. This, in a simple sense, arises from disorder. The origin of the disorder is important information reflected in the location and intensity of diffuse scatter, but it is only necessary to note here that randomness is not necessarily meant. Exact geometric order in much of the fibril may be missing, but covalently defined order is not necessarily absent.
5. There are too few reflections to solve molecular or fibril structure from the diffraction data alone. The data are, however, extremely useful for testing and refining models, or for comparing structural features under different conditions, but the precision of the calculations should not be confused with the accuracy of the model. Intuitive reasoning and independent confirmation play a major role in interpreting collagen diffraction patterns.

With these characteristics in mind, what does the x-ray diffraction pattern say about axial order? Features of molecular order have already been discussed in this context (Section 1.2.1). Information about fibril order is seen in the low angle area on the meridian (Figure 1.13), which shows a series of equally spaced spots or short lines of varying intensity. The spacing is 1/67 nm, so the spots are orders of D, the same major periodicity seen by electron microscopy. Under good conditions there are about 40 orders of D visible, indicative of fine structure within the period to a resolution of 67/40 = 1.7 nm. As in the case of the band pattern seen by electron microscopy, the intensities of the low angle meridional reflections given by type I collagen fibrils can be calculated from the amino acid sequence of the α1(I) and α2(I) chains. This was done by Hulmes et al. (1980). The best agreement with observation is obtained when D is set at 235 residues and it is assumed that the nonhelical ends of the molecules have an average residue spacing about three quarters of the spacing in the triple-helix. Similar calculations from neutron-diffraction patterns yield essentially the same results. Since neutrons are scattered by protons and x-rays by electrons, these are independent measures.

Since the helical portion of the α1(I) chain has 1014 residues and the nonhelical ends have 16 N-terminal and 26 C-terminal residues, and since α2(I) is similar (Chapter 2), the length of the type I collagen molecule can be calculated as $(1014 \times 0.286) + (42 \times 0.22) = 299$ nm. Since D is accurately known to be 67 nm, and 299/67 = 4.46, 0.46 D can be assigned to the overlap and the remainder to the gap region, which is internally consistent with the direct measurement of 0.47 D (Tomlin and Worthington 1956; Hulmes et al. 1980).

Figure 1.13. Medium angle x-ray diffraction patterns of rat tail tendon collagen fibers. *Left:* Normal, slightly stretched fiber showing the sharp meridional reflections arising from the D stagger and equatorial reflections arising from molecular packing. The reflections near 1.3 nm are related to the near-hexagonal packing of molecules. The row line at 3.8 nm is an important dimension of the unit cell. *Right:* Fiber lightly stained with phosphotungstic acid, which intensifies the row lines and shows them to be split and sampled on the 1.0-nm layer line. (Courtesy of R. D. B. Fraser)

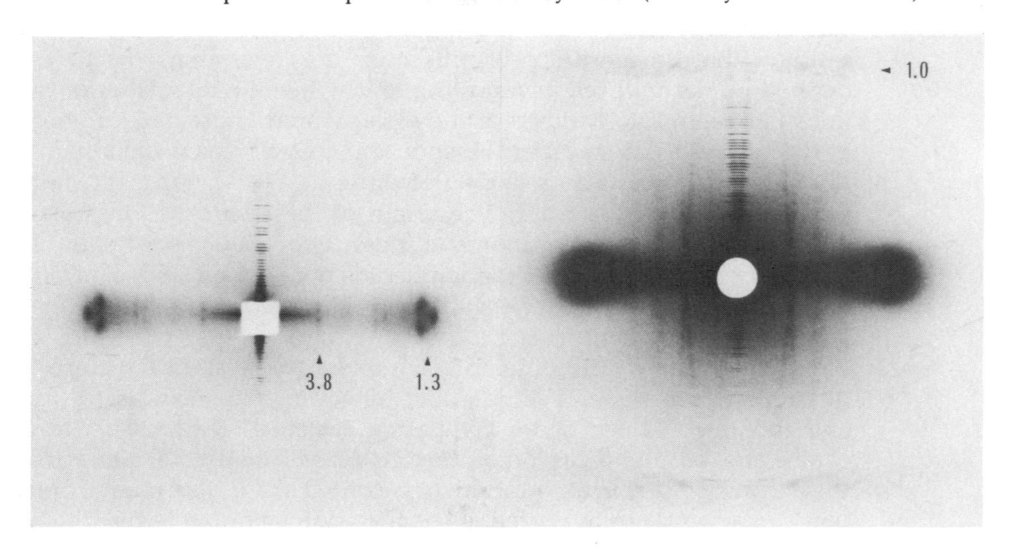

1.3.2. Three-Dimensional Structure

In contrast to the considerable degree of understanding of axial structure of the type I collagen fibril, there is much less understanding about lateral order. There are presently two views. One is that fibrils have an ordered filamentous substructure, with the filaments often being called microfibrils. They have a D period and a specific helical geometry. They are aligned in bundles to form fibrils of various diameters. Molecules in the microfibrils, and perhaps the microfibrils themselves, are supercoiled, resulting in a rope-like structure with several levels of coiling. The microfibril model that has had the best acceptance is shown in Figure 1.14. It was first proposed by Smith (1968). The other view is that a fibril is a three-dimensional crystal in which the collagen molecule is the basic unit without an intermediate substructure. Disorder is common, but quite large domains of coherent crystalline order are present under appropriate conditions. These domains are characterized by near-hexagonal packing of molecules with a straight tilt and a defined stagger pattern. Since models of assembly, stabilization, and function depend on knowing structure, it is worth briefly recounting the arguments and critically judging the conclusions.

Since the early electron microscopic studies of Schmitt and his colleagues (see Bear 1952), it has been recognized that collagen fibrils appear to be composed of filaments. These can be seen in the gap region of negatively-stained, unswollen preparations, as in Figure 1.12, and in fibrils that are swollen or fragmented in a variety of ways. An example of the latter type of preparation is shown in Figure 1.15. Freeze-fracture studies also show thin filaments within fibrils. For example, Ruggeri et al. (1979) found filaments of the diameter predicted for five-fold microfibrils in fibrils from many tissues. In some cases the filaments are straight (for example, rat tail tendon), and in some cases supercoiled (for example, rat skin). Since electron microscopy involves quite drastic procedures during sample preparation (drying and staining, for example) and viewing (exposure to the electron beam), it cannot be ruled out that the filaments are artifacts. However, it is unexpected that such varied treatments would all produce similar artifacts were they not based on a real structural feature.

There is electron microscopic evidence of another kind for an ordered fila-

Figure 1.14. Schematic view of the Smith (1968) five-stranded microfibril model. Collagen molecules (rods) (*a*, *b*, *c*, *d*, and *e*) are staggered by 1 D and rotated by one fifth of a circle to make a closed helix about 4 nm in diameter. The sixth molecule (a') is in line with the first. D = 67 nm. Molecules may be supercoiled to form a rope-like structure. Microfibril bundles form fibrils (Figure 1.17).

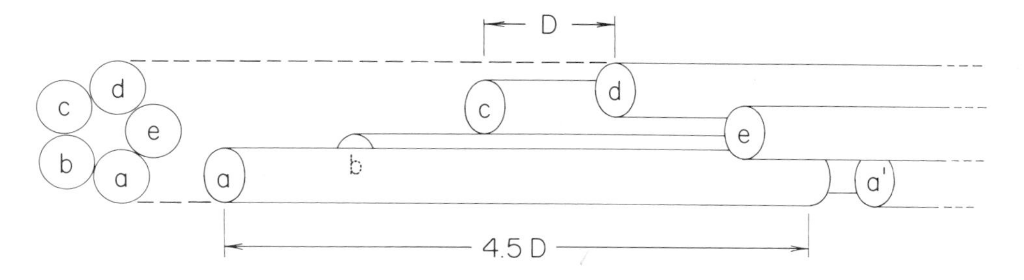

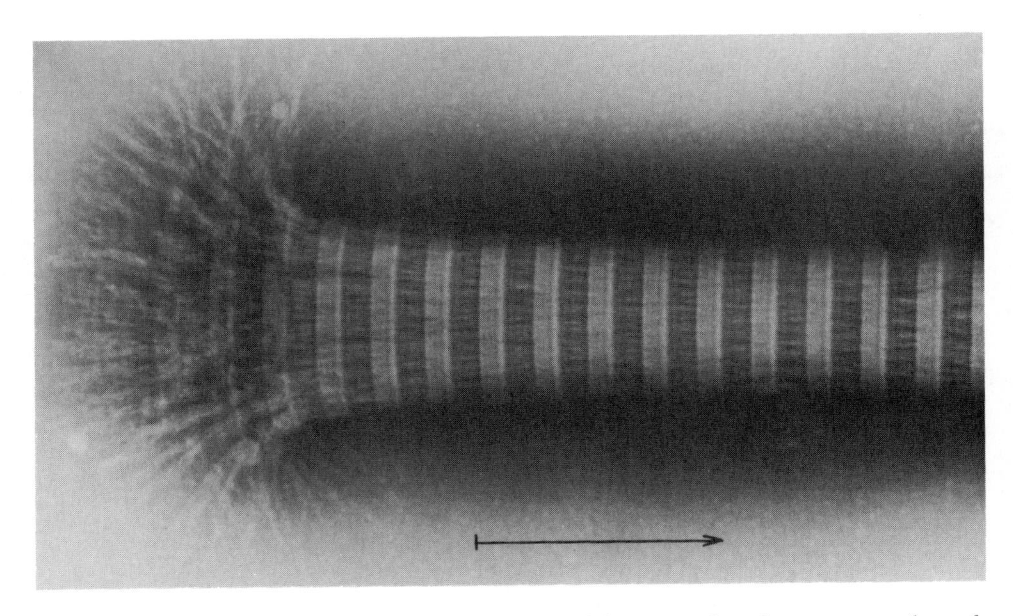

Figure 1.15. An electron micrograph of a collagen fibril from rat rail tendon cut across the end, which has flared showing the filamentous substructure. The arrow represents a collagen molecule, 300 nm long. Negatively stained with phosphotungstic acid.

mentous substructure. There are two polymorphic forms of collagen fibrils, obliquely-banded and D-periodic symmetric, that seem to be composed of a native D-periodic substructure, but not aligned as in the native fibril (Section 1.4). The limiting diameter of the D-periodic substructure is very small, suggesting that it is the five-fold microfibril. At least, it is difficult to imagine that it is not an ordered filament of some type. Also, thin filaments less than 10 nm in diameter and as small as 4 nm have been prepared by Veis et al. (1979) and shown to be D-periodic by x-ray diffraction.

Other evidence provided by electron microscopy is that Parry and Craig (1979) measured the diameters of collagen fibrils in a variety of tissues from a variety of animals and found that diameters are always multiples of 8 nm, beginning at 16 nm (Section 1.3.5). This result implies a unit substructure considerably larger than the molecule.

Independent evidence supporting a microfibril model has been obtained by analyzing regularities in the sequence of the α1(I) and α2(I) chains (Piez and Trus 1978). This analysis shows that charged and large hydrophobic residues can be arranged on the surface of a type I collagen molecule to form different sets of intermolecular interactions on opposite sides. One set seems designed to function within a microfibril; the other between microfibrils as would be predicted by the five-fold microfibril model (Figures 1.14 and 1.17). This is possible because of the different distributions of large hydrophobic residues in α1(I) and α2(I). Details of microfibril symmetry in the original model may not be correct since the x-ray diffraction data on which they were based have been reinterpreted, but the general principle of an ordered microfibril with supercoiled molecules remains as the only existing explanation of the sequence regularities (Piez and Trus 1981).

X-ray diffraction data consisting of reflections on and near the equator (Figure 1.13) that denote three-dimensional crystalline order were originally used to support the five-fold microfibril model. However, difficulties in refinement led to the substitution of a new and very different model, the three-dimensional crystal model (Hulmes and Miller 1979), which has been successfully refined (Miller and Tocchetti 1981; Fraser and MacRae 1981). The model is shown in Figure 1.16. It would apply to domains within large fibrils, as suggested by electron microscopic evidence (Hulmes et al. 1981), but could not explain narrow D-periodic fibrils, since it contains straight-tilted molecules which would extend beyond the boundaries of narrow fibrils (Fraser and MacRae 1981).

It has already been noted that collagen x-ray diffraction data can test models but not derive them. Even though the three-dimensional crystal model has withstood this test far better than any other model, it is not proven. Furthermore, the evidence obtained by electron microscopy and sequence analysis discussed above cannot be easily fit to the new model. Since it is difficult to discard either model, a compromise model in which microfibrils are retained but compressed to put molecules in a unit cell like the Hulmes and Miller (1979) model has been suggested (Trus and Piez 1980). It has been further refined (Piez and Trus 1981) and may offer a solution to the apparent conflict. Its structure is shown in Figure 1.17.

Of course, what is important about fibril structure is what might be called its covalent connectivity—the three-dimensional pattern established by crosslinked polypeptide chains. Once that has been determined, whether crystallinity is present or not is secondary. In fact, lateral crystallinity is seen only in some tendons (notably rat tail tendon) under carefully defined solvent conditions and when stretched to remove the crimp (Section 1.3.5). Disorder is the rule, probably for important functional reasons.

An important new technique has been used in the last few years to give evidence about internal interactions in the native type I collagen fibril, which is

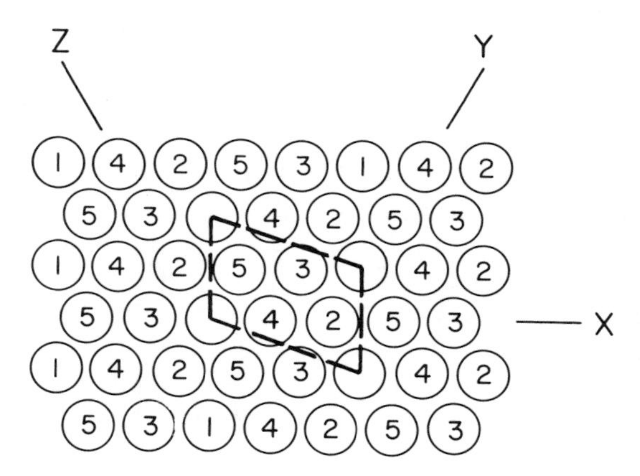

Figure 1.16. A cross-sectional view of molecular packing of type I collagen molecules *(circles)* in the three-dimensional crystal model proposed by Hulmes and Miller (1979). The alternate model of Miller and Tocchetti (1981) is shown. The fibril axis is perpendicular to the page. The lattice is near-hexagonal. The numbers refer to the D segment cut in the cross section. Molecules are tilted from the fibril axis by about 5° in the Y direction. Molecules in the Y and Z planes are staggered by 1 D, and in the X plane by 2 D. Dimensions of the unit cell *(dashed lines)* are: a = 3.90 nm, b = 2.67 nm, γ = 104.58, and height = 1 D.

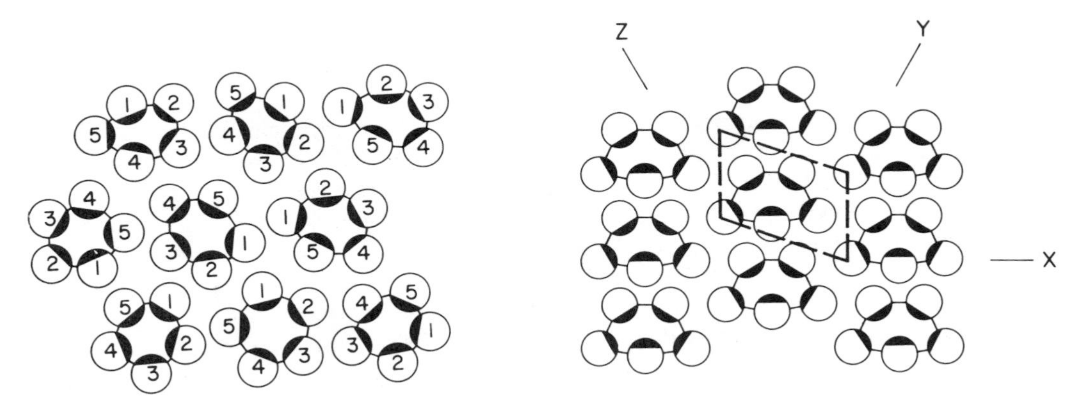

Figure 1.17. Cross-sectional views of packing of type I collagen molecules in the compressed microfibril model proposed by Piez and Trus (1981). *Left:* the normal arrangement with lateral disorder. *Right:* crystalline order under lateral compression. The fibril axis is perpendicular to the page. Molecules in the five-stranded microfibril (Figure 1.14) are supercoiled with a left-handed twist of pitch 400–700 residues (115–200 nm). The numbers indicate the D segment cut in cross section. D-segments are quasi-equivalent. Molecules are two-sided with one side *(filled sector)* directed to the inside of the microfibril. Under lateral compression, molecules become straight-tilted in their overlap regions (Figure 1.11) which, in cross section, lie close to but systematically displaced from a near-hexagonal lattice with the same unit cell as the three-dimensional crystal model (Figure 1.16).

relevant to the question of order. The technique is solid-state nuclear magnetic resonance (nmr) spectroscopy. As normally used, nmr does not give a measurable signal from solid materials like collagen fibrils. However, with the use of decoupling methods and other stratagems, signals can be obtained that give information about molecular motions in the millisecond to nanosecond time range (Torchia 1982). Using type I collagen enriched biosynthetically in ^{13}C or ^{2}H-labeled amino acids and reconstituted into native fibrils, it has been found that collagen molecules are moving about their axes at rates similar to those observed in free solution, while motion in the axial direction is fixed. The motion is probably torsional through an angle of about 30°, with a rotary diffusion coefficient greater than 10^7 s^{-1} (Jelinski et al. 1980). That is, molecules have preferred azimuthal orientations but rapid reorientation occurs about those positions. Superimposed on molecular motion is side-chain motion, again similar to what would be observed in solution. These results mean that contacts between molecules in fibrils are fluid; noncovalent bonds are constantly being made and broken. This view has had a major influence on considerations of order, as already presented, and on fibril stability (Section 1.3.3).

Another approach to studying three-dimensional fibril structure is to calculate interactions between molecules from the amino acid sequence. As already noted, this approach in one dimension was successful in obtaining a value for D (Section 1.3.1). Calculations have been extended to three dimensions, but the results are not convincing because they become model dependent as the model becomes more complex (Piez and Trus 1978). Perhaps this is not surprising in view of the nmr results discussed above, which show that these kinds of interactions are not static.

1.3.3. Stabilization

Charged and hydrophobic interactions between collagen molecules have already been discussed. It seems clear that these and other noncovalent interactions act primarily to form structure, while covalent bonds are the major source of stabilization. Direct evidence comes from studies of lathyrism induced experimentally in animals. If animals are fed a lathyrogen to inhibit the formation of lysine-derived aldehydes and thus covalent crosslinks (Chapter 3), normal-appearing collagen fibrils form, but the connective tissues of the animals are fragile and the collagen is more readily extractable. The extracted collagen behaves normally and can be reconstituted to fibrils by warming in vitro (Section 1.3.4). However, the aldehyde content is low and the fibrils dissociate on cooling. Thus, it is the inability to make covalent crosslinks that is the difference. The chemistry of collagen crosslinks is discussed in Chapter 2. Their location, and its significance in molecular packing, are pertinent here.

There is considerable chemical evidence that the major sites of crosslinking in the α1(I) chain involve lysine (or hydroxylysine) residues in the nonhelical ends at positions 9^N and 16^C and in the helical region at positions 87 and 930 (Chapter 2). Residue 930 is 4 D from 9^N in the N-terminal nonhelical region, and residue 87 is 4 D from 16^C in the C-terminal nonhelical region. (See Chapter 2 for the numbering system used for α chains.) Thus, intermolecular crosslinks 9^N–930 and 16^C–87 will stabilize a 4 D stagger between molecules, as shown in Figure 1.18. Direct evidence for this relationship between native molecules pres-

Figure 1.18. *Top:* a 4 D stagger relationship between two type I collagen molecules. *Bottom:* an enlarged view of the overlap region showing the positions of covalent crosslinks that stabilize the structure. N and C refer to the N- and C-terminal ends of the molecule.

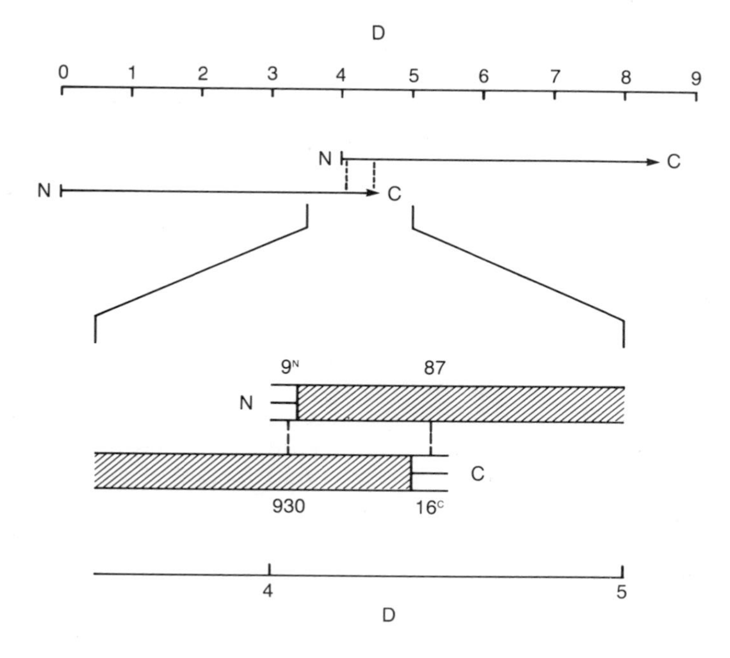

ent as aggregates in soluble type I collagen preparations has been obtained (Zimmerman et al. 1970; Silver and Trelstad 1980).

The α2(I) chain may not participate in exactly the same way as the α1(I) chain, since it apparently lacks a potential crosslinking lysine in its C-terminal nonhelical region (Chapter 2). It does, however, have similar helical crosslink sites and a nonhelical N-terminal lysine-derived aldehyde, and has been shown to participate in at least one crosslink at the 4 D position (Scott 1980). Type II collagen has helical crosslink sites homologous to those in Type I, so it probably has a similar crosslink pattern, but little information is available. Type III collagen has recently been shown to be similar to type I collagen in its crosslink pattern in that a 4 D stagger relationship is stabilized (Nicholls and Bailey 1980). Type IV collagen (Section 1.2.6) is apparently crosslinked both by lysine-derived aldehydes and disulfide bonds (Heathcote et al. 1980), but no details are known about the relationship of crosslinks to structure.

A 4 D stagger is, of course, the corollary of the 1 D stagger in the collagen packing scheme (Figure 1.11). Thus, this relationship must be an important one. Since collagen molecules have three chains, they are polyvalent and crosslinks could join many molecules into a continuous network. This apparently happens, as Light and Bailey (1980) have isolated a polymeric crosslinked material from cyanogen bromide digests of insoluble bovine tendon. Its size and chemistry suggests that neighboring molecules in the fibril must often be related by 0 D as well as 4 D, a result that places constraints on molecular packing schemes (Bailey et al. 1980). It was noted earlier that noncovalent interactions also favor a 0 D relationship as well as other D staggers. (Section 1.3.1). Both models presented above (Section 1.3.2) satisfy these constraints.

1.3.4. Assembly

Formation in vivo of types I, II, and III fibrils is a complex process involving conversion of procollagen to collagen and oxidation of lysines and hydroxylysines to aldehydes as well as assembly of molecules at the right place and time (Chapter 3). It cannot even be assumed that the collagen monomer, as an isolated species, is an intermediate. It has been suggested that a collagen aggregate formed from procollagen bundles is the assembling unit (Bruns et al. 1979) and that the cell surface is intimately involved (Trelstad and Hayashi 1979). However, these ideas are highly speculative at this time.

Because of these complexities, fibril formation in vitro has been widely studied as a means of understanding the basic principles of the self-assembly aspects. Under the right conditions, purified soluble collagen preparations in a defined solvent will re-form fibrils (see Schmitt et al. 1955 for early references). By electron microscopic (Williams et al. 1978) and x-ray diffraction criteria (Eikenberry and Brodsky 1980), these re-formed fibrils are identical to fibrils formed in vivo except that lateral order is generally not as good. Type I collagen has been the most studied. Only morphological results have been reported for types II and III collagen (Section 1.3.1). Thus, the collagen molecule contains all the information necessary to make fibrils. Of course, in vivo, regulation may well require the participation of other components.

Fibril formation in vitro is usually brought about by adding buffer to a cold, acidic solution of collagen to give a pH and ionic strength near physiological values, and then raising the temperature to between about 25 and 37°C. Fibril formation is generally followed by absorbance at a convenient wavelength. As fibrils assemble, a turbid suspension or gel forms in which absorbance at some convenient wavelength is assumed to be proportional to amount of product. A weakness of this assay is that it is not completely clear what aspect of assembly is being measured. A typical turbidity curve is shown in Figure 1.19. It is characterized by a lag phase with no absorbance change, a sigmoidal growth phase, and a plateau which may increase slowly as fibril formation is completed.

Other physical chemical methods have been used to study the lag phase, including viscosity, static and dynamic light scattering, and electron microscopy. A variety of studies (see Comper and Veis 1977a; Williams et al. 1978) have shown that the rate is sensitive to pH, ionic strength, temperature, buffer composition, the presence of added components, and the particular collagen preparation. Differences in the collagen preparations are probably the greatest source of irreproducibility. A further difficulty is that the product of assembly need not be the native fibril; a variety of polymorphic forms exist (Section 1.4). The sensitivity of the system has made it difficult to compare different studies and arrive at general conclusions. However, a beginning understanding of assembly has been achieved, even though details are controversial.

Because of the shape of the turbidity curve, assembly has often been treated formally as a nucleation and growth process (Comper and Veis 1977a). Such a process is characterized by a lag phase during which monomer and a small amount of aggregate are in a state of dynamic exchange. After a time, some aggregates become large enough (nuclei) to grow irreversibly until an equilibrium

Figure 1.19. Assembly of type I collagen in vitro to native fibrils. Rat tail tendon collagen, 0.1 mg/ml in buffer at pH 7.4 and 0.225 ionic strength. *Dashed line:* normal collagen. *Solid line:* reduced collagen. Assembly is initiated by raising the temperature from 4 to 26°C. Cooling to 4°C after assembly is complete does not reverse assembly of normal collagen since it spontaneously forms covalent crosslinks, but does reverse assembly of reduced collagen. Rewarming initiates reassembly without a lag period. *(Source: Redrawn, Gelman, Williams, et al. (1979) J Biol Chem 254:180–186, with permission.)*

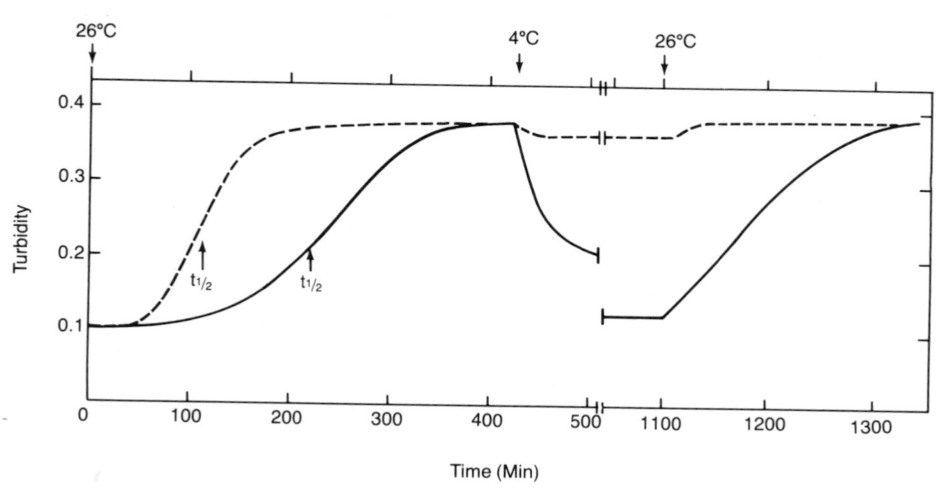

is reached between monomer and, in this case, fibril. However, this mechanism cannot be the entire explanation for several reasons. First of all, the putative nuclei appear to be stable to cooling in that once formed and cooled, rewarming produces fibrils without a lag phase (Figure 1.19). An irreversible conformational change involving the nonhelical ends has been suggested as the explanation for these stable intermediates (Gelman et al. 1979a, 1979b; Helseth and Veis 1981). Second, extensive aggregation appears to occur during the lag phase, but it is not agreed whether the aggregates are small (Silver 1981) or large (Gelman and Piez 1980), or even if they are significant (Helseth and Veis 1981). Third, if the collagen is normal acid-extracted collagen, it will contain aldehydes and will spontaneously crosslink during assembly. This will make at least the final stage irreversible, drive the process to completion, and change the kinetics. Rather than nucleation and growth, in vitro assembly may be better described as a multistep accretion process (Gelman et al. 1979a). Aspects of both mechanisms may be present.

It is agreed that the nonhelical ends of type I collagen are critical to assembly (Comper and Veis 1977b; Gelman et al. 1979b; Helseth and Veis 1981), presumably through noncovalent interactions with the helical body of an adjacent molecule. These interactions may initiate an early 4 D stagger relationship in assembly (Silver 1981) that is later stabilized by covalent crosslinks (Section 1.3.3).

1.3.5. Fibril Organization

Collagen fibrils provide the major source of mechanical strength for all connective tissues in vertebrates. This is obvious for tissues that consist largely (in the dry state) of collagen, such as skin and tendon, but is also true for tissues such as bone, elastic ligament, and cartilage, which contain large amounts of another material. The only exception is the enamel of the teeth, which contains no collagen (Chapter 9). In all cases, it is the tensile strength of differently organized fibrils that is utilized. How these different organizations are achieved is a major unanswered question in biology. In tendon, fibrils are parallel in flexible bundles to transmit force in one direction and around the corners formed by joints. In skin, layers and bundles of fibrils extending in different directions but generally parallel to the surface resist stretch and tear in two directions. In bone, a similar arrangement in three dimensions provides resistance to bending and stretching to an otherwise brittle mineral. In elastic ligaments, a deformable meshwork of collagen fibrils around elastic fibers limits stretch, which would otherwise extend the fiber to its breaking point. In cartilage, single fibrils in random directions maintain form in a proteoglycan matrix that would otherwise swell and deform under osmotic and compressive forces.

Collagen fibrils vary in diameter from 16 nm in teleost cornea and some fetal tissues (Craig and Parry 1981) to more than several hundred nanometers in some tendons (Parry et al. 1978). In some tissues such as cornea, fibril diameter is very regular, while in others diameters cover a wide distribution and may be bimodal. Electron micrographs of cornea and tendon are shown in Figure 1.20.

An often overlooked but significant aspect of fibril organization is that wherever bundles of fibrils occur, they often follow a planar wavy path referred to as a crimp. Rat tail tendon has been studied the most (Kastelic et al. 1978;

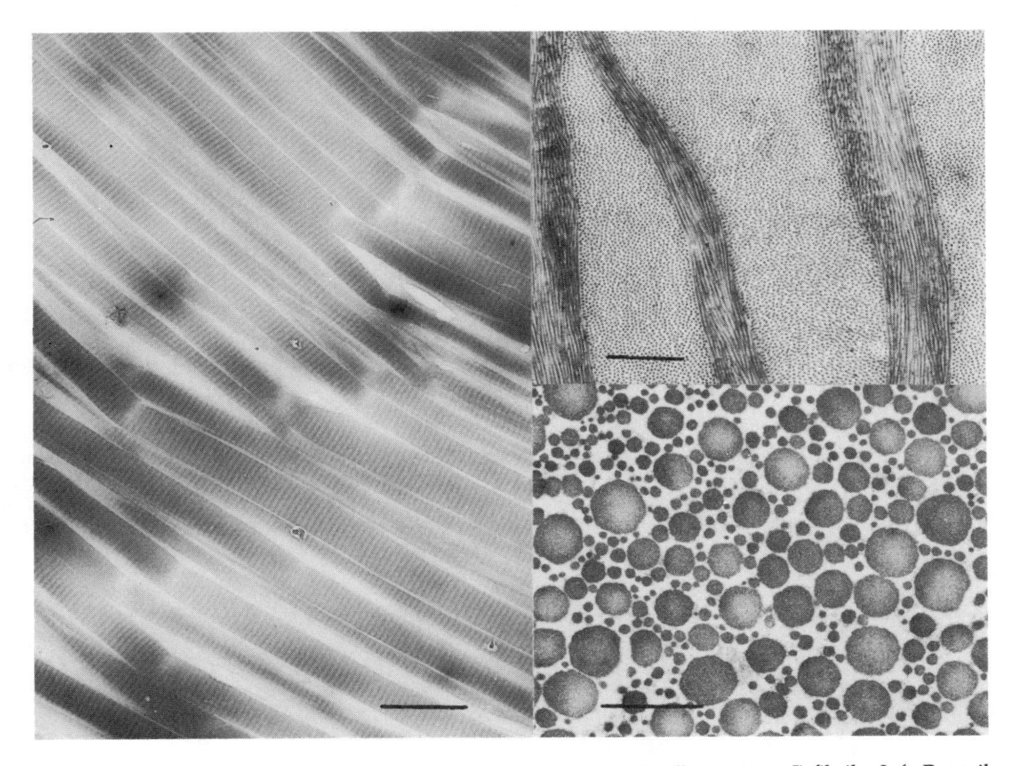

Figure 1.20. Electron micrographs of embedded and sectioned collagen (type I) fibrils. *Left:* Rat tail tendon in the crimp region showing distortion at the bends in the fibrils *(lower left to upper right)*. Uranyl acetate stained. The bar is 1.0 μm. (Courtesy of L. J. Gathercole.) *Upper right:* Magpie cornea showing orthogonal lamellae made up of many fibrils 25 nm in diameter. Uranyl acetate and lead citrate stained. The bar is 1.0 μm. (Courtesy of D. A. D. Parry.) *Lower right:* Horse suspensory ligament in cross section showing a biphasic distribution of fibril diameters. The largest fibrils are 250 nm in diameter. The bar is 0.5 μm. (Courtesy of D. A. D. Parry.)

Gathercole and Keller 1978). The distance between crimps (half-waves) is of the order of 100 μm and has been studied mostly by optical microscopy. Electron micrographs of local regions at the crimp have also been taken (Dlugosz et al. 1978). One is shown in Figure 1.20. The crimp is quite sharp, extending over only a few D periods in individual fibrils and distorting the fibril in that region. The function of the crimp is presumably to take up the initial shock of an applied stress by virtue of the low modulus inherent in such a structure. It may also function to allow a fibril bundle to make gradual bends in any direction without major strain, the crimp angle being reduced on the outside and increased on the inside of the bend.

In some tissues, collagen plays other physical roles of a special nature. Perhaps the most obvious example is the cornea, where the regular size and distribution of the fibrils is related to transparency, although the exact mechanism is uncertain. The filtration properties of the basement membrane is another example. A third possible example is epitactic nucleation of bone mineral (Chapter 9). Collagen fibrils also have pyroelectric, piezoelectric, and optical anisotropic properties that could have important biological consequences (see Roth and Freund 1981).

Beyond the scope of this chapter, but considered in Chapters 7, 10, and 11, are the interactions of collagen with cells. Fibril organization is certainly a part of these important interactions, but in ways not understood.

1.4. Polymorphic Ordered Aggregates

1.4.1. Segment-Long-Spacing Crystallites

As first observed by Schmitt et al. (1953), when soluble collagen is mixed with a polyvalent anion under acidic conditions, a precipitate forms which in the electron microscope shows aggregates about 300 nm long but of variable width. These aggregates were designated segment-long-spacing (SLS) crystallites (Fig-

Figure 1.21. Electron micrographs of segment-long-spacing (SLS) crystallites of type I collagen from rat tail tendon. *Left:* Positively stained with phosphotungstate and uranyl acetate. *Lower right:* Negatively stained with phosphotungstic acid. *Upper right:* Reacted with an osmium reagent to show carbohydrate *(small arrow)*, excess adenosine triphosphate (ATP) giving a light negative stain also. The arrows show the C-terminal direction of collagen molecules, 300 nm long.

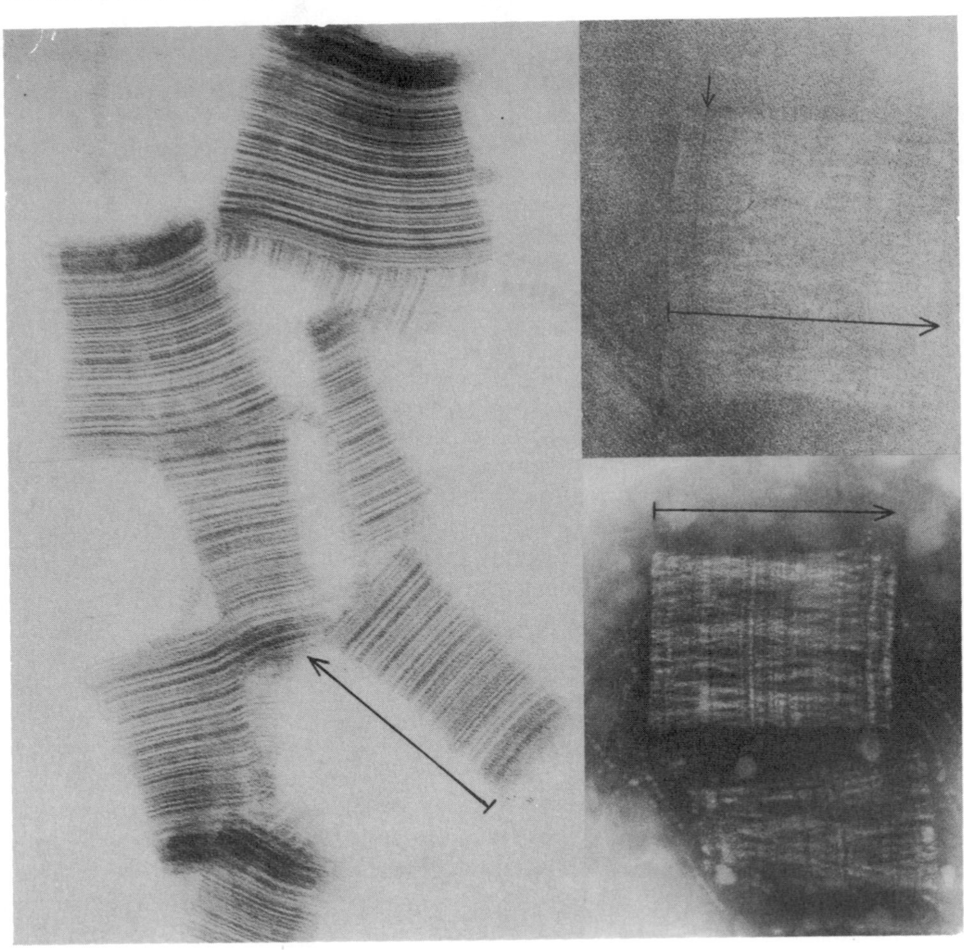

ure 1.21). Positively stained specimens show a characteristic and reproducible band pattern perpendicular to the 300 nm dimension. The interpretation (Schmitt et al. 1955) is that the aggregates are bundles of parallel collagen molecules with ends in register. The bundles may be small or large, the larger ones spreading out to form ribbons. Occasionally, bundles stand on end. The band pattern represents the distribution of charged groups along the molecule which bind heavy metal stain. The charge clusters along individual molecules are seen as a band across the SLS aggregate since molecules are aligned. As noted earlier (Section 1.3.1), SLS aggregates were important in understanding native fibril structure by comparison of the band patterns.

Electron micrographs of positively and negatively stained SLS made from type I collagen are shown in Figure 1.21. It can be easily seen in negatively stained specimens that the bundles are composed of long, thin structures which are molecules. The band pattern of positively stained SLS has been compared to the sequence and to known features of the type I collagen molecule (Bruns and Gross 1973; Piez 1976). As shown in Figure 1.22, the agreement between the sequence and the pattern is excellent.

Figure 1.22. AA segment-long-spacing (SLS) crystallite of type I collagen compared to structural features along the molecule. *From top down:* The distribution of methionine residues in the α1(I) and α2(I) chains; the distribution of charged residues in the α1(I) chain (the α2(I) chain is nearly identical); electron micrograph of a positively stained SLS crystallite (courtesy of K. Kühn); the band numbering system of Bruns and Gross (1973); and a schematic view of the type I collagen molecule, 300 nm long.

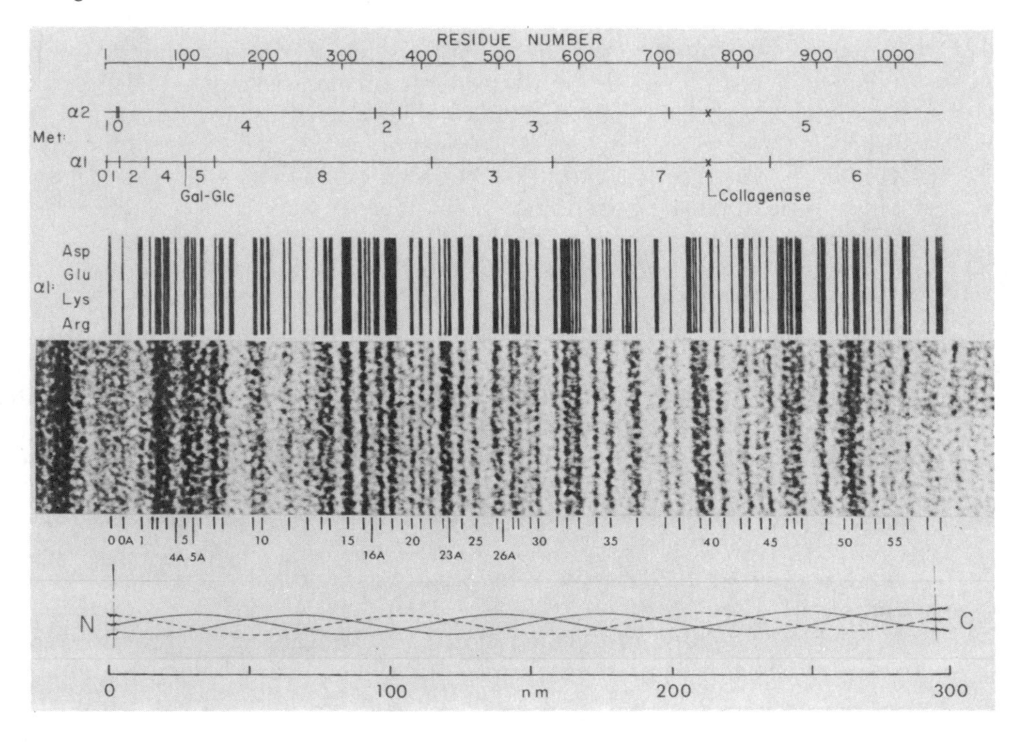

Other features of the collagen molecule can be revealed by stains specific for certain reactive groups. The stain forms a band across the SLS aggregate at each position of the reactive group along the molecule. This has been done for methionine and for carbohydrate (Beer et al. 1979). The methionine band pattern is very similar to prediction from the sequence (Chapter 2). An SLS stained with an osmium reagent for carbohydrate is shown in Figure 1.21. A single dark band locates a major position of hydroxylysine-linked carbohydrate in α1(I) (Figure 1.22 and Chapter 2).

Segment-long-spacing crystallites are usually made by dialyzing adenosine triphosphate (ATP) into a dilute, salt-free acidic solution of collagen. Under these conditions, the collagen is positively charged and the ATP bridges those charges, with maximal neutralization occurring when molecules are in register. Although ATP is convenient to use, other polyvalent anions such as triphosphate serve as well. Native fragments of the collagens will also make short SLS aggregates. This has aided the identification, by matching band patterns, of animal collagenase fragments (Chapter 4) and cyanogen bromide peptides (see Traub and Piez 1971), for example. These and other uses of SLS crystallites have been reviewed by Kühn (1982).

Types I, II, and III collagen make SLS aggregates with very similar band patterns. There are, however, characteristic differences undoubtedly related to differences in sequence.

1.4.2. D-Periodic Symmetric Fibrils

Under various conditions not necessarily very different from those used to reconstitute native fibrils (Section 1.3.3), fibrils may be obtained that are D-periodic but symmetric (DPS). They are sometimes called centrosymmetric fibrils. Four types have been observed by electron microscopy, designated DPS I, II, III, and IV (Doyle et al. 1974b; Bruns 1976; Williams et al. 1978). The former two are reported to be formed by type II collagen and the latter two by type I collagen. However, type II can also form DPS III, or a very similar aggregate (Figure 1.23). Since these fibrils are D-periodic, they can easily be confused with normal fibrils in electron micrographs if resolution is poor or magnification low.

The structures of these fibrils have been explained (Doyle et al. 1974b; Piez and Miller 1974) as arising from normal asymmetric D-periodic substructural units that are oppositely oriented but too small (narrow) to be seen. The pattern is then the sum of the two antiparallel patterns. The different DPS types are explained by different staggers between the substructural units. It is indeed possible to reconstruct the band pattern on this assumption. Examples of DPS III are shown in Figure 1.23. It is not possible to reconstruct any of the patterns simply by reversing the direction of half the molecules in a native fibril. The substructure was believed to be the five-stranded microfibril, but this conclusion is in some doubt in view of the uncertain nature of the structure of the native fibril (Section 1.3.2). However, some kind of native fibrillar substructure is strongly indicated.

Further evidence for this conclusion is that DPS and normal banding are often seen in the same fibril, the transition usually occurring across the fibril. Figure 1.23 shows a transition from DPS III to normal banding in reconstituted type II

Figure 1.23. Electron micrographs of reconstituted fibrils negatively stained with phosphotungstic acid. The large fibril reconstituted from type II collagen (rat chondrosarcoma) shows a transition from DPS III *(lower left side)* to normal banding *(upper right side)*. (Courtesy of S. L. Lee.) *Inset:* A type I collagen fibril (rat tail tendon) also showing the DPS III pattern. (Courtesy of B. W. Williams.) The lines (arrows show the C-terminal direction) show how the DPS III pattern arises from a normal D-periodic substructure arranged antiparallel. Different staggers explain the different DPS forms.

collagen. A DPS III to DPS IV transition has also been seen (Williams et al. 1978). The implication is that the D-periodic substructure prefers the parallel arrangement during assembly, which results in native fibrils, but is occasionally trapped (kinetically) into a slightly less favorable antiparallel relationship. Once started, it continues to form at least in the axial direction.

By sequence analysis, the antiparallel relationship in DPS III has been shown to be related to a maximization of hydrophobic interactions between staggered antiparallel molecules (Doyle et al. 1975). The other DPS forms may arise in a similar way from chance relationships. In some cases, narrow native fibrils of type II collagen (subfibrils) may associate laterally with an alternating antiparallel relationship as in DPS I or DPS II, giving a checkerboard pattern (Doyle et al. 1974a; Bruns 1976).

1.4.3. Obliquely Banded Fibrils

Type II collagen dialyzed first against 0.15 *M* NaCl at acid pH in the cold, and then at neutral pH and room temperature, precipitates as obliquely banded fibrils (Bruns et al. 1973). Careful examination shows that the pattern is actually a step pattern formed by narrow native fibrils (subfibrils) side-by-side and parallel, but axially displaced by about 10 nm (Bruns 1976). The step is slightly variable in height. An example is shown in Figure 1.24 in which the subfibril width is about

Figure 1.24. Reconstituted obliquely-banded fibrils. The major periodicity is 67 nm in the vertical direction. *Left:* A type II collagen fibril with native subfibrils about 20 nm wide and stepped by about 10 nm. Negatively stained with silicotungstate. (Courtesy of R. Bruns.) *Right:* A type I collagen fibril with a similar band pattern composed of subfibrils about 10 nm wide seen best on the left side. Overlying patterns on the right side produce a D/6 periodicity. Negatively stained with phosphotungstate. (Courtesy of B. W. Williams.)

20 nm. The fibrils usually appear to be ribbons; when folded over the angle of the band reverses. Sometimes, oblique bands can be seen in both directions, suggesting folded sheets or flattened helical tubes.

Doyle et al. (1974a) found the subfibril width to be variable from about 4 to 40 nm. The 4 nm width is too narrow to be directly observed, but can be calculated from the angle of the oblique bands using optical diffraction. An earlier example made from type I collagen was found in the literature (Kühn et al. 1964). Since 4 nm is the predicted diameter of a five-fold microfibril and the structure is D-periodic, this observation was used to support the microfibril model, as previously discussed (Section 1.3.2). Although this interpretation is in doubt since the microfibril model is uncertain, an alternative explanation has yet to be made.

Obliquely banded fibrils can also be made from type I collagen from rat tail tendon by dialysis of an acidic solution against a low ionic strength citrate buffer at pH 6.5 (Piez KA, unpublished data). The product is mixed native and obliquely banded fibrils. They may also appear under conditions similar to those used to prepare native fibrils (Williams et al. 1978), as shown in Figure 1.24.

Under some conditions where the subfibril width is small, the negative staining is light, and the axial displacement is an integral submultiple of D, an obliquely banded fibril will be dominated by a periodicity of 10–11 nm (D/6, or possibly D/7). An example is shown in Figure 1.24. Sometimes the oblique bands are not visible, perhaps because the displacements are random multiples of D/6 rather than regular steps. Originally observed by Kühn et al. (1964), these fibrils have been called D/6-periodic fibrils (Doyle et al. 1975).

1.4.4. Nonbanded Fibrils

Under a variety of conditions, fibrils are seen in electron micrographs that show no banding. In principle, this could occur either because periodicity is present but not visible for some reason or because it is truly absent. The true absence of periodicity is obviously difficult to prove and no examples are known. The apparent absence of periodicity is readily explained, for example, if the periodic structure is too narrow for it to be seen in the usual electron micrographs. An example of reconstituted thin filaments without a visible period by electron microscopy but D-periodic by x-ray diffraction has already been cited above (Section 1.3.2). Another possible example is fibrils formed from pepsin-treated collagen (Gelman et al. 1979b). The product is ropy fibrils composed of thin filaments without a visible period. However, an occasional faint D-period suggests that the filaments are D-periodic but filament alignment is too poor for it to be visible. In this sense, at least some nonbanded fibrils fall in the same class as DPS and obliquely banded fibrils in that they also contain a D-periodic substructure abnormally aligned.

1.4.5. Fibrous-Long-Spacing Aggregates

This class of fibril-like aggregates are very different from the forms discussed above that are based on a D-periodic substructure. The fibrous-long-spacing (FLS) aggregates have longer periods and contain other material. They are usually made by mixing an acidic collagen solution (type I) with a glycosaminoglycan preparation and dialyzing against water. They were discovered by Highberger et al. (1950).

There are four forms, designated FLS I, II, III, and IV. Their structures were worked out by Kühn and Zimmer (1961), Chapman and Armitage (1972), and Doyle et al. (1975) from positively and negatively stained electron micrographs by reconstruction of the band pattern from the SLS pattern. Examples of FLS I and IV are shown in Figure 1.25. The formal structural unit is a pair of aligned but oppositely oriented collagen molecules. This unit is overlapped in various ways. The fundamental overlaps are about 50 nm (FLS I) and 130 nm (FLS IV), as illustrated in Figure 1.25. The other FLS forms are combinations of these, as summarized by Doyle et al. (1975). As with the DPS fibrils, FLS fibrils are often seen with transitions from one type to another, but along the fibril, as there is presumably no ordered filamentous substructure.

These solutions are, of course, one-dimensional projected structures. Whether molecules are actually paired, or simply have that projected relationship, is not known. That they can pair, however, is shown by the observation of a symmetrical SLS aggregate (Chapman and Armitage 1972) formed under similar conditions to the FLS aggregates.

An explanation of why these forms occur was suggested by Doyle et al. (1975). While most charges along the type I collagen molecule are paired (Section 1.3.1), the net positive charge of collagen leaves some unpaired positive charges. Calculations of interactions between them, assuming bridging by a polyanion, shows maximal interactions at just the overlaps needed to make the FLS aggregates. Furthermore, the unpaired positive charges have an approximate center of sym-

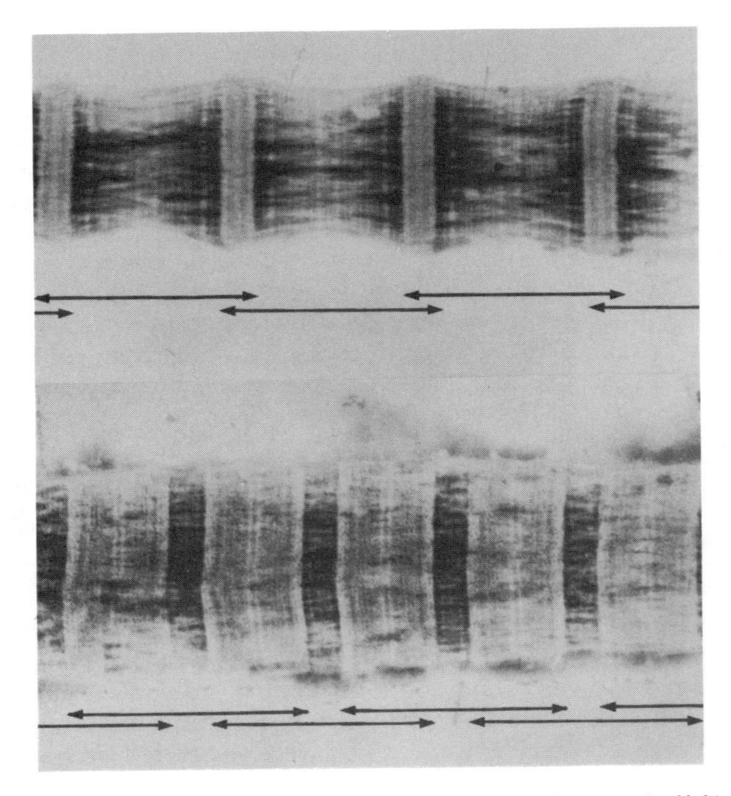

Figure 1.25. Fibrous-long-spacing (FLS) fibrils made from type I calf skin collagen and a chondroitin sulfate glycosaminoglycan showing FLS I *(top)* and FLS IV *(bottom)*, negatively stained with phosphotungstate. The double-headed arrows represent antiparallel pairs of molecules overlapped and bridged with glycosaminoglycan to form the patterns. (Courtesy of J. A. Chapman.)

metry in the collagen molecule, apparently by chance. Thus, the interactions are equivalent regardless of molecular direction; so both are used. Glycosaminoglycan is incorporated into the fibril in an ordered way. These are bridged structures like SLS, but unlike native, DPS, and obliquely banded fibrils, which contain only collagen.

The FLS aggregates, like the DPS fibrils, have not been found in vivo. However, FLS aggregates may form under some culture conditions where a polyanion has been added to the medium (*See* Merker et al. 1978).

1.5. Perspectives

From the review presented here it is evident that the collagen triple helix is one of the fundamental conformations assumed by polypeptide chains akin to the α helix and the β pleated sheet. In particular, its antiquity on the evolutionary scale and broad distribution attest to its importance. However, proteins consisting largely of the triple helix—the collagens—are unique in that they are always a part of extracellular supramolecular structures. No intracellular protein containing triple helix has yet been identified.

The collagens fall into two groups, the fibrillar or interstitial collagens (types I, II, and III) and the basement membrane collagens (type IV). Whether type V should be included with type IV or forms still another group is uncertain. The largely mechanical role of these proteins is understood in a general sense, although some details such as molecular packing and the differences between types I, II, and III collagen remain to be revealed. In contrast, two other topics relating to collagen structure, the control of supramolecular assembly and the interactions of collagenous structures with other connective tissue components, are only beginning to be understood. With the outstanding exception of fibronectin (Chapter 7), it has only been possible to mention these problems here (or elsewhere in this book) because there is so little to say. This is unfortunate since it is becoming increasingly clear that the structure of the extracellular matrix is no less important than intracellular structure for understanding development and disease. Indeed, the two areas increasingly overlap in studies of biological structure.

Thus, continued emphasis on biological structure is warranted. Advances will as always depend on improved or new techniques. Of the major methods referred to in this chapter—x-ray diffraction, electron microscopy, computer model building, and nuclear magnetic resonance—minor improvements can be expected which will slowly advance understanding. While major improvements or completely new techniques do not seem imminent, they are seldom recognized as such until applied. Past experience says they will come.

References

Anglister L, Silman I: (1978) Molecular structure of elongated forms of electric eel acetylcholinesterase. J Mol Biol 125:293–311.

Bächinger HP, Bruckner P, Timpl R, Prockop DJ, Engel J: (1980) Folding mechanism of the triple helix in type-III collagen and type-III pN collagen. Eur J Biochem 106:619–632.

Bailey AJ, Light ND, Atkins EDT: (1980) Chemical cross-linking restrictions on models for the molecular organization of the collagen fibre. Nature 288:408–410.

Bear RS: (1952) The structure of collagen fibrils. Adv Prot Chem 7:69–160.

Beer M, Wiggins JW, Alexander R, Schettino R, Stockert C, Piez K: (1979) Electron microscopy of selectively stained collagen. In: Bailey GW, ed. 37th Annual Proceedings of the Electron Microscopy Society of America. San Antonio: Electron Microscopy Society of America, pp. 28–29.

Berg RA, Prockop DJ: (1973) The thermal transition of a non-hydroxylated form of collagen. Evidence for a role for hydroxyproline in stabilizing the triple-helix of collagen. Biochem Biophys Res Commun 52:115–120.

Bhatnagar RS, Rapaka RS: (1976) Synthetic polypeptide models of collagen: Synthesis and applications. In: Ramachandran GN, Reddi AH, eds. Biochemistry of Collagen. New York: Plenum, pp. 479–523.

Boedtker H, Doty P: (1956) The native and denatured states of soluble collagen. J Am Chem Soc 78:4267–4280.

Bon S, Vigny M, Massoulie J: (1979) Asymmetric and globular forms of acetylcholinesterase in mammals and birds. Proc Natl Acad Sci USA 76:2546–2550.

Bornstein P, Traub W: (1979) The chemistry and biology of collagen. In: Neurath H, Hill RL, eds. The Proteins. Volume IV. New York: Academic Press, pp. 412–632.

Bruns RR: (1976) Supramolecular structure of polymorphic collagen fibrils. J Cell Biol 68:521–538.

Bruns RR, Gross J: (1973) Band pattern of the segment-long-spacing form of collagen. Its use in the analysis of primary structure. Biochemistry 12:808–815.

Bruns RR, Hulmes DJS, Therrien SF, Gross J: (1979) Procollagen segment-long-spacing crystallites: Their role in collagen fibrillogenesis. Proc Natl Acad Sci USA 76:313–317.

Bruns RR, Trelstad RL, Gross J: (1973) Cartilage collagen: A staggered substructure in reconstructed fibrils. Science 181:269–271.

Carver JP, Blout ER: (1967) Polypeptide models for collagen. In: Ramachandran GN, ed. Treatise on Collagen. Volume 1: Chemistry of Collagen. London: Academic Press, pp. 441–523.

Chapman JA, Armitage PM: (1972) An analysis of fibrous long spacing forms of collagen. Conn Tissue Res 1:31–37.

Comper WD, Veis A: (1977a) The mechanism of nucleation for in vitro collagen fibril formation. Biopolymers 16:2113–2131.

Comper WD, Veis A: (1977b) Characterization of nuclei in vitro collagen fibril formation. Biopolymers 16:2133–2142.

Cowan PM, McGavin S, North ACT: (1955) The polypeptide chain configuration of collagen. Nature 176:1062–1064.

Craig AS, Parry DAD: (1981) Collagen fibrils of the vertebrate corneal stroma. J Ultrastruct Res 74:232–239.

Di Blasi R, Verdini AA: (1979) ^{13}C Nuclear magnetic resonance study of the triple helix random-coil transition of $(prolylprolylglycyl)_{10}$. Biopolymers 18:735–738.

Dlugosz J, Gathercole LJ, Keller A: (1978) Transmission electron microscope studies and relation to polarizing optical microscopy in rat tail tendon. Micron 9:71–82.

Doyle BB, Hukins DWL, Hulmes DJS, Miller A, Woodhead-Galloway J: (1975) Collagen polymorphism: Its origins in the amino acid sequence. J Mol Biol 91:79–99.

Doyle BB, Hulmes DJS, Miller A, Parry DAD, Piez KA, Woodhead-Galloway J: (1974a) A D-periodic narrow filament in collagen. Proc R Soc Lond [Biol] 186:67–74.

Doyle BB, Hulmes DJS, Miller A, Parry DAD, Piez KA, Woodhead-Galloway J: (1974b) Axially projected collagen structures. Proc R Soc Lond [Biol] 187:37–46.

Eikenberry EF, Brodsky B: (1980) X-ray diffraction of resconstituted collagen fibers. J Mol Biol 144:397–404.

Engel J, Chen H-T, Prockop DJ, Klump H: (1977) The triple helix-coil conversion of collagen-like polytripeptides in aqueous and nonaqueous solvents. Comparison of the thermodynamic parameters and the binding of water to $(L\text{-}Pro\text{-}L\text{-}Pro\text{-}Gly)_n$ and $(L\text{-}Pro\text{-}L\text{-}Hyp\text{-}Gly)_n$. Biopolymers 16:601–622.

Fraser RDB, MacRae TP: (1973) Conformation in Fibrous Proteins. New York: Academic Press.

Fraser RDB, MacRae TP: (1981) Unit cell and molecular connectivity in tendon collagen. Int J Biol Macromol 3:193–200.

Fraser RDB, MacRae TP, Suzuki E: (1979) Chain conformation in the collagen molecule. J Mol Biol 129:463–481.

Gathercole LJ, Keller A: (1978) Early development of crimping in rat tail tendon collagen: A polarizing optical and SEM study. Micron 9:83–89.

Gelman RA, Piez KA: (1980) Collagen fibril formation in vitro. A quasielastic light-scattering study of early stages. J Biol Chem 255:8098–8102.

Gelman RA, Poppke DC, Piez KA: (1979b) Collagen fibril formation in vitro. The role of the nonhelical terminal regions. J Biol Chem 254:11741–11745.

Gelman RA, Williams BR, Piez KA: (1979a) Collagen fibril formation. Evidence for a multistep process. J Biol Chem 254:180–186.

Grant ME, Heathcote JG, Orkin RW: (1982) Current concepts of membrane structure and function. Biosci Rep 1:819–842.

Grigera JR, Berendsen HJC: (1979) The molecular details of collagen hydration. Biopolymers 18:47–57.

Gustavson KH: (1956) The Chemistry and Reactivity of Collagen. New York: Academic Press.

Hall CE: (1956) Visualization of individual macromolecules with the electron microscope. Proc Natl Acad Sci USA 42:801–804.

Hayashi T, Nagai Y: (1973) Effect of pH on the stability of collagen molecule in solution. J Biochem 73:999–1006.

Heathcote JG, Bailey AJ, Grant ME: (1980) Studies on the assembly of the rat lens capsule. Biochem J 190:229–237.

Helseth DL Jr, Veis A: (1981) Collagen self-assembly in vitro. J Biol Chem 256:7118–7128.

Highberger JH, Gross J, Schmitt FO: (1950) Electron microscopic observations of certain fibrous structures obtained from connective tissue extracts. J Am Chem Soc 72:3321–3322.

Hodge AJ, Petruska JA: (1963) Recent studies with the electron microscope on ordered aggregates of the tropocollagen molecule. In: Ramachandran GN, ed. Aspects of Protein Structure. London: Academic Press, pp. 289–300.

Hodge AJ, Schmitt FO: (1960) Charge profile of the tropocollagen macromolecule and the packing arrangement in native fibrils. Proc Natl Acad Sci USA 46:186–197.

Hofmann H, Fietzek PP, Kühn K: (1980) Comparative analysis of the sequences of the three collagen chains α1(I), α2(I) and α1(III). Functional and genetic aspects. J Mol Biol 141:293–313.

Hulmes DJS, Jesior J-C, Miller A, Berthet-Colominas C, Wolff C: (1981) Electron microscopy shows periodic structure in collagen fibril cross sections. Proc Natl Acad Sci USA 78:3567–3571.

Hulmes DJS, Miller A: (1979) Quasi-hexagonal molecular packing in collagen fibrils. Nature 282:878–880.

Hulmes DJS, Miller A, Parry DAD, Piez KA, Woodhead-Galloway J: (1973) Analysis of the primary structure of collagen for the origins of molecular packing. J Mol Biol 79:137–148.

Hulmes DJS, Miller A, White SW, Timmins PA, Berthet-Colominas C: (1980) Interpretation of the low-angle meridional neutron diffraction patterns from collagen fibres in terms of the amino acid sequence. Int J Biol Macromol 2:338–346.

Jelinski LW, Sullivan CE, Torchia DA: (1980) ^{2}H NMR study of molecular motion in collagen fibrils. Nature 284:531–534.

Kastelic J, Galeski A, Baer E: (1978) The multicomposite structure of tendon. Conn Tissue Res 6:11–23.

Kefalides NA, Alper R, Clark CC: (1979) Biochemistry and metabolism of basement membranes. Int Rev Cytol 61:167–228.

Kühn K: (1982) Segment-long-spacing crystallites, a powerful tool in collagen research. Collagen Rel Res 2:1–20.

Kühn K, Kühn J, Schuppler G: (1964) Kollagen fibrillen mit abnormalen Querstreifungsmuster. Naturwissenschaften 51:337.

Kühn K, von der Mark K: (1978) The influence of proteoglycans on the macromolecular structure of collagen. In: Gastpar H, ed. Collagen–Platelet Interaction. Stuttgart: FK Schattauer Verlag, pp. 123–126.

Kühn K, Wiedemann H, Timpl R, Ristelli J, Dieringer H, Voss T, Glanville RR: (1981) Macromolecular structure of basement membrane collagen. FEBS Lett 125:123–128.

Kühn K, Zimmer E: (1961) Eigenschaften des Tropocollagen-Moleküles und deren Bedeutung für die Fibrillenbildung. Zeit für Naturforschung 16:648–658.

Light ND, Bailey AJ: (1980) Polymeric C-terminal cross-linked material from type-I collagen. Biochem J 189:111–124.

Mays C, Rosenberry TL: (1981) Characterization of pepsin-resistant collagen-like tail subunit fragments of 18S and 14S acetylcholinesterase from *electrophorus electricus*. Biochemistry 20: 2810–2817.

Meek KM, Chapman JA, Hardcastle RA: (1979) The staining pattern of collagen fibrils. J Biol Chem 254:10710–10714.

Merker H-J, Lilja S, Barrach HJ, Gunter Th: (1978) Formation of an atypical collagen and cartilage pattern in limb bud cultures by highly sulfated GAG. Virchows Archiv A 380:11–30.

Miller A: (1976) Molecular packing in collagen fibrils. In: Ramachandran GN, Reddi AH, eds. Biochemistry of Collagen. New York: Plenum, pp. 85–136.

Miller A, Tocchetti D: (1981) Calculated x-ray diffraction pattern from a quasi-hexagonal model for the molecular arrangement in collagen. Int J Biol Macromol 3:9–18.

Miller EJ, Martin GR, Piez KA, Powers MJ: (1967) Characterization of chick bone collagen and compositional changes associated with maturation. J Biol Chem 242:5481–5489.

Nicholls AC, Bailey AJ: (1980) Identification of cyanogen bromide peptides involved in intermolecular cross-linking of bovine type III collagen. Biochem J 185:195–201.

Okuyama K, Okuyama K, Arnott S, Takayanagai M, Kakudo M: (1981) Crystal and molecular structure of a collagen-like polypeptide $(Gly\text{-}Pro\text{-}Pro)_{10}$. J Mol Biol 152:427–443.

Parry DAD, Barnes GRG, Craig AS: (1978) A comparison of the size distribution of collagen fibrils in connective tissues as a function of age and a possible relation between fibril size distribution and mechanical properties. Proc R Soc Lond [Biol] 203:305–321.

Parry DAD, Craig AS: (1979) Electron microscope evidence for an 80 A unit in collagen fibrils. Nature 282:213–215.

Piez KA: (1967) Soluble collagen and the components resulting from its denaturation. In: Ramachandran GN, ed. Treatise on Collagen. Volume 1: Chemistry of Collagen. London: Academic Press, pp. 201–248.

Piez KA: (1976) Primary structure. In: Ramachandran GN, Reddi AH, eds. Biochemistry of Collagen. New York: Plenum, pp. 1–44.

Piez KA, Miller A: (1974) The structure of collagen fibrils. J Supramol Struct 2:121–137.

Piez KA, Sherman MR: (1970) Characterization of the product formed by renaturation of α1-CB2, a small peptide from collagen. Biochemistry 9:4129–4133.

Piez KA, Torchia DA: (1975) Possible contribution of ionic clustering to molecular packing of collagen. Nature 258:87.

Piez KA, Trus BL: (1978) Sequence regularities and packing of collagen molecules. J Mol Biol 122:419–432.

Piez KA, Trus BL: (1981) A new model for molecular packing of type I collagen molecules in native fibrils. Biosci Rep 1:801–810.

Ramachandran GN: (1967) Structure of collagen at the molecular level. In: Ramachandran GN, ed. Treatise on Collagen, Volume 1: Chemistry of Collagen. London: Academic Press, pp. 103–179.

Ramachandran GN: (1976) Molecular structure. In: Ramachandran GN, Reddi AH, eds. Biochemistry of Collagen. New York: Plenum, pp. 45–84.

Ramachandran GN, Kartha G: (1954) Structure of collagen. Nature 174:269–270.

Ramachandran GN, Kartha G: (1955) Structure of collagen. Nature 176:593–595.

Rao NV, Adams E: (1979) Collagen helix stabilization by hydroxyproline in $(\text{Ala-Hyp-Gly})_n$. Biochem Biophys Res Commun 86:654–660.

Reid KBM: (1979) Complete amino acid sequences of the three collagen-like regions present in subcomponent Clq of the first component of human complement. Biochem J 179:367–371.

Reid KBM, Porter RR: (1981) The proteolytic activation systems of complement. Ann Rev Biochem 50:433–464.

Rich A, Crick FHC: (1955) The structure of collagen. Nature 176:915–916.

Rigby BJ: (1968) Thermal transitions in some invertebrate collagens and their relation to amino acid content and environmental temperature. In: Crewther WG, ed. Symposium on Fibrous Proteins. Sydney, Australia: Butterworths, pp. 217–225.

Rosenbloom J, Harsch M, Jimenez S: (1973) Hydroxyproline content determines the denaturation temperature of chick tendon collagen. Arch Biochem Biophys 158:478–484.

Roth S, Freund I: (1981) Optical second-harmonic scattering in rat-tail tendon. Biopolymers 20:1271–1290.

Ruggeri A, Benazzo F, Reale E: (1979) Collagen fibrils with straight and helicoidal microfibrils: A freeze-fracture and thin-section study. J Ultrastruct Res 68:101–108.

Schmitt FO, Gross J, Highberger JH: (1953) A new particle type in certain connective tissue extracts. Proc Natl Acad Sci USA 39:459–470.

Schmitt FO, Gross J, Highberger JH: (1955) Tropocollagen and the properties of fibrous collagen. Exp Cell Res (Suppl) 3:326–334.

Scott PG: (1980) A major intermolecular cross-linking site in bovine dentine collagen involving the α2 chain and stabilizing the 4 D overlap. Biochemistry 19:6118–6124.

Silver FH: (1981) Type I collagen fibrillogenesis in vitro. J Biol Chem 256:4973–4977.

Silver FH, Trelstad RL: (1980) Type I collagen in solution. J Biol Chem 255:9427–9433.

Smith JW: (1968) Molecular pattern in native collagen. Nature 219:157–158.

Timpl R, Martin GR: (1981) Components of basement membranes. In: Furthmayr H, ed. Immunochemistry of the Extracellular Matrix. Boca Raton: CRC Press, pp. 119–150.

Timpl R, Wiedemann H, van Delden V, Furthmayr H, Kühn K: (1981) A network model for the organization of type IV collagen molecules in basement membranes. Eur J Biochem 120:203–211.

Tomlin SG, Worthington CR: (1956) Low-angle x-ray diffraction patterns of collagen. Proc R Soc A 235:189–201.

Torchia DA: (1982) Solid state nmr studies of collagen fibrils. In: Cunningham LW, Frederiksen DF, eds. Methods in Enzymology. Volume 82, Part A. New York: Academic, pp. 174–188.

Traub W, Piez KA: (1971) The chemistry and structure of collagen. Adv Protein Chem 25:243–352.

Trelstad RL, Hayashi K: (1979) Tendon collagen fibrillogenesis: Intracellular subassemblies and cell surface changes associated with fibril growth. Dev Biol 71:228–242.

Trus BL, Piez KA: (1980) Compressed microfibril models of the native collagen fibril. Nature 286:300–301.

Veis A, Miller A, Leibovich SJ, Traub W: (1979) The limiting collagen microfibril. The minimum structure demonstrating native axial periodicity. Biochim Biophys Acta 576:88–98.

von Hippel PH: (1967) Structure and stabilization of the collagen molecule in solution. In: Ramachandran GN, ed. Treatise on Collagen. Volume 1: Chemistry of Collagen. London: Academic Press, pp. 253–335.

Williams BR, Gelman RA, Poppke DC, Piez KA: (1978) Collagen fibril formation. J Biol Chem 253:6578–6585.

Yonath A, Traub W: (1969) Polymers of tripeptides as collagen models. IV. Structure analysis of poly(L-prolyl-glycyl-L-proline). J Mol Biol 43:461–477.

Zimmerman BK, Pikkarainen J, Fietzek PP, Kühn K: (1970) Cross-linkages in collagen. Demonstration of three different intermolecular bonds. Eur J Biochem 16:217–225.

2

Chemistry of the Collagens and Their Distribution

Edward J. Miller

2.1. Introduction

It is now clear that collagen molecules represent a number of closely related, albeit chemically distinct, macromolecules (Miller 1976; Eyre 1980; Bornstein and Sage 1980; Miller and Gay 1982). The individual polypeptide chains (α chains) utilized in forming the different types of collagen molecules are derived from a family of structural genes, a subject which is receiving intensive study (Chapter 3). Moreover, the various molecular species of collagen exhibit some specificity with respect to tissue distribution. In spite of these differences, all of the known types of collagen molecules participate in the formation of fibrous elements or supramolecular structures in extracellular spaces (Chapter 1, Section 1.3). These aggregates function primarily, but not solely, as the major structural components of the various connective tissues. In this regard, then, the different types of collagen molecules are broadly related functionally as well as chemically.

The most obvious chemical similarity is the repeating Gly-X-Y triplet, where X and Y are often proline and hydroxyproline, respectively, which is necessary for the native triple helix (Chapter 1, Section 1.2). However, all the collagens also have nontriplet regions that are presumably globular in structure and are critical to biological function. The amino acid sequence, triplet or nontriplet, is the ultimate chemical information about a protein (plus, of course, posttranslational modifications). Most sequence data presently available on the collagens were determined at the protein level. However, recombinant DNA techniques

From the Department of Biochemistry and Institute of Dental Research, University of Alabama Medical Center, University Station, Birmingham, Alabama.

Work cited from the author's laboratory has been supported by National Institutes of Health grants DE-02670 and HL-11310, as well as by grants from the Osteogenesis Imperfecta and Kroc Foundations.

and the easier sequencing of DNA are already providing, by decoding, amino acid sequences and promise to be the major source of these data in the future. Of course, protein chemistry is needed to characterize the functional forms of proteins and to assign DNA sequences to specific proteins.

This chapter summarizes current information concerning the chemical properties of collagen chains and the molecules formed as a result of the association of three individual chains. Comprehensive discussions and background references will be found in reviews by Traub and Piez (1971); Fietzek and Kühn (1976); Piez (1976); Bornstein and Traub (1979); Miller and Gay (1982); and Miller and Rhodes (1982). Common features of the collagens will be considered first and their distinctive features second. And finally, the distribution of the various collagens in the tissues of vertebrates will be discussed with special reference to the biological significance of the collagen gene family. The abbreviations used for the amino acids, as well as other abbreviations, are found in the Front Matter of the book.

2.2. Collagen Chains, Molecules, and Nomenclature

The term collagen is very old and was originally applied to the crude insoluble material that was the source of gelatin and glue and the major component of leather. While it had a morphological definition in the early 1950s, with increased understanding of the chemistry and structure of collagen, the term is now usually applied to the pure protein in either monomeric or native aggregate form, depending on the context. The term tropocollagen has been used for the monomer but is not often seen today. Until 1969, the protein was always spoken of in the singular since it seemed to be a single protein. It was usually obtained from skin or tendon, which contains largely the collagen now referred to as type I. However, when a cartilage-specific collagen was isolated and shown to differ from type I in chain composition and primary structure, it was called type II since it was clearly a different gene product (Miller and Matukas 1969). Soon after, a second collagen in skin, termed type III, was demonstrated (Miller et al. 1971). Type IV collagen (basement membranes) had always occupied an uncertain position and only recently has it become well enough defined to be assigned a type number with confidence (see Timpl and Martin 1982; Grant et al. 1982). Type V collagen (cell surfaces) was identified by Burgeson et al. (1976) and Chung et al. (1976), and the chains were referred to as αA and αB, or A and B. The identification of a related αC chain (Sage and Bornstein 1979) and the recognition that they were a distinct group led to their assignment as type V and to the α chains being renamed α1(V), α2(V), and α3(V) for B, A, and C, respectively (See Bornstein and Sage 1980; Miller and Gay 1982).

Collectively, collagen types I–V contain nine different α chains that have been characterized, at least with respect to overall compositional features. These α chains are listed in Table 2.1, along with the chain composition of various molecular species and the manner in which individual molecular species or groups of species are currently assigned to types. The molecular species composing types IV and V collagen are still not firmly established. The chains designated α1 are not necessarily any more closely related to one another than they are to chains designated α2 or α3; these assignments are largely arbitrary. In addition,

Table 2.1. Collagen Types and α Chains

Designation	α Chains	Molecular species
Type I	α1(I), α2(I)	$[\alpha1(I)]_2\alpha2(I)$, $[\alpha1(I)]_3$[a]
Type II	α1(II)	$[\alpha1(II)]_3$
Type III	α1(III)	$[\alpha1(III)]_3$
Type IV	α1(IV), α2(IV)	$[\alpha1(IV)]_3$, $[\alpha2(IV)]_3$
Type V	α1(V), α2(V), α3(V)	$[\alpha1(V)]_2\alpha2(V)$, $[\alpha1(V)]_3$, α1(V)α2(V)α3(V)

[a]Sometimes called type I trimer.

a type number may designate one or a group of closely related molecular species, as in the case of type IV (Robey and Martin 1981) and type V (Rhodes and Miller 1981). Assignment of a type number to a collagen or group of collagens not only requires that the α chain or chains be chemically distinct, but also implies that the molecules are a functionally distinct group, although the function may not be strictly known. Thus, the definition of a type has both a chemical and a biological basis, but for lack of information the latter must often be inferred from chemical and localization studies. For this reason, assignment of a type number to a protein believed to be a new collagen should ideally await definitive data about chemistry and some information about function.

Several lines of evidence indicate that the collagens listed in Table 2.1 represent only a partial catalogue of the total number of α chains and molecules actually present in vertebrate species. Direct evidence based on sequence data indicates that there are at least two distinct chains differing in only a few positions within the α1(II) population (Butler et al. 1977). In addition, peptide maps derived from type III collagen synthesized by aortic endothelial cells (Sage et al. 1981), as well as ultrastructural studies on various cell types in tissues from an individual presenting the Ehlers-Danlos type IV syndrome (Chapter 11), have provided preliminary evidence for genetic heterogeneity of the α1(III) chain (Holbrook and Byers 1981). These results suggest that other collagen chains will actually prove to be heterogeneous as well, but these minor differences are so far insufficient to assign new chain or type numbers. They appear to be minor polymorphic forms for which there is as yet no standard nomenclature.

However, additional collagen chains that may require new assignments have been identified and partially characterized. These include three or four new chains present as minor collagenous constituents of cartilage (Burgeson and Hollister 1979; Shimokomaki et al. 1980; Reese and Mayne 1981), an apparently unique chain synthesized by aortic endothelial cells in culture (Sage et al. 1980), and the three chains present in an unusual disulfide-bonded high molecular weight collagenous aggregate found in extracts of vascular tissues (Furuto and Miller 1980). Too little information concerning these components is currently available to discern their relationship to the more well-characterized chains or to decide if they represent constituents of new collagen types.

In view of the data outlined above, it is obvious that the actual size of the collagen gene pool can only be estimated in conservative terms at the present time. Nevertheless, there could be 20 or more nonallelic genes for collagen synthesis normally present in the vertebrate genome. In addition, the triple helix is encoded in other genes, such as those for synthesis of the three chains of the C1q molecule and the chains of the tail region of the acetylcholinesterase molecule (Chapter 1, Section 1.2.7).

2.3. Common Chemical Features

2.3.1. Pro α and α Chains

The most convenient manner of introducing discussions concerning the chemistry of the collagens is to evaluate current information and concepts on the chemistry of individual collagen chains. Since the procollagens are formed by the alignment and folding of three pro α chains and the collagens are formed from the procollagens (Chapter 3), the disposition of chemical features in each chain determines the topological arrangement of these features in molecules. The general structural features of pro α1(I) and its relationship to the α1(I) chain are illustrated diagrammatically in Figure 2.1. The pro α1(I) chain consists of a linear array of 1441 amino acid residues linked exclusively by α-amino, α-carboxy peptide bonds. Although no single pro α1(I) chain has been sequenced in its entirety, the entire primary structure of pro α1(I) is available as a composite of bovine and chick sequences. This is shown in Figure 2.2. Although fewer data are currently available concerning the structural features of other pro α chains, the general characteristics depicted for pro α1(I) are valid for the pro α chains of types I, II, and III collagens. As also indicated in Figures 2.1 and 2.2, pro α chains do not represent the initial translation products of collagen messenger RNA. Convincing evidence has been presented for the existence of prepro α chains that contain an additional N-terminal transient domain or signal sequence. The chemical characteristics of the latter remain largely undetermined, but enough information is available (Palmiter et al. 1979) to show that it resembles the hydrophobic signal sequences of other secretory proteins, which are cleaved off shortly after the beginning of translation (Chapter 3).

As illustrated in Figures 2.1 and 2.2, the pro α1(I) chain may be considered in terms of three major regions, the two precursor peptides pN (N-terminal) and pC (C-terminal) and the α chain. The pN peptide from calf skin collagen contains 139 amino acid residues (Hörlein et al. 1979). This region consists of three unique segments: (1) An initial nontriplet (and thus nonhelical) region of 86 amino acid residues characterized by the presence of relatively high levels of acidic and hydrophobic residues as well as by 10 cysteine residues that form intrachain disulfide bonds; (2) a region composed of 17 contiguous Gly-X-Y triplets which are relatively rich in proline and hydroxyproline residues in the X and Y positions, respectively; and (3) a short sequence of two amino acid residues following the last triplet, which joins the pN peptide to the initial amino acid in the α1(I) chain at a Pro-Gln bond. This bond is cleaved by a specific enzyme, procollagen N-protease, during the conversion of procollagen molecules to collagen (Chapter 3). Information concerning the N-terminal precursor-specific domain of pro α1(I) has been obtained largely from studies on pN collagen type I extracted from the dermis of dermatosparactic calves (Chapter 11). These animals exhibit a genetic defect in the processing of procollagen to

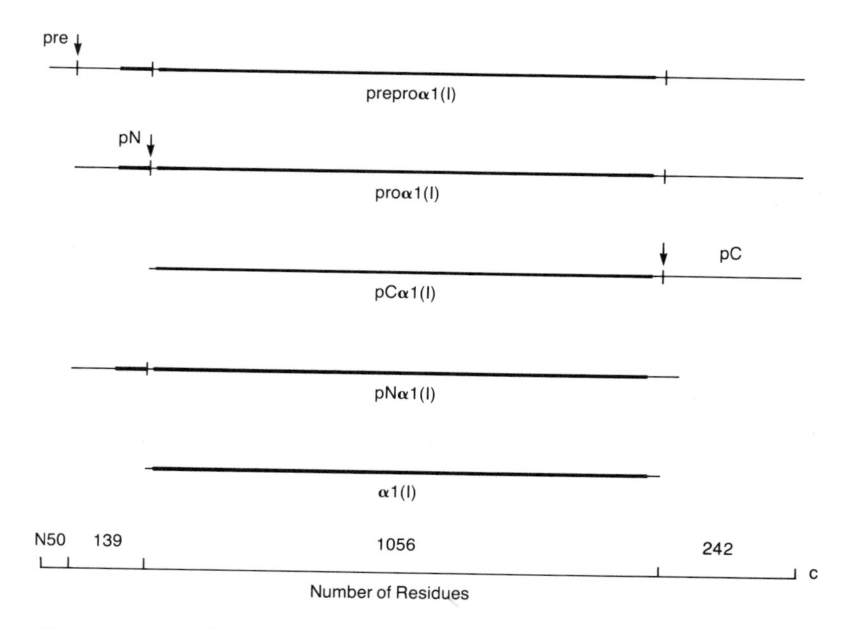

Figure 2.1. A schematic representation of the collagen α1(I) chain and its precursor forms. The heavy line indicates regions of Gly-X-Y triplet repeats, which are helical in the type I collagen molecule; the light line indicates nontriplet regions, which are globular. The arrows and vertical lines indicate positions where proteases cleave during processing. The α2(I), α1(II), and α1(III) chains and their precursors are similar.

collagen, resulting in the accumulation of molecules that retain the pN domain and cannot form normal fibrils.

Following pro α chain association, formation of the triple helix, and removal of the pN and pC domains, three α chain regions as a triple helix with short nonhelical ends constitute a collagen molecule of types I, II, or III. Of course, α chains never exist separately in vivo; they are created by denaturation of soluble collagen in vitro. Sequences are known from studies on the α1(I) chain derived from type I collagen of several species (Fietzek and Kühn 1976; Bornstein and Traub 1979). The α1(I) chain of calf skin collagen has been completely sequenced; it contains 1056 amino acid residues (*See* Hofmann et al. 1978 for references). It consists of three readily distinguishable segments (Figure 2.2): (1) A relatively short N-terminal region containing 16 amino acid residues (seventeen residues in some species), not in triplet form and numbered 1^N to 16^N; (2) a large central region composed of 338[1] contiguous Gly-X-Y triplets in which approximately one third of the total X and Y positions are occupied by proline or hydroxyproline in mammals; and (3) a short terminal segment of 26 amino acid residues numbered 1^C to 26^C which, like the N-terminal segment, is not in triplet form. The

[1]Initial studies reviewed by Fietzek and Kühn (1976) and Bornstein and Traub (1979) suggested the presence of 337 contiguous Gly-X-Y triplets in the triplet region of α1(I). However, subsequent investigations have established that one triplet had been missed in the earlier studies. Hence, residue numbers assigned to amino acids in the sequences presented in the aforementioned reviews for calf α1(I) should be increased by 3, beginning with residue 613 and Gly-Pro-Ala inserted for residues 613–615. For chick α1(I), the missing triplet is Gly-Pro-Hyp, beginning at residue 606.

PRE PEPTIDE:

1 Met-Phe-Ser-Phe-Val-Xxx-Ser-Arg-Leu-Leu-Leu-Leu-Ile-Ala-Ala-Xxx-Xxx-Leu-Leu-(30 or more residues)-

pN PEPTIDE

1 Gln-Glu-Glu-Gly-Gln-Glu-Glu-Gly-Gln-Glu-Glu-Asp-Ile-Pro-Pro-Val-Thr-Cys-Val-Gln-Asp-Gly-Leu-Arg-Tyr-His-Asp-Arg-Asp-Val-
31 Trp-Lys-Pro-Val-Pro-Cys-Gln-Ile-Cys-Val-Cys-Asp-Asn-Gly-Asn-Val-Leu-Cys-Asp-Asp-Val-Ile-Cys-Asp-Gln-Leu-Lys-Asp-Cys-Pro-
61 Asn-Ala-Lys-Val-Pro-Thr-Asp-Glu-Cys-Cys-Pro-Val-Cys-Pro-Glu-Gly-Gln-Glu-Ser-Pro-Thr-Asp-Gln-Glu-Thr-Thr-GLY-Val-Glu-GLY-
91 Pro-Lys-GLY-Asp-Thr-GLY-Pro-Arg-GLY-Pro-Arg-GLY-Pro-Ala-GLY-Pro-Hyp-GLY-Arg-Asp-GLY-Ile-Hyp-GLY-Gln-Pro-GLY-Leu-Hyp-GLY-
121 Pro-Hyp-GLY-Pro-Hyp-GLY-Pro-Hyp-GLY-Pro-Hyp-GLY-Leu-Gly-GLY-Asn-Phe-Ala-Pro-

α CHAIN

1N Gln-Leu-Ser-Tyr-Gly-Tyr-Asp-Glu-LYS-Ser-Thr-Gly-Ile-Ser-Val-Pro-

1 GLY-Pro-Met-GLY-Pro-Ser-GLY-Pro-Arg-GLY-Leu-Hyp-GLY-Pro-Hyp-GLY-Ala-Hyp-GLY-Pro-Gln-GLY-Phe-Gln-GLY-Pro-Hyp-GLY-Glu-Hyp-
31 GLY-Glu-Hyp-GLY-Ala-Ser-GLY-Pro-Met-GLY-Pro-Arg-GLY-Pro-Hyp-GLY-Pro-Hyp-GLY-Lys-Asn-GLY-Asp-Asp-GLY-Glu-Ala-GLY-Lys-Pro-
61 GLY-Arg-Hyp-GLY-Glu-Arg-GLY-Pro-Hyp-GLY-Pro-Gln-GLY-Ala-Arg-GLY-Leu-Hyp-GLY-Thr-Ala-GLY-Leu-Hyp-GLY-Met-Hyl-GLY-His-Arg-
91 GLY-Phe-Ser-GLY-Leu-Asp-GLY-Ala-Lys-GLY-Asp-Ala-GLY-Pro-Ala-GLY-Pro-Lys-GLY-Glu-Hyp-GLY-Ser-Hyp-GLY-Glu-Asn-GLY-Ala-Hyp-
121 GLY-Gln-Met-GLY-Pro-Arg-GLY-Leu-Hyp-GLY-Glu-Arg-GLY-Arg-Hyp-GLY-Ala-Hyp-GLY-Pro-Ala-GLY-Ala-Arg-GLY-Asn-Asp-GLY-Ala-Thr-
151 GLY-Ala-Ala-GLY-Pro-Hyp-GLY-Pro-Thr-GLY-Pro-Ala-GLY-Pro-Hyp-GLY-Phe-Hyp-GLY-Ala-Val-GLY-Ala-Lys-GLY-Glu-Gly-GLY-Pro-Gln-
181 GLY-Ala-Arg-GLY-Ser-Glu-GLY-Pro-Gln-GLY-Val-Arg-GLY-Glu-Hyp-GLY-Pro-Hyp-GLY-Pro-Ala-GLY-Ala-Ala-GLY-Pro-Ala-GLY-Asn-Hyp-
211 GLY-Ala-Asp-GLY-Glu-Hyp-GLY-Ala-Lys-GLY-Ala-Asn-GLY-Ala-Hyp-GLY-Ile-Ala-GLY-Ala-Hyp-GLY-Phe-Hyp-GLY-Ala-Arg-GLY-Pro-Ser-
241 GLY-Pro-Gln-GLY-Pro-Ser-GLY-Pro-Hyp-GLY-Pro-Lys-GLY-Asn-Ser-GLY-Glu-Hyp-GLY-Ala-Hyp-GLY-Asn-Lys-GLY-Asp-Thr-GLY-Ala-Lys-
271 GLY-Glu-Hyp-GLY-Pro-Thr-GLY-Ile-Gln-GLY-Pro-Hyp-GLY-Pro-Ala-GLY-Glu-Glu-GLY-Lys-Arg-GLY-Ala-Arg-GLY-Glu-Hyp-GLY-Pro-Ala-
301 GLY-Leu-Hyp-GLY-Pro-Hyp-GLY-Glu-Arg-GLY-Gly-Hyp-GLY-Ser-Arg-GLY-Phe-Hyp-GLY-Ala-Asp-GLY-Val-Ala-GLY-Pro-Lys-GLY-Pro-Hyp-
331 GLY-Glu-Arg-GLY-Ala-Hyp-GLY-Pro-Ala-GLY-Pro-Lys-GLY-Ser-Hyp-GLY-Glu-Ala-GLY-Arg-Hyp-GLY-Glu-Ala-GLY-Leu-Hyp-GLY-Ala-Lys-
361 GLY-Leu-Thr-GLY-Ser-Hyp-GLY-Ser-Hyp-GLY-Pro-Asp-GLY-Lys-Thr-GLY-Pro-Hyp-GLY-Pro-Ala-GLY-Gln-Asn-GLY-Arg-Hyp-GLY-Pro-Ala-
391 GLY-Pro-Hyp-GLY-Ala-Arg-GLY-Gln-Ala-GLY-Val-Met-GLY-Phe-Hyp-GLY-Pro-Lys-GLY-Ala-Ala-GLY-Glu-Hyp-GLY-Lys-Ala-GLY-Glu-Arg-
421 GLY-Val-Hyp-GLY-Pro-Hyp-GLY-Ala-Val-GLY-Pro-Ala-GLY-Lys-Asp-GLY-Glu-Ala-GLY-Ala-Gln-GLY-Pro-Hyp-GLY-Pro-Ala-GLY-Pro-Ala-
451 GLY-Glu-Arg-GLY-Glu-Gln-GLY-Pro-Ala-GLY-Ser-Hyp-GLY-Phe-Gln-GLY-Leu-Hyp-GLY-Pro-Ala-GLY-Pro-Hyp-GLY-Glu-Ala-GLY-Lys-Hyp-
481 GLY-Glu-Gln-GLY-Val-Hyp-GLY-Asp-Leu-GLY-Ala-Hyp-GLY-Pro-Ser-GLY-Ala-Arg-GLY-Glu-Arg-GLY-Phe-Hyp-GLY-Glu-Arg-GLY-Val-Glu-
511 GLY-Pro-Hyp-GLY-Pro-Ala-GLY-Pro-Arg-GLY-Ala-Asn-GLY-Ala-Hyp-GLY-Asn-Asp-GLY-Ala-Lys-GLY-Asp-Ala-GLY-Ala-Hyp-GLY-Ala-Hyp-
541 GLY-Ser-Gln-GLY-Ala-Hyp-GLY-Leu-Gln-GLY-Met-Hyp-GLY-Glu-Arg-GLY-Ala-Ala-GLY-Leu-Hyp-GLY-Pro-Lys-GLY-Asp-Arg-GLY-Asp-Ala-
571 GLY-Pro-Lys-GLY-Ala-Asp-GLY-Ala-Pro-GLY-Lys-Asp-GLY-Val-Arg-GLY-Leu-Thr-GLY-Pro-Ile-GLY-Pro-Hyp-GLY-Pro-Ala-GLY-Ala-Hyp-
601 GLY-Asp-Lys-GLY-Glu-Ala-GLY-Pro-Ser-GLY-Pro-Ala-GLY-Pro-Thr-GLY-Ala-Arg-GLY-Ala-Hyp-GLY-Asp-Arg-GLY-Glu-Hyp-GLY-Pro-Hyp-
631 GLY-Pro-Ala-GLY-Phe-Ala-GLY-Pro-Hyp-GLY-Ala-Asp-GLY-Gln-Hyp-GLY-Ala-Lys-GLY-Glu-Hyp-GLY-Asp-Ala-GLY-Ala-Lys-GLY-Asp-Ala-
661 GLY-Pro-Hyp-GLY-Pro-Ala-GLY-Pro-Ala-GLY-Pro-Hyp-GLY-Pro-Ile-GLY-Asn-Val-GLY-Ala-Hyp-GLY-Pro-Hyl-GLY-Ala-Arg-GLY-Ser-Ala-
691 GLY-Pro-Hyp-GLY-Ala-Thr-GLY-Phe-Hyp-GLY-Ala-Ala-GLY-Arg-Val-GLY-Pro-Hyp-GLY-Pro-Ser-GLY-Asn-Ala-GLY-Pro-Hyp-GLY-Pro-Hyp-
721 GLY-Pro-Ala-GLY-Lys-Glu-GLY-Ser-Lys-GLY-Pro-Arg-GLY-Glu-Thr-GLY-Pro-Ala-GLY-Arg-Hyp-GLY-Glu-Val-GLY-Pro-Hyp-GLY-Pro-Hyp-
751 GLY-Pro-Ala-GLY-Glu-Lys-GLY-Ala-Hyp-GLY-Ala-Asp-GLY-Pro-Ala-GLY-Ala-Hyp-GLY-Thr-Pro-GLY-Pro-Gln-GLY-Ile-Ala-GLY-Gln-Arg-
781 GLY-Val-Val-GLY-Leu-Hyp-GLY-Gln-Arg-GLY-Glu-Arg-GLY-Phe-Hyp-GLY-Leu-Hyp-GLY-Pro-Ser-GLY-Glu-Hyp-GLY-Lys-Gln-GLY-Pro-Ser-
811 GLY-Ala-Ser-GLY-Glu-Arg-GLY-Pro-Hyp-GLY-Pro-Met-GLY-Pro-Hyp-GLY-Leu-Ala-GLY-Pro-Hyp-GLY-Glu-Ser-GLY-Arg-Glu-GLY-Ala-Hyp-
841 GLY-Ala-Glu-GLY-Ser-Hyp-GLY-Arg-Asp-GLY-Ser-Hyp-GLY-Ala-Lys-GLY-Asp-Arg-GLY-Glu-Thr-GLY-Pro-Ala-GLY-Pro-Hyp-GLY-Ala-Hyp-
871 GLY-Ala-Hyp-GLY-Ala-Hyp-GLY-Pro-Val-GLY-Pro-Ala-GLY-Lys-Ser-GLY-Asp-Arg-GLY-Glu-Thr-GLY-Pro-Ala-GLY-Pro-Ile-GLY-Pro-Val-
901 GLY-Pro-Ala-GLY-Ala-Arg-GLY-Pro-Ala-GLY-Pro-Gln-GLY-Pro-Arg-GLY-Asp-Hyl-GLY-Glu-Thr-GLY-Glu-Glu-GLY-Asp-Arg-GLY-Ile-Hyl-
931 GLY-His-Arg-GLY-Phe-Ser-GLY-Leu-Gln-GLY-Pro-Hyp-GLY-Pro-Hyp-GLY-Ser-Hyp-GLY-Glu-Gln-GLY-Pro-Ser-GLY-Ala-Ser-GLY-Pro-Ala-
961 GLY-Pro-Arg-GLY-Pro-Hyp-GLY-Ser-Ala-GLY-Ser-Hyp-GLY-Lys-Asp-GLY-Leu-Asn-GLY-Leu-Hyp-GLY-Pro-Ile-GLY-Hyp-Hyp-GLY-Pro-Arg-
991 GLY-Arg-Thr-GLY-Asp-Ala-GLY-Pro-Ala-GLY-Pro-Hyp-GLY-Pro-Hyp-GLY-Pro-Hyp-GLY-Pro-Hyp-GLY-Pro-Pro-

1C Ser-Gly-Gly-Phe-Asp-Phe-Ser-Phe-Leu-Pro-Gln-Pro-Pro-Gln-Glu-LYS-Ala-His-Asp-Gly-Gly-Arg-Tyr-Tyr-Arg-Ala-

pC PEPTIDE:

1 Asp-Asp-Ala-Asn-Val-Met-Arg-Asp-Arg-Asp-Leu-Glu-Val-Asp-Thr-Thr-Leu-Lys-Ser-Leu-Ser-Gln-Gln-Ile-Glu-Asn-Ile-Arg-Ser-Pro-
31 Glu-Gly-Thr-Arg-Lys-Asn-Pro-Ala-Arg-Thr-Cys-Arg-Asp-Leu-Lys-Met-Cys-His-Gly-Asp-Trp-Lys-Ser-Gly-Glu-Tyr-Trp-Ile-Asp-Pro-
61 Asn-Gln-Gly-Cys-Asn-Leu-Asp-Ala-Ile-Lys-Val-Tyr-Cys-Asn-Met-Glu-Thr-Gly-Glu-Thr-Cys-Val-Tyr-Pro-Thr-Gln-Ala-Thr-Ile-Ala-
91 Gln-Lys-Asn-Trp-Tyr-Leu-Ser-Lys-Asn-Pro-Lys-Glu-Lys-Lys-His-Val-Trp-Phe-Gly-Glu-Thr-Met-Ser-Asp-Gly-Phe-Gln-Phe-Glu-Tyr-
121 Gly-Gly-Glu-Gly-Ser-Asn-Pro-Ala-Asp-Val-Ala-Ile-Gln-Leu-Thr-Phe-Leu-Arg-Leu-Met-Ser-Thr-Glu-Ala-Thr-Gln-Asn-Val-Thr-Tyr-
151 His-Cys-Lys-Asn-Ser-Val-Ala-Tyr-Met-Asp-His-Asp-Thr-Gly-Asn-Leu-Lys-Lys-Ala-Leu-Leu-Leu-Gln-Gly-Ala-Asn-Glu-Ile-Glu-Ile-
181 Arg-Ala-Glu-Glu-Asn-Ser-Arg-Phe-Thr-Tyr-Gly-Val-Thr-Glu-Asp-Gly-Cys-Thr-Ser-His-Thr-Gly-Ala-Trp-Gly-Lys-Thr-Val-Ile-Glu-
211 Tyr-Lys-Thr-Thr-Lys-Thr-Ser-Arg-Leu-Pro-Ile-Ile-Asp-Leu-Ala-Pro-Met-Asp-Val-Gly-Ala-Pro-Asp-Gln-Glu-Phe-Gly-Ile-Asp-Ile-
241 Gly-Pro-Val-Cys-Phe-Leu

Figure 2.2. A composite sequence of the collagen prepro α1(I) chain divided into precursor and α chain regions. Glycine residues in triplet regions are capitalized (*GLY*). The crosslink-precursor lysine residues in the nonhelical ends are also capitalized (*LYS*). The pre peptide sequence is from translated chick messenger RNA (Palmiter et al. 1979). The pN peptide and α chain sequences are bovine type I collagen except the C-terminal nonhelical end. It and the pC peptide are derived from chick complementary DNA (Fuller and Boedtker 1981). Most of the bovine sequences are published (*See* Hofmann et al. 1980 for references), but some unpublished portions were kindly provided by P. Fietzek and K. Kühn.

α chain terminates at an Ala-Asp bond, which is cleaved by a specific enzyme, procollagen C-protease, during the conversion of procollagen to collagen. The N- and C-terminal segments of the α chain do not possess the structural requirements for triple helix formation, and are sometimes referred to as telopeptides. These segments contain sites for intra- and intermolecular crosslinking and are believed to be of considerable importance in establishing intermolecular contacts during fibril formation (Chapter 1, Section 1.3.4). The long triplet region specifies the triple helical conformation and the rod-like shape of collagen molecules.

The pC peptide constitutes the C-terminal region of pro α1(I). With respect to structural and compositional features, it closely resembles the globular segment of the pN peptide at the opposite end of the chain. However, it is much larger, containing 246 amino acid residues. At least some of the eight cysteine residues in the domain are involved in interchain disulfide bonding. The entire primary structure of the pC peptide has been elucidated as a result of DNA sequence determinations on a series of complementary DNA clones to chick pro α1(I) messenger RNA (Fuller and Boedtker 1981). In addition, the pC peptide contains one mannose-rich oligosaccharide linked to an asparagine residue located presumably at position 147 within the region (Section 2.3.2). Numerous studies indicate that the pC peptide is of considerable importance in promoting the assembly and alignment of newly synthesized pro α chains following their release from polysomes (Chapter 3). The efficient formation of procollagen molecules (and ultimately collagen molecules) is thus largely dependent on the properties of the sequences contained within this domain of the various pro α chains.

Although the general structural features of the pro α1(I) chain may be considered representative of the characteristics of other pro α chains in types I, II, and III collagen, there are some exceptions. The pN peptide of pro α2(I) probably does not contain an extensive globular segment (Becker et al. 1977). Moreover, the nontriplet sequences at the C-terminal end of the α chain region and in the pC peptide of pro α2(I) are somewhat shorter than the homologous sequences in pro α1(I) because of deletions (Fuller and Boedtker 1981). Although detailed features of pro α1(II) remain incomplete, investigations of procollagen type II synthesized in culture (Chapter 3) reveal that they closely resemble procollagen type I molecules with respect to the presence of both pN and pC domains of about the same size. In contrast to pro α1(I) chains, however, pro α1(II) chains appear to contain a mannose-rich oligosaccharide (Section 2.3.2) in both the pN and pC peptides. Somewhat more detailed information is available for the pro α1(III) chain. The pN peptide closely resembles that of pro α1(I) except that it contains additional cysteine residues located in the junction region between the short triplet sequence and the following segment (Bruckner et al. 1978). In the native procollagen type III molecule, these residues participate in interchain disulfide bonding and thereby provide an additional degree of structural stability in this portion of the molecule. The α chain region of pro α1(III) likewise closely resembles that of pro α1(I), but contains a somewhat shorter N-terminal nontriplet region as well as four more contiguous triplets in the protein from calf skin (Fietzek et al. 1979; Allmann et al. 1979). In addition, pro α1(III) contains two cysteine residues in the α chain located between the triplet sequence and the C-terminal nontriplet segment. This placement of cysteine residues is similar to that in the pN peptide and these residues also participate in the formation of interchain disulfide bonds in the native type III procollagen molecule. Relatively little is currently known concerning the pC peptide of pro α1(III), although it is clear that it closely resembles the homologous regions of pro α1(I), pro α2(I), and pro α1(II) chains with respect to size and general chemical properties.

Few details are available concerning the structural features of the chains of type IV collagen but a general outline is emerging (see Timpl and Martin 1982; Grant et al. 1982). On electrophoresis under reducing conditions, newly synthesized type IV molecules dissociate to yield what have been called pro α1(IV) and pro α2(IV) chains with apparent molecular weights of 185,000 and 170,000, respectively. When extracated from tissues, smaller forms appear with apparent molecular weights of about 30,000 less (Tryggvason et al. 1980). However, recent ultrastructural studies on type IV collagen molecules isolated from a mouse tumor or from teratocarcinoma (PYS) cells in culture show the largest forms to consist of apparently triple-helical strands with a length of about 390 nm (accommodating as many as 1350 amino acids in the triplet regions of each chain) plus a globular end (Timpl et al. 1981; See Chapter 1, Section 1.2.6). The globular end is about the same size as the pC domain of procollagen type I. Smaller forms seem to arise from cleavage during isolation rather than being normal tissue components. These data indicate that the chains of type IV collagen are considerably larger than their counterparts in types I, II, and III collagen, and retain some of the features characteristic of the extensive globular domains in types I, II, and III procollagens. However, there is no evidence for any proteolytic processing and thus the existence of pro forms is in doubt. It is clear from chemical studies that there are a number of discontinuities in the triplet sequence (Schuppan et al. 1980), leading to interruptions in triple-helical conformation and explaining the relatively high susceptibility of type IV collagen molecules to cleavage by proteases.

Type V collagen isolated by pepsin digestion contains chains of about the same size as α1(I) chains (Rhodes and Miller 1979), but with a composition more like type IV collagen (Section 2.4.4). Pro α1(V) and pro α2(V) chains, and intermediates apparently analogous to pN or pC α chains, have also been demonstrated (Fessler et al. 1981a, 1981b) but are not yet fully characterized. The fully processed but intact type V molecules have constituent chains with apparent molecular weights considerably larger than α1(I). The structure and chemistry of the type V collagens may be intermediate between types I, II, and III on the one hand and type IV on the other.

Since the collagen types deposited in tissues have varying amounts of globular regions, a simplifying concept is that biological variation has been achieved in part by varying degrees of processing of initially similar gene products. In this regard, types I and II procollagen are rapidly processed to collagen molecules, while in the case of type III processing is quite slow and possibly never complete, and with type IV may not occur at all. Processing is discussed in more detail in Chapter 3.

2.3.2. Carbohydrate

All vertebrate collagens characterized thus far qualify for the designation as glycoproteins, since they contain significant, but highly variable, amounts of covalently linked carbohydrate. With respect to collagen molecules, all carbohydrate units are linked O-glycosidically to hydroxylysine residues in a unique way (Butler and Cunningham 1966). They are among the simplest of all carbohydrate units. They consist either of a single galactose residue, α-D-galactopyranosyl hydroxylysine, or of the disaccharide, α-D-glucopyranosyl-(1 → 2)-β-D-galactopyranosyl hydroxylysine, as shown by Spiro (1967). In both cases, bonding to the protein is achieved through an acetal linkage between C-1 of the pyranose form of galactose and the hydroxy group of hydroxylysine, as shown in Figure 2.3. They are abbreviated Gal-Hyl and Glc-Gal-Hyl, respectively.

The acetal linking groups of Glc-Gal-Hyl are highly stable under alkaline conditions. Therefore, both Glc-Gal-Hyl and Gal-Hyl may be isolated from, or quantitated in, alkaline hydrolysates of collagen. An example of the latter procedure is illustrated in Figure 2.4, which depicts a portion of an amino acid analyzer tracing for an alkaline hydrolysate of α1(V). The chromatogram reveals that the majority of the hydroxylysine residues in this chain are covalently linked to carbohydrate and that the disaccharide derivative is the predominant component.

As noted above, the amount of carbohydrate covalently linked to collagen molecules is highly variable. These differences are apparent even for a given type of collagen isolated from various sources, but are most noticeable when comparing different types of collagen. In general, the chains of types I and III collagen contain the lowest levels of hydroxylysine-linked carbohydrate (0.5–1.0%) whereas the chains of types II, IV, and V collagen contain considerably more, from 6% in α1(II) to nearly 15% in α1(IV). Data on the number of hydroxylysine residues as well as the number of carbohydrate units attached to these residues in the various α chains are summarized in Table 2.2. In highly glycosylated chains, formation of the disaccharide derivative occurs much more frequently than the monosaccharide derivative. Moreover, for certain α chains [(α1(IV), α2(IV), and α1(V)], virtually all hydroxylysine residues contain covalently at-

Figure 2.3. The structure of the collagen-specific carbohydrate Glc-Gal-Hyl in peptide linkage. The carbohydrate unit may also consist only of the Gal portion.

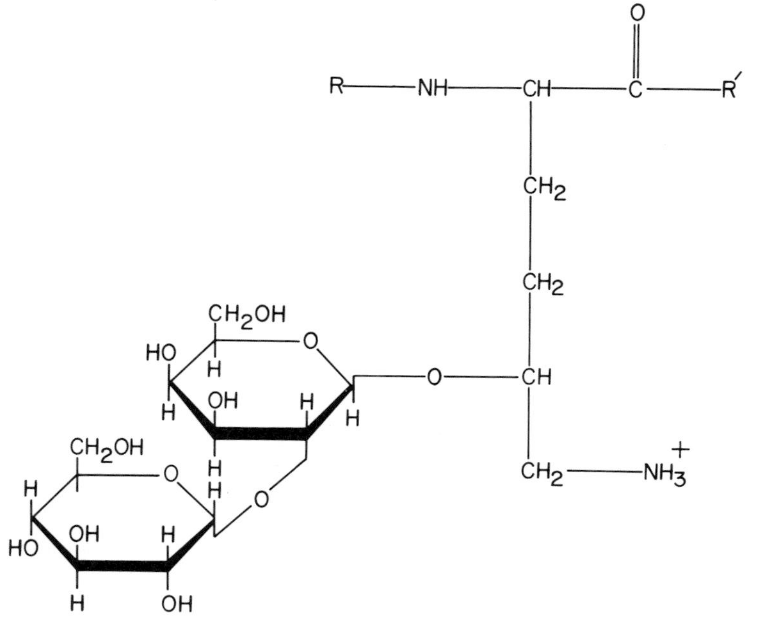

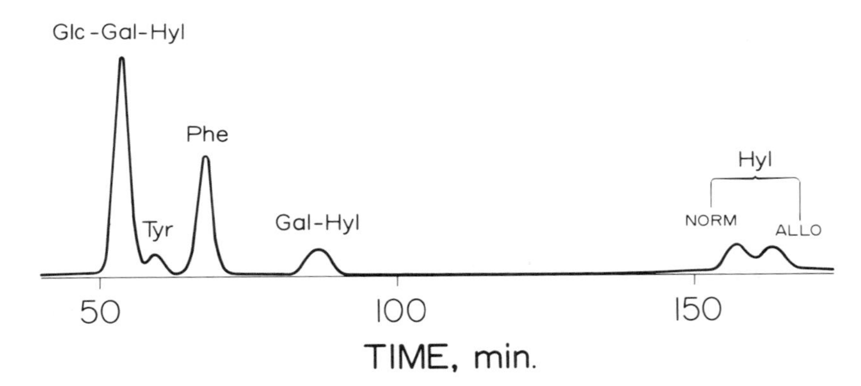

Figure 2.4. A portion of an automatic amino acid analyzer tracing illustrating the elution pattern of Glc-Gal-Hyl and Gal-Hyl in an alkaline hydrolysate of the collagen α1(V) chain. Under the conditions employed for chromatography, most acidic and neutral amino acids are eluted prior to 50 min. Alkaline hydrolysis promotes the epimerization of hydroxylysine and the diastereoisomers partially separate.

tached carbohydrate. This suggests that the presence of a hydroxylysine residue in the Y position in the Gly-X-Y triplet is, in itself, sufficient for glycosylation. A similar situation exists for C1q (Chapter 1, Section 1.2.7). The failure to achieve correspondingly high proportions of glycosylated hydroxylysine residues in some α chains can be attributed to the combination of enzyme activity and the rapidity with which the α chains associate and make triple helix that is no longer a substrate for the glycosylating enzymes (Chapter 3).

In addition to O-glycosidically linked units, procollagen molecules contain N-glycosidically linked units located in the globular segments of the pro domains, as noted in Section 2.3.1. The latter carbohydrate units have been shown to be of the simple mannose-rich type containing approximately eight to nine monosaccharides linked to the protein through an asparagine residue (Clark 1979). See Chapter 7 for a detailed discussion of N-linked oligosaccharides.

Table 2.2. Hydroxylysine (Hyl), Glucosylgalactosylhydroxylysine (Glc-Gal-Hyl), and Galactosylhydroxylysine (Gal-Hyl) Residues in Different Collagen α Chains[a]

Chain	Hyl	Glc-Gal-Hyl	Gal-Hyl
α1(I)[b]	5–9	1	0
α2(I)[b]	8–12	1	1
α1(II)[b]	20–24	12	4
α1(III)	5	1	0
α1(IV)	50	44	2
α2(IV)	36	29	2
α1(V)	36	29	5
α2(V)	23	5	3
α3(V)	43	17	7

[a]Values are expressed as residues/1000 total amino acid residues.
[b]α1(I), α2(I), and α1(II) chains from various sources generally exhibit varying levels of lysyl hydroxylation with somewhat variable levels of glycosylation. In addition, the α1(I) and α2(I) chains of type I collagen in bone contain largely the monosaccharide derivative.

Further studies on the collagens have shown the presence of significant quantities of nonenzymatically linked hexoses (Robins and Bailey 1972). They have been characterized as the aldimine condensation product of glucose, mannose, or galactose with the ε-NH_2 group of either lysine or hydroxylysine. The linkage is stabilized by an Amadori rearrangement to a keto-imine bond (Pape et al. 1981) in the same manner as some of the crosslinks (Section 2.3.3). Since the level of these components in the collagens and in other proteins increases with age as well as in diabetic states, the alteration may play a role in pathological changes.

The functional role of N- or O-linked carbohydrate attached to the procollagens remains undefined. Although glycosylation does not appear to be a prerequisite for transport and secretion of many glycoproteins, there are conflicting reports concerning the role of N-glycosylation in procollagen secretion (Duksin and Bornstein 1977; Housley et al. 1980). In view of the nature of the procollagens and the collagens and the chemical properties of the carbohydrate attached to them in both globular and helical domains, glycosylation at any level would not be expected to alter the molecular conformation. On the other hand, extensive glycosylation would be expected to greatly modify some of the physicochemical properties of the protein. These include hydrophobicity, susceptibility to proteolysis, the ability to form ordered aggregates, and the capacity for interaction with other matrix constituents. For instance, fibrils of type II collagen, a moderately glycosylated collagen, apparently contain about twice as much water as fibrils of type I, as reflected in an increased lateral spacing of molecules (Grynpas et al. 1980). These phenomena have been attributed to the presence of considerably larger amounts of hydrophilic carbohydrate groups in type II molecules and may be of significance in terms of the function of type II fibrils in cartilage. Moreover, it is of interest to note that as a group, the α chains of types IV and V collagen are highly glycosylated (Table 2.2). The bulky carbohydrate groups render the molecules much more hydrophilic than they otherwise would be. It may be significant that each of these collagens apparently requires a specific collagenolytic enzyme for the initiation of degradation in vivo (Chapter 4) and neither of these collagens forms ordered fibrils like types I, II, and III collagen. Indeed, type IV molecules apparently form an extensive open network (Timpl et al. 1981). Aggregate structure of the collagens is discussed in detail in Chapter 1.

2.3.3. Crosslinking

It was known from early physicochemical studies of gelatin that the collagens (type I, at least) were extensively crosslinked by nondisulfide covalent bonds. The chemistry of the crosslinks was controversial until cyanogen bromide peptides were isolated from the N-terminal nonhelical ends of α1(I) and α2(I) and shown to be monomers containing a lysine-derived aldehyde, allysine, or dimers joined by allysine aldol (Bornstein et al. 1966; Bornstein and Piez 1966). It was then found that allysine (and hydroxyallysine derived from hydroxylysine) also formed aldimine condensation products with lysine and hydroxylysine (Bailey and Peach 1968; Tanzer 1968; Mechanic and Tanzer 1970). The aldol crosslinks are apparently largely intramolecular, and the aldimine crosslinks are largely intermolecular. Both eventually participate in complex multichain crosslinks still incompletely understood, as discussed below. Further discussion and references to the early literature can be found in the review by Bornstein and Traub (1979).

It is now clear that the ability of collagen fibrils and other aggregates to serve as the major structural elements in various connective tissues is primarily dependent on this system of covalent crosslinks among molecules. These linkages stabilize the aggregates and confer on them the tensile strength necessary to resist the physical forces normally encountered in different locations of the body. Although disulfide bonding plays an important role in molecular structure of the procollagens and of type III collagen (Section 2.3.1), the available evidence indicates that these linkages play no role in intermolecular bonding, at least not in types I, II, and III collagen. Presumably, the advantage of the aldehyde-derived crosslinks is that they can be multifunctional.

Types I, II, and III collagen molecules contain four major sites involved in crosslinking. In α1(I) (Figure 2.2) these sites consist of: (1) a residue of allysine or hydroxyallysine at position 9^N in the N-terminal nonhelical segment; (2) a residue of hydroxylysine at position 87 in the repetitive Gly-X-Y triplet sequence; (3) a residue of hydroxylysine at position 930 in the repetitive triplet sequence; and (4) a residue of allysine or hydroxyallysine at position 16^C in the C-terminal nonhelical segment. It may be that the lysine and hydroxylysine residues in the nonhelical ends may sometimes escape conversion to aldehydes and participate as amines. Similarly, the hydroxylysine residues at 87 and 930 probably sometimes escape hydroxylation and participate as lysine residues. Nearby histidines in the triplet region may also be involved. The other α chains have homologous sites where the sequence is highly conserved except for the C-terminal region of α2(I). The sequences around these positions are discussed in Section 2.4.6. These are the only known crosslinking sites in these collagens, but others are not ruled out.

As discussed in Chapter 1 (Section 1.3.3), intermolecular crosslinks between 9^N and 930 and between 16^C and 87 stabilize a 4-D stagger relationship between molecules that is important to fibril structure. Crosslinks between aligned molecules also seem to be important. Light and Bailey (1980) have presented evidence for extensive lateral networks of crosslinks across the type I fibril in rat tail tendon joining molecules related by 4-D and 0-D staggers. Position 87 is normally glycosylated and position 930 may sometimes be (Section 2.3.2). What effect this has on crosslink formation or function, if any, is not known. Glycosylation would, of course, prevent the participation of the hydroxy group of hydroxylysine in any other reaction.

The sites involved in crosslinking in type IV collagen have not as yet been determined. The available evidence, however, suggests that type IV molecules possess sites at both ends of the molecule with end overlaps (Kühn et al. 1981; Timpl et al. 1981). Unlike types I, II, and III collagen, the overlaps are between like ends. No information is available about crosslinks in type V collagen. Therefore, the discussion here relates only to types I, II, and III collagen. However, type IV collagen has been reported to contain some of the same crosslinks (Heathcote et al. 1980).

The reactions preliminary to crosslink formation are depicted in Figure 2.5. Lysine and hydroxylysine in peptide linkage are converted to allysine and hy-

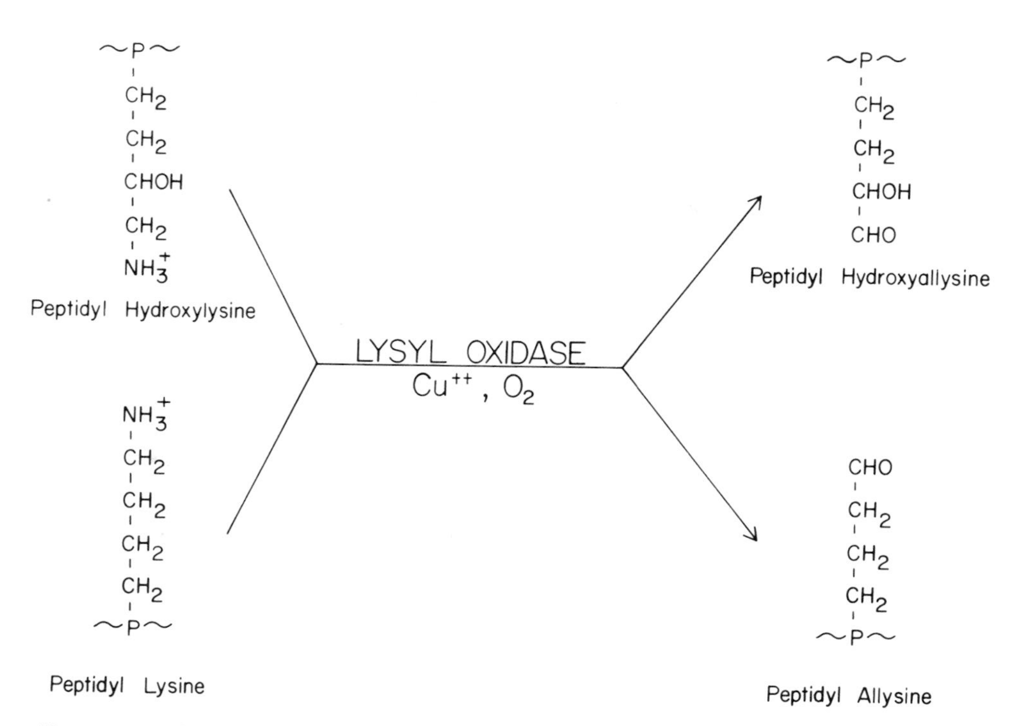

Figure 2.5. The formation of allysine and hydroxyallysine residues in the nonhelical ends of collagen α chains. *P* denotes the polypeptide backbond of an α chain.

droxyallysine, respectively, by the enzyme peptidyl lysine oxidase. Despite major sequence differences between the nonhelical ends of the collagens (Section 2.4.6), there is apparently only one enzyme. Elastin is also a substrate (Chapter 6). Peptidyl lysine oxidase is discussed in more detail in Chapter 3.

Allysine and hydroxyallysine initially react with each other or with lysine and hydroxylysine residues to form aldol and aldimine condensation products. No enzymatic catalysis is required for these condensation reactions. They occur spontaneously, requiring only the physical juxtaposition of appropriate side chains. In types I, II, and III collagen, this requirement is fulfilled through the accuracy with which individual molecules are aligned and packed within the fibrils. In elastin, the situation must be more complex since the polypeptide chains are random (Chapter 6).

In any event, the initial condensation reactions produce difunctional crosslinks capable of linking two separate polypeptide chains. The structures of the difunctional crosslinks found in collagen fibrils are illustrated in Figure 2.6. Allysine aldol (AA) arises as a result of the condensation of two allysine residues. It is isolated, after reduction of the aldehyde, as the α,β-unsaturated form as depicted, although a significant proportion of the compound may be present in the hydrated form in tissues. Formation of AA seems to be largely intramolecular. Covalently linked dimers of α chains (β-components), which are prevalent in the denaturation products of soluble type I collagen preparations, are linked in their N-terminal nontriplet regions by this crosslink. In principle, it could also link chains in aligned molecules, although evidence for that role is lacking.

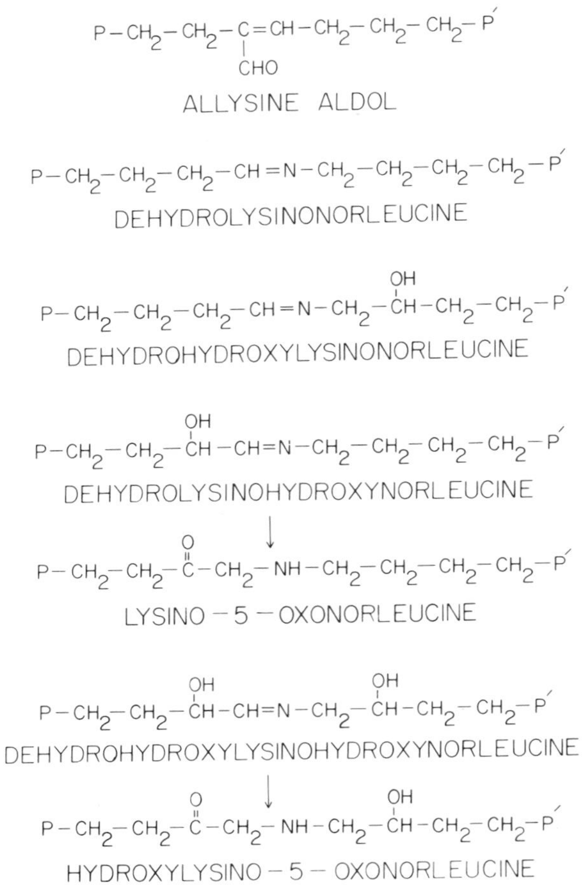

$P-CH_2-CH_2-C(CHO)=CH-CH_2-CH_2-CH_2-P'$

ALLYSINE ALDOL

$P-CH_2-CH_2-CH_2-CH=N-CH_2-CH_2-CH_2-CH_2-P'$

DEHYDROLYSINONORLEUCINE

$P-CH_2-CH_2-CH_2-CH=N-CH_2-CH(OH)-CH_2-CH_2-P'$

DEHYDROHYDROXYLYSINONORLEUCINE

$P-CH_2-CH_2-CH(OH)-CH=N-CH_2-CH_2-CH_2-CH_2-P'$

DEHYDROLYSINOHYDROXYNORLEUCINE

↓

$P-CH_2-CH_2-C(=O)-CH_2-NH-CH_2-CH_2-CH_2-CH_2-P'$

LYSINO − 5 − OXONORLEUCINE

$P-CH_2-CH_2-CH(OH)-CH=N-CH_2-CH(OH)-CH_2-CH_2-P'$

DEHYDROHYDROXYLYSINOHYDROXYNORLEUCINE

↓

$P-CH_2-CH_2-C(=O)-CH_2-NH-CH_2-CH(OH)-CH_2-CH_2-P'$

HYDROXYLYSINO − 5 − OXONORLEUCINE

Figure 2.6. Structures of the difunctional crosslinks found in the fibrillar collagens. *P* and *P′* denote separate α chains.

Dehydrolysinonorleucine (deLN) originates through the condensation of an allysine residue with a lysine residue to form a Schiff base or aldimine bond. It is normally only a minor component in the collagens because at least one of the precursor groups is usually hydroxylysine or hydroxyallysine. Also, deLN is found in elastin where hydroxylysine is absent (Chapter 6).

A more common aldimine crosslink is dehydrohydroxylysinonorleucine (deHLN), which is formed by condensation of an allysine residue with a residue of hydroxylysine. Although this type of linkage would be expected to be stable under physiologic conditions, it is broken by acid and by amines competing for the aldehyde group. This explains the ease of extractability of type I collagen from certain tissues, such as rat tail tendon, by acidic solutions. In terms of function, deLN and deHLN could theoretically serve as intramolecular crosslinks. However, their presence in preparations of highly insoluble collagen as well as their apparent absence in preparations of soluble collagen argue strongly for their role as intermolecular crosslinks. As such, deHLN would be formed as a result of condensation of an allysine residue in the N- or C-terminal nontriplet region, with a hydroxylysine residue located in a helical crosslinking site.

The remaining difunctional intermolecular crosslinks found in collagen fibrils are lysino-5-oxonorleucine (LON) and hydroxylysino-5-oxonorleucine (HLON). The initial step in the formation of LON involves the condensation of a residue of hydroxyallysine with a residue of lysine. The resultant crosslink, dehydrolysinohydroxynorleucine (deLHN), is a structural isomer of deHLN in which the hydroxy group is conjugated with an imminum double bond. This structure readily tautomerizes through an enamine intermediate to produce the much more stable keto-imine form of the crosslink, LON. The same mechanism following condensation of a hydroxyallysine residue with a residue of hydroxylysine forms HLON from dehydrohydroxylysinohydroxynorleucine (deHLHN).

Neither of the crosslink precursors, allysine and hydroxyallysine, nor their condensation products in collagen fibrils, are stable to acid or alkaline hydrolysis. Each of these compounds, however, contains an easily reducible function. In reduced form, all of the components are stable to acid hydrolysis with the exception of the reduced aldol condensation product, which is readily recovered following alkaline hydrolysis. Reduction of collagen preparations with reagents such as tritiated sodium borohydride stabilizes the crosslink precursors and their derivatives to hydrolysis and introduces a tritium atom, which facilitates their detection following hydrolysis. Of course, deHLN, deLHN, and LON give the same reduced products (except for the location of tritium, if used). Resolution and isolation of the reduced components in hydrolysates is commonly achieved by conventional cation-exchange chromatography employing an automatic amino acid analyzer with modified elution programs. A preliminary separation on a molecular sieve column improves the procedure (Housley and Tanzer 1981).

In addition to the difunctional crosslinks described above, five more complex components whose structures suggest a role in crosslinking have been reported to be present in collagen fibrils. Structures proposed for these multifunctional compounds are illustrated in Figure 2.7. One, dehydrohydroxymerodesmosine (deHMD), contains AA and hydroxylysine joined through an aldimine bond, although the order of condensation is unknown. deHMD is related to dehydromerodesmosine, derived from three lysine residues, a crosslink present in elastin (Chapter 6). These crosslinks are trifunctional and could join three chains.

Three others contain histidine. Aldol-histidine (AHis), is clearly a derivative of AA formed through Michael addition of an imidazole nitrogen to the α,β-unsaturated bond of AA. This component could also serve to crosslink three separate chains. Positions 89 and 932 in the repetitive Gly-X-Y sequence of α1(I), α1(II), and α1(III) are occupied by histidine (Section 2.4.6). These residues are just two residues from the helical crosslinking sites occupied by hydroxylysine at positions 87 and 930. Molecules containing an intramolecular aldol condensation product in either the N-terminal or C-terminal nonhelical region could then be linked to an adjacent molecule staggered by 4 D. The crosslink formed would therefore be a combined intra- and intermolecular linkage.

Similar considerations apply to the formation of hydroxyaldol-histidine (HAHis). In this case, however, the imidazole nitrogen is viewed as condensing with the reactive aldehyde group of the hydroxyaldol condensation product. Formation of the conjugated enamine linkage region between the imidazole group and the hydroxyaldol condensation product would require in vivo reduction and would account for the observation that this particular crosslinking component is resistant to borohydride reduction (Housley et al. 1975).

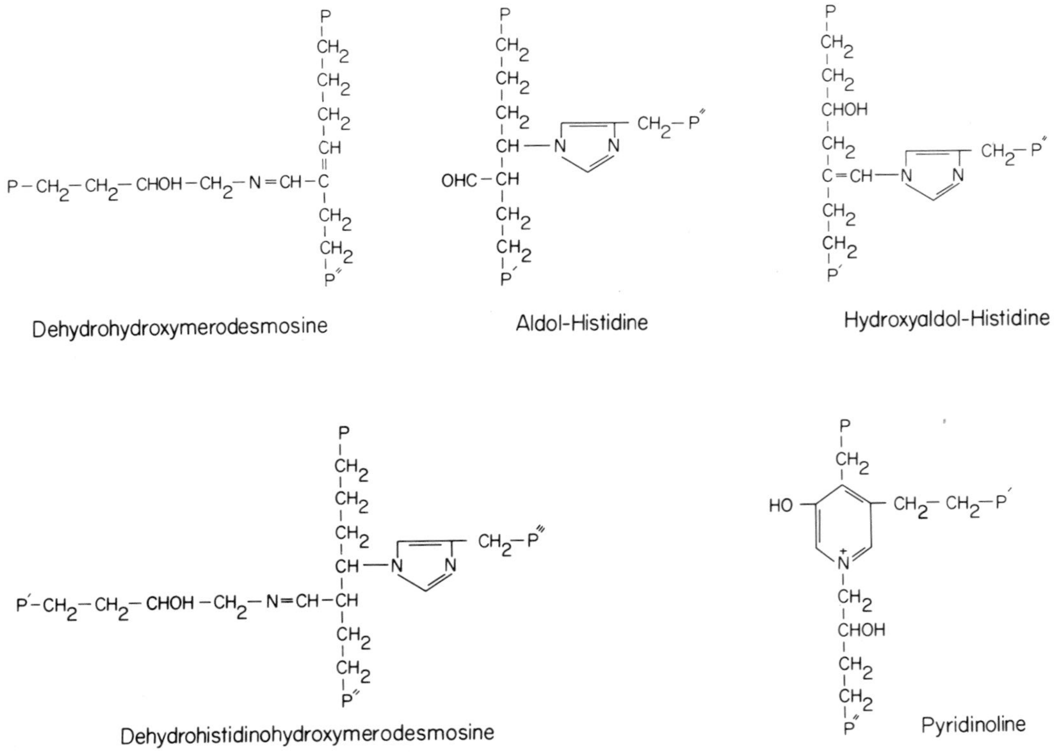

Dehydrohydroxymerodesmosine

Aldol-Histidine

Hydroxyaldol-Histidine

Dehydrohistidinohydroxymerodesmosine

Pyridinoline

Figure 2.7. Proposed structures of the multifunctional crosslinks found in the fibrillar collagens. *P, P′, P″,* and *P‴* denote α chain backbones.

Dehydrohistidinohydroxymerodesmosine (deHisHMD) represents the only tetrafunctional crosslink thus far identified in collagen. It could be formed through Michael addition of an imidazole nitrogen across the ethylene double bond of deHMD or by condensation of a hydroxylysine side chain with the aldehyde group of aldol-histidine. Like many of the other crosslinks, it may stabilize a 4 D stagger between molecules (Bernstein and Mechanic 1980).

The most recently reported multifunctional crosslinking component in collagen fibers is pyridinoline (Fujimoto et al. 1978). Several lines of evidence are consistent with the structure depicted in Figure 2.7. Assembly of the pyridinium ring requires the condensation of three hydroxylysine residues. It has been proposed that it is formed by the interaction of two residues of HLON which would reform a hydroxylysine residue as well as creating one residue of pyridinoline (Eyre and Oguchi 1980).

The crosslinking profile in the collagens of various tissues differs widely depending on the tissue and age of the organism. In dermis, for instance, reducible difunctional crosslinks derived from hydroxylysine residues are prevalent throughout fetal life. Following birth, however, the levels of these crosslinks diminish and there is a corresponding increment in the levels of crosslinks involving lysine residues, as well as some of the multifunctional linkages. In other tissues such as bone, dentin, cartilage and some tendons, reducible crosslinks derived from hydroxylysine residues are the predominant types of linkage throughout fetal life and into the early postnatal years. Regardless of the nature and proportions of the reducible crosslinks in a given tissue, their levels gradually diminish during maturation. Since collagen fibrils normally exhibit an increasing degree of tensile strength over this time span, it is generally assumed that the reducible crosslinks are gradually converted into more stable and perhaps more complex crosslinks during maturation. The conclusion is that the reducible crosslinks represent intermediates.

The issue is somewhat clouded by questions concerning the physiological significance of some of the reducible crosslinks, particularly those involving the Michael addition of an imidazole nitrogen of histidine to previously formed di- or trifunctional crosslinks. It has been suggested that the Michael adducts are formed artifactually in the process of reducing collagen fibers in vitro at relatively high pH. At the same time, there exists persuasive evidence indicating that such components may readily form under physiological conditions (Bernstein and Mechanic 1980). It has also been suggested that pyridinoline arises from noncollagenous components and is not a crosslink (Elsden et al. 1980). Again, there are arguments on both sides (Fujimoto 1980). In any event, HAHis and pyridinoline are both nonreducible trifunctional crosslinks which could, at least for the present, be considered representative of the types of crosslinks present in the mature collagen fibril. Alternatively, the presence of polymeric crosslinked peptides in cyanogen bromide digests of insoluble type I collagen (Light and Bailey 1980) suggests that polymeric crosslinks formed somewhat randomly by continued aldol, aldimine, and perhaps Michael condensations are the answer. It can be imagined that such crosslinks could extend across a fibril at the levels of the nonhelical ends (Chapter 1, Section 1.3.3).

While it is generally agreed that lysine-derived crosslinks are the major crosslinks in the fibrillar collagens, and probably the only crosslinks in newly synthesized fibrils, the literature suggests many other crosslinks not involving lysine. They are too numerous and the evidence is too indirect for presentation here, but they could be important in pathological states and in aging.

2.3.4. Extractability

Under appropriate conditions, collagen molecules present within defined fibrils, as well as those within other types of structural aggregates, can be dissociated and brought into solution in aqueous solvents. The major impediment to dissolution of these structures is, of course, the presence of covalent crosslinks between the molecules. In experimental situations, formation of crosslinks may be dramatically reduced by administration of agents such as β-aminopropionitrile or α-aminoacetonitrile. These agents, termed lathyrogens, irreversibly inhibit peptidyl lysine oxidase, thereby preventing the formation of allysine and hydroxyallysine (Chapter 3). Since the oxidase requires copper as a cofactor, much the same effect can be achieved by feeding a copper-free diet or a copper chelator. Collagens from tissues of lathyritic or copper-deficient animals are thus much more readily extractable than those from tissues of normal animals. Lathyrogens are also often used in cell or tissue culture systems.

Three solvent systems are commonly used for the extraction of native collagen molecules. Specific methods have been recently summarized (Miller and Rhodes 1982). Neutral salt solvents (0.15–1.0 *M* NaCl, pH 7.5) in the cold preferentially extract molecules not yet covalently crosslinked. In some tissues, such as the skin of growing animals, crosslinking is sufficiently slow for this procedure to be effective. However, most tissues have little or no salt-extractable collagen unless a lathyrogen has been used.

Dilute organic acid solvents such as 0.5 *M* acetic acid in the pH range of 2.0–3.0 represent somewhat more potent solvent systems. The efficacy of these solvents may be ascribed largely to their low pH, which promotes swelling of the fibrils and solvation of the component molecules. In addition, as already noted, covalent crosslinks containing an aldimine bond would be expected to be labile at low pH. Since both neutral and acid proteases are present in tissues, which may degrade the nonhelical ends or other regions of the collagens, it is common today to include a mix of appropriate protease inhibitors in extraction and purification solvents (Chandrakasan et al. 1976).

A third solvent system is provided by a dilute organic acid (0.5 *M* acetic acid) to which an acid protease such as pepsin is added, usually at a 1:10 weight ratio of enzyme to dry weight of tissue. The efficacy of this solvent system arises from selective cleavage in the N- and C-terminal nonhelical regions breaking peptide bonds near crosslinks and releasing molecules, which then dissolve in the acid. Covalent crosslinks presumably remain, attaching small peptide remnants to the helical crosslink sites. The collagen is recovered with modified nonhelical ends and is, therefore, not suitable for studies in which intact collagen is critical. Limited proteolysis may also lead to more extensive degradation of the protein, particularly in those instances where the triple helix is not continuous throughout the molecule, as is the case for type IV collagen. However, pepsin-solubilized collagens, including fragments of type IV collagen (Kresina and Miller 1979; Gay and Miller 1979), are useful for characterization and quantitation since recoveries may be quite good (Miller and Rhodes 1982).

Collagen may also be extracted from a variety of sources in denaturing solvents. Although these solvents are often quite useful, the denatured collagen is generally characterized by a continuum of high molecular weight aggregates. In addition, many of the techniques developed for separation of the various types of collagen once they have been brought into solution are useful only when dealing with native, largely monomeric collagen preparations.

2.4. Distinctive Chemical Features

The common chemical features of the collagens described above clearly indicate that this group of proteins represents a unique, closely related set of biological macromolecules. Nevertheless, each α chain, as well as each molecular species composed of the various chains, possesses distinct chemical characteristics which presumably serve as the basis for its biological function and, incidentally, its identification. These characteristics stem largely from the primary structure of the individual α chains. In addition, a given collagen type in a specific location may also exhibit unique chemical properties as a result of posttranslational modification, for example, phosphorylation and the attachment of phosphoprotein molecules to type I collagen in tissues to be mineralized (Chapter 9). These kinds of modifications, however, are not well understood in relation to function and are best considered in discussions of the specific tissue.

2.4.1. Selective Precipitation Behavior

Native collagen molecules dissolved in either a neutral salt solvent or a dilute acidic solvent can be selectively precipitated from solution at discrete salt concentrations (Miller and Rhodes 1982). Although chromatographic techniques have been developed for the resolution of certain combinations of native molecular species (See below), they are not generally applicable. Therefore, selective salt precipitation procedures have been developed. They provide a relatively rapid and convenient method for resolution of the collagen types.

Figure 2.8 depicts a flow diagram illustrating the precipitation behavior of collagens in 1.0 *M* NaCl buffered at pH 7.5. The initial solution is assumed to contain all of the well-characterized collagens and thus represents an idealized mixture that would rarely, if ever, be encountered in practice. However, it represents all possible mixtures. This technique alone does not permit the resolution of all collagen types since types III and IV coprecipitate at 1.8 *M* NaCl and types II, V, and I trimer coprecipitate at the much higher NaCl concentration of 4.0 *M*. In practice, a tissue would be selected that contained collagens that could be separated. For example, mammalian skin contains largely type I, type III, and small amounts of type I trimer, all of which can be resolved from one another. That type I (chain composition, $[\alpha 1(I)]_2\alpha 2(I)$) and type I trimer (chain composition, $[\alpha 1(I)]_3$) are resolved, suggests that the chemical features respon-

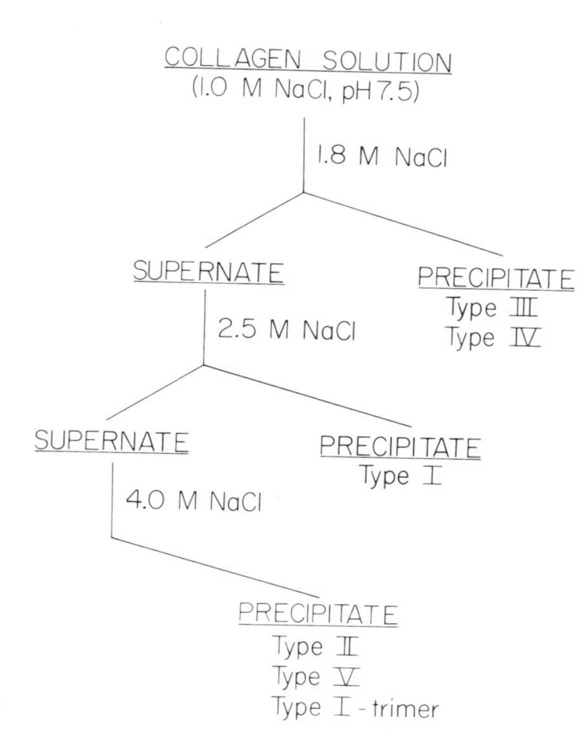

Figure 2.8. A flow diagram illustrating the selective precipitation behavior of different collagens in a cold, neutral salt solvent as the salt concentration is increased.

sible for selective precipitation behavior under these conditions are derived largely from the primary structure of the constituent chains of the different molecular species. This concept is reinforced by the additional observation that the procollagens precipitate selectively from neutral salt solutions at about the same ionic strengths as their respective collagen derivatives.

A much different pattern is observed when the collagens are selectively precipitated from a dilute acid solvent as shown in Figure 2.9. In this case, the collagens precipitate over a lower and narrower range of NaCl concentrations, 0.7–1.2 *M*. In addition, the separations achieved correlate fairly well with the extent of glycosylation of the individual types of collagen. Thus, type I, II, III, and I trimer collagens, which contain low to moderate levels of attached carbohydrate, coprecipitate at 0.7 *M* NaCl, whereas the highly glycosylated type IV and V collagens remain in solution until precipitation at 1.2 *M* NaCl. There is also an apparent correlation with the size of the nonhelical domains, types IV and (almost certainly) V having large globular regions. While precipitation from dilute acid or from neutral salt is not adequate for the resolution of all collagen types, a combination of both procedures will theoretically result in the resolution of most mixtures encountered in the preparation of collagens.

Both of the above procedures assume that the collagens are largely monomeric. If a sample cannot be dissolved in cold neutral 1 *M* NaCl or 0.5 *M* acetic acid, it can be assumed that it is highly aggregated or altered in some other way. Advantage can be taken of this behavior to remove aggregates. For type I collagen in 0.5 *M* acetic acid, a preliminary precipitation at 0.53 *M* NaCl will leave a solution that is largely monomeric (Chandrakasan et al. 1976).

The procedures described above are performed in the cold. The precipitates obtained under these conditions are largely amorphous, indicating random precipitation of unstructured aggregates. At higher temperatures, types IV and V collagen do not precipitate under conditions promoting the formation of native fibrils from solutions of types I, II, and III collagen (e.g., 0.16 *M* potassium

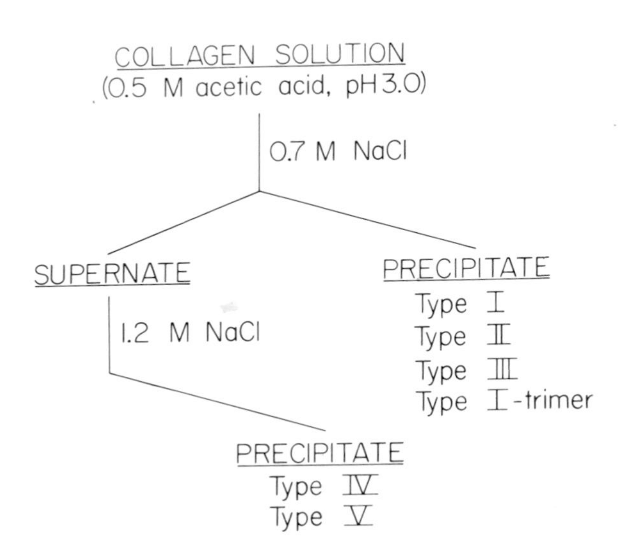

Figure 2.9. A flow diagram illustrating the selective precipitation behavior of different collagens in a cold, dilute acidic solvent as the salt concentration is increased.

phosphate, pH 7.5, 37°C) (Chapter 1, Section 1.3.4). Using this procedure, then, essentially the same resolution can be achieved as from a dilute acid solvent.

2.4.2. Chromatographic Properties

Nondenaturing Conditions. Chromatography of collagens under nondenaturing conditions affords, in principle, an additional means of purifying and resolving collagens that are not completely separated by selective salt precipitation techniques. The range of separations attainable is, however, limited by the high viscosities and nonideal behavior of native collagen solutions. This approach has been utilized largely in the purification of pepsin-solubilized types IV and V collagen preparations (Miller and Rhodes 1982) and the resolution of type I and III procollagens (Byers et al. 1974).

Preparations of type IV collagen obtained by pepsin treatment are conveniently chromatographed at low temperatures (4–8°C) on carboxymethyl (CM)-cellulose in sodium acetate buffer containing 2.0 *M* urea, pH 4.8. Elution is achieved by means of a linear gradient from 0.0 to 0.4 *M* NaCl. Type IV collagen elutes early in the gradient but is well resolved from acidic components that are not retained by the column. Other more basic types of collagen are more strongly retained by the column and elute after type IV collagen. Preparations of pepsin-solubilized type V collagen may likewise be resolved from contaminating collagens by chromatography in slightly alkaline buffers (0.2–0.5 *M* Tris, pH 8.0–8.6) on CM- or diethylaminoethyl (DEAE)-cellulose. When employing CM-cellulose columns under these conditions, the most basic collagen, type V, is retained by the column while other collagen types are not retained. The opposite principles apply when utilizing DEAE-cellulose, with type V collagen eluting earlier than the more acidic components.

Chromatography of type IV and V collagens under nondenaturing conditions has been useful in resolving individual molecular species within a type. Chromatography of pepsin-solubilized murine type IV collagen on DEAE-cellulose in 0.005 *M* Tris containing 2.0 *M* urea, pH 8.6, has led to the resolution and isolation of two molecular species which appear to be homopolymers composed, respectively, of somewhat truncated α1(IV) and α2(IV) chains (Timpl et al. 1979). Chromatography of type V collagen preparations derived by pepsin solubilization from human placental villi on phosphocellulose in 0.04 *M* Tris containing 2.0 *M* urea, pH 8.2, resolves two molecular species (Rhodes and Miller 1981) identified as heteropolymers with the chain compositions α1(V)α2(V)α3(V) and $[\alpha1(V)]_2\alpha2(V)$.

Denaturing Conditions. When the collagens are heated in solution, they denature at characteristic temperatures (Chapter 1, Section 1.2.4) to form random coils. Unlike most proteins, they are soluble under these conditions and behave ideally when chromatographed. The two most useful procedures are molecular sieve chromatography on agarose or crosslinked dextran, and ion exchange chromatography on modified celluloses. The two procedures are particularly powerful when used in combination. They are useful for α chains, crosslinked α chains, and fragments or peptides derived from α chains. While other procedures such as high pressure liquid chromatography of peptides are potentially useful, none are yet in extensive use for α chains or their fragments.

Molecular sieve chromatography separates on the basis of molecular weight and can be used both preparatively and analytically. A chromatogram of denatured type I collagen is shown in Figure 2.10. In addition to monomeric α chains, it shows the presence of crosslinked α chains, primarily dimers with some trimers, and a small amount of higher molecular weight material. The dimers and trimers are sometimes referred to as β- and γ-components, respectively. The solvent, 1 *M* $CaCl_2$ buffered at pH 7.5, is convenient, as it lowers the melting temperature to room temperature for type I collagen and inhibits adsorptive effects (Piez 1968). Solvents containing other denaturing agents, such as sodium dodecyl sulfate or guanidine are also used (Miller and Rhodes 1982). Calibrated columns can be used to determine molecular weights and appear to be free of the anomalous effects observed with gel electrophoresis (Section 2.4.3). However, the latter procedure has much higher resolution.

Since the constituent chains of all known collagens are relatively basic, ion exchange chromatography is commonly performed on CM-cellulose in a sodium acetate buffer, pH 4.8, at a temperature above the melting point of the collagen (Piez et al. 1963). Under these conditions, the random-coil polypeptide chains exhibit a net positive charge and are retained by the cation exchanger. Gradient elution with an increasing concentration of NaCl is then used to displace individual chains as a function of net positive charge. The procedure provides not only a means of purifying and isolating the various chains in preparative applications, but is also a useful analytical procedure in identifying the nature of a given collagen preparation. Current approaches utilized in CM-cellulose chro-

Figure 2.10. Molecular sieve chromatography of denatured type I collagen in 1 *M* $CaCl_2$, pH 7.4 at room temperature. Monomer, dimer, and trimer refer to α chains [α1(I) and α2(I) in this case], and crosslinked dimers and trimers. *(Source: Chandrakasan et al. (1976), with permission.)*

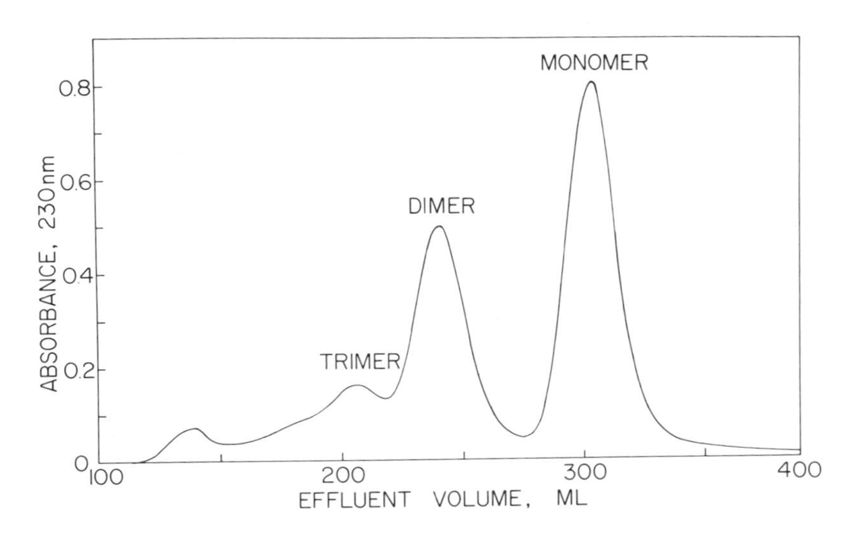

matography, as well as in phosphocellulose chromatography (a stronger cation exchange system) of denatured collagen preparations, have recently been detailed (Miller and Rhodes 1982).

For purposes of illustration, the CM-cellulose elution profile of a preparation of denatured type I collagen is depicted in Figure 2.11. Crosslinked components present in the sample were removed by molecular sieve chromatography (Figure 2.10) prior to CM-cellulose chromatography to more clearly illustrate the elution profile of the primary chains. The latter, α1(I) and α2(I), are present in a 2:1 ratio, which reflects the stoichiometry of type I collagen.

Also indicated in Figure 2.11 are the elution positions observed for the chains derived from other types of collagen when chromatographed under the same conditions. Small species differences may be observed. It is apparent that CM-cellulose chromatography will not resolve all potential combinations of α chains. This approach will, however, generally afford excellent resolution of the α chains present in the denaturation products of individual collagen types. The only exceptions are the chains of the type V collagens which are most readily resolved by sequential chromatographic steps involving initial resolution on phosphocellulose (Rhodes and Miller 1978), followed by rechromatography of fractions containing α1(V) and α3(V) chains on CM-cellulose (Sage and Bornstein 1979).

2.4.3. Electrophoretic Properties

Polyacrylamide gel electrophoresis in sodium dodecyl sulfate (SDS) is the most widely used analytical procedure for the identification of the collagens (Miller and Rhodes 1982). Its utility lies in the relative ease and rapidity with which analyses can be performed, as well as its high sensitivity and resolution. Only microgram quantities of protein are required, whereas conventional chromato-

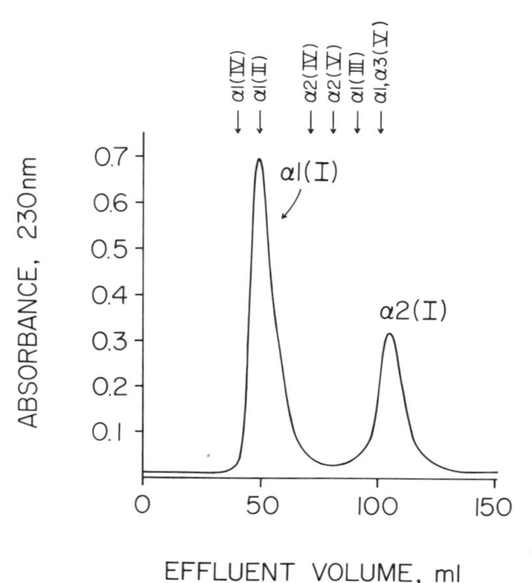

Figure 2.11. CM-cellulose chromatography of the α chains derived from denatured type I collagen. Crosslinked components were removed by molecular sieve chromatography (Figure 2.10). Chromatography was performed at 42°C and pH 4.8 using a linear salt gradient. Arrows at the top of the figure show the positions at which other collagen α chains are eluted under the same conditions.

graphic procedures such as those described above must be performed with milligram quantities. While the constituent chains of some collagens and their procollagens migrate anomalously in SDS gels relative to one another, complicating molecular weight determinations, this factor allows the resolution of some chains with similar molecular weights and thereby further enhances the utility of the method.

All collagen α chains migrate relative to globular proteins as if they were larger as well as having relative differences. Although conformational effects resulting from the high content of proline and hydroxyproline have been used to explain the anomaly, it is likely that the differential effects result from variations in hydrophobicity and SDS binding (Hayashi and Nagai 1980). Other contributing factors, such as the high carbohydrate content of some of the collagen α chains, are not ruled out.

An SDS polyacrylamide gel electrophoretogram of the α chains of types I, II, and III collagen is shown in Figure 2.12. A schematic representation of all the collagen α chains is shown in Figure 2.13. The α chains of type I collagen demonstrate unusual behavior in that α2(I) chains have greater mobility than α1(I) chains of the same molecular weight (95,000). Similar differences are commonly observed for the corresponding Pro α chains. Although α1(II) and α1(III) chains migrate in a manner identical to α1(I) chains, α1(III) is normally present as the trimer and is well separated. Also, α1(II) is unaccompanied by α2(I) if free of type I. The relatively low electrophoretic mobility of the partially truncated α1(IV) and α2(IV) chains undoubtedly reflects the somewhat larger size of these chains (apparent molecular weights 160,000 and 140,000, respectively) relative to other α chains. The α chains of type V collagen are approximately the same size as the other α chains after pepsin digestion (Rhodes and Miller 1979), yet two of the type V chains migrate as apparently larger molecular weight polypeptides. It may be significant that their migration correlates with their carbohydrate content.

Most native collagen preparations will contain intramolecularly crosslinked components, as already noted. This is especially true of type III, which exists largely as the disulfide-bonded γ-component but is readily reduced to α chains, as shown in Figures 2.12 and 2.13. The pattern of α chains and β- and γ-components of type I collagen can also be readily identified. The aldehyde crosslinks are of course unaffected by reducing agents used for electrophoresis but are present in much lower amounts in lathyritic or pepsin-treated samples. Type II collagen preparations show little or no β-component because type II is extractable only with pepsin or from lathyritic animals.

2.4.4. Amino Acid Composition

The amino acid composition of the nine collagen α chains discussed here is presented in Table 2.3. For uniformity, all of the data listed are for human collagens (*See* Miller and Gay 1982 for references). Since species differences are generally minimal, these data may be considered representative of any mammalian species. Extensive species data have been published (Eastoe 1967). Since sequence information is currently available only for the α1(I), α2(I), α1(II), and

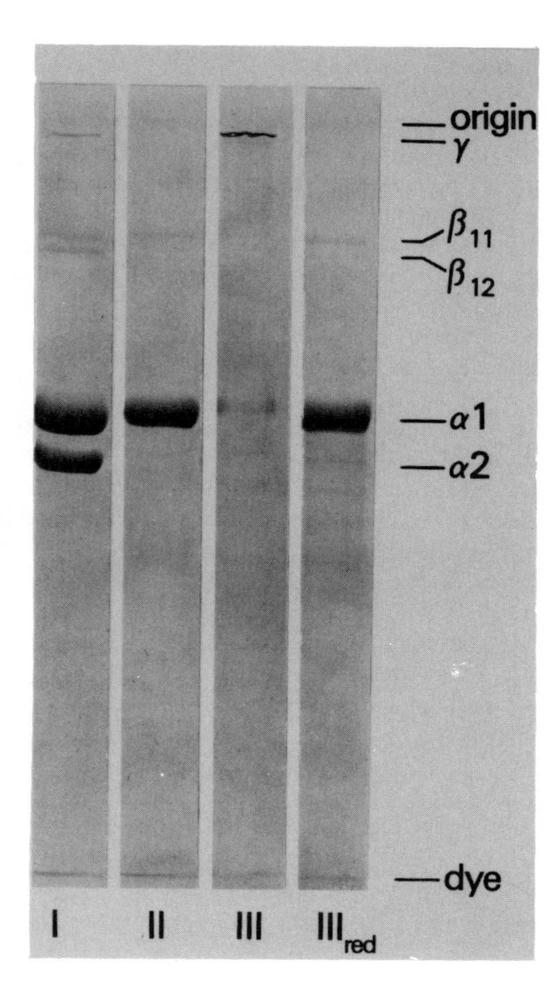

Figure 2.12. Sodium dodecyl sulfate (SDS) gel electrophoresis of different collagen types showing their α chain composition. *(Courtesy of S.L. Lee.)*

Figure 2.13. A diagrammatic presentation of sodium dodecyl sulfate (SDS) gels of the five different collagens (types I–V). The subscripts *U* and *R* denote unreduced and reduced preparations. The symbol γ denotes the disulfide-bonded trimers of denatured types III and IV collagen.

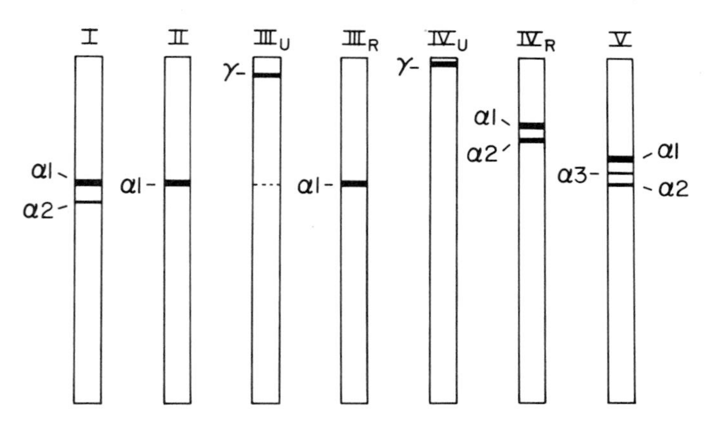

α1(III) chains, the compositional features of the chains constitute the most complete set of data now available for comparisons involving all α chains.

As would be expected, the number of glycine residues in each chain is constant at approximately one-third of the total number of residues. Exceptions to this rule are α1(III), which contains significantly more than one-third of its residues as glycine, and α2(IV), which contains somewhat less than one-third glycine residues. The sequence of bovine α1(III) (*See* Hofmann et al. 1980 for references) contains 15 glycine residues in the X position and two in the Y position of the Gly-X-Y repeat, while α1(I) contains only one extra glycine. The α2(IV) preparation presumably retained some nonhelical portions in spite of the pepsin treatment.

Other compositional differences clearly serve to distinguish individual α chains as well as groups of chains. The α chains of types IV and V collagen all contain fewer alanine residues than the chains of types I, II, and III collagen. In the α chains of type IV collagen, the decreased alanine content is accompanied by an increased number of the larger hydrophobic residues. In the α chains of type V collagen, however, there are fewer of the larger hydrophobic amino acids and a modest increase in the number of basic residues. Other characteristics serve to distinguish certain chains. The proportion of hydroxylated proline and lysine residues to nonhydroxylated residues, for instance, is significantly elevated in the α chains of type IV collagen relative to other α chains. These considerations,

Table 2.3. Amino Acid Composition of Human Collagen α Chains Isolated by Extraction and Purification after Pepsin Digestion[a]

Amino acid	α1(I)	α2(I)	α1(II)	α1(III)	α1(IV)[b]	α2(IV)[b]	α2(V)	α1(V)	α3(V)
3-Hydroxyproline	1	1	2	0	1	1	3	5	1
4-Hydroxyproline	108	93	97	125	122	110	106	110	91
Aspartic acid	42	44	43	42	45	49	50	49	42
Threonine	16	19	23	13	19	30	29	21	19
Serine	34	30	25	39	38	30	34	23	34
Glutamic acid	73	68	89	71	78	65	89	100	98
Proline	124	113	120	107	85	73	107	130	99
Glycine	333	338	333	350	334	324	331	332	332
Alanine	115	102	103	96	30	47	54	39	49
Half-cystine	0	0	0	2	0	2	0	0	1
Valine	21	35	18	14	33	27	27	17	29
Methionine	7	5	10	8	15	14	11	9	8
Isoleucine	6	14	9	13	32	38	15	17	20
Leucine	19	30	26	22	52	56	37	36	56
Tyrosine	1	4	2	3	5	7	2	4	2
Phenylalanine	12	12	13	8	27	36	11	12	9
Hydroxylysine	9	12	20	5	50	36	23	36	43
Lysine	26	18	15	30	6	7	13	14	15
Histidine	3	12	2	6	6	6	10	6	14
Arginine	50	50	50	46	22	42	48	40	42

[a]Values are expressed as residues/1000 total amino acid residues.

[b]Values are for fragments from pepsin solubilized human placental type IV collagen. Both represent largely triple-helical parts of the molecules.

then, suggest that α chains may be classified into three groups: the interstitial α1(I), α2(I), α1(II), and α1(III) chains; the chains derived from type IV collagen; and the chains derived from type V collagen. Structural evidence at the molecular level is also consistent with these distinctions (Chapter 1).

2.4.5. Cyanogen Bromide Peptides

Specific chemical cleavage of collagen α chains at methionine residues with cyanogen bromide (CNBr) has considerably facilitated their further characterization. Since α chains of most species contain relatively few methionine residues (Table 2.3), the cleavage products may generally be isolated and purified by the conventional molecular sieve and ion-exchange chromatographic procedures discussed in Section 2.4.2. The purified cleavage products, which range in size from di- and tripeptides to peptides with molecular weights near 40,000, have proved to be valuable starting materials for primary structure determinations. Moreover, even though the various chains exhibit a great deal of homology, cleavage with CNBr generates a unique set of fragments for a given chain, providing an additional analytical approach to identification (Miller 1976; Bornstein and Sage 1980). In this case, separation by SDS gel electrophoresis is a sensitive means of producing characteristic patterns. A typical set is shown in Figure 2.14. Interpretation is complicated by species differences and by incomplete cleavage yielding extra peptides. However, diagnostic features can be selected for any given system to identify α chains and even for approximate quantitation.

The location of individual CNBr peptides along the length of α1(I), α2(I), α1(II), and α1(III) has been established (*See* Traub and Piez 1971). This information is summarized in Figure 2.15, which depicts each of the human α chains as a linear composite of its CNBr peptides. The illustration clearly reveals a unique placement of methionine residues in each chain. The α1(I) and α1(II) chains appear to be closely related in that each of these chains contains a methionine residue at five common positions. A close relationship between these chains and α1(III) is also apparent since all three chains contain a methionine residue at three of the same positions, residue numbers 123, 402, and 551. The most different is α2(I), which has two methionine residues in common with α1(II), positions 3 and 6, and one in common with α1(I), at position 3.

The distribution of methionine residues along the α chains from most vertebrate species is similar but not identical to that shown for the human chains in Figure 2.15. Additional and missing methionine residues are sometimes observed. This has led to a nomenclature where the CNBr peptides of the first α chain characterized have been numbered consecutively as they appear in chromatographic effluents. Some small peptides originally missed have been designated 0. When placed along the α chain, the numbers then have a random order. If an α chain from another species is found to have a missing methione, the larger peptide is assigned both numbers, reflecting interspecies homology. Thus, α1(I)CB0 and α1(I)CB1 in rat become α1(I)CB0,1 in human. If an extra methionine is found, the designations A and B are used. Thus, α1(I)CB6 in

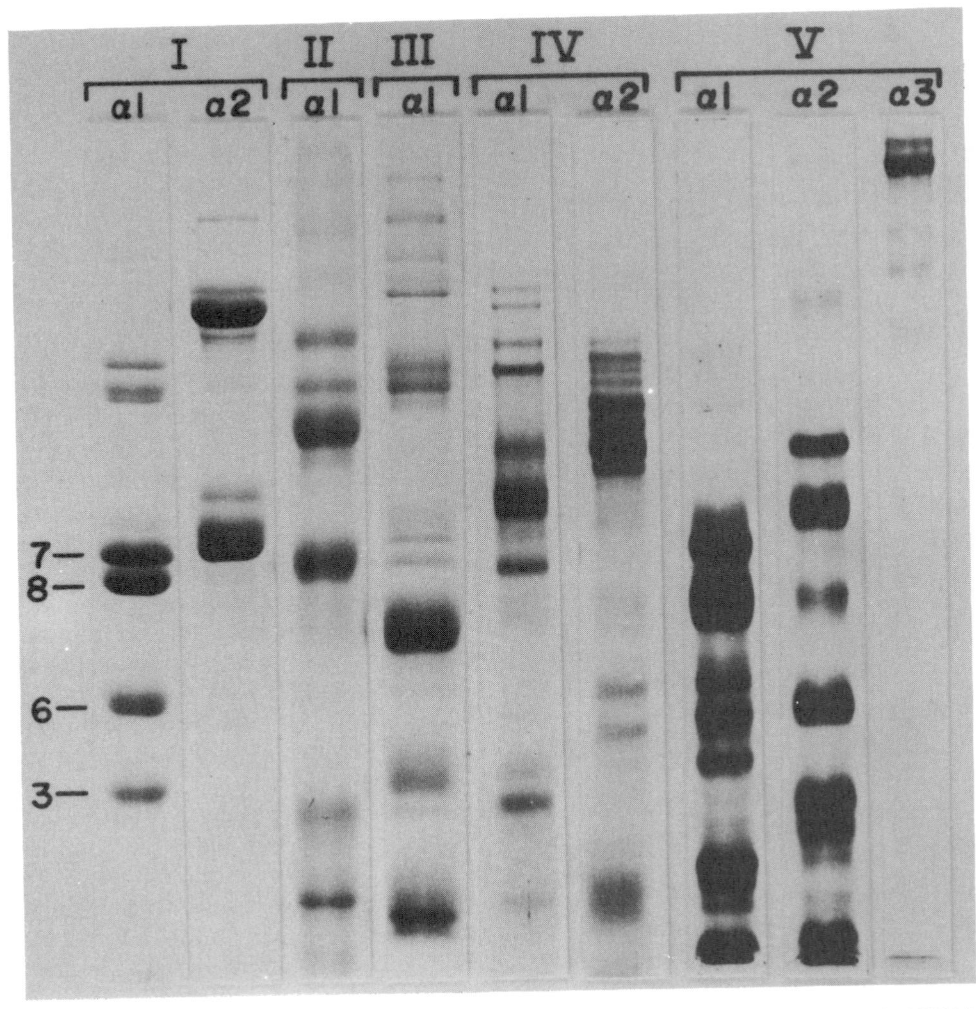

Figure 2.14. Sodium dodecyl sulfate (SDS) gel electrophoresis of the cyanogen bromide (CNBr) peptides of α chains from several collagen types. *(Source: Bornstein and Sage (1980), with permission.)*

human or rat becomes α1(I)CB6A and α1(I)CB6B in chick. Species differences have been summarized by Bornstein and Traub (1979). This nomenclature will not be satisfactory for chains where interspecies homology of methionine residues is low. Also no attempt has been made to show intertype homology. Thus, α1(I)CB3, α1(II)CB8, and α1(III)CB4 are homologous (Figure 2.15) but carry different numbers.

The CNBr peptides derived from α1(V) and α2(V) chains from pepsin-solubilized human type V collagen preparations have recently been isolated and characterized with respect to molecular weight and amino acid composition (Rhodes and Miller 1979; Rhodes et al. 1981). Similar studies have been performed on α1(IV) chains from pepsin-solubilized bovine type IV collagen as well as on α2(IV) chains from pepsin-solubilized porcine type IV collagen. The cleav-

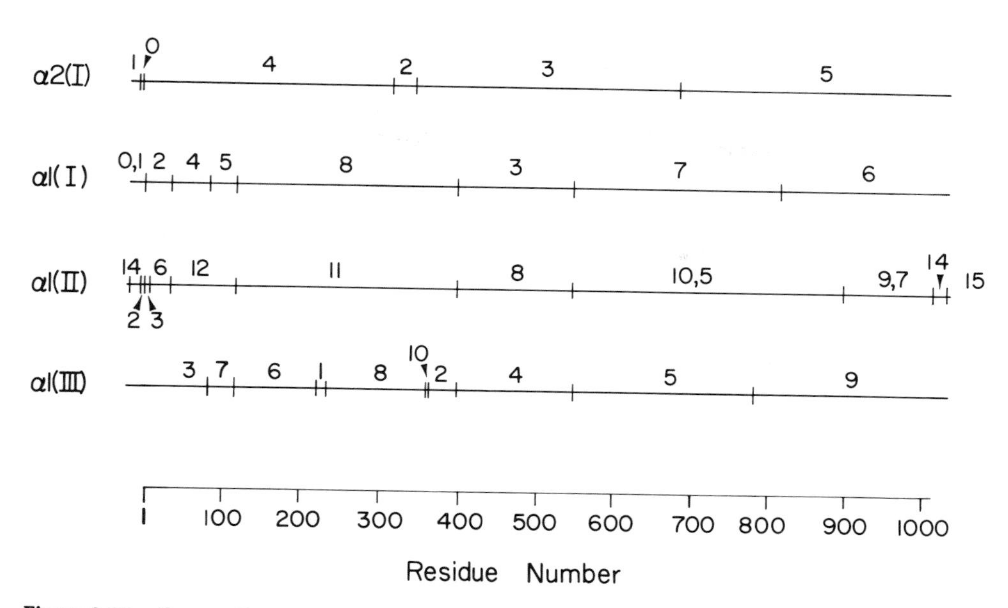

Figure 2.15. Composite representations of human α2(I), α1(I), α1(II), and α1(III) chains illustrating the positions and relative sizes of the cyanogen bromide (CNBr) peptides from each chain. Locations of methionine residues are indicated by vertical lines. Residue number denotes position within the Gly-X-Y repetitive triplet sequence of the α chains. The nonhelical ends are not numbered here.

age products isolated and characterized in these studies have not as yet been ordered. Nevertheless, the data clearly indicate a unique distribution of methionine residues in each chain and confirm that each represents a unique gene product.

It has also been shown that treatment of α1(IV) and α2(IV) chains with either a mast cell protease or *Staphylococcus aureus* V8 protease produces from each chain a unique set of cleavage products which are distinguishable by SDS gel electrophoresis (Bornstein and Sage 1980). Although the fragments have not been isolated and characterized, the use of these enzymes affords an alternative method for obtaining a relatively small number of specific cleavage products and a characteristic peptide pattern.

2.4.6. Primary Structure

Data on the primary structure of individual CNBr peptides together with the position of these sequences within the chains have been used to assemble the complete primary structure of bovine α1(I), α2(I), and α1(III) (*See* Hofmann et al. 1978, 1980), human α1(III) (Seyer and Kang 1981), and a large portion of rat α1(I) and bovine α1(II) (*See* Bornstein and Traub 1979). The sequence of a composite prepro α1(I) chain has been presented (Figure 2.2) as a prototype for the α chains, at least those of types I, II, and III collagen. Only a small amount of sequence data has been reported for type IV (Schuppan et al. 1980) and none as yet for type V. As previously noted, type IV has large noncollagenous domains and the triplet sequence appears to have interruptions in several places. Type V may be like α1(I), at least in the α chain triplet region, as judged by the resistance of the pepsin-solubilized triplet region to further proteolytic cleavage.

Extant information for the pN and pC peptides of types I, II, and III collagens has been presented (Section 2.3.1). The discussion here will concentrate on regions of the α chains of these collagens where something is known about function—the nonhelical ends, the helical crosslink sites, and the animal collagenase site. Some general features of the triplet regions will also be considered. Although data are available for some α chains derived from several species, the data are most complete for bovine α chains. For the most part, therefore, data on bovine chains have been used in the construction of Figures 2.16 through 2.19. Exceptions are noted in the legends.

Figure 2.16 shows sequence data for the N-terminal nonhelical region. A glutamine residue constitutes the initial amino acid in each of the α chains. The procollagen N-proteases (Chapter 3) cleave just before this residue. Cyclization to form pyroglutamic acid either in situ or during extraction of collagen results in a blocked N-terminus. Aside from the initial residue in each chain, very little interchain homology is apparent in the nontriplet sequences except in the region of residues 7^N–10^N. Within this sequence, each chain possesses a lysine or hydroxylysine residue, which is a substrate for peptidyl lysine oxidase and, as the aldehyde, is required for subsequent crosslinking reactions (Section 2.3.3). Although these positions in α2(I), α1(I), and α1(III) are represented by lysine residues in Figure 2.16, they may be partially or fully hydroxylated, depending on the tissue from which the collagen is prepared and its developmental status.

Sequence data for the C-terminal nonhelical region are shown in Figure 2.17 (residues 1^C–26^C). Complete sequences are available only for bovine α1(I) as well as chick α1(I), α1(II), and α2(I), the chick sequences via cDNA. The interchain homology would be very high except for a deletion of 13^C–23^C in α2(I). This is unexpected since the lysine residue at 17^C in α1(I) is believed to be a crosslink precursor. Its absence in α2(I) means that it cannot participate in crosslinking

Figure 2.16. Sequences from the N-terminal nontriplet region (*superscript N*) and the beginning of the repetitive triplet region of several bovine α chains. Gaps have been inserted to improve homology. Commas indicate a suggested sequence from compositional data.

```
             1N                5N                 10N                 15N
α2(I)      Gln-         Phe-        Asp-Ala-Lys-Gly-    Gly-    Gly-    Pro-
α1(I)      Gln-Leu-Ser-Tyr-Gly-Tyr-Asp-Glu-Lys-Ser-Thr-Gly-Ile-Ser-Val-Pro-
α1(II)     Gln-Met,Ala,Phe,Gly,Glx,Asp,Glu,Hyl,Gly,Ala,Gly,Ala,Gly,Met,
α1(III)    Gln-Tyr-Glu-    Ala-Tyr-Asp-Val-Lys-Ser-Gly-Val-Ala-Gly-Gly-

             1               5               10              15
α2(I)      Gly-Pro-Met-Gly-Leu-Met-Gly-Pro-Arg-Gly-Pro-Hyp-Gly-Ala-Ser-
α1(I)      Gly-Pro-Met-Gly-Pro-Ser-Gly-Pro-Arg-Gly-Leu-Hyp-Gly-Pro-Hyp-
α1(II)     Gly-Val-Met-Gly-Pro-Met-Gly-Pro-Arg-Gly-Pro-Hyp-Gly-Pro-Ala-
α1(III)    Gly-Ile-Ala-Gly-Tyr-Hyp-Gly-Pro-Ala-Gly-Pro-Hyp-Gly-Pro-Hyp-
```

	1000 1005 1010 1014
α2(I)	-Gly-Pro-Hyp-Gly-Pro-Hyp-Gly-Pro-Hyp-Gly-Pro-Hyp-Gly-Pro-Pro-
α1(I)	-Gly-Pro-Hyp-Gly-Pro-Hyp-Gly-Pro-Hyp-Gly-Pro-Hyp-Gly-Pro-Pro-
α1(II)	-Gly-Asn-Hyp-Gly-Pro-Hyp-Gly-Pro-Hyp-Gly-Pro-Hyp-Gly-Thr-Gly-
α1(III)	-Gly-His-Hyp-Gly-Gln-Hyp-Gly-Pro-Hyp-Gly-Ala-Hyp-Gly-Pro-Cys-

	1^{C} 5^{C} 10^{C}
α2(I)	Ser-Gly-Gly-Tyr-Glu-Phe-Gly- Phe-Asp-Ala-Glu-
α1(I)	Ser-Gly-Gly-Tyr-Asp-Leu-Ser- Phe-Leu-Pro-Gln-Pro-Pro-
α1(II)	Ile-Asp-Met-Ser-Ala-Phe-Ala-Gly-Leu-Gly-Gln-
α1(III)	Cys-Gly-Ala-Gly-Gly-Val-Ala-Ala-Ile

	15^{C} 20^{C} 25^{C}
α2(I)	Tyr-Tyr-Arg-Ala
α1(I)	Gln-Glu-Lys-Ala-His-Asp-Gly-Gly-Arg-Tyr-Tyr-Arg-Ala
α1(II)	Thr-Glu-Lys-Gly-Pro-Asp-Pro-Ile-Arg-Tyr-Met-Arg-Ala

Figure 2.17. Sequences from the end of the repetitive triplet region and the C-terminal nontriplet region (*superscript C*) of several α chains. The α chains are bovine except for α1(II), which is chick. The sequences represented by 7^{C}–27^{C} in α2(I) and 26^{C}–27^{C} in α1(I) are from chick cDNA. Gaps in the indicated sequences are required to maximize homology. The α1(III) C-terminal sequence is incomplete. The numbers applied to the triplet region of α1(III) should be increased by 12 to compensate for its extra length. The α1(II) sequence was provided by Linda Sandell (unpublished).

in the same way as α1(I). Little can be learned about interchain homology from the partial sequence for α1(III). Procollagen C-protease cleaves just after the alanine residue in α1(I), α2(I), and α1(II).

A much greater degree of interchain homology is observed when comparing the repetitive Gly-X-Y sequences. Examples are residues 1–15 in Figure 2.16 and 1000–1014 in Figure 2.17, the beginning and end of the triplet region. The α1(III) chain is the most dissimilar in the regions shown. Since there are 1026 residues in the triplet region of bovine α1(III)—12 more than in α1(I)—a different alignment to improve homology is permissible but would not greatly change the comparison in the region shown. Comparison of the entire repetitive triplet sequences of α1(I) and α2(I) including the invariant glycine residues reveals an overall homology of 66%. Similar comparisons between α1(I) and α1(II); α1(I) and α1(III); and α1(III) and α2(I) show overall homologies of 75%, 65%, and 58%, respectively. In general, the degree of homology exhibited among the same chains from different species is much higher, 90–95%, depending somewhat on the degree to which the species are related on the evolutionary scale and on the particular chain. With respect to the latter point, interspecies variability tends to be greatest for α2(I) chains, suggesting a somewhat higher rate of substitution in this chain during its evolutionary history. It may be that this occurred because two α1(I) chains were sufficient to carry the minimal required information for molecular interactions; thus α2(I) could evolve more freely and refine function.

A high degree of interregion as well as interchain homology is observed at residues 85–93 and 928–936, as shown in Figure 2.18. The majority of the carbohydrate attached to type I and III collagens is present in the form of a disaccharide bound to hydroxylysine in position 87. This residue is likewise glycosylated in α1(II), although in the latter chain carbohydrate units are bound to several additional hydroxylysine residues throughout the chain, including the one at 930. There is indirect evidence (*See* Traub and Piez 1971) that residue 930 in α1(I) may also be glycosylated in some tissues. Furthermore, the hydroxylysine residues in position 87 and 930 constitute, as described in Section 2.3.3, sites for intermolecular crosslinking within the repetitive Gly-X-Y sequences. Histidine residues which may participate in the formation of multifunctional crosslinks are located at positions 89 and 932 in α1(I), α1(II), and α1(III). In α2(I), the region near 930 is similar to other chains but is shifted three residues farther along the chain. It seems likely, then, that preservation of structure within these sequences can be correlated primarily with their highly important role as sites of intermolecular crosslinking. Since crosslinking can only occur following assembly bringing appropriate sites in adjacent molecules together, these regions may provide unique interaction sites for the nonhelical terminal sequences during fibril formation (Chapter 1, Section 1.3.4). A curious point, however, is that while the two helical crosslink sites are very similar to each other and are conserved among types, the N- and C-terminal nonhelical regions are not similar and are not highly conserved among types, as already discussed.

Residues 772–783 are also of interest with respect to interchain homology. One critical step in the physiologic degradation of native type I and II collagen is collagenase cleavage (Chapter 4) of Gly-Leu and Gly-Ile bonds between residues 775–776, shown in Figure 2.19. Collagenase cleavage of type III apparently involves specific cleavage of a Gly–Ile bond in the same region but in the preceding triplet, as aligned in Figure 2.19 (Seyer et al. 1980). The region immediately following the collagenase cleavage site in each chain is characterized by a relatively hydrophobic, hydroxyproline- and proline-poor sequence. Given

Figure 2.18. Sequences in the helical crosslink sites of several α chains. The Hyl residues at positions 87 and 930 are involved in crosslinking. Carbohydrate is attached at position 87 in all four chains and at residue 930 in α1(II) and possibly others. The α1(II) sequence from 928–936 is chick and was provided by William T. Butler (unpublished). The α2(I) sequence is from chick cDNA (Fuller and Boedtker 1981). Displacement of the α2(I) sequence by three residues maximizes homology. All other sequences are bovine.

```
                  87                 92
α2(I)      -Gly-Phe-Hyl-Gly-Ile-Arg-Gly-His-Asn-
α1(I)      -Gly-Met-Hyl-Gly-His-Arg-Gly-Phe-Ser-
α1(II)     -Gly-Val-Hyl-Gly-His-Arg-Gly-Tyr-Hyp-
α1(III)    -Gly-Met-Hyl-Gly-His-Arg-Gly-Phe-Asp-

                930                 935
α2(I)      -Gly-Leu-Pro-Gly-Leu-Hyl-Gly-His-Asn-
α1(I)      -Gly-Ile-Hyl-Gly-His-Arg-Gly-Phe-Ser-
α1(II)     -Gly-Leu-Hyl-Gly-His-Arg-Gly-Phe-Thr-
α1(III)    -Gly-Ile-Hyl-Gly-His-Arg-Gly-Phe-Hyp-
```

```
                775              780
α2(I)    -Gly-Pro-Gln-Gly-Leu-Leu-Gly-Ala-Hyp-Gly-Phe-Leu-
α1(I)    -Gly-Pro-Gln-Gly-Ile-Ala-Gly-Gln-Arg-Gly-Val-Val-
α1(II)                   -Ile-Ala-Gly-Gln-Arg-Gly-Leu-Val-
α1(III)  -Gly-Ile-Ala-Gly-Leu-Thr-Gly-Ala-Arg-Gly-Leu-Ala-
```

Figure 2.19. Sequences in the collagenase and fibronectin-attachment sites of several α chains. The α chains are bovine except for α1(II) which is human. Collagenase cleaves at the 775–776 bond except in α1(III) where 772–773 is cleaved.

these characteristics, this region may very well represent a segment in which the triple-helical conformation is less stable or well defined than in other portions of the molecule, and may allow specific binding of collagenase and subsequent proteolysis of specific bonds. Of further interest is that fibronectin binds to α1(I) chains at or near the collagenase cleavage site (Chapter 7). Again, the specificity may lie in the conformation as well as the sequence. In the case of type III molecules, this region is susceptible to cleavage by other enzymes in addition to the specific collagenases. These include trypsin, thermolysin, and elastase (See Miller et al. 1976), all of which cleave native type III molecules at susceptible bonds within a few residues of the collagenase site. This broad susceptibility must certainly be conformation dependent.

Glycine, by definition, occupies the initial position in each Gly-X-Y triplet, but is also sometimes encountered in position X in α1(III), as noted above (Section 2.4.4). Peptidyl proline 4-hydroxylase and peptidyl lysine hydroxylase are specific for the Y positon (Chapter 3), accounting for the locations of 4-hydroxyproline and hydroxylysine only in that position. Earthworm cuticle collagen seems to be unique in that a large proportion of its 4-hydroxyproline is in position X (Adams 1978). In types I, II, and III collagen some lysine residues in position Y escape hydroxylation completely and some are partially hydroxylated. The high content of hydroxylysine and attached carbohydrate in types IV and V collagen (Section 2.3.2) suggests that most or all lysine residues in the Y position are hydroxylated, but sequence data are not available. Proline in position Y in types I, II, and III collagen escapes hydroxylation in only two or three positions but may be incompletely hydroxylated in others to give an overall level of 90–95% hydroxylation of eligible proline residues. The level of lysine hydroxylation in type I collagen varies with the tissue and state of development (Barnes et al. 1974).

Also as a result of enzyme specificity, 3-hydroxyproline, which occurs once in α1(I) and α2(I) and twice in α1(II), is found in the X position and always precedes 4-hydroxyproline. 3-Hydroxyproline is present to a much greater extent in type IV collagen preparations derived from glomerular basement membrane and lens capsule, where it likewise appears to be present in position X preceding 4-hydroxyproline (Gryder et al. 1975).

Certain other amino acids exhibit a strong tendency for selective distribution within triplets as discussed in Chapter 1 (Section 1.2.1). Thus, approximately 90% of the glutamine and leucine residues and virtually all phenylalanine and histidine residues are located in X positions. The reason is not known. Other large hydrophobic amino acids such as methionine, isoleucine, and valine exhibit no preference. This is likewise true of aspartic acid. The basic residues, lysine and arginine, exhibit a tendency for localization in Y positions, but to a lesser degree than residues preferring the X position.

The long-range distribution of amino acids along collagen chains is also of interest. In comparing the sequences of different α chains, charged residues—the general locations of which may be visualized in appropriately stained preparations of segment-long-spacing (SLS) cystallites with the electron microscope (Chapter 1, Section 1.4.1)—are the most highly conserved. When changes occur, one charged residue is likely to be replaced by another. The only other amino acid conserved to a similar degree is phenylalanine. Other hydrophobic residues, particularly leucine and valine, exhibit relatively high variability, which can be ascribed largely to differences in α2(I) and may explain its proposed contribution to molecular packing of type I collagen (Chapter 1, Section 1.3.2).

The availability of sequence data for the α chains has prompted several considerations with respect to the evolution of collagen. Analyses of the sequences for α1(I), α2(I), and α1(III) have revealed periodicities of internal similarity ranging from lengths of D (234 residues) to D/13 (18 residues), with the smaller periodicities being most prominent in α1(I) and α1(III) (Hofmann et al. 1980). Of particular interest in this regard are observations that a high proportion of the coding segments (exons) in the chick α2(I) gene have a length of 54 base pairs (bp) which codes for 18 amino acid residues (Yamada et al. 1980; Wozney et al. 1981). Since each of these exons codes for six complete triplets beginning with a glycine residue in triplet 1 and terminating with the Y position in triplet 6, it has been proposed that the first collagen gene arose through duplication of a 54-bp coding sequence (Yamada et al. 1980). This represents an attractive hypothesis which, however, is difficult to relate to protein structure or function. In particular, the 18-residue periodicity in the α1(I) and α1(III) amino acid sequences is a statistical value that is no more, and often less, significant than other periodicities. It has yet to be shown that there is a one-to-one correlation between protein sequence regularities and gene structure. The most that can be said is that 18 is a submultiple of 234, the most important periodicity in the sequence since it is the molecular stagger in the fibril (Chapter 1, Section 1.3.1), and thus the two are consistent.

The point in time at which the first collagen and the different α chains emerged in evolution is likewise open to question. It is likely that all known types of vertebrate α chains exist in even the most primitive vertebrates (Mathews 1980). Indeed, it is known that collagens very similar to the known vertebrate collagens had evolved in early invertebrates (*See* Adams 1978). Species from all multicellular invertebrate phyla, including some sponges, contain at least one collagen as defined by x-ray diffraction and amino acid composition criteria (Chapter 1, Section 1.2.1). In some cases, native banded fibrils, SLS patterns, and bands on SDS gels characteristic of collagens like types I, II, or III have been observed. Although there is insufficient information on chain composition, primary structure and localization to relate invertebrate collagens to individual types as defined for the vertebrates, these observations suggest that at least some of the different collagen chains emerged well before the divergence of vertebrates and invertebrates. It is tempting to relate their emergence to a very early time, about 600 million years ago, since even the most primitive multicelled organisms require some kind of connective tissue. In embryogenesis, at least one collagen appears very early in cell division (Chapter 10). There is no evidence for a collagen or the collagen triple helix in single-celled animals or in plants.

If one assumes that the various collagen α chains arose from a single ancestral gene and equal rates of evolutionary change, the observed interchain homologies for vertebrate α chains suggest initial divergence of α1(III) followed by α2(I), and eventually α1(I) and α1(II). However, earlier suggestions that α2(I) may have been less constrained imply a faster rate of evolution for that chain, and thus a later divergence.

2.5. Distribution

The localization of the collagens within the vertebrate organism is often the best information available about biological function. Although one or more collagens occur in virtually every tissue or organ, there exists a great deal of specificity with respect to the distribution of the collagens among tissues, as well as their localization within tissues. Changes in distribution as a function of development and disease provide important information about mechanisms.

The distribution of the collagens can be studied in two general ways—by tissue isolation and chemical characterization, and by immunochemical localization in tissue sections. Immunology of the collagens with examples of immunofluorescent micrographs is discussed in Chapter 5. Resolution by chemical characterization is limited by the size of the piece of tissue that can be analyzed, generally in the millimeter range. However, quantitative data can be obtained if care is taken to minimize inherent technical problems such as recovery and selection. The usual method is to prepare pepsin digests of dissected tissue and determine their α chain composition from chromatograms or SDS gels of the chains and their cyanogen bromide digests.

Immunochemical methods can, in principle, be applied at both the light and electron microscope levels. At present, the immunostaining methods suitable for electron microscopy, which could achieve resolution to a few tens of namometers, have not been shown to be routinely useful for the collagens. However, they have been successfully applied in some cases (*See* Fleischmajer et al. 1981). Immunofluorescence at the light microscope level has received wide use (*See* Chapter 5) and much of what is known about localization has been obtained this way. Resolution is in the micrometer range. Unfortunately, it is not quantitative and technical problems make it of uncertain reliability as a sole means of identification. It is best used to follow changes in defined systems.

2.5.1. Macroscopic

The gross or macroscopic distribution of the collagens in the major connective tissues is summarized in Table 2.4. When evaluated by chemical methods at this level, it is apparent that types I, II, and IV collagen comprise the overwhelming

Table 2.4. Macroscopic Distribution of the Collagens

Collagen	Tissue[a]
Type I	Bone, cornea, dentin, fibrocartilage, tendon (>95%) Dermis, gingiva, heart valve (85%) Intestinal, large vessel, and uterine wall (50–60%)
Type I-trimer	Dentin, dermis, tendon (<2–3%)
Type II	Notochord, nucleus pulposus, hyaline cartilage, vitreous (>95%)
Type III	Intestinal, large vessel, and uterine wall (35–45%) Dermis, gingiva, heart valve (10–15%)
Type IV	Endothelial, epithelial basement membrane (>95%)
Type V	Cornea, placental membrane, large vessel wall (5%) Bone, gingiva, heart valve, hyaline cartilage (<5%)

[a]Numbers in parentheses indicate approximate percentage of specified collagen type found in tissue's total collagen.

majority of the collagen in selected tissues. There is some correlation between type and gross tissue properties. Thus, the fibrous elements of the less extensible connective tissues—tendon and bone—are composed almost exclusively of type I collagen. In tissues located in regions most likely to encounter compressive forces, such as the nucleus pulposus, hyaline cartilage, and the vitreous humour, structural elements comprised of type II collagen and proteoglycans predominate. Type IV collagen constitutes major units utilized in the structural scaffold of a variety of basement membranes. The more or less open, three-dimensional meshwork formed by aggregates of type IV molecules apparently forms a convenient supporting screen which does not interfere with the passage of small molecules.

In contrast, structural elements containing type III or V collagen are not predominant in any given tissue. However, type III collagen comprises a significant proportion of the collagen in the more distensible connective tissues such as dermis, heart valves and large blood vessel, and intestinal uterine walls. Moreover, the level of type III present in these tissues increases with the degree of distensibility required in normal function with low levels present in the less compliant dermis and heart valves and much higher levels in the more distensible vessel and uterine walls. This apprent correlation implies that type III fibrils play a different role from type I, which is always present in the same tissue. However, type III has not been localized at the ultrastructural level, leaving its role uncertain. It is known that the diminution of type III collagen in individuals presenting the type IV Ehlers-Danlos syndrome (Chapter 11) is accompanied by a marked weakening of the tissues in which this collagen is a major component. It is, however, not easy at this time to dissociate the influences of structure at the molecular, fibril, and tissue level. Type V collagen is the least prevalent of all the collagens in quantitative terms. It is, however, widely distributed.

In addition to the tissues listed in Table 2.4, various collagens are distributed within the connective tissue stroma of all internal organs such as the heart, liver, spleen, muscles, and lungs, and around all structures such as nerves and small blood vessels. Although critical to structure, the amount of any of the collagens is small unless the content has been dramatically elevated as a result of fibrotic processes. Such changes may be dramatic and are an important aspect of many diseases (Chapter 11). Practical considerations relevant to yields and extractability govern the choice of tissue for the isolation and purification of a given collagen type (Miller and Rhodes 1982).

2.5.2. Microscopic

As already noted, the localization of the collagens within tissues has been considerably facilitated by the availability of specific antibodies to different collagens as well as to their respective procollagens (Chapter 5). These approaches have been particularly valuable in ascertaining the disposition and interrelationships of the collagens in a variety of tissues as well as in evaluating the changes which occur during developmental, repair, and pathological processes (Gay and Miller, 1978).

In general, the results of these studies have confirmed and extended observations concerning the macroscopic distribution of collagens. Thus, the fibers in tissues, such as bone, tendon, dermis, and vessel walls, stain avidly with antibodies to type I collagen. Antibodies to type III collagen, however, preferentially stain fine reticular networks that are visible in spaces surrounding larger type I fibers in sections of dermis and vessel walls. Although types I and III collagen apparently form separate and distinct fibrous elements, it has been shown in double-staining experiments with cultured fibroblasts that individual cells are capable of synthesizing both molecular species. In tendon, which is composed predominantly of fibers derived from type I collagen, type III collagen is found localized to the inner and outer tendon sheaths. Antibodies to type III collagen likewise stain the epimysium, perimysium, and endomysium of skeletal muscle.

In accordance with the general distribution of type II collagen, antibodies to this collagen stain only the fibrous elements in selected tissues, such as hyaline cartilage and nucleus pulposus. Thus, reactivity with antibodies to type II collagen is sufficiently specific to warrant its use as a marker for chondrogenesis. Indeed, the whole course of endochondral bone formation may be followed as a function of the appearance of different collagen types (Chapter 10).

Antibodies to type IV collagen preparations specifically stain basement membrane structures. To date, relatively few basement membrane structures have been examined chemically, owing to difficulties in isolating the structures in appropriate quantities. However, since all basement membranes regardless of their location and precise function show an affinity for type IV collagen antibodies, it would appear that all basement membranes are similar in this regard as well as in glycoprotein (Chapter 7) and proteoglycan content (Chapter 8). They may well differ in relative amounts of these components and may contain other components.

Although there was some early confusion, it is now agreed that by immunofluorescence type V collagen is primarily localized within limited regions surrounding the surfaces of several cell types (Gay et al. 1981a, 1981b; Martinez-Hernandez et al. 1982). In these locations, the different molecular species of the type V system apparently constitute what might be designated as the exocytoskeleton or the membranous sheath often associated with mobile cells such as fibroblasts and smooth muscle cells. These results are consistent with the presence of relatively small amounts of type V collagen in a variety of tissues. They are, moreover, entirely consistent with chemical data on isolated basement membranes indicating that these structures do not contain type V collagen.

The author expresses gratitude to Ms. Melody Edwards for skillful assistance in all phases of the preparation of this chapter.

References

Adams E: (1978) Invertebrate collagens. Science 202:591–598.

Allmann H, Fietzek PP, Glanville RW, Kühn K: (1979) The covalent structure of calf skin type III collagen. VI. The amino acid sequence of the carboxyterminal cyanogen bromide peptide α1(III)CB9B (Position 928–1028). Hoppe Seylers Z Physiol Chem 360:861–868.

Bailey AJ, Peach CM: (1968) Isolation and structural identification of a labile intermolecular crosslink in collagen. Biochem Biophys Res Commun 33:812–819.

Barnes MJ, Constable BJ, Morton LF, Royce PM: (1974) Age-related variations in hydroxylation of lysine and proline in collagen. Biochem J 139:461–468.

Becker U, Helle O, Timpl R: (1977) Characterization of the amino-terminal segment in procollagen pα2 chain from dermatosparactic sheep. FEBS Letters 73:197–200.

Bernstein PH, Mechanic GL: (1980) A natural histidine-based imminium cross-link in collagen and its location. J Biol Chem 255:10414–10422.

Bornstein P, Kang AH, Piez KA: (1966) The nature and location of intramolecular cross-links in collagen. Proc Natl Acad Sci USA. 55:417–424.

Bornstein P, Piez KA: (1966) The nature of the intramolecular crosslinks in collagen. The separation and characterization of peptides from the crosslink region of rat skin collagen. Biochemistry 5:3460–3473.

Bornstein P, Sage H: (1980) Structurally distinct collagen types. Ann Rev Biochem 49:957–1003.

Bornstein P, Traub W: (1979) The chemistry and biology of collagen. In: Neurath H, Hill RL, eds. The Proteins. Volume IV. New York: Academic Press.

Bruckner P, Bächinger HP, Timpl R, Engel J: (1978) Three conformationally distinct domains in the amino-terminal segment of type III procollagen and its rapid triple helix ⇌ coil transition. Eur J Biochem 90:595–603.

Burgeson RE, El Adli F, Kaitila II, Hollister DW: (1976) Fetal membrane collagens: Identification of two new collagen alpha chains. Proc Natl Acad Sci USA 73:2579–2583.

Burgeson RE, Hollister DW: (1979) Collagen heterogeneity in human cartilage: Identification of several new collagen chains. Biochem Biophys Res Commun 87:1124–1131.

Butler WT, Cunningham LW: (1966) Evidence for the linkage of a disaccharide to hydroxylysine in tropocollagen. J Biol Chem 241:3882–3888.

Butler WT, Finch JE Jr, Miller EJ: (1977) The covalent structure of cartilage collagen. Evidence for sequence heterogeneity of bovine α1(II) chains. J Biol Chem 252:639–643.

Byers PH, McKenney KH, Lichtenstein JR, Martin GR: (1974) Preparation of type III procollagen and collagen from rat skin. Biochemistry 13:5243–5348.

Chandrakasan G, Torchia DA, Piez KA: (1976) Preparation of intact monomeric collagen from rat tail tendon and skin and the structure of the nonhelical ends in solution. J Biol Chem 251:6062–6067.

Chung E, Rhodes RK, Miller EJ: (1976) Isolation of three collagenous components of probable basement membrane origin from several tissues. Biochem Biophys Res Commun 71:1167–1174.

Clark CC: (1979) The distribution and initial characterization of oligosaccharide units on the COOH-terminal propeptide extensions of the pro-αl and pro-α2 chains of type I procollagen. J Biol Chem 254:10798–10802.

Duksin D, Bornstein P: (1977) Impaired conversion of procollagen to collagen by fibroblasts and bone treated with tunicamycin, an inhibitor of protein glycosylation. J Biol Chem 252:955–962.

Eastoe J: (1967) Composition of collagen and allied proteins. In: Ramachandran GN, ed. Treatise on Collagen. Volume 1: Chemistry of Collagen. London: Academic Press.

Elsden DF, Light ND, Bailey AJ: (1980) An investigation of pyridinoline, a putative collagen cross-link. Biochem J 185:531–534.

Eyre DR: (1980) Collagen: Molecular diversity in the body's protein scaffold. Science 207:1315–1322.

Eyre DR, Oguchi H: (1980) The hydroxypyridinium crosslinks of skeletal collagens: Their measurement, properties and a proposed pathway of formation. Biochem Biophys Res Commun 92:403–410.

Fessler LI, Kumamoto CA, Meis ME, Fessler JH: (1981a) Assembly and processing of procollagen V (AB) in chick blood vessels and other tissues. J Biol Chem 256:9640–9645.

Fessler LI, Robinson WJ, Fessler JH: (1981b) Biosynthesis of procollagen [(proα1(V)$_2$proα2V)] by chick tendon fibroblasts and procollagen (proα1V)$_3$ by hamster lung cell cultures. J Biol Chem 256:9646–9651.

Fietzek PP, Allmann H, Rauterberg J, Henkel W, Wachter E, Kühn K: (1979) The covalent structure of calf skin type III collagen. I. The amino acid sequence of the amino terminal region of the α1(III) chain (position 1–222). Hoppe Seylers Z Physiol Chem 360:809–820.

Fietzek PP, Kühn K: (1976) The primary structure of collagen. Int Rev Connect Tissue Res 7:1–60.

Fleischmajer R, Timpl R, Tuderman L, Raisher L, Wiestner M, Perlishm JS, Graves PN: (1981) Ultrastructural identification of extension aminopropeptides of type I and III collagens in human skin. Proc Natl Acad Sci USA 78:7360–7364.

Fujimoto D: (1980) Evidence for natural existence of pyridinoline crosslink in collagen. Biochem Biophys Res Commun 93:948–953.

Fujimoto D, Moriguchi T, Ishida T, Hayashi H: (1978) The structure of pyridinoline, a collagen cross-link. Biochem Biophys Res Commun 84:52–57.

Fuller F, Boedtker H: (1981) Sequence determination and analysis of the 3′ region of chicken pro-α1(I) and pro-α2(I) collagen messenger ribonucleic acids including the carboxy-terminal propeptide sequences. Biochemistry 20:996–1006.

Furuto DK, Miller EJ: (1980) Isolation of a unique collagenous fraction from limited pepsin digests of human placental tissue. J Biol Chem 255:290–295.

Gay S, Martinez-Hernandez A, Rhodes RK, Miller EJ: (1981a) The collagenous exocytoskeleton of smooth muscle cells. Collagen Rel Res 1:377–384.

Gay S, Miller EJ: (1978) Collagen in the Physiology and Pathology of Connective Tissue. Stuttgart: Fischer Verlag.

Gay S, Miller EJ: (1979) Characterization of lens capsule collagen: Evidence for the presence of two unique chains in molecules derived from major basement membrane structures. Arch Biochem Biophys 198:370–378.

Gay S, Rhodes RK, Gay RE, Miller EJ: (1981b) Collagen molecules comprised of α1(V)-chains (B-chains): An apparent localization in the exocytoskeleton. Collagen Rel Res 1:53–58.

Grant ME, Heathcote JG, Orkin RW: (1982) Current concepts of basement membrane structure and function. Biosci Rep 1:819–842.

Gryder RM, Lamon M, Adams E: (1975) Sequence position of 3-hydroxyproline in basement membrane collagen. J Biol Chem 250:2470–2474.

Grynpas MD, Eyre DR, Kirschner DA: (1980) Collagen type II differs from type I in native molecular packing. Biochim Biophys Acta 626:346–355.

Hayashi T, Nagai Y: (1980) The anomalous behavior of collagen peptides on sodium dodecyl sulfate–polyacrylamide gel electrophoresis is due to the low content of hydrophobic amino acid residues. J Biochem 87:803–808.

Heathcote JG, Bailey AJ, Grant ME: (1980) Studies on the assembly of the rat lens capsule. Biosynthesis of a cross-linked collagenous component of high molecular weight. Biochem J 190:229–237.

Hofmann H, Fietzek PP, Kühn K: (1978) The role of polar and hydrophobic interactions for the molecular packing of type I collagen: A three dimensional evaluation of the amino acid sequence. J Mol Biol 125:137–165.

Hofmann H, Fietzek PP, Kühn K: (1980) Comparative analysis of the sequences of the three collagen chains α1(I), α2 and α1(III). J Mol Biol 141:293–314.

Holbrook KA, Byers PH: (1981) Ultrastructural characteristics of the skin in a form of the Ehlers-Danlos syndrome type IV. Lab Invest 44:342–350.

Hörlein D, Fietzek PP, Wachter E, Lapiere CM, Kühn K: (1979) Amino acid sequence of the aminoterminal segment of dermatosparactic calfskin procollagen type I. Eur J Biochem 99:31–38.

Housley TJ, Rowland FN, Ledger PW, Kaplan J, Tanzer ML: (1980) Effects of tunicamycin on the biosynthesis of procollagen by human fibroblasts. J Biol Chem 255:121–128.

Housley TJ, Tanzer ML: (1981) The separation and amino acid analysis of collagen crosslinks on an extended basic ion-exchange column. Anal Biochem 114:310–315.

Housley TJ, Tanzer ML, Henson E, Gallop PM: (1975) Collagen cross-linking: Isolation of hydroxyaldol-histidine, a naturally-occuring cross-link. Biochem Biophys Res Commun 67:824–830.

Kresina TF, Miller EJ: (1979) Isolation and characterization of basement membrane collagen from human placental tissue. Evidence for the presence of two genetically distinct collagen chains. Biochemistry 18:3089–3097.

Kühn K, Wiedemann H, Timpl R, Risteli J, Dieringer H, Voss T, Glanville RW: (1981) Macromolecular structure of basement membrane collagens. Identification of 7S collagen as a crosslinking domain of type IV collagen. FEBS Letters 125:123–128.

Light ND, Bailey AJ: (1980) Polymeric C-terminal cross-linked material from type I collagen. Biochem J 189:111–124.

Martinez-Hernandez A, Gay S, Miller EJ: (1982) The ultrastructural localization of type V collagen in rat kidney. J Cell Biol 92:108–114.

Mathews MB: (1980) Coevolution of collagen. In: Viidik A, ed. Biology of Collagen. New York: Academic Press.

Mechanic G, Tanzer ML: (1970) Biochemistry of collagen crosslinking. Isolation of a new crosslink, hydroxylysinohydroxynorleucine, and its reduced precursor, dihydroxynorleucine, from bovine tendon. Biochem Biophys Res Commun 41:1597–1604.

Miller EJ: (1976) Biochemical characteristics and biological significance of the genetically-distinct collagens. Mol Cell Biochem 13:165–192.

Miller, EJ, Epstein EH Jr, Piez KA: (1971) Identification of three genetically-distinct collagens by cyanogen bromide cleavage of insoluble human skin and cartilage collagen. Biochem Biophys Res Commun 42:1024–1029.

Miller EJ, Finch JE Jr, Chung E, Butler WT, Robertson PB: (1976) Specific cleavage of the native type III collagen molecule with trypsin. Arch Biochem Biophys 173:631–637.

Miller EJ, Gay S: (1982) Collagen—An overview. Methods Enzymol 82:3–32.

Miller EJ, Matukas VJ: (1969) Chick cartilage collagen: A new type of $\alpha 1$ chain not present in bone or skin of the species. Proc Natl Acad Sci USA 64:1264–1268.

Miller EJ, Rhodes RK: (1982) Preparation and characterization of the different types of collagen. Methods Enzymol 82:33–64.

Palmiter RD, Davidson JM, Gagnon J, Rowe DW, Bornstein P: (1979) NH_2-terminal sequence of the chick pro $\alpha 1(I)$ chain synthesized in the reticulocyte lysate system. J Biol Chem 254:1433–1436.

Pape AL, Muh JP, Bailey AJ: (1981) Characterization of N-glycosylated type I collagen in streptozotocin-induced diabetes. Biochem J 197:405–412.

Piez KA: (1968) Molecular weight determination of random coil polypeptides from collagen by molecular sieve chromatography. Anal Biochem 26:305–312.

Piez KA: (1976) Primary structure. In: Ramachandran GN, Reddi AH, eds. Biochemistry of Collagen. New York: Plenum Press.

Piez KA, Eigner EA, Lewis MS: (1963) The chromatographic separation and amino acid composition of the subunits of several collagens. Biochemistry 2:58–66.

Reese CA, Mayne R: (1981) Minor collagens of chicken hyaline cartilage. Biochemistry 20:5443–5448.

Rhodes RK, Gibson KD, Miller EJ: (1981) Isolation and characterization of the cyanogen bromide peptides from the human $\alpha 2(V)$ collagen chain. Biochemistry 20:3117–3121.

Rhodes RK, Miller EJ: (1978) Physicochemical characterization and molecular organization of the collagen A and B chains. Biochemistry 17:3442–3448.

Rhodes RK, Miller EJ: (1979) The isolation and characterization of the cyanogen bromide peptides from the B chain of human collagen. J Biol Chem 254:12084–12087.

Rhodes RK, Miller EJ: (1981) Evidence for the existence of an α1(V)α2(V)α3(V) collagen molecule in human placental tissue. Collagen Rel Res 1:337–343.

Robey PG, Martin GR: (1981) Type IV collagen contains two distinct chains in separate molecules. Collagen Rel Res 1:27–38.

Robins SP, Bailey AJ: (1972) Age-related changes in collagen: The identification of reducible lysine-carbohydrate condensation products. Biochem Biophys Res Commun 48:76–84.

Sage H, Bornstein P: (1979) Characterization of a novel collagen chain in human placenta and its relation to AB collagen. Biochemistry 18:3815–3822.

Sage H, Pritzl P, Bornstein P: (1980) A unique, pepsin-sensitive collagen synthesized by aortic endothelial cells in culture. Biochemistry 19:5747–5755.

Sage H, Pritzl P, Bornstein P: (1981) Characterization of cell matrix associated collagens synthesized by aortic endothelial cells in culture. Biochemistry 20:436–442.

Schuppan D, Timpl R, Glanville RW: (1980) Discontinuities in the triple helical sequence Gly-X-Y of basement membrane (type IV) collagen. FEBS Letters 115:297–300.

Seyer JM, Kang AH: (1981) Covalent structure of collagen: Amino acid sequence of α1(III)-CB9 from type III collagen of human liver. Biochemistry 20:2621–2627.

Seyer JM, Mainardi C, Kang AH: (1980) Covalent structure of collagen: Amino acid sequence of α1(III)-CB5 from type III collagen of human liver. Biochemistry 19:1583–1589.

Shimokomaki M, Duance VC, Bailey AJ: (1980) Identification of a new disulphide bonded collagen from cartilage. FEBS Letters 121:51–54.

Spiro RG: (1967) Structure of the disaccharide unit of the renal glomerular basement membrane. J Biol Chem 242:4813–4823.

Tanzer ML: (1968) Intermolecular crosslinks in reconstituted collagen fibrils. Evidence for the nature of the covalent bonds. J Biol Chem 243:4045–4054.

Timpl R, Bruckner P, Fietzek P (1979) Characterization of pepsin fragments of basement membrane collagen obtained from a mouse tumor. Eur J Biochem 95:255–263.

Timpl R, Martin GR: (1982) Components of basement membranes. In: Furthmayr H, ed. Immunochemistry of the Extracellular Matrix. Volume 2: Applications. Boca Raton: CRC Press.

Timpl R, Wiedemann H, van Delden V, Furthmayr H, Kühn K: (1981) A network model for the organization of type IV collagen molecules in basement membranes. Eur J Biochem 120:203–211.

Traub W, Piez KA: (1971) The chemistry and structure of collagen. Adv Protein Chem 25:243–352.

Tryggvason K, Robey PG, Martin GR: (1980) Biosynthesis of type IV procollagens. Biochemistry 19:1284–1289.

Wozney J, Hanahan D, Tate V, Boedtker H, Doty P: (1981) Structure of the pro α2(I) collagen gene. Nature 294:129–135.

Yamada Y, Avvedimento VE, Mudryj M, Ohkubo H, Vogeli G, Irani M, Pastan I, de Crombrugghe B: (1980) The collagen gene: Evidence for its evolutionary assembly by amplification of a DNA segment containing an exon of 54 bp. Cell 22:887–892.

3

Biosynthesis of the Collagens

Kari I. Kivirikko and Raili Myllylä

3.1. Introduction

Collagen biosynthesis resembles that of other secretory proteins, but is distinguished by the presence of many unusual reactions. The most characteristic feature is that the initial polypeptide chains are modified by a number of cotranslational and posttranslational events, many of which are unique to the collagens and a few other proteins with collagen-like amino acid sequences. The aim of this chapter is to review the biosynthesis of collagens and its regulation; pathological changes will be described elsewhere (Chapter 11).

Due to the voluminous literature, it is not possible to cite all the original contributions, and therefore the references include only recently published work. For more detailed accounts and more complete references, the reader is referred to the many recent reviews on various aspects of collagen biosynthesis or its cotranslational and posttranslational processing in general (Fessler and Fessler 1978; Bornstein and Traub 1979; Olsen and Berg 1979; Prockop et al. 1979; Bornstein and Sage 1980; Eyre 1980; Minor 1980; Kivirikko and Myllylä 1982b) or on the intracellular modifications (Prockop et al. 1976), hydroxylations (Adams and Frank 1980; Kivirikko and Myllylä 1980), glycosylations (Kivirikko and Myllylä 1979), extracellular modifications (Heathcote and Grant 1980), and lysyl oxidase (Siegel 1979). Recent reviews are also available on the methodology of the preparation and translation of procollagen messenger RNAs (mRNAs) (Haralson 1982; Monson 1982) and the intracellular (Kivirikko and Myllylä 1982a) and extracellular (Prockop and Tuderman 1982) posttranslational enzymes.

Collagen biosynthesis can be regarded as taking place in several stages. The genes for these proteins are first transcribed into corresponding pre-mRNAs,

From the Department of Medical Biochemistry, University of Oulu, Oulu, Finland

Table 3.1. General Steps in the Biosynthesis of Collagen and Their Main Functions

Biosynthetic step	Biological significance
A. Transcription and translation	
1. Biosynthesis and processing of pre-mRNA	Formation of translatable mRNA
2. Translation	Formation of primary structure
B. Intracellular modifications	
1. Removal of prepeptide sequences	Membrane translocation
2. 4-Hydroxylation of peptidyl proline	Essential for triple helix at 37°C
3. 3-Hydroxylation of peptidyl proline	Unknown
4. Hydroxylation of peptidyl lysine	Essential for glycosylation of hydroxylysine; crosslinks
5. Glycosylation of peptidyl hydroxylysine	Unknown
6. Glycosylation of propeptides	Unknown
7. Chain association and disulfide bonding	Essential for triple helix formation
8. Triple helix formation	Essential for secretion and later molecular functions
9. Translocation and secretion of procollagen	Transport
C. Extracellular modifications	
1. Conversion of procollagen to collagen[a]	Essential for normal fibril formation[a]
2. Ordered aggregation	Formation of fibrils or other native structures
3. Crosslink formation	Essential for the stability of native structure

[a]For collagen types I–III; for types IV and V, see text.

which are subsequently processed and translated into precursor forms of various collagen α chains. Intracellular cotranslational and posttranslational modifications are necessary for the formation and secretion of the triple helical procollagen molecules. Extracellular processing converts these molecules into collagens and incorporates them into stable, crosslinked fibrils in the case of types I, II, and III collagen or into other supramolecular structures in the case of types IV and V collagen (Chapter 1). The processing steps are outlined in Table 3.1 and will be considered in the order in which they occur.

3.2. Transcription and Translation

3.2.1. Collagen Genes and Messenger RNAs

Various tissues express different types of collagen molecules which are composed of more than nine distinct α chains (Chapter 2). Therefore, the cells must have more than nine different genes for these chains.

Collagen genes have recently become an object of study using the methods of recombinant DNA, and very rapid progress is now taking place in this field. Complementary DNAs have been synthesized and cloned for mRNAs of the pro α2(I) and pro α1(I) chains of type I procollagen from chick embryos (Lehrach et al. 1978; Sobel et al. 1978; Showalter et al. 1980; Fuller and Boedtker 1981; Vogeli et al. 1981) and human fibroblasts (Chu et al. 1982; Myers et al. 1981a), and for the pro α1(II) chain of type II procollagen from chick cartilage (Vuorio et al. 1982). In addition, genomic clones have been isolated and characterized for sheep (Boyd et al. 1980; Schafer et al. 1980), chick embryo (Vogeli et al. 1980;

Yamada et al. 1980; Wozney et al. 1981a,b) and human (Myers et al. 1981b) pro α2(I) chain, mouse pro α1(I) chain (Monson and McCarthy 1981), and an unidentified human pro α1(I)-like chain (Weiss et al. 1982) (See also Cheah and Grant 1982; de Crombrugghe and Pastan 1982; Tolstoshev and Crystal 1982).

All these genes appear to be about ten times longer than the corresponding translatable mRNAs, with the coding information distributed in more than 50 short coding regions (exons) which are interrupted by noncoding regions (introns) of various sizes (Figure 3.1). All the exons that code for the triple helical regions of the pro α chains begin with a codon for glycine and end with a codon for an amino acid preceding a glycine residue. Most of these exons have a length of 54 or 108 base pairs, thus corresponding to 6 or 12 triplets with amino acid sequences of Gly-X-Y. Some exons have a different length, such as 45 or 99 base pairs, but their base pair numbers are all multiples of nine. Based on these findings, it has been proposed that the ancestral collagen gene has developed by multiple duplication or amplification of a single genetic unit containing an exon of 54 base pairs, and that all collagen genes may follow this rule (Yamada

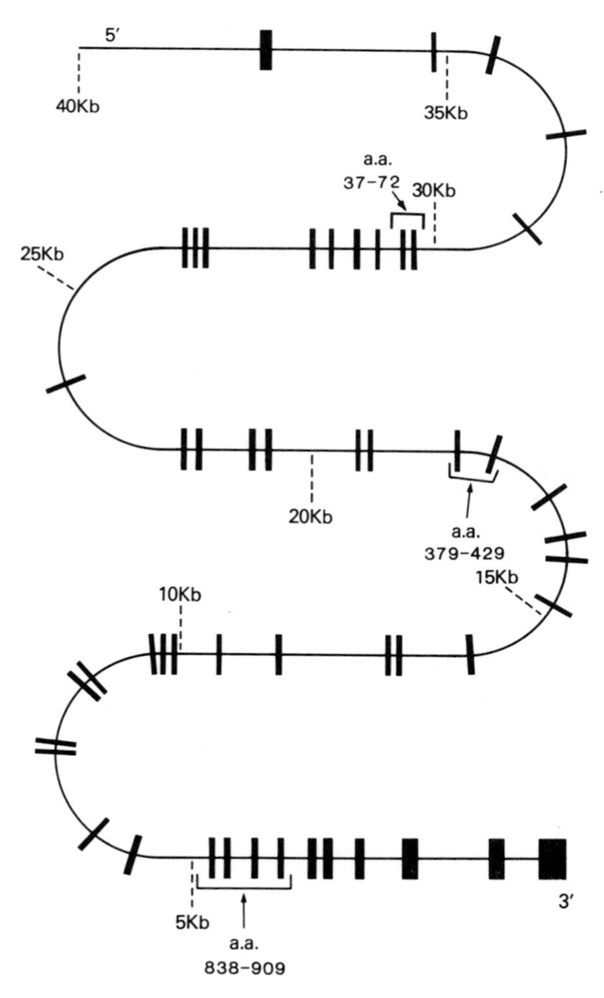

Figure 3.1. Schematic representation of the gene for the chick type I collagen pro α2 chain. The 5′ and 3′ ends of the gene are indicated. The bars indicate the sites of exons (coding sequences) and the connecting lines those of introns (noncoding sequences). The brackets indicate segments of DNA sequenced, the positions of the corresponding amino acid residues also being indicated. *(Source: Yamada, Avvedimento, et al. (1980) Cell 20:887–892, with permission.)*

et al. 1980). Preliminary data on genes for type II and III collagens are consistent with this suggestion (see Tolstoshev and Solomon 1982). Work on other collagen genes is now in progress and will soon further elucidate this hypothesis.

The exons coding for the propeptides of the pro α2(I) chain differ distinctly from those coding for the triple helical region. These propeptide coding exons are often longer and the numbers of their base pairs are not multiples of nine. The DNA sequences of the exons coding for the N-terminal and C-terminal ends of the triple helical region have been determined. In both cases, these exons also code for the nonhelical region of the α chain and part of the propeptide. The exon at the N-terminal end codes for the last four amino acid residues of the propeptide, all 12 residues of the nonhelical region and the first Gly-X-Y triplet. The exon at the C-terminal end codes for the last five Gly-X-Y triplets, all 15 residues of the nonhelical region and the 53 first amino acids of the C-terminal propeptide. These findings that the three structural domains at both ends of the collagen molecule are not separated by introns, may possibly be explained by the fact that they consist of a specific functional domain, namely the procollagen protease cleaving site (Wozney et al. 1981b).

The mRNAs for the pro α chains of various collagen types are synthesized and processed in the same way as those for other proteins in eukaryotes (*See* Brown 1981). The DNA gene is first transcribed into a precursor mRNA that contains copies of both the exons and the introns, and this precursor is then modified by several reactions, including excision and splicing, to form the cytoplasmic translatable mRNA (Figure 3.2). The complex structure of the collagen genes implies that processing of the pre-mRNAs must involve at least 50 excision and splicing events. This number is in fact an underestimate, as the removal of at least some of the introns requires more than one step (Avvedimento et al. 1980). Recent observations of several types of protein mRNAs have indicated that the length of the untranslated region at the 3′-end of the mRNAs for a given protein may vary. Therefore, transcription of a single gene can give rise to two

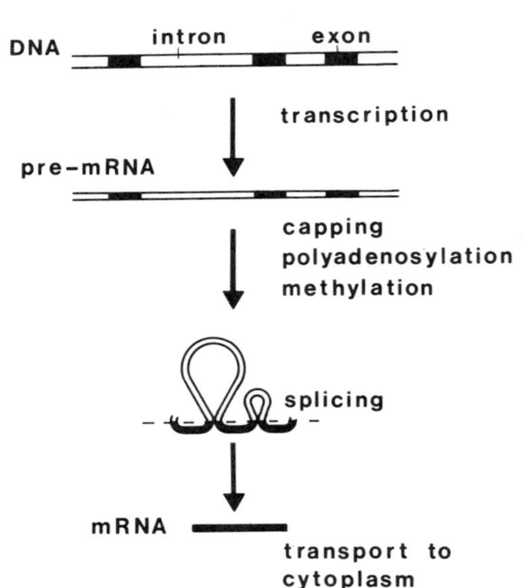

Figure 3.2. Schematic representation of the synthesis of translatable mRNA.

or more mRNAs with different lengths but coding for the same protein. In the case of human type I procollagen, three mRNAs of different sizes have been identified for the pro α2(I) chain (Myers et al. 1981a) and two for the pro α1(I) chain (Chu et al. 1982).

3.2.2. Synthesis of Prepro α Chains

The polypeptide chains are synthesized on membrane-bound ribosomes and pass through the membranes into the cisternae of the rough endoplasmic reticulum while being assembled (*See* Prockop et al. 1976; Harwood 1979). The mRNAs for types I, II, IV, and V procollagens have been translated in cell-free systems in several laboratories, and the primary translation products have been characterized (*See* Cheah et al. 1979; Harwood 1979; Haralson 1982; Kurkinen et al. 1982; Monson 1982). The initially synthesized polypeptide chains have N-terminal hydrophobic pre-peptide (or signal) sequences similar to those in many other secretory proteins. The prepeptide sequences of procollagens may be significantly longer, however, as the prepeptide in the prepro α1(I) chain may contain more than 100 amino acid residues (Olsen and Berg 1979; Palmiter et al. 1979; Sandell and Veis 1980; Graves et al. 1981), in contrast to the 15 to 30 residues usually present. The prepeptide probably binds to the membrane of the rough endoplasmic reticulum and leads the nascent polypeptide chain across the membrane (Blobel et al. 1979; Davis and Tai 1980).

3.3. Intracellular Processing of Procollagen

3.3.1. Cleavage of the Prepeptide Sequences

The prepeptides of secretory proteins are removed during or shortly after translocation across the membrane of the rough endoplasmic reticulum, with a single enzyme probably catalyzing the processing of a number of proteins. It is not known how the specific locus is recognized by the protease involved, as the amino acid sequences of different proteins at the cleavage site vary greatly. Modification of the prepeptide sequences of some proteins by incorporation of the leucine analogue β-hydroxyleucine or the threonine analogue β-hydroxynorvaline inhibits removal of the prepeptide (*See* Hortin and Boime 1981). Some such modified proteins are not translocated into the cisternae of the rough endoplasmic reticulum, but the data clearly indicate that removal of the prepeptide is not an obligatory step in the secretion process.

Proteases that are involved in prepeptide processing have recently been solubilized and in part characterized from dog pancreas membranes (Strauss et al. 1979) and *Escherichia coli* (Zwizinski and Wickner 1980). The dog pancreas enzyme cleaves preplacental lactogen by removing the prepeptide in one step, and is inhibited by high concentrations of chymostatin and by some serine protease inhibitors. The prepeptide of the chick prepro α1(I) chain synthesized in the reticulocyte lysate system is likewise cleaved by the dog pancreas membranes (Palmiter et al. 1979), but it is not known whether this chain is processed by the same protease that is involved in the case of the preplacental lactogen.

3.3.2. Hydroxylation of Proline and Lysine Residues

The hydroxylation of proline and lysine residues is catalyzed by three separate enzymes, peptidyl proline 4-hydroxylase (EC 1.14.11.2, usually termed peptidyl proline hydroxylase), peptidyl proline 3-hydroxylase (EC 1.14.11.?), and peptidyl lysine hydroxylase (EC 1.14.11.4). All three hydroxylate proline or lysine residues in peptide linkage, and they all require Fe^{2+}, 2-oxoglutarate, O_2, and ascorbate (Figure 3.3). The 2-oxoglutarate is stoichiometrically decarboxylated and oxidized during hydroxylation, with one atom of the O_2 molecule being incorporated into succinate while the other is incorporated into the hydroxyl group. The corresponding reaction products, peptidyl 4-hydroxyproline, 3-hydroxyproline, and hydroxylysine, are found in vertebrate proteins, almost exclusively in collagens and a few other proteins with collagen-like amino acid sequences (Chapter 1, Section 1.2.7). For detailed references on the hydroxylation, see Prockop et al. 1976; Kivirikko and Myllylä 1980, 1982a.

Peptidyl proline 4-hydroxylase, peptidyl proline 3-hydroxylase, and peptidyl lysine hydroxylase are probably all located within the cisternae of the rough endoplasmic reticulum. Peptidyl proline 4-hydroxylase is either free within the cisternae or loosely bound to the inner membrane, whereas peptidyl lysine hydroxylase is probably in part free or loosely bound and in part tightly bound. Most of the hydroxylation occurs while the nascent polypeptide chains are grow-

Figure 3.3. Reactions catalyzed by peptidyl proline 4-hydroxylase, peptidyl proline 3-hydroxylase, and peptidyl lysine hydroxylase.

PROLINE + 2-oxoglutarate (COOH–C=O–CH_2–CH_2–COOH) + O_2 —(Ascorbate, Fe^{++})→ 4-HYDROXYPROLINE + succinate (COOH–CH_2–CH_2–COOH) + CO_2

PROLINE + 2-oxoglutarate (COOH–C=O–CH_2–CH_2–COOH) + O_2 —(Ascorbate, Fe^{++})→ 3-HYDROXYPROLINE + succinate (COOH–CH_2–CH_2–COOH) + CO_2

LYSINE (-NH-CH-C(=O)-, CH_2, CH_2, CH_2, CH_2NH_2) + 2-oxoglutarate (COOH–C=O–CH_2–CH_2–COOH) + O_2 —(Ascorbate, Fe^{++})→ HYDROXYLYSINE (-NH-CH-C(=O)-, CH_2, CH_2, CHOH, CH_2NH_2) + succinate (COOH–CH_2–CH_2–COOH) + CO_2

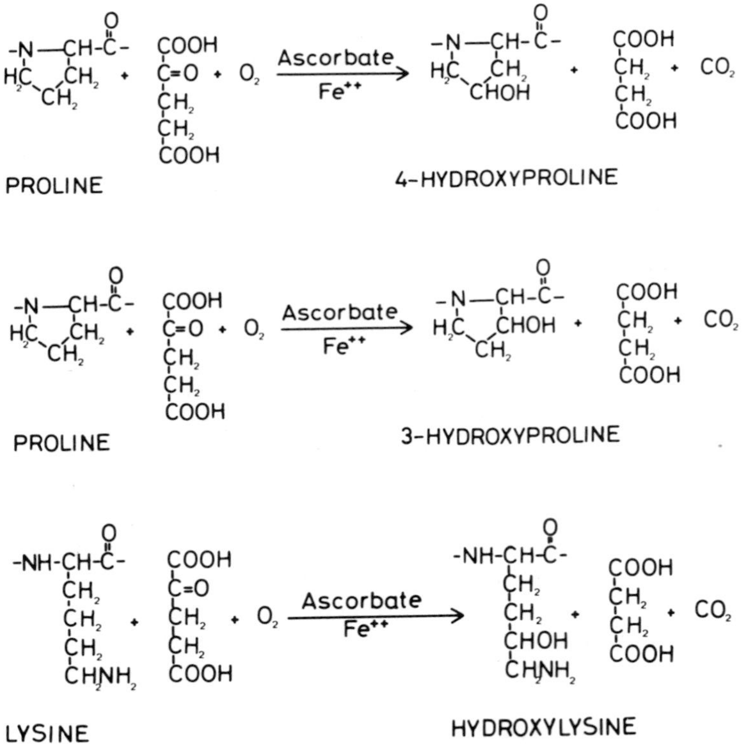

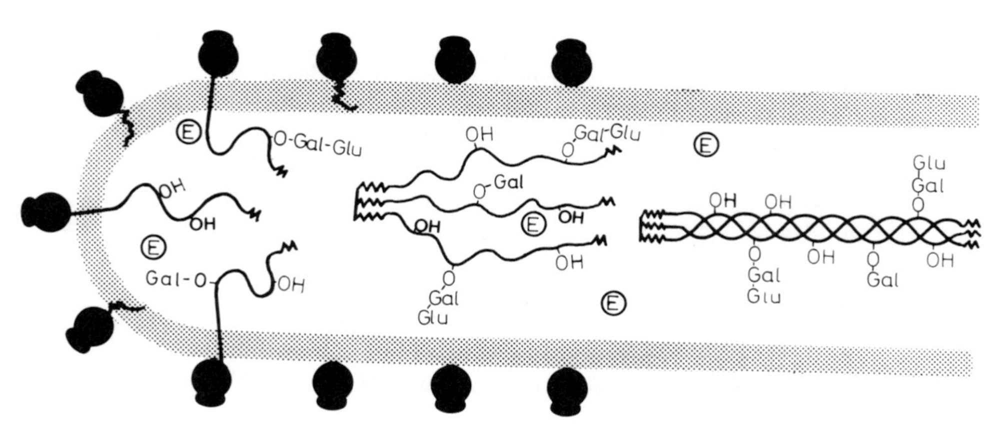

Figure 3.4. Schematic representation of the events of collagen biosynthesis that take place within the cisternae of the rough endoplasmic reticulum. The scheme indicates that the prepeptide is removed during translocation across the membrane and that the hydroxylations of peptidyl proline and lysine residues and the glycosylations of peptidyl hydroxylysine residues are initiated while the polypeptide chains are being assembled on the ribosomes, and continue after the synthesis of complete pro α chains. Formation of the triple helix stops these steps. *E* = enzyme. *(Source: Kivirikko and Myllylä (1979) Int Rev Connect Tissue Res 8:23–72, with permission.)*

ing on the ribosomes (Figure 3.4). These reactions are continued, however, after the release of complete polypeptide chains from the ribosomes (*See* Anttinen and Hulkko 1980; Kivirikko and Myllylä 1980; Pihlajaniemi et al. 1981), until triple helix formation of the pro α chains prevents any further hydroxylation (see below).

Under some conditions hydroxylation occurs only after the release of the completed pro α chains into the cisternae of the rough endoplasmic reticulum. This has been demonstrated by incubating cells or tissues under conditions in which hydroxylation is inhibited and subsequently reversing this inhibition. During the inhibition period the cells synthesize nonhydroxylated polypeptide chains, which accumulate within the cisternae of the rough endoplasmic reticulum and are then hydroxylated during the reversal period.

Peptidyl proline 4-hydroxylase has been purified to homogeneity from several sources, including human tissues (Kivirikko and Myllylä 1980, 1982a). The active enzyme from all these sources is a tetramer ($\alpha_2\beta_2$) with a molecular weight of about 240,000 that consists of two different types of enzymically inactive monomers with molecular weights of about 64,000 (α-subunit) and 60,000 (β-subunit). Distinct differences are found between the two monomers in amino acid and carbohydrate composition, and in the peptide fragments produced by trypsin, chymotrypsin, or *Staphylococcus aureus* V8 protease.

Peptidyl lysine hydroxylase has been isolated as a homogeneous protein from chick embryos and human placentas (Turpeenniemi-Hujanen et al. 1980, 1981). The active enzyme from both sources is probably a dimer with a molecular weight of about 190,000 by gel filtration that consists of only one type of monomer with a molecular weight of about 85,000. Peptidyl proline 3-hydroxylase has been purified about 5000-fold from an ammonium sulphate fraction of chick embryo extract (Kivirikko and Myllylä 1980, 1982a). The molecular weight is about 160,000 by gel filtration, but the subunit structure is unknown.

Some data suggest that peptidyl lysine hydroxylase and peptidyl proline 3-hydroxylase may have collagen type-specific or tissue-specific isoenzymes. Large differences are found in the extent of lysine hydroxylation and proline 3-hydroxylation among genetically distinct collagen types and within the same collagen type from different tissues (*See* Section 3.5 and Chapter 2). The deficiency in peptidyl lysine hydroxylase activity present in the type VI variant of the Ehlers-Danlos syndrome leads to a hydroxylysine deficiency which likewise shows a wide variation in extent from tissue to tissue (Chapter 11). Recent data indicate, however, that the molecular and catalytic properties of peptidyl lysine hydroxylase purified from several sources (Turpeenniemi-Hujanen et al. 1980, 1981) and the catalytic properties determined in crude extracts from cells synthesizing various genetically distinct collagen types (Puistola 1982) are essentially identical. Other data suggest that differences in the extent of lysine hydroxylation and proline 3-hydroxylation between collagen types can be explained by regulation of single peptidyl lysine hydroxylase and peptidyl proline 3-hydroxylase activities, and hence do not require the presence of isoenzymes. Immunological characterization of chick-embryo peptidyl lysine hydroxylase similarly does not support the hypothesis that various tissues in the embryo have collagen type-specific or tissue-specific forms (Turpeenniemi-Hujanen 1981). Accordingly, a number of observations argue against the existence of isoenzymes with different molecular or catalytic properties, but additional studies with other techniques are required to further elucidate this aspect.

The minimum sequence requirement for reaction with a vertebrate peptidyl proline 4-hydroxylase is fulfilled by an -X-Pro-Gly- sequence, and that for reaction with peptidyl lysine hydroxylase by an -X-Lys-Gly- sequence (*See* Prockop et al. 1976; Kivirikko and Myllylä 1980). In some cases, however, both enzymes also seem to hydroxylate certain other triplets. These data agree with the occurrence of 4-hydroxyproline and hydroxylysine in collagens and other vertebrate proteins almost exclusively in the Y positions of X-Y-Gly- sequences. The only exceptions reported are two sequences of X-4Hyp-Ala in the subcomponent C1q of complement, and one sequence of X-Hyl-Ser- or X-Hyl-Ala- in the short nonhelical domains at the ends of the α chains in some collagens. Peptidyl proline 3-hydroxylase appears to require a -Pro-4Hyp-Gly- sequence, whereas the sequence -Pro-Pro-Gly- is probably not hydroxylated. In agreement with this, 3-hydroxyproline has been found in collagens only in the sequence -Gly-3Hyp-4Hyp-Gly-.

The reaction with peptidyl proline 4-hydroxylase and peptidyl lysine hydroxylase is further influenced by the amino acid in the X position of the -X-Y-Gly- sequence, and the reactions with all three enzymes are affected by other nearby amino acids, by the peptide chain length, and by the peptide conformation. In most cases, the structure of the amino acid in the X position of the same triplet probably influences the maximal velocity (V) of the reaction, whereas the main effect of an increase in chain length is a decrease in the Michaelis constant (K_m) of the peptide. The triple helical conformation prevents all three hydroxylations, indicating that these must take place in the cell before triple helix formation of the pro α chains.

The kinetic constants of the three enzymes for their substrates and cosubstrates are very similar, and the catalytic center activities of pure peptidyl proline 4-

hydroxylase and peptidyl lysine hydroxylase are likewise similar (Kivirikko and Myllylä 1980; Turpeenniemi-Hujanen et al. 1980, 1981). Immunological studies suggest that there may be distinct structural similarities close to or at the active sites of peptidyl proline 4-hydroxylase and peptidyl lysine hydroxylase. Several bivalent cations, citric acid cycle intermediates, and certain other compounds compete with some of the cosubstrates to inhibit hydroxylation, and the inhibition constants of these compounds for peptidyl proline 4-hydroxylase and peptidyl lysine hydroxylase are very similar.

The mechanisms of the peptidyl proline 4-hydroxylase and peptidyl lysine hydroxylase reactions have been elucidated in extensive kinetic studies (Kivirikko and Myllylä 1980, 1982a). These and other data (De Jong et al. 1982) are consistent with an ordered binding of Fe^{2+}, 2-oxoglutarate, O_2, and the peptide substrate to the enzyme, in this order; and an ordered release of the hydroxylated peptide, CO_2, succinate, and Fe^{2+}, in which the order of release of the hydroxylated peptide and CO_2 is uncertain and Fe^{2+} is not released between the catalytic cycles (Figure 3.5). The oxygen must obviously be activated, probably to superoxide. Both enzymes catalyze an uncoupled decarboxylation of 2-oxoglutarate at a low rate in the absence of the peptide substrate, but the details of this reaction are unknown.

Ascorbate is not consumed stoichiometrically in the peptidyl proline 4-hydroxylase and peptidyl lysine hydroxylase reactions, and both enzymes can catalyze their reactions for a number of catalytic cycles in the complete absence of this vitamin (Myllylä et al. 1978; Kivirikko and Myllylä 1980, 1982a; Nietfeld and Kemp 1981). These findings, together with the kinetic data, suggest that the reaction with ascorbate is required to prevent occasional oxidation of either the enzyme-bound iron or some other groups in the enzyme molecule (Figure 3.5). The specificity of this vitamin requirement for pure peptidyl proline 4-hydroxylase and highly purified peptidyl lysine hydroxylase is quite high, but ascorbate can be partly replaced by dithiothreitol, L-cysteine and some reduced pteridines in high concentrations (Myllylä et al. 1978; Puistola et al. 1980). A

Figure 3.5. Schematic representation of the mechanism proposed for the peptidyl proline 4-hydroxylase and lysine hydroxylase reactions. The dashed arrows indicate that ascorbate is required to prevent oxidation of either the enzyme-bound iron or some other group during some catalytic cycles, but not the majority. Fe^{2+} need not leave the enzyme between each catalytic cycle, and is shown as remaining enzyme bound. The order of release of the hydroxylated peptide and CO_2 is uncertain. *E* = enzyme; *2-Og* = 2-oxoglutarate; *Pept-OH* = hydroxylated peptide; *Succ* = succinate; *Asc* = ascorbate. *(Source: Modified; Kivirikko and Myllylä (1980) In: Freedman and Hawkins, eds., The Enzymology of Post-Translational Modifications of Proteins. London: Academic, with permission.)*

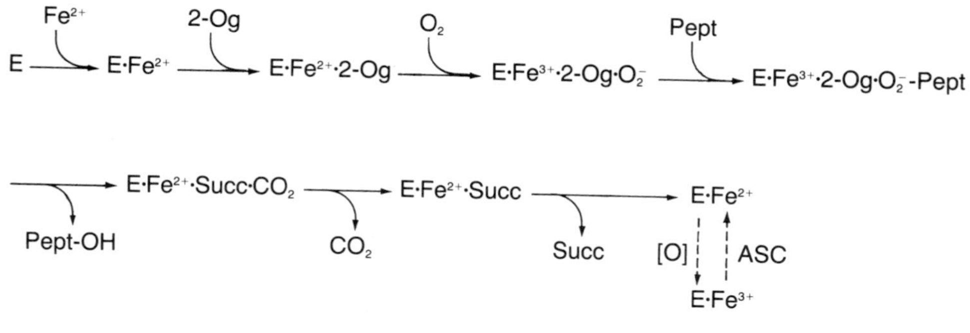

$$E \xrightarrow{Fe^{2+}} E\cdot Fe^{2+} \xrightarrow{2\text{-}Og} E\cdot Fe^{2+}\cdot 2\text{-}Og \xrightarrow{O_2} E\cdot Fe^{3+}\cdot 2\text{-}Og\cdot O_2^- \xrightarrow{Pept} E\cdot Fe^{3+}\cdot 2\text{-}Og\cdot O_2^-\text{-}Pept$$

$$\xrightarrow{-\,Pept\text{-}OH} E\cdot Fe^{2+}\cdot Succ\cdot CO_2 \xrightarrow{-\,CO_2} E\cdot Fe^{2+}\cdot Succ \xrightarrow{-\,Succ} E\cdot Fe^{2+} \underset{ASC}{\overset{[O]}{\rightleftharpoons}} E\cdot Fe^{3+}$$

low hydroxylation rate has also been found in several cultured cells in the complete absence of ascorbate (*See* Prockop et al. 1979; Kivirikko and Myllylä 1980), this being due to an unidentified intramembranous reductant rather than cysteine or tetrahydropteridine (Mata et al. 1981). It is not known, however, if this reductant has any role in vivo.

The hydroxy group of the 4-hydroxyproline residue stabilizes the collagen triple helix under physiological conditions (*See* Prockop et al. 1976; Kivirikko and Myllylä 1980). Pro α chains that contain no hydroxyproline can fold into the triple helix at low temperatures, but the transition temperature (T_m) for the unfolding of such molecules is only about 24°C. Accordingly, nonhydroxylated pro α chains cannot form triple helical molecules in vivo, and an almost complete 4-hydroxylation of proline residues in the Y positions of the Gly-X-Y triplets is necessary for the formation of a molecule that is stable at 37°C. This function is likely to be the same in other proteins with collagen-like amino acid sequences. The mechanism by which 4-hydroxyproline stabilizes the triple helix is not known, however (*See* Chapter 1, Section 1.2.2). The role of the 3-hydroxyproline residues is unknown at present.

The hydroxy group of hydroxylysine has two functions: it acts as the site of attachment for carbohydrate units and it is essential for the stability of the intermolecular collagen crosslinks (*See* Chapter 2). The importance of hydroxylysine is clearly demonstrated by the clinical signs of the type VI variant of the Ehlers-Danlos syndrome, in which a genetic deficiency in peptidyl lysine hydroxylase activity produces a connective tissue disorder with marked changes in the mechanical properties of certain tissues (Chapter 11).

3.3.3. Glycosylation of Hydroxylysine Residues

The formation of hydroxylysine-linked carbohydrate is catalyzed by two specific enzymes (Figure 3.6), first peptidyl hydroxylysine galactosyltransferase (EC 2.4.1.50), transferring galactose to certain hydroxylysines, and then peptidyl galactosylhydroxylysine glucosyltransferase (EC 2.4.9.66), transferring glucose to certain galactosylhydroxylysines. For detailed references on these glycosylations, see Kivirikko and Myllylä 1979. Some of the carbohydrate is present as the monosaccharide galactose and some as the disaccharide glucosylgalactose, these being the only carbohydrate units present in mammalian interstitial collagens (Chapter 2).

Glycosylation takes place largely while the polypeptide chains are still being assembled on the ribosomes, as there is no significant lag between the hydroxylation of lysine and the glycosylation of hydroxylysine. Glycosylation, like hydroxylation, continues after the release of complete pro α chains into the cisternae of the rough endoplasmic reticulum (Figure 3.4), until triple helix formation of the procollagen prevents any further reaction. As in the case of hydroxylation, glycosylation takes place after the release of completed polypeptide chains from the ribosomes if it is blocked earlier.

Peptidyl hydroxylysine galactosyltransferase has been purified to about 1000-fold from an ammonium sulphate fraction of chick-embryo extract (*See* Kivirikko and Myllylä 1979, 1982a). The activity of the partially purified enzyme is found by gel filtration in the form of two major species with apparent molecular weights of about 450,000 and 200,000, respectively, and one minor species with an ap-

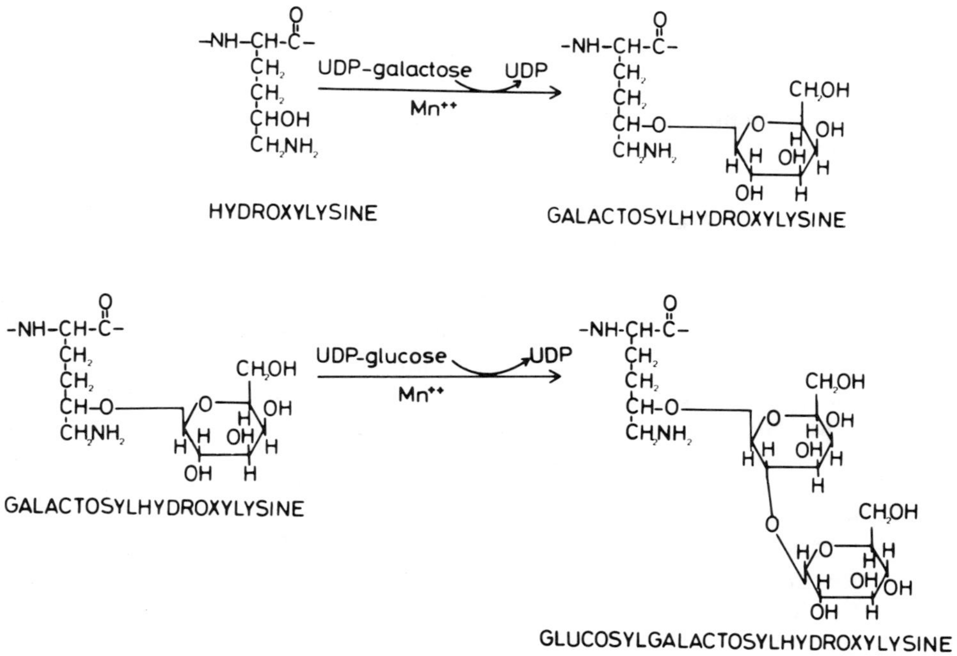

UDP-galactose UDP
Mn^{++}
HYDROXYLYSINE
GALACTOSYLHYDROXYLYSINE

UDP-glucose UDP
Mn^{++}
GALACTOSYLHYDROXYLYSINE
GLUCOSYLGALACTOSYLHYDROXYLYSINE

Figure 3.6. Reactions catalyzed by peptidyl hydroxylysine galactosyltransferase and peptidyl galactosylhydroxylysine glucosyltransferase.

parent molecular weight of about 50,000. Peptidyl galactosylhydroxylysine glucosyltransferase has been purified to homogeneity from chick embryos and to a lesser extent from several other sources. The enzyme consists of one polypeptide chain with a molecular weight of about 70,000. Both transferases, like the three hydroxylases, are probably glycoproteins.

No direct data are available at present to indicate the existence of collagen type-specific or tissue-specific peptidyl hydroxylysine glycosyltransferase isoenzymes. Differences in the extent of hydroxylysine glycosylation among genetically distinct collagen types can be explained by regulation of the amounts of single types of peptidyl hydroxylysine galactosyltransferase and peptidyl galactosylhydroxylysine glucosyltransferase activity. Immunological characterization of chick-embryo peptidyl galactosylhydroxylysine glucosyltransferase likewise does not indicate the presence of isoenzymes with distinctly different specific activities (Myllylä 1981). The possible existence of such isoenzymes nevertheless requires further elucidation.

Hydoxylysine is formed only as a cotranslational and posttranslational modification, and hence it is evident that the carbohydrate units are synthesized in vivo on a peptide substrate. Free hydroxylysine does not act as the carbohydrate acceptor for the galactosyltransferase, whereas free galactosylhydroxylysine does act in vitro as a substrate for the glucosyltransferase. The free ε-amino group of the hydroxylysine residue appears to be an absolute requirement for the two transferases, as its N-acetylation or deamination completely inhibits both reactions (*See* Kivirikko and Myllylä 1979).

The two glycosylations are further affected by the amino acid sequence of the peptide, the peptide chain length, and the peptide conformation (Kivirikko and Myllylä 1979). The number of -X-Hyl-Gly- sequences in a polypeptide chain is an important determinant of its overall glycosylation (Anttinen and Hulkko 1980), whereas the amino acid sequence around the hydroxylysine residue may only be a relative factor. The effect of the peptide chain length is that longer peptides are better substrates than shorter ones. The triple helical conformation of the substrate prevents interaction with both transferases, indicating that glycosylation, like hydroxylation, must occur in the cell before triple helix formation of the newly synthesized polypeptide chains.

The preferential carbohydrate donor for both peptidyl hydroxylysine glycosyltransferases is the corresponding uridine diphosphate (UDP) glycoside, and both reactions require a bivalent cation, this requirement being best fulfilled by manganese. Fe^{2+} and Co^{2+} can also serve as the metal cofactor for the glucosyltransferase in vitro, but Co^{2+} does not show any activity in concentrations comparable to its tissue levels (Myllylä et al. 1979). Several bivalent cations, the reaction product UDP, and certain other nucleotides are inhibitors of both transferases.

The mechanism of the peptidyl galactosylhydroxylysine glucosyltransferase reaction has been studied by analyzing the initial velocity and inhibition kinetics (*See* Kivirikko and Myllylä 1979, 1982a). The data are consistent with an ordered binding of Mn^{2+}, UDP-glucose, and collagen to the enzyme, and an ordered release of the glycosylated collagen, UDP, and Mn^{2+}, in which Mn^{2+} need not leave the enzyme between the catalytic cycles (Figure 3.7). The transferase appears to have at least two Mn^{2+} binding sites with dissociation constants of 3–5 μM (site I) and 50–70 μM (site II) (Myllylä et al. 1979). At high Mn^{2+} concentrations, the enzyme seems to bind two Mn^{2+} ions successively before the binding of UDP-glucose, but due to the relatively high dissociation constant at site II, it is not known whether this site has any significance in vivo. The peptidyl hydroxylysine glycosyltransferase reactions, unlike those involved in the biosynthesis of asparagine-linked oligosaccharide units (see below), do not seem to involve any lipid intermediate.

The role of the hydroxylysine-linked carbohydrate units remains unresolved. As these constitute large groups on the surface of the collagen molecule, it has been suggested that they may have some function in regulating the packing of molecules into the fibrils. This point is discussed in Chapter 2 (Section 2.2.2). Intermolecular collagen crosslinks can be formed from hydroxylysine residues that are glycosylated, but it is not known whether this affects their properties (Chapter 2, Section 2.3.3). The extent of collagen glycosylation may also affect

Figure 3.7. Schematic representation of the mechanism proposed for the peptidyl galactosylhydroxylysine glucosyltransferase reaction. At high Mn^{2+} concentrations, the enzyme probably binds two Mn^{2+} ions successively before the binding of UDP-glucose, but due to the relatively high dissociation constant at the second Mn^{2+} binding site, this situation may not exist in vivo and hence is not shown. Mn^{2+} need not leave the enzyme between each catalytic cycle, and is shown to remain enzyme bound. *E* = enzyme; UDPglc = UDP-glucose; *Glc-pept* = glucosylated peptide. *(Source: Modified, Kivirikko and Myllylä (1979) Int Rev Connect Tissue Res, 8:23–72, with permission.)*

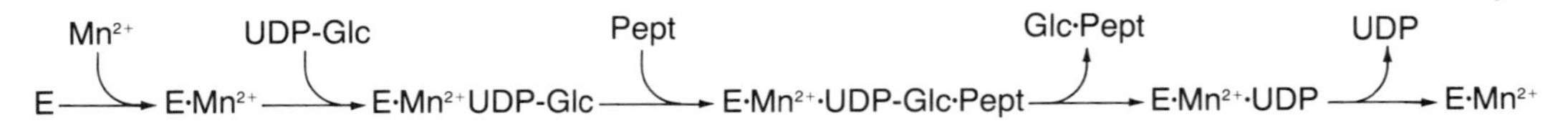

the sensitivity of the protein to cleavage by mammalian collagenases (Chapter 4), and the hydroxylysine-linked carbohydrate units may have a role in the attachment of some cells to type IV collagen (Berman et al. 1980), but little data are currently available to elucidate these possibilities.

3.3.4. Glycosylation of Asparagine Residues in the Propeptides

The propeptides of procollagen contain asparagine-linked carbohydrate units that are not present in the collagen domain of the molecule (*See* Clark 1979, 1982; Kivirikko and Myllylä 1979, 1982b; Chapter 2). In type II procollagen, these units are found in both the N-terminal and C-terminal propeptides, whereas in type I procollagen most or all of the carbohydrate is present in the C-terminal propeptides. The pro α1(I) and pro α2(I) chains probably each contain a single high mannose oligosaccharide unit that consists of two residues of N-acetylglucosamine and about 10 residues of mannose.

The asparagine-linked carbohydrate units of other proteins are first synthesized on a lipid carrier. The preformed oligosaccharide is then transferred as a whole to an asparagine residue in a sequence -Asn-X-Ser/Thr- in the nascent polypeptide chain (*See* Hart et al. 1979; Hubbard and Ivatt 1981). The carbohydrate unit may subsequently be processed further within the rough and smooth endoplasmic reticulum and the Golgi complex. These processes are discussed in more detail in Chapter 8. Procollagen glycosylation occurs by a similar mechanism (*See* Clark 1979, 1982; Kivirikko and Myllylä 1979, 1982b), with the asparagine residue in chick pro α1(I) and pro α2(I) chains present in the sequences -Asn-Val-Thr- (Showalter et al. 1980) and -Asn-Ile-Thr- (Pesciotta et al. 1981), respectively. It is not known, however, whether this oligosaccharide unit is processed further within the smooth endoplasmic reticulum and the Golgi complex.

The functions of the aspargine-linked carbohydrate units of procollagens are unknown. Suggestions that these oligosaccharides are required for a normal rate of procollagen secretion from the cell (Housley et al. 1980), have not been supported by other investigations (*See* Bornstein and Traub 1979; Kivirikko and Myllylä 1982b).

3.3.5. Chain Association, Disulfide Bonding, and Helix Formation

The propeptides of the procollagens contain both intra- and interchain disulfide bonds. Interchain disulfide bonds are found in types I and II procollagens only between the C-terminal propeptides, whereas in type III procollagen, both the N-terminal and C-terminal propeptides are linked by such bonds. Type III procollagen also contains two interchain disulfide bonds at the C-terminal end of the triple helical region in the collagen domain (Chapter 2). The molecules of type IV (*See* Bornstein and Sage 1980) and type V procollagens consist, in both cases, of two or more different polypeptide chains (Fessler et al. 1981a,b; Kumamoto and Fessler 1981). These chains are likewise disulfide bonded, except in cases of molecules consisting of three pro α1(V) chains, which do not form interchain disulfide linkages at all (Fessler et al. 1981a,b; Kumamoto and Fessler 1981; Alitalo et al. 1982).

Interchain disulfide bonds are probably not formed until translation is complete, and perhaps not until after the polypeptide chains have been released from the ribosomes (*See* Prockop et al. 1976, 1979). Bonds between the C-terminal propeptides are formed even when triple helix formation is prevented, whereas bonds between the N-terminal propeptides present in type III procollagen are not formed in such a situation (Fessler and Fessler 1978; Fessler et al. 1981c). Thus the interchain disulfide bonds between the C-terminal propeptides are probably synthesized earlier than those between the N-terminal propeptides, the triple helical conformation appearing to be a requirement for the formation of the latter. The C-terminal propeptides are therefore likely to direct the association between the three pro α chains and to serve as an initiating point for triple helix formation. Experiments in which puromycin was used to produce shortened pro α chains likewise suggest such a function for these propeptides (*See* Bornstein and Traub 1979; Fessler et al. 1981c).

The mechanism by which disulfide bonds are formed in the biosynthesis of proteins is not known. In vitro studies of the formation of disulfide bonds in the reoxidation of reduced proteins have demonstrated that the rate-limiting steps are conformational changes accompanied by rearrangement of random "incorrect" protein disulfide bonds (*See* Freedman and Hillson 1980). The in vivo role of the enzyme known as protein disulfide isomerase (EC 5.3.4.1), which catalyzes such rearrangements (Creighton et al. 1980; Freedman and Hillson 1980), is still not clear, but it has been shown to be widely distributed in animal tissues, to be present in the microsomal fraction, and to have a wide specificity for protein disulfide substrates. Protein disulfide isomerase activity is present in cells actively synthesizing collagen (Freedman and Hillson 1980; Myllylä et al. 1983), and changes in this enzyme activity in the tissues of developing chick embryos (Brockway et al. 1980) and cultured 3T6 fibroblasts (Myllylä et al. 1983) correlate with those in the peptidyl proline 4-hydroxylase and peptidyl lysine hydroxylase activity. There is no direct evidence, however, to demonstrate whether protein disulfide isomerase is involved in disulfide bond formation in the procollagens.

Triple helix formation in vivo is believed to have two requirements: almost all the proline residues present in the Y positions of the Gly-X-Y triplets must have been 4-hydroxylated (*See* function of 4-hydroxyproline, above), and probably the interchain disulfide bonds must have been formed between the C-terminal propeptides. The *cis* → *trans* isomerization of the peptide bonds limits the rate of triple helix formation in vitro, but it is uncertain whether this is a factor in vivo (Bächinger et al. 1978).

Evidence for the interchain disulfide bond requirement for rapid triple helix formation has been obtained in model studies in vitro (Bruckner et al. 1978; Gerard et al. 1981). Studies in vivo suffer from an inadequacy of available methodology (Olsen and Berg 1979), but tentatively suggest that the reduction of disulfide bonds in intact cells prevents triple helix formation in a reversible manner and that a close correlation exists between the time required for disulfide bonding and triple helix formation in cells synthesizing various collagen types (*See* Prockop et al. 1976). The time required for these events varies greatly from one cell type to another. It is shortest—only few minutes—in chick-embryo tendon cells synthesizing type I procollagen; intermediate in chick-embryo cartilage cells synthesizing type II; and longest—about one hour or more—in various cells synthesizing type IV and V procollagens (*See* Prockop et al. 1976; Kivirikko and Myllylä 1979; Fessler et al. 1981a; Kumamoto and Fessler 1981).

The exact intracellular location for triple helix formation is unknown. Most available evidence suggests that it is formed only after the release of the pro α chains into the cisternae of the rough endoplasmic reticulum or as the protein leaves this cell compartment. The possibility is not completely excluded, however, that the triple helix may be formed only as the pro α chains reach the Golgi complex (Prockop et al. 1979).

3.3.6. Structure and Function of the Procollagens

The procollagens of the interstitial types I, II, and III collagen have propeptides at both ends of their three polypeptide chains (Figure 3.8). The N-terminal propeptides are composed of three structural regions: a globular N-terminal portion, a central triple helical collagen-like region, and a short C-terminal noncollagenous sequence (Chapter 2). In the pro α2(I) chain, the N-terminal propeptide of several species lacks the N-terminal globular domain (Figure 3.8), although it may be present in the rat. The N-terminal propeptides in types II and III procollagens are very similar to that in the pro α1(I) chain, except that the propeptide in the pro α1(II) chain is somewhat smaller.

The C-terminal propeptides of types I, II, and III procollagen do not contain any collagenous amino acid sequences (Chapter 2). Type IV collagen has noncollagenous domains at the C-terminal ends of its polypeptide chains, but these

Figure 3.8. Schematic representation of the structure of the procollagen molecule. Glc = glucose; gal = galactose; Man = mannose; GlcNac = N-acetylglucosamine. *(Source: Prockop, Kivirikko, et al. (1979) Reproduced from N Engl J Med 301:13–23, 77–85, with permission.)*

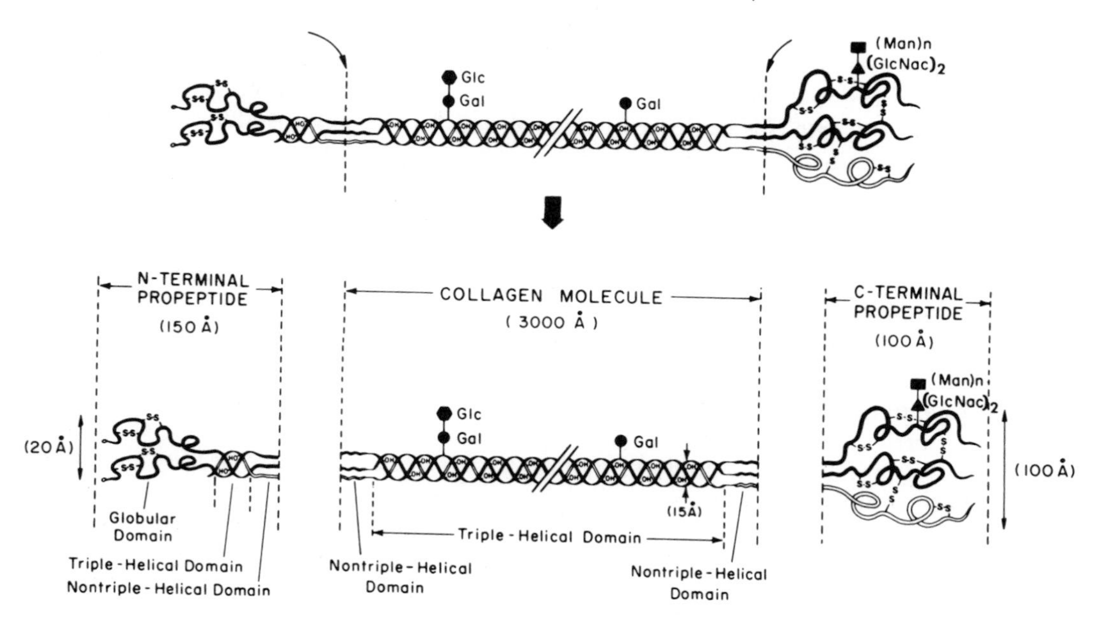

molecules may not be processed (*See* Section 3.4.1). Type V procollagen has large propeptides at both ends of the molecule (Fessler et al. 1981a,b; Kumamoto and Fessler 1981; Bächinger et al. 1982).

The occurrence of asparagine-linked oligosaccharide units and intrachain and interchain disulfide bonds in the propeptides of the interstitial procollagens (Figure 3.8) has been described in Sections 3.3.4 and 3.3.5.

A number of important functions have been proposed for the propeptides (Fessler and Fessler 1978; Bornstein and Traub 1979; Prockop et al. 1979). As discussed above, the C-terminal propeptides probably direct chain association, and their interchain disulfide bonding is probably a requirement for a rapid rate of triple helix formation. An obvious function is to provide properties necessary for transport. They may also play a role in fibril formation and act as feedback inhibitors of procollagen synthesis (*See* Section 3.5.1). Suggestive evidence has been presented to support these latter functions, but additional studies are required to demonstrate them with certainty.

3.3.7. Transport and Secretion of the Procollagens

The secretion of the procollagens is similar to that of other extracellular proteins. The nascent polypeptide chains are translocated across the membrane into the cisternae of the rough endoplasmic reticulum (see above), and the protein passes through the Golgi complex before leaving the cell (Prockop et al. 1976, 1979; Bornstein and Traub 1979).

The conformation of the procollagens has a marked influence on the secretion rate. A number of experiments with isolated cells or tissues have demonstrated that if triple helix formation is prevented, the random coil polypeptide chains first accumulate within the cisternae of the rough endoplasmic reticulum and are then secreted at a delayed rate (*See* Prockop et al. 1976, 1979; Kivirikko and Myllylä 1982b). Helix formation can be prevented by excluding some of the cosubstrates of the hydroxylases or by incubating the cells or tissues with certain proline analogs, such as *cis*-4-hydroxyproline or azetidine-2-carboxylic acid, which are incorporated into protein. The pro α chains containing these proline analogs do not fold into the triple helix. The reason for the decreased secretion rate of nonhelical pro α chains is not known, but it may be due to an interaction between the chains and peptidyl proline 4-hydroxylase or other posttranslational enzymes (*See* Prockop et al. 1979; Kivirikko and Myllylä 1982b). These bind to pro α chains but not to triple helical procollagen (Section 3.3.2), and hence the pro α chains may remain in equilibrium with the enzymes within the cisternae of the rough endoplasmic reticulum.

The secretion of the procollagens is energy dependent and is inhibited by colchicine, vinblastine, and tertiary amine local anesthetics, suggesting the involvement of microtubules and microfilaments (*See* Bornstein and Traub 1979; Eichorn and Peterkofsky 1979; Prockop et al. 1979). Secretion is also inhibited by the monovalent ionophore monensin (Ledger et al. 1980). In the cases of all these inhibitors, the triple helical procollagen accumulates mainly within the Golgi complex. The possible role of the asparagine-linked oligosaccharide units in procollagen secretion is a controversial matter at present (*See* Section 3.3.4).

The form in which procollagen is secreted has not been firmly established. Many observations suggest that procollagen may be packed inside the cell into structured aggregates that are transported to the cell surface and exocytosed (*See* Bruns et al. 1979; Trelstad and Hayashi 1979; Kivirikko and Myllylä 1982b). Aggregation would probably protect procollagen against denaturation at body temperature by increasing the thermal stability of the protein. As the transition temperature (T_m) for the unfolding of type I procollagen is only about 40°C (Hayashi et al. 1979), any increase in this value may be of physiological importance.

3.4. Extracellular Processing

3.4.1. Cleavage of the Pro Domains

Removal of the propeptides requires at least two proteases, one to cleave the N-terminal propeptides and the other to cleave the C-terminal propeptides. The actual number is in fact larger, as at least the N-terminal proteases appear to have collagen type-specific forms.

Kinetic data on intact tissue indicate that processing of types II, III, and V procollagens, and removal of the C-terminal propeptides in type I, clearly take place in the extracellular space subsequent to secretion (*See* Fessler and Fessler 1978; Bornstein and Traub 1979; Prockop et al. 1979; Bornstein and Sage 1980; Heathcote and Grant 1980; Fessler et al. 1981a,b,c; Kivirikko and Myllylä 1982b). These data do not completely exclude the possibility that cleavage of the N-terminal propeptides in type I procollagen may be coincident with secretion, but other findings make such a possibility unlikely. Studies with freshly isolated or cultured fibroblasts indicate that intact types I, II, and III procollagen molecules are secreted by the cells and subsequently processed. Furthermore, colchicine, which inhibits procollagen secretion, also inhibits the removal of the N-terminal propeptides from types I and II procollagens (Fessler and Fessler 1978; Uitto et al. 1979).

There seems to be no obligatory sequence for the cleavage of the two propeptides from a procollagen molecule (Figure 3.9). Nevertheless, the synthesis of type I procollagen in many systems involves removal of the N-terminal propeptides before the C-terminal propeptides (Fessler and Fessler 1978; Bornstein

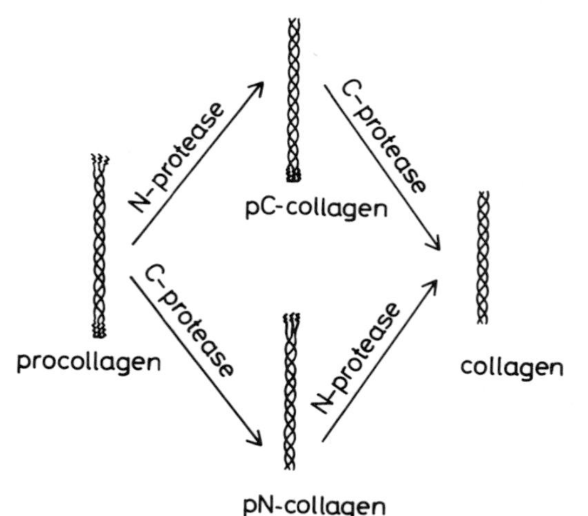

Figure 3.9. Schematic representation of the conversion of procollagen to collagen. There is no obligatory sequence of cleavage for the two propeptides from a procollagen molecule, but most of the processing of types I and II procollagen occurs as shown in the upper pathway, whereas type III procollagen is primarily processed as shown in the lower pathway. *N-protease* = procollagen N-terminal protease; *C-protease* = procollagen C-terminal protease.

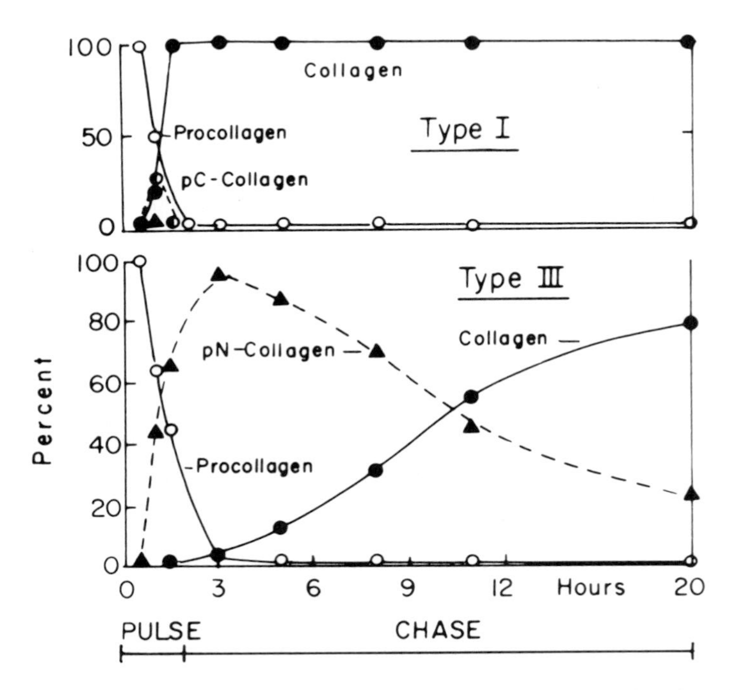

Figure 3.10. Processing of types I and III procollagen in chick-embryo blood vessels. The tissues were incubated with [^{3}H]proline for 2 h, and then chased in a medium supplemented with non-radioactive proline for up to 20 h. The percentage distributions among the types I or III collagenous protein present at each time point are given. The distribution of type I procollagen, pC-collagen, and collagen is shown in the upper panel. The lower panel represents the distribution for the type III proteins. *(Source: Fessler, Timpl, et al. (1981) J Biol Chem 256:2531–2537, with permission.)*

and Traub 1979; Kivirikko and Myllylä 1982b). Some processing takes place in the opposite order in the biosynthesis of type II procollagen (Uitto et al. 1979), and the N-terminal propeptides are removed very slowly in the case of type III procollagen (Figure 3.10) (Fessler and Fessler 1979; Fessler et al. 1981c). This explains the finding that many tissues contain significant amounts of partially processed type III molecules with intact N-terminal propeptides but lacking in C-terminal propeptides (*See* Bornstein and Sage 1980). Type IV collagen is not processed at all in vitro (Crouch and Bornstein 1979; Kefalides et al. 1979; Alitalo et al. 1980; Heathcote et al. 1980; Tryggvason et al. 1980; Pihlajaniemi et al. 1981). Conversion to shorter chains has been demonstrated in vivo in a mouse tumor (EHS sarcoma) (Tryggvason et al. 1982), though it is not known whether any processing takes place in normal tissue. Type V procollagen is processed in at least two steps to final forms that are still larger than the $\alpha1$(V) and $\alpha2$(V) chains obtained from tissues by pepsin extraction (Fessler et al. 1981a,b; Kumamoto and Fessler 1981).

A procollagen N-terminal protease (EC 3.4.24) has been purified about 300 to 4000-fold from chick-embryo tendon extract (Tuderman et al. 1978; Tuderman and Prockop 1982) and to a lesser extent from calf tendons (*See* Fessler and Fessler 1978; Heathcote and Grant 1980). The molecular weight of the chick-embryo tendon enzyme is about 260,000 by gel filtration (Tuderman et al. 1978). This enzyme cleaves both types I and II procollagen, but not types III or IV

(Tuderman et al. 1978; Tuderman and Prockop 1982). Work on N-terminal proteases from other sources likewise indicates that types I and III procollagen are cleaved by separate enzymes (Leung et al. 1979; Nusgens et al. 1980). An N-terminal protease cleaving type III procollagen has been purified in preliminary experiments (Nusgens et al. 1980). This enzyme activity is found by molecular sieve chromatography in the form of two peaks with apparent molecular weights of about 170,000 and 100,000.

A procollagen C-terminal protease with a lower molecular weight (about 80,000) than the N-terminal proteases has been reported, but only partial data are available on its properties (Kessler and Goldberg 1978; Leung et al. 1979; Morris et al. 1979; Prockop et al. 1979). Crude preparations of this C-terminal protease act on types I, III, and V procollagen (Fessler et al. 1981c), but no data are yet available to indicate whether only one enzyme is involved.

The procollagen proteases are endopeptidases, and they all operate at a neutral pH. The type I N-terminal protease and the C-terminal protease have been shown to require a divalent cation such as calcium (Kessler and Goldberg 1978; Tuderman et al. 1978; Leung et al. 1979; Heathcote and Grant 1980) and to act on both monomeric and aggregated procollagen molecules (Leung et al 1979). The triple helical substrate conformation appears to be a requirement for cleavage by type I N-terminal protease whereas the C-terminal protease acts on both the native and the denatured substrate (Tuderman et al. 1978; Morris et al. 1979). The site of cleavage by the N-terminal protease in the pro α1(I) chain is a proline-glutamine bond (Hörlein et al. 1979), and the glutamine subsequently cyclizes to form a pyroglutamate residue (pyrrolidone-carboxylic acid) (Heathcote and Grant 1980). Certain synthetic peptides with sequences similar to the cleavage site in the pro α1(I) chain are inhibitors of this enzyme (Morikawa et al. 1980).

The removal of the propeptides is a prerequisite for the formation of native collagen fibrils (Bornstein and Traub 1979; Prockop et al. 1979). This has been demonstrated especially in animals with dermatosparaxis and patients with the type VII variant of the Ehlers-Danlos syndrome, in which a heritable defect in the cleavage of the N-terminal propeptides causes impaired fibril formation (Chapter 11). The free N-terminal propeptides, once they have been cleaved from the procollagen, may also act as feedback regulators of cellular collagen synthesis (*See* Section 3.5.1).

3.4.2. Fibril Formation

Types I, II, and III collagen molecules produced by removal of the propeptides have a marked ability to self-assemble and spontaneously form fibrils. This event does not need any enzyme or other factors and will occur easily in collagen solutions in vitro. Such fibrils of type I collagen are indistinguishable by criteria based on electron optics and x-ray diffraction from those found in vivo, with the exception of their diameter and length. This has enabled a number of studies to be carried out on fibril formation in vitro, and detailed physicochemical and kinetic data are now available on this process (Chapter 1, Section 1.3.4).

It is not known at present to what extent the in vitro results can be applied to the situation in vivo. Other connective tissue components, e.g., fibronectin (Kleinman et al. 1981) and proteoglycans, are likely to affect the rate of the process and the fibril length and diameter, although these components do not determine the basic structure (Chapter 1). Furthermore, procollagen may be secreted in the form of structured aggregates (*See* Chapter 1, Section 1.3.4), and cleavage of the propeptides may thus lead to aggregated products that are incorporated into the fibrils directly.

Several heritable connective tissue disorders in man and animals involve abnormalities in collagen fibril structure (Chapter 11). Elucidation of the biochemical defects in these patients and animals will probably greatly increase our knowledge of the factors that influence normal collagen fibrillogenesis.

3.4.3. Crosslink Formation

Crosslinking of newly formed collagen fibrils takes place in at least two stages. The initial reaction is oxidative deamination of ε-amino groups in certain lysine and hydroxylysine residues catalyzed by peptidyl lysine oxidase to give reactive aldehydes (Figure 3.11). These aldehydes subsequently participate in the formation of the various crosslinks, as discussed in Chapter 2 (Section 2.3.3). Peptidyl lysine oxidase has been purified to near homogeneity or homogeneity from several sources. For detailed references on peptidyl lysine oxidase, see Siegel 1979; Heathcote and Grant 1980; Kivirikko and Myllylä 1982b. Peptidyl lysine oxidase activity from all tissues studied is found in multiple forms by diethylaminoethyl (DEAE) cellulose chromatography, the molecular weight of most forms being about 30,000 or a multiple thereof. The reason for this heterogeneity is not known, but it may be due to the presence of several enzyme forms, variable aggregation of one enzyme form, or alterations incurred during purification.

Peptidyl lysine oxidase contains tightly bound copper, probably in the cupric form, and requires molecular oxygen and probably pyridoxal. Copper chelators, carbonyl reagents such as phenyl hydrazine, isoniazid, and hydroxylamine, and lathyrogens such as β-aminopropionitrile are inhibitors of this enzyme activity. The mode of inhibition by the lathyrogens is noncompetitive and irreversible, with the inhibitor probably becoming covalently bound to the enzyme.

Figure 3.11. Reactions catalyzed by peptidyl lysine oxidase.

-NH-CH(-C(=O)-)-$(CH_2)_3$-CH_2-NH_2 → (O_2, Cu^{++}) → -NH-CH(-C(=O)-)-$(CH_2)_3$-HC=O

LYSINE ALLYSINE

-NH-CH(-C(=O)-)-$(CH_2)_2$-CHOH-CH_2-NH_2 → (O_2, Cu^{++}) → -NH-CH(-C(=O)-)-$(CH_2)_2$-CHOH-HC=O

HYDROXYLYSINE HYDROXYALLYSINE

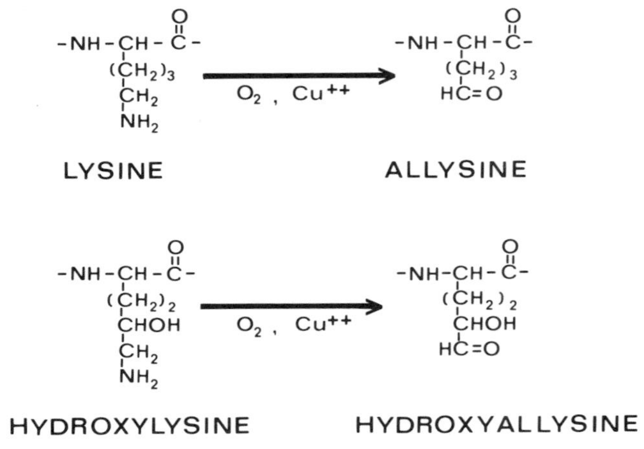

Peptidyl lysine oxidase acts on both lysine and hydroxylysine residues in the nonhelical ends of types I, II, and III collagen, with the activity being greater for hydroxylysine. Highly purified enzyme preparations from several tissues utilize both collagen and elastin as substrates, there being no difference in specificity between the multiple enzyme forms with respect to these two proteins. The highest activity is found with reconstituted collagen fibrils, coacervates of elastin or certain synthetic peptides, and possibly other ordered molecular aggregates of soluble elastin (Chapter 6).

The aldehydes generated by peptidyl lysine oxidase can form two major types of crosslinks, either by aldol condensation of two of the aldehydes or by condensation between one aldehyde and one ε-amino group of a second lysine, hydroxylysine, or glycosylated hydroxylysine residue. The structures of the various collagen crosslinks are described in Chapter 2 (Section 2.3.3) and their localization in Chapter 1 (Section 1.3.3). The crosslinks formed from hydroxylysine-derived aldehyde are known to be more stable than those formed from lysine-derived aldehyde (Chapter 2).

Further stabilization of collagen crosslinks takes place with time. The crosslinks in young connective tissues can be reduced with sodium borohydride in vitro, but such reducible crosslinks gradually disappear from old collagen (Eyre 1980; Heathcote and Grant 1980). This suggests that the reducible crosslinks are long-lived intermediates that can react further. The mechanisms of stabilization in vivo have not been established, however. A detailed discussion is presented in Chapter 2 (Section 2.3.3). There is no evidence that any of the steps beyond the initial oxidation require enzymes. The lysine and hydroxylysine-derived crosslinks have been found in all genetically distinct collagen types except type V, which has not yet been examined. Newly formed fibrils of type I collagen do not have the necessary tensile strength for proper function until the crosslinks have been formed (Chapter 1, Section 1.3.3). This has been demonstrated especially clearly in a number of studies on experimental lathyrism, an acquired molecular disease characterized by a marked fragility of growing connective tissues (Bornstein and Traub 1979; Minor 1980). Lathyrism can be produced by the administration of β-aminopropionitrile or related compounds that inhibit peptidyl lysine oxidase and thus prevent the crosslinking of collagen and elastin. Impaired crosslinking is also found in animals with copper deficiency and in the genetic peptidyl lysine oxidase deficiency present in aneurysm-prone mice and patients with an X-linked form of cutis laxa (Chapter 11).

3.5. Regulation of Collagen Biosynthesis

The regulation of collagen biosynthesis defines the quantity, type, and quality of the protein that is produced. The term type refers to the genetically distinct collagen α chains, and the term quality is used here to indicate that the structure of a single collagen type may vary markedly in the extent to which the polypeptide chains have been modified by the cotranslational and posttranslational enzymes. The type and quantity of the collagen are naturally regulated at the level of transcription, but the quantity may also be regulated at the levels of translation and some of the cotranslational and posttranslational reactions. The quality is primarily determined by regulation of various cotranslational and posttranslational modifications, but a marked change in the rate of procollagen polypeptide chain synthesis can in itself influence the extent to which the α chains are modified.

3.5.1. Transcription and Translation

Various tissues manifest a family of different collagen polypeptide chains, and thus mechanisms must exist for regulation of collagen gene expression. A voluminous and rapidly expanding literature is now available on the assessment of genetically distinct collagen types in various tissues and their synthesis by cultured cells under different conditions. The tissue distribution is described in Chapter 2, the immunological localization in tissues and cells in Chapter 5, and the changes in development in Chapter 10. A recent review (Bornstein and Sage 1980) also deals with collagen types and modulation of their synthesis. Therefore, only a few brief examples on the modulation of collagen types are given below.

Some cells produce only one collagen type: e.g., freshly isolated chick-embryo tendon fibroblasts produce type I and freshly isolated chondrocytes produce type II. A number of cells, however, synthesize more than one collagen type, and there is now firm data to indicate that in many situations a single cell will produce more than one type simultaneously (Bornstein and Sage 1980; Conrad et al. 1980). For example, cultured fibroblasts produce types I and III procollagen, the former usually accounting for about 80% of the total procollagen secreted, and some fibroblasts further synthesize types IV (Alitalo 1980) and/or V (Herrman et al. 1980). Smooth muscle cells in culture likewise synthesize mainly types I and III procollagen, the latter representing the major proportion. As in fibroblasts, small amounts of types IV and V procollagen are also produced (Bornstein and Sage 1980).

Environmental factors influence the collagen gene expression. These factors have been investigated especially with cultured chondrocytes, as these synthesize almost exclusively type II procollagen, and thus a change in the gene expression can be assayed relatively easily. The biosynthetic phenotype of these cells seems to be quite labile, and a number of conditions or factors will cause a switch in the collagen type produced (Table 3.2). The transition involves a loss of type II procollagen synthesis with an onset of production of type I procollagen, type I trimer (*See* Chapter 2), and in some cases also of types III and V. The change in the gene expression of a single cell, at least under some of these conditions,

Table 3.2. Conditions or Factors Causing a Switch in the Collagen Type (II → I) Synthesized by Chondrocytes in Culture[a]

Culturing in monolayer when compared with suspension	Fibronectin
Low cell density	Bromodeoxyuridine
Prolonged time in culture	Ca^{2+}
Embryo extract	Cyclic adenosine monophosphate (cAMP)
	Arrest of growth[b]

[a]For references, see Bornstein and Sage (1980).
[b]Gauss and Müller (1981).

appears to be very abrupt, as only a few cells synthesize types I and II procollagen simultaneously (Bornstein and Sage 1980).

The quantity of procollagen polypeptide chains synthesized differs also markedly between various cell types and even within the same cell type in different situations. For example, about half of the protein synthesized in freshly isolated chick-embryo tendon and cartilage cells is procollagen (Prockop et al. 1979; Risteli et al. 1979), whereas the corresponding value in several types of cultured fibroblast is about 5% (Breul et al. 1980; Myllylä et al. 1981; Pihlajaniemi et al. 1981). In only a few cases have changes in quantity of a collagen produced been ascribed to a specific biosynthetic step.

Cortisol and anti-inflammatory steroids when administered in pharmacological doses, and parathyroid hormone when added to bones in organ culture, reduce the rate of collagen biosynthesis. A specific decrease in the cellular concentration of translatable type I procollagen mRNAs has been demonstrated in these situations (Kream et al. 1980; Oikarinen and Ryhänen 1981), but it is not yet known whether this is due to a decrease in the transcription rate of the corresponding genes or an increase in mRNA degradation. Many other hormones also affect the rate of collagen synthesis (*See* Kivirikko and Risteli 1976), and although these effects have so far not been elucidated in detail, many of the changes are likely to occur at the level of transcription. It should be noted, however, that many hormones also affect the amounts of the cotranslational and posttranslational enzymes (*See* Kivirikko and Myllylä 1979, 1980; Siegel 1979), and the rate of collagen degradation (Kivirikko and Risteli 1976; Chapter 4).

Growth, development, and aging involve major changes in collagen biosynthesis and metabolism (Prockop et al. 1979; Chapter 10). These lead to distinct alterations in collagen types, quantity, and quality. The molecular mechanisms of the changes in collagen quantity have been studied during the development of chick-embryo calvaria and have been found to involve changes in the concentration of type I procollagen mRNAs with an unaltered translational efficiency (Moen et al. 1979). A similar situation has been found during the fetal development of sheep lung and tendon (Tolstoshev et al. 1981b).

Transformation of cultured fibroblasts by RNA tumor viruses, such as the Rous sarcoma virus, DNA viruses, or chemical carcinogens, leads to a rapid, marked reduction in the amount of collagen synthesized and in the levels of type I procollagen mRNAs (Figure 3.12) (Adams et al. 1979; Prockop et al. 1979; Sandmeyer and Bornstein 1979; Sandmeyer et al. 1981b; Sobel et al. 1981). The decreases in the concentrations of the mRNAs coding for the pro α1(I) and pro α2(I) chains are identical (Sandmeyer et al. 1981a; Sobel et al. 1981), these changes being caused primarily by a coordinate decrease in the transcription of the pro α1(I) and pro α2(I) genes (Avvedimento et al. 1981; Sandmeyer et al. 1981a). Each class of transformants appears to possess a characteristic degree of collagen synthesis reduction, correlating inversely with the extent of transformation (Sandmeyer et al. 1981b). It may be noted that transformation affects not only collagen quantity, but also collagen quality and in some cases collagen type (Bornstein and Sage 1980; Myllylä et al. 1981).

Peptide and protein factors that activate or inhibit collagen biosynthesis have been studied from several sources such as silica-treated macrophages, hyperlip-

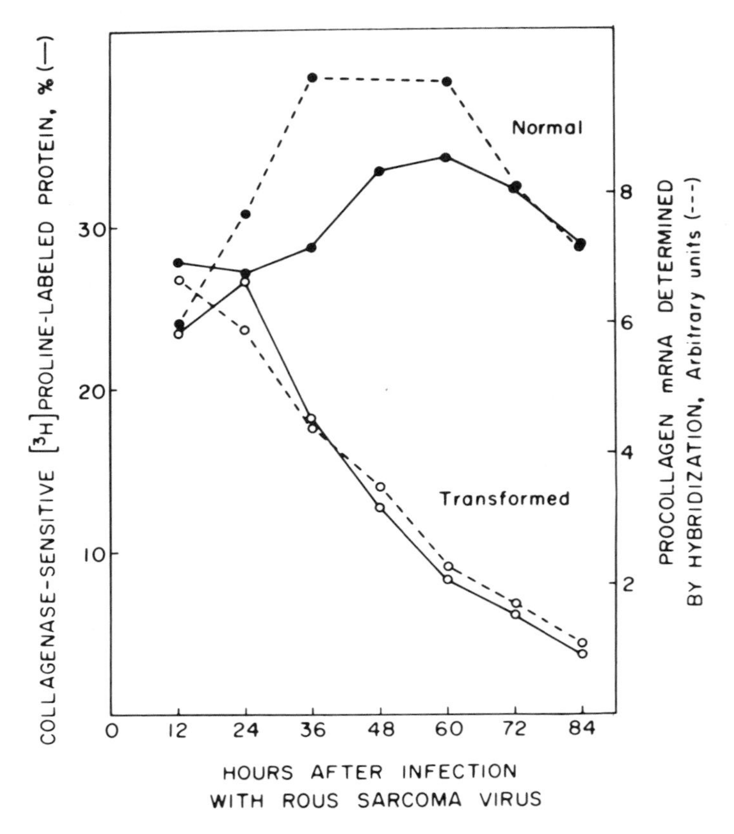

Figure 3.12. Changes in procollagen biosynthesis and procollagen mRNA following transformation of chick-embryo fibroblasts with the Rous sarcoma virus. Procollagen synthesis is indicated by collagenase-sensitive proline-labeled protein in the cell layer *(solid line)*. Procollagen mRNA *(broken line)* was determined by hybridization. *(Source: Sandmeyer and Bornstein (1979) J Biol Chem 254:4950–4953, with permission.)*

idemic serum, regenerating liver, embryonic tissues, experimental granulation tissues, and mononuclear cells (Kivirikko and Risteli 1976; Aalto et al. 1979; Jimenez et al. 1979; Canalis et al. 1980; Schiltz and Ward 1980). Many of these factors are likely to influence transcription, but none of these effects has been identified in any great detail so far.

The detailed mechanisms regulating collagen gene expression are unknown at present. The gene coding for the human pro α1(I) chain has been shown to be located on chromosome 17 (Huerre et al. 1982) and that coding for the human pro α2(I) chain on chromosome 7 (Junien et al. 1982). A gene coding for an unidentified (*See* Chu et al. 1982) human pro α1(I)-like chain is likewise located on chromosome 7 (Weiss et al. 1982), and a gene coding for one polypeptide chain of type IV collagen has preliminarily been assigned to chromosome 17 (Kefalides 1979). No data are available on the location of the other collagen genes. Until the genetic map of the various collagen genes has been elucidated, it will not be known whether any of these genes are arranged in linked clusters or whether they are scattered throughout the genome. A recent study has indicated that there is only one gene for the human pro α2(I) chain, and thus changes in the production of this polypeptide chain are not modulated by differential transcription of multiple α2(I) genes (Dagleish et al. 1982).

The DNA sequence which precedes the start site for transcription of the chick collagen pro α2(I) gene shows several structures that may be interaction sites for regulatory proteins and therefore may play a role in the control of gene expression (Merlino et al. 1981; Vogeli et al. 1981). The pro α2(I) mRNA has three AUG start codons in its initial portion, the first two of them being followed by very short coding sequences. If only the third AUG is recognized as a translation start, the untranslated sequence might play a role in translational control. Alternatively, the two short peptides produced might be involved in the regulation of gene expression.

Although most of the regulation discussed above takes place by transcriptional mechanisms, some observations suggest that collagen synthesis can also be regulated at the level of translation. In cultured human fetal lung fibroblasts the concentrations of type I procollagen mRNAs are twice as high in confluent cells as in logarithmic phase cells, even though the logarithmic phase fibroblasts actually synthesize slightly more type I procollagen per cell than do confluent fibroblasts (Tolstoshev et al. 1981a). This suggests that the logarithmic phase and confluent lung fibroblasts use their type I procollagen mRNAs with different efficiencies. A similar situation has been reported during development of fetal sheep skin (Tolstoshev et al. 1981b). The synthesis of collagen falls approximately 10-fold during the last half of fetal life while the concentrations of type I procollagen mRNAs remain constant. Additional observations suggesting the presence of translational control mechanisms have been made in chondroblasts infected with the Rous sarcoma virus. The transformation of these cells leads to a reduction in the level of type II procollagen mRNA and an induction of type I procollagen mRNAs, but the latter mRNAs are not translated in intact cells (Adams et al. 1982).

It has been suggested that the free N-terminal propeptides of procollagen act as feedback regulators of collagen translation. These peptides are present in significant concentrations in various tissues and biological fluids, such as serum (Rohde et al. 1979). Addition of the N-terminal propeptides of types I or III procollagen to the incubation medium of cultured fibroblasts results in a specific decrease in the rate of collagen synthesis, though no such effect can be demonstrated in cultured chondrocytes (Wiestner et al. 1979). This biological activity resides in the N-terminal globular domain of the propeptide (Figure 3.8), and can also be demonstrated by adding this fragment or the intact N-terminal propeptide to a reticulocyte lysate cell-free translation system of procollagen mRNAs (Paglia et al. 1979, 1981). These findings suggest that the N-terminal propeptides or fragments of them may influence the efficiency with which the procollagen mRNAs are translated, but it is not known at present whether the peptides are actually taken up by the cells and transported to the translation site. It is likewise not known whether this regulatory mechanism exists in vivo, as the peptide concentrations in the extracellular matrix are probably not as high as those required to reduce the rate of collagen synthesis in cultured cells.

3.5.2. Intracellular Processing

The rates of intracellular cotranslational and posttranslational enzyme reactions and the extent to which the collagen is modified are influenced by a large number of different events or factors. In extreme cases, such as when proline 4-hydroxylation is prevented, the formation of fibrillar collagen is inhibited completely. In other situations, the overall rate of collagen synthesis may be unaffected, but there may be marked changes in the properties of the fibrils.

The genetically distinct collagen types show large differences in the extents of proline 3-hydroxylation, lysine hydroxylation, hydroxylysine galactosylation, and galactosylhydroxylysine glucosylation, and distinct though smaller differences in the extent of proline 4-hydroxylation (Chapter 2). Additional variations in the extents of these reactions are found within the same collagen type from different tissues and the same tissue in many physiological and pathological states (Kivirikko and Myllylä 1979, 1980). It should be noted, however, that the content of 4-hydroxyproline in a given collagen type varies only within narrow limits, since an almost complete 4-hydroxylation of proline residues in the Y positions of the Gly-X-Y triplets is required for the formation of a triple helical molecule at physiological temperature (*See* Section 3.3.2).

The amounts of peptidyl proline 4-hydroxylase activity and immunoreactive enzyme protein vary markedly between different cell types and even in the same cell type in different experimental situations (Prockop et al. 1976; Kivirikko and Myllylä 1980, 1982b). A good correlation is found between this enzyme activity and the rate of collagen biosynthesis in various cells and tissues in a large number of physiological and pathological states. On this basis, assays of peptidyl proline 4-hydroxylase activity have been used in numerous studies for estimating the rate of collagen biosynthesis in many experimental situations and in patients with various diseases.

There are some cases, however, in which the amount of peptidyl proline 4-hydroxylase activity does not correlate with the rate of collagen biosynthesis (*See* Kivirikko and Myllylä 1980, 1982b). Some cells possess this activity even though they do not synthesize any collagenous protein, and some have relatively high enzyme activity even though only traces of collagenous polypeptide chains are synthesized. Cultured fibroblasts have relatively low peptidyl proline 4-hydroxylase activity during the logarithmic phase of growth, distinctly lower than during the stationary growth phase, even though the rate of collagen synthesis is higher (Section 3.5.1). In such logarithmic phase cultures, the amount of peptidyl proline 4-hydroxylase activity is clearly rate-limiting for the production of procollagen. The proline residues in the newly synthesized polypeptide chains are under-hydroxylated (Breul et al. 1980), so that the polypeptide chains fail to form triple helical molecules (*See* function of 4-hydroxyproline, Section 3.3.2) and are subject to rapid intracellular degradation (*See* Section 3.6).

The rates of synthesis of the two subunits of peptidyl proline 4-hydroxylase (Section 3.3.2) appear to be regulated differently. The enzyme protein is found in cells not only in the form of active tetramers, but also in an inactive form that corresponds in size and amino acid and peptide map analysis to the smaller subunit (the 60,000-molecular weight β-subunit) (Kivirikko and Myllylä 1980,

1982b). Studies involving radioactive labeling of the enzyme protein in tissues in vivo and in freshly isolated or cultured cells have demonstrated that at least one function of the β-subunit-related protein is to serve as a precursor of the β-subunit in the assembly of the peptidyl proline 4-hydroxylase tetramer (Majamaa et al. 1979; Berg et al. 1980a; Kao and Chou 1980; Kivirikko and Myllylä 1980). The α-subunit appears to be utilized immediately after its synthesis to form the active enzyme, whereas the β-subunit is produced in excess and enters a precursor pool before being incorporated into the tetramer. This precursor probably must be processed in some way before being converted into the final β-subunit, a suggestion supported by some minor but distinct differences found between these two proteins in peptide map analysis. The ratio of active enzyme tetramer to inactive β-subunit-related protein varies markedly between tissues and even in the same cell type or tissue under various conditions, suggesting that synthesis of the α-subunit is regulated more effectively than that of the β-subunit. There are no data currently available to indicate by what mechanism the synthesis of the two peptidyl proline 4-hydroxylase subunits is coupled to that of the collagen polypeptide chains.

The biological significance of the β-subunit-related protein is not yet fully understood. This protein is present in most cells in a very large excess over the active enzyme tetramer, forming up to 99% of total peptidyl proline 4-hydroxylase protein (Kivirikko and Myllylä 1980). The β-subunit-related protein may thus not be exclusively utilized to make enzyme, but may also have some additional functions. It has recently been reported that while the active enzyme tetramer is located only within the cisternae of the rough endoplasmic reticulum, β-subunit-related protein is also found in association with the plasma membrane (Kao and Chou 1980). It has not yet been determined, however, whether this protein is an integral part of the plasma membrane or just associated with the membrane by some unknown mechanism.

Peptidyl proline 3-hydroxylase, peptidyl lysine hydroxylase, and the two peptidyl hydroxylysine glycosyltransferase activities also vary markedly among different cell types and tissues, and in the same cell type or tissue in many experimental and clinical conditions (Kivirikko and Myllylä 1979, 1980). The relative activities are also not constant, but vary markedly. This seems to explain many of the differences in the extent of corresponding modification among the genetically distinct collagen types or in the same collagen type under various conditions (Risteli et al. 1979; Myllylä et al. 1981; Pihlajaniemi et al. 1981).

The rates and/or extents of the intracellular modifications are influenced not only by the amounts of the active enzymes present, but also by the concentrations of the cofactors and cosubstrates required by the enzymes, the rate of procollagen synthesis, and the rate at which the newly synthesized pro α chains fold into the triple helical conformation.

The hydroxylases and glycosyltransferases require several cofactors and cosubstrates, and it is apparent that exclusion of any of these will inhibit the corresponding reactions. In the case of the hydroxylases, this situation has been demonstrated in a number of experiments with isolated cells or tissues in which a nitrogen atmosphere or chelation of the ferrous ions has been used to inhibit these reactions completely (Prockop et al. 1976). There are no direct data to indicate whether the concentrations of the cofactors and cosubstrates are rate-limiting for the corresponding reactions under physiological conditions in vivo, but such a situation is very likely to exist in the case of the two peptidyl hydroxylysine glycosyltransferases (Myllylä et al. 1979; Anttinen and Hulkko 1980), and also in the case of the three hydroxylases in certain pathological states, at least. For example, the K_m of oxygen in the hydroxylation reactions is of about the same magnitude as the oxygen tension measured in several kinds of wounds, suggesting that changes in oxygen tension may affect the rate of these reactions. Scurvy serves as a classic example of the influence of ascorbate concentration. Many symptoms of this disease are believed to result from a reduced rate of collagen synthesis, probably due to a lack of ascorbate for the hydroxylations. Some reports suggest that ascorbate also affects ribosomal synthesis of the procollagen polypeptide chains and the amounts of the active hydroxylase proteins in the cell, but these effects are controversial at present (*See* Booth and Uitto 1981; Murad et al. 1981). It is of interest that a deficiency in this vitamin in cultured cells influences the 4-hydroxylation of proline residues more readily than the hydroxylation of lysine (Quinn and Krane 1979; Anttinen et al. 1981), even though the K_m values for ascorbate in these two reactions are very similar and the two hydroxylases show no differences in the specificity of this vitamin requirement (*See* Section 3.3.2).

The rate of procollagen synthesis affects the ratio of the enzymes to their substrate. A marked decrease in this rate may therefore in itself increase the extent to which the newly synthesized polypeptide chains are modified. Transformed cells in culture provide an example of such a situation. The rate of procollagen synthesis in these cells is markedly decreased (see above) and the extent of modification is distinctly increased, even though the corresponding enzyme activities are unaltered or even decreased (Myllylä et al. 1981).

The rate at which newly synthesized pro α chains fold into the triple helical conformation or the time at which the process is initiated affects the time available for the modifying reactions, as none of the hydroxylases and peptidyl hydroxylysine glycosyltransferases can act on triple helical molecules. Isolated cells can be manipulated so that helix formation is either inhibited or accelerated. Such experiments have demonstrated that when helix formation is inhibited, the extent of modification of the newly synthesized procollagen increases, while opposite changes are observed if helix formation is accelerated (Prockop et al. 1979; Kivirikko and Myllylä 1979, 1980). Large differences are found in the time required for helix formation among cells synthesizing the various genetically distinct collagens (see above), explaining in part differences in the extent of modification among the various collagen types. However, differences in the timing of helix formation are not the main reason for the greater extent of lysine hydroxylation and hydroxylysine glycosylation in types II and IV collagen relative to types I and III (Anttinen and Hulkko 1980; Pihlajaniemi et al. 1981). The main factors appear to be the amounts of the corresponding enzyme activities in the various cells and the availability of lysine residues in the Y positions of the Gly-X-Y triplets in the various pro α chains. The latter is important, especially in the case of the high degree of lysine hydroxylation in type IV collagens, as more than 80% of the lysine residues are hydroxylated (Chapter 2).

3.5.3. Intracellular Degradation of Newly Synthesized Procollagen

A significant fraction of the newly synthesized procollagen is degraded to small peptides within the cell prior to secretion (*See* Baum et al. 1980). The proportion degraded is influenced by the intracellular concentration of cyclic adenosine monophosphate (AMP) and by stimuli that increase this concentration. Elevated levels of cylic AMP enhance the degree of the degradation. Inhibition of lysosomal enzymes suppresses the degradation, suggesting that lysosomes may be the site of this process (Berg et al. 1980b).

The role of intracellular degradation in the regulation of collagen production appears to be to eliminate abnormal or defective pro α chains that are unable to form a stable triple helix. This is suggested by two kinds of data. Firstly, diploid human fibroblasts in the early logarithmic phase of growth degrade about one third of the newly synthesized procollagens, whereas confluent cells degrade only about one tenth (Berg et al. 1980b). As described above, the pro α chains synthesized by the early logarithmic phase cultures are deficient in 4-hydroxyproline and do not form triple helical molecules. Secondly, proline analogs, such as cis-4-hydroxyproline, which are incorporated into procollagen but prevent triple helix formation (*See* Section 3.3.7), significantly enhance the proportion of the newly synthesized pro α chains that are degraded (Berg et al. 1980b). The capacity of fibroblasts to destroy defective collagen by this mechanism appears to be limited, however, as even in the presence of high concentrations of cis-4-hydroxyproline the degree of intracellular degradation does not exceed about one third of the procollagen synthesized.

3.5.4. Extracellular Processing

Little specific information is currently available on the regulation of the extracellular modifications. Peptidyl lysine oxidase activity has been measured in cultured cells and tissues in many physiological and pathological states, but such studies should be evaluated with caution as many technical problems are encountered in the assay of this enzyme activity (Siegel 1979). Both peptidyl lysine oxidase and also procollagen N-terminal protease activity are nevertheless known to vary in the culture medium of human fibroblasts, with maximal values at the stationary phase of growth occuring at about the same time as proline 4-hydroxylase (See Kivirikko and Risteli 1976; Siegel 1979). Peptidyl lysine oxidase activity in implanted sponges correlates with the development of increased wound tensile strength, and with collagen deposition in liver fibrosis. Activity is low in cases of copper deficiency, being restored by administration of trace amounts of this cation, due probably to induction of new enzyme synthesis rather than activation of pre-existing enzyme protein (Rayton and Harris 1979). Hypophysectomy leads to a marked decrease in the amount of peptidyl lysine oxidase activity, in aldehyde content of newly synthesized collagen and in degree of crosslink formation (Siegel 1979).

3.6. Perspectives

The biosynthesis of the collagens as presented in this review reveals the intricate mechanisms underlying the assembly of extracellular collagenous matrices. The triple helical collagen molecule is synthesized by the classical routes of protein synthesis, differing qualitatively only in the large number of co and posttranslational modifications. The transcription of genes, subsequent translation, posttranslational enzymatic modifications, secretion, extracellular processing, and finally assembly into ordered supramolecular aggregates represent the salient steps in this process. The multistep sequential biosynthesis ensures precise biological control. However, any error in this process may lead to deficiencies in the collagenous matrix, and thereby to pathological changes.

The rapid advancement of understanding in this facet of collagen biochemistry has major implications for several heritable and acquired disorders of connective tissue metabolism. It is possible to relate individual biosynthetic steps to individual pathological states, at least in theory and in some cases in fact, as discussed in Chapter 11. Only in the final assembly of the collagen fibril is there a major lack of understanding of mechanisms at the protein level.

Advances in recombinant DNA technology have spawned a new era of study of collagen biosynthesis. The concept of gene families and ideas about transcriptional and translational control apply to the collagens as well as to other proteins. It is this area that will broaden understanding and where there will be continued refinement of answers to long-standing questions about development and disease.

References

Aalto M, Potila M, Kulonen E: (1979) Glycoproteins from experimental granulation tissue and their effects on collagen synthesis in embryonic chick tendon cells. Biochim Biophys Acta 587:606–617.

Adams E, Frank L: (1980) Metabolism of proline and the hydroxyprolines. Annu Rev Biochem 49:1005–1061.

Adams SL, Alwine JC, de Crombrugghe B, Pastan I: (1979) Use of recombinant plasmids to characterize collagen RNAs in normal and transformed chick embryo fibroblasts. J Biol Chem 254:4935–4938.

Adams SL, Boettiger D, Focht RJ, Holtzer H, Pacifici M: (1982) Regulation of the synthesis of extracellular matrix components in chondroblasts transformed by a temperature-sensitive mutant of Rous sarcoma virus. Cell 30:373–384.

Alitalo K: (1980) Production of both basement membrane and interstitial procollagens by fibroblastic WI-38 cells from human embryonic lung. Biochem Biophys Res Commun 93:873–880.

Alitalo K, Myllylä R, Vaheri A, Sage H, Pritzl P, Bornstein P: (1982) Biosynthesis of type V procollagen by A-204, a human rhabdomyosarcoma cell line. J Biol Chem 257:9016–9024.

Alitalo K, Vaheri A, Krieg T, Timpl R: (1980) Biosynthesis of two subunits of type IV procollagen and of other basement membrane proteins by a human tumor cell line. Eur J Biochem 109:247–255.

Anttinen H, Hulkko A: (1980) Regulation of the glycosylations of collagen hydroxylysine in chick embryo tendon and cartilage cells. Biochim Biophys Acta 632:417–427.

Anttinen H, Puistola U, Pihlajaniemi T, Kivirikko KI: (1981) Differences between proline and lysine hydroxylations in their inhibition by zinc or by ascorbate deficiency during collagen synthesis in various cell types. Biochim Biophys Acta 674:336–344.

Avvedimento VE, Vogeli G, Yamada Y, Maizel JV Jr, Pastan I, de Crombrugghe B: (1980) Correlation between splicing sites within an intron and their sequence complementarity with U1 RNA. Cell 21:689–696.

Avvedimento E, Yamada Y, Lovelace E, Vogeli G, de Crombrugghe B, Pastan I: (1981) Decrease in the levels of nuclear RNA precursors for alpha 2 collagen in Rous sarcoma virus transformed fibroblasts. Nucleic Acids Res 9:1123–1131.

Bächinger HP, Bruckner P, Timpl R, Engel J: (1978) The role of *cis–trans* isomerization of peptide bonds in the coil-triple helix conversion of collagen. Eur J Biochem 90:605–613.

Bächinger HP, Doege KJ, Petschek JP, Fessler LJ, Fessler JH: (1982) Structural implications from an electronmicroscopic comparison of procollagen V with procollagen I, pC-collagen I, procollagen IV and a *Drosophila* procollagen. J Biol Chem 257:14590–14592.

Baum BJ, Moss J, Breul SD, Berg RA, Crystal RG: (1980) Effect of cyclic AMP on the intracellular degradation of newly synthesized collagen. J Biol Chem 255:2843–2847.

Berg RA, Kao WW-Y, Kedersha NL: (1980a) The assembly of tetrameric prolyl hydroxylase in tendon fibroblasts from newly synthesized α-subunits and from preformed cross-reacting protein. Biochem J 189:491–499.

Berg RA, Schwartz ML, Crystal RG: (1980b) Regulation of the production of secretory proteins: Intracellular degradation of newly synthesized 'defective' collagen. Proc Natl Acad Sci USA 77:4746–4750.

Berman MD, Waggoner JG, Foidart J-M, Kleinman HK: (1980) Attachment to collagen by isolated hepatocytes from rats with induced hepatic fibrosis. J Lab Clin Med 95:660–671.

Blobel G, Walter P, Chang CN, Goldman BM, Erickson AH, Lingappa VR: (1979) Translocation of proteins across membranes: The signal hypothesis and beyond. Symp Soc Exp Biol 33:9–36.

Booth BA, Uitto J: (1981) The effects of ascorbic acid on procollagen production and prolyl hydroxylase activity. Biochim Biophys Acta 675:117–122.

Bornstein P, Sage H: (1980) Structurally distinct collagen types. Annu Rev Biochem 49:957–1003.

Bornstein P, Traub W: (1979) The chemistry and biology of collagen. In: Neurath H, Hill RL, eds. The Proteins. Volume IV. New York: Academic Press, 411–632.

Boyd CD, Tolstoshev P, Schafer MP, Trapnell BC, Coon HC, Kretschmer PJ, Nienhuis AW, Crystal RG: (1980) Isolation and characterization of a 15-kilobase genomic sequence coding for part of the proα2 chain of sheep type I collagen. J Biol Chem 255:3212–3220.

Breul SD, Bradley KH, Hance AJ, Schafer MP, Berg RA, Crystal RG: (1980) Control of collagen production by human diploid lung fibroblasts. J Biol Chem 255:5250–5260.

Brockway BE, Forster SJ, Freedman RB: (1980) Protein disulphide-isomerase activity in chick-embryo tissues. Correlation with the biosynthesis of procollagen. Biochem J 191:873–876.

Brown DD: (1981) Gene expression in eucaryotes. Science 211:667–674.

Bruckner P, Bächinger P, Timpl R, Engel J: (1978) Three conformationally distinct domains in the amino-terminal segment of type II procollagen and its rapid triple helix $\rightleftharpoons$ coil transition. Eur J Biochem 90:595–603.

Bruns RR, Hulmes DJS, Therrien SF, Gross J: (1979) Procollagen segment-long-spacing crystallites: Their role in collagen fibrillogenesis. Proc Natl Acad Sci USA 76:313–317.

Canalis E, Peck WA, Raisz LG: (1980) Stimulation of DNA and collagen synthesis by autologous growth factor in cultured fetal rat calvaria. Science 210:1021–1023.

Cheah KSE, Grant ME: (1982) Procollagen genes and messenger RNAs, In: Weiss JB, Jayson MIV, eds. Collagen in Health and Disease, Edinburgh: Churchill Livingstone, 73–100.

Cheah KSE, Grant ME, Jackson DS: (1979) Translation of type II procollagen mRNA and hydroxylation of the cell-free product. Biochem Biophys Res Commun 91:1025–1031.

Chu M-L, Myers JC, Bernard MP, Ding J-F, Ramirez F: (1982) Cloning and characterization of five overlapping cDNAs specific for the human proα1(I) collagen chain. Nucleic Acids Res 10:5925–5934.

Clark CC: (1979) The distribution and initial characterization of oligosaccharide units on the COOH-terminal propeptide extensions of the pro-α1 and pro-α2 chains of type I procollagen. J Biol Chem 254:10798–10802.

Clark CC: (1982) Asparagine-linked glycosides. Methods Enzymol 82A:346–360.

Conrad CW, Dessau W, von der Mark K: (1980) Synthesis of type III collagen by fibroblasts from the embryonic chick cornea. J Cell Biol 84:501–512.

Creighton TE, Hillson DA, Freedman RB: (1980) Catalysis by protein-disulphide isomerase of the unfolding and refolding of proteins with disulphide bonds. J Mol Biol 142:43–62.

Crouch E, Bornstein P: (1979) Characterization of a type IV procollagen synthesized by human amniotic fluid cells in culture. J Biol Chem 254:197–202.

Dalgleish R, Trapnell BC, Crystal RG, Tolstoshev P: (1982) Copy number of a human type I α2 collagen gene. J Biol Chem 257:13816–13822.

Davis BD, Tai P-C: (1980) The mechanism of protein secretion across membranes. Nature 283:433–438.

de Crombrugghe B, Pastan I: (1982) Structure and regulation of a collagen gene. Trends Biochem Sci 11-13.

De Jong L, Albracht SPJ, Kemp A: (1982) Prolyl 4-hydroxylase activity in relation to the oxidation state of enzyme-bound iron. The role of ascorbate in peptidyl proline hydroxylation. Biochim Biophys Acta 704:326–332.

Eichorn JH, Peterkofsky B: (1979) Local anesthetic-induced inhibition of collagen secretion in cultured cells under conditions where microtubules are not depolymerized by these agents. J Cell Biol 81:26–42.

Eyre DR: (1980) Molecular diversity in the body's protein scaffold. Science 207:1315–1322.

Fessler JH, Fessler LI: (1978) Biosynthesis of procollagen. Annu Rev Biochem 47:129–162.

Fessler LI, Fessler JH: (1979) Characterization of type III procollagen from chick embryo blood vessels. J Biol Chem 254:233–239.

Fessler LI, Kumamoto C, Meis ME, Fessler JH: (1981a) Assembly and processing of procollagen V (AB) in chick blood vessels and other tissues. J Biol Chem 256:9640–9645.

Fessler LI, Robinson WJ, Fessler JH: (1981b) Biosynthesis of procollagen [$(\text{pro}\alpha 1\text{V})_2\text{pro}\alpha 2\text{V}$] by chick tendon fibroblasts and procollagen $(\text{pro}\alpha 1\text{V})_3$ by hamster lung cell cultures. J Biol Chem 256:9646–9651.

Fessler LI, Timpl R, Fessler JH: (1981c) Assembly and processing of procollagen type III in chick embryo blood vessels. J Biol Chem 256:2531–2537.

Freedman RB, Hillson DA: (1980) Formation of disulphide bonds. In: Freedman RB, Hawkins HC, eds. The Enzymology of Post-translational Modification of Proteins. Volume 1. London: Academic Press, 157–212.

Fuller F, Boedtker H: (1981) Sequence determination and analysis of the 3′ region of chicken pro-α1(I) and pro-α2(I) collagen messenger ribonucleic acids including the carboxy-terminal propeptide sequences. Biochemistry 20:996–1006.

Gauss V, Müller PK: (1981) Change in the expression of collagen genes in dividing and nondividing chondrocytes. Biochim Biophys Acta 652:39–47.

Gerard S, Puett D, Mitchell WM: (1981) Kinetics of collagen fold formation in human type I procollagen and the effect of disulfide bonds. Biochemistry 20:1857–1865.

Graves PN, Olsen BR, Fietzek PP, Prockop DJ, Monson JM: (1981) Comparison of the NH_2-terminal sequences of chick type I preprocollagen chains synthesized in an mRNA-dependent reticulocyte lysate. Eur J Biochem 118:363–369.

Haralson MA: (1982) Cell-free synthesis of non-interstitial (CHL cell) procollagen chains. Methods Enzymol 82A:225–245.

Hart GW, Brew K, Grant GA, Bradshaw RA, Lennarz WJ: (1979) Primary structural requirements for the enzymatic formation of the N-glycosidic bond in glycoproteins. J Biol Chem 254:9747–9753.

Harwood R: (1979) Collagen polymorphism and messenger RNA. Int Rev Connect Tissue Res 8:159–226.

Hayashi T, Curran-Patel S, Prockop DJ: (1979) Thermal stability of the triple helix of type I procollagen and collagen. Precautions for minimizing ultraviolet damage to proteins during circular dichroism studies. Biochemistry 18:4182–4187.

Heathcote JG, Bailey AJ, Grant ME: (1980) Studies on the assembly of rat lens capsule. Biosynthesis of a cross-linked collagenous component of high molecular weight. Biochem J 190:229–237.

Heathcote JG, Grant ME: (1980) Extracellular modification of connective tissue proteins. In: Freedman RB, Hawkins HC, eds. The Enzymology of Post-translational Modifications of Proteins. Volume 1. London: Academic, 457–506.

Herrmann H, Dessau W, Fessler LI, von der Mark K: (1980) Synthesis of type I, III and AB_2 collagen by chick tendon fibroblasts in vitro. Eur J Biochem 105:63–74.

Hörlein D, Fietzek PP, Wachter E, Lapiere CM, Kühn K: (1979) Amino acid sequence of the aminoterminal segment of dermatosparactic calf-skin procollagen type I. Eur J Biochem 99:31–38.

Hortin G, Boime I: (1981) Transport of an uncleaved preprotein into the endoplasmic reticulum of rat pituitary cells. J Biol Chem 256:1491–1494.

Housley TJ, Rowland FN, Ledger PW, Kaplan J, Tanzer ML: (1980) Effects of tunicamycin on the biosynthesis of procollagen by human fibroblasts. J Biol Chem 255:121–125.

Hubbard SC, Ivatt RJ: (1981) Synthesis and processing of asparagine-linked oligosaccharides. Annu Rev Biochem 50:555–584.

Huerre C, Junien C, Weil D, Chu M-L, Morabito M, Van Cong N, Myers JC, Foubert C, Gross M-S, Prockop DJ, Bou A, Kaplan J-C, de la Chapelle A, Ramirez F: (1982) Human type I procollagen genes are located in different chromosomes. Proc Natl Acad Sci USA 79:6627–6630.

Jimenez SA, McArthur W, Rosenbloom J: (1979) Inhibition of collagen synthesis by mononuclear cell supernates. J Exp Med 150:1421–1431.

Junien C, Weil D, Myers JC, Van Cong N, Chu M-L, Foubert C, Gross M-S, Prockop DJ, Kaplan J-C, Ramirez F: (1981) Assignment of the human proα2(I) collagen structural gene (COLIA2) to chromosome 7 by molecular hybridization. Am J Hum Genet 34:381–387.

Kao WW -Y, Chou KK -L: (1980) CRP, immunologically cross-reacting protein of prolyl hydroxylase. Its role in assembly of active prolyl hydroxylase and cellular localization in L-929 fibroblasts. Arch Biochem Biophys 199:147–157.

Kefalides NA: (1979) Persistence of basement membrane collagen phenotype in hybrids of human vascular endothelium and rodent fibroblasts. Fed Proc 38:816.

Kefalides NA, Alper R, Clark CC: (1979) Biochemistry and metabolism of basement membranes. Int Rev Cytol 61:167–228.

Kessler E, Goldberg B: (1978) A method for assaying the activity of the endopeptidase which excises the nonhelical carboxyterminal extensions from type I procollagen. Anal Biochem 86:463–469.

Kivirikko KI, Myllylä R: (1979) Collagen glycosyltransferases. Int Rev Connect Tissue Res 8:23–72.

Kivirikko KI, Myllylä R: (1980) The hydroxylation of prolyl and lysyl residues. In: Freedman RB, Hawkins HC, eds. The Enzymology of Post-translational Modifications of Protein. Volume 1. London: Academic, 53–104.

Kivirikko KI, Myllylä R: (1982a) Posttranslational enzymes in the biosynthesis of collagen: Intracellular enzymes. Methods Enzymol 82A:245–304.

Kivirikko KI, Myllylä R: (1982b) Post-translational modifications. In: Jayson MIV, Weiss JB, eds. Collagen in Health and Diseases. Edinburgh: Churchill Livingstone, 101–120.

Kivirikko KI, Risteli L: (1976) Biosynthesis of collagen and its alterations in pathological states. Med Biol 54:159–186.

Kleinman HK, Wilkes CM, Martin GR: (1981) Interaction of fibronectin with collagen fibrils. Biochemistry 20:2325–2330.

Kream BE, Rowe DW, Gworek SC, Raiz LG: (1980) Parathyroid hormone alters collagen synthesis and procollagen mRNA levels in fetal rat calvaria. Proc Natl Acad Sci USA 77:5654–5658.

Kumamoto CA, Fessler JH: (1981) Propeptides of procollagen V (A,B) in chick embryo crop. J Biol Chem 256:7053–7058.

Kurkinen M, Foster L, Barlow DP, Hogan BLM: (1982) In vitro synthesis of type IV procollagen. J. Biol Chem 257:15151–15155.

Ledger PW, Uchida N, Tanzer ML: (1980) Immunocytochemical localization of procollagen and fibronectin in human fibroblasts: Effects of the monovalent ionophore, monensin. J. Cell Biol 87:664–671.

Lehrach H, Frischauf AM, Hanahan D, Wozney J, Fuller F, Crkvenjakov R, Boedtker H, Doty P: (1978) Construction and characterization of a 2.5-kilobase procollagen clone. Proc Natl Acad Sci USA 75:5417–5421.

Leung MKK, Fessler LI, Greenberg DB, Fessler JH: (1979) Separate amino and carboxyl procollagen peptidases in chick embryo tendon. J Biol Chem 254:224–232.

Majamaa K, Kuutti-Savolainen E-R, Tuderman L, Kivirikko KI: (1979) Turnover of prolyl hydroxylase tetramers and the monomer-size protein in chick-embryo cartilaginous bone and lung in vivo. Biochem J 178:313–322.

Mata JM, Assad R, Peterkofsky B: (1981) An intramembraneous reductant which participates in the proline hydroxylation reaction with intracisternal prolyl hydroxylase and unhydroxylated procollagen in isolated microsomes from L-929 cells. Arch Biochem Biophys 206:93–104.

Merlino GT, Vogeli G, Yamamoto T, de Crombrugghe B, Pastan I: (1981) Accurate in vitro transciptional initiation of the chick α2 (type I) collagen gene. J Biol Chem 256:11251–11258.

Minor RR: (1980) Collagen metabolism. A comparison of diseases of collagen and diseases affecting collagen. Am J Pathol 98:225–280.

Moen RC, Rowe DW, Palmiter RD: (1979) Regulation of procollagen synthesis during the development of chick embryo calvaria. Correlation with procollagen mRNA content. J Biol Chem 254:3526–3530.

Monson JM: (1982) Preparation and translation of interstitial (calvaria) procollagen mRNA. Methods Enzymol 82A:218–225.

Monson JM, McCarthy BJ: (1981) Identification of a Balb/c mouse proα1(I) procollagen gene: Evidence for insertions or deletions in gene coding sequences. DNA 1:59–69.

Morikawa T, Tuderman L, Prockop DJ: (1980) Inhibitors of procollagen N-protease. Synthetic peptides with sequences similar to the cleavage site in the proα1(I) chain. Biochemistry 19:2646–2650.

Morris NP, Fessler LI, Fessler JH: (1979) Procollagen propeptide release by procollagen peptidases and bacterial collagenase. J Biol Chem 254:11024–11032.

Murad S, Grove D, Lindberg KA, Reynolds G, Sivarajah A, Pinnell SR: (1981) Regulation of collagen synthesis by ascorbic acid. Proc Natl Acad Sci USA 78:2879–2882.

Myllylä R: (1981) Preparation of antibodies to chick-embryo galactosylhydroxylysyl glucosyltransferase and their use for an immunological characterization of the enzyme of collagen synthesis. Biochim Biophys Acta 658:299–307.

Myllylä R, Alitalo K, Vaheri A, Kivirikko KI: (1981) Regulation of collagen post-translational modification in transformed human and chick-embryo cells. Biochem J 196:683–692.

Myllylä R, Anttinen H, Kivirikko KI: (1979) Metal activation of galactosylhydroxylysyl glucosyltransferase, an intracellular enzyme of collagen biosynthesis. Eur J Biochem 101:261–269.

Myllylä R, Koivu J, Pihlajaniemi T, Kivirikko KI: (1983) Protein disulphide-isomerase activity in various cells synthesizing collagen. Eur J Biochem 134:7–11.

Myllylä R, Kuutti-Savolainen E-R, Kivirikko KI: (1978) The role of ascorbate in the prolyl hydroxylase reaction. Biochem Biophys Res Commun 83:441–448.

Myers JC, Chu M-L, Faro SH, Clark WJ, Prockop DJ, Ramirez F: (1981a) Cloning a cDNA for the proα2 chain of human type I collagen. Proc Natl Acad Sci USA 78:3516–3520.

Myers JC, Sangiorgi FO, Chu ML, Ding G, Prockop DJ, Ramirez F: (1981b) Isolation and characterization of part of the human proα2 collagen gene. Fed Proc 40:1650.

Nietfeld JJ, Kemp A: (1981) The function of ascorbate with respect to prolyl 4-hydroxylase activity. Biochim Biophys Acta 657:159–167.

Nusgens BV, Goebels Y, Shinkai H, Lapiere CM: (1980) Procollagen type III N-terminal endopeptidase in fibroblast culture. Biochem J 194:699–706.

Oikarinen J, Ryhänen L: (1981) Cortisol decreases the concentration of translatable type-I procollagen mRNA species in the developing chick-embryo calvaria. Biochem J 198:519–524.

Olsen B, Berg RA: (1979) Post-translational processing and secretion of procollagen in fibroblasts. In: Hopkins CR, Duncan CJ, eds. Secretory Mechanisms. Cambridge, England: Cambridge University, 57–78.

Paglia LM, Wiestner M, Duchene M, Ouellette LA, Hörlein D, Martin GR, Müller P: (1981) Effects of procollagen peptides on the translation of type II collagen messenger ribonucleic acid and on collagen biosynthesis in chondrocytes. Biochemistry 20:3525–3527.

Paglia LM, Wilczek J, de Leon LD, Martin GR, Hörlein D, Müller P: (1979) Inhibition of procollagen cell-free synthesis by amino-terminal extension peptides. Biochemistry 18:5030–5034.

Palmiter RD, Davidson JM, Gagnon J, Rowe DW, Bornstein P: (1979) NH_2-terminal sequence of the chick-proα1(I) chain synthesized in the reticulocyte lysate system. J Biol Chem 254:1433–1436.

Pesciotta DM, Dickson LA, Showalter AM, Eikenberry EF, de Crombrugghe B, Fietzek PP, Olsen BR: (1981) Primary structure of the carbohydrate-containing regions of the carboxyl propeptides of type I procollagen. FEBS Letters 125:170–174.

Pihlajaniemi T, Myllylä R, Alitalo K, Vaheri A, Kivirikko KI: (1981) Post-translational modifications in the biosynthesis of type IV collagen by a human tumor cell line. Biochemistry 20:7409–7415.

Prockop DJ, Berg RA, Kivirikko KI, Uitto J: (1976) Intracellular steps in the biosynthesis of collagen. In: Ramachandran GN, Reddi AH, eds. Biochemistry of Collagen. New York: Plenum, 163–273.

Prockop DJ, Kivirikko KI, Tuderman L, Guzman NA: (1979) The biosynthesis of collagen and its disorders. N Engl J Med 301:13–23, 77–85.

Prockop DJ, Tuderman L: (1982) Posttranslational enzymes in the biosynthesis of collagens: Extracellular enzymes. Methods Enzymol 82A:305–319.

Puistola U: (1982) Catalytic properties of lysyl hydroxylase from cells synthesizing genetically different collagen types. Biochem J 201:215–219.

Puistola U, Turpeenniemi-Hujanen TM, Myllylä R, Kivirikko KI: (1980) Studies on the lysyl hydroxylase reaction. I. Initial velocity kinetics and related aspects. Biochim Biophys Acta 611:40–50.

Quinn RS, Krane SM: (1979) Collagen synthesis by cultured skin fibroblasts from siblings with hydroxylysine-deficient collagen. Biochim Biophys Acta 585:589–598.

Rayton JK, Harris ED: (1979) Induction of lysyl oxidase with copper. Properties of an in vitro system. J Biol Chem 254:621–626.

Risteli L, Risteli J, Salo L, Kivirikko KI: (1979) Intracellular enzymes of collagen biosynthesis in 3T6 fibroblasts and chick-embryo tendon and cartilage cells. Eur J Biochem 97:297–303.

Rohde H, Vargas L, Hahn E, Kalbfleisch H, Bruguera M, Timpl R: (1979) Radioimmunoassay for type III procollagen peptide and its application to human liver disease. Eur J Clin Invest 9:451–459.

Sandell L, Veis A: (1980) The molecular weight of the cell-free translation product of α1(I) procollagen mRNA. Biochem Biophys Res Commun 92:554–562.

Sandmeyer S, Bornstein P: (1979) Declining procollagen mRNA sequences in chick embryo fibroblasts infected with Rous sarcoma virus. Correlation with procollagen synthesis. J Biol Chem 254:4950–4953.

Sandmeyer S, Gallis B, Bornstein P: (1981a) Coordinate transcriptional regulation of type I procollagen genes by Rous sarcoma virus. J Biol Chem 256:5022–5028.

Sandmeyer S, Smith R, Kiehn D, Bornstein P: (1981b) Correlation of collagen synthesis and procollagen messenger RNA levels with transformation in rat embryo fibroblasts. Cancer Res 41:830–838.

Schafer MP, Boyd CD, Tolstoshev P, Crystal RG: (1980) Structural organization of a 17 KB segment of the α2 collagen gene: Evaluation by R loop mapping. Nucleic Acids Res 8:2241–2253.

Schiltz JR, Ward S: (1980) Effects of chick embryo extract fractions on collagen and glycosaminoglycan metabolism by chick chondroblasts. Biochim Biophys Acta 628:343–354.

Showalter AM, Pesciotta DM, Eikenberry EF, Yamamoto T, Pastan I, de Crombrugghe B, Fietzek PP, Olsen BR: (1980) Nucleotide sequence of a collagen cDNA-fragment coding for the carboxyl end of proα1(I)-chains. FEBS Letters 111:61–65.

Siegel RC: (1979) Lysyl oxidase. Int Rev Conn Tissue Res 8:73–118.

Sobel ME, Yamamoto T, Adams SL, DiLauro R, Avvedimento VE, de Crombrugghe B, Pastan I: (1978) Construction of a recombinant bacterial plasmid containing a chick pro-α2 collagen gene sequence. Proc Natl Acad Sci USA 75:5846–5850.

Sobel ME, Yamamoto T, de Crombrugghe B, Pastan I: (1981) Regulation of procollagen messenger ribonucleic acid levels in Rous sarcoma virus transformed chick embryo fibroblasts. Biochemistry 20:2678–2684.

Strauss AW, Zimmerman M, Boime I, Ashe B, Mumford RA, Alberts AW: (1979) Characterization of an endopeptidase involved in preprotein processing. Proc Natl Acad Sci USA 76:4225–4229.

Tolstoshev P, Berg RA, Rennard SJ, Bradley KH, Trapnell BC, Crystal RG: (1981a) Procollagen production and procollagen messenger RNA levels and activity in human lung fibroblasts during periods of rapid and stationary growth. J Biol Chem 256:3135–3140.

Tolstoshev P, Crystal R: (1982) The collagen alpha-2 chain gene. J Invest Dermatol 79:60s–64s.

Tolstoshev P, Haber R, Trapnell BC, Crystal RG: (1981b) Procollagen messenger RNA levels and activity and collagen synthesis during the fetal development of sheep lung, tendon and skin. J Biol Chem 256:9672–9679.

Tolstoshev P, Solomon E: (1982) Collagen genes. Nature 300:581–582.

Trelstad RL, Hayashi K: (1979) Tendon collagen fibrillogenesis: Intracellular subassemblies and cell surface changes associated with fibril growth. Dev Biol 71:228–242.

Tryggvason K, Gehron-Robey P, Martin GR: (1980) Biosynthesis of type IV procollagens. Biochemistry 19:1284–1289.

Tryggvason K, Pihlajaniemi T, Liotta LA, Salo T, Kivirikko KI: (1982) Studies on the biosynthesis and degradation of type IV procollagen. In: Kühn K, Schoene H-H, Timpl R, eds. New Trends in Basement Membrane Research, New York: Raven Press, 187–193.

Tuderman L, Kivirikko KI, Prockop DJ: (1978) Partial purification and characterization of a neutral protease which cleaves the N-terminal propeptides from procollagen. Biochemistry 17:2948–2954.

Tuderman L, Prockop DJ: (1982) Procollagen N-proteinase: Properties of the enzyme purified from chick embryo tendons. Eur J Biochem 125:545–549.

Turpeenniemi-Hujanen TM: (1981) Immunological characterization of lysyl hydroxylase, an enzyme of collagen synthesis. Biochem J 195:669–676.

Turpeenniemi-Hujanen TM, Puistola U, Kivirikko KI: (1980) Isolation of lysyl hydroxylase, an enzyme of collagen synthesis, from chick embryos as a homogeneous protein. Biochem J 189:247–253.

Turpeenniemi-Hujanen TM, Puistola U, Kivirikko KI: (1981) Human lysyl hydroxylase: Purification to homogeneity, partial characterization and comparison of catalytic properties with those of a mutant enzyme from Ehlers-Danlos syndrome type VI fibroblasts. Coll Rel Res 1:355–366.

Uitto J, Allan RE, Polak KL: (1979) Conversion of type II procollagen to collagen. Extracellular removal of the amino-terminal and carboxy-terminal extensions. Eur J Biochem 99:97–103.

Vogeli G, Avvedimento EV, Sullivan M, Maizel JV Jr, Lozano G, Adams SL, Pastan I, de Crombrugghe B: (1980) Isolation and characterization of genomic DNA for α2 type I collagen. Nucleic Acids Res 8:1823–1837.

Vogeli G, Ohkubo H, Sobel ME, Yamada Y, Pastan I, de Crombrugghe B: (1981) Structure of the promoter for the chick alpha 2 type I collagen gene. Proc Natl Acad Sci USA 78:5334–5338.

Vuorio E, Sandell L, Kravis D, Sheffield VC, Vuorio T, Dorfman A, Upholt WB: (1982) Construction and partial characterization of two recombinant cDNA clones for procollagen from chicken cartilage. Nucleic Acids Res 10:1175–1192.

Weiss EH, Cheah KSE, Grosveld FG, Dahl HHM, Solomon E, Flavell RA: (1982) Isolation and characterization of a human collagen α1(I)-like gene from a cosmid library. Nucleic Acids Res 10:1981–1994.

Wiestner M, Krieg T, Hörlein D, Glanville RW, Fietzek P, Müller PK: (1979) Inhibiting effect of procollagen peptides on collagen biosynthesis in fibroblast cultures. J Biol Chem 254:7016–7023.

Wozney J, Hanahan D, Morimoto R, Boedtker H, Doty P: (1981a) Fine structural analysis of the chicken pro α2 collagen gene. Proc Natl Acad Sci USA 78:712–716.

Wozney J, Hanahan D, Tate V, Boedtker H, Doty P: (1981b) Structure of the pro α2(I) collagen gene. Nature 294:129–135.

Yamada Y, Avvedimento VE, Mudryj M, Ohkubo H, Vogeli G, Irani M, Pastan I, de Crombrugghe B: (1980) The collagen gene: Evidence for its evolutionary assembly by amplification of a DNA segment containing an exon of 54 bases. Cell 22:887–892.

Zwizinski C, Wickner W: (1980) Purification and characterization of leader (signal) peptidase from *Escherichia coli*. J Biol Chem 255:7973–7977.

4

Mammalian Collagenases

David E. Woolley

4.1. Introduction

The collagens as a group represent the major structural proteins of all connective tissues. Their degradation in both healthy and diseased states has received much attention in recent years. Normal growth and development, as well as the repair of tissues, requires as precise regulation of collagen catabolism as of synthesis. Several collagenolytic enzymes are now recognized that attack collagen fibrils in vitro. In some cases, this collagenolysis reflects a depolymerization of the fibrils by enzymatic attack in the nonhelical regions, releasing truncated molecules (Chapter 2, Section 2.3.4), in contrast to the selective and specific transection of the triple helical portion of the molecule by what may be called the true collagenases. There are also the bacterial collagenases which cleave α chains at sites preceding most glycine residues. They are primarily important as laboratory tools because of their specificity (Mandl 1972).

Since the first description of a vertebrate neutral collagenase from cultures of tadpole tail fins (Gross and Lapière 1962), many similar enzymes have been isolated from a variety of cells and tissues. (For earlier reviews, see Harris and Krane 1974; Gross 1976; Weiss 1976; Harris and Cartwright 1977; Perez-Tamayo 1978; Harper 1980.) A key event in the first demonstration of a true collagenase was establishing criteria of its identity. It was defined as a protease that acts in a specific manner on native collagen at neutral pH (Gross and Lapiere 1962). The key part of this definition is "native," as the collagens are readily denatured in vitro and become substrates for many proteases. This chapter is largely restricted to the true collagenases, which are thought to have a unique, rate-limiting role in many examples of collagen degradation in vivo. Most of what

---

From the Department of Medicine, University Hospital of South Manchester, Manchester, United Kingdom.

is described here relates to the fibrillar collagens, types I, II, and III, which are similar in chemistry (Chapter 2) and structure (Chapter 1) and therefore present similar but not identical (Section 4.5.4) substrates. This report summarizes our present knowledge of collagenase physiology, properties, interactions with the collagens, and regulation and distribution in various tissues.

4.2. Pathways of Collagenolysis

The extracellular matrix is generally composed of a meshwork of tightly or loosely packed collagen fibrils. There are two main pathways for their degradation—extracellular and intracellular—as illustrated in Figure 4.1. There are many instances where fragments of collagen fibrils have been observed within cells, especially at sites of rapid collagen resorption, and it is thought that this phagocytosed collagen is degraded by the lysosomal enzyme system (*See* Etherington 1980). The cathepsins B and N have been shown to have collagenolytic properties, but only at the lower pH values found within phagolysosomes. However, for collagen fibrils to be ingested by cells, some preliminary extracellular degradation to produce fragments of suitable size for phagocytosis would be expected.

The extracellular enzymes responsible for the degradation of collagen fibrils operate at neutral pH. A specific sequence of enzymatic events leading to complete degradation has been proposed by several authors (*See* Harris and Krane 1974; Weiss 1976; Harris and Cartwright 1977; Woessner 1980). However, it is likely that several variations of a general scheme exist in vivo, and the enzymes involved probably depend on the tissue, the type of cells, rate of collagen breakdown, and disease state.

Figure 4.1. Pathways of collagen degradation.

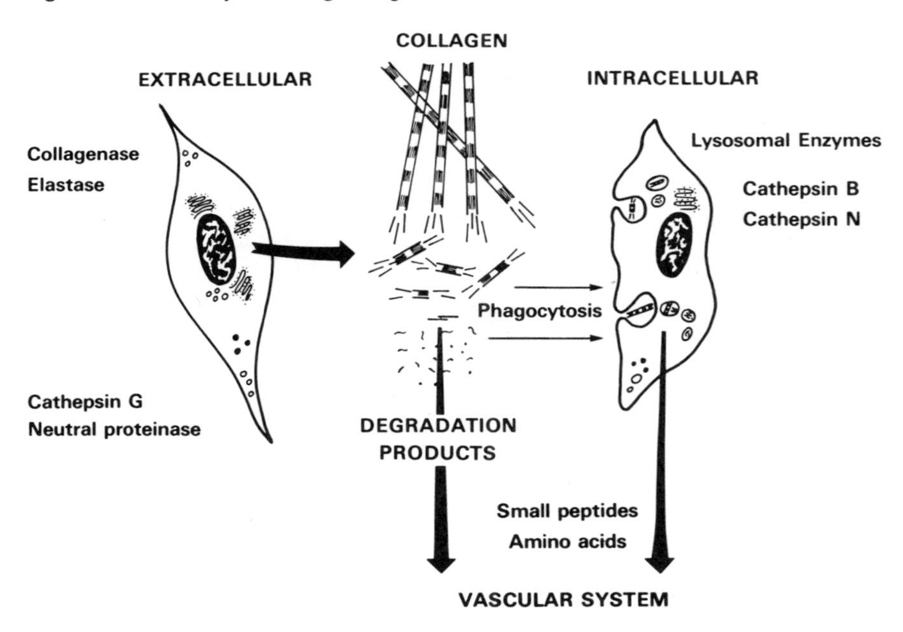

Depolymerization of collagen fibrils has often been suggested as a primary step in degradation, and elastase and cathepsin G isolated from human polymorphonuclear cells can solubilize fibrillar type II collagen, probably by attack on the nonhelical ends (Starkey et al. 1977). However, the demonstration that a highly purified collagenase, degrades crosslinked collagen fibrils by itself (Section 4.5.2), and immunolocalization studies demonstrating the enzyme at sites of collagen breakdown in various tissues (Section 4.9.2), strongly suggest that the true collagenases, with their high degree of selectivity for collagen substrates, are physiologically crucial for collagen degradation. Other proteases may facilitate their action, and a concerted action of several enzymes may accelerate the rate of collagen breakdown, but the action and role of such enzymes has yet to be demonstrated in vivo.

4.3. Sources of Collagenase

Collagenases have been reported from a variety of mammalian cells and tissues. Usually the enzyme has been detected and isolated from explant cultures using serum-free culture media, but small amounts of collagenase have been extracted directly from some tissues. Most tissue explants in primary culture show a lag phase of 1 to 3 days before active enzyme is detected in the medium. However, the recognition of latent collagenase and its activation by trypsin or organomercurials (Section 4.7) has revealed collagenase production during even the early stages of culture.

The addition of fetal calf serum to cell cultures, although essential for cell adhesion and growth, also provides various inhibitors such as α_2-macroglobulin, which explains why only latent enzyme is obtainable from such cultures. This problem may be partially overcome by using acid-treated fetal calf serum, a process which denatures the inhibitory capacity, or by replacing the serum with 0.2% lactalbumin hydrolysate solution for short-term cultures.

As a result of recent improvements in techniques, cell cultures have become useful sources of collagenase. Rabbit synovial fibroblasts (Werb and Burleigh 1974), rheumatoid synovial cells (Dayer et al. 1980a), human skin fibroblasts (Stricklin et al. 1977), macrophages (Wahl et al. 1974; Werb 1978a; Passwell et al. 1980), rat uterine cells (Halme et al. 1980), endothelial cells (Moscatelli et al. 1980), BALB/c 3T3 cells (Kluve et al. 1981), bone cells (Puzas and Brand 1979), and chondrocytes (Ridge et al. 1980) have all been reported to produce collagenase in culture. Quite often such cell cultures may have relatively low or negligible levels of enzyme production, but the recognition that activating factors such as mononuclear cell factor (Dayer et al. 1980a), cytochalasin B (Harris et al. 1975), or phorbol (Moscatelli et al. 1980; Brinckerhoff et al. 1979) markedly stimulate collagenase production suggests that these agents may be useful in generating much higher yields of enzyme.

The collagenases from these sources all have the characteristic properties originally described for the tadpole collagenase by Gross and colleagues (Nagai et al. 1966; Sakai and Gross 1967). They are all extracellular enzymes with neutral pH optima that attack the native triple helical collagen molecule at a specific locus, are generally synthesized de novo, and are secreted quite quickly rather than being stored or packaged within the cell. One exception to this rule is the polymorphonuclear leucocyte collagenase, which is stored within the specific rather than the azurophil (lysosomal) granules (Murphy et al. 1980).

4.4. Methodology

4.4.1. Assay

There are numerous techniques available for the measurement of collagenolytic activity (*See* Harris and Vater 1980), but the most common assay, which utilizes radioactively labeled reconstituted collagen fibrils, has changed little since its introduction nearly 20 years ago (Lapière and Gross 1963; Nagai et al. 1966). Because of the high specificity of collagenase for collagen, and the fact that other proteases have little effect on the triple helical collagen molecule at neutral pH, most assays usually utilize purified collagen as substrate.

Radioactively labeled collagen is prepared either by in vivo labeling of collagen using ^{14}C-glycine or ^{14}C-proline, or by in vitro labeling of purified collagen by acetylation with ^{3}H- or ^{14}C-acetic anhydride, or by ^{3}H-sodium borohydride. The in vitro methods, despite being easier and less expensive than the in vivo method, also label the nonhelical, protease-sensitive regions as well as the triple helical portion of the collagen molecule. As most assays monitor the solubilization of label, this approach sometimes results in misleading data, as contaminating nonspecific proteases can preferentially attack the nonhelical regions of the collagen molecule and contribute to the solubilized radioactivity although individual collagen monomers remain intact. The in vitro labeled collagen substrates also tend to give slightly higher values for buffer and trypsin controls (as a percent of the total count) than in vivo labeled substrate. However, both methods of labeling provide a sensitive and reasonably specific method for measuring collagenase if these reservations are kept in mind.

The choice of collagenase assay depends upon several factors, but in many laboratories the availability of very low levels of enzyme activity is the major problem to be overcome. Therefore, it is of interest that the radioactive soluble collagen assay of Terato et al. (1976) has almost a ten-fold increase in sensitivity over the conventional reconstituted collagen fibril assay, and even higher sensitivity when a substrate of high specific activity is available (Ishikawa and Nimni 1979).

Methods for extracting collagenase activity from homogenates of involuting rat uterine tissue (Weeks et al. 1976) and from Graafian follicles of rat ovary (Morales et al. 1978) have been reported. However, it seems likely that the collagenolytic activity obtained quite probably reflects the concerted action of several proteases rather than collagenase per se. In general, it has been difficult to extract collagenase activity from tissue homogenates because of the presence of inhibitors and the high affinity of the enzyme to its substrate.

Recent interest in the degradation of types IV and V collagen has depended upon the development of assay systems specific for these substrates, as they are very different from the fibrillar collagens. Garbisa et al. (1980a, 1980b) have reported an assay system based on biosynthetically ^{14}C-proline-labeled type IV collagen. Despite having relatively high trypsin blanks, the assay has helped in the identification of a tumor-derived neutral proteinase that cleaves type IV collagen (Liotta et al. 1981a).

As many experimental systems often produce an inactive, latent, or masked form of collagenase, it is appropriate to mention here the various methods of activation required prior to assay. Activation of latent collagenase is usually achieved either by mild proteolysis with trypsin or plasmin, followed by soybean trypsin inhibitor, or by exposure to various mercurials such as 1 m*M* 4-aminophenylmercuric acetate (APMA) or mersalyl. Full details of these procedures are summarized in the review by Harris and Vater (1980).

Synthetic peptides such as the PZ-peptide (4-phenylazogenzyloxycarbonyl-Pro-Leu-Gly-Pro-D-Arg), the DNP-peptide (2,4-dinitrophenyl-Pro-Leu-Gly-Ile-Ala-Gly-Arg-amide), and the coumarin peptide (succinyl-Gly-Pro-Leu-Gly-Pro-4-methylcoumaryl-7-amide) have been used to measure collagen peptidase activity. Unfortunately these substrates are not specific for true collagenase activity (Harris and Cartwright 1977; Woessner 1979a; Kojima et al. 1979), but it is hoped that eventually a synthetic substrate specific for collagenase will be found. Such a substrate would offer many advantages, especially for the determination of specific activities and provision of comparative data between different laboratories.

4.4.2. Purification

Because of variations in the chemistry of vertebrate collagenases (Section 4.5), it is not possible to use a single method of purification for all enzymes. For example, differences in four human collagenases derived from different tissues have been demonstrated by molecular sieve and ion exchange chromatography (Woolley 1980). Perhaps the most detailed purification procedures are those for the collagenases from rabbit synovial fibroblast (Werb and Reynolds 1975a), rheumatoid synovium (Woolley et al. 1975), rabbit V_2-carcinoma (McCroskery et al. 1975), human skin (Stricklin et al. 1977; Woolley et al. 1978a), and pig synovium (Cawston and Tyler 1979). Different techniques were used in these procedures; they usually produced only a few micrograms of pure protein.

Concentration of collagenase activity has often been a problem in many purification procedures. Many laboratories now use ultrafiltration as the method of choice, and although this may be a time-consuming process for crude culture media, it provides superior recoveries in comparison to ammonium sulfate precipitation or lyophilization techniques.

In addition to the usual molecular sieve materials, which fractionate proteins according to size, and the ion exchange resins, which separate proteins by their charge properties, several new materials for use in collagenase purification have recently become available. They are based on affinity groups linked to Sepharose (Sakamoto et al. 1978a; Cawston and Tyler 1979; Cawston et al. 1981a).

Affinity chromatography using collagen linked to Sepharose has received mixed reports. The major disadvantage is that the collagen apparently interacts with a variety of proteins, especially certain gammaglobulins, and the procedure is therefore relatively inefficient with crude enzyme preparations, but may well be useful later in the purification scheme. McCroskery et al. (1975) effectively used the cyanogen bromide collagen peptide αl-CB7, which contains the specific cleavage site (Section 4.5.2), in an affinity column for the purification of the rabbit tumor enzyme. Although mouse bone procollagenase is effectively bound and eluted by affinity chromatography on collagen-Sepharose (Gillet et al. 1977), the human skin procollagenases do not bind to collagen fibrils (Stricklin et al. 1978), and the latent form of synovial collagenase has a much reduced ability to bind to collagen (Vater et al. 1978a).

With the availability of monospecific antisera to certain collagenases, the use of affinity columns with purified immunoglobulins has become an important method of purification. Though problems of retaining enzyme activity following elution from such columns remain, the method theoretically provides highly purified enzyme protein for structural and composition studies.

Specific activities (usually expressed as μg collagen degraded at 37°C/min/μg enzyme) of purified collagenases are useful only inasmuch as they provide comparative values for each step of the purification procedure. They cannot be compared directly with published values for other enzymes because of variations in methods and conditions of assay, and because of problems of reproducibility even with the same method. Purified enzymes are also subject to loss of activity in the final stages of purification. As yet no means of preserving the activity of highly purified preparations have been reported. Human collagenases are usually stable for about one week at most when kept at 0–2°C, and freezing is often deleterious (Woolley DE unpublished observation).

Claims for a highly purified collagenase have usually been based on the demonstration of a single homogeneous protein band on polyacrylamide gel electrophoresis, with and without sodium dodecyl sulfate (SDS). This evidence must be complemented by elution of enzyme activity from the gel in a position which corresponds to the stained protein band. This criterion has been achieved for only a few enzymes (Werb and Reynolds 1975a; Woolley et al. 1975, 1978a; Tyler

Figure 4.2. Diagrammatic illustration showing cleavage of the type I collagen molecule (TC) by collagenase at 25°C. Two helical fragments, TC_A and TC_B, three quarters and one quarter the length of the original molecule, respectively, are produced.

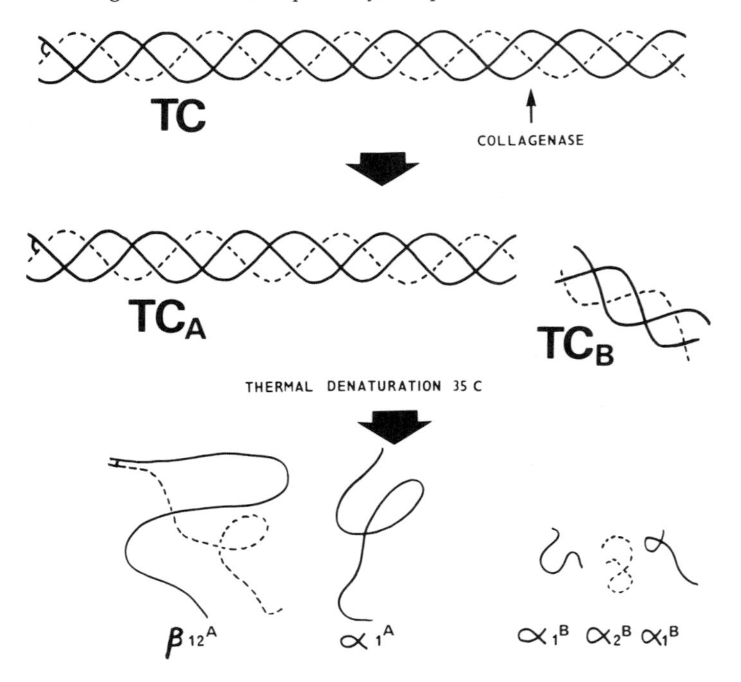

and Cawston 1980), but would seem to be essential before producing antisera to the final enzyme preparation.

4.5. Biological Properties

4.5.1. Action on Soluble Collagen

Once collagenolytic activity has been recognized by an assay, its identification as a specific collagenase is based upon an analysis of the reaction products brought about by incubation of the enzyme with collagen in solution at 25°C. The true neutral collagenases attack the collagen molecule at one specific locus, producing two characteristic products—the N-terminal three-quarter fragment and the C-terminal one-quarter fragment. These remain as triple helical rods at 25°C, but at temperatures greater than about 33°C, they spontaneously denature to random coil fragments, as illustrated in Figure 4.2. Since type I collagen contains nonidentical chains, $\alpha1(I)$ and $\alpha2(I)$, and most soluble type I collagen molecules contain intramolecular crosslinks at the N-terminal end, these dimers of the α chains, designated β_{11} and β_{12}, yield crosslinked three-quarter fragments of β_{12} and β_{11}, as illustrated in Figures 4.2 and 4.3. The various fragments may then be degraded further either by the collagenase itself, albeit at a much slower rate than the initial cleavage, or by other proteases if present.

Figure 4.3. Polyacryamide gel electrophoresis in sodium dodecyl sulfate (SDS) of the reaction products produced by incubation of solutions of types I, II, and III collagen with purified human lung fibroblast collagenase at 25°C. (Courtesy of R. G. Crystal.)

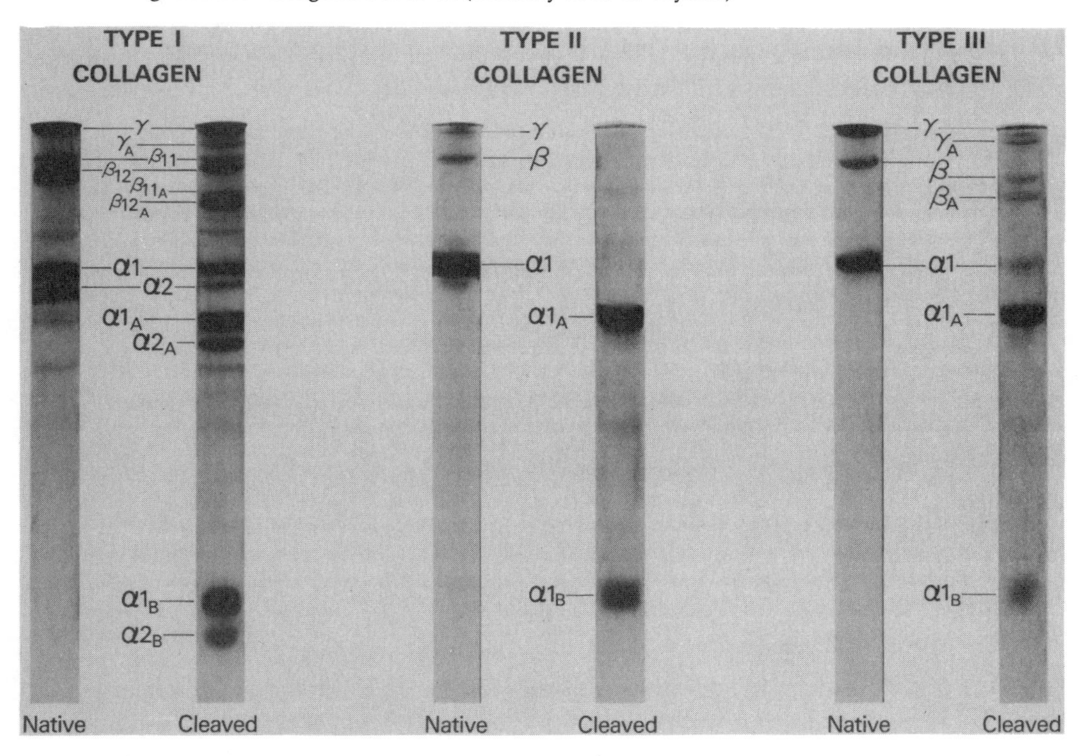

There are three methods of studying the products resulting from collagenase cleavage of types I, II, and III collagen. However, because type III collagen is susceptible to attack by other proteases such as trypsin and elastase, type I collagen should be used for collagenase identification studies. The three methods, a legacy from the initial characterization of the tadpole collagenase by Gross and colleagues (Nagai et al. 1966; Sakai and Gross 1967), are summarized as follows:

1. Viscometry. The specific viscosity of a solution of collagen at 25°C is reduced by approximately 58% upon incubation with collagenase. In contrast, bacterial or Clostridial collagenase brings about a complete loss of viscosity on account of its total destruction of the collagen molecule, as illustrated in Figure 4.4.
2. Electrophoresis. Polyacrylamide gel electrophoresis using 5% acrylamide containing SDS is the method of choice for identification of the products shown in Figure 4.3. The reaction mixture at 25°C is inactivated by the addition of 1 m*M* o-phenanthrolene and then heated at 60°C for 30 min under suitable reducing conditions using either dithiothreitol or 2-mercaptoethanol. This treatment brings about denaturation of the helical fragments to their characteristic α chain fragments.
3. Electron microscopy. Study of segment-long-spacing (SLS) aggregates with the electron microscope has proved very useful in determining the site of cleavage along the collagen molecule, as shown in Figure 4.5. Collagen molecules and the products resulting from collagenase attack at 25°C form crystallites, the latter being three quarters and one quarter of the original length, respectively. The preparation and staining characteristics of SLS crystallites are described in Chapter 1, Section 1.4.1. Collagenases cleave the collagen molecule between bands 41 and 42 (Bruns and Gross 1973), although there are a few exceptions where cleavage products of different sizes have been reported (Harris and Cartwright 1977; Harris and Vater 1980). These variations are as yet unexplained.

Figure 4.4. Viscometric response of a type I collagen solution at 25°C to human collagenase, bacterial collagenase and trypsin. *(Source: Woolley, Glanville, et al. (1975) Eur J Biochem 54:611–622, with permission.)*

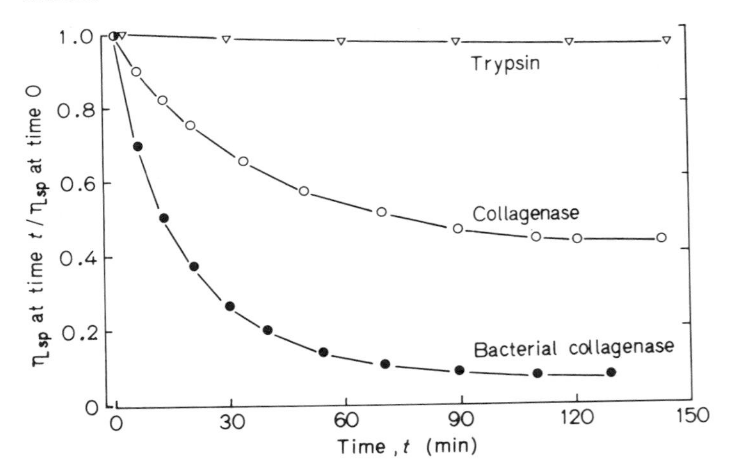

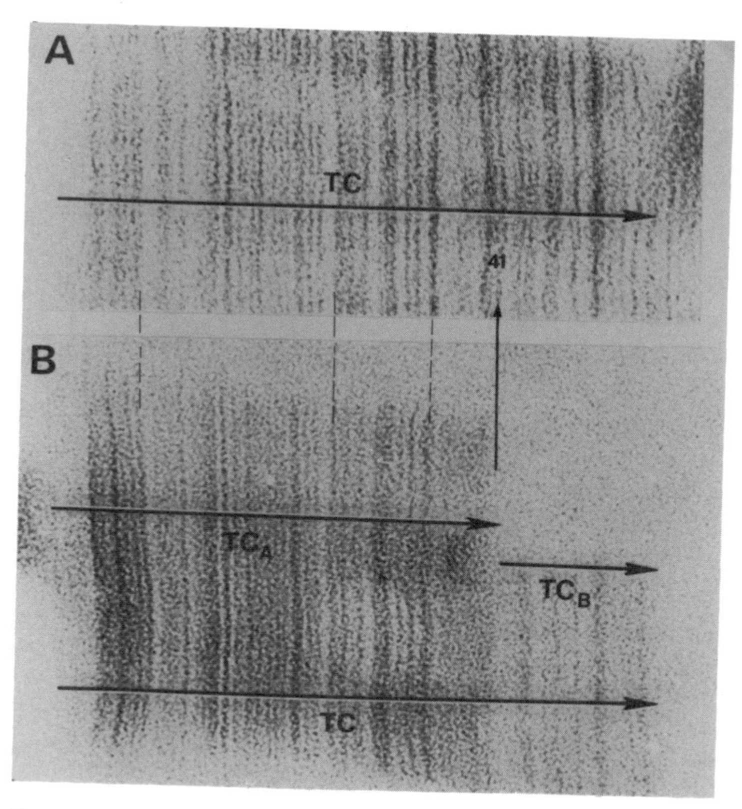

Figure 4.5. Determination of collagenase cleavage locus of type I collagen at 25°C from electron micrographs of segment-long-spacing (SLS) crystallites of the reaction products. A = control; B = products. Alignment of the banding patterns indicates that the cleavage locus is between bands 41 and 42 (*See* Figure 1.22, Chapter 1) for most mammalian collagenases. *(Source: Woolley, Akroyd, et al. (1978b) Biochim Biophys Acta 522:205–217, with permission.)*

4.5.2. Mechanism of Action

The triple helical structure of the collagen molecule renders it highly resistant to the action of most proteases, although recent studies with type III collagen have shown it to be susceptible to trypsin and elastase (Miller et al. 1976a; Gadek et al. 1980). What then are the special properties of the collagen molecule that make it susceptible to specific cleavage by collagenase?

Collagenase action on the type I collagen molecule results in cleavage of a Gly-Ile peptide bond (residues 775–776) of the $\alpha1(I)$ chains, and a Gly–Leu bond in the $\alpha2(I)$ chain (Figure 4.6). Although the same bonds occur elsewhere in the α chains, they are not attacked. Amino acid sequence determinations in the region of the cleavage locus have shown that though it contains the necessary requirements for a triple helical configuration, it is a region of low helix stability in relation to the rest of the molecule (Gross et al. 1980; Highberger et al. 1979; Weiss 1976). Consistent with this hypothesis are studies using a 36-residue peptide containing the region around the collagenase-susceptible Gly-Ile bond, which show that this peptide cannot renature to the triple helical configuration.

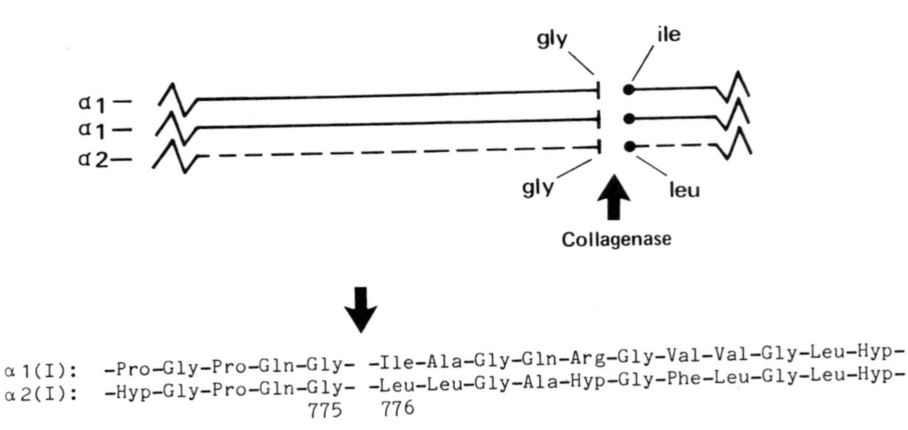

Figure 4.6. Schematic view of the cleavage site and amino acid sequences in the region for chick skin α1(I) and calf skin α2(I) chains of collagen. *(Source: Highberger, Gross, et al. (1980) In: Woolley and Evanson, eds. Collagenase in Normal and Pathological Connective Tissues. Chichester, UK: Wiley 11–35, with permission.)*

In contrast, the 36-residue peptide α1-CB2, which is not cleaved by the enzyme, does form a stable triple helix as judged by circular dichroism and molecular sieve chromatography (Highberger et al. 1979).

Sequence studies show that the α1(II) chains of cartilage collagen are attacked at a Gly-Ile bond, whereas α1(III) chains are cleaved at a Gly-Leu bond (Miller et al. 1976b; Gross et al. 1980). From these studies it appears that the minimum specific sequence required for collagen cleavage is Gly-Ile-Ala or Gly-Leu-Ala. The sequences around these regions are shown in Chapter 2, Figure 2.19. Thus, the main feature of the collagenase-susceptible region of collagen molecules is the presence of a particular sequence plus conditions that lead to low helix stability.

4.5.3. Action on Collagen Fibrils

As collagen in vivo is in the form of crosslinked fibrils, these fibrils must represent the natural substrate for the specific collagenases. Since early studies suggested that purified collagenase had no effect on native fibrillar collagen, it was claimed that degradation of such a substrate required an initial attack on the nonhelical regions of the collagen molecules by some other protease, thereby releasing collagen monomers which were then susceptible to collagenase. The introduction of chemical crosslinks into reconstituted collagen fibrils, either by formaldehyde or by peptidyl lysine oxidase, produces a substrate more resistant to collagenase attack (Vater et al. 1979b). Indeed, the incorporation of approximately 0.1 Schiff-base crosslink per collagen molecule endows a two- to three-fold resistance to collagenase when compared with reconstituted fibrils lacking crosslinks.

Similarly, studies on the action of highly purified rheumatoid synovial and human skin collagenases, each with no detectable nonspecific proteolytic activity, show that highly crosslinked, native collagen fibrils from human tendon and skin are susceptible to degradation (Woolley et al. 1975, 1978a). Although

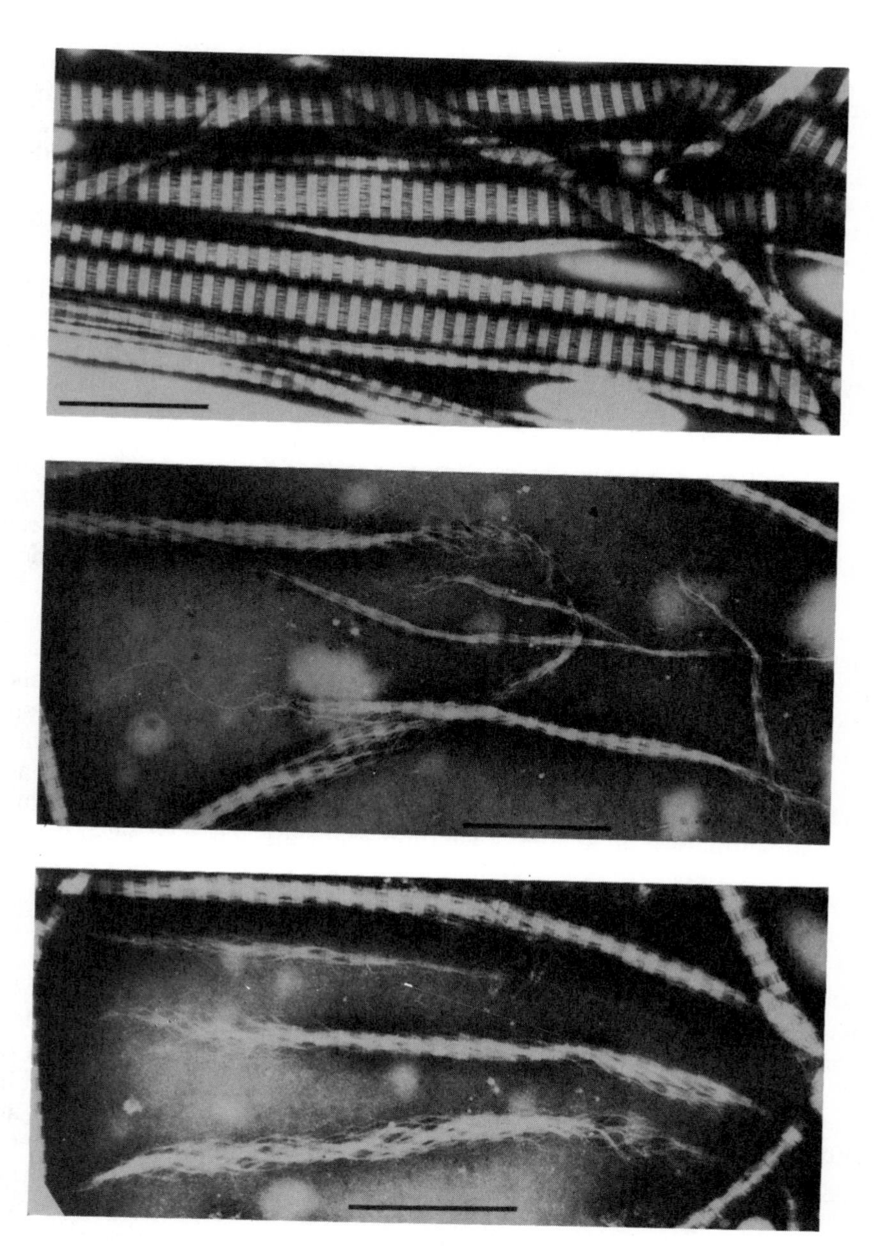

Figure 4.7. Electron micrographs of human insoluble type I collagen fibrils before and after exposure to purified human collagenase at 37°C. *Top:* Insoluble collagen fibrils from human skin without enzyme. *Middle* and *bottom:* Collagen fibrils in various stages of degradation after exposure to collagenase. All preparations negatively stained with phosphotungstic acid. Bars are 0.5 μm long. *(Source: Woolley, Glanville, et al. (1978a) Biochem J 169:265–276, with permission.)*

these fibrils are more than 20 times less susceptible to collagenase than reconstituted collagen fibrils lacking crosslinks, the important observation is that collagenase can bring about complete degradation of the fibrils without the intervention of other proteases. This is demonstrated in Figure 4.7, where exposure of collagen fibrils to purified human collagenase has resulted in disintegration and fragmentation at 37°C. Vater et al. (1979b) suggested that the extent of intermolecular crosslinking among collagen fibrils may provide a mechanism for regulating the rate of collagen catabolism in certain in vivo conditions. This may be a factor, but recent immunolocalization studies (Section 4.9) show that collagenase in vivo is often pericellular in nature. It seems likely that local variations in enzyme concentration are probably more important in determining the rate of collagen breakdown than the degree of crosslinking (Section 4.8.3).

The probable explanation of the observation that crosslinks confer resistance to collagenase attack is that since solubilization of the substrate is used to measure degradation, the component collagen monomers of the fibrils may well be cleaved but the fragments are held in place by crosslinks (Gross et al. 1980). Welgus et al. (1980) studied the binding characteristics of human skin collagenase with fibrillar collagen and concluded that the enzyme bound tightly to its substrate at 25°C or 37°C. Moreover, it appeared that the enzyme bound initially to those collagen molecules at or near the surface of each fibril (approximately 10% of the total), the remaining molecules becoming accessible to enzyme at later times. Steven (1976) also reported that the collagen molecules on the outer surface of the fibril are attacked first and the general understanding is that at physiological temperature these would then spontaneously denature to expose intact collagen molecules deeper in the fibril.

It also seems likely that other proteases may complete collagenolysis once initial clips are made by a collagenase. The intervention of enzymes such as elastase, gelatinase, and neutral proteinase could obviously supplement collagenase activity and bring about more rapid collagen degradation by concerted enzyme action. For example, the rapid loss of collagen in uterine involution or severe inflammatory disease may well represent the action of several enzymes working together.

4.5.4. Collagen Type-Specificity

Most of the specific neutral collagenases attack types I, II, and III collagen in a similar manner at the characteristic locus at 25°C, although type II collagen is degraded five to six times more slowly than types I and III collagen. The reason for this is uncertain, but as type II collagen appears to have a greater binding affinity (approximately four-fold) for tadpole collagenase than types I and III (Sundala et al. 1981), it would seem that the rate-limiting factors for type II collagen degradation reside in the mechanism of cleavage that follows formation of the enzyme-substrate complex. Differences in the amino acid sequence and degree of glycosylation around the cleavage locus may be important factors for explaining the relative resistance of type II collagen to collagenase attack. Alternatively it is possible that some collagenase specificity exists that is dependent upon the source of enzyme. For example, Horwitz et al. (1977) showed that preparations of human leucocyte collagenase cleave type I collagen 15 times faster than type III collagen. It would therefore be of interest to know whether chondrocyte collagenase might degrade type II collagen (presumably its natural substrate) at a faster rate than types I and III. However, such comparative studies designed to examine the action of specific tissue enzymes on their natural substrates have yet to be reported.

Several groups have demonstrated the resistance of types IV and V collagen to degradation by the specific mammalian collagenases (Woolley et al. 1978a; Sage et al. 1979; Sage and Bornstein 1979; Liotta et al. 1979) that readily attack types I, II, and III collagen. Such observations suggest that specific types IV and V collagenase may exist, and a variety of serine and metalloproteinases have since been reported to degrade these two substrates.

Sage and coworkers (1979) showed that a chymotrypsin-like enzyme from mast cells can degrade the various native components of type IV placenta collagen at 22°C, but not type V collagen. Similarly, a serine esterase isolated from extracts of polymorphonuclear leucocytes and thought to be elastase can degrade native lens capsule type IV collagen at 37°C, but type V collagen is unaffected (Mainardi et al. 1980a). Davies et al. (1977) reported that elastase and cathepsins G and B are able to solubilize hydroxyproline from isolated renal glomeruli, a source of type IV collagen.

A neutral protease extracted from human leucocytes degrades the native basement membrane collagen of bovine lens capsules into specific triple helical fragments at 26°C (Uitto et al. 1980). This activity is not accounted for by the leucocyte collagenase that attacks types I, II, and III collagen. Different metalloproteases separated by gel filtration of rabbit bone culture medium also have the capacity to degrade types IV and V collagens (Murphy et al. 1981a).

Perhaps the best-characterized enzyme with activity against type IV collagen is that described by Liotta et al. (1981a). This neutral metalloprotease, molecular weight approximately 70,000–80,000, is obtained from cultured metastatic murine tumor cells and has no activity against type I collagen or fibronectin, but produces specific cleavage products for both chains of mouse type IV collagen. The existence of a latent metalloprotease secreted by cultured malignant macrophages that produces limited cleavage products from soluble type V collagen at 25°C (Liotta et al. 1981b), and a similar type V collagen-degrading enzyme from macrophages (Mainardi et al. 1980b) suggests that a family of metalloproteases exists having different substrate specificities with respect to collagen type. Such enzymes derived from tumor cells could have an important role in the penetration of basement membranes during tumor invasion; other cell types may have the ability to elaborate similar enzymes (Liotta et al. 1980).

At present, the question as to whether the enzymes described above are specific for their substrates remains unclear. Another cautionary note with respect to such studies is that since the substrates are not yet well characterized, it is not possible to be certain that they are in their native state. A normal physiological role for the enzymes remains to be demonstrated.

4.6. Physicochemical Properties

4.6.1. Neutral Metalloproteases

Mammalian collagenases operate over the neutral pH range from 5.2–9.6, with optimal activity at pH 7.5–8.0. Enzyme activity is lost below pH 5.2. They are metalloproteases requiring both Ca^{2+} and Zn^{2+}. The metal-chelating agents ethylenediaminetetraacetate (EDTA) and 1,10-phenanthroline produce total inhibition of collagenase activity, and various thiol reagents such as cysteine, dithiothreitol, and D-penicillamine (1–10 m*M* concentrations) produce partial inhibition.

The removal of Ca^{2+} from various mammalian collagenases, either by a chelating resin or by molecular sieve chromatography, results in loss of enzyme activity that can be restored by the subsequent addition of calcium ions. However, this restoration becomes less effective with increasing periods of Ca^{2+}-free incubation at physiological temperature and pH (Seltzer et al. 1976). Collagenases from various sources are half-maximally activated at concentrations around 0.5 m*M* Ca^{2+} and are optimally active between 5–20 m*M* Ca^{2+}. Calcium is not required for enzyme–substrate binding, and Ba^{2+} and Sr^{2+} are effective substitutes for Ca^{2+} in human skin collagenase.

The inhibition of collagenases by the chelating agents EDTA and 1,10-phenanthroline in the presence of excess Ca^{2+} ions was the first suggestion that another metal ion is involved in enzyme activity. The finding that Zn^{2+} ions reverse and even prevent such inhibition by EDTA and 1,10-phenanthroline demonstrates an essential role for this divalent cation (Berman and Manabe 1973; Seltzer et al. 1977; Swann et al. 1981). The biosynthetic incorporation of $^{65}Zn^{2+}$ into corneal and human skin collagenases, as judged by copurification procedures using column chromatography and atomic absorption analysis, confirms that Zn^{2+} is an integral part of the enzyme protein (Berman 1980; Seltzer et al. 1977). Furthermore, it has recently been demonstrated by Swann et al. (1981) that rabbit bone collagenase incorporates $^{65}Zn^{2+}$ by exchange when incubated with 1,10-phenanthroline and the Zn^{2+} isotope. Moreover, the bound $^{65}Zn^{2+}$ is removed from its binding site in the enzyme by 1,10-phenanthroline or EDTA, and once collagenase is rendered irreversibly inactive by these inhibitors, it is no longer able to incorporate $^{65}Zn^{2+}$.

4.6.2. Molecular Weight

The molecular weights of the vertebrate collagenases show wide variation, even when they are derived from tissues of the same species. Usually the values fall within the range of 30,000–80,000. One comparative study of four human collagenases using identical molecular sieve conditions for each enzyme preparation reported values of 33,000, 38,000, 54,000, and 63,000 for the rheumatoid synovial, gastric mucosal, polymorphonuclear leucoctye, and skin collagenases, respectively (Woolley 1980). The values for the synovial and skin enzymes were confirmed by SDS polyacrylamide gel electrophoresis under reducing conditions, an observation that suggests that the larger enzyme is not a polymer of the smaller. A likely explanation for obtaining several enzyme peaks from molecular sieve studies of crude enzyme preparations is the formation of enzyme–inhibitor or enzyme–collagen peptide complexes (Fiedler-Nagy et al. 1977), especially when separations are performed at 2–4°C. Another consideration is the presence of precursor forms of the enzyme. Studies using cultures of human skin fibroblasts (Stricklin et al. 1977), mouse embryo calvaria (Vaes 1980), and others have usually shown the inactive form of the enzyme to have a molecular weight 10,000–20,000 larger than the active form (Section 4.7).

Molecular weights determined for collagenases derived from tumor tissues or cells also show variations, and it is uncertain whether a given tumor collagenase has the same physicochemical properties as those produced by the normal cells from which the tumor cells were derived. The molecular weight of the tumor enzyme that attacks basement membrane collagen has been reported as 70,000–80,000 (Liotta et al. 1981a), this being somewhat larger than most of the enzymes active against types I, II, and III collagen.

Another example of heterogeneity is the collagenase from human uterine cervix, where extraction by homogenization reveals one active and two latent forms of enzyme (Kitamura et al. 1980). Such studies suggest that the human uterine cervix contains three forms of collagenase. However, as homogenization of any tissue permits a variety of protein interactions to occur, some of which are probably artifactual, caution is required in interpreting such data. For instance, it is not known what proportion of enzyme is intra- or extracellular, whether or not enzyme is complexed with the large molecular weight serum inhibitor α_2-macroglobulin prior to homogenization, or what is the distribution or availability of other tissue inhibitors. In general, direct tissue extraction techniques have produced relatively little active enzyme. One exception is the rabbit V_2-carcinoma, where significant amounts of active enzyme were available for effective purification and characterization (McCroskery et al. 1975).

Stricklin et al. (1977) have pointed out the variability in molecular weight determined for human skin collagenase when using different molecular sieve materials. Another consideration for molecular sieve analysis is that it has always been assumed that the collagenases are globular proteins, but no information is as yet available on shape. This, and possible heterogeneity in carbohydrate content of the enzymes might well affect their behavior on molecular sieves.

In addition to the variations listed above, there may also be genetically distinct collagenases. This is not yet known because of the absence of data on primary structure. The presence of a gene family of collagenases could explain differences in enzymatic properties and susceptibility to certain natural inhibitors.

4.7. Latent Collagenase

Latent collagenase is the term used to describe the inactive form of the enzyme found either in the culture media of certain cells or tissue explants, or in tissue homogenates. This latent form may be activated or unmasked by a variety of proteases and by certain chemical compounds such as organic mercurials or chaotropic agents. Such observations would normally be interpreted as evidence for two forms of latent enzyme, one being a true precursor (procollagenase or zymogen) and the other an enzyme–inhibitor complex.

An interesting debate has ensued as to the exact nature of latent collagenases, knowledge of which is essential for a complete understanding of the mechanism of collagenase regulation in vivo. The fundamental question, which has yet to be answered, is what form of collagenase do cells secrete? Figure 4.8 summarizes three possibilities that are proposed in the literature.

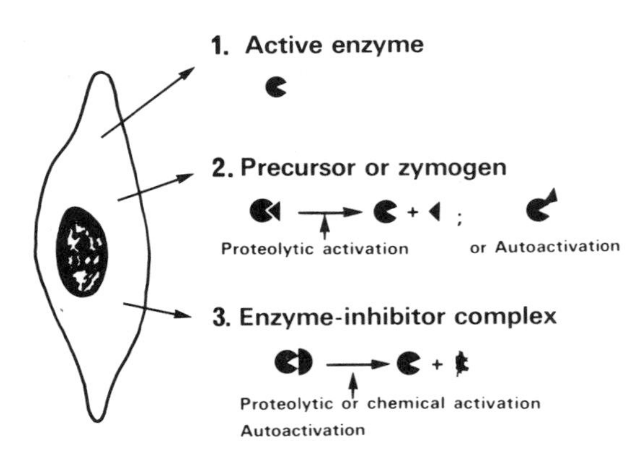

Figure 4.8. Summary of possible forms of collagenase released from cells.

4.7.1. Precursor or Proenzyme

Since the first reports of a procollagenase or zymogen form of collagenase from tadpole tissue (Harper et al. 1971) and mouse bone cultures (Vaes 1972)—as evidenced by trypsin's activation of previously inactive culture media—similar findings for other tissues have been reported by several laboratories (Birkedal-Hansen et al. 1976a; Woessner 1977). Activation can be achieved by a variety of proteases including trypsin, chymotrypsin, cathepsin B, kallikrein, plasmin (Vaes 1980), and mast cell proteases (Birkedal-Hansen et al. 1976b), but elastase and cathepsin G are not as effective as trypsin or plasmin for the activation of latent pig synovial collagenase (Cawston et al. 1981a). Plasminogen activator has also been shown to activate latent collagenase via the production of plasmin (Werb et al. 1977; Paranjpe et al. 1980).

Stricklin et al. (1977, 1978) purified and examined the properties of a procollagenase derived from human skin fibroblasts and also raised a monospecific antiserum that has been used effectively for both qualitative and quantitative studies of collagenase from various pathological skin conditions (Section 4.10). The unusual finding from these skin fibroblast or explant cultures is the apparent presence of two procollagenase proteins, with molecular weights of 60,000 and 55,000, respectively, each of which is reduced by 10,000 upon trypsin activation. Autoactivation of the procollagenases without a reduction in molecular weight also occurs. Analysis of the purification data (Stricklin et al. 1977) shows that the pure procollagenases apparently account for approximately 10% of the total protein produced by the cultured fibroblasts, an exceptionally high proportion for one enzyme.

Rabbit synovial fibroblasts in culture also secrete a doublet form of procollagenase with molecular weights of 61,000 and 57,000, but these forms are preceded by the synthesis of a pre-procollagenase of molecular weight 59,000. This latter form is thought to be processed via the microsomal system into a polypeptide chain of molecular weight 57,000, some of which is post-translationally glycosylated to the 61,000 molecular weight form. The ratio of glycosylated to nonglycosylated forms within cells was about 1 : 4, but in the culture medium this ratio was 0.07 : 1 (Nagase et al., 1981, 1983). The significance of the glycosylation in terms of secretory events and eventual activation has yet to be reported.

Fibroblast cultures and pulse-chase labeling experiments have demonstrated that intracellular procollagenase first appears after 15 minutes, whereas 30–35 minutes are required for secretion of the proenzyme (Valle and Bauer, 1979; Nagase et al., 1983). If cells do secrete a true precursor or procollagenase, then regulation of collagenase activity in the extracellular matrix must depend on the availability of some activating system that has not as yet been clearly defined. Alternatively, the autoactivation process, also poorly understood at present, may be important for in vivo collagenolysis. One explanation relates to the observation that latent leucocyte collagenase is activated via a disulfide-thiol exchange mechanism (Tschesche and Macartney 1981).

4.7.2. Enzyme–Inhibitor Complexes

That latent collagenase preparations can be activated chemically, especially by certain organic mercurials or chaotropic agents, suggests that latency is the result of enzyme–inhibitor complexes rather than precursor forms of the enzyme (For reviews, see Harris and Cartwright 1977; Murphy and Sellers 1980). Sodium thiocyanate releases collagenase activity from enzyme–α_2-macroglobulin complexes in rheumatoid synovial fluids. A similar recovery from bovine gingival collagenase–α_2-macroglobulin complexes can be achieved by brief exposure to trypsin or dialysis against thiocyanate.

With the recognition that tissues and cells in culture often produce a collagenase inhibitor, especially during the early stages, it has been shown that 4-hydroxymercuribenzoate, 4-aminophenylmercuric acetate, phenylmercuric chloride, and mersalyl can either dissociate or prevent the formation of such enzyme–inhibitor complexes.

What is uncertain from some studies is whether or not this enzyme–inhibitor complex is a pre- or postsecretion event. Do cells or tissues secrete a latent collagenase in the form of an enzyme–inhibitor complex into the extracellular environment, or does complex formation with inhibitors occur after secretion of active enzyme? Once the complex is formed, can it be activated, or is it destined for removal from the tissue? Reports of significant quantities of free inhibitor being secreted in various cell cultures partly support the view that the complex is destined for removal from the tissue (Section 4.8.4).

Although tissue culture techniques provided the initial breakthrough in the study of collagenase and are generally indispensable, it is possible that the latent collagenase story represents an artifact of this system. While it is possible to study the various forms of collagenase that accumulate in culture, it is very difficult to interpret the interrelationships in vivo. The same reservation applies to tissue extractions. It is possible for a small amount of active enzyme to complex with a variety of inhibitors that would not normally occur together in vivo.

It seems most likely that cells secrete an active enzyme in response to various environmental stimuli. Although precursor forms of the enzyme are secreted by cells in vitro, confirmation of this in vivo is lacking. Moreover, there is only indirect evidence for the existence of natural tissue activators, these being proteases which generally lack specificity for procollagenase. As collagenase activity appears restricted to pericellular locations (Section 4.9.2) it is probable that activation of collagenase precursors occurs either during or immediately after secretion, but as yet there is little information on the existence of activation mechanisms associated with the plasma membrane. Enzyme-inhibitor complexes are

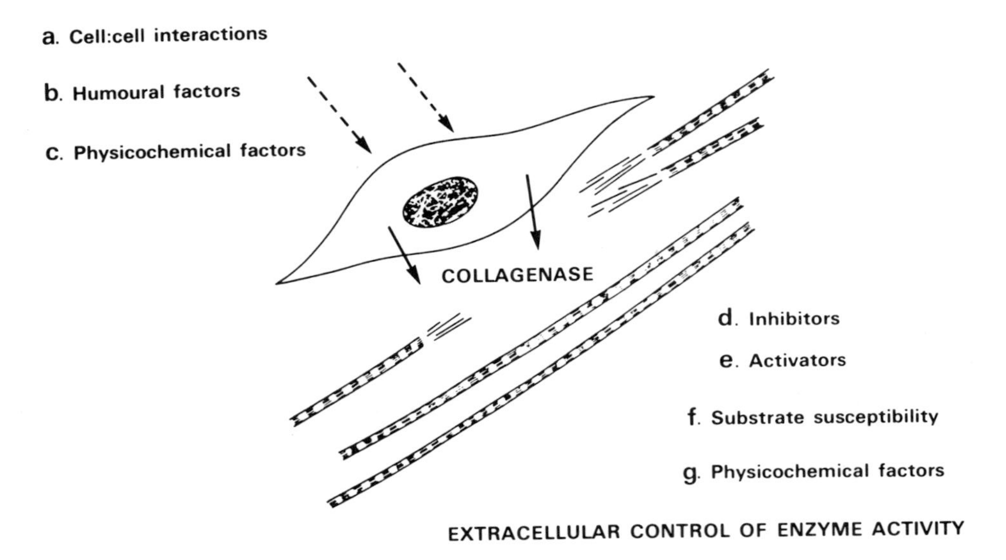

Figure 4.9. Regulation of collagenase activity. Collagenase regulation in vivo involves both cellular events and environmental factors in a multifactorial process.

probably terminating reactions. Unless other proteases capable of reactivating such complexes are available, it is expected that they would be disseminated or eliminated from the tissue quite quickly. However, there may well be exceptions to these views, and a unitary explanation of collagen degradation mediated by collagenase may well be inappropriate for all tissues and pathologies.

4.8. Regulation of Activity

Much attention is being focused on the vivo modulation of collagenase activity, and this is proving to be an exciting area of study. Regulation of collagenase activity may be conveniently divided into two aspects—the induction or *intra*-cellular phase of enzyme synthesis and release, and the *extra*cellular phase of inhibition, activation, and substrate susceptibility. Stimulation or suppression of collagenase activity in tissues is a multifactorial process. Some of the avenues of investigation that have been considered are summarized in Figure 4.9.

4.8.1. Cell–Cell Interactions

Grillo and Gross (1967) were the first to recognize that collagenase production in tissue culture can be drastically modified by the introduction of another cell type. They showed that the addition of separated epithelium to cultures of granulation tissue derived from wounded guinea pig skin results in enhanced collagenolytic activity to a much greater extent than that produced by the epithelial or mesenchymal tissue alone.

Other studies have also provided indirect evidence that cell–cell interactions are important in regulating collagenase production. Fibroblasts cultured from basal cell carcinoma demonstrate an increased capacity for collagenase production, suggesting that this represents a stimulation engendered by the neoplastic process (Bauer et al. 1979). Similarly, explants of mouse fibrosarcoma when cultured together with mouse calvaria stimulate the production of bone collagenase (Matsumoto et al. 1979).

Thus, it has become apparent not only that certain cells (such as fibroblasts and adherent synovial cells) are more proficient than others in producing collagenase, but also that these cells are sensitive to the presence of other non-collagenase-producing cells (such as epithelial, lymphocytic, and blood mononuclear cells), resulting in increased rates of collagenase production. These views are based on studies involving cell culture techniques, where medium harvested from one cell type is added to a culture of different effector cells, or the cells of one type are introduced into another and cocultivated for several days (e.g. Dayer et al. 1980a; Gross et al. 1980).

Recently, important and significant contributions to our understanding of cell–cell interactions and collagenase production in inflammatory tissues have been made by Dayer et al. (1980a, 1980b). They found that cultured adherent synovial cells obtained by proteolytic dispersion of minced synovium from patients with rheumatoid arthritis produce high levels of collagenase in primary culture. After passage of this adherent cell population, the enzyme levels decrease, but are stimulated up to several hundred-fold by a soluble factor released in vitro from cultured human blood mononuclear cells. This factor is known as mononuclear cell factor (MCF) and is produced only by monocyte-macrophages. Its production is moderated by T-lymphocytes, thus linking the immune response to altered collagenase activity in inflammatory disease.

A variety of other studies using corneal stromal cells (Newsome and Gross 1979), fibroblasts (Huybrechts-Godin et al. 1979), and chondrocytes (Deshmukh-Phadke et al. 1978; Ridge et al. 1980) have all suggested that macrophages produce a factor or factors that stimulate collagenase production by effector cells. Though macrophages are capable of producing small amounts of collagenase after activation by endotoxins or mitogens (Wahl et al. 1974; Werb 1978a; McCarthy et al. 1980), their more important role in collagenolysis appears to be the induction of other cells. Epithelial cells appear to have a similar function, producing factors that modulate collagenase production by corneal stromal cells. Condi-

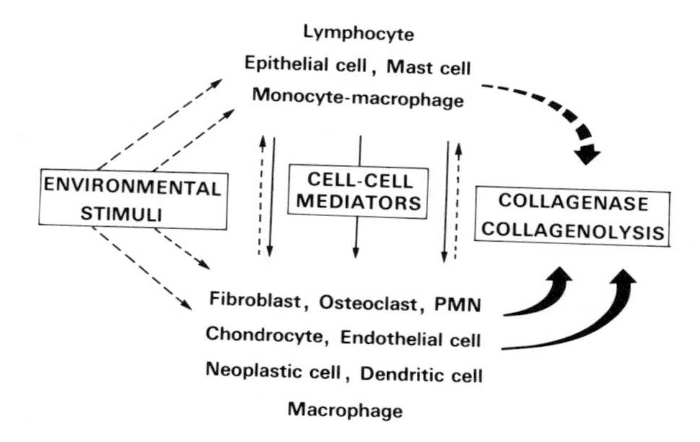

Figure 4.10. Cell–cell interactions and collagenase production. Summary of the various effector and mediator cell types involved in collagenase production. Environmental stimuli may induce mediator cells to release factors that stimulate effector cells to produce collagenase, or may act directly on effector cells. The nature of the mediators and the cell types involved at specific sites have yet to be determined in most cases.

tioned medium from cultured corneal epithelial cells is capable of either stimulating or inhibiting the production of collagenase by stromal cells, this depending upon the density of the epithelial cell cultures (Johnson-Muller and Gross 1978; Johnson-Wint 1980).

As yet, little is known about the function of the mast cell in the process of collagenolysis. Mast cell products such as heparin and histamine may have the ability to stimulate collagenase production by effector cells, but we do not as yet know if mast cells themselves produce collagenase. Figure 4.10 provides a generalized summary of cell–cell interactions. Although it is recognized that direct cell contact may be important in interactions between cells, the existence of soluble mediators that regulate cell behavior is now an established fact, and these in turn are probably produced in response to various environmental stimuli.

4.8.2. Humoral Factors

In addition to cell–cell mediators, hormones and related intracellular messengers also play a role in regulation of collagenase activity. Uterine involution after parturition, and bone resorption in response to parathyroid hormone, are believed to be examples of the influence of hormones on collagen degradation.

The sequential changes in levels of estradiol and progesterone during pregnancy and immediately after parturition suggest that these hormones are important in the regulation of collagenolytic activity. Progesterone (.25 μM) or medroxyprogesterone (1 μM) inhibit collagenase production in culture of postpartum uterine explants or uterine deciduoma (Koob and Jeffrey 1974; Jeffrey 1981). The inhibition of collagenase production by these two hormones is apparently concentration dependent. However, Tyree et al. (1980), using similar concentrations of progesterone and medroxyprogesterone in rat uterine explant cultures, found that total collagenase production, i.e., latent plus active enzyme, is little changed, and that reduced levels of active collagenase may be explained by a decrease in the amount of a proteolytic activator.

Estradiol has no inhibitory or stimulatory effect on collagenase production when added to cultures of involuting rat uterus (Woessner 1979b, 1980; Tyree et al. 1980). When estradiol-17β is administered intraperitoneally to rats shortly after parturition, an inhibition of collagen breakdown, which parallels a decrease in total collagenase activity, is observed (Woessner 1979b, 1980). This effect is most likely explained by the prevention of an influx of such cells as macrophages, heterophils, and eosinophils, which may be responsible for the production or stimulation of collagenase in uterine involution. However, Wahl (1977) reported that endotoxin-stimulated guinea pig peritoneal macrophages, which produce collagenase activity in culture, are directly inhibited when the cells are exposed to progesterone or estrogen (5 μM), either in vitro or in vivo. In contrast, collagenase production by thioglycollate-stimulated mouse peritoneal macrophages is unaffected by progesterone or estradiol (Werb 1978b).

A relationship between cyclic adenosine monophosphate (cyclic-AMP) levels and collagenase production has been suggested by several studies. Exogenous dibutyryl cyclic-AMP suppresses collagenase activity and explant degradation in cultures of ulcerated rabbit cornea, and 5′-adenosine monophosphate is even more effective than cyclic-AMP (Berman et al. 1976, 1977). In uterine or human skin explants, dibutyryl cyclic-AMP (1 m*M*) also partially inhibits collagenase production (Koob et al. 1980; Jeffrey 1981). However, cyclic-AMP is generally regarded primarily as an intracellular messenger and elevated cellular levels of this nucleotide are produced upon exposure either to prostaglandins (McCarthy et al. 1980) or to MCF (Dayer et al. 1980a).

The role of collagenase in bone resorption has recently been reviewed in detail, including the effects of heparin, parathyroid hormone, and calcitonin (Vaes 1980). The addition of heparin to the culture medium of bone explants results in increased collagenase production (Sakamoto et al. 1975a; (Lenaers-Claeys and Vaes 1979). Similarly, parathyroid hormone stimulates collagenase production in bone cultures (Sakamoto et al. 1975b) and also in isolated bone cells cultured on collagen gels (Puzas and Brand 1979), but cannot always be demonstrated (Lenaers-Claeys and Vaes 1979) even though increased bone resorption may occur. Calcitonin, while showing no effect on collagenase production, inhibits the bone-resorbing action of parathyroid hormone (Vaes 1980).

Glucocorticoids have an inhibitory effect on collagenase production by various cells and tissues. Dexamethasone at 10-100 nM completely inhibits collagenase production in rat uterine or human skin explant cultures (Koob et al. 1980). Similarly, enzyme production by stimulated mouse macrophages is inhibited by dexamethasone, cortisol, and corticosterone (Werb 1978a, 1978b). Collagenase production in cultures of rheumatoid synovial cells is inhibited when cultured with all-*trans* and 13-*cis*-retinoic acid (Brinckerhoff et al. 1980).

Humoral factors such as hormones, prostaglandins, and activating or inhibiting proteins clearly have important regulatory functions in collagenase production by various effector cells. At present many of these responses appear to be tissue-specific and the hierarchy of control factors has yet to be determined.

4.8.3. Physicochemical and Substrate Factors

A wide range of physicochemical factors are reported to stimulate collagenase production in many experimental systems. Cartilage collagen (Fisher et al. 1978) and the cyanogen bromide peptides of type II cartilage collagen (Golds et al. 1979) stimulate rheumatoid synovial explants in culture to produce more collagenase. Similarly, when collagen types I, II, or III, either in native or denatured form, are added separately to cultured human skin fibroblasts or synovial cells, each markedly stimulates collagenase production (Biswas and Dayer 1979). Immune complexes, Fc fragments of immunoglobulins, and lectins such as concanavalin A may indirectly stimulate collagenase production by their interaction with mononuclear cells with subsequent production of MCF (Dayer et al. 1980a, 1980b).

Phagocytosis of poorly digestible substances such as polystyrene latex particles or dextran sulfate stimulates collagenase production in rabbit synovial fibroblasts (Werb and Reynolds 1974) and alkali-burned corneal fibroblasts (Berman 1980). When rabbit synovial fibroblasts are exposed to various neutral proteinases such as plasmin, trypsin, and elastase, they subsequently secrete elevated levels of collagenase (Werb and Aggeler 1978).

Several chemical treatments of cultured cells or tissues also stimulate collagenase production, but as yet the mechanisms of induction are poorly understood. Colchicine produces a two- to ten-fold stimulation of enzyme release in synovial explant cultures (Harris and Krane 1971). Cytocholasin B increases collagenase production in human skin fibroblast cultures (Harris et al. 1975) and also in cocultures of rabbit corneal epithelial and stromal cells (Johnson-Muller and Gross 1978, Johnson-Wint 1980). Addition of the tumor promoter 12-0-tetradecanoyl phorbol-13-acetate to cultures of human endothelial cells produces a 5- to 30-fold stimulation of enzyme secretion (Moscatelli et al. 1980) and phorbol myristate acetate similarly induces collagenase production in rabbit synovial fibroblasts (Brinckerhoff et al. 1979). Ionic iron was also found to increase collagenase production by rabbit synovial fibroblasts (Okazaki et al. 1981).

Collagenase production by BALB/C 3T3 cells is greatly enhanced when they are cultured in the presence of mycoplasmas (Kluve et al. 1981). This observation may well be relevant to disease states where increased levels of collagenase are often associated with infection. The nature of the collagenase-stimulating activity from mycoplasma infection is unknown, but clearly this represents a direct effect on effector cells.

Human fibroblast cultures show a five- to ten-fold increase in collagenase production shortly after confluence is established, suggesting that cell culture density modulates collagenase synthesis and release (Bauer 1977). Environmental pH also affects fibroblast enzyme production (Busiek and Bauer 1979) and an increase in the O_2 tension of culture medium increases enzyme production by porcine gingival explants (Pettigrew et al. 1978). The application of a continued tensile mechanical stress upon cultured rabbit coronal sutures produces a small increase in collagenase expression (Meikle et al. 1980).

In addition to the factors stimulating collagenase synthesis and release there are those affecting extracellular enzyme activity. In vitro studies have shown that the rate of collagen degradation depends on the collagen type, its state of aggregation, the amount of crosslinking (Section 4.5), and on the presence of other macromolecules associated with native collagen, such as elastin and proteoglycan. The different susceptibilities of various collagen substrates to collagenase attack suggests that this could be an important factor in the regulation of collagen degradation (Vater et al. 1979a; Gross et al. 1980). Indeed, Perez-Tamayo et al. (1980) have suggested that collagen breakdown in vivo is primarily regulated by the substrate's susceptibility to collagenase. However, the presence and concentration of the enzyme are the first essentials to be considered.

As the distribution of different collagens varies from tissue to tissue, and also with age or disease, it is likely that the different susceptibilities of the various collagen types to a given tissue collagenase would represent an important secondary feature of collagen catabolism. With this in mind, it has often been suggested that poorly crosslinked or newly synthesized collagen would be more susceptible to degradation than highly crosslinked collagen fibrils of the same tissue. This would be true if the poorly crosslinked fibrils were able to attract and bind more collagenase molecules, but the enzyme is unlikely to be that selective and probably binds tightly to the first available substrate with which it makes contact. Hayashi et al. (1980a, 1980b) reported that the degradation rates of types II and III collagens by tadpole collagenase increased and decreased, respectively, when mixed together, as compared to the rates when incubated alone. Apparently, the binding of collagenase to type II collagen is much stronger than to type III.

Temperature and pH are two factors known to affect the rate of collagen degradation. Harris and McCroskery (1974), using native cartilage collagen fibrils as a substrate for synovial collagenase, report a ten-fold increase in degradation rate on raising the temperature from 30 to 36°C. Increases in temperature may occur during severe inflammation of tissues and may well enhance the rate of collagenolysis in vivo, probably by reducing fibril stability (Chapter 1, Section 1.2.4). On the other hand, a small reduction in extracellular pH brought about by lactate accumulation may well reduce the rate of collagen degradation by decreasing enzyme activity.

Proteases represent another factor that might have a regulatory effect on extracellular collagenase activity. Activation of latent collagenase by several proteases has been discussed (Section 4.7), but the possibility exists that high levels of these enzymes might destroy the collagenase protein. The question to be answered is whether or not endogenous latent collagenase becomes exposed to activating levels of proteolytic activity in vivo. The report that a neutral protease capable of activating a collagenase precursor was found on Ehrlich ascites cell surfaces is of particular interest in relation to this question (Steven et al. 1980).

4.8.4. Natural Inhibitors

An understanding of tissue collagenolysis requires an appreciation of the inhibitory factors that regulate extracellular collagenase activity. Many tissues extracted by homogenization techniques have been found to contain an excess of collagenase inhibitors. In most cases, the inhibitory activity is derived from the plasma antiproteases, especially α_2-macroglobulin. Other collagenase inhibitors have recently been identified and isolated from either explant or cell cultures; some of these are included in the review by Murphy and Sellers (1980). Thus, there are two major sources of inhibitory activity—the plasma inhibitors of the vascular system and those produced by certain tissue cells (Table 4.1).

The two collagenase inhibitors present in plasma are α_2-macroglobulin and β_1-anticollagenase. α_1-Antiprotease (α_1-antitrypsin) does not inhibit collagenase. α_2-Macroglobulin has long been known to inhibit collagenase as well as other proteases and accounts for more than 95% of the total collagenase inhibitory activity of whole serum. Several researchers have ascribed a regulatory role for α_2-macroglobulin in collagenase physiology, but because of its large molecular weight (725,000) it is probably confined to the vascular system and is likely to be ineffective at extravascular tissue sites, except perhaps in severe inflammation with increased capillary permeability. β_1-Anticollagenase is a small molecular weight (40,000) inhibitor with high specificity for the tissue collagenases, and although it represents less than 5% of the inhibitory activity of whole serum, it is likely to have access to most extravascular locations (Woolley et al. 1976a).

Reynolds and colleagues have characterized several inhibitors obtained from cultured rabbit tissues (Reynolds et al. 1977; Murphy and Sellers 1980). These inhibit three rabbit bone metalloproteases, have molecular weights of 28,000, and are reportedly inactivated by 4-aminophenylmercuric acetate. They have been described as tissue inhibitors of metalloprotease (TIMP) (Murphy and Sellers 1980; Cawston et al. 1981b). However, the properties of the various inhibitors derived from tissue and cell cultures, or from extracts of various tissues, show variability. Some inhibitor preparations are inactivated by exposure to organomercurials or trypsin, but not the inhibitors derived from aorta (Nolan et al.

Table 4.1. Natural Inhibitors of Collagenase[a]

Source	Molecular weight	Reference
Plasma		
α_2-Macroglobulin	725,000	Werb et al. (1974)
β_1-Anticollagenase	40,000	Woolley et al. (1976a, 1978b)
Cultured mammalian tissues		
Rabbit bone, skin, uterus (TIMP)	28,000	Murphy and Sellers (1980); Cawston et al. (1981b)
Human tendon	25,000	Vater et al. (1979a)
Bovine aorta	45,000	Nolan et al. (1980)
Chick cartilage, bone	25,000	Yasui et al. (1981)
Cultured cells		
Human skin fibroblasts	33,000	Welgus et al. (1979)
Rheumatoid synovial cells	12,000–35,000	Vater et al. (1978b)
Smooth muscle cells	25,000	Kerwar et al. (1980)
Mammalian fibroblast, epithelial cells	10,000–50,000	Pettigrew et al. (1980)
Human gingival fibroblasts	27,000–30,000	Simpson and Mailman (1981)
Extracted tissues		
Bovine cartilage, aorta	11,000	Kuettner et al. (1976)
Rabbit V_2 carcinoma	40,000–50,000	McCroskery et al. (1975)
Physiological fluids		
Human amniotic	28,000	Murphy et al. (1981b)

[a]Includes only the well-characterized inhibitors of collagenase and those most recently reported. See Murphy and Sellers (1980) for a more detailed list of inhibitors.
TIMP = tissue inhibitors of metalloprotease.

1980), smooth muscle cells (Kerwar et al. 1980), and periodontal ligament fibroblasts (Pettigrew et al. 1980). Most inhibitors are inactivated by reduction and alkylation but show variability to heat inactivation. Table 4.1 illustrates the differences in molecular weights of some recently isolated natural inhibitors of collagenase. Such variations in physicochemical properties suggest the existence of a family of small molecular weight inhibitors. As yet it is uncertain whether these variations are species- or tissue-specific.

Purified inhibitor preparations from human skin fibroblasts (Welgus et al. 1979) and rabbit bone (Cawston et al. 1981b) interact with purified collagenase with a 1 : 1 stoichiometry. Kinetic data suggest a similar relationship for two cationic collagenase inhibitors derived from growth cartilage of chick bone (Yasui et al. 1981). However, the different susceptibilities of four human collagenases to β_1-anticollagenase (Woolley et al. 1976a), and the different results found for the human amniotic fluid inhibitor against enzymes from three species (Murphy et al. 1981b), suggest that enzyme–inhibitor interaction is a complex process. Some reports suggest the formation of tightly bound complexes (Cawston et al. 1981b) while other complexes may be relatively weak and possibly require the presence of collagen (Welgus et al. 1979).

Natural inhibitors are likely to be important factors controlling local extracellular collagenase activity in connective tissue. The identification of several tissue inhibitors, together with plasma β_1-anticollagenase, supports the hypothesis that collagen catabolism at any one site depends upon the balance between the local concentrations of collagenase and its inhibitor(s), and the susceptibility of the particular enzyme to such inhibition (Woolley et al. 1976a; Woolley and Evanson 1977; Reynolds et al. 1977; Murphy and Sellers 1980).

4.9. Immunolocalization Studies

4.9.1. Antibody Preparations

The availability of monospecific antibody to collagenase provides an important tool with which to obtain a more exact interpretation of the enzyme's functional role in physiological and pathological conditions. The use of such an antibody, either as a specific collagenase inhibitor in biological studies or as a specific label in immunolocalization studies, has been the aim of several research groups (Table 4.2). Early work was fraught with problems of monospecificity, and although antisera to a variety of purified collagenases have been reported, only a few have been characterized and examined in detail (Bauer et al. 1970, 1972; Werb and Reynolds 1975b; Woolley et al. 1976b, 1980a; Sakamoto et al. 1978a; Vater et al. 1981). From such studies it is possible to summarize some common properties of anticollagenase immunoglobulin (IgG) as follows:

1. Anticollagenase IgG generally reacts with immunologic identity to collagenase derived from different tissues of the same species. One exception to this rule is the polymorphonuclear leucocyte collagenase, which fails to react with antibodies raised against rabbit fibroblast collagenase or human synovial collagenase (Werb and Reynolds 1975b; Woolley et al. 1976b) but apparently does react with the antibody to human skin collagenase (Bauer et al. 1972).
2. Most antibodies are essentially species-specific and do not precipitate or inhibit collagenases from other sources. When they occasionally show some crossreactivity with enzymes from other species, the response, whether determined by immunoprecipitation, enzyme inhibition, or radioimmunoassay, is generally much reduced (Bauer et al. 1972; Woolley et al. 1976b; Vater et al. 1981).
3. All antibody preparations inhibit collagenase activity. While early studies indicated that inhibition was related to the extent of immunoprecipitation, the use of nonprecipitating Fab immunoglobulin fragments confirms that immunoprecipitation is not essential for inhibition (Werb and Reynolds 1975b; Woolley et al. 1976b). Moreover, as antibody IgG or Fab preparations fail to inhibit the general proteolytic activity of crude collagenase samples, it is apparent that this is a means for the specific inhibition of collagenase (Werb and Reynolds 1975b; Woolley et al. 1976b), a finding that has yet to be applied to biological experiments.
4. Most antibody preparations fail to distinguish between latent, inhibited, or active forms of collagenase. One exception is the antihuman synovial collagenase preparation, which reacts primarily with active enzyme (Woolley et al. 1977, 1979, 1980a).

Antisera to various purified collagenases obtained by several research groups and their subsequent immunolocalization findings are summarized in Table 4.2. Unfortunately, the available antibodies have not all been scrutinized in sufficient depth for monospecificity and it is therefore difficult to assess the credibility of some localization studies. Claims for monospecificity based on homogeneity of

Table 4.2. Summary of Collagenase Immunolocalization Studies

Antibody	Tissue	Summary	Reference
Anti-human skin collagenase	Human skin, skin fibroblasts	Enzyme localized to collagen fibers of upper or papillary dermis and cytoplasm of skin fibroblasts.	Reddick et al. (1974)
Anti-human skin collagenase	Basal cell carcinoma	Enzyme demonstrated in stromal elements surrounding tumor cells. No staining of epithelial cells.	Bauer et al. (1977)
Anti-human skin collagenase	Middle ear cholesteotoma	Enzyme appeared in granulation tissue and dermis of canal skin; also within fibroblasts, macrophages and endothelial cells.	Abramson & Huang (1977)
Anti-human skin collagenase	Human cornea	Enzyme demonstrated on stromal structures of ulcerating corneas; not detected in normal corneas. No staining of epithelial cells.	Gordon et al. (1980)
Anti-human rheumatoid synovial collagenase	Rheumatoid joint tissue	Enzyme confined to cartilage–pannus junction; generally very little enzyme observed.	Woolley et al. (1977, 1980a)
Anti-human rheumatoid synovial collagenase	Adherent rheumatoid synovial cells	Enzyme associated with a relatively small population of dendritic cells in vitro; also intracellular localization with synovial fibroblasts.	Woolley et al. (1978c, 1979, 1980a)
Anti-human skin, anti-guinea pig skin collagenase	Bone tissues—cholesteatoma & chronic otitis media	Enzyme associated with mononuclear cells of granulation tissue and around osteocytes adjacent to areas of resorption.	Gantz et al. (1979)
		microenvironmental in nature.	

the antigen and single precipitin responses in agarose diffusion plates are insufficient in themselves.

All anticollagenase immunoglobulins must be exhaustively examined using a variety of immunoelectrophoretic techniques, precipitation–inhibition assays, and crossreactivity checks with collagenases derived from different tissues. Selective absorption of antibody IgG by affinity chromatography using purified collagenase coupled to Sepharose has several advantages. These include the exclusion of any contaminating antibody and the use of pure anticollagenase immunoglobulins at concentrations of 10–50 μg IgG/ml, which minimizes the possibility of nonspecific staining.

As collagenase probably exists in a variety of forms in tissues, e.g., as a precursor, complexed with various inhibitors, active or denatured, it is essential to examine the reactivity of each antibody against all available forms of the enzyme protein. Accurate interpretation of immunolocalization data therefore depends not only on the antibody's purity and specificity, but also on knowledge of its behavior with all natural forms of the enzyme.

Table 4.2. *(continued)*

Antibody	Tissue	Summary	Reference
Anti-human rheumatoid synovial collagenase	Skin melanomas, normal skin	Enzyme rarely detected in normal skin when fixed at time of sampling. More frequently observed in tumors, mainly on connective tissue but variable distribution. Epithelial cells negative for enzyme.	Woolley & Grafton (1980)
Anti-human rheumatoid synovial collagenase	Gingival tissues (periodontal disease)	Enzyme associated with inflammatory lesions of diseased gingival tissue.	Woolley & Davies (1981)
Anti-rat uterus collagenase	Involuting rat uterus, normal rat tissues	Enzyme demonstrated on connective tissue structures of all organs examined.	Montfort & Perez-Tamayo (1975a, 1975b)
Anti-rat uterus collagenase	Rat liver cirrhosis	Enzymes associated with connective tissue septums of cirrhotic liver. Cell origin not indicated.	Montfort & Perez-Tamayo (1978)
Anti-mouse bone collagenase	Mouse bone & normal tissues	Enzyme associated with all connective tissues and within fibroblasts, endothelial cells, chondrocytes, and all bone cells.	Sakamoto et al. (1978b)
Anti-human skin, anti-guinea pig skin collagenase	Bone tissues—cholesteatoma & chronic otitis media	Enzyme associated with mononuclear cells of granulation tissue and around osteocytes adjacent to areas of resorption.	Gantz et al. (1979)
Anti-mouse bone collagenase	Mouse peritoneal macrophages	Intracellular granular enzyme in activated macrophages; absent in nonstimulated control cells.	Sakamoto et al. (1981).

4.9.2. Normal and Pathological Specimens

The general conclusion from studies of rheumatoid, tumor, inflamed gingival, and normal tissues is that most healthy tissue contains relatively little immunoreactive enzyme whereas pathological specimens more frequently demonstrate an increase in both the distribution of immunoreactive enzyme and the intensity of immunofluorescence.

While specimens of normal skin tissue examined for immunoreactive collagenase are usually devoid of enzyme when frozen within minutes of sampling, explants subjected to tissue culture in serum-free medium for 2–4 days consistently show an increased distribution of immunoreactive enzyme (Woolley 1982). This finding has previously been explained as an increased production of collagenase in response to the wounding effects of explantation, the adjustment of the tissue cells to the new in vitro environment, and the lack of certain in vivo restraints of collagenase synthesis, such as hormones and plasma factors. Another explanation is that the increase in immunoreactive enzyme in cultured tissue represents an accumulation, rather than increased production per se, with the released collagenase being bound and retained by the collagenous structures of the explant during the period of culture. Similar observations have been made with a variety of cultured human tissues. Thus, to obtain a true assessment of the in vivo distribution of collagenase in any specimen, it is essential to fix tissues as soon as possible after excision.

There are many pathological conditions in which degradation and loss of collagen is a common feature. Collagenase immunolocalization techniques have been used to examine cartilage erosion in the rheumatoid joint and breakdown of connective tissues associated with tumor invasion, periodontal disease, and corneal ulceration. Some of the results can be summarized as follows:

1. Rheumatoid tissues. Immunoreactive collagenase has been demonstrated at sites of cartilage erosion in 5 of 11 rheumatoid specimens of cartilage–pannus junctions. Enzyme is not present in chondrocytes or in the cartilage matrix distant from the site of erosion, and surprisingly, very little enzyme is observed in the hypertrophied synovial membrane overlying the cartilage (Figure 4.11). Most of the immunoreactive enzyme is restricted to the resorbing edge of the cartilage and is often microenvironmental in nature. Such findings provide strong direct evidence that collagenase has an important role in cartilage degradation of the rheumatoid joint (Woolley et al. 1977, 1980a).
2. Tumor. The hypothesis that certain neoplasms and tumor cells invade normal tissues by their ability to secrete matrix-degrading or collagenolytic enzymes has been proposed by several groups. Immunolocalization studies of human melanomas or gastric adenocarcinomas show that about 40% of primary and secondary tumors examined contain immunoreactive collagenase (Woolley and Grafton 1980; Woolley 1982). Usually, enzyme is associated with stromal connective tissue elements surrounding small groups of cells; only occasionally is enzyme detected intracellularly (Figure 4.11). As yet, it has been difficult to identify which cells are responsible for collagenase production in vivo. The evidence to date suggests that both tumor and certain normal cells can elaborate enzyme, but this may depend upon the type and/or location of the tumor. These findings, together with those of Bauer et al. (1977), who reported similar observations for basal cell carcinomas, support the idea that collagenase facilitates the collagen breakdown that is commonly associated with tumor growth and invasiveness.
3. Corneal ulceration. Gordon et al. (1980) demonstrated immunoreactive collagenase associated with the stromal tissue of ulcerated corneas, whereas no immunologic evidence of the enzyme is seen in nonulcerated corneas. Moreover, when either ulcerated or nonulcerated corneal tissues are subjected to tissue culture, all explants manifest an increased distribution of immunoreactive enzyme.

4.9.3. Cell Origin

It has usually been difficult to determine the cellular source of collagenase in tissue sections, especially in those containing a heterogeneous cell population, and much of our knowledge on this aspect has derived from cell cultures. Human skin fibroblasts are known to produce collagenase in culture and localization

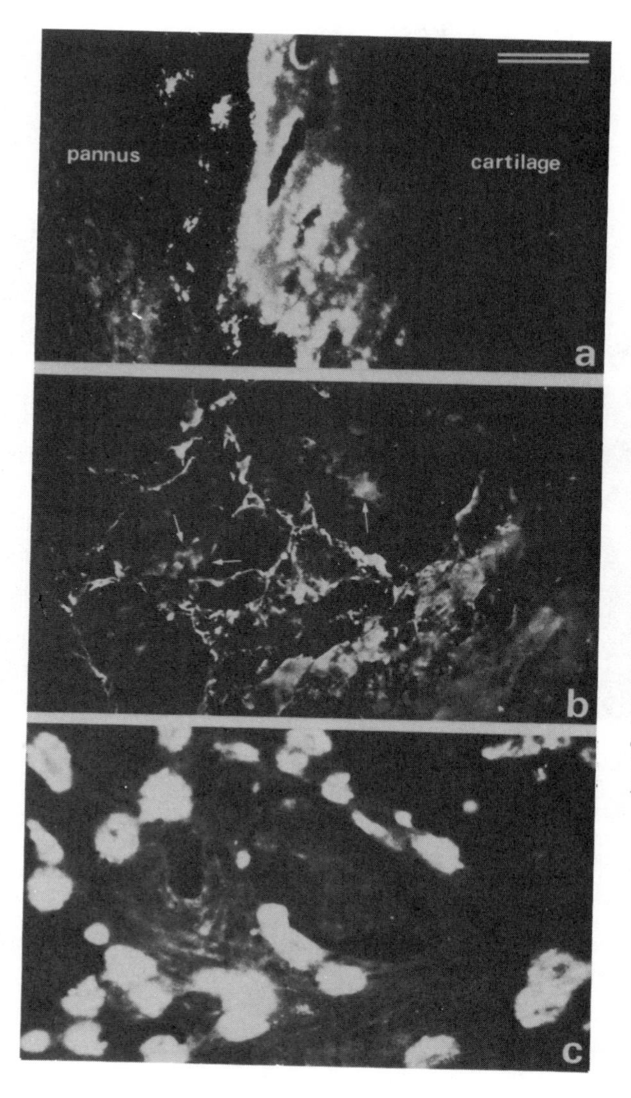

Figure 4.11. Immunolocalization of collagenase in human tissues. **(a)** Collagenase occurs at sites of cartilage erosion in a rheumatoid arthritic joint. Enzyme is absent in deeper regions of cartilage and synovial pannus tissue remote from junction. **(b)** Enzyme in a primary breast melanoma is associated with residual collagenous structures and adjacent stromal tissue, and a few cells are positive for enzyme *(arrows)*. **(c)** Collagenase surrounds individual cells in a gastric adenocarcinoma suggesting microenvironmental rather than widespread collagenase activity. Bar is 50 μm long and applies to *a, b,* and *c*. *(Source:* (Figure 4.11.a): *Woolley, Crossley, et al. (1977) Arthritis Rheum 20:1231–1239, with permission.)*

studies demonstrate the enzyme in a granular or reticulated pattern, often in a perinuclear region of the cells (Reddick et al. 1974). Similar observations for cultured rheumatoid synovial fibroblasts (Figure 4.12d) show the enzyme to be restricted to a reticulated perinuclear region of the cytoplasm (Woolley et al. 1979).

Immunolocalization studies carried out on cultures of adherent rheumatoid synovial cells are shown in Figure 4.12. The mixed synovial cell population contains fibroblasts, macrophages, and dendritic cells, as characterized morphologically. When attached to a collagenous substratum, the dendritic cells can be shown to be those mainly responsible for collagenase production in vitro. Whereas immunoreactive enzyme can be demonstrated both intra- and extracellularly around the cytoplasmic extensions of dendritic cells, relatively few fibroblasts and macrophages are positive for collagenase (Woolley et al. 1978c, 1979). An interesting observation is that not all dendritic cells of the same culture

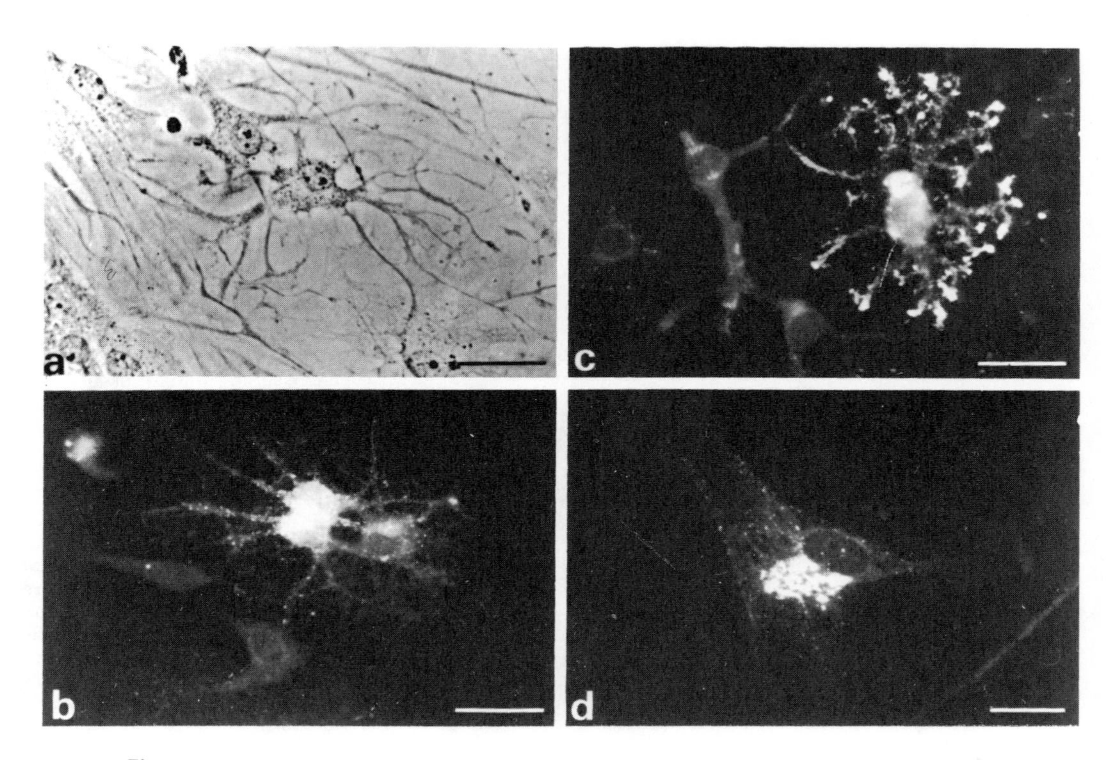

Figure 4.12. Immunolocalization of collagenase in cultured human adherent synovial cells. **(a)** Mixed population of adherent cells plated on a thin collagen substratum. Bar is 50 μm long. **(b)** Cells with dendritic morphology are positive to enzyme; note adjacent negative cells. Bar is 30 μm long. **(c)** As in *(b)*, enzyme is associated with one synovial dendritic cell but not with similar adjacent cells, suggesting either transient production of enzyme or differences in phenotypic expression by cells of similar morphology under identical culture conditions. Bar is 30 μm long. **(d)** Collagenase in a synovial fibroblast showing perinuclear distribution. Bar is 15 μm long. *(Source:* (Figure 4.12b): *Woolley, Brinckerhoff, et al. (1979) Ann Rheum Dis 38:262–230, with permission.* (Figure 4.12c): *Woolley, Harris, et al. (1978c) Science 200:773–775, 19 May 1978, Copyright 1978 by the American Association for the Advancement of Science, with permission.)*

produce enzyme at the same time, despite presumably being exposed to identical environmental stimuli (Figure 4.12c). Such an observation might be explained by the existence of subpopulations of dendritic cells, a collagenase physiology dependent upon the life cycle of these cells, or the possibility that collagenase production is a transient phase of cell activity, even in vitro.

4.9.4. Interpretation

From the observations summarized in Table 4.2., it is evident that no coherent pattern emerges from the reported findings of various research groups. The following account therefore represents one view, which may have to be modified as more information becomes available.

The general lack of immunoreactive collagenase in most specimens fixed immediately after excision, especially normal tissues, suggests that collagenase production and collagen catabolism are not continuous processes in vivo. The failure to detect significant intracellular enzyme in normal cells also supports the idea that, with the notable exception of the polymorphonuclear leucocyte collagenase, collagenase is not stored or packaged within the cell, but is probably synthesized and released as required. Furthermore, collagenase is restricted to microenvironmental locations rather than having widespread distribution and is usually associated with substrate rather than cells. It can be concluded that whereas the limited production of collagenase by cells in normal tissues suggests a defined role in the collagen remodeling processes, its more frequent occurrence in diseased tissues is consistent with an important and exaggerated role in pathological collagen resorption.

4.10. Future Trends

Anticipation of future trends in collagenase research depends to a large extent on the development of new techniques. Improvements in cell cultures (especially cocultivation of different cell types), synthetic substrates, assay systems for newly identified collagens, ultrastructural immunolocalization and microenvironmental probes, and the application of molecular biology techniques would all provide new approaches toward obtaining a better understanding of collagenase physiology. But what are the questions to be examined? Some of these are summarized in Figure 4.13, which illustrates the main avenues of research currently in progress.

Two of the most important questions relate to (a) identification of the factors that stimulate cells to produce collagenase and their hierarchy in the induction process; and (b) the roles of the different inhibitors and whether or not any of these are deficient in disease states. Both these questions have a bearing on the development of rational therapeutic approaches to those clinical conditions in which connective tissue degradation plays a central role. In many cases, a selective interruption of different tissue collagenases would appear to be the prime

Figure 4.13. Summary illustration of some of the questions that have to be considered in understanding regulation of collagenase activity in vivo. Questions include: Which cells produce collagenase? Do they release active or latent enzyme? Are proteolytic activators of latent collagenase required (or available) for collagenase activity in vivo? What is the distribution of collagenase in various tissues in vivo? What environmental stimuli induce collagenase production? What is the nature of the cell–cell interactions required for collagenase production at specific tissue locations? What inhibitors are available for modulation of extracellular enzyme activity? How, and at what rate, are enzyme–inhibitor complexes removed from tissues?

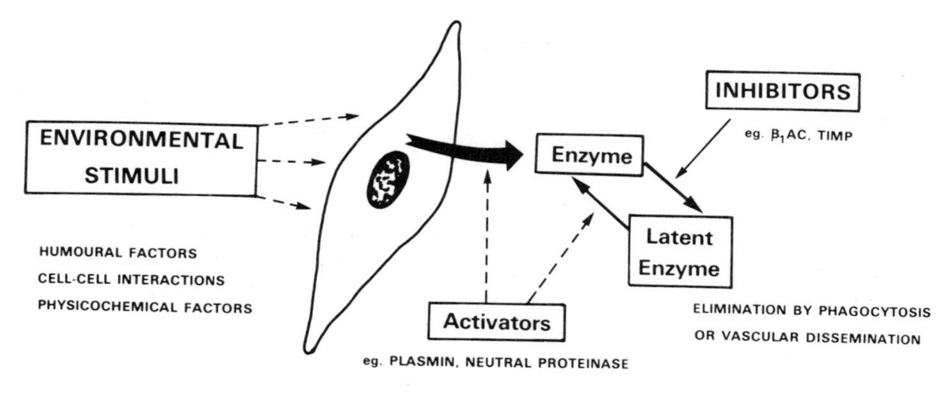

goal, for it seems unlikely that total inhibition of collagenase synthesis or activity throughout the body would prove beneficial (Woolley and Evanson 1980).

There is now good evidence from immunolocalization and biochemical studies that collagenase is present in increased quantities at sites of pathological collagen resorption (Section 4.9). In many cases, these sites are restrictive or microenvironmental in nature; for example the cartilage–pannus junction in rheumatoid arthritis, the inflamed gingival lesions in periodontal disease and the host–tumor junction as demonstrated in melanomas and basal cell carcinoma. Such tissue sites may well be amenable to the targeting of specific drugs or agents that could modify collagenase physiology, but this probably depends to some extent on the identification of the cells responsible for collagenase production and those that release induction factors.

Studies to elucidate the interrelationships between the various activating or inhibitory factors, prostaglandins, cyclic-AMP and collagenase production are essential for a better understanding of in vivo collagenolysis. The nature, characterization, and cellular origin of natural activator or suppressor factors is an exciting avenue of research, and may provide alternative targets for the pharmacological intervention of collagenase production.

Corticosteroids have provided encouraging results in various experimental systems. Topical application of medroxyprogesterone can prevent perforation of the alkali-burned rabbit cornea (Newsome and Gross 1977). Gross et al. (1981) have also reported that medroxyprogesterone and dexamethasone applied to rabbit V_2 carcinoma implanted in the rabbit cornea subsequently prevents both tumor growth and neovascularization—observations probably related to the suppression of collagenase production. Enhanced production of collagenase in skin fibroblast cultures derived from patients with recessive dystrophic epidermolysis bullosa suggests that this is pathogenetically determined (Bauer and Eisen 1978). Phenytoin (diphenylhydantoin) reduces collagenase production in human skin explant or fibroblast cultures, and when administered to patients with recessive dystrophic epidermolysis bullosa, blistering is reduced and a favorable clinical response is observed (Bauer et al. 1980). These examples suggest that modification of collagenase production in tissues may well have therapeutic advantages.

One question which has not yet been answered conclusively is whether or not collagenase is required for the cellular invasion of connective tissue. Do tumor cells, or other infiltrative cells, such as polymorphonuclear leucocytes or macrophages, secrete collagenase during their migration through connective tissue matrices, and do cells that traverse such organized collagenous structures as basement membranes secrete specific enzymes for this purpose? Similarly, is collagenase released by capillary buds during the process of neovascularization? These questions may well be answered by immunolocalization studies, especially at the ultrastructural level.

This review is not meant to leave the reader with the impression that collagenase is primarily a "bad" enzyme—one only occurring at sites of pathological collagen resorption—for it is obvious that normal collagen metabolism requires precise regulation of catabolic activity, which is probably mediated through the neutral collagenases. The excessive collagen breakdown that occurs in disease may well reflect exaggerated collagenase activity, possibly in association with other proteases. This lack of control may relate either to increased production of the enzyme or to a deficiency in available collagenase inhibitors.

The author thanks all his colleagues for their contributions, especially John Evanson and Susan Davies for their support and assistance in the preparation of the manuscript.

References

Abramson M, Huang C-C: (1977) Localization of collagenase in human middle ear cholesteatoma. Laryngoscope 87:771–791.

Bauer EA: (1977) Cell culture density as a modulator of collagenase expression in normal human fibroblast cultures. Exp Cell Res 107:269–276.

Bauer EA, Cooper TW, Tucker DR, Esterly NB: (1980) Phenytoin therapy of recessive dystrophic epidermolysis bullosa. Clinical trial and proposed mechanism of action on collagenase. N Engl J Med 303:776–781.

Bauer EA, Eisen AZ: (1978) Recessive dystrophic epidermolysis bullosa. Evidence for increased collagenase as a genetic characteristic in cell culture. J Exp Med 148:1378–1387.

Bauer EA, Eisen AZ, Jeffrey JJ: (1970) Immunologic relationship of a purified human skin collagenase to other human and animal collagenases. Biochim Biophys Acta 206:152–160.

Bauer EA, Eisen AZ, Jeffrey JJ: (1972) Radioimmunoassay of human collagenase. I. Specificity of the assay and quantitative determination of in vivo and in vitro human skin collagenase. J Biol Chem 247:6679–6685.

Bauer EA, Gordon JM, Reddick ME, Eisen AZ: (1977) Quantitation and immunocytochemical localization of human skin collagenase in basal cell carcinoma. J Invest Derm 69:363–367.

Bauer EA, Uitto J, Walters RC, Eisen AZ: (1979) Enhanced collagenase production by fibroblasts derived from human basal cell carcinomas. Cancer Res 39:4594–4599.

Berman MB: (1980) Collagenase and corneal ulceration. In: Woolley DE, Evanson JM, eds. Collagenase in Normal and Pathological Connective Tissues. Chichester, UK: Wiley, 141–174.

Berman MB, Cavanagh HD, Gage J: (1976) Regulation of collagenase activity in ulcerating cornea by cyclic AMP. Exp Eye Res 22:209–218.

Berman MB, Cavanagh HD, Gage J: (1977) 5′Adenosine monophosphate prevents collagen degradation in culture but does not prevent corneal ulceration. Exp Eye Res 24:391–397.

Berman MB, Manabe R: (1973) Corneal collagenases: Evidence for zinc metalloenzymes. Ann Ophthalmol 5:1193–1209.

Birkedal-Hansen H, Cobb CM, Taylor RE, Fullmer HM: (1976a) Synthesis and release of procollagenase by cultured fibroblasts. J Biol Chem 251:3162–3168.

Birkedal-Hansen H, Cobb CM, Taylor RE, Fullmer HM: (1976b) Activation of fibroblast procollagenase by mast cell proteases. Biochim Biophys Acta 438:273–286.

Biswas C, Dayer JM: (1979) Stimulation of collagenase production by collagen in mammalian cell cultures. Cell 18:1035–1041.

Brinckerhoff CE, McMillan RM, Dayer JM, Harris ED Jr: (1980) Inhibition by retinoic acid of collagenase production in rheumatoid synovial cells. N Engl J Med 303:432–436.

Brinckerhoff CE, McMillan RM, Fahey JV, Harris ED Jr: (1979) Collagenase production by synovial fibroblasts treated with phorbol myristate acetate. Arthritis Rheum 22:1109–1116.

Bruns RR, Gross J: (1973) Band pattern of the segment-long-spacing form of collagen. Its use in the analysis of primary structure. Biochemistry 12:808–815.

Busiek DF, Bauer EA: (1979) Environmental pH modulation of collagenase in normal human fibroblast cultures. Biochem Biophys Acta 585:389–397.

Cawston TE, Tyler JA: (1979) Purification of pig synovial collagenase to high specific activity. Biochem J 183:647–656.

Cawston TE, Mercer E, Tyler JA: (1981a) The activation of latent pig synovial collagenase. Biochim Biophys Acta 657:73–83.

Cawston TE, Galloway WA, Mercer E, Murphy G, Reynolds JJ: (1981b) Purification of rabbit bone inhibitor of collagenase. Biochem J 195:159–165.

Davies M, Barrett AJ, Travis J, Sanders E, Coles GA: (1977) The degradation of human glomerular basement membrane with purified lysosomal proteinases. Clin Sci Mol Med 54:233–240.

Dayer JM, Goldring SR, Robinson DR, Krane SM: (1980a) Cell–cell interactions and collagenase production. In: Woolley DE, Evanson JM, eds. Collagenase in Normal and Pathological Connective Tissues. Chichester, UK: Wiley, 83–104.

Dayer JM, Passwell JH, Schneeberger EE, Krane SM: (1980b) Interactions among rheumatoid synovial cells and monocyte-macrophages: Production of collagenase-stimulating factor by human monocytes exposed to concanavalin A or immunoglobulin Fc fragments. J Immunol 124:1712–1720.

Deshmukh-Phadke K, Lawrence M, Nanda S: (1978) Synthesis of collagenase and neutral proteases by articular chondrocytes: Stimulation by a macrophage-derived factor. Biochem Biophys Res Commun 85:490–496.

Etherington DJ: (1980) Proteinases in connective tissue breakdown. Ciba Found Symp 75:87–103.

Fiedler-Nagy C, Coffey JW, Salvador RA: (1977) Factors influencing the apparent molecular weight of collagenase produced by human-skin explants. Eur J Biochem 76:291–297.

Fisher WD, Lyons H, Van Der Rest M, Cooke TDV, Poole AR: (1978) Synovial inflammation in rheumatoid arthritis: Stimulation of collagenase secretion by type II cartilage collagen. Arthritis Rheum 21:556–560.

Gadek JE, Fells GA, Wright DG, Crystal RG: (1980) Human neutrophil elastase functions as a type III collagen "collagenase." Biochem Biophys Res Commun 95:1815–1822.

Gantz BJ, Maynard J, Burnstead RM, Huang CC, Abramson M: (1979) Bone resorption in chronic otitis media. Ann Otol Rhinol Laryngol 88:693–700.

Garbisa S, Kniska K, Tryggvason K, Foltz C, Liotta LA: (1980a) Quantitation of basement membrane collagen degradation by living tumour cells in vitro. Cancer Letters 9:359–366.

Garbisa S, Tryggvason K, Foidart JM, Liotta LA: (1980b) Assay for radiolabelled type IV collagen in the presence of other proteins using a specific collagenase. Anal Biochem 107:187–192.

Gillet C, Eeckhout Y, Vaes G: (1977) Purification of procollagenase and collagenase by affinity chromatography on Sepharose-collagen. FEBS Letters 74:126–128.

Golds EE, Lyons H, Cooke TDV, Van Der Rest M, Poole AR: (1979) Rheumatoid lymphocytes stimulated with collagen peptides produce a lymphokine which enhances collagenase secretion from synovial cell cultures. Arthritis Rheum 22:614.

Gordon JM, Bauer EA, Eisen AZ: (1980) Collagenase in human cornea: Immunologic localisation. Arch Opthalmol 98:341–345.

Grillo HC, Gross J: (1967) Collagenolytic activity during mammalian wound repair. Dev Biol 15:300–317.

Gross J: (1976) Aspects of animal collagenases. In: Ramachandran GN, Reddi AH, eds. Biochemistry of Collagen. New York: Plenum, 275–317.

Gross J, Azizkhan RG, Biswas C, Bruns RR, Hsieh DS, Folkman J: (1981) Inhibition of tumor growth, vascularization, and collagenolysis in the rabbit cornea by medroxyprogesterone. Proc Natl Acad Sci USA 78:1176–1180.

Gross J, Highberger JH, Johnson-Wint B, Biswas C: (1980) Mode of action and regulation of tissue collagenases. In: Woolley DE, Evanson JM, eds. Collagenase in Normal and Pathological Connective Tissues. Chichester, UK: Wiley, 11–35.

Gross J, Lapière CM: (1962) Collagenolytic activity in amphibian tissues: A tissue culture assay. Proc Natl Acad Sci USA 48:1014–1022.

Halme J, Tyree B, Jeffrey JJ: (1980) Collagenase production by primary cultures of rat uterine cells. Partial purification and characterisation of the enzyme. Arch Biochem Biophys 199:51–60.

Harper E: (1980) Collagenases. Ann Rev Biochem 49:1063–1078.

Harper E, Bloch KJ, Gross J: (1971) The zymogen of tadpole collagenase. Biochemistry 10:3035–3041.

Harris ED Jr, Cartwright EC: (1977) Mammalian collagenases. In: Barrett AJ, ed. Proteinases in Mammalian Cells and Tissues. Amsterdam: Elsevier, 249–283.

Harris ED Jr, Krane SM: (1971) Effects of colchicine on collagenase in cultures of rheumatoid synovium. Arthritis Rheum 13:83–94.

Harris ED Jr, Krane SM: (1974) Collagenases. N Engl J Med 291:557–563, 605–609, 652–661.

Harris ED Jr, McCroskery PA: (1974) Influence of temperature and fibril stability on degradation of cartilage collagen by rheumatoid synovial collagenase. N Engl J Med 290:1–6.

Harris ED Jr, Reynolds JJ, Werb Z: (1975) Cytochalasin B increases collagenase production by cells in vitro. Nature (Lond) 257:243–244.

Harris ED Jr, Vater CA: (1980) Methodology of collagenase research: Substrate preparation, enzyme activation and purification. In: Woolley DE, Evanson JM, eds. Collagenase in Normal and Pathological Connective Tissues. Chichester, UK: Wiley, 37–63.

Hayashi T, Nakamura T, Hori H, Nagai Y: (1980a) The degradation rates of type I, II and III collagens by tadpole collagenase. J Biochem 87:809–815.

Hayashi T, Nakamura T, Hori H, Nagai Y: (1980b) Degradation rates of type II and III collagens by tadpole collagenase are modulated by mutual presence. J Biochem 87:993–995.

Highberger JH, Corbett C, Gross J: (1979) Isolation and characterisation of a peptide containing the site of cleavage of the chick skin collagenα_1 (I) chain by animal collagenases. Biochim Biophys Res Commun 89:202–208.

Horwitz AL, Hance AJ, Crystal RG: (1977) Granulocyte collagenase: Selective digestion of type I relative to type III collagen. Proc Natl Acad Sci USA 74:897–901.

Huybrechts-Godin G, Hauser P, Vaes G: (1979) Macrophage-fibroblast interactions in collagenase production and cartilage degradation. Biochem J 184:643–650.

Ishikawa T, Nimni ME: (1979) A modified collagenase assay method based on the use of p-dioxane. Anal Biochem 92:136–143.

Jeffrey JJ: (1981) Collagen synthesis and degradation in the uterine deciduoma: Regulation of collagenase activity by progesterone. Collagen Rel Res 1:257–268.

Johnson-Muller B, Gross J: (1978) Regulation of corneal collagenase production: Epithelial-stromal cell interactions. Proc Natl Acad Sci USA 75:4417–4421.

Johnson-Wint B: (1980) Regulation of stromal cell collagenase production in adult rabbit cornea: In vitro stimulation and inhibition by epithelial cell products. Proc Natl Acad Sci USA 77:5331–5335.

Kerwar SS, Nolan JC, Ridge SC, Oronsky AL, Slakey LL: (1980) Properties of a collagenase inhibitor partially purified from cultures of smooth muscle cells. Biochim Biophys Acta 632:183–191.

Kitamura K, Ito A, Mori Y: (1980) The existing forms of collagenase in the human uterine cervix. J Biochem (Tokyo) 87:753–760.

Kluve B, Merrick WC, Stanbridge EJ, Gershman H: (1981) Mycoplasmas induce collagenase in BALB/c 3T3 cells. Nature 292:855–857.

Kojima K, Kinoshita H, Kato T, Nagatsu T, Takada K, Sakakibara S: (1979) A new and highly sensitive fluorescence assay for collagenase-like peptidase activity. Anal Biochem 100:43–50.

Koob TJ, Jeffrey JJ: (1974) Hormonal regulation of collagen degradation in the uterus: Inhibition of collagenase expression by progesterone and cyclic AMP. Biochim Biophys Acta 354:61–70.

Koob TJ, Jeffrey JJ, Eisen AZ, Bauer EA; (1980) Hormonal interactions in mammalian collagenase regulation. Comparative studies in human skin and rat uterus. Biochim Biophys Acta 629:13–23.

Kuettner KE, Hiti J, Eisenstein R, Harper E: (1976) Collagenase inhibition by cationic proteins derived from cartilage and aorta. Biochem Biophys Res Commun 72:40–46.

Lapière CM, Gross J: (1963) Animal collagenase and collagen metabolism. In: Sognnaes RF, ed. Mechanisms of Hard Tissue Destruction. Washington: American Association for the Advancement of Science, 663–694.

Lenaers-Claeys G, Vaes G: (1979) Collagenase, procollagenase and bone resorption. Effects of heparin, parathyroid hormone and calcitonin. Biochim Biophys Acta 584:375–388.

Liotta LA, Abe S, Robey PG, Martin GR: (1979) Preferential digestion of basement membrane collagen by an enzyme derived from a metastatic murine tumor. Proc Natl Acad Sci USA 76:2268–2272.

Liotta LA, Tryggvason K, Garbisa S, Hart I, Foltz CM, Shafie S: (1980) Metastatic potential correlates with enzymatic degradation of basement membrane collagen. Nature 284:67–68.

Liotta LA, Tryggvason K, Garbisa S, Robey PG, Abe S: (1981a) Partial purification and characterisation of a neutral protease which cleaves type IV collagen. Biochemistry 20:100–104.

Liotta LA, Lanzer WL, Garbisa S: (1981b) Identification of a type V collagenolytic enzyme. Biochem Biophys Res Commun 98:184–190.

McCarthy JB, Wahl SM, Rees JC, Olsen CE, Sandberg L, Wahl LM: (1980) Mediation of macropahge collagenase production by 3′-5′ cyclic adenosine monophosphate. J Immunol 124:2405–2409.

McCroskery PA, Richards JF, Harris ED Jr: (1975) Purification and characterisation of collagenase extracted from rabbit tumours. Biochem J 152:131–142.

Mainardi CL, Dixit SN, Kang AH: (1980a) Degradation of type IV (basement membrane) collagen by a proteinase isolated from human polymorphonuclear leucocytes granules. J Biol Chem 255:5435–5441.

Mainardi CL, Seyer JM, Kang AH: (1980b) Type specific collagenolysis: A type V collagen-degrading enzyme from macrophages. Biochem Biophys Res Commun 97:1108–1115.

Mandl I: (1972) Collagenase. New York: Gordon and Breach, 215.

Matsumoto A, Sakamoto S, Sakamoto M: (1979) Stimulation of bone collagenase synthesis by mouse fibrosarcoma in resorbing bone in vitro cultures. Arch Oral Biol 24:403–405.

Meikle MC, Sellers A, Reynolds JJ: (1980) Effect of tensile mechanical stress on the synthesis of metalloproteinases by rabbit coronal sutures in vitro. Calcif Tissue Int 30:77–82.

Miller EJ, Finch JE Jr, Chung E, Butler WT, Robertson PB: (1976a) Specific cleavage of the native type III collagen molecule with trypsin. Similarity of the cleavage products to collagenase-produced fragments and primary structure at the cleavage site. Arch Biochem Biophys 173:631–637.

Miller EJ, Harris ED Jr, Chung E, Finch JE Jr, McCroskery PA, Butler WT: (1976b) Cleavage of type II and III collagens with mammalian collagenase: Site of cleavage and primary structure at the NH_2-terminal portion of the smaller fragment released from both collagens. Biochemistry 15:787–792.

Montfort I, Perez-Tamayo R: (1975a) Distribution of collagenase in rat uterus during postpartum involution. Connect Tiss Res 3:245–252.

Montfort I, Perez-Tamayo R: (1975b) The distribution of collagenase in normal rat tissues. J Histochem Cytochem 23:910–920.

Montfort I, Perez-Tamayo R: (1978) Collagenase in experimental carbon tetrachloride cirrhosis of liver. Am J Pathol 92:411–420.

Morales TI, Woessner JF, Howell DS, Marsh JM, LeMaire WJ: (1978) A microassay for the direct demonstration of collagenolytic activity in Graafian follicles of the rat. Biochim Biophys Acta 524:428–434.

Moscatelli D, Jaffe E, Rifkin DB: (1980) Tetradecanoyl phorbol acetate stimulates latent collagenase production by cultured human endothelial cells. Cell 20:343–351.

Murphy G, Bretz U, Baggiolini M, Reynolds JJ: (1980) The latent collagenase and gelatinase of human polymorphonuclear neutrophil leucocytes. Biochem J 192:517–525.

Murphy G, Cawston TE, Galloway WA, Barnes MJ, Bunning RAD, Mercer E, Reynolds JJ, Burgeson RE: (1981a) Metalloproteinases from rabbit bone culture medium degrades type IV and V collagens, laminin and fibronectin. Biochem J 199:807–811.

Murphy G, Cawston TE, Reynolds JJ: (1981b) An inhibitor of collagenase from human amniotic fluid. Purification, characterization and action on metalloproteinases. Biochem J 195:167–170.

Murphy G, Sellers A: (1980) The extracellular regulation of collagenase activity. In: Woolley DE, Evanson JM, eds. Collagenase in Normal and Pathological Connective Tissues. Chichester, UK: Wiley, 65–81.

Nagai Y, Lapière CM, Gross J: (1966) Tadpole collagenase. Preparation and purification. Biochemistry 5:3123–3130.

Nagase H, Brinckerhoff E, Vater CA, Harris ED Jr: (1983) Biosynthesis and secretion of procollagenase by rabbit synovial fibroblasts. Biochem J 214:281–288.

Nagase H, Jackson RC, Brinckerhoff CE, Vater CA, Harris ED Jr: (1981) A precursor form of latent collagenase produced in a cell-free system with mRNA from rabbit synovial cells. J Biol Chem 256:11951–11954.

Newsome DA, Gross J: (1977) Prevention by medroxyprogesterone of perforation in alkali-burned rabbit cornea: Inhibition of collagenolytic activity. Invest Ophthalmol 16:21–31.

Newsome DA, Gross J: (1979) Regulation of corneal collagenase production: Stimulation of serially passaged stromal cells by blood mononuclear cells. Cell 16:895–900.

Nolan JC, Ridge SC, Oronsky AL, Kerwar SS: (1980) Purification and properties of a collagenase inhibitor from cultures of bovine aorta. Atherosclerosis 35:93–102.

Okazaki I, Brinckerhoff CE, Sinclair JF, Sinclair PR, Bonkowsky HL, Harris ED Jr: (1981) Iron increases collagenase production by rabbit synovial fibroblasts. J Lab Clin Med 97:396–402.

Paranjpe M, Eugel L, Young N, Liotta LA: (1980) Activation of human breast carcinoma collagenase through plasminogen activator. Life Sci [II] 26:1223–1231.

Passwell JH, Dayer JM, Gass K, Edelson PJ: (1980) Regulation by Fc fragments of the secretion of collagenase, PGE_2 and lysozyme by mouse peritoneal macrophages. J Immunol 125:910–913.

Perez-Tamayo R: (1978) Pathology of collagen degradation: Review. Am J Pathol 92:509–566.

Perez-Tamayo R, Montfort I, Pardo A: (1980) What controls collagen resorption in vivo? Med Hypothesis 6:711–726.

Pettigrew DW, Ho GH, Sodek J, Brunette DM, Wang HM: (1978) Effect of oxygen tension and indomethacin on production of collagenase and neutral proteinase enzymes and their latent forms of porcine gingival explants in culture. Arch Oral Biol 23:767–777.

Pettigrew DW, Sodek J, Wang HM, Brunette DM: (1980) Inhibitors of collagenolytic enzymes synthesised by fibroblasts. Arch Oral Biol 25:269–274.

Puzas JE, Brand JS: (1979) Parathyroid hormone stimulation of collagenase secretion by isolated bone cells. Endocrinology 104:559–562.

Reddick ME, Bauer EA, Eisen AZ: (1974) Immunocytochemical localisation of collagenase in human skin and fibroblasts in monolayer culture. J Invest Dermatol 62:361–366.

Reynolds JJ, Sellers A, Murphy G, Cartwright E: (1977) New factor that may control collagen resorption. Lancet II:333–335.

Ridge SC, Oronsky AL, Kerwar SS: (1980) Induction of the synthesis of latent collagenase and latent neutral protease in chondrocytes by a factor synthesised by activated macrophages. Arthritis Rheum 23:448–454.

Sage H, Bornstein P: (1979) Characterisation of a novel collagen chain in human placenta and its relation to AB collagen. Biochemistry 18:3815–3822.

Sage H, Crouch E, Bornstein P: (1979) Collagen synthesis by bovine aortic endothelial cells in culture. Biochemistry 18:5433–5442.

Sakai T, Gross J: (1967) Some properties of the products of reaction of tadpole collagenase with collagen. Biochemistry 6:518–528.

Sakamoto M, Alfant M, Sakamoto S: (1981) Isolation and immunochemical localisation of collagenase in mouse peritoneal macrophages. J Biochem 90:715–720.

Sakamoto S, Sakamoto M, Goldhaber P, Glimcher MJ: (1975a) Studies on the interaction between heparin and mouse bone collagenase. Biochim Biophys Acta 385:41–50.

Sakamoto S, Sakamoto M, Goldhaber P, Glimcher MJ: (1975b) Collagenase and bone resorption: Isolation of collagenase from culture medium containing serum after stimulation of bone resorption by addition of parathyroid hormone extract. Biochem Biophys Res Commun 63:172–178.

Sakamoto S, Sakamoto M, Goldhaber P, Glimcher MJ: (1978a) Mouse bone collagenase. Purification of the enzyme by heparin-substituted Sepharose 4B affinity chromatography and preparation of specific antibody to the enzyme. Arch Biochem Biophys 188:438–449.

Sakamoto M, Sakamoto S, Goldhaber P, Glimcher MJ: (1978b) Immunocytochemical studies of collagenase localization in normal mouse tissues. In: Horton JE, Tarpley TM, Davis WF, eds. Mechanisms of Localized Bone Loss. Washington DC: Information Retrieval 441.

Seltzer JL, Jeffrey JJ, Eisen AZ: (1977) Evidence for mammalian collagenases as zinc ion metalloenzymes. Biochim Biophys Acta 484:179–187.

Seltzer JL, Welgus HG, Jeffrey JJ, Eisen AZ: (1976) The function of calcium ion in the action of mammalian collagenase. Arch Biochem Biophys 173:355–361.

Simpson JW, Mailman ML: (1981) Synthesis of a collagenase inhibitor by gingival fibroblasts in culture. Biochem Biophys Acta 673:279–285.

Starkey PM, Barratt AJ, Burleigh MC: (1977) The degradation of articular collagen by neutrophil proteinases. Biochim Biophys Acta 483:386–397.

Steven F: (1976) Observations on the different behaviour of tropocollagen molecules in solution and intermolecularly cross-linked tropocollagen within insoluble polymeric collagen fibrils. Biochem J 155:391–400.

Steven FS, Griffin MM, Itzhaki S, Al-Habib A: (1980) A trypsin-like neutral proteinase on Ehrlich ascites cell surfaces: Its role in the activation of tumour-cell zymogen of collagenase. Brit J Cancer 42:712–721.

Stricklin GP, Bauer EA, Jeffrey JJ, Eisen AZ: (1977) Human skin collagenase: Isolation of precursor and active forms from both fibroblast and organ cultures. Biochemistry 16:1607–1615.

Stricklin GP, Eisen AZ, Bauer EA, Jeffrey JJ: (1978) Human skin fibroblast collagenase: Chemical properties of precursor and active forms. Biochemistry 17:2331–2337.

Sundala H, Hayashi T, Hori H, Nagai Y: (1981) Dual effects of type II collagen on the degradation of type I collagen by tadpole collagenase. Collagen Rel Res 1:177–185.

Swann JC, Reynolds JJ, Galloway WA: (1981) Zinc metalloenzyme properties of active and latent collagenase from rabbit bone. Biochem J 195:41–49.

Terato K, Nagai Y, Kawanishi K, Yamamoto S: (1976) A rapid assay method of collagenase activity using ^{14}C-labelled soluble collagen as substrate. Biochim Biophys Acta 445:753–762.

Tschesche H, MacCartney HW: (1981) A new principle of regulation of enzymic activity. Activation and regulation of human polymorphonuclear leucocyte collagenase via disulphide-thiol exchange as catalysed by the glutathione cycle in a peroxidase-coupled reaction to glucose metabolism. Eur J Biochem 120:183–190.

Tyler JA, Cawston TE: (1980) Properties of pig synovial collagenase. Biochem J 189:349–357.

Tyree B, Halme J, Jeffrey JJ: (1980) Latent and active forms of collagenase in rat uterine explant cultures: Regulation of conversion by progestational steroids. Arch Biochem Biophys 202:314–317.

Uitto VJ, Schwartz D, Veis A: (1980) Degradation of basement-membrane collagen by neutral proteases from human leucocytes. Eur J Biochem 105:409–417.

Vaes G: (1980) Collagenase, lysosomes and osteoclastic bone resorption. In: Woolley DE, Evanson JM, eds. Collagenase in Normal and Pathological Connective Tissues. Chichester, UK: Wiley, 185–207.

Valle KJ, Bauer EA: (1979) Biosynthesis of collagenase by human skin fibroblasts in monolayer culture. J Biol Chem 254:10115–10122.

Vater CA, Mainardi CL, Harris ED Jr: (1978a) Binding of latent rheumatoid synovial collagenase to collagen fibrils. Biochim Biophys Acta 539:238–247.

Vater CA, Mainardi CL, Harris ED Jr: (1978b) Activation in vitro of rheumatoid synovial collagenase from cell cultures. J Clin Invest 62:987–992.

Vater CA, Mainardi CL, Harris ED Jr: (1979a) Inhibitor of human collagenase from cultures of human tendon. J Biol Chem 254:3045–3053.

Vater CA, Harris ED Jr, and Siegel RC: (1979b) Native cross-links in collagen fibrils induce resistance to human synovial collagenase. Biochem J 181:639–645.

Vater CA, Hahn JL, Harris ED Jr: (1981) Preparation of a monospecific antibody to purified rabbit synovial fibroblast collagenase. Collagen Rel Res 1:527–542.

Wahl LM: (1977) Hormonal regulation of macrophage collagenase activity. Biochem Biophys Res Commun 74:838–845.

Wahl LM, Wahl SM, Mergenhagen SE, Martin GR: (1974) Collagenase production by endotoxin-activated macrophages. Proc Natl Acad Sci

Weeks JG, Halme J, Woessner JF Jr: (1976) Extraction of collagenase from the involuting rat uterus. Biochim Biophys Acta 445:205–214.

Weiss JB: (1976) Enzymatic degradation of collagen. In: Hall DA, Jackson DS, eds. International Review of Connective Tissue Research. Volume 7. New York: Academic Press 101–157.

Welgus HG, Jeffrey JJ, Stricklin GP, Roswit WT, Eisen AZ: (1980) Characteristics of the action of human skin fibroblast collagenase on fibrillar collagen. J Biol Chem 255:6806–6813.

Welgus HG, Stricklin GP, Eisen AZ, Bauer EA, Cooney RV, Jeffrey JJ: (1979) A specific inhibitor of vertebrate collagenase produced by human skin fibroblasts. J Biol Chem 254:1938–1943.

Werb Z: (1978a) Pathways for the modulation of macrophage collagenase activity. In: Horton JE, Tarpley TM, Davis WF, eds. Mechanisms of Localized Bone Loss. Washington, DC: Information Retrieval, 213–228.

Werb Z: (1978b) Biochemical actions of glucocorticoids on macrophages in culture. Specific inhibition of elastase, collagenase, and plasminogen activator secretion and effects on other metabolic functions. J Exp Med 147:1695–1712.

Werb Z, Aggeler J: (1978) Proteases induce secretion of collagenase and plasminogen activator by fibroblasts. Proc Natl Acad Sci USA 75:1839–1843.

Werb Z, Burleigh MC: (1974) A specific collagenase from rabbit fibroblasts in monolayer culture. Biochem J 137:373–385.

Werb Z, Burleigh MC, Barrett AJ, Starkey PM: (1974) The interaction of α_2-macroglobulin with proteinases. Binding and inhibition of mammalian collagenases and other metal proteinases. Biochem J 139:359–368.

Werb Z, Mainardi CL, Vater CA, Harris ED Jr: (1977) Endogenous activation of latent collagenase by rheumatoid synovial cells. N Engl J Med 296:1017–1023.

Werb Z, Reynolds JJ: (1974) Stimulation by endocytosis in controlling the secretion of collagenase and neutral proteinase from rabbit synovial fibroblasts. J Exp Med 140:1482–1497.

Werb Z, Reynolds JJ: (1975a) Purification and properties of a specific collagenase from rabbit synovial fibroblasts. Biochem J 151:645–653.

Werb Z, Reynolds JJ: (1975b) Immunochemical studies with a specific antiserum to rabbit fibroblast collagenase. Biochem J 151:655–663.

Woessner JF Jr: (1977) A latent form of collagenase in the involuting rat uterus and its activation by a serine proteinase. Biochem J 161:535–542.

Woessner JF Jr: (1979a) Separation of collagenase and a metal-dependent endopeptidase of rat uterus that hydrolyzes a heptapeptide related to collagen. Biochem Biophys Acta 571:313–320.

Woessner JF Jr: (1979b) Total, latent and active collagenase during the courses of post-partum involution of the rat uterus. Biochem J 180:95–102.

Woessner JF Jr: (1980) Collagenase in uterine resorption. In: Woolley DE, Evanson JM, eds. Collagenase in Normal and Pathological Connective Tissues. Chichester, UK: Wiley, 223–239.

Woolley DE: (1980) Human Collagenases: Comparative and immunolocalization studies. Ciba Found Symp 75:69–86.

Woolley DE: (1982) Collagenase immunolocalization studies of human tumors. In: Liotta LA, Hart I, eds. Tumor Invasion and Metastasis. Boston: Martinus Nijhoff, 391–404.

Woolley DE, Davies RM: (1981) Immunolocalisation of collagenase in periodontal disease. J Periodont Res 16:292–297.

Woolley DE, Evanson JM: (1977) Collagenase and its natural inhibitors in relation to the rheumatoid joint. Connect Tissue Res 5:31–35.

Woolley DE, Evanson JM: (1980) Present status and future prospects in collagenase research. In: Woolley DE, Evanson JM, eds. Collagenase in Normal and Pathological Connective Tissues. Chichester, UK: Wiley, 241–250.

Woolley DE, Grafton CA: (1980) Collagenase immunolocalisation studies of cutaneous secondary melanomas. Br J Cancer 42:260–265.

Woolley DE, Glanville RW, Crossley MJ, Evanson JM: (1975) Purification of rheumatoid synovial collagenase and its action on soluble and insoluble collagen. Eur J Biochem 54:611–622.

Woolley DE, Roberts DR, Evanson JM: (1976a) Small molecular-weight B_1 serum protein which specifically inhibits human collagenases. Nature (Lond) 261:325–327.

Woolley DE, Crossley MJ, Evanson JM: (1976b) Antibody to rheumatoid synovial collagenase. Its characterization, specificity and immunological cross-reactivity. Eur J Biochem 69:421–428.

Woolley DE, Crossley MJ, Evanson JM: (1977) Collagenase at sites of cartilage erosion in the rheumatoid joint. Arthritis Rheum 20:1231–1239.

Woolley DE, Glanville RW, Roberts DR, Evanson JM: (1978a) Purification, characterisation and inhibition of human skin collagenase. Biochem J 169:265–276.

Woolley DE, Akroyd C, Evanson JM, Soames JV, Davies RM: (1978b) Characterisation and serum inhibition of neutral collagenase from cultured dog gingival tissue. Biochem Biophys Acta 522:205–217.

Woolley DE, Harris ED Jr, Mainardi CL, Brinckerhoff CE: (1978c) Collagenase immunolocalization in cultures of rheumatoid synovial cells. Science 200:773–775.

Woolley DE, Brinckerhoff CE, Mainardi CL, Vater CA, Evanson JM, Harris ED Jr: (1979) Collagenase production by rheumatoid synovial cells: Morphological and immunohistochemical studies of the dendritic cell. Ann Rheum Dis 38:262–230.

Woolley DE, Tetlow LC, Evanson JM: (1980a) Collagenase immunolocalisation studies of rheumatoid and malignant tissues. In: Woolley DE, Evanson JM, eds. Collagenase in Normal and Pathological Connective Tissues. Chichester, UK: Wiley, 105–125.

Woolley DE, Tetlow LC, Mooney CJ, Evanson JM: (1980b) Human collagenase and its extracellular inhibitors in relation to tumor invasion. In: Strauli P, Barrett AJ, Baici A, eds. Proteinases in Tumor Invasion. New York: Raven Press, 97–115.

Yasui N, Hori H, Nagai Y: (1981) Production of collagenase inhibitor by the growth cartilage of embryonic chick bone: Isolation and partial characterisation. Collagen Rel Res 1:59–72.

5

Immunology of the Collagens

Rupert Timpl

5.1. Nature and Recognition of Antigenic Determinants in Proteins

Most proteins possess multiple sites designated antigenic determinants that are capable of inducing an antibody response when the protein comes in contact with cells of the immune system. This contact can be easily produced by injection of the protein (antigen) into a variety of laboratory animals (immunization). Antigenic determinants are also capable of interacting noncovalently with the induced antibodies to form tightly bound antigen–antibody complexes with affinity constants in the range 10^6 to 10^{12} $1 \cdot mol^{-1}$. These determinants are in most cases located on the surface of the proteins, where they occupy regions composed of 4 to 15 different amino acid residues. Each protein usually possesses several different antigenic determinants. Their number is determined by the structural complexity and the size of the protein antigen. Small proteins such as myoglobin (molecular weight 17,000) contain as many as five distinct determinants. Multiple copies of a single determinant may exist in proteins with repeating sequences or with identical subunits (Crumpton 1974; Atassi 1977).

Another factor contributing to the number and nature of antigenic determinants is foreignness. This term vaguely defines the relationship between structures of the antigen and of the proteins of the animal species used for immunization. Normally, only those structures which are foreign to the immunized animal are recognized as antigenic determinants. Shared structures are antigenically silent due to a delicate control of the immune system (self-tolerance) which prevents antibody responses against structures of one's own body. Foreignness, however, is not the only factor involved in triggering immune responses and

From the Max-Planck-Institut für Biochemie, Martinsried, Federal Republic of Germany

can be circumvented under certain pathologic or experimental conditions (autoimmunity).

The activity of the majority of antigenic determinants not only depends on the linear array of amino acids in a particular sequence but also on the secondary, tertiary, or quarternary structure of this sequence within the protein molecule (conformational antigenic determinants). Modification of the structure by denaturation, proteolysis, or chemical cleavage abolishes the ability of these determinants to interact with antibodies and may create new determinants. Since these treatments are common approaches in identifying a particular structure within a protein, the precise characterization of conformational antigenic determinants is a rather difficult task. Certain segments of a number of proteins exhibit a less rigid conformational structure. If such segments participate in triggering an immune response, they often survive extensive degradation of the protein without loss of antigenic activity (sequential antigenic determinants). Examples of sequential antigenic determinants include linear arrays of amino acids within a protein chain. Characterization of such determinants has indicated that they may be as small as pentapeptides (Sela et al. 1967).

Distinct antigenic determinants are unique for a particular protein and determine its antigenic specificity. Antibodies raised against a protein usually fail to react with other proteins, except for those possessing similar determinants due to a homologous amino acid sequence and similar conformation. These reactions are commonly observed between identical proteins obtained from different animal species (interspecies crossreaction). Crossreactions can also exist between related proteins obtained from a single animal source, such as between various isozymes or other proteins sharing certain structural elements (for example, immunoglobulin isotypes). Crossreactions are usually manifested for some but not all antigenic determinants, and this property permits the preparation of specific antibodies for almost all proteins.

The presence of potential antigenic determinants in a protein is usually not sufficient for inducing an antibody response but requires another quality, referred to as immunogenicity. Immunogenic activity does not necessarily depend on the same structures as antigenic specificity. This has been known for more than 50 years from studies with immunogens composed of small organic compounds such as sulfanilic acid (hapten) covalently attached to a protein (carrier). Immunization with the conjugate but not with the hapten alone elicits large amounts of antibodies against the hapten. This observation is currently interpreted as being due to two functionally different sets of structures, haptenic and carrier determinants (Figure 5.1), being present on each T cell dependent antigen (*See* below), which also includes all natural protein antigens. The former determinants interact directly with antibody-producing cells while the latter modulate this process by interaction with other immune cells not directly involved in antibody production. Both sets of determinants are frequently localized in different regions of a protein antigen (Crumpton 1974; Atassi 1977).

Antibody-producing cells belong to the B cell lineage of small lymphocytes. They possess antibody molecules inserted into their plasma membrane, which are able to interact with haptenic determinants. This is presumably one of the final steps in antibody induction and this interaction results in proliferation and differentiation of B cells. Some of the cells eventually become plasma cells that

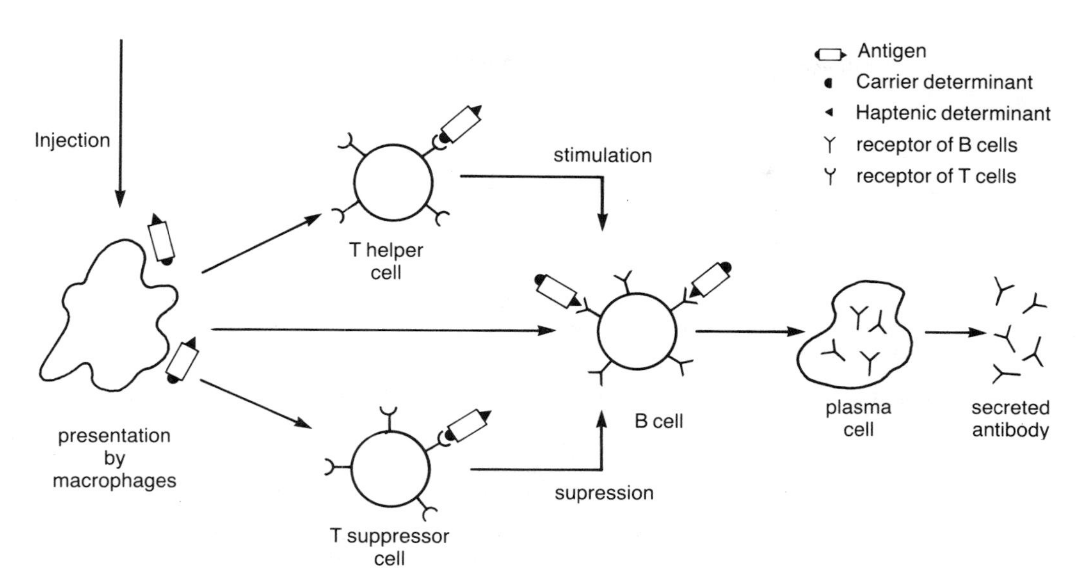

Figure 5.1. Scheme of cellular cooperation involved in the antibody response against a typical protein antigen. The initial step consists of antigen uptake by macrophages followed by presentation to T and/or B cells. Different regulatory T cells then either enhance or suppress the antigen-induced proliferation of B cells. Eventually, B cells develop into plasma cells that secrete abundant amounts of antibody. T and B cells react with the antigen via receptors that do not necessarily have the same specificity.

produce up to 1000-fold more antibody than nonstimulated cells. Most of the antibody is secreted and then found in extracellular fluids, for example, blood plasma. It is thought that the membrane-bound receptor and the secreted antibody are molecules with identical specificity. Excessive antibody production is controlled by the short life span of plasma cells and, in addition, by production of additional antibodies directed against the primary antibody (Eisen 1981; Klein 1982).

Several steps precede antibody production and involve cooperation between different immune cells (Figure 5.1). These include presentation of the antigen to immune cells in membrane-bound form by macrophages. Another set of small lymphocytes (regulatory T cells) are able to react via specific receptors with carrier determinants of protein antigens. These interactions can either enhance (T helper cells) or abolish (T suppressor cells) the production of antibody by B cells. The chemical nature of T cell receptors, as well as the mechanism of T and B cell cooperation, are not well understood. Several possibilities, such as the bridging of T and B cells by the antigen or the release of soluble mediators, are currently discussed as events involved in stimulating the proliferation of B cells. A few antigens are known to be independent of T cell control, but these usually do not include protein antigens. The repertoire of receptor specificities of B and T cells is incompletely known, but is presumably overlapping. Thus it is possible that the same structure of a protein may function both as a carrier and a haptenic determinant. An exclusive T cell response against an antigen can also exist (effector T cells) and is often related to pathologic phenomena (cytotoxic effects, transplantation rejection).

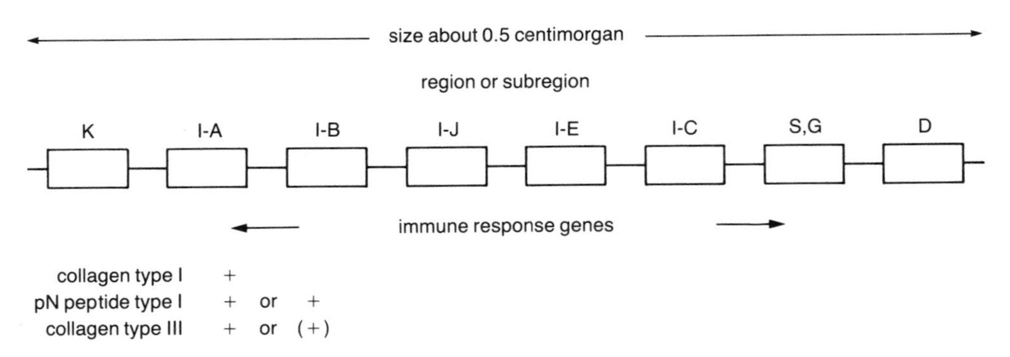

Figure 5.2. Map of the major histocompatibility complex H-2 at chromosome 17 of the mouse. Loci K and D code for major transplantation antigens that show some similarity to immunoglobulins and occur on almost every cell. The I locus codes for immune response genes, which control antibody response at the level of regulatory T cells, and for membrane antigens on B cells and certain T cells. Immune response genes for types I and III collagens and type I procollagen, among many other antigens, have been mapped to certain subregions of the I locus. This is accomplished by studies in particular inbred strains that show recombinations between segments of high and low responder alleles within the I locus. (For reviews, see Dorf 1981; Klein 1982.)

Immune cell receptors are characterized by the enormous diversity of their combining sites, which to a large extent is genetically controlled. Recent research has elucidated the DNA recombinational events underlying this diversity and the reader is referred to reviews for a more detailed discussion (Eisen 1981; Gottlieb 1980; Klein 1982). The B cell population is divided into many groups of clones, which differ from each other in receptor specificity. Antibody production is manifested by the selective stimulation of those clones possessing sufficient receptor affinity for the determinants present on the injected antigen. The repertoire of B cell receptors is usually large enough to allow the stimulation of many clones by a single protein antigen (polyclonal antibody response). Monoclonal antibody responses are a rather rare event, even against a single antigenic determinant. T cells apparently exist also in many clones, but the structural and genetic basis of this diversity is less well established. Genetic control of T cell activity can be studied by characterizing the strength of the antibody response against T cell-dependent antigens, which is usually done in inbred strains of mice. This control is due to immune response genes that are closely linked to the major histocompatibility complex coding for transplantation antigens (Figure 5.2). Frequently, a strong antibody response depends on two or more immune response genes located in this region that complement each other. The nature of the immune response gene products that are involved in T cell receptor activity is not known (Dorf 1981; Klein 1982).

5.2. Production of Antibodies Against Collagens and Their Immunochemical Analysis

Collagenous proteins were considered nonimmunogenic for a long time. This opinion changed because of studies done during the past 20 years, but it has been established that most of the collagens are poor immunogens as compared with many other proteins (Timpl 1976, 1982; Furthmayr and Timpl 1976; Beard et al. 1977; Michaeli 1977; Furthmayr 1982). Successful production of antibodies usually requires immunization with a solubilized collagen together with complete Freund's adjuvant (a mixture of mineral oil and heat-killed mlycobacteria), which augments antibody responses in a general way. Antibody levels achieved with the interstitial collagens (types I, II, and III) are in the range 100–500 μg/ml antiserum. They exceed 1 mg/ml after immunization with a procollagen or basement membrane collagen (type IV), indicating that these components are better immunogens. Injection of insoluble, crosslinked collagen usually fails to elicit antibodies. Immunogenicity is also decreased after denaturation which, however, may expose new antigenic determinants. The low immunogenicity of collagen increases the risk of antibodies against impurities (serum proteins, procollagen peptides, fibronectin) present in the materials used for immunization. This possibility requires careful purification of antibodies for certain applied studies (Timpl et al. 1977). The reasons for the low antibody response against the collagens are not known, but could be due to a low level of foreignness or to certain structural properties (rigidity, low solubility, protease resistance) that interfere with proper handling of the antigen by immune cells.

The collagens and procollagens belong to the class of antigens that require T cell help for optimal antibody production (Table 5.1). This was shown by two different sets of experiments (Nowack et al. 1976). In the first, the immunologic capacity of mice was destroyed by thymectomy and whole body x-ray irradiation, which eliminates all B and T cells. Injection of sufficient numbers of B and T cells into the animals, but not of B cells alone, restored the full antibody response against collagen. In the second model it was shown that athymic mice lacking functionally active T cells do not respond against the interstitial collagens. Similar data were obtained for the procollagens. The T cell dependence of the antibody response against denatured collagen is still controversial.

Another way of demonstrating the involvement of T cells in the antibody response against the collagens and procollagens is the identification of immune response control linked to the H-2 locus of the mouse genome (Table 5.2). These studies are usually done in congenic resistant strains, which differ only at this gene locus. The experiments demonstrated that high and low antibody responses

Table 5.1. Experimental Evidence That Bovine Type I Collagen Behaves as T Cell-Dependent Antigen in Inbred Strains of Mice[a]

Group of mice	Antibody titer[b] (mean ± SD)
Control	8.0 ± 2.4
Thymectomized; B cells	1.5 ± 1.2
Thymectomized; B and T cells	8.5 ± 2.6
Athymic nude mice (nu/nu)	<2.0
Heterozygous control (nu/+)	7.8 ± 1.6

[a]Data from Nowack et al. 1976. In one experiment, adult mice were thymectomized and irradiated to destroy immune cells. Immune capacity was then restored prior to immunization by injection of B cells alone or a mixture of B and T cells. The second model used nude mice homozygous for the nu gene with a low level of functionally active T cells. This is a recessive trait and heterozygotes possess an almost unimpaired immunologic capacity.

[b]Determined by passive hemagglutination and expressed as highest antiserum dilution ($-\log_2$) showing a positive reaction.

against types I, II, III, and V collagen and against the pN peptides of types I and III procollagen are governed by different alleles of this gene region. This indicates that various types of collagen can also be distinguished at the level of T cell activity. The observation of other strains with intermediate antibody responses, however, and of two gene complementation also indicates that immune recognition of the collagens is a more complex phenomenon not controlled by only genes of the H-2 locus (Nowack et al. 1975b; Kemp and Madri 1981). The H-2 associated immune response region is rather large and may account for several hundred individual genes. Studies of appropriate recombinant strains allows the subdivision into five different segments (Figure 5.2). Control of the antibody response against type I collagen appears to be associated with the I-A subregion and against its pN peptide or against type III collagen, either with I-A or I-B (Nowack et al. 1975b, 1977; Kemp and Madri 1981).

Several different laboratory animals have been used for the production of antibodies against the collagens (Timpl 1982). Rabbits respond mainly against the terminal regions of the molecule and the reaction is not dependent on an intact triple helical conformation. Most other animals (guinea pigs, rats, mice, sheep, and chickens) recognize conformational antigenic determinants located within the triple helix. Chickens also show a strong antibody response against denatured collagen, which is a rare observation in other animals. All antisera react with a variety of antigenic determinants and are thus of polyclonal nature.

Heterogeneity of the induced antibodies can also be demonstrated by analyzing antibody affinity. It is convenient to determine affinity constants by radioimmunoassay. Here a constant amount of antiserum is reacted with labeled antigen over a wide range of concentration. After equilibrium has been reached, the relative proportions of bound and free antigen are determined. Analysis of the data by a Scatchard plot (Figure 5.3a) usually shows a curved line for polyclonal antibodies, indicating heterogeneity of binding sites. Such plots still permit calculation of approximate affinity constants, which are of the order of 10^{11} $1 \cdot mol^{-1}$ for antibodies against the pN peptides of the procollagens. Affinity

Table 5.2. Control of Antibody Production Against Collagens and Procollagens by Different Immune Response Genes Associated with the Major Histocompatibility Locus H-2 of the Mouse[a]

	H-2 allele responsible	
Immunogen	High response	Low response
Bovine type I collagen	b, f, s	a, d, k, q
Bovine type II collagen	q, s	b, k, d
Chick type II collagen	b, d, q, s	k, f
Human type III collagen	s	a, b, d
Human type V collagen	k (b,d)[b]	b, d, q
Bovine type I pN peptide	b, k, a	d, q, s
Bovine type III pN peptide	b, k, s	d, q

[a]Data from Nowack et al. 1975a, 1975b, 1977; Kemp and Madri 1981; Wooley et al. 1981; and unpublished observations.
[b]Only certain strains, due to gene complementation.

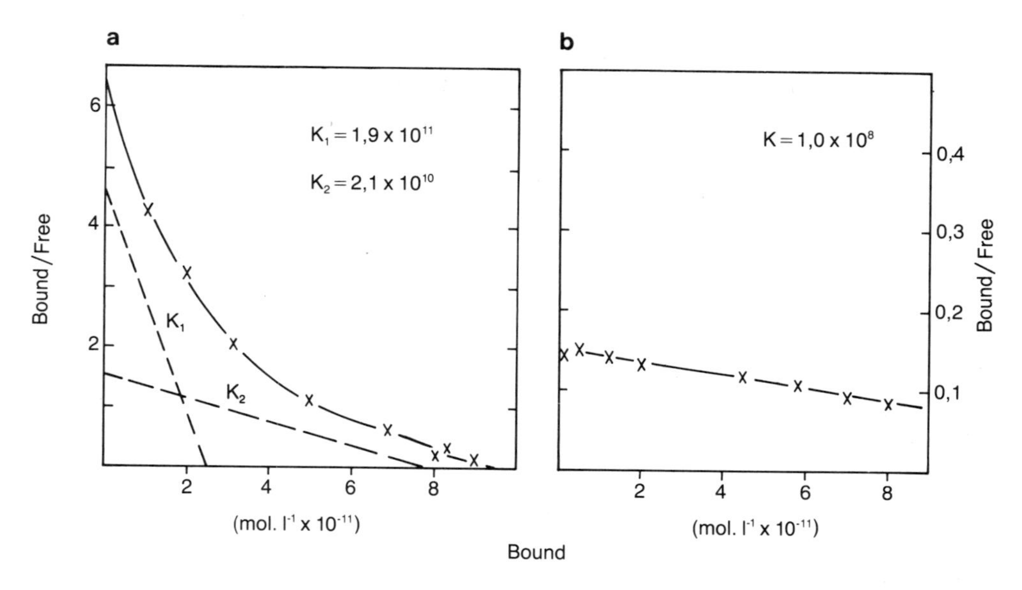

Figure 5.3. Determination of affinity constants (K) of polyclonal rabbit antibodies (**a**) and monoclonal mouse antibodies (**b**) against the pN peptide of type I procollagen by radioimmunoassay. The test was performed with constant amounts of antibodies incubated under equilibrium conditions, with various amounts of labeled pN peptide followed by separation of the bound (B) from free (F) peptide. The data were then analyzed by Scatchard plots. A plot of B/F against B gives a straight line for systems with homogeneous binding sites and K can be calculated from the slope of the curve, as shown for the monoclonal antibody. A curved plot is observed for the polyclonal antibodies indicating a considerable heterogeneity. The data were analyzed by assuming two different populations of antibodies with affinity constants K_1 and K_2. K is defined as $K = \frac{[AbAg]}{[Ab]\,[Ag]}\ 1 \cdot mol^{-1}$. (*Source: Timpl and Risteli (1982) In: Furthmayr H, ed. Immunochemistry of the Extracellular Matrix. Boca Raton, Florida: CRC Press, with permission; and unpublished data, with permission.)*

constants of antibodies against the helical regions of the collagens are usually lower by several orders of magnitude.

Monoclonal antibodies (Galfre and Milstein 1981) with a homogeneous binding site for a single antigenic determinant can be produced from triggered immune cells by a special cell culture technique (Table 5.3). The procedure starts with a polyclonal population of antibody-producing cells from spleen, which are fused with enzyme-deficient tumor cells of B cell origin (myeloma cells). Successful spleen tumor cell hybrids are then selected in a special culture medium which eliminates both parental cell types. Hybrid cells preserve two useful properties. They continue to produce the immunoglobulins of the parental spleen cells and, as tumor cells, they have an indefinite life span. Those hybridomas which produce antibody to the original immunogen are then selected by appropriate cell cloning. Since each hybrid cell produces only a single antibody and large numbers of hybrid cells may be grown, this technique allows the production of homogeneous antibody on a large scale (several hundred milligrams) in a reproducible manner (Linsenmayer et al. 1979). Homogeneity of monoclonal antibodies can be assessed in several ways including the determination of their affinity. Scatchard analysis of a monoclonal antibody directed against type I procollagen (Figure 3b) shows a straight line, as expected for a single homogeneous antibody binding site and a single antigenic determinant.

Table 5.3. Procedure for the Production of Monoclonal Antibodies by the Hybridoma Technique[a]

1. Mice are immunized with the antigen by a conventional method. Spleen cells containing antibody-producing clones are harvested 1 to 10 days after the last antigen injection.
2. Spleen cells are fused with a mutant mouse myeloma cell line by treatment, for example, with polyethylene glycol. The myeloma cells are usually nonantibody-producing malignant immune B cells, deficient in the enzyme hypoxanthine-guanine phosphoribosyl transferase.
3. Fused cells are cloned and grown in a selective culture medium (HAT medium) in many small wells of a multi-well culture plate. Only hybridoma cells (hybrid between a spleen and a myeloma cell) will survive and proliferate.
4. Antibody-producing hybrids are identified by immunochemical analysis of the culture medium of each well.
5. Large amounts of monoclonal antibodies are produced by growth of a selected cloned cell in the ascites fluid of nude mice, or, alternatively, by mass culture in large tissue culture plates.

[a]For a review, see Galfre and Milstein 1981.

Even though the advantages of monoclonal antibodies are obvious, they may also have a few disadvantages as compared with polyclonal antibodies. For unknown reasons, monoclonal antibodies may have a low affinity constant (example shown in Figure 5.3) which may not allow sensitive assays capable of detecting antigen in the nanogram range. Very often monoclonal antibodies exhibit an extraordinarily fine specificity and distinguish between antigens that are considered homogeneous by other criteria (Linsenmayer et al. 1979). This is conveniently shown by binding analysis of radiolabeled antigen (Figure 5.4). Excess amounts of polyclonal antibodies usually bind 90–100% of the labeled antigen. Binding curves with monoclonal antibodies frequently level off at a lower plateau, indicating that they are able to react with only 5–30% of the labeled antigen.

The structural basis for partial binding is not yet known. It could be due to

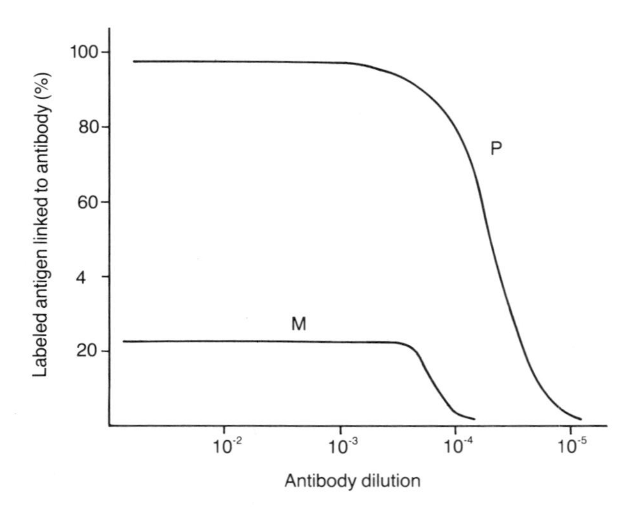

Figure 5.4. Typical binding profiles in radioimmunoassays obtained with polyclonal antibodies (*P*) and with monoclonal antibodies (*M*) showing specificity for a site not present on all antigen molecules. The test is carried out with various dilutions of the antibody that were incubated with a constant amount of ^{125}I-labeled antigen. Bound and free antigen are then separated by a suitable precipitation reaction.

posttranslational modifications of collagen (hydroxylated or glycosylated amino acids), which are known to vary among different molecules. Another possibility could be microheterogeneity, which may normally exist in a collagen or may be introduced during purification (denaturation, proteolytic modification). Monoclonal antibodies against the collagens often show the phenomenon of partial binding. This is, however, not necessarily a typical property, since other monoclonal antibodies, such as one against type I procollagen, (Figure 5.3b) bind the antigen like a polyclonal antibody (Figure 5.4).

Radioimmunoassays (Timpl and Risteli 1982) and enzyme immunoassays (Rennard et al. 1980) are the most commonly used methods for determining antibody titers and immunochemical properties of the collagens and procollagens. The former assay is usually done with antigen (1–5 ng) labeled with ^{125}I, which allows a more sensitive assay than labeling with tritium or ^{14}C. Binding of the labeled antigen to antiserum is usually carried out in a titration assay (Figure 5.4) and involves separation of free and bound antigen by a suitable precipitation reaction (usually second antibody against immunoglobulin). Inhibition assays are commonly used variants of radioimmunoassays. They are based on the competition of unlabeled antigen (inhibitor) and labeled antigen for the same antibody binding sites (Figure 5.5). The particular shape of an inhibition profile is determined by the affinity of the inhibitor for the antibody

Figure 5.5 Schematic protocol of an inhibition radioimmunoassay. This particular variant is referred to as sequential saturation assay since it includes preincubation of nonlabeled inhibitor with antibody to achieve equilibrium prior to the addition of labeled antigen. Simultaneous incubation of antibody, inhibitor, and labeled antigen (equilibrium assay) is usually a less sensitive variant. Sensitivity is also increased by using an excess of labeled antigen so that only 50% is bound in the binding control.

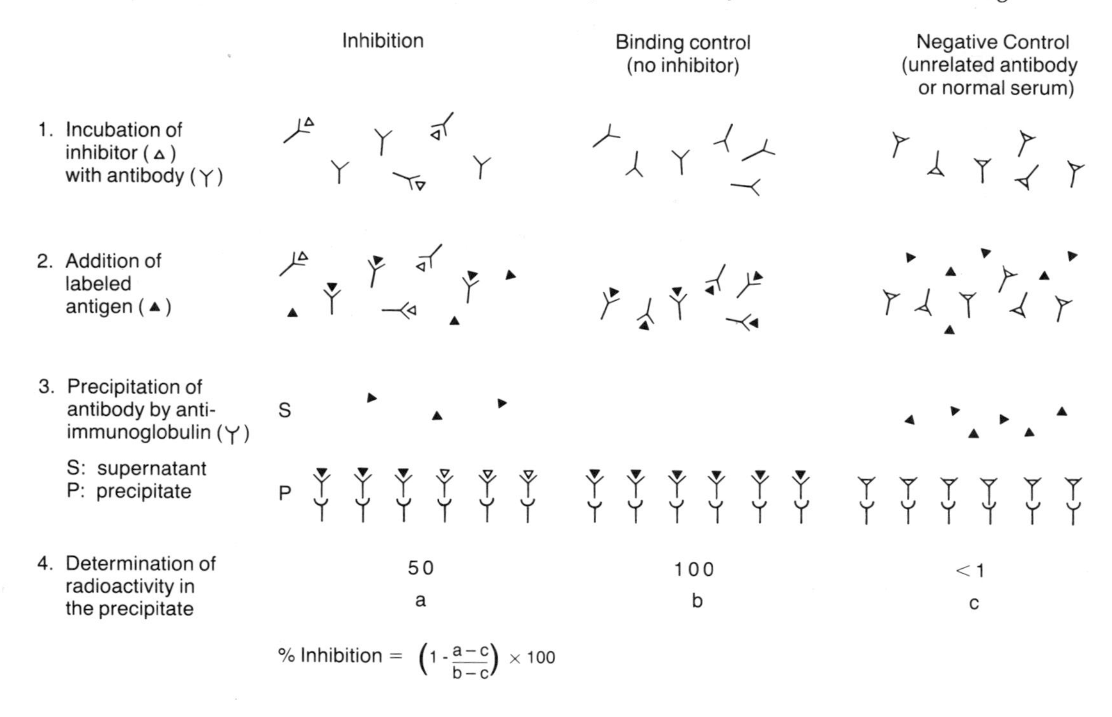

and allows the identification of partial or complete crossreactivity. Radioimmunoassays are also well suited for the quantitative determination of minute amounts of antigen, even in complex biological samples.

Enzyme immunoassays are in principle similar to radioimmunoassays except for a few technical differences. The antigen is used in unlabeled form as an insoluble coat on the surface of plastic wells. Uptake of antibody by the absorbed antigen is measured by reaction with a second antibody against immunoglobulin conjugated with an enzyme. Binding of the enzyme to the antibody layer is then analyzed by an appropriate colorimetric reaction. These assays are frequently referred to as enzyme-linked immunosorbent assays or ELISA. Inhibition enzyme immunoassays follow basically the same protocol as inhibition radioimmunoassays (Figure 5.5). Both assays are usually of comparable sensitivity.

Passive hemagglutination (Timpl 1982) is another assay used in immunologic studies of the collagens. The assay exploits the agglutination reaction by antibodies to erythrocytes which are coated with the antigen. Reaction patterns are read visually, so they are semiquantitative at best. Despite this disadvantage, the assay has been successfully used for analyzing the antigenic determinants of a collagen.

5.3. Antigenic Determinants of the Collagens

A highly diverse antigenic structure is likely for molecules with the size and structural complexity of the collagens (Chapters 1 and 2), and has in fact been demonstrated in studies over the past 20 years (Figure 5.6). The triple helical

Figure 5.6. Distribution of different antigenic determinants in type I collagen. Helical determinants depend on an intact triple helical conformation and their precise number and positions are insufficiently known. Terminal and central determinants can also be detected on unfolded α1(I) and α2(I) chains and are mapped on different cyanogen bromide peptides along the chains. The data refer to studies with calf and rat collagen and are more comprehensively discussed in previous reviews (Timpl 1976; Furthmayr and Timpl 1976).

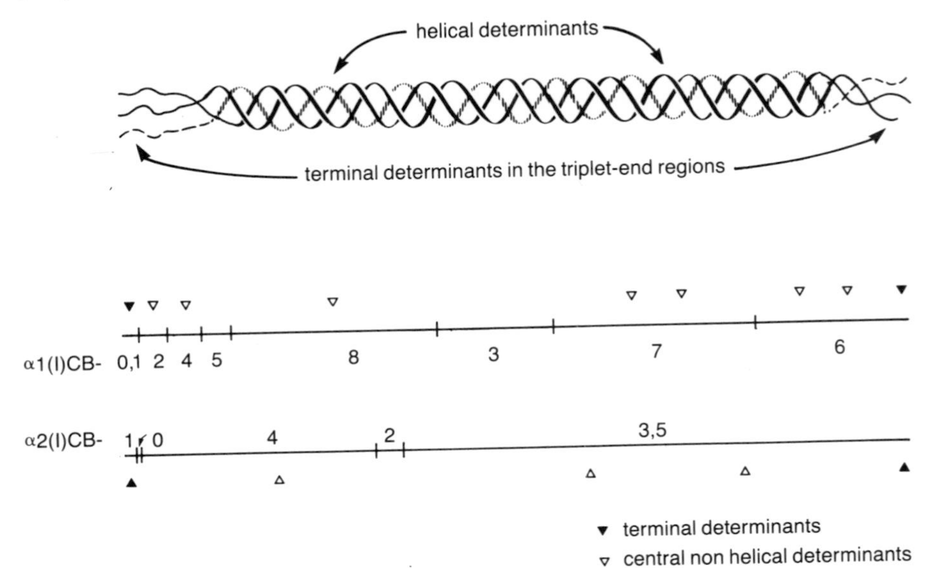

conformation endows the collagens with a unique feature: all amino acid side chains are directed to the outside of the molecule and thus represent potential antigenic determinants. Studies with types I, II, and III collagen have identified three groups of antigenic determinants with unique properties (For earlier reviews, see Timpl 1976; Furthmayr and Timpl 1976; Michaeli 1977):

1. Helical determinants are localized in the triple helix and are destroyed by unfolding the triple helix.
2. Central nonhelical determinants are exposed by unfolding the triple helix.
3. Terminal determinants are restricted to the nonhelical end regions of the α chains showing the highest structural variability among different collagens. They can be detected both on triple helical molecules and unfolded α chains.

Antibody responses against the helical determinants of collagen are the most commonly observed pattern in animals other than the rabbit (Beil et al. 1973). This is particularly strong if protease-solubilized collagens are used as immunogens since the terminal determinants are absent. Due to their conformational nature, little is known about their precise localization and structure. Antisera raised in rats against type I collagen clearly distinguish among molecules of the chain composition $[\alpha1(I)]_2\alpha2(I)$, $[\alpha1(I)]_3$, and $[\alpha2(I)]_3$ (Hahn and Timpl 1973). The latter two materials are formed by renaturation of purified α chains but $[\alpha1(I)]_3$ also exists in situ (*See* Chapter 2). The data indicate the presence of antigenic determinants (Crumpton 1974) composed of segments from different polypeptide chains. Current (Linsenmayer et al. 1979) and forthcoming studies of monoclonal antibodies will provide new sets of immunologic reagents specific for the triple helix, since they are usually derived from mouse and rat spleen cells. These antibodies may allow a comprehensive characterization and mapping of helical determinants, for example, by electron microscopy (*See* Figure 5.10).

Antibody responses against central nonhelical (denatured) determinants are rather rare and have so far only been observed in chickens and some rabbits (Furthmayr et al. 1972). The strength of the response may be increased by using a denatured collagen as immunogen, which, however, often lacks substantial immunogenicity. Central determinants have been localized on several cyanogen bromide (CNBr) peptides of the α1(I) and α2(I) chains (Figure 5.6) and presumably include many different sequences. The amino acid sequence of a single determinant that is localized in the triplet region of the pN peptide of type I procollagen has been elucidated (Figure 5.7, residues 93–98). Antibodies against central determinants usually do not crossreact with triple helical collagen, even though they may have been raised by immunization with triple helical immunogens. This paradoxic observation has not been adequately explained, but similar data have been obtained with noncollagenous protein antigens.

The importance of the nonhelical terminal regions for the antigenicity of collagen was originally recognized by demonstrating their protease sensitivity (Schmitt et al. 1964) and by localizing antigenic determinants on terminal CNBr peptides (Michaeli et al. 1969). At present these terminal determinants are the best characterized antigenic regions of the collagens (Fig. 5.7). The principal way to elucidate their antigenic structure involves localization of a distinct determinant to a particular CNBr peptide (Figure 5.6) followed by appropriate proteolytic digestion of the peptide to smaller fragments that still possess full antigenic activity in inhibition assays. Table 5.4 illustrates that approach for the 14-residue

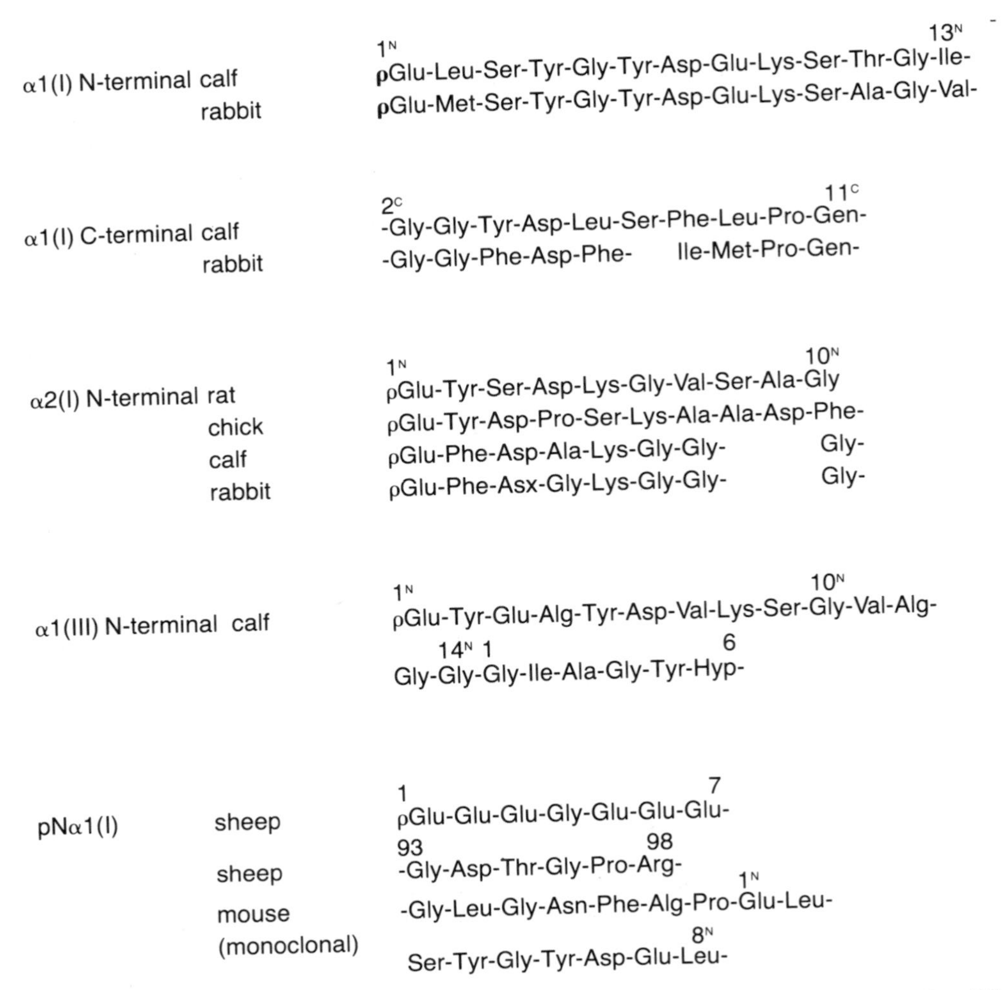

Figure 5.7. Amino acid sequences involved in sequential antigenic determinants of types I and III collagen and the pN peptide of type I procollagen. The determinants in the α chains are located in the N- and C-terminal nontriplet regions of calf, rat, and chick collagen. They are compared with homologous sequences of rabbit collagen, since the antigenic determinants were found in studies with rabbit antibodies except for one mouse monoclonal antibody as indicated. Lines above the sequence outline the minimum size of a particular antigenic determinant (See text and Table 5.4). Sequences without horizontal lines correspond to the size of the smallest fragment studied possessing full antigenic activity. The structure of the actual antigenic determinant within these sequences could be smaller. The site reacting with a monoclonal mouse antibody in the pro-α1(I) chain includes the Pro-Gln bond cleaved by procollagen N protease *(arrow)*. Data from reviews (Furthmayr and Timpl 1976; Timpl 1976; Michaeli 1977), original references (Becker et al. 1976; Rohde and Timpl 1979), and unpublished results.

N-terminal peptide α2(I)-CB1 of the rat collagen α2(I) chain (Furthmayr and Timpl 1972). One particular rabbit antiserum (No. 290) could be fully inhibited by a hexapeptide corresponding to positions 1^N-6^N of α2(I)-CB1, but essentially not by the pentapeptide 1^N-5^N or an overlapping peptide 3^N-14^N. The data indicate that the minimal size of the antigenic determinant includes positions 2^N-6^N of the sequence, leaving open (due to the lack of appropriate fragments) whether or not residue 1 is also essential for antigenicity. Recognition of a particular sequence by individual animals can show considerable diversity, as illustrated for a second antiserum (Table 5.4). Other examples appear in Figure 5.7.

Terminal determinants have the peculiarity of being recognized only by rabbits and not, as far as we know, by other animals. Owing to their sequential nature, it is feasible to compare their structure with those sequences present in homologous regions of rabbit collagen (Figure 5.7). It has been shown that each antigenic determinant contains at least one (but usually more than one) amino acid residue that is different from the rabbit sequence. Identical or nearly identical sequences (compared to the rabbit) such as the N-terminal peptides of rat α1(I) chain or calf α2(I) chain lack antigenicity. This emphasizes foreignness as an important factor in antigenic recognition. On the other hand, considerable sequence differences exist between the nontriplet peptides of rat and chick α2(I) chains (Figure 5.7), but the chicken fails to respond against this region after immunization with rat collagen (Furthmayr et al. 1972). The reasons for such observations remain entirely obscure. Lack of appropriate cooperation between certain haptenic and carrier determinants or proteolytic degradation of peptides by macrophages could be possible explanations.

The type IV collagens, which are unique to basement membranes, and possibly the type V collagens possess a more complex multidomain structure, including large noncollagenous segments and are therefore expected to differ from the interstitial collagens in certain antigenic properties. A detailed antigenic structure has been recently established for type IV collagen indicating the presence of several major and minor antigenic determinants (Figures 5.8, 5.9). The globular domain NC1 as well as a fragment containing the other domains of type IV collagen each show only partial inhibition of the reaction. Both inhibiting activities were found to be additive, demonstrating different antibody populations in the antiserum, with each being specific for a particular fragment. Another major set of antigenic determinants was identified in the 7S collagen segment

Table 5.4. Localization of Different Antigenic Determinants in the N-Terminal Nontriplet Region of Rat Collagen α2(I) Chain by Passive Hemagglutination Inhibition[a]

Inhibitor peptide position[b]	Minimal concentration (μM) of inhibitor blocking the reaction of antiserum	
	No. 290	No. 317
1^N-14^N	0.025	0.1
1^N-6^N	0.025	>12.8
7^N-14^N	>12.8	>12.8
1^N-7^N	0.05	>12.8
8^N-14^N	0.8	6.4
3^N-14^N	>12.8	0.05
1^N-5^N + 6^N-14^N	>12.8	>12.8

Source: Furthmayr, Timpl (1972) Biochem Biophys Res Commun 47:944–950, with permission.

[a]The reacting partners were erythrocytes coated with α2(I) chains and two individual hyperimmune rabbit antisera against rat type I collagen. These antisera contain mainly antibodies reacting with the cyanogen bromide peptide α2(I)-CB1, which occupies sequence position 1^N-14^N in the α2(I) chain.

[b]Reference inhibitor was α2(I)-CB1 with the sequence PCA-Tyr-Ser-Asp-Lys-Gly-Val-Ser-Ala-Gly-Pro-Gly-Pro-Hse (1^N-14^N) which was as active as the whole α2(I) chain. Further inhibitors were prepared by digesting α2(I)-CB1 with trypsin, chymotrypsin, elastase, or thermolysin, which all produced a single cleavage at different positions of the sequence.

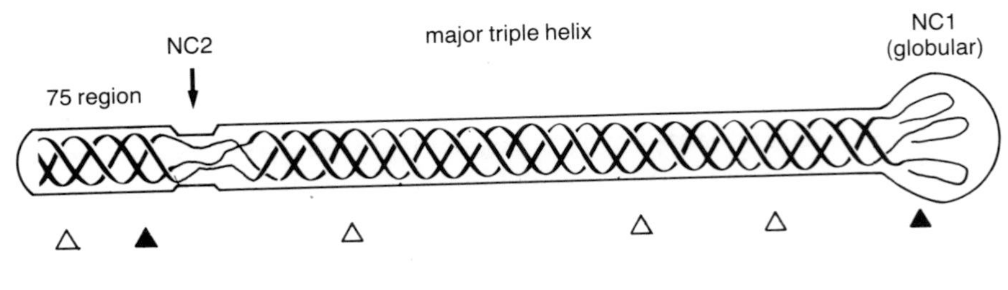

Figure 5.8 Localization of different antigenic determinants within the four domains of type IV (basement membrane) collagen. NC1 and NC2 denote two large noncollagenous segments interrupting or terminating the triple helix. Experimental data (Risteli et al. 1980, 1981a; Timpl et al. 1981) that allowed the identification of major and/or minor antigenic determinants are illustrated in Figure 5.9.

(Risteli et al. 1981b), which is a short triple helix involved in the crosslinking region of the type IV collagens. Studies with pepsin fragments indicate additional, usually weak determinants in the major triple helix, which show cross-reactivity in triple helical and unfolded forms. This may be due to frequent interruptions of the triple helix by short noncollagenous segments that endow the type IV collagens with rather unique antigenic properties. Little is known about the immunochemistry of type V collagen except that some data indicate a major antibody response against helical determinants.

The structure and precise localization at the peptide level of individual antigenic determinants has not yet been established for type IV collagen. This will be a considerable task considering the complexity of the molecule and the still incomplete knowledge of its primary structure. Other methods, such as electron microscopy, may be helpful in this approach. The rotary shadowing technique (Chapter 1, Figure 1.1) visualizes triple helical molecules as thin, extended rods.

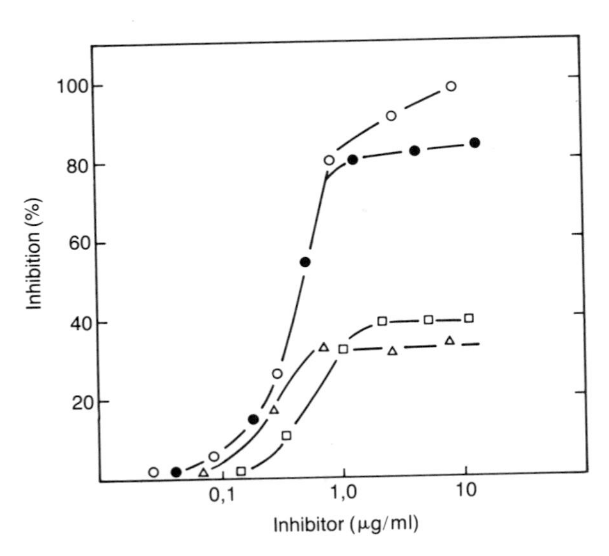

Figure 5.9. Analysis of the antigenic structure of type IV collagen by inhibition radioimmunoassay. The test system consisted of ^{125}I-labeled type IV collagen (structure illustrated in Figure 5.8) and rabbit antibodies against type IV collagen. Inhibitors of the reaction were type IV collagen *(open circles)*, pepsin-treated type IV collagen that had lost the domain NC1 *(squares)*, the isolated domain NC1 *(triangles)*, and a mixture (5 : 1, w/w) of the latter two inhibitors *(filled circles)*. Note that the individual fragments possess only partial inhibiting activities, which were additive as shown by the mixture. (*Source: Timpl, Wiedemann, et al. (1981) Eur J Biochem 120:203–211, with permission.)*

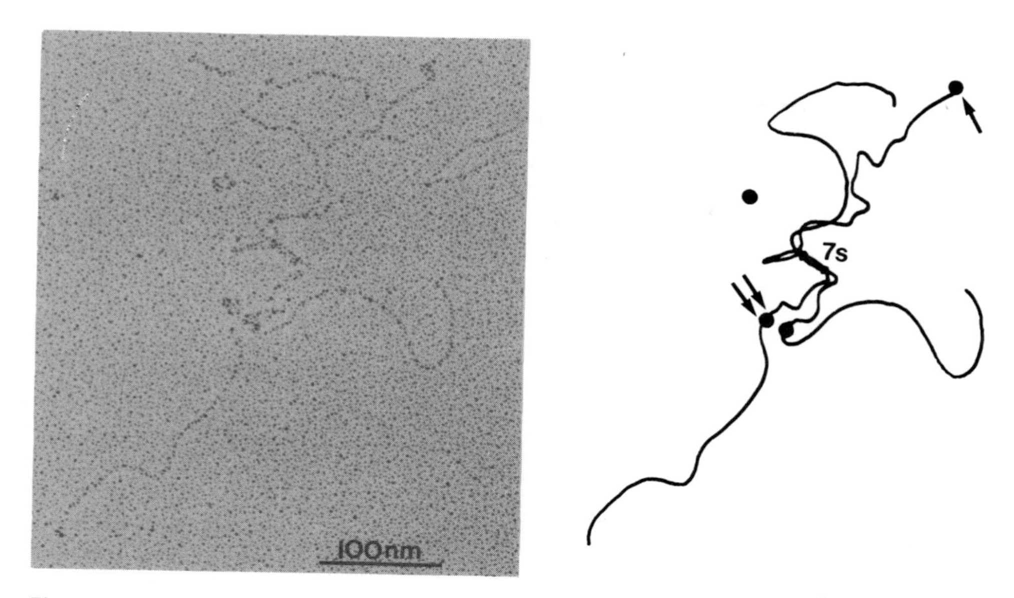

Figure 5.10. Electron microscopic visualization of a particular antigenic determinant detected by a monoclonal antibody on human type IV collagen using the rotary shadowing technique. *Left:* original electron micrograph. *Right:* interpretation of data. The antigen was solubilized by limited pepsin digestion and consists of four type IV collagen molecules connected to each other by the 7S domains. The NC1 regions have been lost by proteolysis. The monoclonal antibodies react primarily with a site located in the strands 60 nm from the 7S domain *(double arrow)*. Nonspecific interaction with the free end of the triple helix *(single arrow)* and free antibody (globule) are less frequently observed. (Courtesy of David W. Hollister and Hanna Wiedemann.)

After reaction with antibody, binding to particular sites can be demonstrated by the appearance of globular immunoglobulin molecules along the rods. Figure 5.10 shows an application of this method to a monoclonal antibody reacting with type IV collagen (Hollister et al. 1983).

5.4. Procollagens are Stronger Antigens

The precursor forms of the interstitial collagens are about 50% larger than the collagens, owing to the presence of precursor regions that have in large part globular structures, (*See* Chapter 2). It was therefore anticipated that the procollagens should possess additional antigenic determinants and presumably a stronger immunogenicity compared to the collagens derived from them. This was confirmed by the first immunologic studies with the procollagens (von der Mark et al. 1973; Timpl et al. 1973). Due to limitations in obtaining substantial amounts of pure antigenic material, the immunochemical knowledge is still restricted to type I procollagen obtained from cell and organ cultures and to types I and III pN collagens, which lack the pC peptides and can be isolated from pathologic or fetal tissues.

The precursor-specific structures possess unique antigenic determinants located in the noncollagenous portions of the pN and pC peptides (Figure 5.11). Antibodies against these determinants show a low level of crossreactivity with other procollagen peptides as well as with structures located in the collagenous

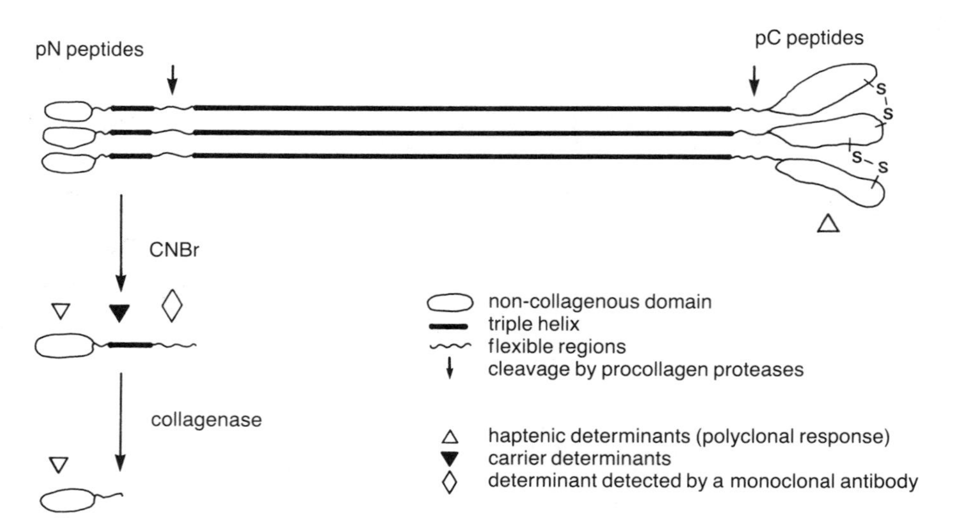

Figure 5.11. Schematic structure of an interstitial procollagen and identification of various segments involved in different immunologic activities of procollagen. Polyclonal antibodies produced in rabbits and mice react with a variety of conformational antigenic determinants located in noncollagenous portions of the pC and pN peptides. Carrier determinants, as identified in studies with inbred mice (Table 5.5), can be localized within or close to the triple helix of the pN peptide. The sequence of the antigenic determinant detected by a single monoclonal antibody is shown in Figure 5.7.

triple helix or in the nonhelical regions that remain with the collagen. Antigenicity is usually destroyed by reduction of inter- and/or intrachain disulfide bonds in the pN and pC peptide segments, indicating the conformational nature of the major antigenic determinants (Rohde et al. 1976). This has hampered their precise structural characterization. A minor antibody response can be detected against supposedly flexible regions within the pN peptides and some of these antigenic determinants (Rohde and Timpl 1979) are known with respect to their amino acid sequence (Figure 5.7). Triple helical sequences within the pN peptides lack substantial antigenicity. A more general interpretation of the immunochemistry of the procollagens still suffers from the lack of a comprehensive approach, including the use of animal species other than rabbit.

Characterization of monoclonal mouse antibodies against the pN peptides is still preliminary. A particular clone reacting with an antigenic determinant located adjacent to the cleavage site of procollagen N protease in type I procollagen has been accidentally identified (Figure 5.7). This would be an atypical reaction pattern for a polyclonal antibody response. This monoclonal antibody can block the proteolytic release of the peptide (Figure 5.12), either because the peptide bond to be cleaved is part of the antigenic determinant or by a more remote steric hindrance of enzyme action by the bulky antibody (Foellmer et al. 1983). Since the removal of the pN peptides is presumably associated with a variety of biological activities (Chapter 3), the importance of such specific reagents for experimental studies is obvious. In this context it should, however, be emphasized that production of monoclonal antibodies is dictated by the arbitrary selection of clones during and after cell fusion (*See* Table 5.3), and may require an enormous experimental investment before a reagent of a desired specificity can be obtained.

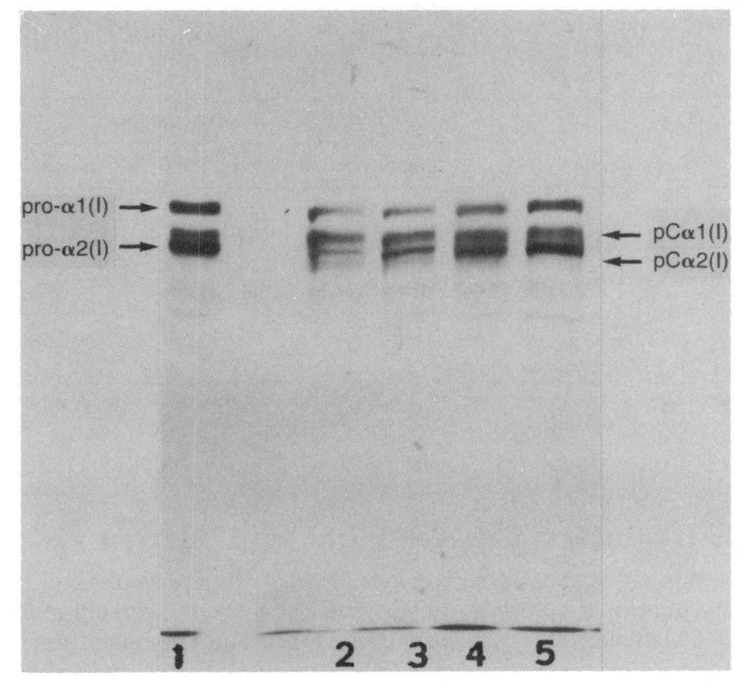

Figure 5.12. Blocking of procollagen conversion by a monoclonal antibody reacting with a determinant located close to the cleavage site of procollagen N protease. The substrate *(lane 1)* was metabolically labeled chick type I procollagen, which consists of intact pro α chains and some pC α chains (see Chapter 2). These chains were separated after reduction by electrophoresis and visualized by fluorography. Treatment with purified N protease converted most of the pro α chains to pC α chains *(lane 2)*. Preincubation of the substrate with increasing amounts of monoclonal antibody (molar ratios 1 : 1, 2 : 1, and 3 : 1, lanes 3 to 5, respectively) produced increasing inhibition of the conversion, which was complete at a ratio of 3 : 1. (Courtesy of Leena Tudermann.)

When used in inbred strains of mice, procollagen antigens also offer a way to study the immunologic carrier properties of collagenous structures (Table 5.5). A genetically determined low antibody response against the collagen triple helix can be converted to a high responder state after immunization with procollagen. Owing to the complexity of the immunogens used, it is not possible to establish the structural requirements for these effects. A better understanding has been achieved with the pN peptides, which, depending on the isolation procedure (Figure 5.11), either lack or possess the precursor-specific triple helical sequence. The latter peptide is a good immunogen in low responder strains, indicating that the triple helix possesses potent carrier determinants reacting with helper T cells that stimulate a B cell response against haptenic determinants located in the nontriple helical segment of the pN peptides (Nowack et al. 1977; Rohde et al. 1978). The data also indicate that the antibody response is regulated by suppressor T cells. Such regulatory events very likely contribute to the immune reaction in noninbred animals and account for the quantitative and qualitative variations in the antibody response patterns observed in different individuals and different species.

Table 5.5. Carrier Effects in the Low Responder Mouse Strain B10.D2 (H-2^d Haplotype) Augmenting the Antibody Response Against Type I Collagen or the Noncollagenous Segment of Procollagen

		Antibody reaction	
Immunogen	Test antigen	Hemagglutination ($-\log_2$)	^{125}I-Antigen bound (%)
Collagen	Collagen	2.3 ± 0.5	
Procollagen	Collagen	6.0 ± 2.8	
pN Peptide Col 1, collagenase treated	pN Peptide Col 1		0.3 ± 0.2
pN Peptide, intact	pN Peptide Col 1		86.0 ± 3.0

The intact pN peptide was isolated by cyanogen bromide cleavage and contained the noncollagenous and collagenous segments. Treatment with collagenase removed the collagenous portion (See Figure 5.11).
Data compiled from Nowack et al. (1975b) and Rohde et al. (1978).

5.5. Immunologic Distinction Between Different Collagens

One of the major aims of immunologic studies of the collagens has been the production of antibodies capable of distinguishing between closely related proteins. This is of considerable practical importance since collagenous proteins share many macromolecular properties that may prevent the application of biochemical methods, particularly in the analysis of small amounts of material in complex biological specimens. The efficient use of immunologic reagents is largely dependent on a precise characterization of antibody specificity and their potential crossreactions. Studies of tissues may require the purification of antibodies by affinity chromatography (immunoadsorption) to ensure specificity (see Figure 5.13).

Polyclonal antisera raised against helical and terminal determinants of the interstitial collagens usually possess a strong antibody component with an exclusive specificity for the immunogen. They often contain additional (minor) antibodies which crossreact with other types of collagens owing to the presence of shared antigenic determinants. The production of a type-specific antibody, i.e., for type I collagen, therefore usually requires crossabsorptions on types II and III collagen prior to the final purification on a type I collagen absorbent (Figure 5.13). Purified antibodies with a rather high specificity for types I, II, or III collagen have been prepared in many laboratories (reviewed by Timpl et al. 1977; Gay and Miller 1978; von der Mark 1980, 1981; Timpl 1982). Similar immunoadsorption procedures have been used for the preparation of specific antibodies against type I pC peptides and types I and III pN peptides. It is also possible to prepare antibodies specific for α1(I) and α2(I) chains or even for a single antigenic determinant located within the chains. A few monoclonal antibodies are available which have an exclusive specificity for a particular collagen type and in principle do not require purification. Since these antibodies are homogeneous proteins, it would be theoretically impossible to remove a preexisting crossreacting activity by immunoadsorption.

Antisera against types IV and V collagen normally show only negligible crossreaction with the interstitial collagens, as expected in view of the large structural

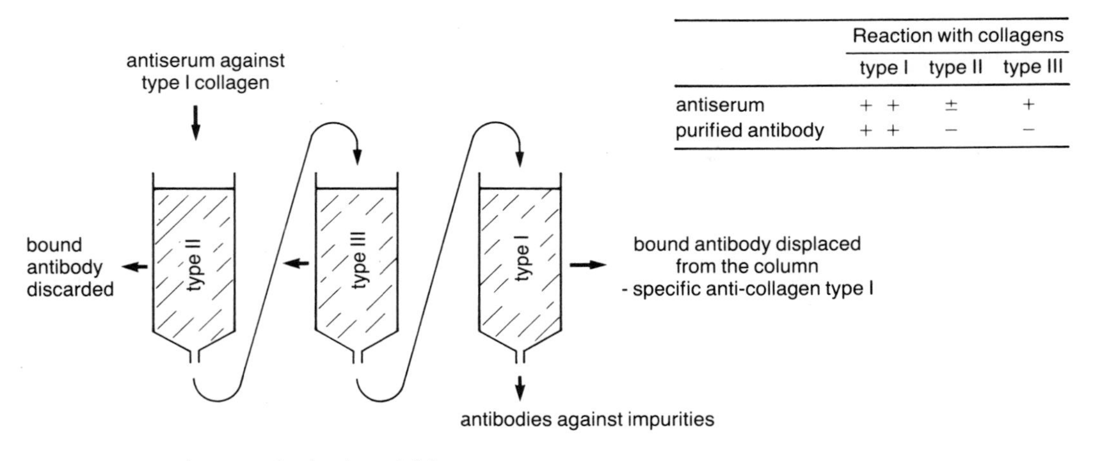

	Reaction with collagens		
	type I	type II	type III
antiserum	+ +	±	+
purified antibody	+ +	−	−

Figure 5.13. Schema of the preparation of purified antibodies against type I collagen by sequential immunoadsorption. Immunoadsorbents are usually prepared from a solubilized collagen and cyanogen bromide-activated agarose as insoluble support and retain the ability to react with stoichiometric amounts of antibodies. These antibodies are bound to columns of such adsorbents by passing a crude antiserum over them. Column-bound antibodies are then displaced by elution with acidic buffer or 3 *M* KSCN. Antibodies crossreacting with other collagen types are removed by prior passage over appropriate adsorbents. The purified antibodies then show a highly type-specific reaction (See insert) and are also low in antibodies against impurities, provided such impurities are a minor component of the column-bound collagen. (*Source: Modified Furthmayr and Timpl (1976) Int Rev Connect Tissue Res 7:61–99, with permission.)*

differences between the two groups of proteins. Purified antibodies with a strict specificity for types IV or V collagen have been obtained by appropriate immunoadsorption or are available in the form of monoclonal reagents. The panel of characterized antibodies also includes some specific for α1(IV), α2(IV), and α1(V) chains and others reacting exclusively with terminal domains of the type IV collagens.

Crossreactions between the same collagen type obtained from different mammalian species are usually much stronger than crossreactions between different types of collagen from one species. This is apparently due to a high degree of homology of the nontriplet terminal regions (Figure 5.7) as well as the triplet sequences (Chapter 2). These interspecies crossreactions are often only of a partial nature, pointing to the possibility of the preparation of species-specific antibodies by immunoadsorption. Such reagents, including monoclonal antibodies, have been used in studies aimed at the localization of collagen genes (Chapter 3) and will be of value in studies of hybrid cells. Antibodies with a broad interspecies crossreaction are, however, also often versatile reagents, particularly in studies of human material since the antigen may be difficult to obtain. Crossreactivity is in part determined by the nature of the antigenic determinants, with increases found from helical to terminal to central determinants. Antibodies against central determinants frequently react with α chains obtained from the animal species used for immunization (i.e., rabbit, chicken). This presumably results from lack of self-tolerance to structures present on denatured collagen.

Crossreactions between collagens of less related species (for example, between mammals and fish or invertebrates) are, as expected, weak or absent. There have also been attempts to produce antibodies against several synthetic polypeptides of the general formula $(Gly\text{-}X\text{-}Y)_n$. Even though such polymers may resemble certain structural features of collagens, they are usually poor immunogens and rarely show distinct crossreactions with authentic collagen (Maoz et al. 1973; Beard et al., 1977). Apparently, this approach is only of very limited value in characterizing the complex antigenic structure of the collagens.

Antisera against collagenous proteins normally fail to react with structurally unrelated proteins. Exceptions include the complement component C1q and acetylcholinesterase, which both possess a short collagenous triple helix (Chapter 1). These crossreactions are rather weak and still remain to be explained in terms of shared amino acid sequences.

5.6. Immunohistology of Collagens

Several but not all collagenous proteins appear in the tissue as distinct fibrils (Chapter 1), but their morphology does not necessarily reveal the chemical nature of the components. During the last ten years immunohistologic techniques have been used increasingly for the localization of the collagens in different tissues of the body (Timpl et al. 1977; Gay and Miller 1978; von der Mark 1980, 1981; Timpl 1982). The most commonly used light microscopic method is immunofluorescence (Wick et al. 1978), in which the reaction of antibodies with a particular tissue structure is visualized by a fluorescent label. The label can be either on the primary antibody (direct variant) or on a second antibody, which reacts with the primary antibody (indirect variant). The indirect variant is the more versatile method and is outlined in detail in Figure 5.14. Another convenient label is peroxidase, which is able to oxidize a small substrate to a large dark polymer deposited close to the antibody. Double staining is a variant of the indirect immunofluorescence method based on two different fluorescent markers, permitting the simultaneous detection of two antigens on the same tissue section (Figure 5.15).

Immunohistologic techniques have allowed a comprehensive analysis of collagens in a large variety of tissues (Chapter 2). Because of the qualitative nature of these methods, staining intensity does not necessarily correlate with the amount of antigen in a discrete place. Other factors contributing to the reaction are the quality and specificity of the primary anticollagen antibody and its ability to penetrate certain tissue regions. The accessibility of the collagen is particularly low in mineralized tissue and hyaline cartilage, which require chemical or enzymatic pretreatments to assure good antibody penetration. The brilliant staining of the bone section shown in Figures 5.15a, 5.15b was obtained after removal of hydroxyapatite with a chelating agent (EDTA) and after proteoglycan degradation by hyaluronidase. In addition, such studies need careful negative controls with unrelated antibodies to exclude nonspecific binding of immunoglobulin by tissue components. Immunofluorescence is also 10–100-fold less sensitive in detecting antibodies than, for example, radioimmunoassay.

Despite these disadvantages, the indirect immunofluorescent method is very useful, particularly for comparing the anatomic distribution of various types of collagenous proteins. Large differences in localization have been observed for

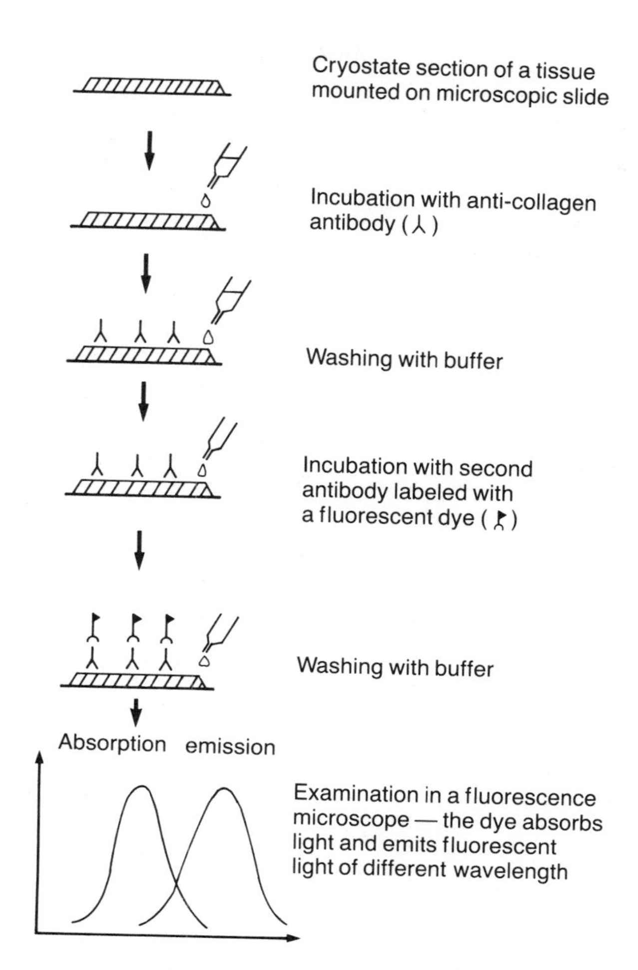

Figure 5.14. Outline of the indirect immunofluorescence procedure on a tissue section. The final reaction pattern is visualized by fluorescent light, which is characteristic for the particular dye used (fluorescein: green; rhodamine: red). Further details can be found in a review (Wick et al., 1978).

types I and II collagen, as illustrated in Figures 5.15a, 5.15b. Other collagens, such as types I and III, often show extensive codistribution. Owing to insufficient resolution by light microscopy (200–500 nm), this does not necessarily demonstrate that both proteins coexist in the same collagen fibrils. As expected, type IV collagens have been found mainly in basement membrane zones and rarely in interstitial connective tissue. Immunofluorescence also provides a rapid means for detecting compositional changes of connective tissue during development, even in small tissue samples such as biopsies that usually are of insufficient size for biochemical analyses. Studies on embryos have demonstrated the appearance of type IV collagen as early as the 32-cell stage (Leivo et al. 1980), and dramatic changes in the deposition and removal of discrete collagen types during further development (von der Mark 1980). Certain, normally more limited changes are also observed under pathologic conditions (*See* Chapter 11, Gay and Miller 1978).

Immunofluorescence is also useful for detecting collagens and procollagens in the intra- and extracellular compartments of cultured cells (Figures 5.15c, 5.15d). For this purpose, cells are grown on small slides and are then dried or treated with organic solvents to fracture the plasma membrane. Extracellular material often appears as fine fibrils. Intracellular collagens are usually visualized

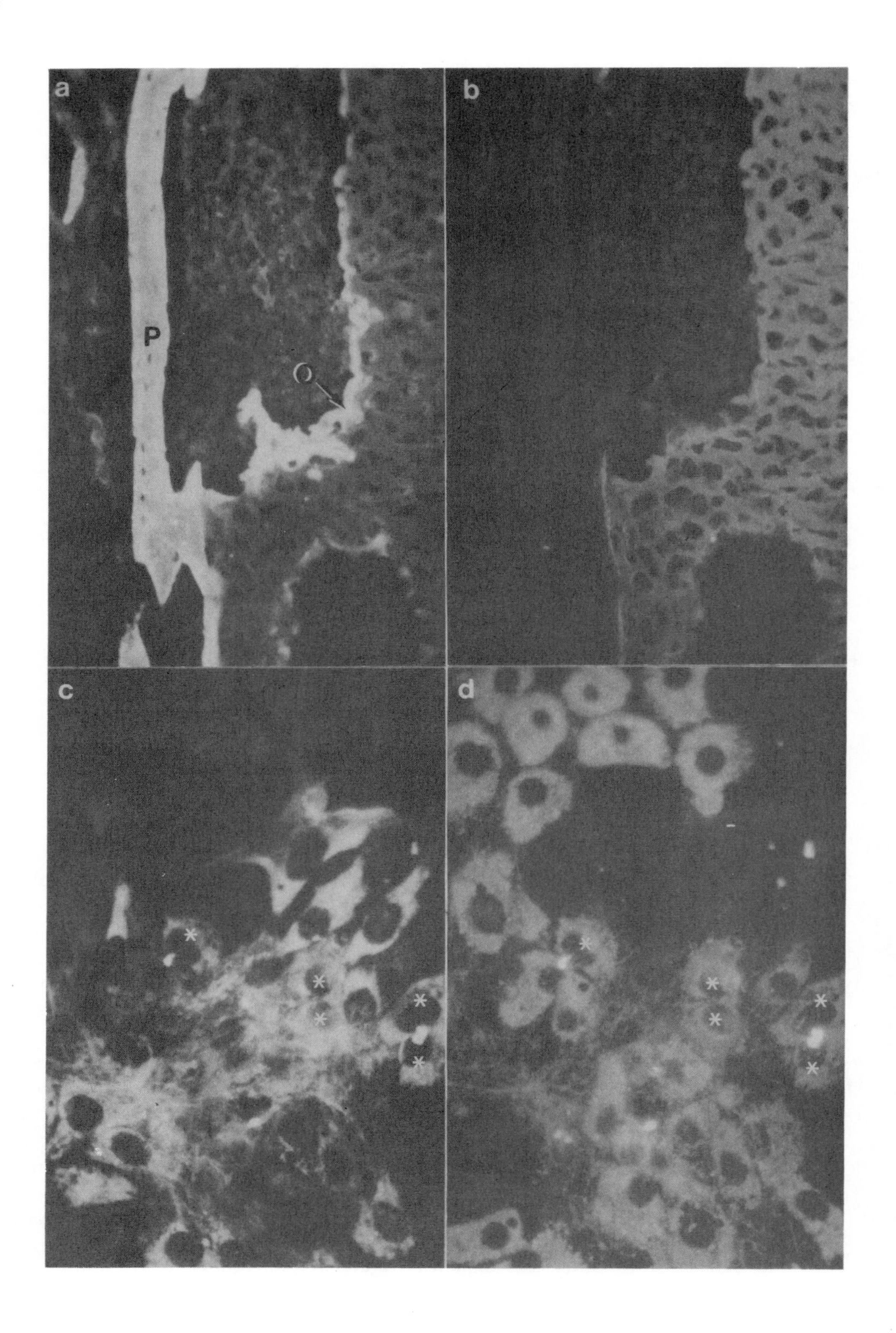
a
b
P
O
c
d
*
*
*
*
*
*
*
*
*
*
*

as granular deposits around the nucleus, suggesting a localization within the rough endoplasmatic reticulum (Gay et al. 1976). This pattern is commonly interpreted as indicating production of collagen by this individual cell. Presumably, due to the lack of penetration and/or a low antigen content, it is usually not possible to stain intracellular collagens of cells still present in intact tissues. This is a considerable handicap, as it would allow the identification of cellular sites of collagen production in situ. Hydrophilic embedding media allowing more sensitive immunofluorescence assays are currently being developed and may help to approach these goals.

An immunohistologic analysis of the collagens at the molecular level is feasible by immunoelectron microscopy (Figure 5.16). The procedures resemble in many technical details immunofluorescence and are also normally carried out in the form of an indirect variant. Ferritin, an iron-containing protein, is still the most frequently used label because it appears optically more dense than most other tissue structures. Peroxidase is another label that allows the study of antigens by both light and electron microscopy. Labeling with colloidal gold particles of uniform diameter (range, 5–15 nm) has been recently developed and can be used for double staining by varying the particle size. The bulky nature of the labels causes additional technical problems related to penetration of tissue and nonspecific association with matrix proteins.

Antibodies to types I and III collagen localize quite heavily on cross-striated collagen fibrils, as shown by the ferritin technique (Figure 5.16a). It has also been possible to identify the pN peptide of type I procollagen along thin fibrils, indicating that the peptide is present, at least in part, during fibril formation (Figure 5.16b). Other studies indicate that types I and III collagen are assembled in different fibrils (Fleischmajer et al. 1981). It is also possible to locate type IV collagen on the lamina densa of basement membranes and type V collagen in close vicinity to certain cells (Roll et al. 1980). Because of the considerable technical difficulties of the immunoelectron microscopic technique, there are still many questions that might be answered by the technique but have not, such as the association between collagen and fibronectin in situ. Penetration of cells presents another problem. However, it has been possible to demonstrate procollagens within the rough endoplasmatic reticulum and Golgi vesicles (Olsen

Figure 5.15. Visualization of collagen types I (a and c) and II (b and d) in a section of developing chick bone (a and b) and in cultured chondrocytes (c and d) by the double staining immunofluorescent technique. The sections were first exposed to purified rabbit antibodies against collagen type II followed by anti-rabbit immunoglobulin rhodamine conjugate. Then the same sections were exposed to purified guinea pig antibodies against collagen type I followed by anti-guinea pig immunoglobulin-fluorescein conjugate. Visualization of the differently colored fluorescent conjugates was accomplished by a set of appropriate filters in the microscope. In the black and white rendition shown here, the colors appear as light areas on a dark background. The bone section (a and b) contains collagen type I in the periostal bone (p) and newly synthesized osteoid (o). Collagen type II, however, is mainly found in the central regions consisting of hyaline cartilage. In the chondrocytes (c and d) collagens are mainly detected in the form of granular staining patterns. These resemble intracellular deposits most likely within the rough endoplasmatic reticulum demonstrating that a particular cell is able to produce a collagen. Most cells either synthesize collagen type I or type II. A few cells (marked by asterisks) stain for both collagens indicating that they are in a transient phase switching from type II to type I collagen production. Photographs were kindly supplied by Dr. Klaus von der Mark and reproduced with permission from von der Mark (1981).

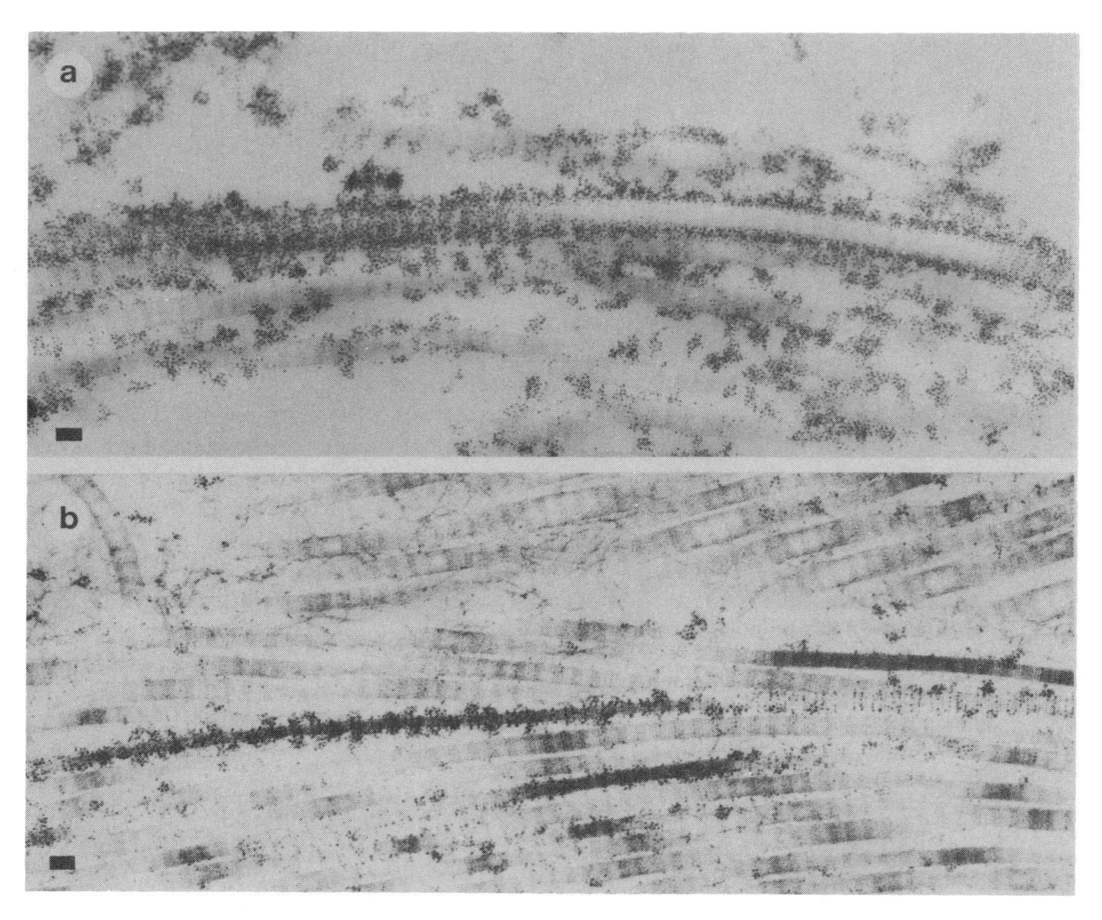

Figure 5.16. Ultrastructural localization of type I collagen and procollagen in a section of human skin by immunoelectron microscopy. A small piece of tissue was first exposed to purified antibody against either type I collagen (**a**) or against the pN peptide of type I (**b**), and then followed by anti-immunoglobulin antibodies conjugated to ferritin. The slice was then washed to remove nonspecifically bound reagents, cut into ultrathin sections, and examined under the electron microscope after poststaining with uranyl acetate to visualize cross-striated collagen fibrils. Type I collagen antigen is visualized by small, dense dots (mainly single ferritin particles) occurring in abundance along longitudinally or cross-sectioned collagen fibrils varying in diameter from 20 to 80 nm. Almost no reaction was observed in the nonfibrillar regions. Antibodies against the pN peptide react mainly with thin collagen fibrils (20–40 nm) in a periodic fashion associated with the D-stagger (67 nm). The bar denotes a length of 100 nm. (Courtesy Raul Fleischmajer.) (*Source: Fleischmajer, Timpl, et al. (1981) Proc Natl Acad Sci USA 78:7360–7364, with permission.*)

and Prockop 1974), defining an intracellular pathway similar to that of other secretory proteins.

5.7. Antibodies as Tools in Biosynthetic Studies

Studies of collagen synthesis are faced with the problems of a complex biological situation and small amounts of available material. Immunofluorescence is an extremely useful method for characterizing the biosynthetic repertoire of individual cells. This has an impact on the important question of whether the coordinated production of matrix proteins is mainly regulated within cells or requires cellular cooperation. Double staining studies have clearly shown that most cultured fibroblasts are engaged in the simultaneous production of types I and III collagen (Gay et al. 1976). Cultured chondrocytes may produce both types I and II collagen, by different cells as shown by immunofluorescence (von der Mark et al. 1977). A few cells appear to be double producers (Figures 5.15c, 5.15d), which forces the interpretation of a switch in the production of a particular collagen within single cells. Many more examples of the biosynthesis of collagenous and noncollagenous matrix proteins, including basement membrane components, have been analyzed during the past few years by immunologic methods (reviewed by Timpl and Risteli 1982; Timpl 1982). As already emphasized, these studies need to be extended to cells living within intact tissues.

Specific antibodies are also valuable tools for identifying metabolically labeled collagens and procollagens obtained from cell and organ cultures. This is conveniently carried out with proteins secreted into the culture medium. Immunoprecipitation of the medium proteins by an appropriate panel of antibodies followed by electrophoretic analysis of the precipitates is a common way to characterize newly synthesized proteins (Figure 5.17). The precipitation reaction basically resembles a radioimmunoassay and can therefore be used to determine the relative amounts of radioactivity incorporated into a particular protein during culture. This is usually done by incubating aliquots of the medium with increasing amounts of antibody until no further radioactivity is precipitated. Alterna-

Figure 5.17. Identification of types III and IV procollagen and of fibronectin (*F*) produced by a neoplastic muscle cell line (rhabdomyosarcoma) by immunoprecipitation. Medium proteins (*M*) were obtained from ^{14}C-proline labeled cultured cells and showed five major polypeptide bands by electrophoretic analysis. Different aliquots of the medium were then incubated with a variety of specific antibodies and radioactive material bound to the antibodies was removed by reaction with anti-immunoglobulin or staphylococcal protein A coupled to agarose. Polypeptides were then solubilized from these complexes under reducing conditions and examined by electrophoresis followed by fluorography. Antibodies against the pN peptide of type III collagen precipitated two chains identified as pro α1(III) and pN α1(III) (see Chapter 2). Antibodies against collagen type IV precipitated both the pro α1(IV) *(upper band)* and pro α2 *(IV)* chain *(lower band)*. Antibodies against fibronectin reacted only with a single chain. (Courtesy Antti Vaheri.) (*Source: Krieg, Timpl, et al. (1979) FEBS Letters 104:405–409, with permission.)*

tively, absolute amounts of newly synthesized proteins may be determined by inhibition radioimmunoassays, which must be sensitive enough to cope with the concentrations commonly observed in culture media (range, 1 ng–1 μg/ml). Specific immunologic assays have also been useful in analyzing the conversion of procollagens to collagens either by immunoprecipitation or radioimmunoassays. Other applications include antibody-adsorbents (prepared from purified antibodies and agarose), allowing the isolation of minor components from complex samples. This has aided in the analysis of discrete steps involved in the assembly of pro α1(III) chains into type III procollagen (Bächinger et al. 1981).

There are several other applications where the potential of specific antibodies can be exploited. These include the identification of polypeptides produced in cell-free translation systems of mRNA obtained either directly or by the recombinant DNA technology (Chapter 3). The antigenic materials obtained usually lack a distinct conformation, requiring that the immunologic assay should be designed for sequential determinants of the antigen. Many adherent cells possess the ability to interact with collagenous substrates directly or by the help of adhesive proteins (see Chapter 7). These properties are very likely restricted to certain regions of the collagen, which might be localized by specific antibodies. The aggregation of platelets induced by the fibrillar collagens can in fact be inhibited by polyclonal antibodies (Balleisen et al. 1979). It is obvious that monoclonal antibodies with a restricted specificity will be enormously helpful in studies aimed at mapping biological activities that depend on the quarternary or fibrillar structure of the collagens.

5.8. Quantitative Immunologic Assays for Biological and Clinical Studies

Radioimmunoassays and enzyme immunoassays in the form of an inhibition test allow the quantitative determination of antigens with a precision up to 5–15%. Depending on the affinity of the antibody, it is feasible to analyze the antigens in the pico- to nanogram range (Figure 5.18), even when present in very complex biological samples. Radioimmunoassays were developed for this particular purpose about 1960. The quality of radioimmunoassay data is also determined by the nature of the inhibition curve given by the biological sample, which should have a superimposable shape compared to the curve produced by the authentic inhibitor (Figure 5.18). Frequently, the unknown antigens show nonparallel inhibition curves. This can be due to a modified antigenic structure that reacts with different affinity or due to a number of other technical parameters (interference by high protein concentrations, protease damage, etc.). Thus, not every radioimmunoassay is suited for precise quantitative analysis, but may still be used for relative comparisons (Timpl and Risteli 1982).

Sensitive radioimmunoassays offer the unique possibility of measuring connective tissue proteins when they are present at very low concentrations, for example, in small amounts in serum. These circulating antigens presumably reflect the steady turnover of connective tissue elsewhere in the body and a change in their concentration may indicate a pathologic change in production or degradation. Radioimmunoassays for biosynthetic precursors are particularly useful and have been developed for pC and pN peptides of the procollagens

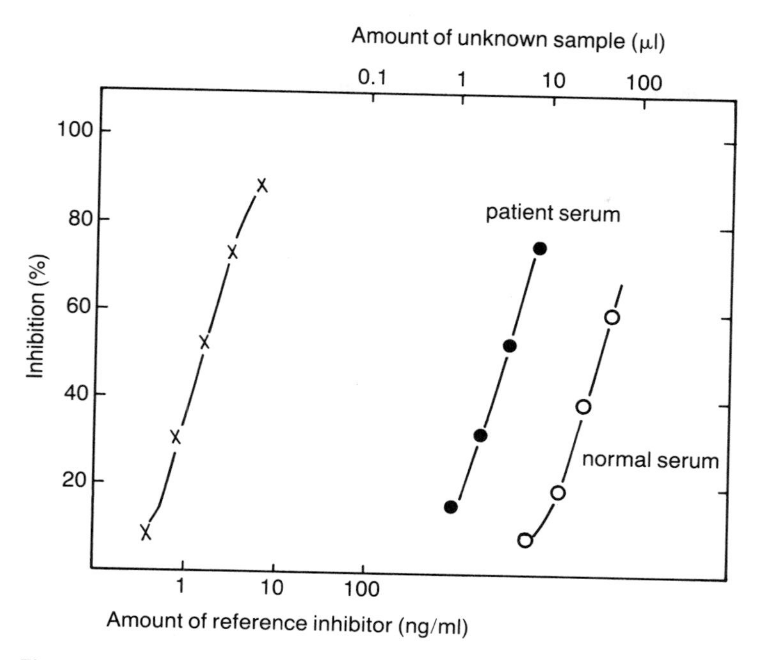

Figure 5.18. Application of a radioimmunoassay for the pN peptide of type III collagen to measure circulating forms of the peptide in human serum. The assay consisted of ^{125}I-labeled pN peptide, antiserum against type III procollagen, and unlabeled pN peptide as reference inhibitor. Inhibitors of the reaction were pN-peptide (x), normal human serum (○) and a patient serum (●). Note the high sensitivity of the assay and the parallel inhibition curves produced by the serum samples, allowing quantitative determination of the antigen in the nanogram range. Reproduced with permission from Rohde et al. (1979).

(Taubman et al. 1976; Rohde et al. 1979). A circulating form of the pN peptide of type III collagen can be detected in normal individuals and increases in patients with inflammatory or fibrotic reactions of the liver (Table 5.6). Other studies have demonstrated an increase of the serum concentration of pC peptide type I in various bone diseases (Chapter 11). Such assays have been developed rather recently and their full diagnostic and prognostic potential are not yet established.

Table 5.6. Serum Concentrations of the pN Peptide of Type III Procollagen in Patients with Liver Disease

Patients	No.	Mean ± SD (ng/ml)
Alcoholic liver disease	29	44 ± 34
Acute hepatitis	17	33 ± 17
Chronic active hepatitis	14	21 ± 13
Other liver diseases	14	13 ± 3
Rheumatoid arthritis	17	10 ± 3
Normal controls	24	7 ± 2

Data from Rohde et al. 1979.

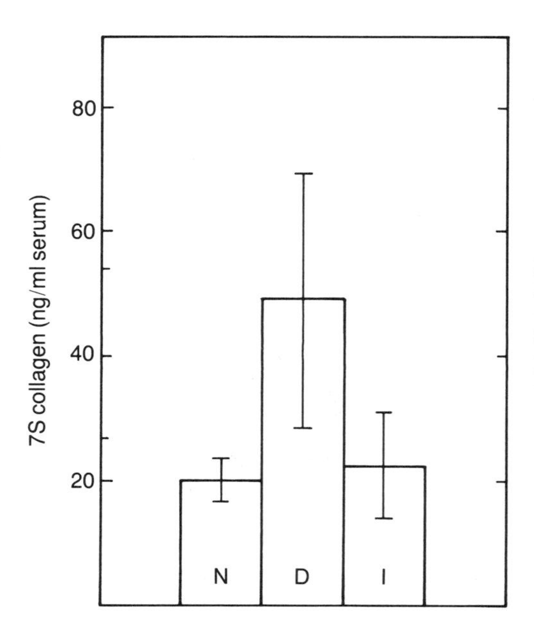

Figure 5.19. Serum concentrations of the 7S fragment from type IV collagen in normal rats (N), diabetic rats (D) and diabetic rats under insulin therapy (I) determined by radioimmunoassay. Reproduced with permission from Risteli et al. (1982).

Quantitative assays are of course useful for many other purposes and have been applied in studying collagen production by cultured cells (Rennard et al. 1980) and experimental animal models of disease. The latter approach can help in the exploration of possible clinical applications and in the study of pathogenesis. For example, thickening of basement membranes is a common and fatal late complication of the metabolic disease diabetes mellitus. Rats made diabetic by injection of streptozotocin rapidly develop glucose intolerance, while basement membrane changes in the kidney are first detected several months later. A radioimmunoassay for 7S collagen (the crosslinking domain of type IV collagen) (*See* Figure 5.8) has demonstrated an increase of the circulating antigen within two weeks, which could be prevented by insulin treatment (Figure 5.19). The choice of an assay for 7S collagen instead of for type IV collagen was determined by the low sensitivity of the latter assay and the observation of nonparallel inhibition curves produced by serum samples. Radioimmunoassays based on stable antigenic fragments will also be useful in the analysis of small tissue samples and, for example, allow an estimate of the amount of type IV collagen in kidneys by determining the 7S collagen content in a pepsin digest (Risteli et al. 1981b).

5.9. Autoimmunity of the Collagens

Loss of self-tolerance against self-proteins is an occasionally observed event and often the ascribed reason for a variety of autoimmune diseases. The initial steps in these processes are not well understood, but may include lack of efficient elimination of autoreactive immune cell clones or the generation of new autoantigens that the immune system has not previously encountered. Autoantibodies against matrix proteins including the collagens are found in several connective tissue diseases. Antibodies reacting particularly with denatured collagens (types I, II, and III) are frequently detected in patients suffering from rheumatoid arthritis (Andriopolous et al. 1976), an inflammatory chronic joint disease. One model postulates that such antibodies are responsible for the chronic progression of the disease (Steffen 1970). Patients with polychondritis, a disease characterized by destruction of cartilage, often possess autoantibodies reacting with native type II collagen (Foidart et al. 1978).

Other autoimmune reactions against collagen can be due to effector T cells (delayed type hypersensitivity) and, for example, have been observed in patients with progressive systemic sclerosis (Stuart et al. 1976) and rheumatoid arthritis. Such reactivities are usually detected by skin tests, by the binding of labeled antigens to lymphocytes, or by the release of soluble mediator substances inhibiting macrophage migration after stimulation by antigen. Experimental models of T cell reactivity against the collagens are known, for example, after immunization of guinea pigs, and allow a more precise study of the parameters of these reactions (Adelmann and Kirrane 1973; Senyk and Michaeli 1973). Thus, autoimmune responses against collagenous proteins can be due to stimulation of B cells, of regulatory T cells, and/or of effector T cells.

The detection of an autoimmune response does not necessarily imply that this reaction is of pathogenetic significance in the development of the disease. In fact, many data on the autoimmune nature of rheumatoid arthritis are still controversial, particularly with respect to the nature and role of anticollagen autoantibodies. A pathologic role, however, is well established for autoantibodies against basement membranes found in a few rare diseases. They include Goodpasture's syndrome (manifested by lung hemorrhages and renal failure) and bullous pemphigoid, a blistering skin disease. The autoantigens involved, however, are very likely noncollagenous matrix proteins rather than collagen. Antibodies raised in rabbits against mouse type IV collagen injected into mice bind to several basement membranes and immediately produce clinical symptoms (for example, respiratory distress) and histologic findings similar to those found in patients with Goodpasture's syndrome (Wick et al. 1982). This illustrates that pathogenetic potentials of antibodies are not necessarily connected to their autoimmune nature. More recent data, however, indicate that the Goodpasture antigen is a structure associated with some but not all NC1 globular domains of type IV collagen (*See* Figure 5.8).

The successful generation of autoimmune disease has recently been reported for rats and mice after injection of type II collagen. The animals produce a strong antibody response even against rat type II collagen and several weeks later develop joint lesions somewhat similar to those found in rheumatoid arthritis (Trentham et al. 1977; Stuart et al. 1982). The mechanism underlying this experimental autoimmune model is still controversial. It has been reported that transfer of both serum (antibodies) or lymphocytes (B and T cells) from sick into healthy animals produces arthritic symptoms. Studies with chick type II collagen in inbred strains of mice demonstrate control of antibody response by several alleles (Woolley et al. 1981) of the immune response region (Table 5.2). Arthritis, however, can only be induced in animals of the H-2^q haplotype. This observation is important in two ways. It demonstrates again that generation of an autoimmune disease is under more complex genetic control than the production of antibodies capable of reacting with autoantigens. The genetically restricted arthritogenic potential of type II collagen apparently resembles in some way the observations that several human diseases, including rheumatoid arthritis, juvenile diabetes, and ankylosing spondylitis (degenerative deformation of the spine), appear to be associated with certain antigens of the major histocompatibility complex, designated HLA in the human (Dorf et al. 1981; Klein 1982). This could indicate control by immune response genes and emphasizes the value of experimental studies in inbred animals.

The author is greatly indebted to Raul Fleischmajer, David W. Hollister, Leena Tudermann, Antti Vaheri, Klaus von der Mark, and Hanna Wiedemann for providing the photographs used here and for many helpful suggestions and comments. The excellent help of Hanna Kirchisner in preparing the manuscript is also appreciated.

References

Adelmann BC, Kirrane J: (1973) The structural basis of cell-mediated immunological reactions of collagen. The species specificity of the cutaneous delayed hypersensitivity reaction. Immunology 25:123–130.

Andriopolous NA, Mestecky J, Miller EJ, Bradley EL: (1976) Antibodies to native and denatured collagens in sera of patients with rheumatoid arthritis. Arthritis Rheum 19:613–617.

Atassi MZ, ed.: (1977) Immunochemistry of Proteins. Volumes I and II. New York: Plenum.

Bächinger HP, Fessler LI, Timpl R, Fessler JH: (1981) Chain assembly intermediate in the biosynthesis of type III procollagen in chick embryo blood vessels. J Biol Chem 256:13,193–13,199.

Balleisen L, Nowack H, Gay S, Timpl R: (1979) Inhibition of collagen-induced platelet aggregation by antibodies to distinct types of collagens. Biochem J 184:683–687.

Beard HK, Faulk WP, Conochie LB, Glynn LE: (1977) Some immunological aspects of collagen. Prog Allergy 22:45–106.

Becker U, Nowack H, Gay S, Timpl R: (1976) Production and specificity of antibodies against the aminoterminal region in type III collagen. Immunology 31:57–65.

Beil W, Timpl R, Furthmayr H: (1973) Conformation dependence of antigenic determinants on the collagen molecule. Immunology 24:13–24.

Crumpton MJ: (1974) Protein antigens: The molecular bases of antigenicity and immunogenicity. In: Sela M, ed. The Antigens. Volume II. New York: Academic, 1–78.

Dorf ME, ed.: (1981) The role of the major histocompatibility complex in immunobiology. New York: Garland STPM Press.

Eisen HN: (1981) Immunology. In: Davis BD, Dulbecco R, Eisen HN, Ginsberg W, eds. Microbiology. Hagerstown: Harper & Row, 290–547.

Fleischmajer R, Timpl R, Tuderman L, Raisher L, Wiestner M, Perlish JS, Graves PN: (1981) Ultrastructural identification of extension aminopropeptides of type I and III collagens in human skin. Proc Natl Acad Sci USA 78:7360–7364.

Foellmer HG, Kawahara K, Madri JA, Furthmayr H, Timpl R, Tuderman L: (1983) A monoclonal antibody specific for the aminoterminal cleavage site of procollagen type I. Eur J Biochem 134:183–189.

Foidart JM, Shigeto A, Martin GR, Zizic TM, Barnett EV, Lawley TJ, Katz SI: (1978) Antibodies to type II collagen in relapsing polychondritis. N Engl J Med 299:1203–1207.

Furthmayr H, ed.: (1982) Immunochemistry of the Extracellular Matrix. Volumes I and II. Boca Raton, Florida: CRC Press.

Furthmayr H, Stoltz M, Becker U, Beil W, Timpl R: (1972) Chicken antibodies to soluble rat collagen—II. Specificity of the reactions with individual polypeptide chains and cyanogen bromide peptides of collagen. Immunochemistry 9:789–798.

Furthmayr H, Timpl R: (1972) Structural requirements of antigenic determinants in the aminoterminal region of the rat collagen α2 chain. Biochem Biophys Res Commun 47:944–950.

Furthmayr H, Timpl R: (1976) Immunochemistry of collagens and procollagens. Int Rev Connect Tissue Res 7:61–99.

Galfre G, Milstein C: (1981) Preparation of monoclonal antibodies: Strategies and procedures. Methods Enzymol 73:3–46.

Gay S, Martin GR, Müller PK, Timpl R, Kühn K: (1976) Simultaneous synthesis of types I and III collagen by fibroblasts in culture. Proc Natl Acad Sci USA 73:4037–4040.

Gay S, Miller EJ: (1978) Collagen in the Physiology and Pathology of Connective Tissue. Stuttgart: Fischer.

Gottlieb PD: (1980) Immunoglobulin genes. Mol Immunol 17:1423–1435.

Hahn E, Timpl R: (1973) Involvement of more than a single polypeptide chain in the helical antigenic determinants of collagen. Eur J Immunol 3:442–446.

Hollister DW, Dieringer H, Wiedemann H, Timpl R, Sakai LY, Kühn K: (1983) Anti-human basement membrane collagen monoclonal antibody: em localization of a unique antigenic binding site. Clin Res 31:118a.

Kemp JD, Madri JA: (1981) The immune response to human type III and type V (AB_2) collagen: Antigenic determinants and genetic control in mice. Eur J Immunol 11:90–94.

Klein J: (1982) Immunology. The science of self-nonself discrimination. J. Wiley, New York.

Krieg T, Timpl R, Alitalo K, Kurkinen M, Vaheri A: (1979) Type III procollagen is the major collagenous component produced by a continuous rhabdomyosarcoma cell line. FEBS Letters 104:405–409.

Leivo I, Vaheri A, Timpl R, Wartiovaara J: (1980) Appearance and distribution of collagens and laminin in the early mouse embryo. Dev Biol 76:100–114.

Linsenmayer TF, Hendrix MJC, Little CD: (1979) Production and characterization of a monoclonal antibody to chicken type I collagen. Proc Natl Acad Sci USA 76:3703–3707.

Maoz A, Fuchs S, Sela M: (1973) Immune response to the collagen-like synthetic ordered polypeptide $(L\text{-}Pro\text{-}Gly\text{-}L\text{-}Pro)_n$. Biochemistry 12:4238–4245.

Michaeli D: (1977) Immunochemistry of collagen. In: Atassi MZ, ed. Immunochemistry of Proteins. Volume I. New York: Plenum, 371–399.

Michaeli D, Martin GR, Kettman J, Benjamini E, Leung DYK, Blatt BA: (1969) Localization of antigenic determinants in the polypeptide chains of collagen. Science 166:1522–1524.

Nowack H, Hahn E, Timpl R: (1975a) Specificity of the antibody response in inbred mice to bovine type I and type II collagen. Immunology 29:621–628.

Nowack H, Hahn E, David CS, Timpl R, Götze, D: (1975b) Immune response to calf collagen type I in mice: A combined control of Ir-1A and non H-2 linked genes. Immunogenetics 2:331–335.

Nowack H, Hahn E, Timpl R: (1976) Requirement for T cells in the antibody response of mice to calf skin collagen. Immunology 30:29–32.

Nowack H, Rohde H, Götze D, Timpl R: (1977) Genetic control and carrier and suppressor effects in the antibody response of mice to procollagen. Immunogenetics 4:117–125.

Olsen BR, Prockop DJ: (1974) Ferritin-conjugated antibodies used for labeling of organelles involved in the cellular synthesis and transport of procollagen. Proc Natl Acad Sci USA 71:2033–2037.

Rennard SI, Berg R, Martin GR, Foidart JM, Gehron Robey P: (1980) Enzyme-linked immunoassay (ELISA) for connective tissue components. Anal Biochem 104:205–214.

Risteli J, Schuppan D, Glanville RW, Timpl R: (1980) Immunochemical distinction between two different chains of type IV collagen. Biochem J 191:517–522.

Risteli J, Wick G, Timpl R: (1981a) Immunological characterization of the 7S domain of type IV collagens. Coll Rel Res 1:419–432.

Risteli J, Rohde H, Timpl R: (1981b) Sensitive radioimmunoassays for 7S collagen and laminin: Application to serum and tissue studies of basement membranes. Anal Biochem 113:372–378.

Risteli J, Draeger KE, Regitz G, Neubauer HP: (1982) Increase in circulating basement membrane proteins in diabetic rats and effects of insulin treatment. Diabetologia 23:266–269.

Rohde H, Becker U, Nowack H, Timpl R: (1976) Antigenic structure of the aminoterminal region in type I procollagen. Characterization of sequential and conformational determinants. Immunochemistry 13:967–974.

Rohde H, Nowack H, Timpl R: (1978) Localization of antigenic activity and immunogenic capacity in different conformational domains of procollagen peptide. Eur J Immunol 8:141–143.

Rohde H, Timpl R: (1979) Structure of antigenic determinants in the N-terminal region of dermatosparactic sheep procollagen type I. Biochem J 179:643–647.

Rohde H, Vargas L, Hahn E, Kalbfleisch H, Bruguera M, Timpl R: (1979) Radioimmunoassay for type III procollagen peptide and its application to human liver disease. Eur J Clin Invest 9:451–459.

Roll FJ, Madri JA, Albert J, Furthmayr H: (1980) Codistribution of collagen types IV and AB_2 in basement membranes and mesangium of the kidney. An immunoferritin study of ultrathin frozen sections. J Cell Biol 85:597–616.

Schmitt FO, Levine L, Drake MP, Rubin AL, Pfahl D, Davison PF: (1964) The antigenicity of tropocollagen. Proc Natl Acad Sci USA 51:493–497.

Sela M, Schechter B, Schechter M, Borek F: (1967) Antibodies to sequential and conformational antigenic determinants. Cold Spring Harbor Symp Quant Biol 32:537–545.

Senyk G, Michaeli D: (1973) Induction of cell-mediated immunity and tolerance to homologous collagen in guinea pigs: Demonstration of antigen reactive cells for a self-antigen. J Immunol 111:1381–1388.

Steffen C: (1970) Consideration of pathogenesis of rheumatoid arthritis as collagen autoimmunity. Z Immunitaetsforsch 139:219–227.

Stuart JM, Cremer MA, Townes AS, Kang AH: (1982) Type II collagen-induced arthritis in rats. Passive transfer with serum and evidence that IgG anticollagen antibodies can cause arthritis. J Exp Med 155:1–16.

Stuart JM, Postlethwaite AE, Kang AH: (1976) Evidence for cell-mediated immunity to collagen in progressive systemic sclerosis. J Lab Clin Med 88:601–607.

Taubman MB, Kammerman S, Goldberg B: (1976) Radioimmunoassay of procollagen in serum of patients with Paget's disease of bone. Proc Soc Exp Biol Med 152:284–287.

Timpl R: (1976) Immunological studies on collagen. In: Ramachandran GN, Reddi AH, eds. Biochemistry of Collagen. New York: Plenum, 319–375.

Timpl R: (1982) Antibodies to collagens and procollagens. Methods Enzymol 82:472–498.

Timpl R, Risteli L: (1982) Radioimmunoassays in studies of connective tissue proteins. In: Furthmayr H, ed. Immunochemistry of the Extracellular Matrix. Boca Raton, Florida: CRC Press pp. 199–235.

Timpl R, Wick G, Furthmayr H, Lapiere CM, Kühn K: (1973) Immunochemical properties of procollagen from dermatosparactic calves. Eur J Biochem 32:584–591.

Timpl R, Wick G, Gay S: (1977) Antibodies to distinct types of collagens and procollagens and their application in immunohistology. J Immunol Meth 18:165–182.

Timpl R, Wiedemann H, von Delden V, Furthmayr H, Kühn K: (1981) A network model for the organization of type IV collagen molecules in basement membranes. Eur J Biochem 120:203–211.

Trentham DE, Townes AS, Kang AH: (1977) Autoimmunity to type II collagen: An experimental model of arthritis. J Exp Med 146:857–868.

von der Mark K: (1980) Immunological studies on collagen type transition in chondrogenesis. Curr Top Dev Biol 14:199–225.

von der Mark K: (1981) Localization of collagen types in tissues. Int Rev Connect Tissue Res 9:265–324.

von der Mark K, Click EM, Bornstein P: (1973) The immunology of procollagen. I. Development of antibodies to determinants unique to the proα1 chain. Arch Biochem Biophys 156:356–364.

von der Mark K, Gauss V, von der Mark H, Müller P: (1977) Relationship between cell shape and type of collagen synthesized as chondrocytes lose their cartilage phenotype in culture. Nature 267:531–532.

Wick G, Baudner S, Herzog F: (1978) Immunofluorescence. 2nd Edition. Marburg/Lahn: Germany Med Verlagsgesellschaft.

Wick G, Müller PK, Timpl R: (1982) In vivo localization and pathological effects of passively transferred antibodies to type IV collagen and laminin in mice. Clin Immunol Immunopathol 23:656–665.

Wooley PH, Luthra HS, Stuart JM, David CS: (1981) Type II collagen-induced arthritis in mice. I. Major histocompatibility complex (I region) linkage and antibody correlates. J Exp Med 154:688–700.

6

Elastin

John M. Gosline and Joel Rosenbloom

6.1. Introduction

Within the connective tissues of the vertebrate body, rigid materials (bone) and materials of high tensile strength (tendon and ligaments) are prominent components. In addition to these components, there is an obvious need for pliant materials that can stretch, twist, and bend with normal movements, as well as serve certain specialized functions. Thus, the vertebrate body is encased in a tough, deformable skin, and large arteries and lung tissue possess the property of elastic recoil. The elastic properties of these tissues are due in large part to the presence of elastic fibers in the extracellular matrix. In this chapter we shall be concerned with the chemical and physical structure of these fibers and try to account for their elasticity in terms of molecular properties.

While the elastic fibers may quantitatively compose only a relatively small but important proportion of the total weight of some tissues such as the skin, in others, such as the large arteries and certain specialized ligaments, they are major components comprising more than 50% of the dry weight. With the light microscope, the elastic fibers are highly refractive and have been identified by characteristic staining reactions, chiefly the orcein stain and reaction with resorcin-fuchsin, tetraphenyl porphonate, or other multicomponent stains (*See* Figure 6.1). While these stains are relatively specific for elastic fibers, they do not provide unequivocal identification, and ultrastructural analysis may be necessary. Electron microscopic examination of elastic fibers has revealed that they are comprised of two morphologically distinguishable components (Figure 6.2). The amorphous component, so named because it usually does not possess any

From the Department of Zoology, University of British Columbia, Vancouver, British Columbia, Canada; and the Department of Anatomy and Histology, School of Dental Medicine, University of Pennsylvania, Philadelphia, Pennsylvania

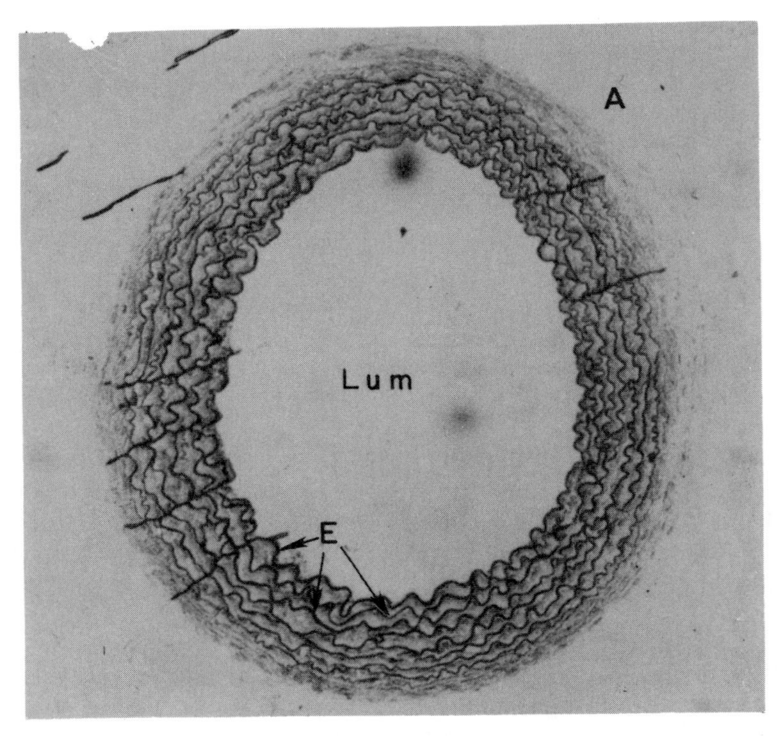
A
Lum
E

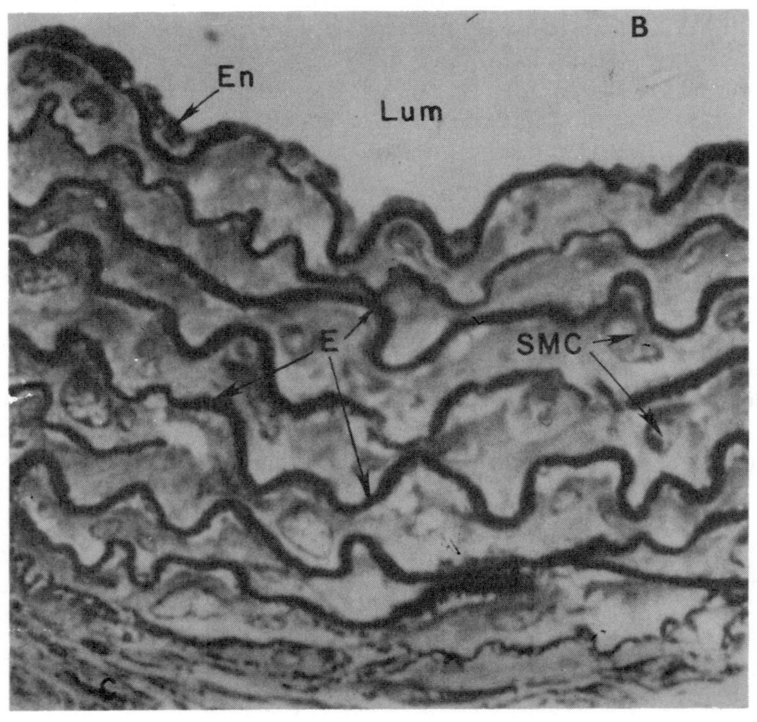
B
En
Lum
E
SMC

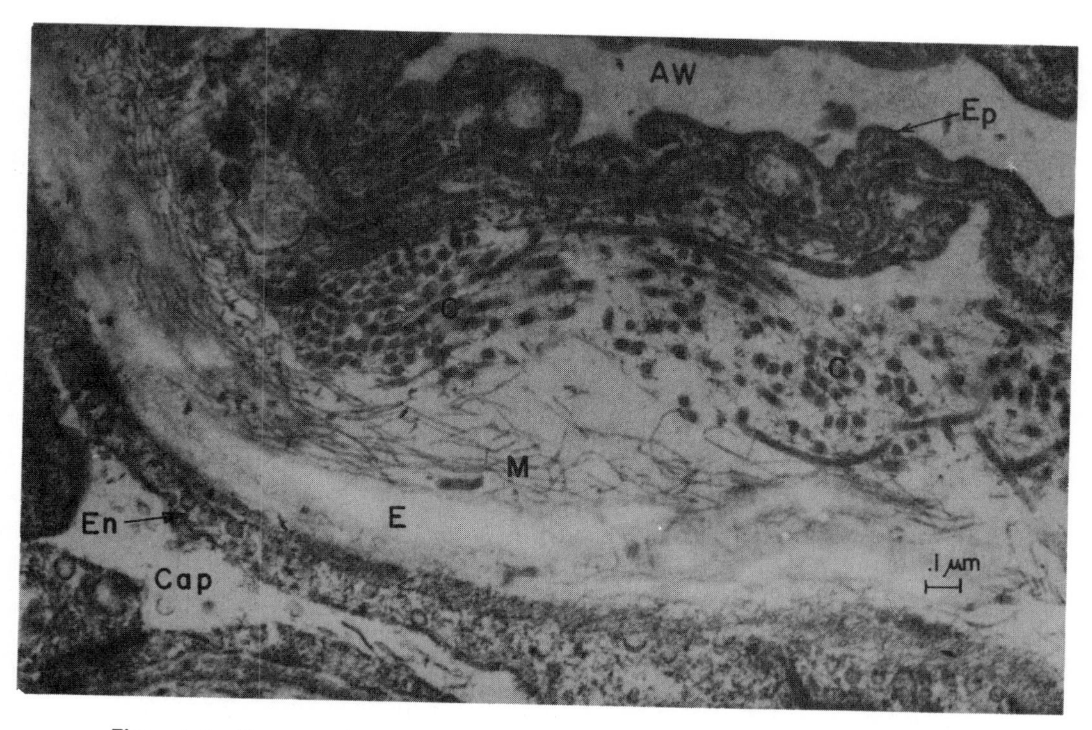

Figure 6.2. Electron micrograph of alveolar wall of dog lung. The section illustrates the various cellular and extracellular matrix components. Note the prominent number of microfibrils on the surface of the elastin fiber. Most of the collagen fibers are cut in cross-section. Stained with uranyl acetate and lead citrate. Magnification, ×60,000. *E* = elastin; *M* = microfibrils; *C* = collagen; *AW* = airway; *Cap* = capillary; *En* = endothelium; *Ep* = epithelium.

regular repeating structure or banding pattern, is quantitatively the largest portion, composing upwards of 90% of the mature fiber. The microfibrillar component, as its name implies, is composed of small fibrils roughly 10–12 nm in diameter, which are found primarily around the periphery of the amorphous component but to some extent are interspersed within it. It is now clear that these two components, in addition to being distinguishable morphologically, are chemically distinct. The name elastin has been reserved for the protein that makes up the amorphous portion of the elastic fiber and is responsible for the elastic properties. Early work on the histochemistry and chemical and physical properties of elastin has been reviewed by Partridge (1962). More recent work on structure and biosynthesis is contained in the proceedings of a conference held in 1976 (Sandberg et al. 1977). The involvement of elastin in blood vessel disease is discussed by Rucker and Tinker (1977).

Figure 6.1. Light microscopic sections of thoracic aorta of 12-day-old hamster. The tissue was fixed in formaldehyde-glutaraldehyde, embedded in Epon, and 1 μm sections were cut. The sections were then stained with a mixture of methylene blue, azure II, and basic fuchsin (Humphrey's stain). The bands of elastin, alternating with layers of cells, are well visualized. Magnification, (A) ×235; (B) ×1400. *Lum* = lumen; *E* = elastin; *C* = collagen; *SMC* = smooth muscle cell; *En* = endothelium.

6.2. Chemical Structure of the Elastic Fiber

6.2.1. Isolation of Elastin

The purification of elastin has relied largely on its great insolubility under all conditions in which there is no appreciable cleavage of peptide bonds. Early methods employed relatively nonspecific and fairly harsh extraction conditions. For example, tissues were extracted with 0.1 *M* NaOH at 98° for 30 to 60 minutes (Lansing et al. 1952) or repeatedly autoclaved in water until no further protein was solubilized and the insoluble residue was operationally taken to be elastin. These procedures are reasonably satisfactory for those tissues containing fairly large amounts of elastin, such as major arteries and the ligamentum nuchae of herbivores, in that they result in preparations with a characteristic amino acid composition and appearance with the electron microscope. In cases where the elastin content is small, such as the parenchyma of fetal lungs, these methods of isolation are less satisfactory and the insoluble fraction is contaminated by other proteins. In addition to these limitations, the vigorous extraction procedures result in elastin preparations in which there is significant cleavage of peptide bonds as well as destruction of the microfibrillar component.

Milder methods have been developed that yield preparations in which the elastin chains are largely intact and that yield in addition a protein fraction clearly derived from the microfibrillar component. The first of these procedures involves an extraction of the tissue with 5 *M* guanidine which removes the soluble collagen, glycoproteins, and proteoglycans, after which the remaining contaminating insoluble collagen is digested with highly purified clostridial collagenase (Ross and Bornstein 1969). Electron microscopic observation of the preparation at this stage reveals elastic fibers containing amorphous and microfibrillar components while no collagen or other proteins are visualized. A second extraction with 5 *M* guanidine containing β-mercaptoethanol then solubilizes the microfibrils from the periphery of the fibers, leaving an insoluble elastin residue. Amino acid analyses of the soluble fraction are rich in acidic and other hydrophilic amino acids as well as unusually rich in cystine/cysteine. The microfibrillar protein is, of course, denatured at this point and no further characterizations of such preparations in terms of molecular weight or other properties have been reported. The amino acid composition of insoluble ligamentum nuchae elastin prepared in this fashion is similar to that prepared by hot alkali extraction. However, this procedure may not be sufficient to remove all contaminating proteins and glycoprotein from tissues containing smaller proportions of elastin, and additional steps have been added in other purification schemes (Starcher and Galione 1976). These have included extraction with detergent solutions, digestion with other proteolytic enzymes such as trypsin against which elastin is resistant, and reaction with cyanogen bromide in formic acid, as it is generally agreed that most elastins contain no methionine (for exceptions, see Section 6.2.2). Depending on the starting tissue, one or the other of these modified procedures should be suitable for preparing acceptable elastin in the absence of alkali treatment.

6.2.2. Comparative Amino Acid Composition of Elastin

A large proportion of the early work on the analysis and characterization of elastin was carried out on samples isolated from ligamentum nuchae or the aortas of mammals. It was readily apparent from these analyses that, consistent with its unique physical properties, elastin is peculiar in its chemical composition. Approximately one third of the amino acids are glycine, 10–13% are proline, over 40% are other amino acids with hydrophobic side chains, and there are very small amounts of hydrophilic amino acids. The amino acid composition is sufficiently similar to that of collagen to suggest that it might be an unusual form of collagen. While structural work has clearly shown that this is not the case (Section 6.5.1), there is some evidence for a homologous relationship at the amino acid sequence level (Section 6.2.3). Elastin preparations from different tissues of the same species have very similar amino acid compositions, with the exception of that from elastic cartilage. Based upon this difference, it has been suggested that the elastin in elastic cartilage may be composed of a genetically distinct polypeptide chain (Keith et al. 1977). However, because of the particular difficulty in removing contaminating protein in cartilage, further work is necessary to substantiate this point.

A very extensive survey of the occurrence and amino acid composition of elastin throughout the animal kingdom was carried out by Sage and Gray (1979), who performed analyses on representative species from all vertebrate classes and a number of invertebrate phyla. Although they prepared and analyzed several tissues in a number of species, most of the assays were carried out on samples of aorta and related vessels. They utilized a relatively nondegradative procedure involving defatting of the tissue, extraction with a concentrated guanidine solution containing β-mercaptoethanol, and finally, autoclaving to purify the elastin. Where they could be compared, the resulting preparations had amino acid compositions very similar to those prepared using alkali extraction, cleavage

Table 6.1. Amino Acid Composition of Elastin from Aortas of Mammals and Birds[a]

	Pig		Dog		Human	Dolphin	Kangaroo	Chicken	Goose
Lys	5.2	6[b]	5.2	3[b]	4[b]	7.0	6.3	2[b]	4.8
His	1.0	0.7	1.9	0.7	0.5	2.0	1.0	0.7	1.6
Arg	7.9	6	9.1	8	9	ND	13	5	6.9
Asx	6.4	7	7.6	6	6	13	12	7	3.7
Thr	15	14	24	19	12	11	16	9	11
Ser	12	11	16	10	8	12	14	6	5.5
Glx	19	18	22	17	18	25	19	13	17
Hyp	8.7	9	11	12	10	2.5	10	26	11
Pro	113	110	107	111	131	124	115	131	136
Gly	313	326	314	352	295	311	301	338	338
Ala	244	233	249	232	233	232	235	179	190
Val	128	132	99	100	143	114	108	173	154
Ile	18	19	28	29	23	23	25	20	21
Leu	54	59	46	48	58	62	63	56	53
Tyr	19	15	29	28	23	20	31	13	15
Phe	33	30	25	21	22	38	26	19	24
Cys	<1		2.3			<1	<1		2.3
Met	<1		2.1			1.5	3.2		0.9
Ide	1.9	1.7	1.8	1.5	2.2	1.7	2.0	1.2	1.4
Des	1.3	2.3	1.3	1.9	2.8	1.2	1.4	1.4	1.0

Data from Sage and Gray (1979) except where indicated. ND = not determined.
[a]Values are expressed as residues/1000.
[b]Data from Starcher and Galione 1976.

with cyanogen bromide, or digestion with proteolytic enyzmes. Their results are very clean cut with respect to the distribution within the vertebrate phyla. Elastin is found in every vertebrate species examined except for those in the Agnatha. Although the amino acid compositions of all the elastins have similar general characteristics in that they contain desmosines (for crosslinks characteristic of elastin, see Section 6.2.4), are rich in glycine, proline, and hydrophobic amino acids, and poor in hydrophilic ones, there is significant variation among species within a class and considerable variation among classes. Representative compositions are given in Tables 6.1 and 6.2, which also contain some analyses of Starcher and Galione (1976), who used a different isolation method. In contrast to the previous results with mammalian and bird elastin in which histidine, methionine, and cysteine were found to be absent, these amino acids are found in the elastin of many reptiles, amphibians, and fish. Consideration of the changes in composition during evolution suggested that the earliest elastin, which arose after the divergence of the cyclostome and gnathosome lines, was similar in amino acid composition and crosslinking to that of mammalian elastin although there has been a progressive increase in hydrophobicity with time. This trend in hydrophobic residues may be related to a parallel change in systolic blood pressure, which also increases from a low of 30 mm of mercury in fish and amphibians to 120–150 mm in mammals and birds. Hydrophobic effects may play a major role in the mechanism of elastin fiber formation and rubber elasticity, as discussed in Section 6.5.5.

In no case did Sage and Gray find elastin in invertebrate species either by chemical or histologic methods, and these findings suggest that the earlier his-

Table 6.2. Amino Acid Compositions of Elastin from Aortas of Reptiles, Amphibians, and Fish[a]

	Turtle	Bullfrog	Salamander	Black grouper	Yellow fin tuna	Longnose gar	African lungfish	Hornshark
Lys	6.8	7.4	4.8	7.9	15	18	12	14
His	3.6	2.0	2.6	2.3	2.6	4.5	5.3	6.1
Arg	7.6	8.6	9.3	18	16	21	15	27
Asx	3.4	12	11	11	26	25	14	26
Thr	18	23	23	67	63	48	33	33
Ser	11	15	14	25	34	23	23	29
Glx	24	29	40	27	43	67	35	70
Hyp	16	5.1	7.5	5.9	5.5	7.8	12	5.4
Pro	130	104	121	101	101	117	112	109
Gly	319	402	373	439	391	297	351	266
Ala	184	154	121	120	103	104	105	128
Val	151	83	118	57	64	98	110	123
Ile	17	33	28	10	19	29	43	36
Leu	58	60	60	36	45	62	80	45
Tyr	34	42	44	34	37	49	27	49
Phe	13	15	14	34	28	22	18	21
Cys	<0.9	3.0	3.9	<0.8	<1	<0.5	<0.8	1.6
Met	2.0	3.6	2.4	3.2	7.1	5.9	3.5	8.5
Ide	1.5	1.0	0.9	0.6	0.3	1.0	0.7	1.5
Des	1.2	1.0	0.8	0.7	0.3	1.0	0.5	1.3

Data from Sage and Gray 1979.
[a]Values are expressed as residues/1000.

tologic identifications of elastin based upon characteristic staining reactions are probably in error. It is true that other rubber-like proteins such as resilin and abductin occur in the invertebrates, but they are not related to elastin.

6.2.3. Primary Structure

Because of crosslinking and consequent insolubility, chemical work on the structure of elastin has been greatly hampered and has progressed slowly. Although elastin can be partially solubilized by nonspecific hydrolysis with weak acids or alkali, the resulting mixture of peptides (called α-elastin when oxalic acid is used) is very heterogeneous and difficult to resolve because of the similarity in amino acid composition and chemical properties of the peptides. Surprisingly, as far back as the early 1900s, amino acid and sequence analyses were performed by Emil Fisher and co-workers and the dipeptides Ala-Leu and Leu-Ala were identified in bovine ligamentum nuchae elastin (Fischer and Abderhalden 1907). Much later, in the 1960s, other dipeptides, Val-Val, Leu-Val, and Val-Pro, were identified but no extensive sequences were determined. The major significant achievement of that time was the elucidation of the structure of the desmosine crosslinks by Thomas et al. (1963), and the demonstration that these were derived from lysine residues (Miller et al. 1964; Partridge et al. 1964), as described in Section 6.2.4. The availability of highly purified preparations of pancreatic elastase and other proteases enabled elastin to be digested in a way that resulted in a complex but somewhat less heterogeneous collection of peptides. Anwar (1977) purified peptides containing desmosines from such a mixture and used preparative Edman degradation to release peptides that were C-terminal to these crosslinks. Fractionation of the released peptides by ion exchange chromatography and limited sequence determination by analytical Edman degradation yielded sequences of which the following are representative:

Porcine Aorta	Bovine Aorta	Human Aorta
Tyr-Gly-Ala-Pro-Gly-Ala-Gly	Phe-Gly-Ala-Ala	Tyr-Gly-Ala-Ala
Ala-Pro-Gly-Gly-Gly-Gly-Ala	Ala-Gly-Tyr-Pro-Thr	Ala-Gly-Tyr-Pro-Thr
Leu-Gly-Ala-Ala	Leu-Gly-Ala-Gly-Gly-Ala	Phe-Gly-Ala-Gly

Thus, tyrosine is found frequently on the carboxyl side of crosslinks (often phenylalanine in bovine elastin). Other sequences often begin with alanine and sometimes with leucine or isoleucine. Two peptides containing desmosine were isolated from a subtilisin digest of ligamentum nuchae α-elastin by Foster et al. (1974). Sequence analysis of these peptides has suggested that each desmosine crosslinks two chains (Section 6.2.4), but because the two chains in the crosslinked peptides were sequenced simultaneously, the deduced sequences shown in Figure 6.3 must be viewed as provisional. However, these analyses as well as others point out the clustering of alanine residues surrounding the lysine residues.

A major advance in understanding elastin chemistry came about with the isolation from the tissues of copper deficient animals of a soluble protein that is clearly related to elastin. It had been observed in nutritional studies involving

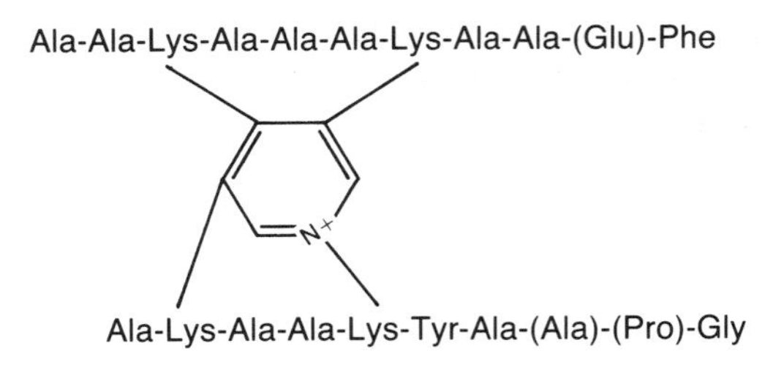

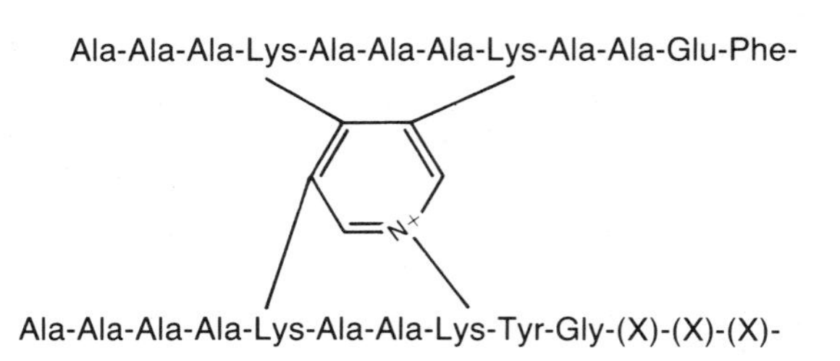

Figure 6.3. Amino acid sequences of desmosine crosslinked peptides isolated from bovine ligamentum nuchae. The peptides were isolated from a subtilisin digest of α-elastin, as discussed in the text. Lys connected to the pyridinium ring is used to denote the desmosine crosslinks. See Figure 6.8 for desmosine structure. The orientation of the pyridinium ring relative to the four lysines from which it is formed is hypothetical. Tentative assignments are given in parentheses and (*X*) signifies that the residue is unknown. Redrawn from Foster, Rubin, et al. (1974) J Biol Chem 249:6191–6196.

trace metals that animals on a copper-deficient diet suffered aneurysms of the aorta and other defects that could be attributed to a decreased content of the amorphous component in their elastic fibers. This led to the isolation by Smith et al. (1968) of a soluble polypeptide from the aorta of copper deficient pigs. The relationship of this protein, which has come to be called tropoelastin, to insoluble elastin was solidified by the work of Sandberg et al., (1969) who showed that the amino acid composition was very similar to that of insoluble elastin except for the absence of crosslinks and a corresponding increase in lysine residues. The total lysine content is 38 residues/mole in tropoelastin compared to about 6 residues/mole in mature elastin. Peptide maps obtained by pancreatic elastase digestion demonstrate that insoluble elastin and the soluble protein share a number of peptides. Although there are peptides that are unique to each, these differences are due to the retention in tropoelastin of many lysine residues. Some of the properties of tropoelastin are illustrated in Figure 6.4. Tropoelastin has been isolated from copper-deficient chicks and calves and from lathyritic animals. (Lathyrism is induced by feeding animals β-aminopropioni-

Figure 6.4. Properties of tropoelastin, the biosynthetic intermediate of the elastin fiber. (*A*) illustrates the similarity of the peptide maps resulting from the digestion of porcine tropoelastin isolated from copper-deficient pigs and insoluble aortic elastin with highly purified pancreatic elastase. Digestion was for 4 hours at 37°C at an enzyme–substrate ratio of 1 : 100. The resulting peptide mixture was then spotted on Whatman filter paper and resolved by high voltage electrophoresis in one direction and chromatography in the other. The positions of free valine and lysine are indicated. Data from Sandberg et al. 1979. (*B*) illustrates the unusual property of coacervation exhibited by solutions of tropoelastin as well as α-elastin. When the temperature of a cool solution of tropoelastin is raised to 25°C or greater, a phase separation occurs in which the tropoelastin molecules are found in an oily liquid. (*C*) illustrates the crosslinking of tropoelastin into the isoluble fiber mediated by the Cu^{++}-requiring enzyme, peptidyl lysine oxidase.

A Tropoelastin And Insoluble Elastin Yield Similar Peptide Maps Upon Digestion With Pancreatic Elastase

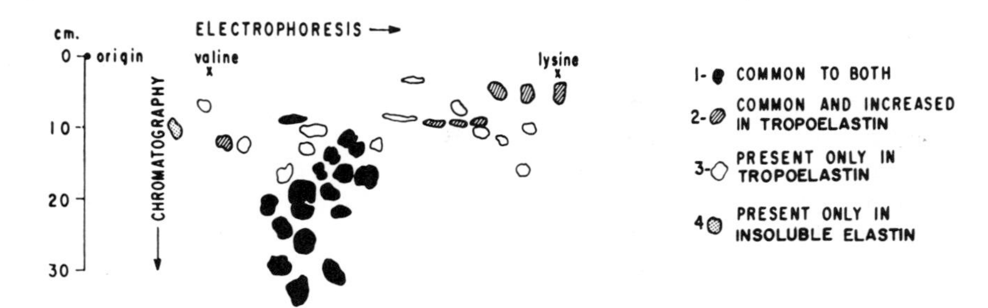

B Solutions Of Tropoelastin Undergo A Phase Separation Or Coacervate Upon Raising The Temperature From 4° to >25°

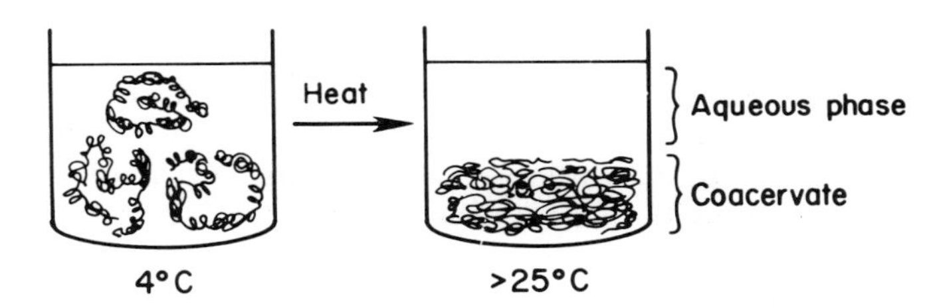

C 72,000 Dalton Tropoelastin Is Crosslinked Into An Insoluble Fiber By The Action Of Lysyl Oxidase

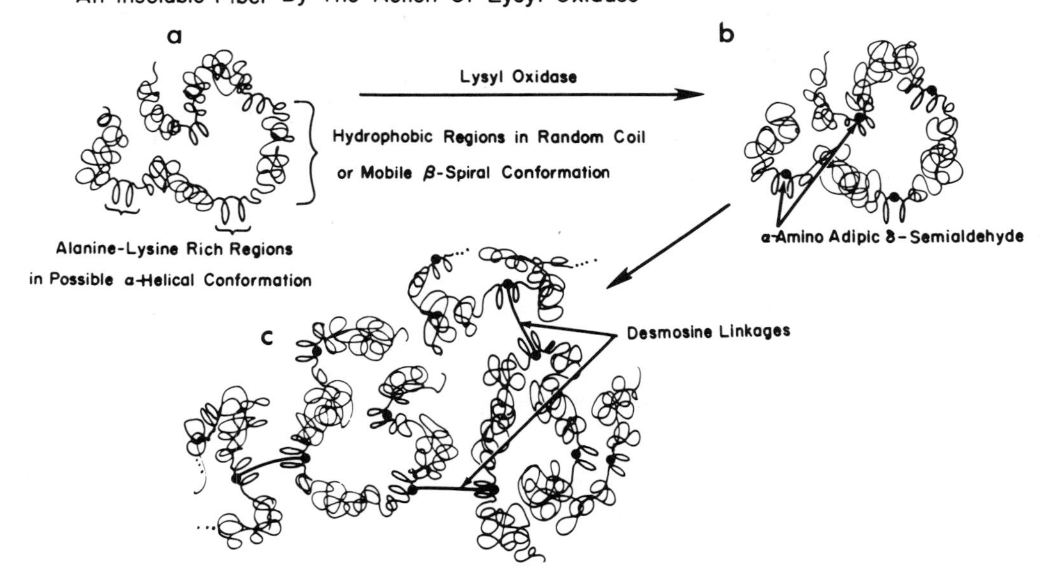

trile, an inhibitor of peptidyl lysine oxidase.) The tropoelastins from all species share a number of features in addition to their similarity in amino acid composition, including a molecular weight of 72,000–74,000, unusually high solubility in concentrated solutions of short chain alcohols, and a negative temperature coefficient of solubility in salt solutions. The last property leads to the phenomenon of coacervation or phase separation of tropoelastin from a cold solution when the temperature is raised to greater than 25°C (Figure 6.4).

Porcine tropoelastin has been digested with trypsin, the peptides fractionated by ion exchange chromatography, and a number of these purified peptides sequenced by automated Edman degradation (Sandberg et al. 1971; Foster et al. 1973). Representative sequences are illustrated in Tables 6.3 and 6.4. These analyses have corroborated and extended the limited sequence data on peptides from insoluble elastin. Some general conclusions can be drawn from these data as well as from other results in which tropoelastin was succinylated prior to tryptic digestion, which confines the cleavage to arginine residues from the intervening regions. Two classes of tryptic peptides are found: small ones rich in alanine (Table 6.3) are derived from regions that will form the crosslinks and other larger peptides rich in hydrophobic residues (Table 6.4). Two of the small peptides, Ala-Ala-Ala-Lys and Ala-Ala-Lys, are repeated six times per mole of tropoelastin, while several others are repeated twice. Undoubtedly these small peptides are spaced throughout the tropoelastin separated by the larger tryptic peptides. Within these larger peptides, smaller limited repeats may be discerned. Such repeating sequences have raised the possibility of a secondary helical structure peculiar to elastin (Section 6.5.1). The sequences of the larger porcine peptides also show that unlike collagen, even though glycine constitutes approximately one third of the residues in elastin, glycine is not found regularly as every third amino acid and there are numerous occurrences of glycine residues adjacent to one another. However, chick tropoelastin contains a stretch of 17 Gly-X-Y triplets, of which most are Gly-Val-Pro (Table 6.4). Presumably this region cannot form a collagen triple helix because proline is usually in the wrong position (Chapter 1, Section 1.2.2). However, this finding raises the possibility that the collagens and elastin have a distant homologous relationship at the primary structure level. Furthermore, the apparent marked difference between chick and porcine tropoelastin, also seen in the amino acid composition of many elastins (Tables 6.1, 6.2), suggests that interspecies homology is low. This may occur because there are presumably many ways to make a random coil, the

Table 6.3. Small Tryptic Peptides Rich in Alanine

Sequence	Moles/mole protein
-Ala-Ala-Ala-Lys-	6
-Ala-Ala-Lys-	6
-Ser-Ala-Lys	2
-Ala-Pro-Gly-Lys-	2
-Ala-Lys-	1
-Tyr-Gly-Ala-Lys-	2

Table 6.4. Larger Tryptic Peptide Rich in Hydrophobic Residues

Repeat	Sequence
Tripeptide	- Tyr - Val - Ala - Gly - Val - Pro - Gly - Val - Gly - Val - Pro - Gly - Val * Gly - Ile - Gly * Gly - Val - Pro * Gly - Val - Pro * Gly - Val - Pro * Gly - Val - Pro * Gly - Val - Pro * Gly - Val - Pro * Gly - Val - Pro * Gly - Val - Pro * Gly - Val - Pro * Gly - Val - Pro * Gly - Val - Pro * Gly - Val - Pro * Gly - Val - Val * Gly - Gly - Val * Gly - Pro - Val * Gly - Val - Ala * Ala - Ala - Ala - Ala - Ala - Ala - Ala - Ala -
Tetrapeptide	* Gly - Gly - Val - Pro * Gly - Ala - Val - Pro * Gly - Gly - Val - Pro * Gly - Gly - Val - Phe - Phe - Pro - Gly - Ala - Gly - Leu - Gly - Gly - Leu - Gly -
Pentapeptide	- Tyr - Gly - Ala - Ala - Gly - Gly - Leu - Val - Pro - Gly - Ala - Pro - Gly - Glu - Gly * Pro - Gly - Val - Gly - Val * Pro - Gly - Val - Gly - Val * Pro - Gly - Val - Gly - Val * Pro - Gly - Xxx - Gly - Val * Xxx - Gly - Val - Xxx - Val * Pro - Gly - Val - Xxx - Xxx *
Hexapeptide	- Ala - Ala - Gln - Phe - Gly - Leu - Phe * Pro - Gly - Ile - Gly - Val - Ala * Pro - Gly - Val - Gly - Val - Ala * Pro - Gly - Val - Xxx - Val - Ala * Pro - Gly - Val - Gly - Val - Xxx * Pro - Gly - Val - Gly - Val - Ala - Pro - Xxx - Ile -
None	- Gly - Gly - Val - Gly - Val - Gly - Gly - Ile - Pro - Thr - Phe - Gly - Val - Gly - Ala - Gly - Gly - Phe - Pro - Phe - Gly - Gly - Val - Gly - Val - Gly - Gly - Val - Val - Pro - Gly - Val - Gly - Leu - Pro - Gly - Gly - Val - Tyr - Pro - Gly -

*Similar or identical repeats.

major factor in the properties of elastin (Section 6.5.1), and selective constraints are less than for ordered proteins. There is a preferential occurrence of proline residues on the amino side of glycine, with proline only found rarely on the carboxyl side. The prolines on the amino side may be partially hydroxylated in elastin. In collagen, hydroxyproline has been shown to stabilize the triple helix (Chapter 1, Section 1.2.2) but no known function has been found for hydroxyproline in elastin. Porcine tropoelastin contains four arginines per mole and these are relatively evenly spaced throughout the molecule, as five large peptides of similar molecular weight have been recovered after tryptic digestion of succinylated tropoelastin.

6.2.4 Crosslinking

The crosslinking of elastin is mediated by the enzyme peptidyl lysine oxidase which oxidizes selective lysine residues in peptide linkage to α-amino adipic semialdehyde (trivial name, allysine). This is the same enzyme that appears to be involved in collagen crosslinking and its isolation, purification, assay, and properties are discussed in detail in Chapter 3. (For a review of peptidyl lysine oxidase, see Siegel 1980.) The activity of peptidyl lysine oxidase was first observed using a substrate of [6-^{3}H] lysine-labeled protein from chick aorta. The substrate was prepared by incubating the aortas with β-amino propionitrile to inhibit endogenous peptidyl lysine oxidase and consisted of an insoluble mixture containing largely labeled elastin with some collagen. When this substrate was incubated with an extract of embryonic chick bones, tritiated water was released and allysine was formed. In these experiments, it was not possible to determine what proportion of the total allysine was being formed in each of the two possible substrates, collagen and elastin. Subsequent experiments using more highly purified preparations of peptidyl lysine oxidase and very pure collagen or elastin have demonstrated that the enzyme functions better on insoluble forms of the substrates. Thus, preincubation of soluble collagen to allow aggregation, or of tropoelastin to form a coacervate, results in increased oxidation by a given amount of enzyme.

The above observation is consistent with the finding that a substantial amount of peptidyl lysine oxidase is associated with insoluble fibers in the connective tissue matrix. The enzyme can be extracted with high concentrations of urea with retention of enzymatic activity upon removal of the urea by dialysis. This enzyme–substrate complex can be visualized using fluorescent tagged antibody against lysyl oxidase. Purification of the enzyme and fractionation by either ion exchange or molecular sieve chromatography have demonstrated a number of forms of the enzyme differing in charge and having molecular weights ranging from 3×10^4 to greater than 10^6. The higher molecular weight forms may be composed of similar subunits, but the significance of this heterogeneity is not clear because each of the components exhibits similar enzymatic activity. The finding of enzyme associated with matrix fibers substantiates the idea that it is secreted and acts in the extracellular space. When fibroblasts are grown in culture, peptidyl lysine oxidase is found largely in the incubation medium and comparatively little is associated with the cell layer, again suggesting that it is secreted and acts extracellularly.

Crosslink formation in elastin follows the same course as in the collagens (Chapter 2, Section 2.3.3) with three major exceptions: (1) Hydroxylysine is not involved; there is none in elastin. (2) Histidine is not involved; there is little or none in elastin. (3) The final products of the series of aldol and aldimine condensations—desmosine and isodesmosine, which are absent from the collagens—are known. Thus, the picture is relatively much simpler for elastin.

There are two difunctional crosslinks: dehydrolysinonorleucine (deLNL), formed from one residue of allysine and one of lysine; and allysine aldol (AA), formed from two residues of allysine. Their structures are shown at the top of page 203. Also deLNL occurs as the reduced, and thus stabilized, secondary amine, lysinonorleucine. Allysine aldol, shown as the dehydrated aldol, may be hydrated in vivo. A trifunctional crosslink, dehydromerodesmosine (deMD, shown), may also be present in the reduced form in vivo. Desmosine and isodesmosine, whose structures are shown in Figure 6.5, are tetrafunctional but the evidence discussed earlier (Section 6.2.3) suggests that they normally join only two chains, with two nearby lysine residues from each chain contributing to their formation (Figure 6.3). The desmosines could be formed from AA and deLNL as illustrated in Figure 6.6. Alternatively, they could form by several routes by the repeated addition of single residues of lysine or allysine; deMD could be an intermediate. Overall, desmosine formation requires oxidation (loss of two protons). Perhaps there is a balance in the reduction of deLNL and deMD.

$$P{-}CH_2{-}CH_2{-}CH_2{-}CH{=}N{-}CH_2{-}CH_2{-}CH_2{-}CH_2{-}P'$$

DEHYDROLYSINONORLEUCINE

$$P{-}CH_2{-}CH_2{-}\underset{\displaystyle CHO}{\underset{|}{C}}{=}CH{-}CH_2{-}CH_2{-}CH_2{-}P'$$

ALLYSINE ALDOL

$$P{-}CH_2{-}CH_2{-}\underset{\displaystyle N{-}CH_2{-}CH_2{-}CH_2{-}CH_2{-}P''}{\underset{|}{\underset{\displaystyle CH}{\underset{|}{C}}}}{=}CH{-}CH_2{-}CH_2{-}CH_2{-}P'$$

DEHYDROMERODESMOSINE

Insofar as is known, all of these reactions are spontaneous. They presumably occur essentially randomly between residues in the highly mobile elastin chains (Section 6.5.1). Early models in which elastin is ordered and assembles in an ordered manner remain to be proven. Table 6.5 shows an approximate balance of the various crosslinks and crosslink precursors in mature elastin. About 10% of the lysine residues found in tropoelastin remain unaccounted for in elastin, either because of losses during analysis or the presence of other unidentified derivatives. The net result of the crosslinking is a highly insoluble polymer in which some type of interchain link occurs frequently along the original tropo-

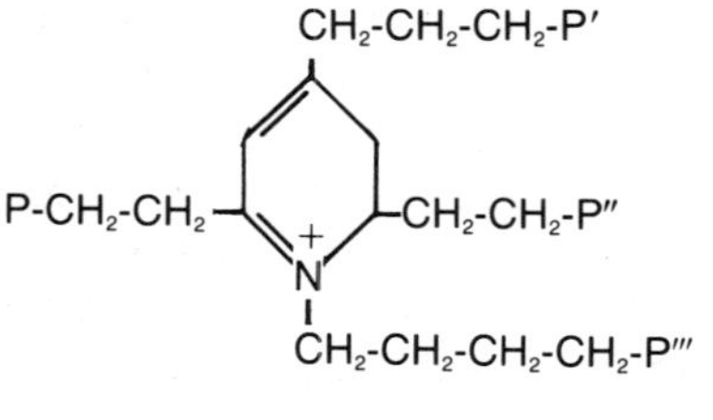

CH_2-CH_2-CH_2-P′

P-CH_2-CH_2 — [pyridinium ring, N^+] — CH_2-CH_2-P″

CH_2-CH_2-CH_2-CH_2-P‴

DESMOSINE

Figure 6.5. The structures of desmosine and isodesmosine, the final crosslink products in elastin.

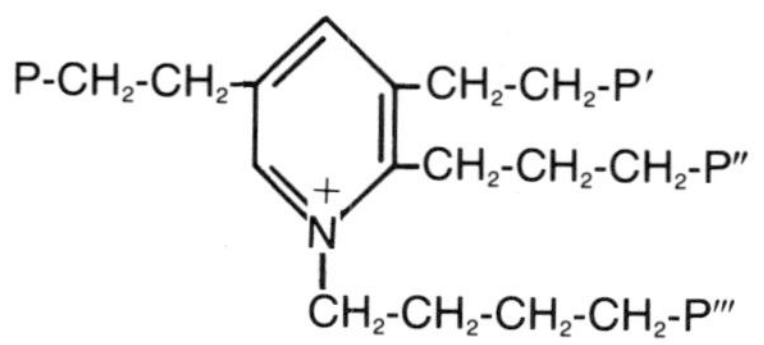

P-CH_2-CH_2 — [pyridinium ring, N^+] — CH_2-CH_2-P′

CH_2-CH_2-CH_2-P″

CH_2-CH_2-CH_2-CH_2-P‴

ISODESMOSINE

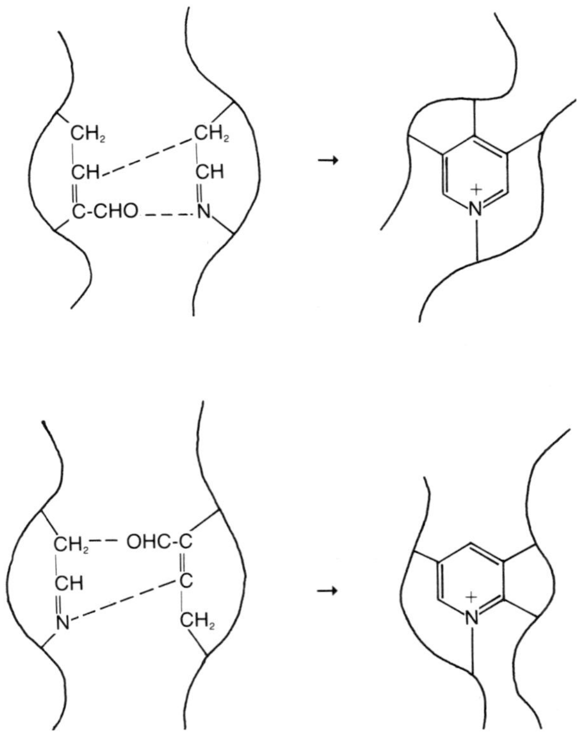

Figure 6.6. Schematic illustration of the possible formation of desmosine and isodesmosine via the reaction of intrachain allysine aldol with intrachain dehydrolysinonorleucine. Other pathways are equally possible.

elastin chains. The frequency can be calculated from the crosslink content. The average distance in amino acid residues between lysine-derived crosslink points is: $2M_r/R_wX$, where M_r is the molecular weight of tropoelastin (72,000), R_w is the average residue weight (85), X is the sum of crosslink points (26, Table 6.5) and the factor 2 arises from the fact each crosslink probably involves two nearby lysine residues functioning as a single point. The value of 65 obtained agrees well with the value of 70 calculated from the mechanochemical properties of elastin (Section 6.5.6).

Table 6.5. Crosslink and Crosslink Precursors in Mature Elastin

Crosslink	Lysine equivalents[a]
Desmosine + Isodesmosine	15
Merodesmosine	2
Lysinonorleucine	3
Allysine aldol	6
Allysine	2
Lysine	6
Total	34

[a]Residues/mol tropoelastin.

6.2.5. The Microfibrillar Component

The above discussion focused on elastin itself, but at least in higher vertebrate tissues with the possible exception of elastic cartilage, the elastic fiber contains microfibrils that must be presumed to have an important functional role. This microfibrillar component consists of 10–12 nm microfibrils located largely on the surface of mature fibers and to a smaller extent enmeshed and scattered throughout the amorphous component (Figure 6.2). Definition of the molecular composition of the microfibrillar component has progressed slowly, largely because of difficulties in solubilizing the microfibrils in any solvents other than powerful denaturants or the classic Lansing method (Lansing et al. 1952). The Lansing method consists of preparing the amorphous component by extraction of tissues with hot 0.1 *M* sodium hydroxide, which hydrolyzes the microfibrillar protein. Milder isolation methods using sequential extraction with a concentrated guanidine solution, digestion with purified collagenase, followed by guanidine extraction under reducing conditions have permitted isolation of a protein fraction with an amino acid composition distinct from that of the amorphous component (Ross and Bornstein 1969). It contains large amounts of acidic amino acids as well as serine and threonine. An unusual feature of this preparation is the large amounts of cystine/cysteine, which probably explains the high degree of insolubility of the microfibrils in the absence of reducing agents. Incubation of the microfibrillar protein with trypsin or chymotrypsin releases soluble peptides while these enzymes do not solubilize the amorphous component to any significant extent.

Though the extraction procedure described above differentiates the amorphous and microfibrillar components, the microfibrillar polypeptide fraction is poly-disperse and no satisfactory characterization has been reported. Another view of the chemical nature of the protein has come through use of cultured fibroblasts from bovine ligamentum nuchae. Electron microscopic observations of developing bovine fetuses have demonstrated that the microfibrillar component appears in the extracellular matrix prior to amorphous elastin, which becomes prominent after 180 days' gestation (Greenlee et al. 1966). Cultured fibroblasts isolated from 45 to 135-day-old fetuses synthesize microfibrils resembling those in the intact tissue (Sear et al. 1981). Isolation of protein from the medium by immunoprecipitation with antibody against a microfibrillar protein preparation revealed two proteins. One is partly collagenous in nature, as shown by sensitivity to bacterial collagenase and incorporation of labeled proline into hydroxyproline. While these results are interesting, the conclusion that these proteins are indeed related to the microfibrillar component depends critically on the monospecificity of the antibody, a technically difficult property to prove. Results such as these must be confirmed by independent methods.

6.3. Biosynthesis and Degradation of Elastin

6.3.1. Biosynthesis

The isolation and characterization of tropoelastin from copper-deficient animals suggested that this 72,000 dalton polypeptide is a soluble intermediate in the biosynthesis of insoluble elastin. However, the possibility remained that the occurrence of tropoelastin is the result of the abnormal nutritional status of the animal and that tropoelastin is an artifact, albeit a useful and informative one. To delineate the status of tropoelastin in the biosynthetic pathway, other types of experiments have been performed. (Figure 6.7 illustrates the putative biosynthetic pathway.) Tissues and cells that rapidly synthesize elastin were incubated with radioactive amino acids and the soluble labeled proteins extracted and characterized by a number of techniques (Murphy et al. 1972; Uitto et al. 1976). Tropoelastin was identified by polyacrylamide gel electrophoresis in sodium dodecyl sulfate, by its solubility in aqueous alcohol solutions, by its labeling pattern with various amino acids including glycine, proline, alanine and valine, and by its resistance to cyanogen bromide degradation and immunoprecipitation with affinity-purified antibody to insoluble elastin. It was found that approximately 20 minutes are required to synthesize the tropoelastin molecule and secrete it into the extracellular matrix. Unlike the case of procollagen, inhibition of peptidyl proline hydroxylation does not inhibit the rate of tropoelastin secretion. Also, in contrast to procollagen, the incorporation of cis-hydroxyproline in place of proline does not alter secretion. However, agents that depolymerize

Figure 6.7. Schematic illustration of the elastin biosynthetic pathway. The intranuclear reactions of mRNA transcription and processing have not been specifically studied in the case of elastin and are presumed to be similar to that found in other eukaryotic mRNAs. The biosynthetic intermediate, tropoelastin, has been shown to contain a signal sequence by in vitro translation of elastin mRNA and to contain hydroxyproline from sequence analysis. The postulate that tropoelastin is secreted in membrane-bound vesicles is based largely on ultrastructural studies using elastin-specific antibody or relatively specific chemical stains, and needs to be confirmed by biochemical means. There is strong evidence that peptidyl lysine oxidase acts extracellularly to crosslink the tropoelastin as discussed in the text.

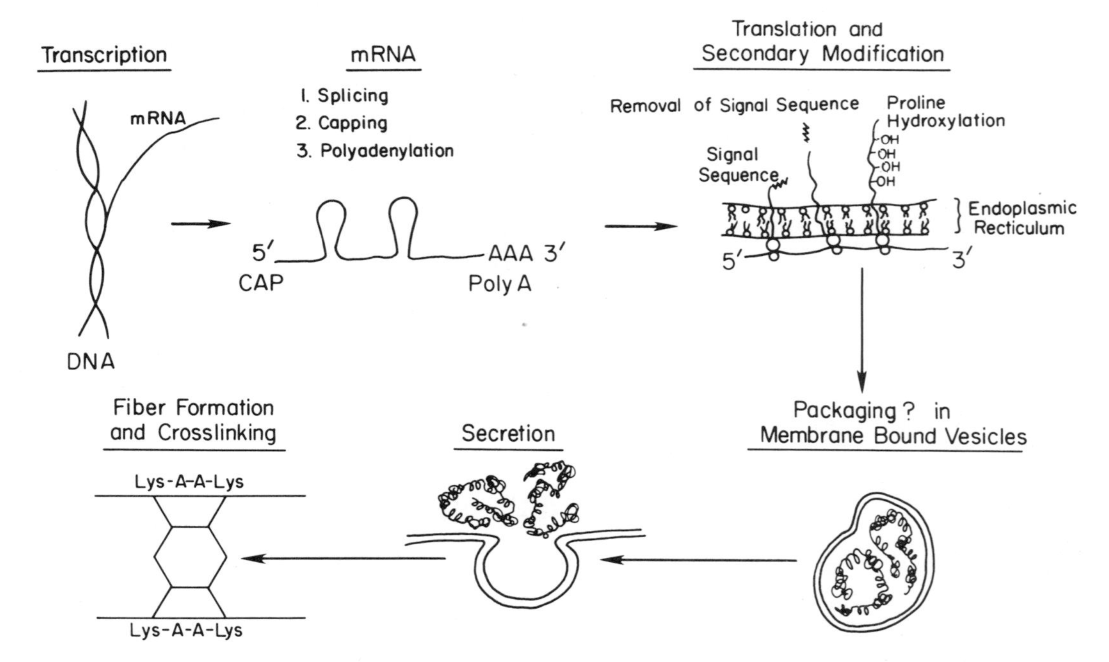

microtubules, such as colchicine, significantly inhibit secretion. Pulse-chase experiments have demonstrated that tropoelastin is incorporated into insoluble elastin and experiments designed to detect cleavage prior to incorporation have not revealed any such cleavage (Narayanan and Page 1976). In all likelihood, tropoelastin is incorporated intact as single molecules onto the surface of forming fibers. Additional evidence for the legitimate role of tropoelastin as a biosynthetic intermediate has come from the observation that when isolated [^{14}C] lysine-labeled porcine tropoelastin is incubated with fresh preparations of normal new-born pig aorta, labeled desmosines as well as lysinonorleucine and merodesmosine can be recovered (Smith et al. 1975).

Although it was clear from the experiments described above that tropoelastin could be incorporated into insoluble fibers, the question remained whether it was the primary translation product, as other higher molecular weight intermediates in the 110,000–120,000 range have been reported. However, when polyribosomes synthesizing elastin were isolated from embryonic chick aortas and the completion of nascent chains carried out in vitro, the predominant polypeptide found had a molecular weight of 72,000. Similarly when messenger RNA (mRNA) isolated from chick embryo aortas was translated in a nuclease-treated rabbit reticulocyte lysate system and the labeled products precipitated with elastin specific antibody, the largest polypeptide found was slightly larger than a tropoelastin standard and no significant amounts of a higher molecular weight species were detected (Burnett et al. 1980). These experiments strongly suggest that tropoelastin is the primary translation product. The slightly larger size of the in vitro translation product is due to an N-terminal prepeptide or signal sequence of 20–25 amino acids.

Control of elastin synthesis has been investigated in the developing embryonic chick aorta and sheep nuchal ligament. In both these systems, measurements of functional elastin mRNA by translation in the nuclease-treated reticulocyte lysate demonstrate a strong correlation between messenger levels and rate of elastin synthesis in the tissue, suggesting that the rate of synthesis is controlled by the mRNA level (Burnett et al. 1980). Recombinant cDNA clones have been obtained for chick aorta elastin mRNA and used to quantitate relative elastin mRNA concentrations in various age chick aortas. These results show that the functional mRNA levels are a reflection of the actual levels as measured by hybridization techniques and confirm the conclusion that the mRNA level controls the rate of elastin synthesis (Burnett et al. 1982). Sheep genomic elastin clones have been obtained, and these along with the cDNA clones should prove useful for studying further the regulation of elastin synthesis as well as for facilitating the elucidation of the primary sequence of the protein through rapid DNA sequencing techniques.

6.3.2. Fiber Structure and Formation

Consistent with its chemical properties, elastin binds the usual electron microscopic stains very poorly, and often not at all. Newly formed elastin, presumably because it still contains a number of unmodified lysine residues, does take up phosphotungstate, while mature elastin appears darkened only after prolonged osmification. In contrast, the microfibrillar component stains heavily with uranyl acetate and lead citrate. Under some conditions and with negative staining, amorphous elastin fibers are reported to exhibit a filamentous structure when examined at high magnification (Gotte et al. 1974; Cleary and Cliff 1978). These 3–5 nm filaments may also have a beaded appearance and appear to intertwine. It is still controversial whether these filaments represent the state of the native fiber or are artifacts caused by the drying and infiltration procedure used in negative staining.

A fundamental question arises as to the role of each of the major components in the organization and construction of the fiber. This problem has been approached by systematic studies of fetal and newborn animals in which qualitative changes have been observed in the appearance of the elastic fiber during development. These electron microscopic observations include studies on the developing rat and chick aorta, bovine ligamentum nuchae, and rat flexor tendon (Greenlee et al. 1966; Karrer and Cox 1961). A general observation made in all of these studies is that the appearance of the microfibrillar component in the extracellular matrix precedes that of elastin itself. This finding has led to the hypothesis that the morphologic structure of the microfibrils determines the pattern of the mature fiber. While this is an attractive idea, the supporting evidence is purely circumstantial. Microfibrils are found consistently interspersed throughout the fiber as if they were acting as nucleating sites for the packing of soluble elastin molecules. The function of the microfibrils on the outside of mature fibers is not known.

The role of the cell in elastin fibrogenesis is also speculative. Based again upon electron microscopic observations, newly forming elastic fibers appear to lie within folds or crevices of the plasma membrane of the synthesizing cells. This has suggested that the cell can control the orientation of the fiber by secreting elastin and/or microfibrillar protein over a limited surface of the cell in a preferred way. However, the details of the secretory pathway have not been determined and it is not even known whether secretion takes place as single molecules or in the form of packets, either of pure elastin or microfibrillar protein or a mixture of the two components. Cells synthesizing elastin contain numerous coated vesicles that appear, based upon characteristic staining reactions, to contain elastin (Thyberg et al. 1979). These vesicles fuse with the plasma membrane and discharge their contents by exocytosis. Even though some experiments using elastin-specific antibodies have identified reactive material within intracellular vesicles, further experiments are necessary to delineate the secretory pathway.

6.3.3. Degradation

While the microfibrillar component is readily degraded by a variety of proteases, elastin is susceptible only to a few enzymes, which have been designated elastases because of this capability. However, unlike the collagenases, which are highly specific (Chapter 4), the elastases are general and powerful proteases attacking a wide variety of proteins. The first of the elastases to be identified and best characterized is readily isolated from pig pancreas. It is one of the family of serine proteases and shares their general properties, including similar reaction mechanism and inhibition by α-1 antiprotease and α-2 macroglobulin.

Pancreatic elastase preferentially cleaves on the carboxyl side of alanine residues in elastin, but also hydrolyzes peptide bonds of other residues. The rate of elastin degradation is markedly increased by anionic detergents such as sodium dodecyl sulfate, probably by increasing the binding of elastase to the insoluble fiber. It is unlikely, under normal circumstances, that the pancreatic enzyme has a functional role other than as a digestive enzyme in the intestinal tract, because plasma α-1 antiprotease effectively inhibits it. When the enzyme is released in overwhelming amounts during pancreatitis, it contributes to the localized tissue destruction.

The elastase synthesized and secreted by polymorphonuclear leukocytes readily degrades elastin, releasing peptides that differ from those released by pancreatic elastase. The neutrophil enzyme is also a serine protease and is inhibited by α-1 antiprotease. Recently an elastase has been prepared from macrophages that appears to be distinct from the neutrophil enzyme. It is a metalloprotease inhibited by α-2 macroglobulin but not α-1 antiprotease, and the cleavage products differ from those generated by the other two elastases. There is considerable interest in the leukocyte enzymes as they may be involved in the pathogenesis of several diseases, such as emphysema.

6.4. Immunologic Studies

6.4.1. Generation and Characterization of Antielastin Antibodies

Elastin has been regarded for many years as a relatively poor antigen. This may in part be attributed to the techniques used to test the antibodies or to the poor response of the species immunized. Early studies utilized as antigens peptides obtained by digestion of elastin with either alcoholic potassium hydroxide or oxalic acid. In general, the resulting antisera were of low titer and contained only weakly precipitating antibodies, although they did react well in hemagglutination reactions. More recently, antibodies have been generated against insoluble chicken elastin, which was administered subcutaneously to sheep as a fine suspension in complete Freund's adjuvant (Sykes and Chidlow 1974). These antibody preparations crossreacted, as observed by immunodiffusion with purified tropoelastin prepared from lathyritic chick, although no precipitin line could be observed with chick α-elastin. Antibodies to insoluble human and dog elastin have also been prepared in sheep. These antibodies, however, were observed by passive hemagglutination to crossreact with α-elastin prepared from the insoluble elastin used as the immunogen. Similarly, antibodies against insoluble ligamentum nuchae react with α-elastin labeled with ^{125}I in a radioimmunoassay (Mecham and Lange 1980). Thus the technique of observation and the sensitivity of the method may determine the ability to detect the reactivity of antibody preparations with various forms of antigen. In general, precipitating antibodies are not obtained and appropriate techniques for detecting soluble antigen–antibody complexes must be used.

Various forms of soluble elastin have been used to generate antibodies, including tropoelastin, α-elastin, and mixtures of soluble peptides obtained from purified insoluble elastin by digestion with pancreatic or neutrophil elastase (Kucich et al. 1981). Antisera against pig tropoelastin crossreact strongly with pig α-elastin. Though the antibodies generated against peptides obtained by leukocyte elastase digestion crossreact with those generated by pancreatic elastase and vice versa, they do not react at all with α-elastin. Antibodies to α-elastin crossreact poorly with the elastase digests. Thus, the antigenic sites in the peptides from elastase digestion are not represented in α-elastin, and the antigenic sites in α-elastin are poorly represented in the elastase derived peptides. As desmosine and other crosslinks are found in all of these solubilized forms of elastin, this implies that at least some, and perhaps a substantial proportion, of the antibody molecules are directed against sites other than the crosslinks. This conclusion is reinforced by the observation that antibodies to insoluble elastin crossreact well with tropoelastin, which contains no crosslinks. These considerations do not preclude formation of antibodies against antigenic sites containing desmosines or other crosslinks, and in some cases these antibodies may even be the predominate species. Interestingly, attempts to elicit antibodies against the free desmosines have been unsuccessful, but when desmosine has been covalently linked to serum albumin as carrier, very high titer antisera have been obtained. These antisera to desmosine react only with free desmosine and not with desmosine in peptide linkage or with isodesmosine in any form (King et al. 1980).

In general, antielastin antibodies do not show tissue specificity with respect to the source of elastin within a given vertebrate species. Thus, analysis with antibodies has not provided any evidence for the presence of tissue-specific forms of elastin, although recent evidence suggests that more than one type of elastin polypeptide chain may occur in a single species (Foster et al. 1980). Although antibody preparations frequently appear to react only with the elastin from the species that was the source of immunogen, this may again be a reflection of the test procedures used. For example, antibody against bovine ligamentum nuchae appears only to react with forms of bovine α-elastin as the test antigen. However, when labeled α-elastin from several other mammalian species was tested, crossreaction was easily demonstrated. In this particular case, two classes of antibody molecules appeared to be present: a class that was bovine-specific corresponding to that of the immunogenic source, and a second class that reacted with antigenic sites shared by several species.

6.4.2. Applications of Antielastin Antibodies

The development of antielastin antibodies has provided a useful tool in the investigation of the biosynthesis, localization, and degradation of elastin. Thus antibodies have been used to identify labeled and unlabeled tropoelastin synthesized by organ and cell cultures. They have also been used to immunoprecipitate polypeptides synthesized by in vitro systems in response to mRNA addition. Because of the lack of a characteristic ultrastructure in the elastic fiber, identification and localization of elastin cannot always be readily made. Specific double antibody techniques have been developed for identification of elastin in the extracellular matrix. Antibodies have also been used to show that the same cell can simultaneously synthesize elastin and collagen. As many of the antibody preparations react with degradation products of insoluble elastin, they are potentially useful in problems where biological degradation may be an important feature.

6.5. Elastin Mechanochemistry

One of the main features of elastin-containing tissues is their ability to be deformed to large extensions with very small forces and then to return to their original dimensions. These properties of low stiffness and high, reversible extensibility are characteristics of rubber-like materials, and they arise from fundamentally different molecular conformations than are seen for stiff, relatively inextensible materials like collagen. In addition, in some circumstances these elastic tissues are required to store elastic energy efficiently and pay back this energy in elastic recoil. The following sections will discuss the features of molecular structure and physical chemistry that account for these functionally important mechanical properties in elastin and in elastin-containing connective tissues. A general discussion of the molecular and mechanical design of biological materials can be found in Wainwright et al. (1976), and more specific reviews on the mechanochemistry of elastin can be found in Gosline (1976, 1980).

Most of what is known about the mechanochemistry of elastin is based on work with elastic ligaments and arteries. These tissues are so rich in elastin that it is difficult to distinguish them from solid rubber on casual observation. They are, however, composite materials that contain collagen, glycosaminoglycans, and in the case of artery large amounts of smooth muscle, in addition to elastin. The mechanical properties of these composites clearly reflect the contribution of at least two of the major components, elastin and collagen. Figure 6.8A shows typical mechanical properties for a large artery, where the properties are represented by plotting the force in the circumferential direction due to an internal pressure versus the circumferential expansion of the artery. Actually, the force and the extension are converted into stress (σ) and strain (ε) by normalizing these values to the initial dimensions of the test specimen. Thus, stress is the force (F) divided by the area (A) over which the force is applied ($\sigma = F/A$), and the strain is the change in the circumferential length (L) divided by the initial circumference ($\varepsilon = \Delta L/L_0$). The slope of the resulting stress-strain curve is a measure of the ability of the arterial wall to resist change in diameter due to increases in internal pressure. However, because the actual forces and extensions are normalized, the stiffness refers to the properties of the material from which the artery is constructed and not to the individual test sample. The stiffness of the material, called its modulus of elasticity (E_m), can be obtained from the slope of the stress-strain curve.

The nonlinear or **J**-shaped stress-strain curve seen for artery is quite characteristic of all elastin-containing connective tissues. Roach and Burton (1957) demonstrated that this kind of stress-strain curve arises from the parallel arrangement of collagen and elastic fibers. Initially the tissue is very extensible because small forces can easily extend the soft elastic fibers. The collagen fibril bundles do not resist extension in this initial region because they are coiled or kinked (Chapter 1, Section 1.3.5). As the tissue extends, however, the collagen fibrils become straightened and aligned in the direction of extension. At about 50% extension, the load is transferred to the collagen fibers, and the tissue becomes very much stiffer. Functionally, this arrangement allows soft tissues to deform easily but provides a set of limits, as determined by the dimensions of the collagen fiber network.

For comparison, Figures 6.8B and C show typical stress-strain curves for purified arterial elastin and for isolated tendon collagen. Note that the modulus of elastin is about 10^6 Nm^{-2} while that of collagen is about 1000 times greater at about 10^9 Nm^{-2}. Elastin can be extended by 100–150% before it breaks, but collagen from tendon breaks at extensions of about 5–8%. (The fibrils in tendon are nearly straight to begin with.) The modulus values shown in Figure 6.8A for arterial wall material in the elastin-dominated region of the curve and in the collagen-dominated region are much lower than those for the pure substances because only a portion of the total tissue volume is occupied by elastin or collagen.

The two types of stress-strain diagrams with high and low moduli can serve as a starting point for the discussion of the molecular basis for the elastic properties of fibrous biopolymers. As discussed in Chapter 1, the helical polypeptide chains in the collagen molecule are arranged essentially parallel to the long axis of the fiber, and the elastic properties of collagen arise primarily from the distortion of the bonding forces that stabilize this highly ordered system. In fact, the basic elastic mechanism exhibited by collagen, or for that matter stiff materials like steel or concrete, is often called energy elasticity because the elasticity arises

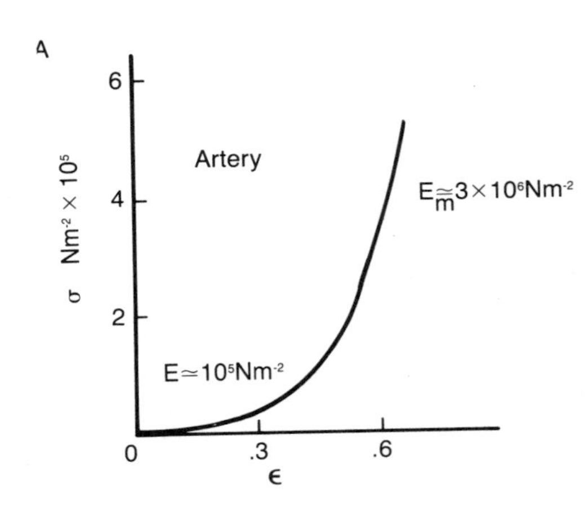

Figure 6.8. (*A*) The mechanical properties of a typical vertebrate artery. The tensile modulus values (E_m, Neutrons per square meter, Nm^{-2}) given correspond roughly to the lower and upper range of physiologic pressures. (*B*) The stress-strain properties of purified arterial elastin. The modulus value given corresponds to the initial portion of the stress-strain curve (*C*) Typical stress-strain properties for tendon collagen. The modulus value corresponds to the linear portion of the curve.

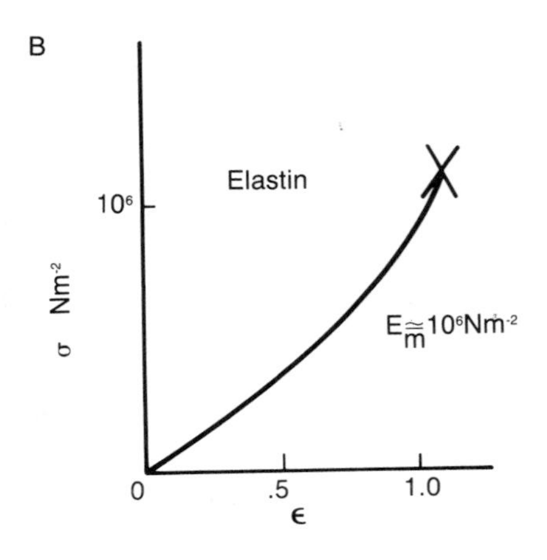

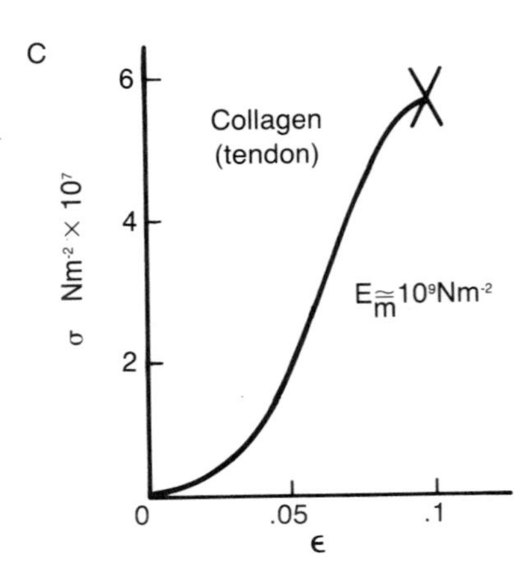

from changes in bond energy levels. Energy elastic materials are stiff because the various bonding forces are quite strong, and thus difficult to deform. The limited extensibility arises primarily because bonding forces drop off quite rapidly with increasing atomic separation.

The low stiffness and high extensibility of rubbery materials arise from a completely different molecular phenomenon, and a good conceptual model for a biochemist to use when thinking about rubber-like elasticity is that of a globular protein that has been denatured by treatment with agents such as guanidine hydrochloride or urea. These denaturing agents weaken the secondary bonding forces that stabilize the globular structure of a protein and cause the polypeptide chain to take on what is known as a random-coil conformation. Such denatured proteins while swollen with solvent are highly folded like the original globular molecule, but the random coil differs in two very important ways from its stable precursor: (1) The random coil lacks short- and long-range order (i.e., it is totally disordered); and (2) random-coil chains are kinetically free and therefore have a constantly changing conformation due to thermal agitation.

When random-coil chains are crosslinked to form a macroscopic network, a rubber-like material is created. The great extensibility of rubbers is due to the coiled nature of the constituent molecules. In fact, some lightly crosslinked rubbers can be stretched 6–8 times their original length before the highly coiled chains are elongated to the point where the applied load actually distorts the covalent bonds in the chain backbone. How then do these rubbery materials store elastic energy? The answer lies in the entropy of the random-chains, and rubber-like elasticity is often called entropy elasticity for this reason. An unperturbed rubber network is in a state of maximum entropy (i.e., maximum disorder), and an external force that acts to extend the network will increase the order of the network (i.e., decrease its entropy) by imposing a small degree of preferred orientation in the direction of extension on each of the random chains in the network. The free energy of a system can be increased and elastic energy can be stored if its entropy is decreased, and this is exactly what occurs in the distortion of a rubber network. The laws of thermodynamics say that systems proceed spontaneously to a state of maximum entropy. Thus, when the external force is removed, elastic recoil will occur because the partially ordered chains will return spontaneously to their initial, disordered state.

With this brief introduction to the mechanochemical principles of biopolymers, a more detailed analysis of the relationship between the structure and mechanical properties of elastin can be considered. First, the structure and conformation of elastin as they relate to the requirements for rubber-like elasticity will be discussed. The elastic mechanism will then be examined in detail and the role of hydrophobic interactions considered. Finally, the effect of solvent environment on the ability of elastin to store elastic energy under dynamic loading conditions will be discussed.

6.5.1. Elastin Structure and Conformation

Extracellular crosslinking reactions are able to link the individual tropoelastin molecules into a macroscopic network (Section 6.2.4), and thus the only other requirement for rubber-like behavior in elastin is a molecular conformation approaching that of a random coil. All proteins become random coils when denatured but the vast majority of proteins have a single or very limited number of stable conformations under cellular conditions (i.e., in dilute aqueous salt solutions). As it is the amino acid sequence that determines the conformation of a protein, some aspect of sequence design allows the elastin protein chains to maintain a degree of mobility and disorder that is quite atypical of other proteins. Although there is no firm understanding of the design of sequences that give rise to random-coil conformations, it is clear that elastin achieves this state to a remarkable degree.

The features expected for a random-coil network are: (1) structural isotropy due to the lack of short- and long-range order; and (2) kinetic mobility of the protein chains due to the lack of stable bond interactions between the various groups along the peptide chains. Structural isotropy has been demonstrated for elastin in a number of ways. Polarized light microscopy of single elastin fibers reveals that elastin is optically isotropic. Transmission electron microscopy of elastin cut in thin sections generally reveals a homogeneous material, and x-ray diffraction studies of unstretched and stretched elastin indicates that there is virtually no short-range order in elastin. Nuclear magnetic resonance (NMR) studies on elastin clearly indicate that the backbone chain of the elastin protein is isotropically highly mobile, with individual residues in the chain able to tumble freely in three dimensions over a time scale on the order of 10^{-7} seconds (Torchia and Piez 1973; Fleming et al. 1980). Thus, elastin appears to meet the general requirements for a kinetically free, random-coil network. In contrast, collagen fibers are highly birefringent when observed with a polarized light microscope, and they can be seen as axially ordered fibrils in the electron microscope. X-ray diffraction patterns of collagen also reveal crystalline organization, and NMR studies indicate that the backbone chains in collagen are axially immobile. (Details of these observations are discussed in Chapter 1.) The differences in mechanical properties between elastin and collagen shown in Figure 6.8 thus clearly reflect differences in molecular structure and conformation.

Although the dominant molecular pattern of elastin is that of a kinetically free, random-coil network, there are some features of elastin structure that indicate regions of local order. For example, circular dichroic spectra indicate the presence of helical secondary structure associated with the alanine-rich sequences in the crosslink regions (Foster et al. 1976). Freeze fracture electron microscopy of elastin reveals an isotropic structure for unstretched samples, but reveals apparently ordered filaments (approximately 13 nm in diameter) in samples that have been stretched 150–200% (Pasquali-Ronchetti et al. 1979). In addition, negatively stained fragments of insoluble elastin reveal a filamentous organization, with the filaments having a diameter of the order of 3–5 nm (Gotte et al. 1974) or even less (Cleary and Cliff 1978). However, negative staining involves the drying of elastin fragments in solutions of heavy metal ions, and the final structure that is observed with the electron microscope is one that has formed from the deposition of stains (e.g., phosphotungstate salt) from a supersaturated solution. For a highly mobile system like elastin, it is possible that such extreme conditions might create artifacts. Very thin filaments have not been observed in elastin by any other technique and the observation needs corroboration, for such filaments seem to be inconsistent with isotropically mobile chains.

A

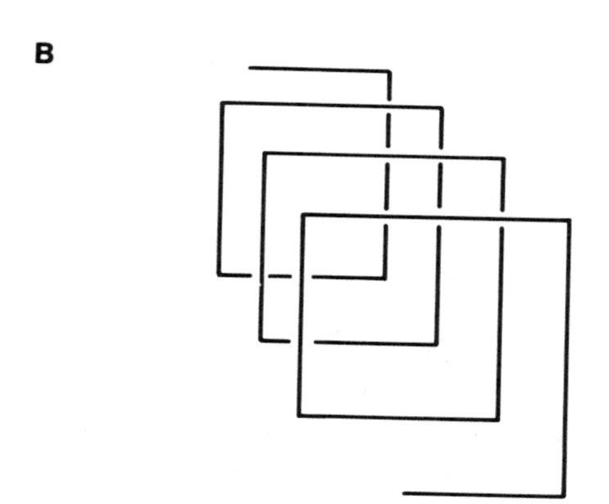

Figure 6.9. (*A*) A possible beta-turn structure for the elastin repeat tetrapeptide, poly(Val-Pro-Gly-Gly), based on the work of Urry and Long (1976). (*B*) Diagrammatic representation of a beta-spiral formed from a series of beta-turns. Each turn of the beta-spiral corresponds roughly to two beta-turns.

Some interesting studies of elastin structure have been carried out by Urry and his co-workers (Urry and Long 1976). The amino acid sequences obtained for elastin suggest possible repeating sequences that might provide a basis for elastin's rubber-like properties, and Urry has constructed polymers of these repeating units and studied their conformational properties using NMR. The repeat peptides show preferred conformations called beta-spirals, which are created from a number of beta-turns in succession (Figure 6.9). Although the beta-spirals represent preferred conformations, the NMR data indicate that even for synthetic polypeptides containing many of these sequence repeats, the spirals are not stable structures like the collagen helix, but are highly mobile, dynamic structures. Since it now appears that the repeating units in the elastin sequence are limited and variable (Section 6.2.3), it is unlikely that beta-spirals play a dominant role in the mechanical properties of the elastin network.

6.5.2. The Elastic Mechanism—Elasticity of a Random Chain

A quantitative theory of rubber-like elasticity has been developed from the notion that elasticity can arise from changes in entropy associated with the distortion of kinetically free, random-coil molecules. A full discussion of this theory can be found in Flory (1953) or Treloar (1975). The theory is based on a random-

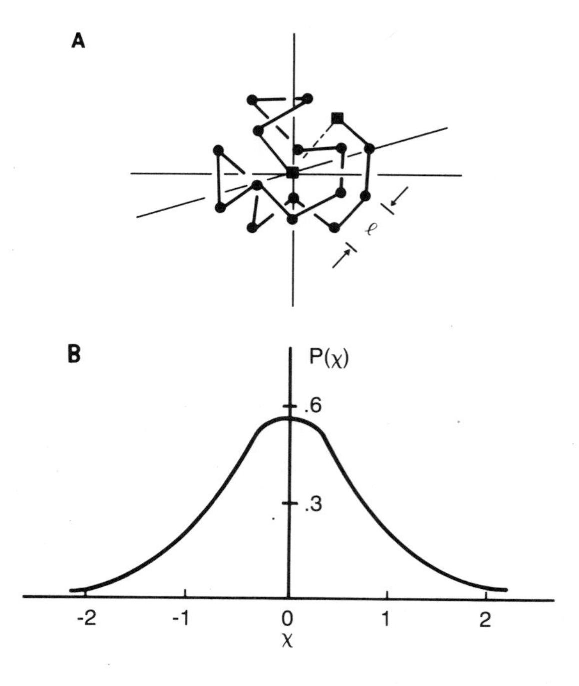

Figure 6.10. (*A*) Diagrammatic representation of an ideal random-coil molecule made up of a linear series of segments of length l. The points where individual segments meet are assumed to allow complete rotational freedom, and the segments orient randomly with respect to all other segments in the molecule. The chain-end separation (r) is represented by the dotted line. (*B*) Plot of the Gaussian probability distribution function for the projected chain-end separation of a random-coil molecule, in the x-direction.

walk model for the network chains, in which each chain is visualized as a series of n ideal segments, each of length l. The segments are ideal in the sense that they are joined to one another by junction points that allow complete freedom of rotation in all directions, and each segment orients randomly with respect to all other segments in the polymer molecule (Figure 6.10A). The system is then set into motion by thermal agitation. Because the model has no fixed form, the dimensions of this hypothetical random walk can be described only in statistical terms, and the normal way of describing such a model is in terms of the separation (r) between the ends of the randomly coiled molecule. Taking the simplest case (i.e., the one dimensional case), the probability of finding a chain-end separation projected onto the x-axis is

$$P(x) = C\,e^{-b^2x^2} \tag{1}$$

where x is the end-to-end separation in the x-direction, C is a constant and $b = 3/2\ nl^2$. This distribution has the form of a Gaussian probability function, and is plotted in Figure 6.10B. The main feature of the chain-end distribution function is that the most probable chain-end separation for this, and for any other one dimensional random walk, is zero.

On the basis of this probability function it is possible to compute the conformational entropy associated with a random-coil molecule. Statistical thermodynamics tells us that entropy (S) is really just a probability statement; specifically,

$$S = k \ln P \tag{2}$$

where k is Boltzmann's constant and P is a probability function. Substituting the chain probability function (Equation 1) into Equation 2 gives

$$S = D - k\,b^2\,x^2 \tag{3}$$

where $D = k \ln C$.

The mechanical properties of a random-coil molecule can be determined by assuming that the change in free energy associated with the distortion of the random coil is due entirely to a change in conformational entropy. Thus,

$$G = H - TS, \text{ or } G = E + PV - TS \tag{4}$$

where G is the Gibbs free energy, H is the enthalpy, E is the internal energy, P is the pressure, V is the volume, T is the absolute temperature, and S is the entropy. If the random-coil molecule is deformed at constant temperature and pressure without change in volume and without change in internal energy, then

$$G = -T\,\Delta\,S \tag{5}$$

The lack of change in internal energy is equivalent to assuming that the random-coil molecule can change its shape without distorting any of the chemical bonds in the polymer backbone (i.e., there is no energy elasticity). If these conditions apply, then the change in free energy is equal to the stored elastic energy (W) and thus

$$W = -T\,\Delta\,S. \tag{6}$$

Since the elastic energy is

$$W = \int f\,dx \tag{7}$$

where f is the elastic force and x is the extension, then

$$f = dW/dx \tag{8}$$

or from Equation 6,

$$f = -T\,(dS/dx). \tag{9}$$

Now, substituting the first derivative of Equation 3 into Equation 9 gives

$$f = 2\,k\,T\,b^2\,x. \tag{10}$$

This means that if it were somehow possible to grab hold of the two ends of a single random-coil molecule and pull them apart by a distance x (Figure 6.11A), then the force required to separate the ends would increase linearly with the separation x. In other words, a random-coil molecule is a molecular spring with a spring constant (i.e., stiffness) proportional to kT or thermal agitation. In fact, in rubber-like elasticity the resistance to deformation is simply the tendency of random thermal motion to drive the flexible molecules back to their most random or most highly coiled state.

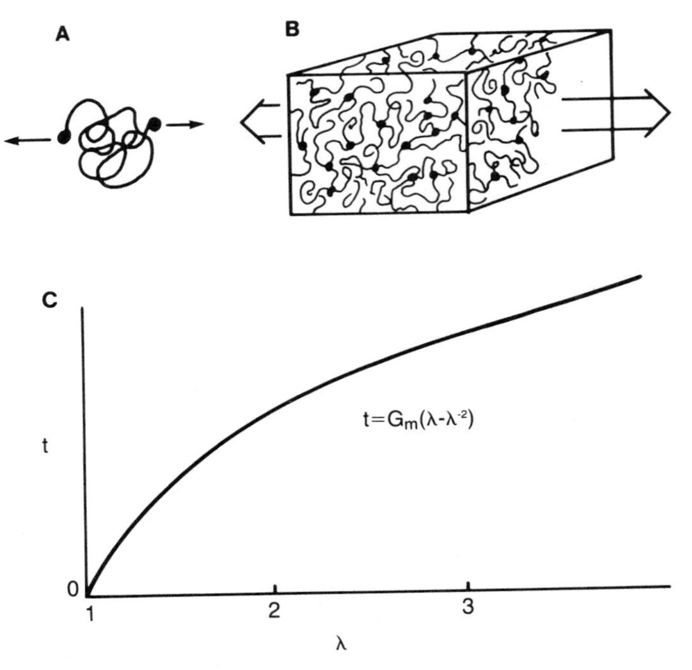

Figure 6.11. (*A*) The elasticity of a single random coil arises when an external force acts to increase the chain-end separation and thus force the molecule to take on a less probable set of conformations. (*B*) A macroscopic network can be formed if many random-coil molecules are joined together by permanent junction points. The junction points are usually covalent crosslinks, and in this diagram are represented by dots. When a crosslinked network is deformed, the macroscopic deformation of the whole network is imposed on all of the constituent chains. (*C*) Plot of the force-extension ($t - \lambda$) curve expected for a crosslinked network of random chains. G_m is the elastic modulus.

6.5.3. Random Networks

Because it is generally not feasible to deform single random-coil molecules, these principles must be transferred to a macroscopic level. This can be done by visualizing many random-coil molecules packed densely together and physically linked to one another by permanent crosslinks. Through this process, the many individual molecules become a single, huge molecule of essentially infinite molecular weight (Figure 6.11B). Crosslinking in network formation is absolutely essential because it allows macroscopic distortions of the whole network to be transferred into conformational changes at the level of the individual protein chains between the network junction points (the crosslinks). If crosslinks were not present, an external force would simply cause the random chains to flow past one another, and the material would behave as a viscous fluid. As random chains in a crosslinked network deform in proportion to the deformation of the whole macroscopic network, it is possible to apply the theory for single chains to the whole network. Network theory predicts that the force required to stretch a crosslinked network will be

$$t = N\,k\,T\,(\lambda - \lambda^{-2}) \tag{11}$$

where the force is now expressed as a tension (t), which is defined as the force per unit initial cross-section area. N is the number of random chains per unit volume, and λ (the extension ratio) is defined as the stretched length divided by the initial length. The elastic force now no longer increases linearly with extension (Figure 6.11C) because the relationship takes into account both the decrease in entropy associated with the extension of the network in the x-direction and the decrease in entropy due to the contraction of the network in the y- and z-directions.

One very important conclusion of network theory is that the stiffness or elastic modulus of any crosslinked network is determined simply by the number of network chains per unit volume and by the intensity of thermal energy (kT), or in other words, the elastic modulus $G_m = N\,k\,T$. In this case the elastic modulus (G_m) is defined slightly differently than the tensile modulus (E_m) used previously, but in principle it provides the same information. As the elasticity of a network is determined by the number of chains, if the density of the polymer from which the network is constructed is known, then it is possible to calculate an average value for the molecular weight of the chains between crosslinks in the network. Thus

$$G_m = \rho\,R\,T/M_c \tag{12}$$

where ρ is the density of the polymer, R is the gas constant, and M_c is the average molecular weight of the random chains between crosslinks.

6.5.4. Thermoelasticity—Testing of Network Theory

Thermoelastic measurements provide a simple protocol that can be used to experimentally verify the hypothesis that elastic energy in a rubber network is stored as a decrease in conformational entropy rather than as a change in internal energy due to the distortion of chemical bonds. For any elastic solid, the elastic force (f) can be divided into two components, one associated with changes in internal energy and the other associated with changes in entropy. Thus

$$f = (\partial E/\partial L)_{T,V} - T\,(\partial S/\partial L)_{T,V} \tag{13}$$

where L is the length of the elastic material. Because the stiffness of an entropy elastic system will increase in direct proportion to absolute temperature (*See* Equations 10, 11), and that of an energy elastic system will be temperature independent, it is possible to assess the relative importance of energy and entropy mechanisms by measuring the temperature dependence of the elastic force. Thus

$$f = (\partial E/\partial L)_{T,V} + T\,(\partial F/\partial T)_{L,V}. \tag{14}$$

Equation 14 provides the basis for thermoelastic measurements. An elastic material is first stretched to a length L, and then the force required to maintain this length is measured as a function of temperature. The slope of the force-temperature plot for the sample is equal to the entropy component of the elastic force: $-(\partial F/\partial T)_{L,V} = (\partial S/\partial L)_{T,V}$. The intercept of the force-temperature plot at zero degrees absolute is equal to the energy component of the elastic force, $(\partial E/\partial L)_{T,V}$. Note that the subscripts for all these partial derivatives indicate that the experiment is carried out with no change in volume. It is generally quite difficult to maintain constant volume during the deformation of a solid, but when this condition has been achieved for elastin and for a variety of other rubbers, it is observed that virtually all of the elastic force is due to changes in entropy (*See* Figure 6.12, *line A*). In this case there are no volume changes, so the observed change in entropy can be attributed unequivocally to changes in the conformation of the polymer chains in the rubber network. Thus, thermoelastic measurements substantiate the major prediction of the theory of rubber elasticity, that elasticity arises from changes in the conformational entropy of kinetically free, coiled molecules. There are, however, some small but significant changes in internal energy associated with the extension of elastin and other rubbers. These changes arise because there are restrictions to free rotation around chemical bonds, and therefore real polymer molecules can approach, but not fully achieve, the rotational freedom assumed for the ideal segments in the random walk.

6.5.5. The Role of Hydrophobic Interactions

One very interesting aspect of the elasticity of elastin is that elastic fibers in living animals do not deform at constant volume. Rather, they exist in swelling equilibrium with a dilute aqueous salt solution. When these fibers are stretched, the volume of the fiber increases because water is absorbed by the rubber network. As mentioned earlier, elastin is a very hydrophobic protein, and the water absorbed during the extension of elastin is forced to associate with some of the

Figure 6.12. Thermoelastic measurements provide a simple way to experimentally assess the relative importance of entropy and internal energy changes to the elastic mechanism of any elastic material. In the examples given, line A represents the pattern expected for a rubber-like material. The elastic force increases with increased temperature, indicating a large entropy contribution. The intercept of the force-temperature plot at zero degrees absolute is small, indicating that the internal energy contribution to the elastic force is quite small. Line B represents a typical energy elastic material, where the energy component (i.e., the intercept) accounts for virtually all of the elastic force.

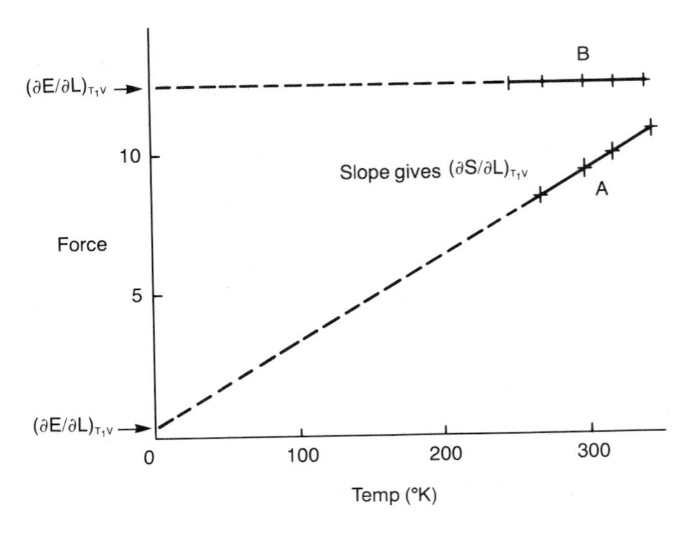

numerous nonpolar amino acid side chains, disrupting hydrophobic interactions. Elastic energy can actually be stored in this disruption of hydrophobic forces because the nonpolar groups will tend to aggregate and expel the absorbed water and thus give rise to elastic recoil (Gosline 1978). Although this hydrophobic elastic mechanism may sound quite different from the conformational entropy mechanisms described above, it is really only a special case of the random network model that occurs when the network is swollen in a small molecular weight solvent and then deformed while in swelling equilibrium with that solvent. The network chains go through exactly the same kind of conformational changes, and the elastic energy is still stored as a decrease in entropy. But due to the rather unusual nature of the interaction between nonpolar compounds and water (i.e., the hydrophobic effect), the origin of the entropy change is quite different.

A hydrophobic interaction is disrupted when nonpolar groups (e.g., several nonpolar amino acid side chains), which have aggregated to form a hydrophobic region, are forced apart and must therefore interact more completely with water. In thermodynamic terms this process can be viewed as the transfer of a nonpolar molecule from an organic solvent into aqueous solution, and this transfer process is accompanied by a large increase in free energy. The large positive free energy change is simply a reflection of the very low solubility of nonpolar compounds in water, and of course, provides the basis for the formation of hydrophobic regions in the first place. As there are no real bonds involved in the formation of a hydrophobic interaction between two nonpolar groups, but rather just a tendency to avoid water, the hydrophobic interactions in the elastin network do not greatly restrict the dynamic mobility of the protein chains. Interestingly, the large positive free energy change for the disruption of a hydrophobic interaction does not arise from a large positive heat of reaction (in fact, the process is exothermic), but from an extremely large decrease in entropy. For example, the free energy change for the transfer of ethane from hydrocarbon solution to water at 25°C is +3.9 kcal/mole. The enthalpy change for this process is $\Delta H = -2.5$ kcal/mole, and the entropy change is $T \Delta S = -6.4$ kcal/mole (*See* Tanford 1980). It is generally believed that the transfer of a nonpolar group into water results in the ordering of the water molecules that form the hydration shell around the nonpolar molecule. The negative enthalpy change probably reflects an increase in hydrogen bonding between the water molecules in this hydration shell, and this increase in water "structure" accounts for the large decrease in entropy. Relating this process to elastin, when an elastin fiber is stretched and water is forced to interact with nonpolar groups in the elastin network, this water is "ordered," and the resulting decrease in the entropy of the water contributes to the restoring force that will allow the fiber to recoil elastically.

6.5.6. Conformational Properties of the Elastin Network

The force extension curve for a random network rubber should follow Equation 11 and should appear as shown in Figure 6.11C. The form of this curve is based on a Gaussian distribution function for the chain-end separation of the random-walk model, and this equation provides a very accurate description of the mechanical properties of rubbery materials up to quite large extensions. However, as the coiled molecules are drawn out at large extensions, the Gaussian distribution function becomes inaccurate and dramatically underestimates the changes in entropy that accompany extension. In fact, all rubbery networks become much stiffer at large extensions, and the form of the force-extension curve at high extensions can be described by more exact, non-Gaussian distribution functions (See Treloar 1975). If the number of steps in the random walk, or equivalently, the number of ideal segments in the random chain, becomes small, the non-Gaussian elastic properties become important at quite low extensions (Figure 6.13), and it becomes possible to use the non-Gaussian properties to obtain some feeling for the conformational properties of the polymer chain.

A recent study of elastin (Aaron and Gosline 1981) indicates that the elastic modulus for single elastin fibres is 4.1×10^5 Nm^{-2}. By modifying Equation 12 to take the swelling of elastin by water and the presence of chain-ends into account, the average molecular weight of the effective random chains between crosslinks (M_c) can be computed to be about 6000 g/mole. From the amino acid composition, the average residue weight of elastin is 85 g/mole, and thus, the random chains between crosslinks contain, on average, about 70 amino acid residues (i.e., $6000/85 \simeq 70$). An analysis of the non-Gaussian elastic properties of elastin (Figure 6.13) indicates that there are 9 to 10 ideal segments per random chain in the network, and therefore roughly 7 to 8 amino acid residues constitute one ideal segment of the random chain. Each amino acid in a polypeptide contains two single bonds that allow some rotational freedom in the backbone,

Figure 6.13. The non-Gaussian elastic properties of random networks. The dotted line is the Gaussian approximation from Equation 11. It essentially assumes an infinite number of segments in the random chain. The broken lines are calculations based on non-Gaussian probability functions, with different numbers of steps (n) in each random chain. The solid line shows the form of the force-extension curve obtained from single elastin fibers by Aaron and Gosline (1981). The curve for elastin is best described by a non-Gaussian curve with n equal to 9 or 10.

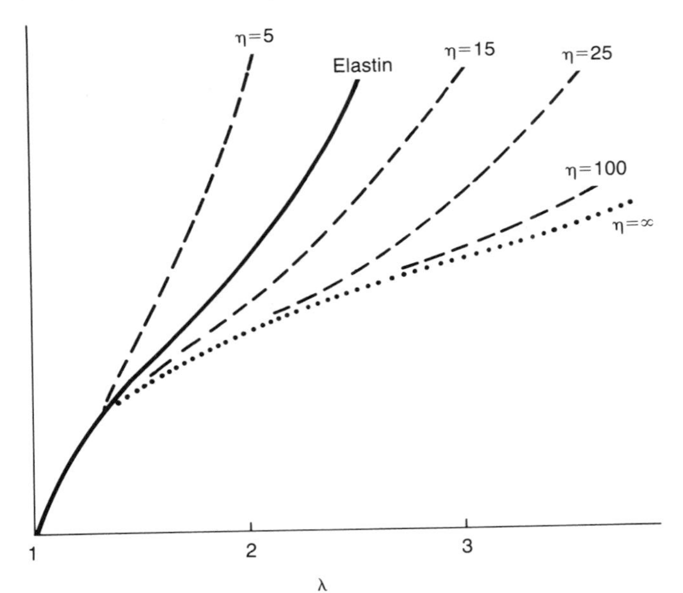

and the analysis presented here suggests that about 14 to 16 of these points of partial freedom are required for the backbone to approach the complete rotational freedom of the ideal segment. The impression obtained from this analysis is that peptide chains, even ones like elastin that have apparently evolved to exhibit a great deal of rotational freedom, are in fact relatively stiff. In comparison, a similar analysis of natural rubber indicates that the ideal segment of this polymer contains only about 5 points of partial freedom. Interestingly, the segment length obtained by very different methods for a variety of globular proteins that had been denatured in 6 *M* guanidine hydrochloride is virtually identical to that obtained here for elastin in water. This observation confirms the previous conclusion (Section 6.5.1) that there must be something special about the amino acid sequence of elastin that allows the elastin chains to exhibit such dynamic mobility without the benefit of strong denaturing agents.

6.5.7. Dynamic Elasticity

One final aspect of elastin mechanochemistry that can be related directly to function is the ability of elastin to effectively store elastic energy under dynamic loading conditions. The analysis presented here has so far made no mention of the rate at which elastin samples are deformed. In fact, the thermoelastic experiments that have been used to verify the basic concept of entropy elasticity are carried out very slowly to ensure that the process of deformation is as close to an equilibrium one as is possible. One of the most important functional roles of elastin in the vertebrate body is in the circulatory system, primarily in the large arteries. Blood is circulated by a pulsatile pump—the heart—and the rubber-like elasticity of the arteries provides a mechanism for storing some of the energy released by the heart when it contracts. This stored energy can then be used to maintain blood flow to the tissues during the refilling phase of the cardiac cycle. For arteries to function in this manner, they must not only stretch and retract, but they must be able to store and release elastic energy efficiently over the time scale of the cardiac cycle. In exercising humans this means storing and releasing energy at up to 3 Hz (cycles per second), and in smaller mammals up to 10 Hz. In comparison, a thermoelastic experiment, as drawn in Figure 6.12, may take several hours to complete.

When temporal restrictions are imposed on the deformation of a rubber network, it is necessary to consider not only the difference in entropy between the relaxed and extended equilibrium states, but also the rate at which the limited mobility of the chains will allow the network chains to attain these changes in conformation. If the rate of external loading approaches the rate at which thermal agitation can bring about changes in conformation at the molecular level, then some energy must be put into overcoming frictional forces between the chain segments. As the energy required to overcome internal friction is lost and not recoverable during elastic recoil, limited chain mobility will reduce the efficiency of elastic energy storage, or in other words, will reduce the rebound resilience. It is not necessary to understand fully the basis of molecular friction in order to appreciate the significance of rebound resilience. Resilience is defined as the ratio of the energy recovered from the deformation of an elastic solid divided by the energy required to deform the solid in the first place. In more familiar terms, when a tennis ball is dropped and the height of its bounce is observed, its rebound resilience is being tested. The energies put into and recovered from the ball are proportional to the square of the height. Thus $R = h_1^2/h_0^2$, where h_0 and h_1 represent the height from which the ball is dropped and the height of the first bounce. Very resilient rubbers, like Super-Balls, have resiliences around 90%. That is, they bounce to about 95% of the starting height.

One can obtain this kind of information for elastin over a broad time scale by applying sinusoidal extensions at different frequencies. Figure 6.14 shows the results of this type of analysis. For elastin in equilibrium with distilled water at 37°C (curve labeled H_2O) the resilience is above the 90% level over the entire biological range of frequencies, but the resilience drops off very rapidly at frequencies above about 10 Hz. This drop in resilience is the first indication of a shift in properties, called the glass transition, that is seen for all rubber-like systems. The fact that elastin appears to enter this transition at such low frequencies is yet another indication of the limited mobility of the protein chains in elastin. The time scale (t) associated with a sinusoidal deformation at a frequency (ν) is usually expressed as $t = \frac{1}{2\pi\nu}$. Thus, at times of the order of 10^{-2} seconds, (i.e., at about 10 Hz) the limited mobility of the segments in the elastin network begins to affect the storage of elastic energy. However, at frequencies well above the biological range (e.g., at t about 10^{-7}–10^{-8} seconds) the chains are virtually unable to change their conformation at all in response to an applied load, and the material behaves like a rigid glass.

One interesting and potentially important aspect is the interaction between water content and chain mobility. Although elastin is a very hydrophobic protein, it binds a reasonable amount of water. At 37°C elastin binds about 0.55 gram water per gram of protein, but if elastin is dehydrated at a relative humidity as high as 97%, almost half of the bound water is removed. Thus, about half of the bound water is very loosely associated with elastin. The remaining water is

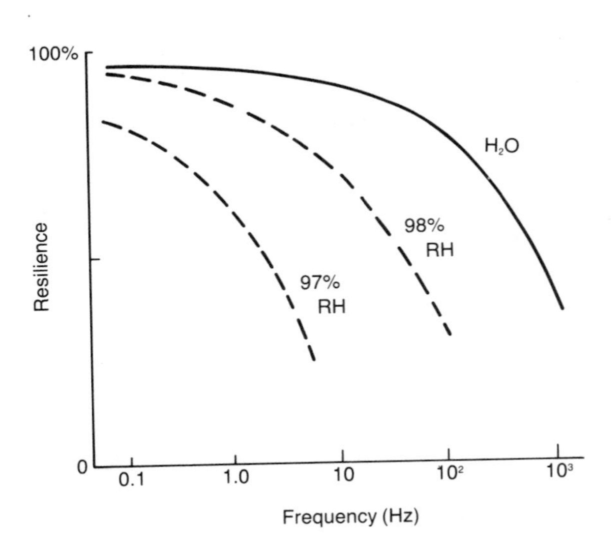

Figure 6.14. The dynamic elastic properties of elastin. The curve labeled H_2O is for elastin swollen in water, while the curves labelled 98% RH and 97% RH show the properties of elastin that has been partially dehydrated at the relative humidities indicated. Data from Gosline and French (1979).

more tightly bound and can be removed only by reducing the relative humidity to much lower values. As there are very few polar side chains in elastin, most of the tightly bound water must be associated with the backbone peptide groups. The large fraction of very loosely bound water apparently associates with the numerous nonpolar amino acid side chains. The interesting thing is that the removal of the loosely bound water dramatically reduces the mobility of the elastin chains. As indicated by the curves labeled 98% and 97% RH in Figure 6.14, when the loosely bound water is removed, elastin enters its glass transition at much lower frequencies. Although it is not known if the reduction in mobility arises from a stabilization of peptide–peptide hydrogen bonds (as in beta-turns) or from the formation of larger and more stable associations between the nonpolar side chains, it is clear that the removal of this loosely bound water can alter the time scale of the transition by several orders of magnitude, and move it into the biological range of frequencies. On a more practical level, it is possible that the accumulation of extracellular materials like glycosaminoglycans, lipids, and calcium deposits, seen in many diseased arteries, may well alter the state of the tissue water and bring about an indirect dehydration of elastin. The shift in mechanical properties that would then result from the removal of this loosely bound water may contribute to the mechanical degradation of the vascular tissues. This aspect of elastin mechanochemistry is now receiving considerable attention.

6.6. Conclusions

The chemical and physical structures of elastin have been determined to a large extent. Elastin is rich in glycine, proline, and hydrophobic amino acids and has an unusual primary amino acid sequence that leads to a conformation approaching that of a random coil. This highly mobile structure results in the property of elasticity. Elastin is found throughout the vertebrate kingdom, but apparently not in invertebrates, in which other proteins serve a similar elastic function. These proteins include abductin in mollusk hinge and resilin in insect cuticle. The crosslinks in elastin, like those in collagen, are derived by the oxidation of lysine residues by the enzyme peptidyl lysine oxidase. However, the final products, particularly the desmosines, differ from those in collagen, and the causes of these differences are not clear. In the case of collagen, crosslinking occurs in an ordered environment with only short range mobility, while elastin crosslinking occurs in a largely disordered, isotropically mobile environment. These differences, in addition to the clustering of lysine residues in elastin, may foster different reaction pathways.

Insoluble elastin is synthesized via a 72,000 dalton soluble intermediate designated tropoelastin, which is secreted from the cell into the extracellular matrix. Control of elastin synthesis in developing systems is largely regulated by the level of tropoelastin mRNA. The details of the mechanism of secretion as well as the assembly of the fiber remain to be elucidated.

The elastic properties of elastin are carefully regulated to achieve maximum efficiency under physiologic conditions. Because the theory of rubber-like polymers is highly developed, structure–function relationships in elastin are understood to a degree unusual for a protein. For this reason, the mechanochemistry is presented here in some detail. The physical chemist should take considerable pride in having solved this problem, and the biologist may take elastin as an example of the degree of understanding that should be sought for all systems.

References

Aaron BB, Gosline JM: (1981) Elastin as a random network elastomer. Biopolymers 20:1247–1260.

Anwar RA: (1977) Desmosine peptides : amino acid sequences and the role of those sequences in crosslink formation. In: Sandberg LB, Gray WR, Franzblau C, eds. Elastin and Elastic Tissue. Pages 329–342 New York: Plenum.

Burnett W, Eichner R, Rosenbloom J: (1980) Correlation of functional elastin messenger ribonucleic acid levels and rate of elastin synthesis in the developing chick aorta. Biochemistry 19:1106–1111.

Burnett W, Finnigan-Bunick A, Yoon K, Rosenbloom J: (1982) Analysis of elastin gene expression in the developing chick aorta using cloned elastin cDNA. J Biol Chem 257:1569–1572.

Cleary EG, Cliff WJ: (1978) The substructure of elastin: Exptl Mol Pathol 28:227–246.

Fisher E, Abderhalden E: (1907) Bildung von polypeptiden bei der hydrolyse der proteine. Ber der Chem Ges 40:3544–3562.

Fleming WW, Sullivan CE, Torchia DA: (1980) Characterization of molecular motions in ^{13}C-labelled elastin by ^{13}C-^{1}H magnetic double resonance. Biopolymers 19:597–618.

Flory PJ: (1953) Principles of Polymers Chemistry. Ithaca: Cornell Univ Press.

Foster JA, Bruenger E, Gray WR, Sandberg LB: (1973) Isolation and amino acid sequences of tropoelastin peptides. J Biol Chem 248:2876–2879.

Foster JA, Bruenger E, Rubin L, Imberman M, Kagan H, Mecham R, Franzblau C: (1976) Circular dichroism studies of an elastin cross-linked peptide. Biopolymers 15:833–841.

Foster JA, Rubin L, Kagan HM, Franzblau C, Bruenger E, Sandberg LB: (1974) Isolation and characterization of cross-linked peptides from elastin. J Biol Chem 249:6191–6196.

Foster JA, Rich CB, Fletcher S, Karr SR, Przybyla A: (1980) Translation of chick aortic elastin messenger ribonucleic acid. Comparison to elastin synthesis in chick aorta organ culture. Biochemistry 19:857–864.

Gosline JM: (1976) Physical properties of elastic tissues. Int Rev Conn Tissue Res 7:211–249.

Gosline JM: (1978) Hydrophobic interaction and a model for the elasticity of elastin. Biopolymers 17:677–695.

Gosline JM: (1980) Elastic properties of rubber-like proteins and highly extensible tissues. In: Vincent JFV, Currey JD, eds. Mechanical Properties of Biological Materials. Symposium of the Society for Experimental Biology, Volume 34 Cambridge: Cambridge Univ Press. 33–357.

Gosline JM, French CJ: (1979) Dynamic mechanical properties of elastin. Biopolymers 18:2091–2103.

Gotte L, Giro MG, Volpin D, Horne RW: (1974) The ultrastructural organization of elastin. J Ultrastruct Res 46:23–33.

Greenlee TK Jr, Ross R, Hartman JL: (1966) The fine structure of elastic fibers. J Cell Biol 30:59–71.

Karrer HE, Cox J: (1961) Electronmicroscope study of developing chick embryo aorta. J Ultrastruct Res 4:420–454.

Keith DA, Paz MA, Gallop PM, Glimcher MJ: (1977) Histologic and biochemical identification and characterization of an elastin in cartilage. J Histochem Cytochem 25:1154–1162.

King GS, Mohan VS, Starcher BC: (1980) Radioimmunoassay for desmosine. Conn Tissue Res 7:263–267.

Kucich U, Christner P, Rosenbloom J, Winbaum G: (1981) An analysis of the organ and species immunospecificity of elastin. Conn Tissue Res 8:121–126.

Lansing AI, Rosenthal TB, Alex M, Dempsey EW: (1952) The structure and chemical characterization of elastic fibers as revealed by elastase and electron microscopy. Anat Rec 114:555–575.

Mecham RP, Lange B: (1980) Measurement by radioimmunoassay of soluble elastins from different animal species. Conn Tissue Res 7:247–252.

Miller EJ, Martin GR, Piez KA: (1964) The utilizaton of lysine in the biosynthesis of elastin crosslinks. Biochem Biophys Res Commun 17:248–253.

Murphy L, Harsch M, Mori T, Rosenbloom J: (1972) Identification of a soluble intermediate during synthesis of elastin by embryonic chick aortae. FEBS Letters 21:113–117.

Narayanan AS, Page RC: (1976) Demonstration of a precursor-product relationship between soluble and cross-linked elastin, and the biosynthesis of the desmosines in vitro. J Biol Chem 251:1125–1130.

Partridge SM: (1962) Elastin. In: Anfinsen CB Jr, Anson ML, Bailey KJ, Edsall JT, eds. Advances in Protein Chemistry. New York: Academic.

Partridge SM, Elsden DF, Thomas J: (1964) Biosynthesis of the desmosine and isodesmosine cross-bridges in elastin. Biochem J 93:30c–33c.

Pasquali-Ronchetti I, Fornieri C, Baccarani-Contri M, Volpin D: (1979) Ultrastructure of elastin revealed by freeze fracture electron microscope. Micron 10:89–99.

Roach MR, Burton AC: (1957) Reason for the shape of the distensibility curve for arteries. Can J Biochem Physiol 35:681–690.

Ross R, Bornstein P: (1969) The elastic fiber I. The separation and partial characterizaton of its macromolecular components. J Cell Biol 40:366–381.

Rucker RB, Tinker D: (1977) Structure and metabolism of arterial elastin. Int Rev Exp Pathol 17:1–47.

Sage H, Gray WR: (1979) Studies on the evolution of elastin—I. Phylogenetic distribution. Comp Biochem Physiol 64B:313–317.

Sandberg LB, Gray WR, Franzblau C, eds: (1977) Elastin and Elastic Tissue. New York: Plenum.

Sandberg LB, Weissman N, Gray WR: (1971) Structural features of tropoelastin related to the sites of cross-links in aortic elastin. Biochemistry 10:52–56.

Sandberg LB, Weissman N, Smith DW: (1969) The purification and partial characterization of a soluble elastin-like protein from copper-deficient porcine aorta. Biochemistry 8:2940–2945.

Sear CHJ, Grant ME, Jackson DS: (1981) The nature of the microfibrillar glycoprotein of elastic fibers. A biosynthetic study. Biochem J 194:587–598.

Siegel RC: (1980) Lysyl oxidase. Int Rev Conn Tissue Res 8:73–118.

Smith DW, Abraham PA, Carnes WH: (1975) Crosslinkage of salt-soluble elastin in vitro. Biochem Biophys Res Commun 66:893–899.

Smith DW, Sandberg LB, Leslie BH, Wolt TB, Minton ST, Myers B, Rucker RB: (1981) Primary structure of a chick tropoelastin peptide: Evidence for a collagen-like amino acid sequence. Biochem Biophys Res Commun 103:880–885.

Smith DW, Weissman N, Carnes WH: (1968) Cardiovascular studies on copper-deficient swine. XII. Partial purification of a soluble protein resembling elastin. Biochem Biophys Res Commun 31:309–315.

Starcher BC, Galione MJ: (1976) Purification and comparison of elastins from different animal species. Anal Biochem 74:441–447.

Sykes B, Chidlow JW: (1974) Precipitating antibodies directed against soluble elastin–The basis of a sensitive assay. FEBS Letters 47:222–224.

Tanford C: (1980) The Hydrophobic Effect, 2nd ed. New York: Wiley.

Thomas J, Elsden DF, Partridge SM: (1963) Partial structure of two major degradation products from the cross-linkages in elastin. Nature 200:651–652.

Thyberg J, Hinek A, Nilsson J, Friberg U: (1979) Electron microscope and cytochemical studies of rat aorta. Intracellular vesicles containing elastin- and collagen-like material. Histochem J 11:1–17.

Torchia DA, Piez KA: (1973) Mobility of elastin chains as determined by ^{13}C nuclear magnetic resonance. J Mol Biol 76:419–424.

Treloar LRG: (1975) Physics of Rubber Elasticity, 2nd Ed. Oxford: Claredon.

Uitto J, Hoffman H-P, Prockop DJ: (1976) Synthesis of elastin and procollagen by cells from embryonic aorta. Arch Bioch Biophys 173:187–200.

Urry DW, Long MM: (1976) Conformations of the repeat peptides of elastin in solution. Crit Rev Biochem 4:1–45.

Wainwright SA, Biggs WD, Currey JD, Gosline JM: (1976) Mechanical Design in Organisms. London: Edward Arnold.

7

Fibronectin, Laminin, and Other Extracellular Glycoproteins

S. Hakomori, M. Fukuda,* K. Sekiguchi, and W.G. Carter

7.1. Introduction

Cells are surrounded by a complex of glycoproteins, a part of the connective tissue matrix. The matrix of mesenchymal cells may be divided into that part close to the cell surface, and perhaps interacting specifically with it, and that part more distant. These are often referred to as the pericellular and intercellular matrix, respectively, but generally there is no sharp demarcation between them. The term extracellular matrix usually implies both. These matrix glycoproteins persist in cultured fibroblasts in vitro; they become particularly conspicuous when cells attain confluency. In addition, epithelial as well as endothelial cells adhere to the glycoprotein matrix of basement membranes, which have a different chemical composition and organization than mesenchymal matrix. Both extracellular matrices and basement membranes have a role in binding cells in an organized state but also influence and regulate genetic expression of these cells. Striking effects of matrices on gene expression have been demonstrated during development. Early biological information is compiled in the monograph edited by Slavkin and Greulich (1975). It is appropriate to quote from the introductory remarks of the above monograph, clearly addressed to this concept, by Grobstein (1975).

> These essential aspects are: first, that intercellular matrix is a complex macromolecular product of the cells imbedded in it; second, that matrix materials not only provide mechanical cohesiveness but act upon and alter the activities of the imbed-

*Present address: La Jolla Cancer Research Foundation, La Jolla, Calif. 92037

From the Division of Biochemical Oncology, Fred Hutchinson Cancer Research Center, and University of Washington, Seattle, Washington

Work by the authors cited in this review has been supported by National Institutes of Health grant CA-23907, and a New Investigator Research Award to W.G.C. from the National Institutes of Health, R23-CA-29172.

ded cells; third, that among the influences exerted by matrix is modification of the gene expression of its associated cells; fourth, that such modification of gene expression is particularly striking when groups of cells having different developmental history come secondarily into association; fifth, that cell-matrix interactions cannot only "turn on" new gene expressions but can act to maintain these expressions in the adult organism.

One can clearly distinguish the effect of matrices on cells from the effects of growth factors and nutrients. Malignancy may also be greatly influenced by pericellular environment. In fact, recent experiments clearly address this problem. For example, cell growth behavior and various malignant phenotypes are modified to resemble normal cells when malignant cells are grown in contact with the pericellular matrices obtained from normal corneal endothelial cells (Vlodavsky et al. 1980). The addition of fibronectin (Section 7.3) to the culture medium of transformed cells induces enrichment of pericellular matrices and restores a part of normal cell phenotype, although growth control persists similar to malignancy (Yamada et al. 1976). Pericellular enrichment of fibronectin is also induced by addition of dextran sulfate to the culture medium (Gahmberg and Hakomori 1973), whereby cell growth is arrested at G1 phase (Goto et al. 1972). It is now known that dextran sulfate interacts with fibronectin (Ruoslahti et al. 1979). The environmental effect on genetic expression of cellular differentiation is most impressive in the remarkable experiment performed by Mintz and Illmensee (1975). When the inner cell mass of blastoid body teratocarcinoma routinely grown in ascites of a 129 white mouse was injected into a blastocyst of a C57 black mouse, the teratocarcinoma differentiated into a perfectly normal hybrid individual, a black and white chimera mouse. The malignant phenotype of the teratocarcinoma was obviously lost in the environment of a normal C57 blastocyst, resulting in normal differentiation of the teratocarcinoma into a normal mouse. This can be partially explained by contact with a normal pericellular environment. It is unnecessary to stress further the significance of pericellular matrices in genetic expression of individual cells and tissues. Unfortunately, our knowledge of the chemistry of matrices is still fragmentary, particularly their molecular organization and function, except possibly for the fibrillar collagens.

The following glycoproteins have been isolated and partially characterized from such matrices and basement membranes: (a) various kinds of glycosaminoglycans and their protein complexes (Chapter 8); (b) collagens of various types and their procollagens (Chapters 1, 2, and 3); (c) fibronectin; (d) laminin; and (e) other as yet poorly defined adhesive glycoproteins and matrix components with molecular weights of 140,000, 170,000, 190,000, and 250,000. This chapter is limited to discussion of fibronectin, laminin, and other adhesive glycoproteins. In addition, the known chemistry of glycoprotein carbohydrate chains and their possible functions will be described. The abbreviations used for carbohydrate residues are given in the frontmatter of the book.

7.2. Glycoprotein Carbohydrate

7.2.1. Chemical Properties and Structure

Most extracellular proteins, either structural or functional, are glycosylated and therefore are technically glycoproteins. Four types of carbohydrate linkage to protein can be distinguished among vertebrates (Table 7.1). These are the proteoglycans, which contain polysaccharide (glycosaminoglycan) O-glycosidically linked through xylose to serine or threonine (Chapter 8); the collagens, which contain glucose and galactose O-glycosidically linked to hydroxylysine (Chapter 2, Section 2.3.2); the mucins, which contain oligosaccharide O-glycosidically linked to serine or threonine; and the large body of glycoproteins, which contain oligosaccharide with a mannose core N-glycosidically linked to asparagine. The last two types are the major objects of attention here, particularly the N-asparagine type, and are generally what is meant by the term glycoprotein. The carbohydrate will be referred to here as O-linked or serine/threonine-linked for the mucins, and N-linked or asparagine-linked for the mannose-core type N-glycosidically linked to asparagine.

Some proteins contain more than one type of linkage. The procollagens contain asparagine-linked carbohydrate in the precursor position of the molecule as well as hydroxylsine-linked carbohydrate in the triple helix. The proteoglycans frequently contain N-linked and sometimes O-linked carbohydrate as well as their characteristic glycosaminoglycan chains. N-linked and O-linked carbohydrate can also occur together in other glycoproteins (Winzler et al. 1967).

Our knowledge of the carbohydrate structure of glycoproteins, particularly those N-linked to asparagine, has been reviewed extensively by Montreuil (1980) and briefly by Kornfeld and Kornfeld (1980). Progress in this area has been made by improvement in various analytical methods such as:

1. Separation of glycopeptides after exhaustive digestion of glycoprotein by pronase and passage through a lectin affinity column, which appears to separate according to the peripheral (antennary) structure (Ogata et al. 1975; Finne et al. 1980).
2. Sugar sequences and the linkages between sugars have been determined by methylation (Hakomori 1964) followed by direct probe mass spectrometry (Karlsson et al. 1974), and by gas chromatography-mass spectrometric identification of partially O-methylated sugars and amino sugars (Björndal et al. 1970; Stellner et al. 1973).
3. New exo- and endo-glycosylhydrolases (*See* Kobata 1979) with well-defined specificities have been isolated and applied to elucidation of the chemistry of carbohydrate chains.
4. Hydrazinolysis (Fukuda et al. 1976) and a new glycopeptidase (Takahashi and Nishibe 1981) have been used for cleavage of the GlcNAc→asparagine linkage.

The major events in establishing the structural chemistry of asparagine-linked carbohydrate chains are listed in Table 7.2. Based on the major core structures and peripheral variations in N-linked carbohydrate chains (Table 7.1, Part I), a few structural rules can be described as follows:

1. Branching of mannose residues in high-mannose type chains occurs through $\alpha 1 \rightarrow 3$ or $\alpha 1 \rightarrow 6$ linkage; the second and third branches always occur on the chain linked $\alpha 1 \rightarrow 6$.
2. Extension of $\alpha 1 \rightarrow 2$ linked mannosyl residues in high mannose type chains takes place on the $\alpha 1 \rightarrow 3$ linked core mannose.
3. The addition of a GlcNAc residue to the core mannosyl residue to form a biantennary complex type takes place through $\beta 1 \rightarrow 2$ linkage.

Table 7.1. Basic Structure of Glycoprotein Carbohydrate

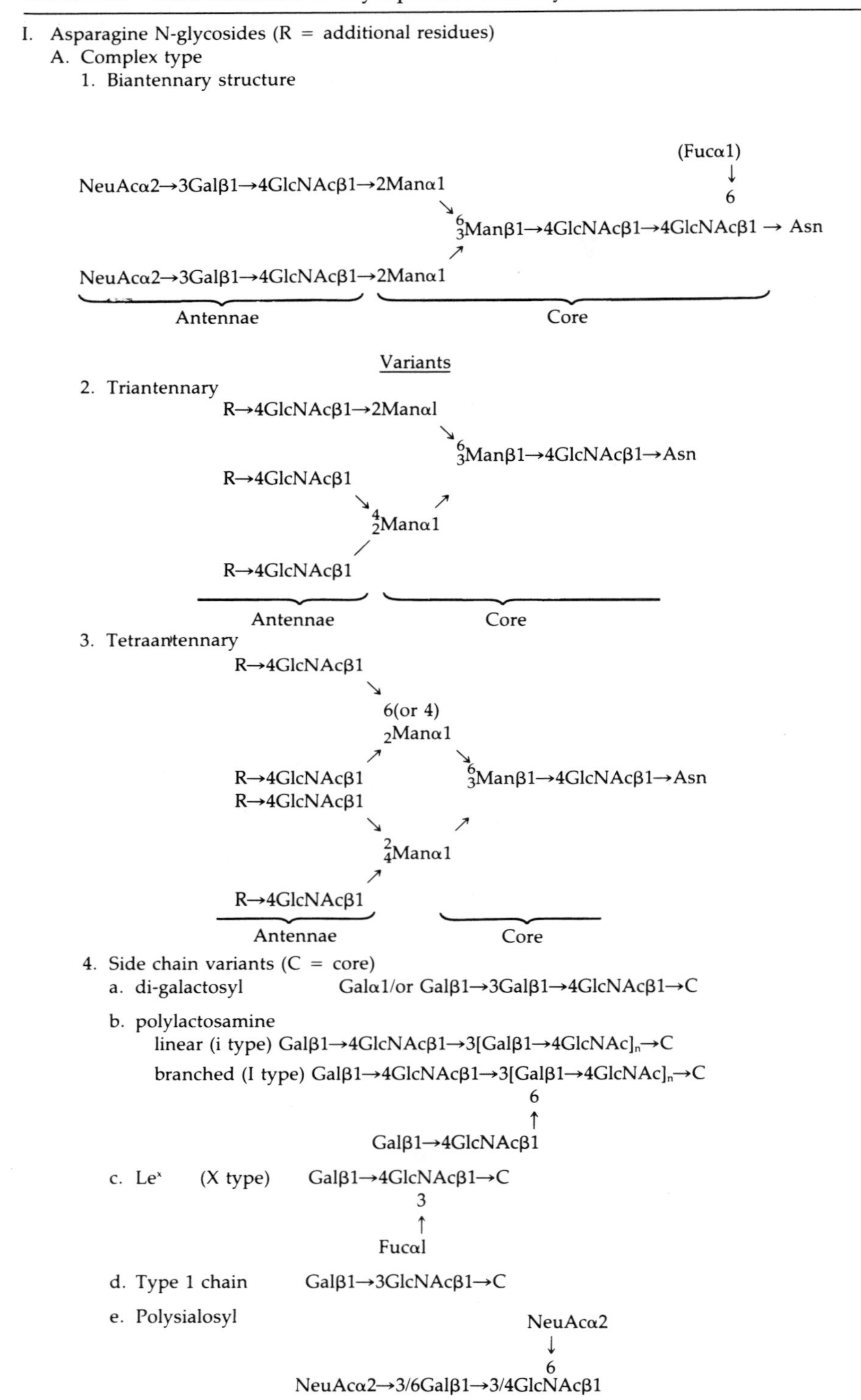

I. Asparagine N-glycosides (R = additional residues)
   A. Complex type
      1. Biantennary structure

(Fucα1)
↓
6
NeuAcα2→3Galβ1→4GlcNAcβ1→2Manα1
↘
${}^{6}_{3}$Manβ1→4GlcNAcβ1→4GlcNAcβ1 → Asn
↗
NeuAcα2→3Galβ1→4GlcNAcβ1→2Manα1

Antennae | Core

Variants

2. Triantennary

R→4GlcNAcβ1→2Manαl
↘
${}^{6}_{3}$Manβ1→4GlcNAcβ1→Asn
R→4GlcNAcβ1
↘ ↗
${}^{4}_{2}$Manα1
/
R→4GlcNAcβ1

Antennae | Core

3. Tetraantennary

R→4GlcNAcβ1
↘
6(or 4)
${}_{2}$Manα1
↗ ↘
R→4GlcNAcβ1 ${}^{6}_{3}$Manβ1→4GlcNAcβ1→Asn
R→4GlcNAcβ1
↘ ↗
${}^{2}_{4}$Manα1
↗
R→4GlcNAcβ1

Antennae | Core

4. Side chain variants (C = core)

a. di-galactosyl Galα1/or Galβ1→3Galβ1→4GlcNAcβ1→C

b. polylactosamine
   linear (i type) Galβ1→4GlcNAcβ1→3[Galβ1→4GlcNAc]$_n$→C
   branched (I type) Galβ1→4GlcNAcβ1→3[Galβ1→4GlcNAc]$_n$→C
   6
   ↑
   Galβ1→4GlcNAcβ1

c. Le^x (X type) Galβ1→4GlcNAcβ1→C
   3
   ↑
   Fucαl

d. Type 1 chain Galβ1→3GlcNAcβ1→C

e. Polysialosyl
   NeuAcα2
   ↓
   6
   NeuAcα2→3/6Galβ1→3/4GlcNAcβ1

Table 7.1 *(continued)*

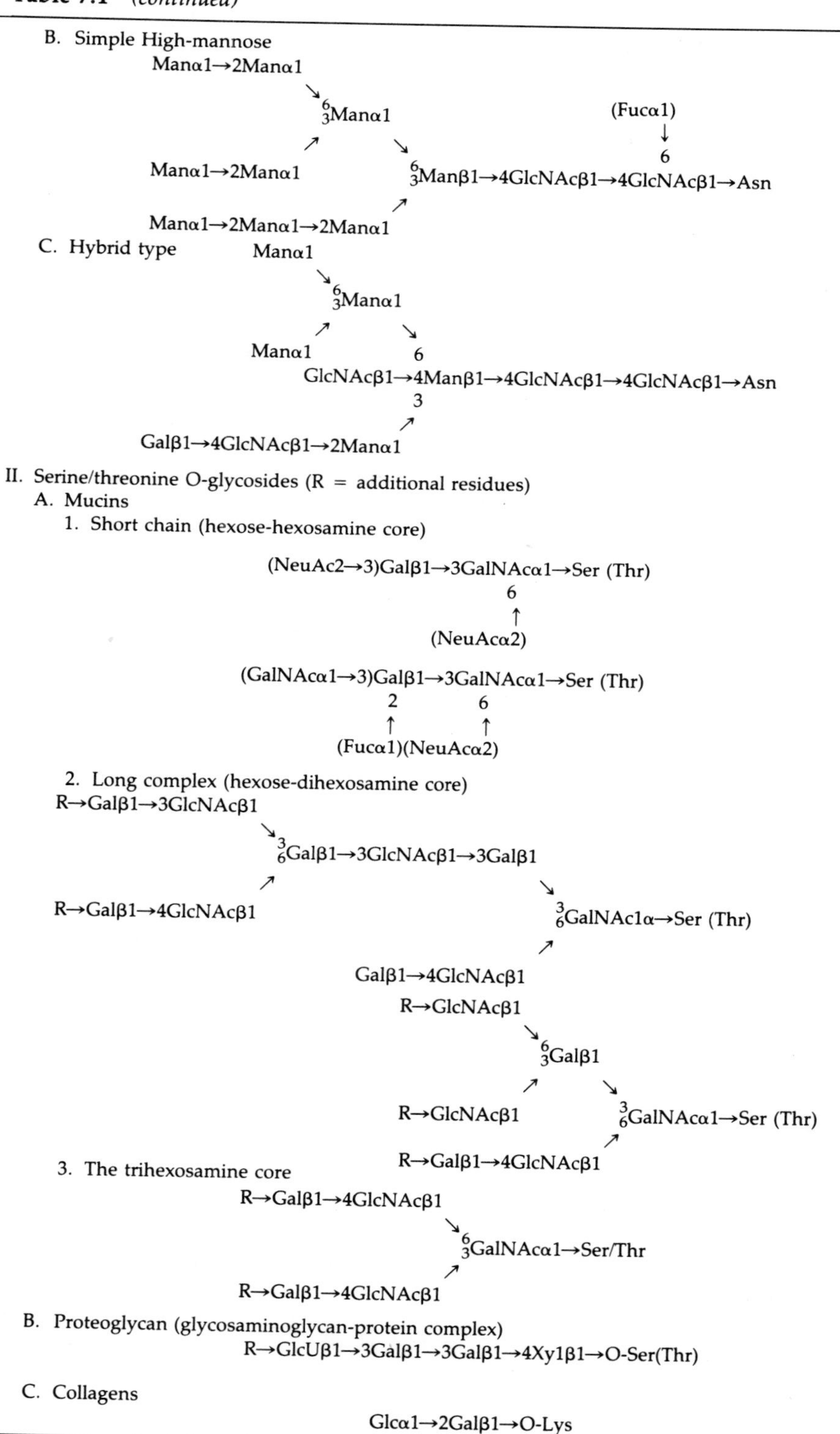

B. Simple High-mannose

Manα1→2Manα1 ↘ ${}^{6}_{3}$Manα1

Manα1→2Manα1 ↗ ${}^{6}_{3}$Manα1 ↘ ${}^{6}_{3}$Manβ1→4GlcNAcβ1→4GlcNAcβ1→Asn (with (Fucα1) → 6 on the first GlcNAc)

Manα1→2Manα1→2Manα1 ↗ ${}^{6}_{3}$Manβ1

C. Hybrid type

Manα1 ↘ ${}^{6}_{3}$Manα1

Manα1 ↗ ${}^{6}_{3}$Manα1 ↘ 6

GlcNAcβ1→4Manβ1→4GlcNAcβ1→4GlcNAcβ1→Asn

3 ↗ Galβ1→4GlcNAcβ1→2Manα1

II. Serine/threonine O-glycosides (R = additional residues)

A. Mucins

1. Short chain (hexose-hexosamine core)

(NeuAc2→3)Galβ1→3GalNAcα1→Ser (Thr)
6 ↑ (NeuAcα2)

(GalNAcα1→3)Galβ1→3GalNAcα1→Ser (Thr)
2 ↑ (Fucα1) 6 ↑ (NeuAcα2)

2. Long complex (hexose-dihexosamine core)

R→Galβ1→3GlcNAcβ1 ↘ ${}^{3}_{6}$Galβ1→3GlcNAcβ1→3Galβ1

R→Galβ1→4GlcNAcβ1 ↗ ${}^{3}_{6}$Galβ1

Galβ1→3GlcNAcβ1→3Galβ1 ↘ ${}^{3}_{6}$GalNAc1α→Ser (Thr)

Galβ1→4GlcNAcβ1 ↗ ${}^{3}_{6}$GalNAc1α

R→GlcNAcβ1 ↘ ${}^{6}_{3}$Galβ1

R→GlcNAcβ1 ↗ ${}^{6}_{3}$Galβ1

${}^{6}_{3}$Galβ1 ↘ ${}^{3}_{6}$GalNAcα1→Ser (Thr)

R→Galβ1→4GlcNAcβ1 ↗ ${}^{3}_{6}$GalNAcα1

3. The trihexosamine core

R→Galβ1→4GlcNAcβ1 ↘ ${}^{6}_{3}$GalNAcα1→Ser/Thr

R→Galβ1→4GlcNAcβ1 ↗ ${}^{6}_{3}$GalNAcα1

B. Proteoglycan (glycosaminoglycan-protein complex)

R→GlcUβ1→3Galβ1→3Galβ1→4Xy1β1→O-Ser(Thr)

C. Collagens

Glcα1→2Galβ1→O-Lys

Table 7.2. Important Events in Structural Chemistry of Asparagine-linked Carbohydrate Chains

Event	Reference (Major area of work)
Establishment of GlcNAc Asparagine N-glycoside linkage	Neuberger, Cunningham, Yamashima, and associates—See Neuberger et al. 1966
Distinction of two types of oligosaccharide chains, high mannose type and complex type	Lee et al. 1965 (ovalbumin) Wagh et al. 1969 (α_1-acid glycoprotein) Spiro 1965 (separation of two types from thyroglobulin)
Presence of N-acetylchitobiose structure at the core (GlcNAcβ1 → 4GlcNAcβ1 → Asn)	Tarentino et al. 1970 (ribonuclease B)
Establishment of high mannose type structure	Yamaguchi et al. 1971 (aspergillus orizae α-amylase)
Discovery of β-mannosyl structure linked to GlcNAc (Manβ1 → 4GlcNAcβ1 → 4GlcNAcβ1 → Asn)	Sukeno et al. 1972; Toyoshima et al. 1973 (complex type) Sukeno et al. 1972; Sugahara et al. 1972; Lee and Scocca 1972 (high mannose type)
Establishment of common core structure between high mannose type and complex type, and the presence of Fuc at the innermost GlcNAc residue through 1 → 6 linkage Manα1 ↘ $^{6}_{3}$Manβ1 → 4GlcNAcβ1 → 4GlcNAcβ1 → Asn; Manα1 ↗; Fucα1 → 6 (innermost GlcNAc)	Baenziger and Kornfeld 1974
Microheterogeneity and structure of carbohydrate chain in ovalbumin and discovery of hybrid type glycopeptide	Yamashita et al. 1978
The correlation of antennary structure and concanavalin A reactivity	Ogata et al. 1975; Baenziger and Fiete 1979
Discovery of new type of side chain structure	
Polylactosaminoglycan	Järnefelt et al 1978; Krusius et al. 1978; Fukuda et al. 1979
Le^x (or X-hapten)	Krusius and Finne 1978
Galβ1 → 3Gal	Kornfeld 1978
NeuAcα2 → 3Galβ1 → 4GlcNAcβ1	Mizuochi et al. 1979
Galβ1 → 3GlcNAc (type 1 chain)	Mizuochi et al. 1979
Atypical core structure Manα1 ↘ $^{6}_{2}$Manβ1 → 4GlcNAcβ1 → 4GlcNAc; Xylα1 ↗; Fucα1 → 3 (GlcNAc)	Fukuda et al. 1976; Ishihara et al. 1979

4. Branching by addition of a GlcNAc residue in the complex type takes place first through β1 → 4 linkage on the α1 → 3-linked core mannose to form the triantennary structure; further branching takes place through β1 → 6 linkage on the α1 → 6-linked core mannose to form the tetraantennary structure.
5. The third GlcNAc residue dissecting the biantennary mannose core structure (Manα1 → 6[Manα1 → 3]Man) always occurs as GlcNAcβ1 → 4Man to form

a hybrid type structure. A similar GlcNAcβ1 → 4Man linkage added to complex type structures results in penta- and hexaantennary complex type structures.

In contrast to the extensive understanding of the chemistry of asparagine-linked N-glycoside carbohydrate chains, carbohydrate chains O-linked to serine or threonine need further characterization. The base-catalyzed β-elimination cleavage of O-glycosidic linkage to serine or threonine residues was initially applied to chondromucoprotein (Anderson et al. 1963), submaxillary mucin (Bhavanandan et al. 1964; Tanaka et al. 1964), and blood group glycoprotein (Schiffman et al. 1964). With protective reduction (Carlson 1968; Iyer and Carlson 1971) the procedure has been used to characterize a number of oligosaccharides, as reviewed by Kornfeld and Kornfeld (1980). However, microheterogeneity of O-linked carbohydrate seems to be much greater than N-linked carbohydrate. For example, more than 30 kinds of oligosaccharide are produced from sheep gastric mucin glycoprotein even if reducing terminals are carefully protected by reduction (Wood et al. 1981)—some of them have a trihexosamine-type core structure (Hounsell et al. 1981). The major variations in the core structure of O-linked carbohydrate chains are shown in Table 7.1, Part IIA.

7.2.2. Function

Possible functions of carbohydrate chains linked to protein are to modify physical properties; alter susceptibility to proteases; regulate metabolic behavior of the protein; and influence indirectly the mobility, localization, and organization of protein within cells, in membranes, and in pericellular tissues. Table 7.3 lists the major known functions of glycoprotein carbohydrate chains, with examples and references.

The biological activity of certain proteins, such as hormones, enzymes, and lectins, has been studied by modification of carbohydrate chains by exo- and endoglycosidase treatment. Enzyme activities are usually unchanged, but some exceptions are reported that are believed due to an influence on tertiary structure (Chu et al. 1978; Wang and Hirs 1977). Lectin activity is consistently unchanged. Loss of glycoprotein hormone activity by sialidase treatment was reported in the early literature; these results are now ascribable to the elimination of the hormone from the serum as a result of altered recognition. Some carbohydrate chains at the cell surface are obviously antigenic, particularly for the Ii determinants of blood group ABH (*See* Watkins 1980; Hakomori 1981). However, loss of antigenicity by modification of carbohydrate chains may not always mean that the antigenic determinants are carbohydrates. Sialosyl residues may have a role in stabilizing polypeptide conformation and thus play a role in maintaining polypeptide antigenicity. Examples are seen in blood group MN determinant glycoprotein (Lisowska and Kordowicz 1977; Furthmayr 1978; Sadler et al. 1979) and in the Paul-Bunnell antigen directed to antibodies in sera of patients with infectious mononucleosis (Watanabe et al. 1980).

The obvious contribution of carbohydrate chains to the viscosity of mucins, the antifreeze activity of serum glycoproteins of antarctic fish, and resistance to denaturation are regarded as modifications of the physicochemical properties of the glycoprotein.

Table 7.3. Function of Glycoprotein Carbohydrate Chains

A. Effects of carbohydrate chains on catalytic or binding activities of proteins, enzymes, hormones, lectins, and antigens
   1. Many enzyme activities not affected; some enzyme activities modified in their kinetics, pH optima, and denaturation resistance, etc.
   2. Lectin activities not affected (Biroc and Etzler 1978; Lotan et al. 1975)
   3. In vivo activities of glycoprotein hormones (follicle stimulating hormone, luteinizing hormone, and erythropoietin) are lost by sialidase treatment (*See* Warren et al. 1978), but in vitro activities not changed (Goldwasser et al. 1974)
   4. Stabilizes polypeptide conformational antigenic determinants, e.g., blood group MN antigen (Lisowska and Kordowicz 1977; Furthmayr 1978; Blumenfeld and Adamany 1978)

B. Effects of carbohydrate chains on physicochemical properties of proteins
   1. Mucin viscosity (Gottschalk and Thomas 1961)
   2. Antifreeze activity (Feeney and Yeh 1978)
   3. Resistance to denaturation and protease digestability (Vegarud and Christensen 1975; Chu et al. 1978)

C. Glycosylation inhibition by tunicamycin
   1. General: embryogenesis (Heifetz and Lennarz 1979; Surani 1979), virus budding and migration (Schwarz et al. 1976; Morrison and McQuain 1978) cell adhesion (Duksin et al. 1978a, 1978b), and metastasis (Irimura et al. 1981)
   2. Specific:
      a. Inhibition of immunoglobulin mobilization and secretion (Hickman and Kornfeld 1978)
      b. Enhanced degradation of fibronectin (Olden et al. 1978), inhibited release of fibronectin (Housely et al. 1980)
      c. Blocked processing of procollagen to collagen (Duksin and Bornstein 1977) due to the altered processing protease (Duksin et al. 1978a)

D. Recognition of carbohydrate chain by cell surface protein
   1. Recognition of serum glycoprotein carbohydrate chain by liver cell surface glycoprotein (*See* Ashwell and Morrel 1974)
   2. Recognition of mannose 6-phosphate residue of lysosomal enzymes by cell surface protein (*See* Sly and Stahl 1979; Neufeld and Ashwell 1980).

E. Interaction and recognition of cell surface carbohydrates by carbohydrate binding proteins (lectins, glycosidases, and glycosyltransferases)
   1. Plant lectin interaction and mitogenesis (*See* Goldstein and Hayes 1978)
   2. Cell surface lectin and its possible role in cell–cell recognition (*See* Simpson et al. 1978; Barondes 1981)
   3. Interaction of cell surface carbohydrates and microbial lectins (*See* Ofek et al. 1978)
   4. Cell adhesion induced by glycosylhydrolases (Rauvala et al. 1981; Carter et al. 1981; Rauvala and Hakomori 1981)

F. Nonspecific cell adhesion mediated by adhesive proteins fibronectin, laminin, and others

In 1971 a new antiviral antibiotic, tunicamycin, was isolated (Takatsuki et al. 1971). Subsequent studies revealed that it blocks the conversion of dolichol phosphate to dolichol phosphate N-acetylglucosamine (Tkacz and Lampen 1975), intermediates in the biosynthesis of asparagine-linked carbohydrate. A number of studies have been carried out to examine alteration or modification of function as a result of blocked glycosylation. The studies are summarized in Table 7.3, Part C. Notably, fibronectin degradation is enhanced if synthesized in the presence of tunicamycin due to loss of the protective action of carbohydrate on protease digestion (Olden et al. 1979). In addition, fibronectin secretion is reduced by inhibition of glycosylation (Housley et al. 1980). Conversion of type I procollagen to collagen is also inhibited in vitro in the presence of tunicamycin (Duksin and Bornstein 1977). Therefore, it was first postulated that the carbohydrate chain on procollagen has a role in conversion of procollagen to collagen. However, further studies indicated that glycosylation of a processing protease is also inhibited, and the overall inhibition of procollagen processing to collagen can be explained by inhibition of the processing protease (Duksin et al. 1978a, 1978b). As tunicamycin may inhibit not only glycosylation but also protein synthesis, its overall effect should be carefully interpreted, particularly because the observed effect may be very general. Examples are inhibition of embryo development, inhibition of cell adhesion, and alteration of metastatic properties of tumor cells.

The basic mechanisms supporting cell social life are the recognition by cells of exogenous ligands, pericellular matrices, and the surface structures of adjacent cells. Intercellular recognition may be based on the reaction between a carbohydrate structure at the cell surface and its recognition protein, or conversely, its recognition protein at the cell surface and the carbohydrate of a glycoprotein. For the latter type of cell recognition, two systems have been well characterized (Table 7.3, Part D). The first system is the reaction between cell surface protein of hepatocytes and serum glycoprotein carbohydrate originally described by Ashwell and Morrel (1974). The second is the recognition and uptake of lysosomal glycoprotein enzymes, discovered in correlation with the pathogenesis of the mucopolysaccharidoses, the genetic disorders of glycosaminoglycan (mucopolysaccharide) catabolism (Hickman et al. 1974; Neufeld et al. 1975). Subsequently, Kaplan et al. (1977), Ullrich et al. (1977), and Distler et al. (1979) observed that the uptake of various glycosylhydrolase glycoproteins by the cells is inhibited by mannose-6-phosphate, which suggests that mannose-6-phosphate residues in lysosomal glycoproteins are recognized by cell surface protein receptors. Further studies indicate that recognition through mannose-6-phosphate is regulated by an α-N-acetylglucosaminyl residue linked to the phosphate group of the mannose phosphate (Hasilik et al. 1980; Tabas and Kornfeld 1980). In both examples, the interaction is followed by a series of cellular events including endocytosis and recycling of receptor proteins (*See* Schlessinger 1980; Pearse 1980; Neufeld and Ashwell 1980).

In the other type of interaction, cell surface carbohydrate is recognized by various carbohydrate-binding proteins, collectively called lectins (For reviews, see Simpson et al. 1978; Barondes 1981). Examples are given in Table 7.3, Part E. However, the evidence so far presented for the presence of cell-surface lectins is weak, and the basis for claiming that such a cell-surface lectin is indeed functional in cell recognition is even weaker. Reactivity at the cell surface has been described for several systems including fluorescein-labeled oligosaccharides (Kieda et al. 1979); glycolipid liposome including fluorescein (Huang 1978); interaction of hepatocytes with galactosyl-S-glycoside-Sepharose complex (Weigel et al. 1979); inhibition of teratocarcinoma-rabbit erythrocyte aggregation by mannose or fucose (Grabel et al. 1979); and asialo-fetuin-dependent cell aggregation and its inhibition by lactose (Raz and Lotan 1981). However, the same phenomenon can also be observed with the haptenic substrates of glycosyltransferases (Roseman 1970) and hydrolases. Recently, a role of glycosylhydrolases in cell recognition and cell adhesion has been fully documented (Rauvala et al. 1981; Carter et al. 1981; Rauvala and Hakomori 1981).

7.3. Fibronectin

The fibronectins are a class of multifunctional high molecular weight glycoproteins present at the cell surface and in the pericellular and intercellular matrix, basement membranes, as well as in a large variety of body fluids. Their chemical properties and structure vary significantly depending on the source and species, but they share common properties that will be defined later. It is not known whether the heterogeneity arises solely from posttranslational chemical differences or is also present at the gene level. A number of reviews on fibronectin have been published (Hynes 1976, Mosesson 1977; Culp 1978; Grinnell 1978; Vaheri and Mosher 1978; Yamada and Olden 1978; Vaheri et al. 1978; Pearlstein et al. 1980; Ruoslahti et al. 1981; Mosher 1980; Mosesson and Amrani 1980; Saba and Jaffe 1980; Chen 1981; Hynes 1982; Kleinman et al. 1981; Hay 1981; Hynes and Yamada 1982; Yamada 1983). They should be consulted for details and more complete references.

7.3.1. Discovery and Definition

Interest in the biology and biochemistry of fibronectin was initially aroused by its presence as a cell surface component and its decrease associated with oncogenic transformation (Hynes 1973; Gahmberg and Hakomori 1973; Ruoslahti et al. 1973). However, it has often been said that fibronectin, in the form called cold insoluble globulin, was discovered by Morrison et al. (1948), and isolated by Mosesson and Umfleet (1970) from plasma. In fact, cold insoluble globulin, the plasma glycoprotein that interacts with fibrin and fibrinogen, was known only to a few blood coagulation specialists until Ruoslahti and Vaheri (1975) discovered its antigenic identity with a new fibroblast cell surface glycoprotein sensitive to oncogenic transformation. It was discovered by several investigators in 1973 (loc. cit) and is now called fibronectin.

Research to discriminate cell surface components of malignant cells from normal cells has existed since glycolipid (Hakomori and Murakami 1968) and glycopeptide (Warren et al. 1973) differences were disclosed. Systematic protein analyses were unsatisfactory until sodium dodecyl sulfate-polyacrylamide gel electrophoresis (SDS-PAGE) was introduced by Weber and Osborn (1969) and adapted to slab gels by Laemmli (1970). Sufficient sensitivity to discriminate the pattern of transformed cells from normal cells was achieved by $[^{125}I]$-iodination with lactoperoxidase, introduced by Phillips and Morrison (1970), and by labeling surface carbohydrate with galactose oxidase and $NaB[^{3}H]_4$, devised by Gahmberg and Hakomori (1973). A transformation-sensitive high molecular weight glycoprotein was described in two papers in 1973 (Hynes 1973; Gahmberg and Hakomori 1973) and several papers in 1974 (Hogg 1974; Stone et al. 1974; Gahmberg et al. 1974; Robbins et al. 1974; Yamada and Weston 1974). Ruoslahti et al. (1973) also recognized a cell surface fibroblast (SF) antigen secreted into the medium. The antigen was also deleted from transformed cells (Vaheri and Ruoslahti 1974), as detected by cell surface labeling. The glycoprotein was soon identified as the same or similar to factors observed previously to cause biological phenomena related to cell adhesion and phagocytosis called cell attachment protein (Klebe 1974), cell adhesion factor (Pearlstein 1976), cell spreading factor (Grinnell 1976), antigelatin factor (Wolff et al. 1967; Dessau et al. 1978), and opsonic factor (Allen et al. 1973). A collective designation—fibronectin—was proposed by Kuusela et al. (1976). It is now used for the class of compounds, although cell surface fibronectin is significantly different from plasma fibronectin, and the fibronectins in various tissues, stages of development, and species may not be identical.

Fibronectins can be defined as sharing three properties: (a) two disulfide bonded subunits that may or may not be identical, with molecular weights near 220,000; (b) they bind to collagen, heparin and fibrin, and induce cell adhesion and spreading; and (c) they are antigenically crossreactive irrespective of source and species. For example, a glycoprotein of molecular weight 45,000 from sperm that binds to collagen and shows immunofluorescence with antifibronectin antibodies (Koehler et al. 1980), or a glycoprotein of molecular weight 140,000 from human fibroblasts (Section 7.5.2) that promotes cell adhesion but does not react with antifibronectin antibody, cannot be called fibronectin. Presumably the sperm protein is closely related to fibronectin with a similar function but only part of the structure, while the other protein is related in function but not structure.

7.3.2. Distribution

The distribution of fibronectin in cells, tissues, and body fluids has been studied largely by immunofluorescence or radioimmunoassay, rather than by chemical identification and characterization. Therefore our knowledge on this topic is tentative in some cases because immunologic reactivity does not necessarily mean that the same molecule is present (as noted in the previous section), nor does it supply adequate quantitative data. The data for distribution in various cells and tissues is listed in Table 7.4.

In general, fibronectin is synthesized by and is present around fibroblasts, endothelial cells, chondrocytes, glial cells, and myocytes. Fibronectin is abundant in the connective tissue matrix and in basement membranes. The distribution is quite characteristic for each cell and tissue. Epithelial cells, in general, are associated with much less fibronectin than the other cells listed above—it appears as spotty areas between cells. However, abundant fibronectin is normally found at the interphase between epithelial cells and basement membranes to which epithelial cells adhere. Blood cells, such as lymphocytes, myelocytes, and erythrocytes, lack fibronectin. Platelets and macrophages contain some but not a large quantity of fibronectin. Although the role of fibronectin in platelets is unknown, in macrophages it is important as a nonspecific opsonin for their phagocytotic activity (*See* Mosher 1980; Saba and Jaffe 1980). The fibronectin concentration in various body fluids is given in Table 7.4. Fibronectin exogenously transfused into the blood stream is rapidly deposited in the matrix of various tissues. It is postulated that tissue fibronectin could be derived from plasma fibronectin (Ruoslahti et al. 1981), but that many cell types produce it.

Table 7.4. Distribution of Fibronectin in Cells, Tissues, and Body Fluids[a]

Large amounts		Absent or trace amounts
Fibroblasts of all sources		Epithelial cells (except the attachment site between cells and between basement membranes)
Astroglial cells and Schwann cells		Transformed fibroblasts (see text for exceptions)
Amniotic cells		Lymphocytes, myelocytes, and erythrocytes
Endothelial cells		
Chondrocytes, osteocytes, myocytes		
Connective tissues and glial tissue		
Basement membranes		
Platelets		
Macrophages		
Presence in body fluids (μg/ml)		
Plasma	300	
Amniotic fluid	170	
Cerebrospinal fluid	1–3	
Seminal fluid	100	
Urine	10	

[a]Data from Ruoslahti et al. 1981.

7.3.3. Transformation, Cell Cycle, and Cell Contact

Loss or decrease of fibronectin on the cell surface was observed initially in virally transformed cells. The finding was further extended to chemically transformed cells (Clarke and Fink 1977) and spontaneous tumor cells (Pearlstein et al. 1976). Deletion is also induced by tumor promoters (Blumberg et al. 1976).

Expression of fibronectin in fibroblasts transformed with temperature-sensitive mutants of oncogenic viruses shows a good correlation with the expression of oncogenic phenotypes dependent on permissive and nonpermissive temperatures (Gahmberg et al. 1974; Vaheri and Ruoslahti 1974; Stone et al. 1974; Hynes and Wyke 1975; Chen et al. 1976). The malignant phenotype of Chinese hamster ovary cells is suppressed in the presence of cyclic AMP; at the same time synthesis and pericellular organization of fibronectin are restored to the same state as in normal cells (Nielson and Puck 1980). Similarly, fibronectin synthesis has been found to be restored in various malignant cells whose growth behavior was induced to be normal (reverse transformants) by dexamethazone (Furcht et al. 1979), sodium butyrate (Hayman et al. 1980), and interferon (Pfeffer et al. 1980).

Changes in pericellular fibronectin also correlate with cell cycle and cell contact. Hynes and Bye (1974) and Gahmberg and Hakomori (1975) initially observed that synchronized hamster embryonic fibroblast cells express fibronectin maximally during G1 phase, at a decreased rate progressively through S and G2 phases, and almost not at all during M phase. A large reduction of pericellular fibronectin at M phase has also been observed by immunofluorescence (Stenman et al. 1977). Loss or reduction of fibronectin is therefore a common phenomenon with oncogenic transformants and normal cells during mitosis, although the fibronectin of transformants is consistently reduced at all phases of the cell cycle.

On the other hand, fibronectin accumulates in the intercellular space when cells contact. Several to ten-fold accumulation of fibronectin may occur, a much larger change than observed for glycolipids or other glycoproteins. The changes may well be related to contact inhibition of cell growth (Gahmberg and Hakomori 1975; Carter and Hakomori 1981). Figure 7.1 shows the results of an experiment illustrating these changes.

Decrease of fibronectin synthesis associated with oncogenic transformation and its increase on cell contact may not be specific for fibronectin, but also may be observed for procollagen and other glycoproteins (Levinson et al. 1975; Hata and Peterkovsky 1977; Vaheri et al. 1978; Carter and Hakomori 1978, 1981; Carter 1982). Therefore the deletion mechanism is important for understanding the regulation of extracellular matrix in general. Several possibilities have been considered: (a) suppression of fibronectin synthesis at the transcription level initially proposed by Olden and Yamada (1977) and confirmed through the study of the cloned fibronectin cDNA (Fagan et al. 1981); (b) enhanced degradation of fibronectin by a protease specific for fibronectin (Keski-Oja and Todaro 1980); (c) modification of fibronectin by phosphorylation (Ali and Hunter 1981), aberrant glycosylation (Wagner et al. 1981), or sulfhydryl modification (Wagner and Hynes 1979), inhibiting incorporation of fibronectin into the pericellular matrix. It is possible that multiple factors are cooperating in loss or removal of fibronectin and other pericellular proteins.

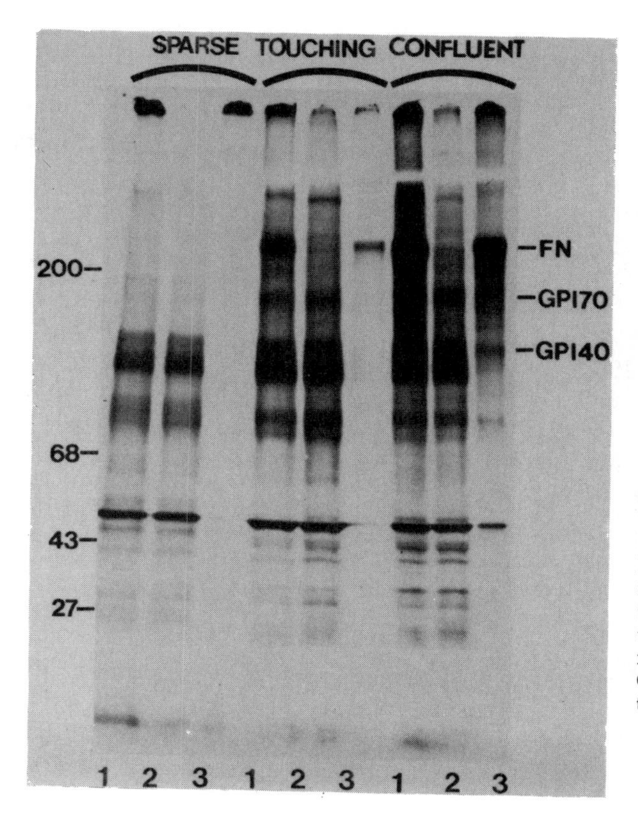

Figure 7.1. Cell density-dependent accumulation of matrix glycoprotein. Hamster embryonic fibroblasts (NIL) cells were plated at sparse, touching, and confluent densities and grown for 29 hours. The three cell populations were cell surface labeled utilizing the galactose oxidase-NaB(^{3}H)$_4$ method, and then each cell sample was extracted with detergent (Empigen BB). Whole cells (gel 1), the detergent soluble material (gel 2), and the insoluble matrix (gel 3) were analyzed on polyacrylamide gels in the presence of sodium dodecyl sulfate (SDS) followed by fluorography. Numbers in the left margin represent migration positions of standard proteins of known molecular weight in thousands. The accumulation of the matrix glycoproteins fibronectin *(FN)*, GP 170, and GP 140 is dependent on the density of the cell population.

7.3.4. Tumorigenicity and Metastatic Properties

Although the in vitro phenotype of transformed cells as measured by properties such as contact inhibition, serum requirement, cell saturation density, and morphology is well correlated with the level of pericellular fibronectin, the question remains whether pericellular fibronectin is related to tumorigenicity in vivo. Initially, a strong positive correlation was found between various rat tumor cells transformed by adenovirus and their tumorigenicity in rats. The greater the tumorigenicity, the greater the reduction of pericellular fibronectin (Chen et al. 1976). Consequently, the tumorigenicity of 28 cell lines in nude mice was compared by Kahn and Shin (1979). They found a number of cases where tumorigenic cells were still associated with high levels of fibronectin. Hybrid cell lines between normal human fibroblasts and HeLa cells having different degrees of tumorigenicity show no correlation with loss of pericellular fibronectin in vitro (Der and Stanbridge 1978). However, further studies (Der and Stanbridge 1980) indicate a clear difference in fibronectin organization and localization. Extensive studies by Chen et al. (1979) with more than 250 tumor cells screened for fibronectin reactivity by immunofluorescence at the cell surface revealed a strong correlation between tumorigenicity and fibronectin staining (correlation coefficient, 0.92). Although a few cell lines showed high tumorigenicity and were associated with high levels of fibronectin, tumors formed from these cells were negative for fibronectin in nude mice. The complexity of the relationship is shown by the fact that cultured cells from the tumors in the nude mice regained the ability to synthesize fibronectin.

A correlation between the metastatic properties of virally transformed cells and their fibronectin content has been studied by Chen et al. (1978), who found that highly metastatic cells have a reduced level of pericellular fibronectin. In colonic and other human carcinomas, the fibronectin levels of metastases and the original tumor have been compared. The metastases show lower levels of fibronectin (Smith et al. 1979). In an independent study of breast cancer, a lower concentration of fibronectin was also found in the metastatic lesion (Labat-Robert et al. 1980). However, the fibronectin level is related more strongly to whether the tumor is carcinoma or sarcoma than to whether it is primary or secondary (Stenman and Vaheri 1981). Fibronectin immunofluorescence and released fibronectin of mammary carcinoma and its high- and low-metastatic variants in vitro and in vivo have been compared, but no simple correlation was found (Neri et al. 1981). Because there are many other adhesive glycoproteins besides fibronectin (Section 7.5), metastatic properties should be correlated with the total profile of pericellular adhesive matrix rather than a single component.

7.3.5. Function

Various functions of fibronectin have been suggested and are summarized with references in Table 7.5. It should be remembered that activities observed in vitro or in cell culture may not always be related to biological function in vivo. Important points are as follows:

Functions Related to Plasma. Fibronectin promotes cell attachment and spreading; the assumption is that most cells make it as needed. However, some cells

Table 7.5. Proposed Functions of Fibronectin

I. At cell surfaces
   A. Cell adhesion and spreading (Grinnell et al. 1977; Höök et al. 1977; Grinnell and Feld 1979; Ruoslahti and Hayman 1979)
   B. Cell proliferation and cell movement (Orly and Sato 1979; Ali and Hynes 1978)
   C. Phagocytic activity of reticuloendothelial cells; nonspecific opsonic activity (Blumenstock et al. 1977, 1978)

II. In the matrix
   A. Binding to collagen and mediation of cell adhesion on collagen-coated surfaces
      1. Cell adhesion on collagen-coated plates mediated by fibronectin
      2. Specific binding of collagen to fibronectin, and application of this property to affinity purification of fibronectin (Engvall and Ruoslahti 1977)
      3. Preferential binding to type III collagen (Engvall and Ruoslahti 1978; Jilek and Hörmann 1979)
   B. Binding to proteoglycan, particularly heparin (Stathakis and Mosesson 1977; Jilek and Hörmann 1979; Yamada et al. 1980)
   C. Binding to cell surface receptor, possibly a 47K glycoprotein (Hughes et al. 1982)
   D. Disulfide-stabilized aggregation (Hynes and Destree 1977)
   E. Crosslinking to collagen and fibrin by transglutaminase (Mosher 1975; Keski-Oja et al. 1976)

III. Clearance of debris and promotion of wound healing and homeostasis
   a. Interaction with fibrin and fibrinogen (Ruoslahti and Vaheri 1975; Stemberger and Hörmann 1976) promoting clearance of blood clots and replacement by cells and matrix (*See* Mosher 1980; Saba and Jaffe 1980)
   B. Nonspecific opsonic activity of macrophages, monocytes, and reticuloendothelial cells (Blumenstock et al. 1978; *See* Saba and Jaffe 1980)
   C. Binding to bacteria (Kuusela 1978; Mosher and Proctor 1980) and intracellular components, e.g., actin, DNA (*See* Ruoslahti et al. 1981)

do not synthesize fibronectin but require it as an essential component for proliferation—the mechanism by which they acquire it is not known. Cell movement is also greatly enhanced by addition of fibronectin. Promotion of phagocytosis of reticuloendothelial cells, such as macrophages, monocytes, and liver Kupfer cells, is classically known as nonspecific opsonic activity. These activities are mediated by interaction of fibronectin and cell surface receptors, which activate appropriate cytoskeletal or cytomotal proteins. The specific domain of fibronectin involved has been isolated as a fragment of about 150,000 molecular weight (Hahn and Yamada 1979; Sekiguchi and Hakomori 1980). A monoclonal antibody has been prepared that is directed to the locus of cell adhesion and spreading of fibronectin that blocks these activities. The locus is represented by a polypeptide with a molecular weight of about 15,000 (Pierschbacher et al. 1981). Polysialo-gangliosides (GD_{1a}, GT) have been proposed as part of the cell surface receptor because they inhibit cell attachment (Kleinman et al. 1979). However, these gangliosides also inhibit cell attachment to lectin (Rauvala et al. 1981). More recently, Aplin et al. (1981) and Hughes et al. (1982) have detected a glycoprotein of about 47,000 molecular weight that is selectively crosslinked to fibronectin when cells are attached to fibronectin on a glass surface. Antibody directed to this glycoprotein inhibits cell attachment, further indicating that this glycoprotein is the receptor for cell-surface fibronectin.

Functions Related to Organization of Matrix. Another class of fibronectin activity is directing and mediating the organization of pericellular and intercellular matrices and basement membranes. These matrices not only bind cell to cell but also may regulate function of cells, as mentioned in Section 7.1. The role of fibronectin in matrix organization is not understood, but it presumably is mediated by the major binding capabilities of fibronectin to collagen (Klebe 1974; Kleinman et al. 1976; Pearlstein 1976; Engvall and Ruoslahti 1977); to proteoglycan, particularly heparin, heparan sulfate, and hyaluronic acid (Stathakis and Mosesson 1977; Jilek and Hörmann 1979; Yamada et al. 1980; Laterra and Culp 1982; Oldberg and Ruoslahti 1982); and to the cell surface. In addition, fibronectin forms a disulfide-stabilized self-aggregate that may also involve other macromolecules (Hynes and Destree 1977; Carter and Hakomori 1981). It is able to covalently crosslink to collagen, fibrin, and other macromolecules by the action of a transglutaminase (*See* Mosher 1980). Fibronectin contains a susceptible glutamine residue in its N-terminal region (Jilek and Hörmann 1977, 1979; McDonald and Kelley 1980; Mosher et al. 1980; Sekiguchi et al. 1981), which may form an ε-amino-γ-glutamyl bond to a lysine residue in another molecule by the action of transglutaminase.

Functions Related to Homeostasis and Wound Healing. The third major class of fibronectin activities is to direct clearance of cell debris, blood clots, and other material, thus promoting wound healing and homeostasis. Again, the mechanism by which this occurs is unknown but presumably involves the ability of fibronectin to interact with fibrin (Ruoslahti and Vaheri 1975; Stemberger and Hörmann 1976) and to be crosslinked by transglutaminase. In addition, the ability of fibronectin to interact with C1q component of complement (Menzel et al. 1981; Pearlstein et al. 1982; Bing et al. 1982), certain bacteria (Kuusela 1978; Courtney et al. 1983), actin (Keski-Oja et al. 1980), and deoxyribonucleic acid (Zardi et al. 1979) may be important. These interactions are followed by phagocytotic clearance by the reticuloendothelial system (*See* Saba and Jaffe 1980). These functions are part of the original concept of constancy of the *"milieu interieur,"* proposed by Claude Bernard about 100 years ago and later elaborated as "homeostasis" by Cannon (1932).

Adhesion Plaque and Fibronectin. The substrate loci to which cell pseudopodia are tightly attached are seen as black spots by interference reflection microscopy and are called adhesion plaques. A proposed organization of an adhesion plaque is shown in Figure 7.2. Electron microscopy reveals that many actin cables terminate at the plasma membrane locus corresponding to the plaque, where a specific actin-binding protein—vinculin—is located (Geiger 1979). Vinculin may regulate actin attachment and organization at the junction between actin and a plasma membrane protein that is directly involved in cell adhesion. It is not yet known whether vinculin directly connects actin to the as yet unidentified plasma membrane protein (*See* Birchmeier 1981). Interestingly, the src gene product ($pp60^{src}$) may phosphorylate a tyrosine residue of vinculin (Sefton et al. 1981). It is suspected that there is a close correlation between the organization of actin and vinculin phosphorylation. A great deal of interest has also been focused on whether or not the adhesion plaques contain fibronectin. Recent studies with interference reflection microscopy clearly indicate that fibronectin is not located exactly at the locus of the adhesion plaque, but is distributed in its close neigh-

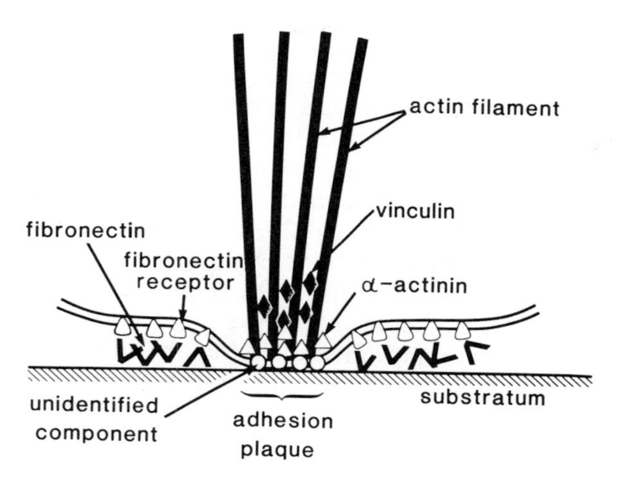

Figure 7.2. Schematic model of the architecture of an adhesion plaque. Adhesion plaques are specialized sites on the dorsal surface of cells spread out on the substratum where the cell membrane is less than 10 nm from the substratum. Actin filament bundles terminate there. Immunofluorescence studies reveal that both vinculin and α-actinin are also localized there. These proteins are considered to bundle and anchor actin filaments to the cytoplasmic surface of the cell membrane.

borhood (Birchmeier et al. 1980; Chen and Singer 1980). Fibronectin may be selectively eliminated from the adhesion site during the process of cell adhesion to form the adhesion plaque (Avnur and Geiger 1981). Consistent with this model is the observation that the kinetic curve of cell adhesion on a fibronectin coat is sigmoid, with the lag period suggesting preliminary steps before cell adhesion actively begins. This is in striking contrast to the adhesion kinetic curve on lectin-coated surfaces (Rauvala et al. 1981; Carter et al. 1981).

7.3.6. Structure

Functional Domains. Multifunctional molecules can usually be thought of as consisting of multiple domains, each of which is specialized to perform a specific function. Each domain is an independently folded portion of the polypeptide chain interconnected through a flexible region. The connecting regions are often more susceptible to protease digestion than the functional domains. Therefore, a selection of proteolytic conditions that preferentially hydrolyze the connecting regions leaving the functional domains intact is a powerful tool for elucidating the domain structure. Our present knowledge of the domain structure of fibronectin is mainly based on the plasma form, as the cellular form is difficult to obtain in an unaggregated state. Other species and variants of fibronectin, which may differ in their domain structures, have also not been well characterized as yet. What information is available is discussed here.

Plasma fibronectin consists of two unequal subunits, α and β, which are held together by disulfide bonds. The subunits may have slightly different molecular weights in different species. In the hamster, the molecular weights of the α and β subunits are 230,000 and 210,000, respectively. It is convenient in this section to refer to them as 230K and 210K subunits and to use the same kind of nomenclature for their fragments.

Various procedures have been tried to obtain quantitative yields of fragments representing each domain structure. The most successful has been a combination of trypsin followed by thermolysin, or a direct hydrolysis with thermolysin (Sekiguchi and Hakomori 1980; Sekiguchi et al. 1981). The fragmentation pattern obtained by these hydrolytic conditions is shown in Figure 7.3. After each cleavage, fragments are separated on affinity columns containing gelatin, fibrin, or heparin. Trypsin treatment of intact fibronectin gives three fragments, 32K, 200K, and 180K. Time course analysis indicates that the α and β subunits are converted into the larger fragments upon release of the 32K fragment. Both the 200K and 180K fragments bind to a gelatin column, but the 32K fragment does not. The collagen-binding domain present in the 200K and 180K fragments can be isolated as a smaller 40K fragment after thermolysin digestion of the 200K and 180K fragments (*See* below). The 200K and 180K fragments are separable on a fibrin column; only the 200K fragment is retained.

The 32K fragment, released by trypsin, can also bind to fibrin; thus, there are two distinct fibrin-binding domains—one in the 32K fragment and another in the 200K fragment. The 200K fragment is further cleaved into 150K, 40K, and 21K fragments by thermolysin, whereas the 180K fragment is cleaved into 140K and 40K fragments. The 21K fragment, produced only from the 200K tryptic fragment and therefore considered to be derived from the 230K subunit, is capable of binding to fibrin (Sekiguchi et al. 1981).

Five fragments are obtained from intact fibronectin by direct thermolysin treatment: 24K, 40K, 140K, 150K, and 21K. Of these, the 40K binds to collagen; the 24K binds to fibrin, heparin, *Staphylococcus aureus*, and interacts with transglutaminase; the 140K and 150K fragments bind to heparin; and the 21K binds to fibrin. Cell adhesion followed by spreading is induced only by the 140K and

Figure 7.3. Controlled fragmentation of hamster plasma fibronectin by trypsin and thermolysin. Trypsin cleaves α and β subunits into 200K and 180K fragments, repectively, by releasing 32K fragments from both of them. It also releases a small C-terminal segment that contains one or more interchain disulfide bonds. The tryptic 200K and 180K fragments are further cleaved into 150K + 40K + 21K, and 140K + 40K fragments, respectively. The tryptic 32K fragment is also degraded to 24K by thermolysin. Direct thermolysin digestion of intact fibronectin also gives 150K, 140K, 40K, 24K, and 21K fragments.

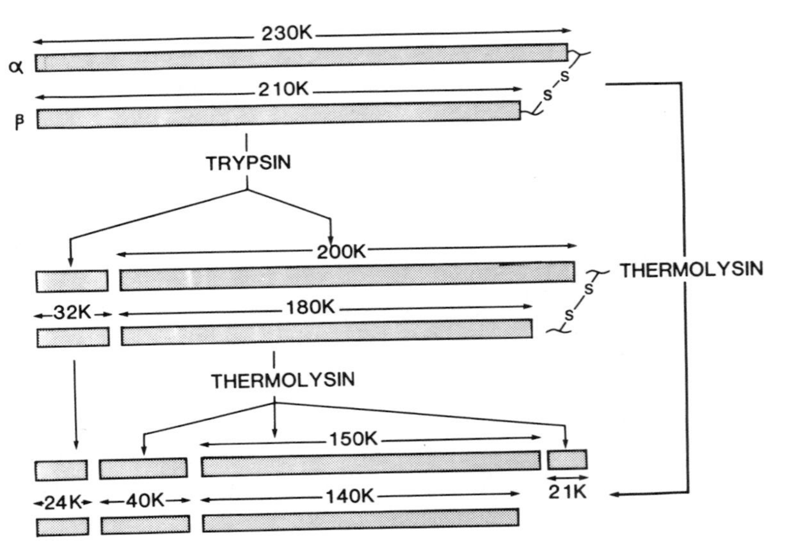

150K fragments when cells are added to plastic surfaces coated with these fragments. No other fragments induce cell attachment and spreading. Gelatin binding in vitro is generally equated to collagen binding. The apparent weaker binding to fibrillar collagen may be a reflection of steric effects.

There are two heparin-binding (i.e., 24K and 150K–140K) and two fibrin-binding (i.e., 24K and 21K) domains in intact fibronectin, but their binding affinities and specificities may vary (Richter et al. 1981; Hayashi and Yamada 1982). For example, the 150K–140K domain exhibits higher affinity to heparin than the 24K domain, whereas the 150K–140K domain has much less affinity to low-sulfated heparin sulfate, chondroitin sulfates, and hyaluronic acid than the 24K domain (Sekiguchi, et al. 1983).

The alignment of these domains has been studied by chemical cleavage of the intact protein and its proteolytic fragments at cysteine residues by S-cyanylation. Plasma fibronectin contains 1–2 cysteine residues per subunit, which are localized on the 150K and 140K domains (Fukuda and Hakomori 1979a; McDonald and Kelley 1980; Sekiguchi and Hakomori 1983a). The N-terminal half of both the α and β subunits is released as a 145K fragment by S-cyanylation of intact fibronectin. The cyanylation-cleaved 145K fragment consists of three domains: 24K, 40K, and 55-58K, the last being the N-terminal portion of the 150K–140K domain. The 21K domain is recovered within the fragments derived from the C-terminal half of the intact molecule. Upon S-cyanylation, the tryptic 200K and 180K fragments give a 120K fragment. This and other evidence indicates that the alignment of the domains (from N- to C-terminal) is 24K–40K–150K–21K for the larger α subunit, and 24K–40K–140K for the smaller β subunit, where interchain disulfide bonding is located near the C-terminal (Sekiguchi and Hakomori 1983a). A possible arrangement of the domain structure of hamster plasma fibronectin is shown in Figure 7.4A. Both α and β subunits contain one cysteine residue at about 145K from the N-terminal.

Most of the carbohydrate is located in the collagen-binding domain, but a small amount (possibly only one oligosaccharide/fragment) is present in the 150K–140K domain.

The unequal distribution of the C-terminal 21K domain found in hamster plasma fibronectin may not be generalized to other types of fibronectins because recent analysis of the domain structure of human plasma fibronectin suggests that the 21K domain is present in both α and β subunits (Sekiguchi and Hakomori 1983b). In fact, the molecular weight difference between α and β subunits in human fibronectin is much smaller than that of hamster fibronectin. The basic domain structure of human fibronectin, nevertheless, is almost the same as hamster fibronectin except for the symmetric distribution of the 21K domain (Figure 7.4B). Similar results have been obtained with not only human plasma fibronectin (McDonald and Kelley 1980; Ruoslahti et al. 1981), but also chicken cellular and plasma fibronectins (Hayashi and Yamada 1981).

Carbohydrate. The carbohydrate structure of fibronectin has been studied in the plasma and cellular forms, both attached and released into the medium. As shown in Table 7.6, cellular fibronectin has a fucose residue in the core structure, whereas plasma fibronectin does not contain fucose. More sialosyl substitution can be seen in plasma fibronectin than in the cellular form. The position of sialic

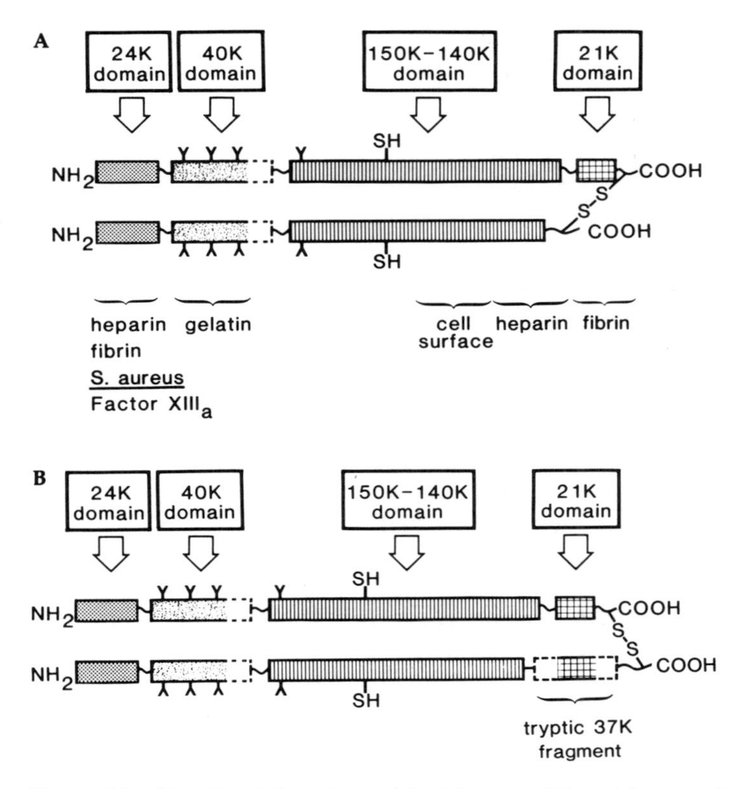

Figure 7.4. Functional domain model of hamster (**A**) and human plasma fibronectin (**B**). Both fibronectins are composed of at least four distinct structural domains. Each domain exhibits different biological activity, as listed below **A**. *Y* represents a carbohydrate unit. Three domains (150K–140K, 40K, and 24K) are present in both large (α) and small (β) subunits, but the 21K domain is considered to only be present in the large subunit. In human plasma fibronectin (**B**), all four domains are equally present in both large and small subunits, except that the 21K domain in the large subunit appears to be larger than that in the small subunit and is recovered as a 37K fragment by mild trypsin digestion.

acid in bovine plasma fibronectin is different from that in hamster plasma fibronectin. In bovine plasma fibronectin, sialic acid is located not only at the terminal galactose, but also at the subterminal N-acetylglucosamine. However, fucose is absent in the core region, which could be a general pattern of plasma fibronectin in contrast to cellular fibronectin. The carbohydrate structure of fibronectin released from transformed fibroblasts seems to be different from that found in fibronectin from normal cells (Wagner et al. 1981). Similar differences have been demonstrated for other glycoproteins (Warren et al. 1973).

Amino Acid Sequence. Partial primary structure of bovine plasma fibronectin has been reported by Petersen et al. (1983). They identified three types of internal sequence homologies (designated I, II, and III), suggesting that fibronectin evolved by duplication of several prototype genes. Studies of genomic DNA clones of fibronectin also indicate that fibronectin arose by multiple gene duplications of

Table 7.6. Glycosylation Patterns of Fibronectin

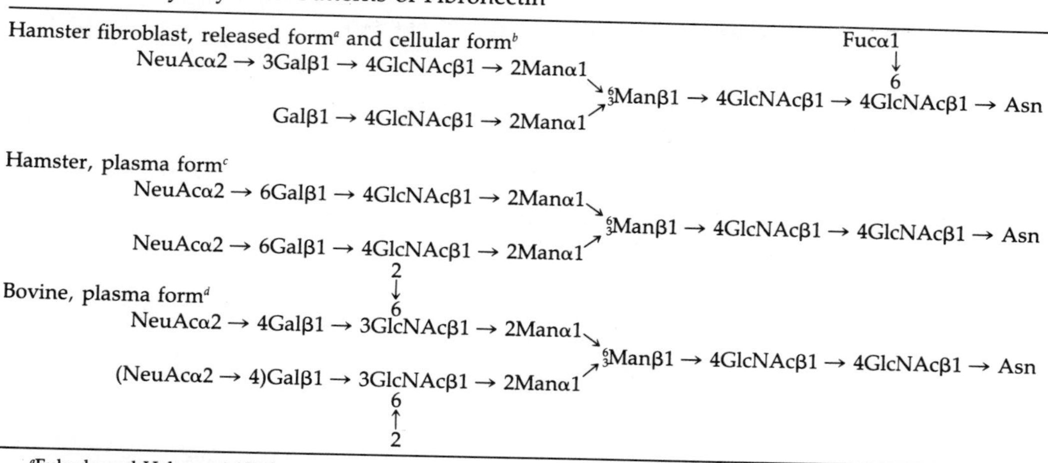

[a]Fukuda and Hakomori 1979b.
[b]Carter and Hakomori 1979.
[c]Fukuda et al. 1982.
[d]Takasaki et al. 1979.

a primordial gene(s) (Hirano et al. 1983). The N-terminal domain, isolated as a 29K plasmin fragment, consists of five type I homologies. The gelatin-binding domain, isolated as a 45K fragment by sequential digestion with plasmin and chymotrypsin, contains at least one type I homology and two type II homologies. The type III homology is localized in one of the heparin-binding domains, which appears to correspond to the C-terminal region of the 150K–140K domain. The C-terminal domain, recovered as a 23K plasmin fragment, also contains three type I homologies. The extreme C-terminal segment is obtained as a 6K fragment containing interchain disulfide bond(s). The 6K fragment also contains a phosphate group at the serine residue, which is the third amino acid from the C-terminal.

Molecular Shape. Fibronectin and laminin (Section 7.4.4) have been investigated by the rotary shadowing technique developed by Shotton et al. (1979) and used for some of the collagens (Chapter 1, Figure 1.5). Electron micrographs of fibronectin molecules reveal a characteristic V-shaped configuration as shown in Figure 7.5. Two flexible arms about 60 nm long, often with a fixed angle between them, may well be the α and β subunits interconnected at one end through a disulfide linkage (Engel et al. 1981). The linear nature of the molecule supports the linear arrangement of domains deduced from cleavage studies.

However, Koteliansky et al. (1981) reported that fibronectin that had been frozen and lyophilized on grids displayed a rather compact configuration with an axial ratio of approximately 2:1. Measurement of hydrodynamic properties of fibronectin indicates that fibronectin assumes an elongated and flexible configuration under physiologic conditions, but it further unfolds at high pH and/or ionic strength (Williams et al. 1982). Thus, the elongated, V-shaped config-

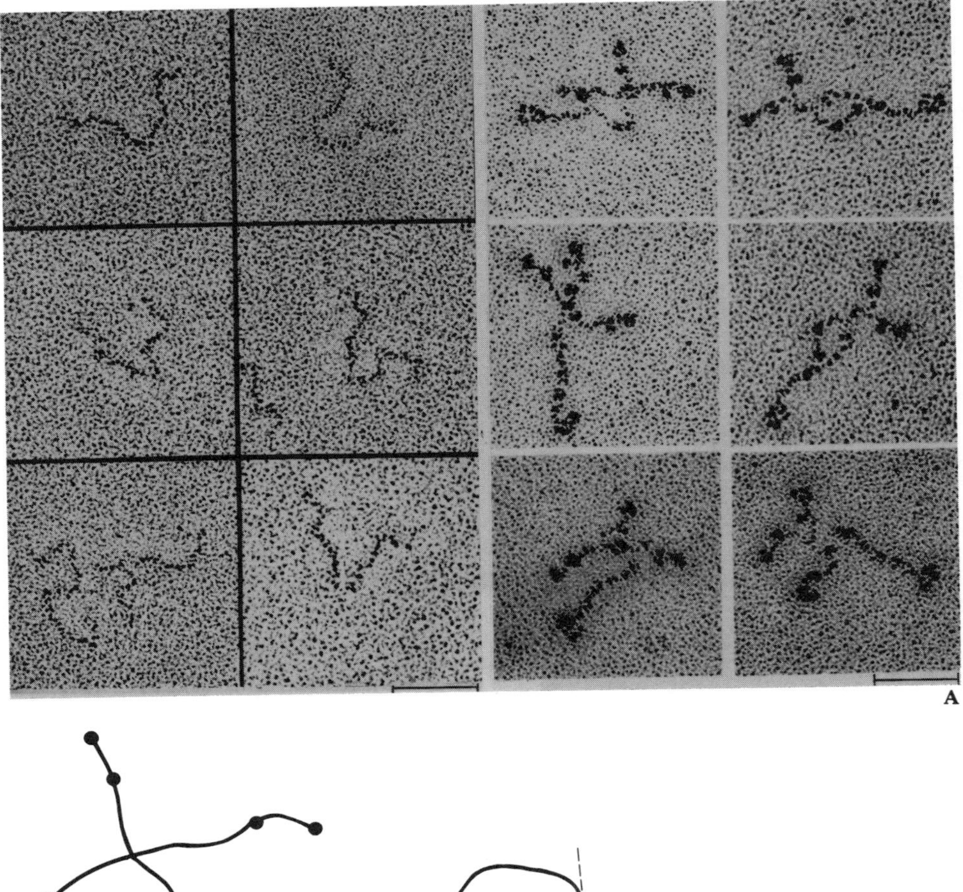

Figure 7.5. A: Electron micrographs of fibronectin *(left)* and laminin *(right)* obtained from rotary shadowed specimens (Engel et al. 1981). The bars correspond to 50 nm. **B:** Schematic representation of fibronectin *(right)* and laminin *(left)* as seen in the electron micrographs. *(Courtesy of Jurgen Engel.)*

uration observed by the rotary shadowing technique may in part be an artifact of the preparation procedure.

Subclasses and Species Differences. Differences in molecular weight, solubility, electrophoretic behavior on SDS gel electrophoresis, and carbohydrate composition and structure have been observed between plasma fibronectin and cellular fibronectin. Comparisons of the domain and carbohydrate structures of cellular and plasma fibronectins from hamster have been done. The cellular form seems to be a homodimer present largely as aggregates in contrast to the plasma form, which is a heterodimer as shown in Figure 7.4. The culture medium of fibroblasts accumulates a relatively large amount of soluble fibronectin. The released fibronectin has essentially the same chemical properties and composition as the cellular form, but is more soluble in aqueous solvents. It tends to aggregate, but the basis of the association is not understood. The carbohydrate structure of the released form is essentially the same as the cellular form, but is distinctively different from the plasma form (Table 7.7).

An extensive study of the differences in domain structures of chick cellular and plasma fibronectins has been performed (Hayashi and Yamada 1981). The collagen-binding domain (44–56K) of the cellular form is about 1K smaller in molecular weight than the plasma form, whereas the cell-binding domain (120K–140K) of the cellular form is about 10K larger in molecular weight than the plasma form. These differences seem to be too large to be explained by posttranslational modifications and suggest differences in primary structure.

Differences among fibronectins according to species and sources are essentially related to differences in subunits and their domains. Detailed information about

Table 7.7. Differences Among Fibronectins

	Cellular form	Released form	Plasma form
Subunit organization	Multimeric; dimeric	Essentially dimeric, α unit = β unit	Dimeric, α unit > β unit
Solubility	Insoluble	Soluble; tends to become insoluble after isolation	Soluble
Binding ability to gelatin, heparin, fibrin, etc.	+	+	+
Domain structure	Gelatin-binding domain is 1K smaller, but cell-binding domain is 10K larger in cellular form than in plasma form.[b]	Cell-binding domain is more sensitive to protease attack in cellular form than in plasma form.[c]	N-terminal half obtained by S-cyanylation is smaller, but C-terminal half is larger in cellular form than in plasma form.[c]
Carbohydrate structure			
Core	→GlcNAcβ1 → 4GlcNAc → Asn 6 ↑ Fuc1		→GlcNAcβ1 → 4GlcNAc → Asn
Peripheral	Galβ1 → 4GlcNAc ↑ ±(NeuAcα2)		Galβ1 → 4GlcNAc 6 6 ↑ ↑ NeuAcα2 NeuAcα2

[a]+ = present; [b]Hayashi and Yamada 1981; [c]Sakiguchi et al. 1983b.

the subunit structures (α and β) and their domains of various fibronectins from various sources and species is not available at present. The apparent molecular weights of α and β subunits in plasma fibronectin are different in various species. The difference between plasma fibronectin hamsters and humans is now correlated to differences in domains in each subunit (Sekiguchi and Hakomori 1983b; also see Fig. 7.4). It is likely that the size of each domain may be variable, as well as the structure of the carbohydrate chains and their distribution along the polypeptide chain (Table 7.2).

7.4. Laminin

7.4.1. Isolation and Localization

Chung et al. (1977) first observed that an extracellular membranous structure is synthesized by and released upon cytochalasin B treatment from a mouse embryonal carcinoma cell line, M1536-B3, in suspension culture. The membranes were purified by repeated centrifugation or by adsorption on a concanavalin A-agarose. On solubilization of the purified membrane in SDS and 2-mercaptoethanol, two major glycoproteins were found with apparent molecular weights of 320,000 and 230,000, respectively, which were called GP-1 and GP-2.

Further studies (Chung et al. 1979) on amino acid and carbohydrate composition of purified GP-1 and GP-2 showed that these glycoproteins are rich in half-cystine residues, but do not contain hydroxy amino acids, indicating that they are not collagenous proteins. These glycoproteins contain 12–15% carbohydrate, which is characterized by the presence of a relatively large amount of galactosamine. Small amounts of xylose and glucose are also present. However, the type of carbohydrate is uncertain because contamination with proteoglycan was not excluded. Although GP-2 and fibronectin have similar molecular weights, GP-2 is distinguishable from fibronectin in amino acid and carbohydrate composition. Antibodies directed to GP-2 do not crossreact with GP-1 or fibronectin. The antibodies specifically stain basement membranes of a variety of mouse and human tissues, including kidney, placenta, skin, and blood vessels.

A similar high molecular weight glycoprotein was extracted and purified by Timpl et al. (1979) from the Engelbreth-Holm-Swarn (EHS) sarcoma that produces an extracellular matrix of basement membranes (Orkin et al. 1977). The name laminin has come to be accepted for this glycoprotein. Laminin migrates essentially as a single band on SDS-polyacrylamide gel electrophoresis with an apparent molecular weight of more than 600,000. It can be resolved into two subunits with molecular weights of 440,000 and 220,000, respectively, under reducing conditions. These subunits appear to correspond to GP-1 and GP-2, respectively. Antibody directed to laminin stains by immunofluorescence the extracellular matrix of the tumor and the basement membranes of normal epidermis and various other tissues. Further studies at the ultrastructural level have revealed that laminin is concentrated in the lamina rara, but is absent in the lamina densa (Farquhar 1981). The amino acid composition of laminin is different from fibronectin and from type IV collagen, the basement membrane collagen.

A typical purification procedure of laminin consists of the following steps (Timpl et al. 1982): (a) homogenization of laminin-producing tumors in high-salt buffer to remove soluble proteins; (b) extraction with neutral buffer containing 0.5 *M* NaCl; and (c) DEAE-cellulose and molecular sieve chromatography. Affinity chromatography on immobilized heparin can be included because laminin binds to heparin (Sakashita et al. 1980). Laminin can also be purified from crude extracts by affinity chromatography on immobilized *Griffonia simplicifolia* I, a lectin that recognizes terminal α-D-galactose (Shibata et al. 1982). This lectin binds specifically to basement membranes in a variety of murine tissues (Peters and Goldstein 1979).

7.4.2. Cell Adhesion

Localization of laminin in the lamina rara suggests that laminin may play a role in the attachment of epithelial cells to type IV collagen, a major constituent of the lamina densa. Indeed, several lines of evidence indicate that laminin is an adhesive protein that preferentially mediates attachment of epithelial cells to type IV collagen. For example, an epithelial cell line PAM-212 adheres preferentially to type IV collagen-coated substrates and cell adhesion is specifically promoted by the presence of a small quantity of laminin (1-5 μg/mL), but not by fibronectin (Terranova et al. 1980). Laminin does not promote cell attachment to a substrate coated with types I, II, III, and V collagen. Attachment of fibroblasts on collagen-coated substrates is not affected by laminin.

Vlodavsky and Gospodarowicz (1981) observed that laminin promotes adhesion of epithelial tumor cells (carcinoma), but not sarcoma cells, on type IV collagen. Carcinoma cells adhere well to matrices elaborated by endothelial cells that contain laminin, fibronectin, and type IV collagen. The growth behavior of these malignant cells on corneal endothelial basement membrane mimics that of normal cells (Vlodavsky et al. 1980) (For discussion, see Section 7.1).

There is some evidence that ability of tumor cells to attach to type IV collagen through laminin correlates with the metastatic potential of tumor cells. Terranova et al. (1982) demonstrated that metastatic fibrosarcoma cells bind preferentially to type IV collagen, and laminin promotes their attachment, whereas nonmetastatic fibrosarcoma cells bind preferentially to type I collagen, and fibronectin specifically promotes their attachment. A subpopulation of metastatic cells selected for laminin-mediated attachment on type IV collagen produces more pulmonary metastasis than cells that did not attach when these variants are injected into mice through the tail vein. Pretreatment of metastatic cells with antilaminin antibodies greatly reduces the number of pulmonary metastases. Varani et al. (1983) reported that highly metastatic murine sarcoma cells readily attach to type IV collagen, whereas lowly metastatic cells do not. The addition of laminin enhances the attachment of the lowly metastatic cells.

The specificity in the requirement for laminin is clear, but not exclusive, among different types of cells. Hepatocytes have been shown to attach to both laminin- and fibronectin-coated substrates (Johansson et al. 1981). The attachment on these adhesive proteins appears to be mediated by independent receptors. Human embryonic skin fibroblasts also attach and spread on laminin-coated surfaces as well as on fibronectin-coated surfaces (Couchman et al. 1983). As clear specificity for these adhesive proteins has been observed in cell attachment on collagen-coated substrates, but not on substrates directly coated with laminin or fibronectin, the assay method may affect the apparent specificity.

A surface receptor for laminin on murine fibrosarcoma cells has been isolated by affinity chromatography on immobilized laminin (Malinoff and Wicha 1983). Two distinct proteins are adsorbed on the affinity column after solubilization of the cells with a detergent. One is type IV collagen and the other is a protein with an apparent molecular weight of 69,000 after reduction. The 69K protein has a high affinity for laminin with $K_d = 2 \times 10^{-9} M$. This is in striking contrast to the poorly characterized surface receptor for fibronectin that is not retained on immobilized fibronectin due to its low affinity.

7.4.3. Interaction with Other Matrix Components and Bacteria

Many extracellular matrix components have been shown to bind to one another. The binding of laminin to other components has been examined because laminin is one of the major components of the basement membrane, along with type IV collagen, heparan sulfate proteoglycan, and fibronectin (Farquhar 1981). Laminin binds to heparin-agarose and the binding is inhibited by a large excess of heparan sulfate (Sakashita et al. 1980). Systematic studies on the binding of laminin to various types of glycosaminoglycans immobilized on agarose indicate that laminin binds to heparin with highest affinity; to heparan sulfate, dermatan sulfate, and chondroitin sulfate C with moderate affinity; and to hyaluronic acid and chondroitin sulfate, A with minimal affinity, if at all (Sekiguchi, et al. 1983a). The binding to heparan sulfate may be physiologically important because both laminin and heparan sulfate proteoglycan are localized in the lamina rara (Farquhar 1981).

Among different types of collagen, laminin preferentially binds to type IV collagen and mediates the cell attachment on the collagen-coated substrate (Terranova et al. 1980, 1983). Laminin does not bind to fibronectin with appreciable affinity (Sakashita et al. 1980).

Laminin has been shown to bind to uropathogenic *Escherichia coli*, but not to *Staphylococcus aureus* to which fibronectin can bind (Speziale et al. 1982). The binding of laminin to *E. coli* appears to occur through its carbohydrate residues, because the hair-like structures of pathogenic *E. coli*, referred to as pilli, possess a lectin-like activity specific to terminal α-D-galactose (Köllenius et al. 1981). Indeed, laminin binds to *Griffonia simplicifolia* I lectin that is also specific to terminal α-D-galactose (Shibata et al. 1982).

7.4.4. Molecular Properties

Several physicochemical properties of laminin have been studied. A sedimentation profile of freshly prepared laminin exhibits a major sharp boundary with a sedimentation constant of $s^{\circ}_{20,w} = 11.5$ S (Engel et al. 1981). An aggregated form can be detected as a faster migrating broad boundary upon prolonged storage. The frictional coefficient (f/f_m) of 2.9 is calculated by using a monomeric molecular weight of 900,000 (Engel et al. 1981), indicating that laminin in neutral solution has a very elongated shape. Circular dichroism indicates that laminin contains 30% α helix, 15% β structure, and 55% aperiodic structures (Otto et al.

1982). The α helix is readily destroyed by protease treatment. An abrupt change in secondary structure is observed near 58°C when laminin is heated in solution (Otto et al. 1982). This change is predominantly due to destruction of α helix.

The molecular shape of laminin has been examined with the electron microscope by the rotary shadowing technique. Engel et al. (1981) found that laminin has the shape of an asymmetric cross with three short (36 nm) arms and a long (77 nm) arm (Figure 7.5). Seven globular structures, two on each short arm and one on the long arm, are also distinguishable. The long and short arms are considered to represent the large and small subunits, respectively, because laminin depleted of the large subunit by thrombin digestion retains three intact short arms, but lacks the long arm (Rao et al. 1982a).

7.4.5. Domain Structure

Currently, laminin is considered to consist of three small subunits of molecular weight 200,000–220,000 and one large subunit of molecular weight 400,000–440,000 that are held together by interchain disulfide bonds (Timpl et al. 1979; Engel et al. 1981; Rao et al. 1982a). Large and small subunits are often referred to as β and α subunits, respectively (Rao et al. 1982a). These subunits are biochemically and immunologically distinct, and display quite different peptide mapping patterns after *Staphylococcus aureus* V8 protease digestion (Cooper et al. 1981). The large subunit is highly susceptible to proteolytic digestion whereas the small subunit is more resistant (Otto et al. 1982; Rao et al. 1982a, 1982b). For example, thrombin selectively removes the large subunit without affecting the size and interchain disulfide bonding of the small subunits (Rao et al. 1982a). Antibodies against the small subunits do not crossreact with the large subunit (Chung et al. 1979; Cooper et al. 1981).

A series of fragments have been generated upon limited proteolysis of laminin with various proteases (Rohde et al. 1980; Otto et al. 1982; Rao et al. 1982a, 1982b). The major fragments have been isolated and partially characterized (Table 7.8). As described above, thrombin digestion selectively degrades the large subunit, yielding a fragment designated α3 (Rao et al. 1982a). Electron microscopic studies show that fragment α3 consists of three intact short arms, but lacks the long arm, although a short residual stump of the long arm is often detected. Fragment α3 binds to type IV collagen and mediates cell attachment on the collagen-coated surface (Terranova et al. 1983).

Prolonged digestion of laminin with other proteases, including pepsin, elastase, trypsin, subtilisin, *S. aureus* V8 protease, cathepsin G, and plasmin, produces a major, large fragment of molecular weight 260,000–350,000, irrespective of the type of protease (Rohde et al. 1980; Otto et al. 1982; Rao et al. 1982b). This fragment is referred to as fragment 1 (Otto et al. 1982). Fragment 1 gives two major fragments of molecular weight 160,000–180,000 and 130,000–140,000, respectively, after reduction (Rao et al. 1982b). Fragment 1 is considered to be derived from the small subunits because a similar fragment is also generated from fragment α3. Electron microscopic studies reveal that fragment 1 consists of three rod-like elements 26–32 nm long connected to each other at one end, indicating that it represents the T-shaped core region of the three short arms.

Table 7.8. Molecular and Biological Properties of Proteolytic Fragments of Laminin

Fragment	Protease	Molecular shape	Molecular weight	Binding activity	Reference
α3	Thrombin	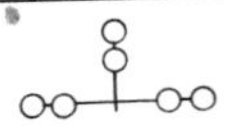	600K (nonreduced); 200K (reduced)	Cell surface; Type IV collagen	Rao et al. (1982a, 1982b)
1	Pepsin, elastase, trypsin, plasmin, cathepsin G, subtilisin, *S. aureus* V8	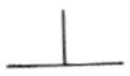	260K–350K (nonreduced); 160K–180K and 130K–140K (reduced)	Cell surface	Rohde et al. (1980) Otto et al. (1982) Rao et al.(1982a, 1982b)
2	Pepsin, trypsin, elastase, subtilisin		50K (reduced)	Unknown	Otto et al. (1982)
3	Elastase, trypsin, *S. aureus* V8		50K (reduced)	Heparin	Otto et al. (1982)
4	Elastase, *S. aureus* V8	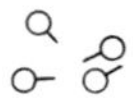	75K (reduced)	Unknown	Otto et al. (1982)

Incomplete digestion often generates a series of fragments consisting of fragment 1 with a variable number of the globular domains (Otto et al. 1982).

Fragment 1 is highly enriched in half-cystine (i.e., 12%) and does not contain detectable amounts of α helix or β structure (Otto et al. 1982). Strong resistance of this fragment to protease attack may be due to the extensive intra- and interchain disulfide bonds. Fragment 1 does not bind to type IV collagen. Instead, it inhibits laminin-mediated cell attachment to type IV collagen in a competitive manner, indicating that it binds to the cell surface receptor for laminin (Terranova et al. 1983). Since fragment α3 binds to type IV collagen, the globular domains of the small subunit are considered to be capable of binding to type IV collagen.

Three additional fragments have been isolated from the proteolytic digests of laminin by heparin-agarose and subsequent molecular sieve chromatography (Otto et al. 1982). Fragment 3, of molecular weight 50,000, is the major fragment retained on immobilized heparin. This fragment shows a globular appearance upon electron microscopic examination. Circular dichroism indicates that this fragment contains predominantly β-pleated sheet structure. The molar ratio of fragment 3 to fragment 1 is estimated to be 0.8–0.9. Fragment 3 may be derived from the globular domain of the large subunit.

Fragments 2 and 4 are structurally and antigenically related. Their amino acid composition and circular dichroic spectra are very similar. Antibodies directed to either fragment crossreact with the other, but not with fragments 1 or 3 (Otto et al. 1982). Under an electron microscope, fragments 2 and 4 appear as short rods and globules connected to short rods, respectively.

Several lines of evidence indicate that fragment 4 is derived from the terminal globular domain of the short arms. Digestion of intact laminin with *S. aureus* V8 protease generates a large fragment of molecular weight 400,000–440,000.

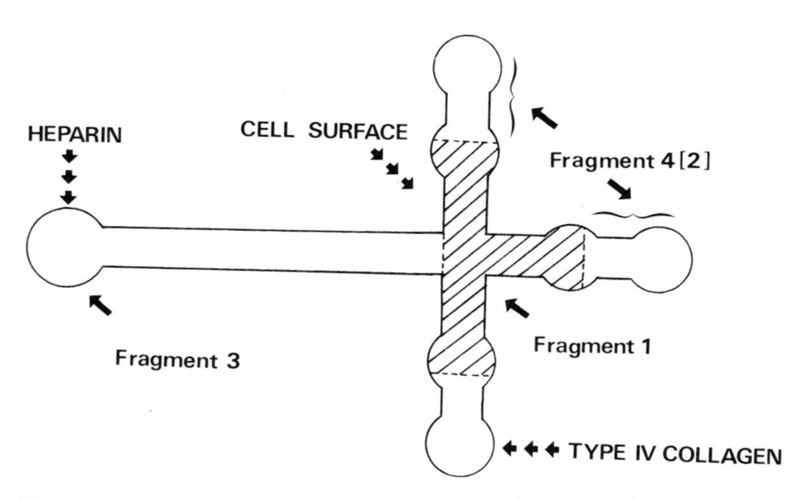

Figure 7.6. Functional domain structure of laminin, as proposed by Otto et al. (1982). The area of the internal portion of the short arms containing the intersecting point *(hatched)* is highly resistant to protease attack.

This fragment, when examined by electron microscopy, consists of three short arms with most of the globular domains. It crossreacts with antibodies directed to fragments 1 and 4 respectively (Otto et al. 1982), indicating that fragments 1 and 4 are in proximity to each other in the large fragment. In support of this analysis, the molar ratio of fragment 4 to fragment 1 in the early stage of elastase digestion is 2.5.

The functional domain structure of laminin is schematically shown in Figure 7.6. The collagen-binding region has not yet been identified, but fragment 4, which represents the terminal globular domain of the short arms, may bind to type IV collagen. The rod-like portion of the long subunit is highly sensitive to protease attack. This portion may be rich in α helix because plasmin digestion of laminin selectively degrades the large subunit and concordantly destroys most of the α helix in the intact molecule. Location of fragments 2, 3, and 4 is still tentative and must be confirmed by detailed investigation.

7.4.6. Properties of Carbohydrates in Laminin

No detailed information is available on the composition and structure of the carbohydrate in laminin, although the initial study by Chung et al. (1979) suggested that the carbohydrate composition of laminin is uniquely different from that of fibronectin in its high content of galactosamine. Some carbohydrate side chains of laminin may have a terminal sequence Galα1 → 3Galβ1 → 4GlcNAcβ1 → R based on the following evidence: (a) laminin has an α-D-Gal residue at the terminus because it interacts with, and can be purified on a column of immobilized *Griffonia simplicifolia* I (*See* Section 7.4.1); and (b) Ehrlich ascites carcinoma has a cell surface glycoprotein containing a terminal sugar sequence Galα1 → 3Galβ1 → 4GlcNAcβ1 → R (Blake and Goldstein 1981), and this glycoprotein was found to be a laminin (Goldstein, personal communication). The variety of carbohydrate chains, their location, and their structures largely remain to be studied.

7.4.7. Transformation Dependency

Laminin of normal rat kidney (NRK) cells decreases on oncogenic transformation, just as fibronectin displays an acute decrease on transformation (Hayman et al. 1981). The decrease of laminin and fibronectin was observed concomitant with a disappearance of a network of heparan sulfate at the surface of NRK cells associated with oncogenic transformation, whereas chondroitin sulfate proteoglycan does not decrease. Cultivation of the transformed cells in the presence of sodium butyrate reverses the transformed morphology and restores surface matrix of fibronectin, laminin, and heparan sulfate proteoglycan (Hayman et al. 1982). As heparan sulfate binds to both fibronectin and laminin and codistributes with these matrix proteins, it has been assumed that heparan sulfate proteoglycan may play a central role in the assembly of the pericellular complex and in their transformation-dependent changes.

It should be noted, however, that many transformed cells, including various types of fibrosarcoma, melanoma, teratocarcinoma, and Ehrlich ascites carcinoma, all contain laminin at the cell surface, and its higher quantity correlates with higher metastatic property (*See* Section 7.4.2). Changes in laminin associated with oncogenic transformation in these cells are inconsistent and need further study.

7.4.8. Development and Differentiation: Role of Laminin in the Biomatrix

As laminin apparently mediates the adhesion of epithelial and endothelial cells to basement membranes, it also may influence expression of various functions of these cells. For example, it may play a role in differentiation of epithelial cells such as kidney tubules (Ekblom et al. 1980). Changes in laminin expression have also been reported in tooth development (Thesleff et al. 1981) and endochondral bone differentiation (Foidart and Reddi 1980).

During mouse embryogenesis, laminin is the first extracellular matrix component that is detectable in the sixteen-cell compacted morulae (Leivo et al. 1980). Type IV collagen, another basement membrane component, becomes detectable in the inner cell mass of the blastocyst. After implantation, laminin and type IV collagen are present in all embryonic and extraembryonic basement membranes. Laminin is a major component of Reichert's membrane of mouse embryo and is synthesized and secreted by parietal endoderm cells attached to Reichert's membrane (Leivo et al. 1980; Hogan et al. 1980). Mouse embryonal carcinoma cells, F9, are shown to synthesize laminin upon differentiation induced by retinoid acid and dibutyryl cyclic AMP (Strickland et al. 1980). Thus, laminin can be used as a marker for studies on differentiation and embryogenesis, but its physiologic role in these phenomena awaits further investigation.

Many types of cellular differentiation proceed through functional cell adhesion upon contact with a biomatrix system such as basement membrane. A complex biomatrix consists of not only macromolecules (collagen, glycosaminoglycans, fibronectin, laminin, etc.), but also low molecular components (cations, biogenic amines, etc.). Laminin is an essential component of some types of biomatrix, and its function for promoting cellular differentiation may be expressed only through an organized state. Isolated laminin may have no affect on differentiation. Biomatrix prepared from rat liver and added to a culture of rat hepatocytes promotes the differentiation and long-term survival of rat hepatocytes (Rojkind et al. 1980). Similarly, the contact of mammary epithelial cells with the biomatrix of pregnant rat mammary glands promotes long-term growth and differentiation of rat mammary epithelium as expressed by lactoglobin synthesis (Wicha et al. 1982).

7.5. Other Matrix Glycoproteins

7.5.1. Organization of Pericellular Adhesive Glycoproteins

Cells are surrounded by, or imbedded in, a complex matrix system composed not only of collagen, fibronectin, and laminin, but also of several other adhesive glycoproteins. Although the composition of these glycoproteins differs from one type of cell to another, there are common features that are summarized in this section.

Extraction of confluent cell cultures of hamster or human fibroblasts with dilute aqueous nonionic or zwitterionic detergents solubilizes most membranes and cytoplasmic components, leaving an insoluble residue consisting of a fibrillar meshwork structure that crosses over and under nuclei (Carter and Hakomori 1977, 1979, 1981; Osborn and Weber 1977; Lehto et al. 1980; Gonen et al. 1979; Keski-Oja et al. 1981). A typical example of such a matrix revealed by scanning electron microscopy is shown in Figure 7.7. The matrix consists of cytoskeletal and nuclear components, as well as a few major collagenous and noncollagenous glycoproteins including fibronectin. Studies with cell surface labeling, followed by sequential extraction with various solutions containing detergents, with or without a reducing reagent, strongly suggest that these glycoproteins are organized in the matrix and display extensive interactions. They are, in general, disulfide-bonded, forming a multimeric structure insoluble in nonionic or zwitterionic detergents. However, the components are readily soluble under reducing conditions using dithiothreitol or 2-mercaptoethanol (Hynes et al. 1976; Carter and Hakomori 1977, 1981). The relative quantities of the various glycoproteins present in such a pericellular matrix differ depending on the type of cells. However, five major cell surface glycoproteins have been detected in human fibroblasts: fibronectin, Gp 250, Gp 190, Gp 170, and Gp 140. The numbers refer to their apparent molecular weight in thousands after denaturation and reduction. In hamster fibroblasts, the quantity of Gp 250 and Gp 140 is much lower than in human fibroblasts. In NRK and mouse 3T3 fibroblasts, fibronectin is a minor component and Gp 170 is the major component (*See* below).

Fibronectin and Gp 170 are protease-sensitive and can be eliminated rapidly from the cell surface and from the detergent-insoluble matrix by mild trypsin treatment. All of these glycoproteins can be cell-surface labeled and can be

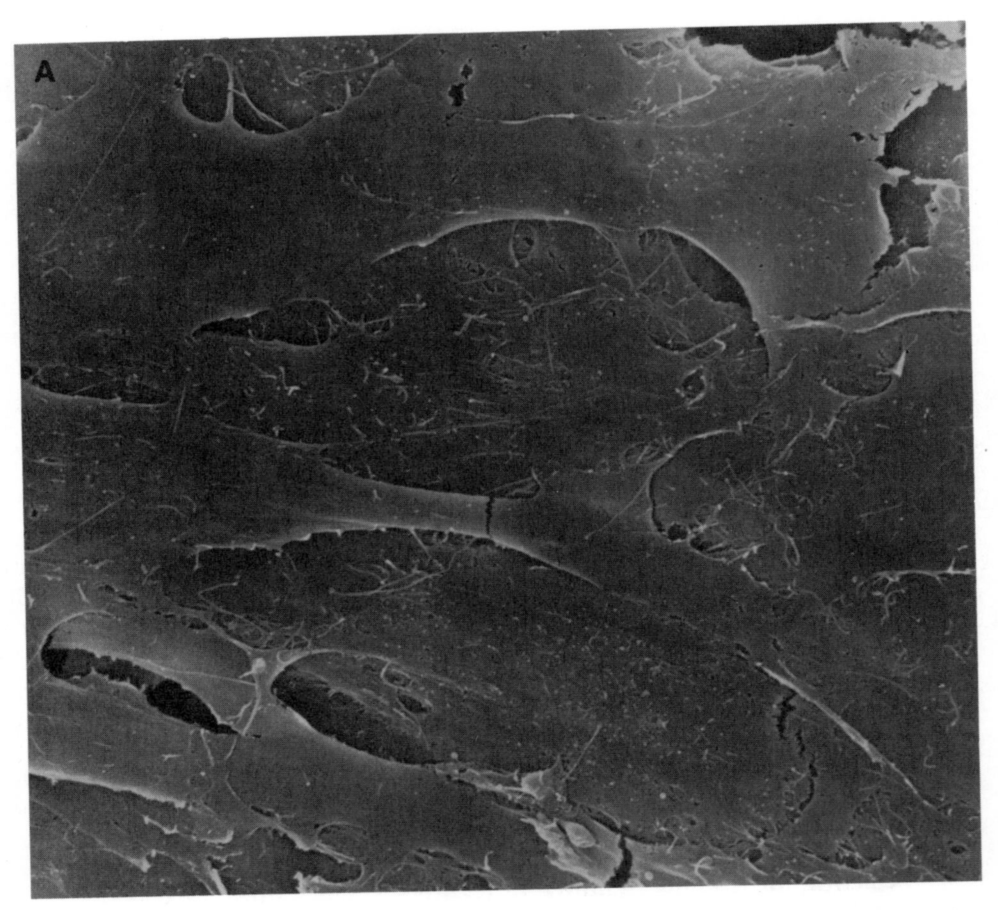

Figure 7.7. Scanning electron micrographs of confluent NIL cells before (**A**) and after (**B**) extraction with detergent (Empigen BB). The prominent convoluted nuclei are surrounded by a filamentous structure arising largely from the pericellular matrix. Magnification, 80×.

secreted or released into the culture medium. However, the susceptibility of these glycoproteins at the cell surface to trypsin and collagenase is different from that of the same glycoproteins secreted into the culture medium.

Although matrix proteins remain insoluble in nonionic and zwitterionic detergents, they can be solubilized by extraction with a hot solution containing sodium dodecyl sulfate. Essentially all the matrix proteins and approximately half of the fibronectin can be solubilized. The solubilized matrix complex behaves as an aggregated macromolecular complex under nonreducing conditions, which can be fractionated by chromatography on Sepharose 2B. The macromolecular complex can be resolved into individual components under reducing conditions. The complex has been demonstrated to contain fibronectin, Gp 250, Gp 140, Gp 190, and Gp 170. The last two are less aggregated than the others, as shown in Figure 7.8.

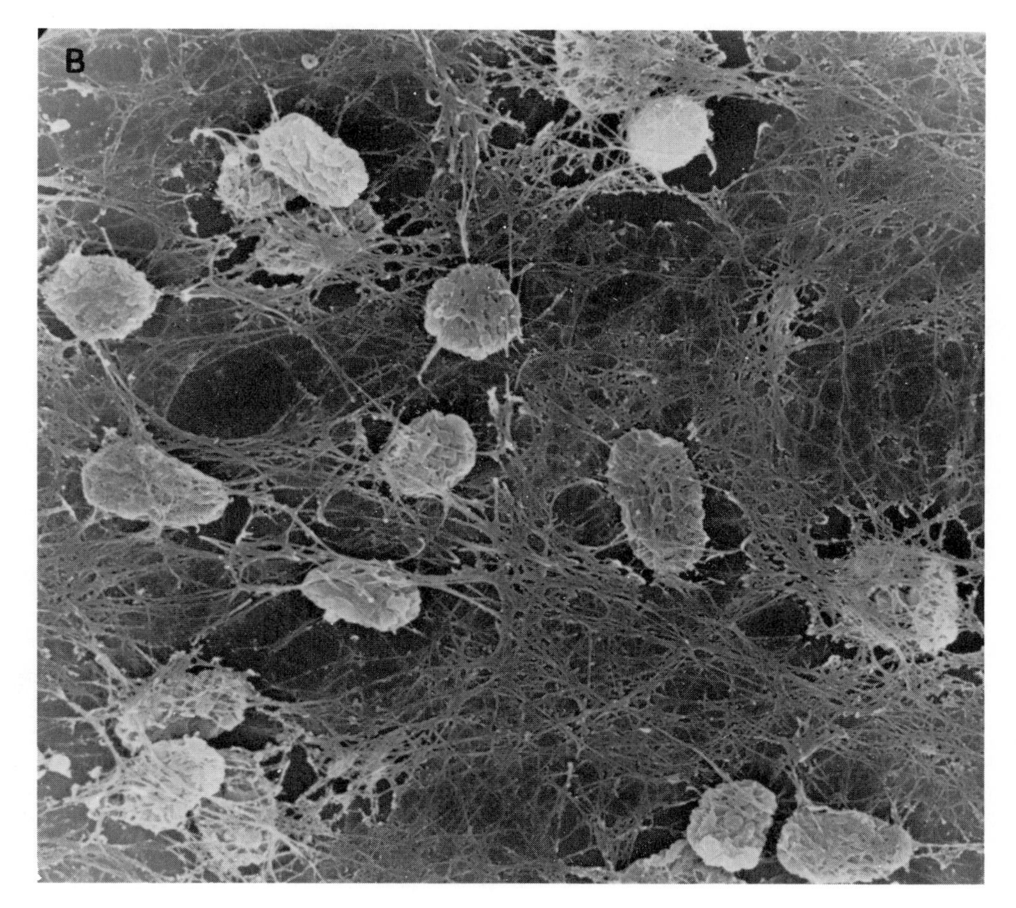

Figure 7.7 *(continued)*

7.5.2. Properties of Adhesive Glycoproteins and Oncogenic Transformation

Gp 140. The Gp 140 subunit has been isolated from the matrix by solubilization in sodium trichloroacetate after trypsin treatment of the matrix to remove fibronectin. The glycoprotein thus isolated adheres and spreads cells in a cooperative fashion with fibronectin. The purified Gp 140 contains mannose, galactose, and N-acetylglucosamine totaling 2.7% of the molecule. In addition, mild periodate oxidation followed by reduction with NaB^3H_4 strongly labelled Gp 140 with tritium. This is not due to periodate-sensitive carbohydrates, but is due to the presence of hydroxylysine residues (Carter 1982a). Since hydroxylysine is present only in proteins containing the collagen helix, a close relationship between Gp 140 and a collagen or procollagen is suspected. The Gp 140 isolated under nonreducing conditions induces stable cell attachment and cell spreading when

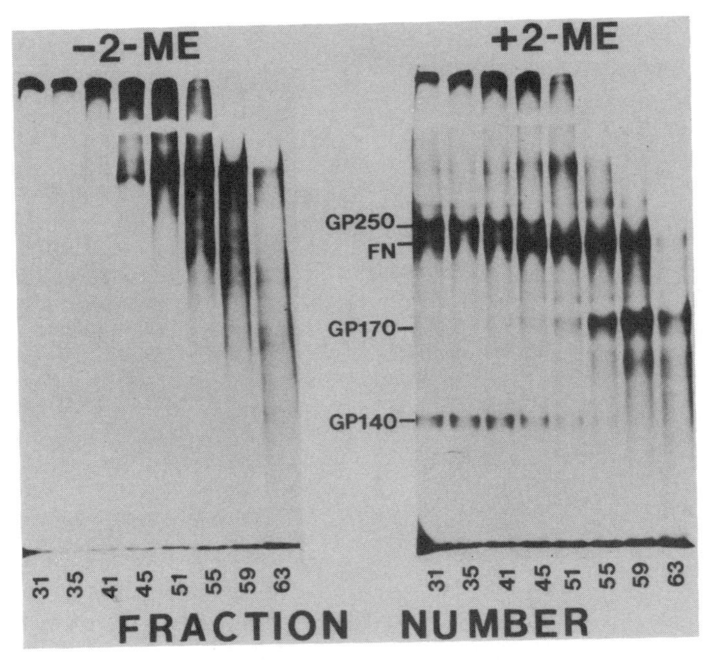

Figure 7.8. Polyacrylamide SDS gel electrophoresis and fluorography of fractionated pericellular glycoproteins fibronectin *(FN)*, GP 250, GP 170 from human fibroblasts. Aliquots from the indicated fractions from a molecular sieve column were analyzed under reducing (mercaptoethanol, *right*) and nonreducing *(left)* conditions. The high molecular weight aggregates that do not enter the gel under nonreducing conditions can be resolved into four components under reducing conditions. Comparing the two patterns, it can be seen that before reduction the disulfide-dependent multimers of fibronectin have a very broad size distribution; GP 170 exists primarily as a relatively small aggregate; and GP 140 exists as a relatively large aggregate.

coated on plastic surfaces. Although Gp 140 at the cell surface is insensitive to collagenase digestion, the same protein secreted in the culture medium has been found to be sensitive. Gp 140 interacts with fibronectin, although to a lesser degree than other collagen molecules. From various other criteria, it appears that Gp 140 could be a new type of pericellular collagen (Carter 1982b).

Gp 250. This component has been clearly demonstrated in the matrix of human fibroblasts, but is barely detectable in other types of fibroblasts. Antibodies directed to Gp 250 do not crossreact with fibronectin, and antibodies directed to fibronectin do not react with Gp 250, showing that Gp 250 and fibronectin are different molecules. This glycoprotein is therefore a part of the complex macromolecule which is stabilized by disulfide bonds.

Gp 170 and Gp 190. Some cells, such as NRK and mouse 3T3, contain relatively small quantities of fibronectin, Gp 140, and Gp 250, but a large quantity of glycoproteins in the molecular weight range of 170,000–190,000. The latter group increases greatly on cell confluency and decreases or is lost on transformation in the same way as fibronectin and Gp 140 (Carter and Hakomori 1978, 1981;

Carter 1982b). Although the composition of glycoproteins within this molecular weight range, some of the major components expressed at the cell surface may be procollagens (Keski-Oja et al. 1981) as their loss on oncogenic transformation has been observed by many investigators (Vaheri et al. 1978; Levinson et al. 1975; Sandmeyer et al. 1981; Hata and Peterkofsky 1977). Both Gp 170 and Gp 190 secreted in the medium are sensitive to bacterial collagenase. These molecules at the cell surface appear to be organized in a manner resistant to collagenase. Gp 170 and Gp 190 of human fibroblasts WI38 have been identified as pro α1(I) and pro α2(I), respectively (Carter 1982b).

Fernandez-Pol (1979) observed a peculiar iron-dependent change of the transformation-sensitive glycoprotein of molecular weight 170,000: the glycoprotein decreased slightly in molecular weight as a result of iron starvation. While a procollagen was suggested at first, in a subsequent report it was concluded that the 170K glycoprotein is a receptor for transferrin (Fernandez-Pol and Klos 1980). The component may be different from matrix-associated procollagen because it is soluble in detergent. In another study, hamster embryonic fibroblast NIL cells accumulated a glycoprotein of molecular weight 180,000 when their growth was

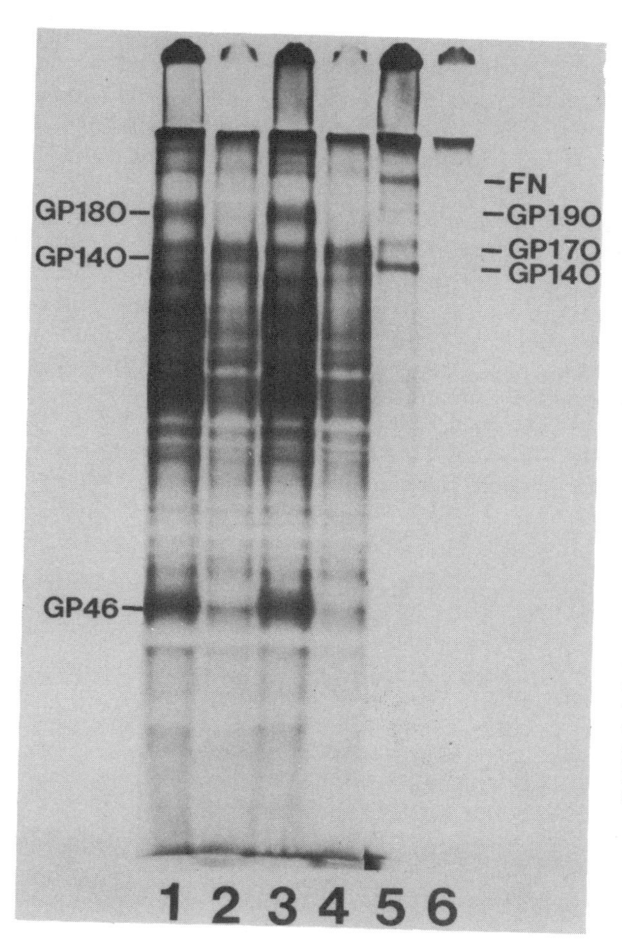

Figure 7.9. Transformation sensitivity of detergent-insoluble matrix glycoproteins of normal human fibroblasts (WI-38) and human fibroblasts transformed with SV40 virus (WI-26 VA4) metabolically labeled with [^{3}H]-glucosamine. Fractions were analyzed by SDS gel electrophoresis followed by fluorography. Gel 1: WI-38, whole cells. Gel 2: WI-26 VA4, whole cells. Gel 3: WI-38, detergent soluble extract. Gel 4: WI-26 VA4, detergent soluble matrix. Gel 5: WI-38, detergent insoluble matrix. Gel 6: WI-26 VA4 detergent insoluble matrix. A number of small quantitative differences in glycoproteins can be seen by comparison of whole or detergent soluble extracts of normal and transformed cells. These are labeled in the left-hand margin as GP 46, GP 140, and GP 180. However, the glycoproteins labeled fibronectin *(FN)* GP 190, GP 170, and GP 140 *(right margin)* are present in the detergent-insoluble matrix of normal cells but are absent from the transformed cells.

inhibited in the presence of retinoic acid (Patt et al. 1978). The glycoprotein is highly dependent on cell growth rate and is suspected to be a type of collagen.

All the extracellular matrix glycoproteins—Gp 250, Gp 190, Gp 170, and Gp 140—of various human fibroblasts are deleted in transformed cells when tested not only by cell surface labeling and metabolic labeling, but also with antibody staining after gel electrophoresis, followed by blotting on nitrocellulose sheets. A typical example of a simultaneous deletion of all the matrix glycoproteins of human fibroblasts by oncogenic transformation is shown in Figure 7.9.

7.5.3. Entactin and Chondronectin

In 1981, Carlin et al. reported that the extramembranous sac of M1536 cells contains a previously undemonstrated noncollagenous glycoprotein with a molecular weight of 158,000, in addition to the laminin components GP-1 and GP-2. This glycoprotein can be metabolically labeled by $^{35}SO_4$. This sulfated glycoprotein is referred to as entactin (Carlin et al. 1981). Entactin is also localized in subendothelial and vascular basement membranes (Bender et al. 1981). A similar glycoprotein is synthesized by mouse parietal endoderm cells. This glycoprotein, referred to as PYS-C, can be immunoprecipitated by antilaminin antibodies together with the large and small subunits of laminin (Hogan et al. 1980). Antibodies raised against PYS-C glycoprotein crossreact with the small subunit, but not the large subunit, of laminin (Cooper et al. 1981). Although PYS-C and the small subunit of laminin give quite different peptide mapping patterns with *S. aureus* V8 protease, these two glycoproteins share some antigenic determinants.

A serum glycoprotein that promotes the interaction of cartilage cells or chondroblasts with cartilage-specific collagen has been found and termed chondronectin (Hewitt et al. 1982). This glycoprotein is also secreted by chondroblasts together with type II collagen and the cartilage-specific proteoglycans. Chondronectin, isolated from serum by ion exchange chromatography, has a molecular weight of 180,000 and consists of subunits with molecular weights of 80,000 joined by disulfide bonds. This new glycoprotein has been found only in cartilage and the vitrous body.

The authors are grateful to Dr. Virginia Richmond for assembling the references.

References

Ali IU, Hunter T: (1981) Structural comparison of fibronectins from normal and transformed cells. J Biol Chem 256:7671–7677.

Ali IU, Hynes RO: (1978) Effects of LETS glycoprotein on cell motility. Cell 14:439–446.

Allen C, Saba TM, Molnar J: (1973) Isolation, purification and characterization of opsonic protein. J Reticuloendothel Soc 13:410–423.

Anderson B, Hoffman P, Meyer K: (1963) A serine-linked peptide chondroitin sulfate. Biochim Biophys Acta 74:309–311.

Aplin JD, Hughes RC, Jaffe CL, Sharon N: (1981) Reversible cross-linking of cellular components of adherent fibroblasts to fibronectin and lectin-coated substrata. Exp Cell Res 134:488–494.

Ashwell G, Morrel AG: (1974) The role of surface carbohydrates in the hepatic recognition and transport of circulating glycoproteins. Adv Enzymol 41:99–128.

Avnur Z, Geiger B: (1981) The removal of extracellular fibronectin from areas of cell-substrate contact. Cell 25:121–132.

Baenziger J, Fiete D: (1979) Structural determinants of concanavalin-A specificity for oligosaccharides. J Biol Chem 254:2400–2407.

Baenziger J, Kornfeld S: (1974) Structure of carbohydrate units of IgA_1 immunoglobulin. I. Composition, glycopeptide isolation, and structure of the asparagine-linked oligosaccharide units. J Biol Chem 249:7260–7269.

Barondes SH: (1981) Lectins: Their multiple endogenous cellular functions. Ann Rev Biochem 50:207–231.

Bender BL, Jaff R, Carlin B, Chung AE: (1981) Immunolocalization of entactin, a sulfated basement membrane component, in rodent tissues, and comparison with GP-2 (laminin). Am J Pathol 103:419–426.

Bhavanandan VP, Buddecke E, Carubelli R, Gottschalk A: (1964) The complete enzymic degradation of glycopeptides containing O-seryl and O-threonyl linked carbohydrate. Biochem Biophys Res Commun 16:353–357.

Bing DH, Almeda S, Isliker H, Lahav J, Hynes RO: (1982) Fibronectin binds to the C1q component of complement. Proc Natl Acad Sci USA 79:4198–4201.

Birchmeier C, Kreis TE, Eppenberger HM, Winterhalter KH, Birchmeier W: (1980) Corrugated attachment membrane in WI-38 fibroblasts: Alternating fibronectin fibers and actin-containing focal contacts. Proc Natl Acad Sci USA 77:4108–4112.

Birchmeier W: (1981) Fibroblast's focal contacts. Trends in Biochem Sci. 6:234–237.

Biroc SL, Etzler ME: (1978) The effect of periodate oxidation and α-mannosidase treatment of *Dolichos biflorus* lectin. Biochim Biophys Acta 544:85–92.

Björndal H, Hellerqvist CG, Lindberg B, Svensson S: (1970) Gas-liquid chromatography and mass spectrometry in methylation analysis of polysaccharides. Angew Chem (Engl) 9:610–619.

Blake DA, Goldstein IJ: (1981) An α-D-galactosyltransferase activity in Ehrlich ascites tumor cells: Biosynthesis and characterization of a trisaccharide (α-D-galactose 1 → 3 N-acetyllactosamine). J Biol Chem 256:5387–5393.

Blumberg PM, Driedger PE, Rossow PW: (1976) Effect of a phorbol ester on a transformation-sensitive surface protein of chick fibroblasts. Nature 264:446–447.

Blumenfeld OO, Adamany AM: (1978) Structural polymorphism within the amino-terminal region of MM, NN, and MN glycoproteins (glycophorins) of the human erythrocyte membrane. Proc Natl Acad Sci USA 75:2727–2731.

Blumenstock F, Weber P, Saba TM: (1977) Isolation and biochemical characterization of α-2-opsonic glycoprotein from rat serum. J Biol Chem 252:7156–7162.

Blumenstock FA, Saba TM, Weber P, Laffin R: (1978) Biochemical and immunological characterization of human opsonic α_2-SB glycoprotein: Its identity with cold-insoluble globulin. J Biol Chem 253:4287–4291.

Cannon WB: (1932) The Wisdom of the Body. New York: Norton. 312.

Carlin B, Jaff R, Bender B, Chung AE: (1981) Entactin, a novel basal lamina-associated sulfated glycoprotein. J Biol Chem 256:5209–5214.

Carlson DM: (1968) Structures and immunochemical properties of oligosaccharides isolated from pig submaxillary mucins. J Biol Chem 243:616–626.

Carter WG: (1982a) The cooperative role of the transformation sensitive glycoproteins, GP140 and fibronectin, in cell attachment and spreading. J Biol Chem 257:3249–3259.

Carter WG: (1982b) Transformation-dependent alterations in glycoproteins of the extracellular matrix of human fibroblasts: Characterization of GP250 and the collagen-like GP140. J Biol Chem 257:13805–13815.

Carter WG, Hakomori S: (1977) Isolation and partial characterization of "galactoprotein a" (LETS) and "galactoprotein b" from hamster embryo fibroblasts. Biochem Biophys Res Commun 76:299–308.

Carter WG, Hakomori S: (1978) A protease resistant, transformation-sensitive membrane glycoprotein ("170K Gp") and an intermediate filament-forming protein (IFP) of hamster fibroblasts. J Biol Chem 253:2867–2878.

Carter WG, Hakomori S: (1979) Isolation of galactoprotein a from hamster embryo fibroblasts and characterization of the carbohydrate unit. Biochemistry 18:730–738.

Carter WG, Hakomori S: (1981) A new cell surface, detergent-insoluble glycoprotein matrix of human and hamster fibroblasts. J Biol Chem 256:6953–6960.

Carter WG, Rauvala H, Hakomori S: (1981) Studies on cell adhesion and recognition. J Cell Biol 88:138–148.

Chen LB: (1981) Fibronectin and related transformation-sensitive cell surface proteins. In: Oncology Overview. US Department of Health and Human Services, Public Health Service, National Institutes of Health, National Cancer Institute, International Cancer Research Data Bank.

Chen LB, Burridge K, Murray A, Walsh MC, Copple CP, Bushnell A, McDougall JK, Gallimore PH: (1978) Modulation of cell surface glycocalyx: Studies on large, external, transformation-sensitive protein. Ann NY Acad Sci 212:366–381.

Chen LB, Gallimore PH, McDougall JK: (1976) Correlation between tumor induction and the large external transformation-sensitive protein on the cell surface. Proc Natl Acad Sci USA 73:3570–3574.

Chen LB, Summerhayes I, Hsieh P, Gallimore PH: (1979) Possible role of fibronectin in malignancy. J Supra Str 12:139–150.

Chen WT, Singer SJ: (1980) Fibronectin is not present in the focal adhesions formed between normal cultured fibroblasts and their substrata. Proc Natl Acad Sci USA 77:7318–7322.

Chu FK, Trimble RB, Maley F: (1978) The effect of carbohydrate depletion on the properties of yeast external invertase. J Biol Chem 253:8691–8693.

Chung AE, Freeman IL, Braginsky JE: (1977) A novel extracellular membrane elaborated by a mouse-embryonal carcinoma-derived cell line. Biochem Biophys Res Commun 79:859–868.

Chung AE, Jaffe R, Freeman EL, Vergenes J-P, Braginsky JE, Carlin B: (1979) Properties of basement membrane-related glycoprotein synthesized in culture by a mouse embryonal carcinoma-derived cell line. Cell 16:277–287.

Clarke SM, Fink LM: (1977) The effects of specific antibody-complement-mediated cytotoxicity on transformed Syrian hamster cells. Cancer Res. 37:2985–2992.

Cooper AR, Kurkinen M, Taylor A, Hogan BLM: (1981) Studies on the biosynthesis of laminin by murine parietal endoderm cells. Eur J Biochem 119:189–197.

Couchman JR, Höök M, Rees DA, Timpl R: (1983) Adhesion, growth, and matrix production by fibroblasts on laminin substrates. J Cell Biol 96:177–183.

Courtney HS, Simpson WA, Beachey EH: (1983) Binding of Streptococcal lipoteichoic acid to fatty acid-binding sites on human plasma fibronectin. J Bacteriol 153:763–770.

Culp LA: (1978) Biochemical determinants of cell adhesions. Curr Top Membranes Transport 11:327–396.

Der CJ, Stanbridge EJ: (1978) Lack of correlation between the decreased expression of cell surface LETS protein and tumorigenicity in human cell hybrids. Cell 15:1241–1251.

Der CJ, Stanbridge EJ: (1980) Alterations in the extracellular matrix organization associated with the re-expression of tumorigenicity in human cell hybrids. Int J Cancer 26:451–459.

Dessau W, Adelmann BC, Timpl R, Martin G: (1978) Identification of the sites in collagen α-chains that bind serum antigelatin factor (cold-insoluble globulin). Biochem J 169:55–59.

Distler J, Hieber V, Sahagian G, Schmickel R, Jourdian GW: (1979) Identification of mannose 6-phosphate in glycoproteins that inhibit the assimilation of beta-galactosidase by fibroblasts. Proc Natl Acad Sci USA 76:4235–4239.

Duksin D, Bornstein P: (1977) Impaired conversion of procollagen to collagen by fibroblasts and bone transfer with tunicamycin, an inhibitor of protein glycosylation. J Biol Chem 252:955–962.

Duksin D, Davidson JM, Bornstein P: (1978a) The role of glycosylation in the enzymatic conversion of procollagen to collagen: Studies using tunicamycin and ConA. Arch Biochem Biophys 185:326–332.

Duksin D, Holbrook K, Williams K, Bornstein P: (1978b) Cell surface morphology and adhesive properties of normal and virally transformed cells treated with tunicamycin, an inhibitor of protein glycosylation. Exp Cell Res 116:153–165.

Ekblom P, Alitalo K, Vaheri A, Timpl R, Saxen L: (1980) Induction of basement membrane in embryonic kidney possible role of laminin in morphogenesis. Proc Natl Acad Sci USA 77:485–489.

Engel J, Odermatt E, Engel A, Madori J, Furthmayr H, Rohde H, Timpl R: (1981) Shapes, domain organizations and flexibility of laminin and fibronectin, two multifunctional proteins of the extracellular matrix. J Mol Biol 150:97–120.

Engvall E, Ruoslahti E: (1977) Binding of soluble form of fibroblast surface protein, fibronectin to collagen. Int J Cancer 20:1–5.

Engvall E, Ruoslahti E, Miller EJ: (1978) Affinity of fibronectin to collagens of different genetic types and to fibrinogen. J Exp Med 147:1584–1595.

Fagan JB, Sobel ME, Yamada JM, Decrombrugghe B, Pastan I: (1981) Effects of transformation on fibronectin gene expression using cloned fibronectin cDNA. J Biol Chem 256:520–525.

Farquhar MG: (1981) The glomerular basement membrane: A selective molecular filter: In: Hay ED, ed. Cell Biology of Extracellular Matrix. New York: Plenum. 335–378.

Feeney RE, Yeh Y: (1978) Antifreeze proteins from fish bloods. Adv Protein Chem 32:191–282.

Fernandez-Pol JA: (1979) Isolation and immunological characterization of an iron-regulated transformation-sensitive cell surface protein of normal rat kidney cells. J Supra Str 11:371–390.

Fernandez-Pol JA, Klos DJ: (1980) Isolation and characterization of normal rat kidney cell membrane proteins with affinity for transferrin. Biochemistry 19:3904–3912.

Finne J, Krusius T, Järnefelt J: (1980) Fractionation of glycopeptides. In: Varmavuori A, ed. 27th International Congress of Pure and Applied Chemistry. Oxford: Pergamon. 147–159.

Foidart J-M, Reddi AH: (1980) Immunofluorescent localization of type IV collagen and laminin during endochondral bond differentiation and regulation by pituitary growth hormone. Dev Biol 75:130–136.

Fukuda M, Fukuda MN, Hakomori S: (1979) Developmental change and genetic defect in the carbohydrate structure of band 3 glycoprotein of human erythrocyte membranes. J Biol Chem 254:3700–3703.

Fukuda M, Hakomori S: (1979a) Proteolytic and chemical fragmentation of galactoprotein a, a major transformation-sensitive glycoprotein released from hamster embryo fibroblasts. J Biol Chem 254:5442–5450.

Fukuda M, Hakomori S: (1979b) Carbohydrate structure of galactoprotein a, a major transformation-sensitive glycoprotein released from hamster embryo fibroblasts. J Biol Chem 254:5451–5457.

Fukuda M, Kondo T, Osawa T: (1976) Studies on the hydrazinolysis of glycoproteins. Core structures of oligosaccharides obtained from porcine thyroglobulin and pineapple stem bromelin. J Biochem 80:1223–1232.

Fukuda M, Leverey S, Hakomori S: (1982) Carbohydrate structure of hamster plasmaform fibronectin. J Biol Chem 257:6856–6860.

Fukuda M, Matsumura G: (1976) Endo-β-galactosidase of *Escherichia freundii*. Purification and endoglycosidic action on keratan sulfates, oligosaccharides and blood group active glycoprotein. J Biol Chem 251:6218–6225.

Furcht LT, Mosher DF, Wendelschafer-Crabb G, Fiodart J-M: (1979) Reversal by glucocorticoid hormones of the loss of a fibronectin and procollagen matrix around transformed human cells. Cancer Res 39:2077–2083.

Furthmayr H: (1978) Structural comparison of glycophorins and immunochemical analysis of genetic variants. Nature 271:519–524.

Gahmberg CG, Hakomori S: (1973) Altered growth behavior of malignant cells associated with changes in externally labelled glycoprotein and glycolipid. Proc Natl Acad Sci USA 70:3329–3333.

Gahmberg CG, Hakomori S: (1975) Surface carbohydrates of hamster fibroblasts. I. Chemical characterization of surface-labelled glycosphingolipids and a specific ceramide tetrasaccharide fortransformants. J Biol Chem 250:2438–2446.

Gahmberg CG, Kiehn D, Hakomori S: (1974) Changes in a surface-labelled galactoprotein and in glycolipid concentrations in cells transformed by a temperature-sensitive polyoma virus mutant. Nature 248:413–415.

Geiger B: (1979) A 130K protein from chicken gizzard: Its localization at the termini of microfilament bundles in cultured chicken cells. Cell 18:193–205.

Goldstein IJ, Hayes CE: (1978) The lectins: Carbohydrate-binding proteins of plants and animals. Adv Carbohyd Chem Biochem 35:127–340.

Goldwasser E, Kung CKH, Eliason J: (1974) On the mechanism of erythropoietin-induced differentiation. 13. The role of sialic acid in erythropoietin action. J Biol Chem 249:4204–4206.

Gonen A, Weisman-Shomer P, Fry M: (1979) Cell adhesion and acquisition of detergent resistance by the cytoskeleton of cultured chick fibroblasts. Biochim Biophys Acta 552:307–321.

Goto M, Kataoka Y, Sato H: (1972) Decrease of saturation density in cultured tumor cells by dextran sulfate. Gann 63:371–374.

Gottschalk A, Thomas MAW: (1961) Studies on mucoproteins. V. The significance of N-acetylneuraminic acid for the viscosity of ovine submaxillary gland mucoprotein. Biochim Biophys Acta 46:91–98.

Grabel LB, Rosen SD, Martin GR: (1979) Teratocarcinoma stem cells have a cell surface carbohydrate-binding component implicated in cell-cell adhesion. Cell 17:477–484.

Grinnell F: (1976) Cell spreading factor. Occurrence and specificity of action. Exp Cell Res 102:51–62.

Grinnell F: (1978) Cellular adhesiveness and extracellular substrata. Int Rev Cytol 53:65–144.

Grinnell F, Feld MK: (1979) Initial adhesion of human fibroblasts in serum-free medium: Possible role of secreted fibronectin. Cell 17:117–129.

Grinnell F, Hays DG, Minter D: (1977) Cell adhesion and spreading factor: Partial purification and properties. Exp Cell Res 110:175–190.

Grobstein C: (1975) Development role of intercellular matrix: Retrospection and prospection. In: Slavkin HC, Greulisch RC, eds. Extracellular Matrix Influences on Gene Expression. New York: Academic. 9–16.

Hahn L-HE, Yamada KM: (1979) Isolation and biological characterization of active fragments of the adhesive glycoprotein fibronectin. Cell 18:1043–1051.

Hakomori S: (1964) Rapid methylation of glycolipid and polysaccharide as catalyzed by methylsulfinylcarbanion in dimethylsulfoxide. J Biochem (Tokyo) 55:205–208.

Hakomori S: (1981) Blood group ABH and Ii antigens of human erythrocytes: Chemistry, polymorphism, and their developmental change. Semin Hematol 18:39–62.

Hakomori S, Murakami WT: (1968) Glycolipids of hamster fibroblasts and derived malignant transformed cell lines. Proc Natl Acad Sci USA 59:254–261.

Hasilik A, Kelin U, Wahee A, Strecker G, von Figura K: (1980) Phosphorylated oligosaccharides in lysosomal enzymes: Identification of α-N-acetylglucosamine-1-phospho-6-mannose diester groups. Proc Natl Acad Sci USA 77:7074–7078.

Hata R, Peterkofsky V: (1977) Specific changes in the collagen phenotype of Balb 3T3 cells as a result of transformation by sarcoma viruses or a chemical carcinogen. Proc Natl Acad Sci USA 74:2933–2937.

Hay ED: (1981) Extracellular matrix. J Cell Biol 91:205–223.

Hayashi M, Yamada KM: (1981) Differences in domain structure between plasma and cellular fibronectin. J Biol Chem 256:11,292–11,300.

Hayashi M, Yamada KM: (1982) Divalent cation modulation of fibronectin binding to heparin and DNA. J Biol Chem 257:5263–5267.

Hayman EG, Engvall E, Ruoslahti E: (1980) Butyrate restores fibronectin at cell surface of transformed cells. Exp Cell Res 127:478–481.

Hayman EG, Engvall E, Ruoslahti E: (1981) Concomitant loss of cell surface fibronectin and laminin from transformed rat kidney cells. J Cell Biol 88:352–357.

Hayman EG, Oldberg A, Martin RG, Ruoslahti E: (1982) Codistribution of heparan sulfate proteoglycan; laminin, and fibronectin in the extracellular matrix of normal rat kidney cells and their coordinate absence in transformed cells. J Cell Biol 94:28–35.

Heifetz A, Lennarz WJ: (1979) Biosynthesis of N-glycosidically linked glycoproteins during gastrulation of sea urchin embryos. J Biol Chem 254:6119–6127.

Hewitt AT, Varner HH, Silver MH, Dessau W, Wilkes CM, Martin GR: (1982) The isolation and partial characterization of chondronectin, an attachment factor for chondrocytes. J Biol Chem 257:2330–2334.

Hickman S, Kornfeld S: (1978) Effect of tunicamycin on IgM, IgA and IgG secretion by mouse plasmacytoma cells. J Immunol 121:990–996.

Hickman S, Shapiro LJ, Neufeld EF: (1974) A recognition marker required for uptake of a lysosomal enzyme by cultured fibroblast. Biochem Biophys Res Commun 57:55–61.

Hirano H, Yamada Y, Sullivan M, de Crombrugghe B, Pastan I, Yamada KM: (1983) Isolation of genomic DNA clones spanning the entire fibronectin gene. Proc Natl Acad Sci USA 80:46–50.

Hogan BLM, Cooper AR, Kurkinen M: (1980) Incorporation into Reichert's membrane of laminin-like extracellular proteins synthesized by parietal endoderm cells of the mouse embryo. Dev Biol 80:289–300.

Hogg NM: (1974) A comparison of membrane proteins of normal and transformed cells by lactoperoxidase labeling. Proc Natl Acad Sci USA 71:489–492.

Höök M, Rubin K, Oldberg A, Vaheri A: (1977) Cold-insoluble globulin mediates the adhesion of rat liver cells to plastic petri dishes. Biochem Biophys Res Commun 79:726–733.

Hounsell EF, Wood E, Feizi T, Fukuda M, Powell ME, Hakomori S: (1981) Structural analysis of hexa- to octa-saccharide fractions isolated from sheep gastric-glycoproteins having blood group I and i activities. Carbohyd Res 90:283–307.

Housley TJ, Rowland LN, Ledger PW, Kaplan J, Tanzer ML: (1980) Effects of tunicamycin on the biosynthesis of procollagens by human fibroblasts. J Biol Chem 255:121–128.

Huang RTC: (1978) Cell adhesion mediated by glycolipids. Nature 276:624–626.

Hughes RC, Butters TD, Aplin JD: (1981) Cell surface molecules involved in fibronectin-mediated adhesion. A study using specific anti-sera, Eur J Cell Biol 26:198–207.

Hynes RO: (1973) Alteration of cell surface proteins by viral transformation and proteolysis. Proc Natl Acad Sci USA 70:3170–3174.

Hynes RO: (1976) Cell surface proteins and malignant transformation. Biochim Biophys Acta 458:73–107.

Hynes RO: (1981) Relationship between fibronectin and the cytoskeleton. In: Poste G, Nicolson GL, eds. Cell Surface Reviews 7:97–106. Amsterdam: Elsevier/North Holland.

Hynes RO, Bye JM: (1974) Density and cell cycle dependence of cell surface proteins in hamster fibroblasts. Cell 3:113–120.

Hynes RO, Destree A: (1977) Extensive disulfide bonding at the mammalian cell surface. Proc Natl Acad Sci USA 74:2844–2859.

Hynes RO, Destree AT, Mautner V: (1976) Spatial organization at the cell surface. In: Marchesi VT, ed. Progress in Chemical and Biological Research. New York: Alan R Liss. 189–201.

Hynes RO, Wyke JA: (1975) Alterations in surface proteins in chicken cells transformed by temperature-sensitive mutants of Rous sarcoma virus. Virology 64:492–504.

Hynes RO, Yamada KM: (1982) Fibronectins: Multifunctional modular glycoproteins. J Cell Biol 95:369–377.

Irimura T, Gonzalez R, Nicolson GL: (1981) Effects of tunicamycin on B16 metastatic melanoma cell surface glycoproteins and blood-borne arrest and survival properties. Cancer Res 41:4311–4318.

Ishihara H, Takahashi N, Oguir S, Tejira S: (1979) Complete structure of the carbohydrate moiety of stem bromelin: An application of the almond glycopeptidase for structural studies of glycopeptides. J Biol Chem 254:10,715–10,719.

Iyer RMN, Carlson DM: (1971) Alkaline borohydride degradation of blood group H substance. Arch Biochem Biophys 142:101–105.

Järnefelt J, Rush J, Li Y-T, Laine RA: (1978) Erythroglycan, a high molecular weight glycopeptide with the repeating structure {galactosyl-(β1 → 4)-2-deoxy-2-acetamido-glycosyl (1 → 3)} comprising more than one-third of the protein-bound carbohydrate of human erythrocyte stroma. J Biol Chem 254:8006–8009.

Jilek F, Hörmann H: (1977) Cold-insoluble globulin, III. Cyanogen bromide and plasminolysis fragments. Hoppe Seylers Z Physiol Chem 358:1165–1168.

Jilek F, Hörmann H: (1979) Fibronectin (cold-insoluble globulin), VI: Influence of heparin and hyaluronic acid on the binding of native collagen. Hoppe Seylers Z Physiol Chem 360:597–603.

Johansson S, Kjellén L, Höök M, Timpl R: (1981) Substrate adhesion of rat hepatocytes: A comparison of laminin and fibronectin as attachment proteins. J Cell Biol 90:260–264.

Kahn P, Shin SI: (1979) Cellular tumorigenicity in nude mice: Test of associations among loss of cell-surface fibronectin, anchorage independence, and tumor-forming ability. J Cell Biol 82:1–16.

Kaplan A, Achord DT, Sly W: (1977) Phosphohexosyl components of a lysosomal enzyme are recognized by pinocytosis receptors on human fibroblasts. Proc Natl Acad Sci USA 74:2026–2030.

Karlsson K-A, Pascher I, Pimlott W, Samuelsson BE: (1974) Use of mass spectrometry for the carbohydrate composition and sequence analysis of glycosphingolipids. Biomed Mass Spectro 1:49–56.

Keski-Oja J, Mosher DF, Vaheri A: (1976) Cross-linking of a major fibroblast surface-associated glycoprotein (fibronectin) catalyzed by blood coagulation factor XIII. Cell 9:29–35.

Keski-Oja J, Sen A, Todaro GJ: (1980) Direct association of fibronectin and actin molecules in vitro. J Cell Biol 85:527–533.

Keski-Oja J, Todaro GJ: (1980) Specific effects of fibronectin-releasing peptides on the extracellular matrices of culture human fibroblasts. Cancer Res 40:4722–4727.

Keski-Oja J, Todaro GJ, Vaheri A: (1981) Thrombin affects fibronectin and procollagen in the pericellular matrix of culture human fibroblasts. Biochim Biophys Acta 673:323–331.

Kieda C, Roche A-C, Delmotte F, Monsigny M: (1979) Lymphocyte membrane lectins: Direct visualization by the use of fluoresceinyl-glycosylated cytochemical markers. FEBS Letters 99:329–332.

Klebe RU: (1974) Isolation of a collagen-dependent cell attachment factor. Nature 250:248–251.

Kleinman HK, Klebe RJ, Martin GR: (1981) Role of collagenous matrices in adhesion and growth of cells. J Cell Biol 88:473–485.

Kleinman HK, McGoodwin EB, Klebe RJ: (1976) Localization of the cell attachment region in types I and II collagens. Biochem Biophys Res Commun 72:426–432.

Kleinman HK, Martin GR, Fishman PH: (1979) Ganglioside inhibitor of fibronectin-mediated cell adhesion to collagen. Proc Natl Acad Sci USA 76:3367–3371.

Kobata A: (1979) Use of endo- and exoglycosidases for structural studies of glycoconjugates. Anal Biochem 100:1–14.

Koehler JK, Nudelman ED, Hakomori S: (1980) A collagen-binding protein on the surface of ejaculated rabbit spermatozoa. J Cell Biol 86:529–536.

Köllenius G, Möllby R, Svenson SB, Winberg J: (1981) Microbial adhesion and the urinary tract. Lancet ii:866.

Kornfeld R: (1978) Structure of the oligosaccharides of three glycopeptides from calf thymocyte plasma membranes. Biochemistry 17:1415–1423.

Kornfeld R, Kornfeld S: (1980) Structure of glycoproteins and their oligosaccharide unit. In: Lennarz WJ, ed. Biochemistry of Glycoproteins and Proteoglycans. New York: Plenum. 1–34.

Koteliansky VE, Glukhova MA, Bejanian MV, Smirnov VN, Filimonov VV, Zalite OM, Venyaminov SY: (1981) A study of the structure of fibronectin. Eur J Biochem 119:619–624.

Krusius T, Finne J: (1978) Characterization of a novel sugar sequence for rat brain glycoprotein containing fucose and sialic acid. Eur J Biochem 84:395–403.

Krusius T, Finne J, Rauvala H: (1978) The poly(glycosyl) chains of glycoprotein characterization of a novel type of glycoprotein saccharides from human erythrocyte membrane. Eur J Biochem 92:284–300.

Kuusela P: (1978) Fibronectin binds to *Staphylococcus aureus*. Nature 276:718–720.

Kuusela P, Ruoslahti E, Engvall E, Vaheri A: (1976) Immunological interspecies cross-reactions of fibroblast surface antigen (fibronectin). Immunochemistry 1:639–642.

Labat-Robert J, Birembaut P, Adnet JJ, Mercantini F, Robert L: (1980) Loss of fibronectin in human breast cancer. Cell Biol Int Rep 4:609–616.

Laemmli UK: (1970) Cleavage of structural proteins during the assembly of the head of bacteriophage T4. Nature 227:680–685.

Laterra J, Culp LA: (1982) Differences in hyaluronate binding to plasma and cell surface fibronectins. J Biol Chem 257:719–726.

Lee YC, Ballou CE, Montgomery R, Wu YC, Lee Y-T: (1965) Periodate oxidation of glycopeptide from ovalbumin. Biochemistry 4:578–587.

Lee YC, Scocca JR: (1972) A common structural unit in asparagine-oligosaccharides of several glycoproteins from different sources. J Biol Chem 247:5753–5758.

Lehto V-P, Vartico T, Virtanen I: (1980) Enrichment of a 140K surface glycoprotein in adherent, detergent-resistant cytoskeleton of cultured human fibroblast. Biochem Biophys Res Commun 95:909–916.

Leivo I, Vaheri A, Timpl R, Wartiovaara J: (1980) Appearance and distribution of collagens and laminin in the early mouse embryo. Dev Biol 76:100–114.

Levinson W, Bhatnagar RS, Liu TA: (1975) Loss of ability to synthesize collagen in fibroblasts transformed by Rous-sarcoma virus. J Natl Cancer Inst 55:807–810.

Lisowska E, Kordowicz M: (1977) Immunochemical properties of M and N blood group antigens and their degradation products. In: Human Blood Groups, 5th International Convocation on Immunology. Basel: Karger. 188–196.

Lotan R, Debray H, Cacan M, Cacan R, Sharon N: (1975) Labeling of soybean agglutinin by oxidation with sodium periodate followed by reduction with sodium [^{3}H-]borohydride. J Biol Chem 250:1955–1957.

Malinoff HL, Wicha MS: (1983) Isolation of a cell surface receptor protein for laminin from murine fibrosarcoma cells. J Cell Biol 96:1475–1479.

McDonald JA, Kelley DG: (1980) Degradation of fibronectin by human leukocyte elastase. J Biol Chem 255:8848–8858.

Menzel EJ, Smolen JS, Liotta L, Reid KBM: (1981) Interaction of fibronectin with C1q and its collagen-like fragment. FEBS Letters 129:188–192.

Mintz B, Illmensee K: (1975) Normal genetically mosaic mice produced from malignant teratocarcinoma cells. Proc Natl Acad Sci USA 72:3585–3589.

Mizuochi T, Yamashita K, Fujikawa K, Kisiel W, Kobata A: (1979) The carbohydrate of bovine prothrombin occurrence of Gal 1 3GlcNAc grouping in asparagine-linked sugar chains. J Biol Chem 254:6419–6525.

Montreuil J: (1980) Primary structure of glycoprotein glycans: Basis for the molecular biology of glycoproteins. Adv Carbohyd Chem Biochem 37:157–223.

Morrison PR, Edsall JT, Miller SG: (1948) Preparation and properties of serum and plasma proteins XVIII. The separation of purified fibrinogen from fraction I of human plasma. J Am Chem Soc 70:3103–3108.

Morrison TA, McQuain CO: (1978) Assembly of viral membranes. Maturation of the vesicular stomatitis virus glycoprotein in the presence of tunicamycin. J Virol 28:368–374.

Mosesson MW: (1977) Cold-insoluble globulin (Clg): A circulating cell surface protein. Thromb Haemostas 38:742–750.

Mosesson MW, Amrani DL: (1980) The structure and biologic activities of plasma fibronectin. Blood 56:145–158.

Mosesson MW, Umfleet RA: (1970) The cold-insoluble globulin of human plasma. J Biol Chem 245:5728–5736.

Mosher DF: (1975) Cross-linking of cold-insoluble globulin by fibrin-stabilizing factor. J Biol Chem 250:6614–6621.

Mosher DF: (1980) Fibronectin. In Spaet TH, ed. Progress in Hemostasis and Thrombosis 5. New York: Grune and Stratton. 111–151.

Mosher DF, Proctor RA: (1980) Binding and factor XIIIa-mediated cross-linking of a 27-kilodalton fragment of fibronectin to *Staphylococcus aureus*. Science 209:927–929.

Mosher DF, Schad PE, Vann JM: (1980) Cross-linking of collagen and fibronectin by factor XIIIa. J Biol Chem 255:1181–1188.

Neri A, Ruoslahti E, Nicolsen GL: (1981) Distribution of fibronectin on clonal cell lines of rat mammary adenocarcinoma growing in vitro and in vivo at primary and metastatic sites. Cancer Res 41:5082–5093.

Neuberger A, Gottschalk A, Marshall RD: (1966) Carbohydrate-peptide linkages in glycoproteins and methods for their elucidation. In: Gottschalk A, ed. Glycoprotein. Amsterdam: Elsevier. 273–294.

Neufeld EF, Ashwell G: (1980) Carbohydrate recognition systems for receptor-mediated pinocytosis. In: Lennarz WJ, ed. The Biochemistry of Glycoproteins and Proteoglycans. New York: Plenum. 241–262.

Neufeld EF, Lim TW, Shapiro LJ: (1975) Inherited disorders of lysosomal metabolism. Ann Rev Biochem 44:357–376.

Nielson SE, Puck TT: (1980) Deposition of fibronectin in the course of reverse transformation of Chinese hamster ovary cells by cyclic AMP. Proc Natl Acad Sci USA 77:985–989.

Ofek I, Beachey EH, Sharon N: (1978) Surface sugars of animal cells as determinants of recognition in bacterial adherence. Trends Biochem Sci 3:159–160.

Ogata S, Muramatsu T, Kobata A: (1975) Fractionation of glycopeptides by affinity column chromatography on ConA-sepharose. J Biochem 78:687–696.

Oldberg A, Ruoslahti E: (1982) Interactions between chondroitin sulfate proteoglycan, fibronectin, and collagen. J Biol Chem 257:4859–4863.

Olden K, Pratt RM, Yamada KM: (1978) Role of carbohydrates in protein secretion and turnover: Effects of tunicamycin on the major cell surface glycoprotein of chick embryo fibroblasts. Cell 13:461–473.

Olden K, Pratt RM, Yamada KM, (1979) Role of carbohydrate in biological function of the adhesive glycoprotein fibronectin. Proc Natl Acad Sci USA 76:3343–3347.

Olden K, Yamada KM: (1977) Mechanism of the decrease in the major cell surface protein of chick embryo fibroblasts after transformation. Cell 11:957–969.

Orkin RW, Gehron P, McGoodwin EG, Martin GR, Valentine T, Swarm R: (1977) A murine tumor producing matrix of basement membrane. J Exp Med 145:204–220.

Orly J, Sato G: (1979) Fibronectin mediates cytokinesis and growth of rat follicular cells in serum-free medium. Cell 17:295–305.

Osborn M, Weber K: (1977) The detergent-resistant cytoskeleton of tissue culture cells includes the nucleus and the microfilament bundles. Exp Cell Res 106:339–349.

Otto U, Odermatt E, Engel J, Furthmayr H, Timpl R: (1982) Protease resistance and conformation of laminin. Eur J Biochem 123:63–72.

Patt LM, Itaya K, Hakomori S: (1978) Retinol induces densely dependent growth inhibition and changes in glycolipids and LETS. Nature 273:379–381.

Pearlstein E: (1976) Plasma membrane glycoprotein which mediates adhesion of fibroblasts to collagen. Nature 262:497–500.

Pearlstein E, Gold LI, Garcia-Pardo A: (1980) Fibronectin: A review of its structure and biological activity. Molec Cell Biochem 29:103–128.

Pearlstein E, Hynes RO, Franks LM, Hemmings V: (1976) Surface proteins and fibrinolytic activity of cultured mammalian cells. Cancer Res 36:1475–1480.

Pearlstein E, Sorvillo J, Giglii I: (1982) The interaction of human plasma fibronectin with a subset of the first component of complement, C1q. J Immunol 128:2036–2039.

Pearse B: (1980) Coated vesicles. Trends Biochem Sci 5:131–134.

Peters BP, Goldstein IJ: (1979) The use of fluorescein-conjugated *Bandeiraea simplicifolia* B_4-isolectin as a histochemical reagent for the detection of α-D-galactopyranosyl groups: The occurrence in basement membranes. Exp Cell Res 120:321–334.

Petersen TE, Thogersen HC, Skorstengaard K, Vibe-Pedersen K, Sahl P, Sottrup-Jensen L, Magnusson S: (1983) Partial primary structure of bovine plasma fibronectin: Three types of internal homology. Proc Natl Acad Sci USA 80:137–141.

Pfeffer LM, Wang E, Tamm I: (1980) Interferon effects on microfilament organization, cellular fibronectin distribution, and cell motility in human fibroblasts. J Cell Biol 85:9–17.

Phillips DR, Morrison M: (1970) The arrangement of proteins in the human erythrocyte membrane. Biochem Biophys Res Commun 40:284–289.

Pierschbacher MD, Hayman EG, Ruoslahti E: (1981) Locations of the cell-attachment site in fibronectin with monoclonal antibodies and proteolytic fragments of the molecule. Cell 26:250–267.

Rao CN, Margulies IMK, Goldfarb RH, Madri JA, Woodley DT, Liotta LA: (1982b) Differential proteolytic susceptibility of laminin alpha and beta subunits. Arch Biochem Biophys 219:65–70.

Rao CN, Margulies IMK, Tralka TS, Terranova VP, Madri JA, Liotta LA: (1982a) Isolation of a subunit of laminin and its role in molecular structure and tumor cell attachment. J Biol Chem 257:9740–9744.

Rauvala H, Carter WG, Hakomori S: (1981) Studies on cell adhesion and recognition. I. Extent and specificity of cell adhesion triggered by carbohydrate-reactive proteins (glycosidases and lectins) and fibronectin. J Cell Biol 88:127–137.

Rauvala H, Hakomori S: (1981) Studies on cell adhesion and recognition III. The occurrence of α-mannosidase at the fibroblast cell surface and its possible role in cell recognition. J Cell Biol 88:149–159.

Raz A, Lotan R: (1981) Lectin-like activities associated with human and murine neoplastic cells. Cancer Res 41:3642–3647.

Richter H, Seidl M, Hörmann H: (1981) Location of heparin-binding sites of fibronectin: Detection of a hitherto unrecognized transaminidase sensitive site. Hoppe-Seylers Z Physiol Chem 362:399–408.

Robbins PW, Wickus GG, Branton PE, Gaffney BJ, Hirschberg CB, Fuchs P, Blumberg PM: (1974) The chick fibroblast cell surface after transformation by Rous sarcoma virus. Cold Spring Harbor Symp Quant Biol 39:1173–1180.

Rohde H, Bächinger HP, Timpl R: (1980) Characterization of pepsin fragments of laminin in a tumor basement membrane: Evidence for the existence of related proteins. Hopp-Seylers Z Physiol Chem 361:1651–1660.

Rojkind M, Gatmaitan Z, Mackensen S, Giambrone M, Ponce P, Reid LM: (1980) Connective tissue biomatrix: Its isolation and utilization for long-term cultures of normal rat hepatocytes. J Cell Biol 87:255–263.

Roseman S: (1970) The synthesis of complex carbohydrates by multi-glycosyltransferase systems and their potential function in intercellular adhesion. Chem Phys Lipids 5:270–297.

Ruoslahti E, Engvall E, Hayman E: (1981) Fibronectin: Current concepts of its structure and function. Collagen Res 1:95–128.

Ruoslahti E, Hayman EG: (1979) Two active sites with different characteristics in fibronectin. FEBS Letters 97:221–224.

Ruoslahti E, Pekkala A, Engvall E: (1979) Effects of dextran sulfate on fibronectin collagen interaction. FEBS Letters 107:51–54.

Ruoslahti E, Vaheri A: (1975) Interaction of soluble fibroblast surface antigen with fibrinogen and fibrin. Identity with cold insoluble globulin of human plasma. J Exp Med 141:497–501.

Ruoslahti E, Vaheri A, Kuusela P, Linder E: (1973) Fibroblast surface antigen: A new serum protein. Biochim Biophys Acta 322:352–358.

Saba TM, Jaffe E: (1980) Plasma fibronectin (opsonic glycoprotein): Its synthesis by vascular endothelial cells and role in cardiopulmonary integrity after trauma as related to reticuloendothelial function. Am J Med 68:577–594.

Sadler JE, Paulson JC, Hill RL: (1979) The role of sialic acid in the expression of human MN blood group antigens. J Biol Chem 254:2112–2119.

Sakashita S, Engvall E, Ruoslahti E: (1980) Basement membrane glycoprotein laminin binds to heparin. FEBS Letters 116:243–246.

Sandmeyer S, Smith R, Kiehn D, Bornstein P: (1981) Correlation of collagen-synthesis and procollagen messenger-RNA levels with transformation in rat embryo fibroblasts. Cancer Res 4:830–838.

Schiffman G, Kabat EA, Thompson W: (1964) Immunochemical studies of blood groups XXX. Cleavage of A,B, and H blood group substances by alkali. Biochemistry 3:113–120.

Schlessinger J: (1980) The mechanism and role of hormone-induced clustering of membrane receptors. Trends Biochem Sci 5:210–215.

Schwarz RT, Rhorschneider JM, Schmidt MFG: (1976) Suppression of glycoprotein formation of Semliki Forest, influenza and avian sarcoma virus by tunicamycin. J Virol 19:782–791.

Sefton BM, Hunter T, Ball BH, Singer SJ: (1981) Vinculin: A cytoskeletal target of the transforming protein of Rous sarcoma virus. Cell 24:165–174.

Sekiguchi K, Fukuda M, Hakomori S: (1981) Domain structure of hamster plasma fibronectin: Isolation and characterization of four functionally distinct domains and their unequal distribution between two subunit polypeptides. J Biol Chem 256:6452–6462.

Sekiguchi K, Hakomori S: (1980) Functional domain structure of fibronectin (thermolysin; gelatin binding; heparin binding; cell adhesion). Proc Natl Acad Sci USA 77:2661–2665.

Sekiguchi K, Hakomori S: (1983a) Topological arrangement of four functionally distinct domains in hamster plasma fibronectin: A study with combination of S-cyanylation and limited proteolysis. Biochemistry 22:1415–1422.

Sekiguchi K, Hakomori S: (1983b) Domain structure of human plasma fibronectin: Differences and similarities between human and hamster fibronectins. J Biol Chem 258:3967–3973.

Sekiguchi K, Hakomori S, Funahashi M, Matsumoto I, Seno N: (1983a) Binding of fibronectin and its proteolytic fragments to glycosaminoglycans: Exposure of cryptic glycosaminoglycan-binding domains upon limited proteolysis. J Biol Chem 258 (in press).

Sekiguchi K, Siri A, Zardi L, Hakomori S: (1983b) Differences in domain structure between pericellular matrix and plasma fibronectins as revealed by domain-specific antibodies combined with limited proteolysis and S-cyanylation: A preliminary note. Biochem Biophys Res Commun 116:534–540.

Shibata S, Peters BP, Roberts DD, Goldstein IJ, Liotta LA: (1982) Isolation of laminin by affinity chromatography on immobilized *Griffonia simplicifolia* I lectin. FEBS Letters 142:194–198.

Shotton DM, Burke B, Branton D: (1979) The molecular structure of human erythrocyte spectrin. Biophysical and electron microscopic studies. J Molec Biol 131:303–329.

Simpson DL, Thorne DR, Loh HH: (1978) Lectins: Endogenous carbohydrate-binding proteins from certebrate tissues: Functional role in recognition process. Life Sci 22:727–748.

Slavkin HC, Greulich RC: (1975) Extracellular matrix influences on gene expression. New York: Academic. 833.

Sly WS, Stahl P: (1979) Receptor mediated uptake of lysosomal enzymes. In: Silverman S, ed. Transport of Molecules in Cellular Systems. Berlin: Dahlem Conferenzen. 229–244.

Smith HS, Riggs JL, Mosesson MW: (1979) Production of fibronectin by human epithelial cells in culture. Cancer Res 39:4138–4144.

Speziale P, Höök M, Wadström T, Timpl R: (1982) Binding of the basement membrane protein laminin to *Escherichia coli*. FEBS Letters 146:55–58.

Spiro RG: (1965) The carbohydrate units of thyroglobulin. J Biol Chem 240:1603.

Stathakis NE, Mosesson MW: (1977) Interactions among heparin, cold-insoluble globulin, and fibrinogen in formation on the heparin-precipitable fraction of plasma. J Clin Invest 60:855–865.

Stemberger A, Hörmann H: (1976) Affinity chromatography on immobilized fibrinogen and fibrin monomer, II: The behavior of cold-insoluble globulin (1). Hoppe Seylers Z Physiol Chem 357:1003–1005.

Stellner K, Saito H, Hakomori S: (1973) Determination of aminosugar linkages in glycolipids by methylation: Aminosugar linkage of ceramide pentasaccharide of rabbit erythrocytes and of Forssman antigen. Arch Biochem Biophys 155:464–472.

Stenman S, Vaheri A: (1981) Fibronectin in human solid tumors. Int J Cancer 27:427–435.

Stenman S, Wartivaara J, Vaheri A: (1977) Changes in the distribution of a major fibroblast protein, fibronectin, during mitosis and interphase. J Cell Biol 74:453–467.

Stone KR, Smith RE, Joklik WK: (1974) Changes in membrane polypeptides that occur when chick embryo fibroblasts and NRK cells are transformed with avian sarcoma viruses. Virology 58:86–100.

Strickland S, Smith KK, Marotti KR: (1980) Hormonal induction of differentiation in teratocarcinoma stem cells: Generation of parietal endoderm by retinoic acid and dibutyryl cAMP. Cell 21:347–355.

Sugahara K, Okmura T, Yamashima I: (1972) Purification of β-mannosidase from a snail, *Achatina fulica*, and its action on glycopeptides. Biochem Biophys Acta 268:488–496.

Sukeno T, Tarentino AL, Plummer TH Jr, Maley F: (1972) Purification and properties of α-D- and β-D-mannosidases from hen oviduct. Biochemistry 11:1493–1401.

Surani MAH: (1979) Glycoprotein synthesis and inhibition of glycosylation by tunicamycin in preimplantation mouse embryos: Compaction and trophoblast adhesion. Cell 18:217–227.

Tabas I, Kornfeld S: (1980) Biosynthetic intermediates of β-glucuronidase contain high mannose oligosaccharides with blocked phosphate residues. J Biol Chem 255:6633–6639.

Takahashi N, Nishibe H: (1981) Almond glycopeptidase acting on aspartylglycosylamine linkages. Biochem Biophys Acta 657:457–467.

Takasaki S, Yamashita K, Suzuki K, Iwanaya S, Kobata A: (1979) The sugar chains of cold insoluble globulin, a protein related to fibronectin. J Biol Chem 254:8548–8553.

Takatsuki A, Arima K, Tamura G: (1971) Tunicamycin, a new antibiotic. I. Isolation and characterization. J Antibiot 24:215.

Tanaka K, Bertonlini M, Pigman W: (1964) Serine and threonine glycosidic linkages in bovine submaxillary mucin. Biochem Biophys Res Commun 16:404–409.

Tarentino A, Plummer TH Jr, Maley F: (1970) Studies on the oligosaccharide sequence of ribonuclease B. J Biol Chem 245:4250–4157.

Terranova VP, Liotta LA, Russo RG, Martin GR: (1982) Role of laminin in the attachment and metastasis of murine tumor cells. Cancer Res 42:2265–2269.

Terranova VP, Rao CN, Kalebic T, Margulies IM, Liotta LA: (1983) Laminin receptor on human breast carcinoma cells. Proc Natl Acad Sci USA 80:444–448.

Terranova VP, Rohrbach DA, Martin GR: (1980) Role of laminin in the attachement of PAM 212 (epithelial) cells to basement membrane collagen. Cell 22:719–726.

Thesleff I, Barrach HJ, Foidart JM, Vaheri A, Pratt RM, Martin GR: (1981) Changes in the distribution of type IV collagen, laminin, proteoglycan, and fibronectin during mouse tooth development. Dev Biol 81:182–192.

Timpl R, Rohde H, Risteli L, Otto U, Robey PG, Martin GR: (1982) Laminin. Methods Enzymol 82:831–838.

Timpl R, Rohde H, Robey PG, Rennard SI, Foidart J-M, Martin GR: (1979) Laminin–A glycoprotein from basement membranes. J Biol Chem 254:9933–9937.

Tkacz JS, Lampen JO: (1975) Tunicamycin inhibition of polyisoprenyl N-acetyl glucosaminyl pyrophosphate formation in calf-liver microsomes. Biochem Biophys Res Commun 65:248–257.

Toyoshima S, Fukuda M, Osawa T: (1973) The presence of β-mannoside linkage in acidic glycopeptide from porcine thyroglobulin. Biochem Biophys Res Commun 51:945–959.

Ullrich K, Mersmann G, Weber E, van Figura K: (1977) Evidence for lysosomal enzyme recognition by human fibroblasts via a phosphorylated carbohydrate moiety. Biochem J 170:643–650.

Vaheri A, Kurkinen M, Lehto V-P, Linder E, Timpl R: (1978) Pericellular matrix proteins, fibronectin and (pro)collagen: Codistribution in cultured fibroblasts and loss in transformation. Proc Natl Acad Sci USA 75:4944–4948.

Vaheri A, Mosher DF: (1978) High molecular weight, cell surface-associated glycoprotein (fibronectin) lost in malignant transformation. Biochim Biophys Acta 516:1–25.

Vaheri A, Ruoslahti E: (1974) Disappearance of a major cell-type specific surface glycoprotein antigen (SF) after transformation of fibroblasts by Rous sarcoma virus. Int J Cancer 13:579–586.

Vaheri A, Ruoslahti E, Mosher DF (eds.): (1978) Fibroblast surface protein conference. Ann NY Acad Sci 312:453.

Varani J, Lovett EJ, McCoy JP, Shibata S, Maddox DE, Goldstein IJ, Wicha M: (1983) Differential expression of a laminin-like substance by high and low metastatic tumor cells. Am J Pathol 111: 27–34.

Vegarud G, Christensen TB: (1975) The resistance of glycoproteins to proteolytic inactivation. Acta Chem Scand (B) 29:887–888.

Vlodavsky I, Gospodarowicz D: (1981) Respective roles of laminin and fibronectin in adhesion of human carcinoma and sarcoma cells. Nature 289:304–306.

Vlodavsky I, Lui GM, Gospodarowicz D: (1980) Morphological appearance, growth behavior, and migratory activity of human tumor cells maintained in extracellular matrix versus plastic. Cell 19:607–616.

Wagh IV, Bornstein I, Winzler RJ: (1979) The structure of a glycopeptide from human orosomucoid (α_1-acid glycoprotein). J Biol Chem 244:658–665.

Wagner DD, Hynes RO: (1979) Domain structure of fibronectin and its relation to function. Disulfides and sulfhydryl groups. J Biol Chem 254:6746–6754.

Wagner DD, Ivatt R, Destree AT, Hynes R: (1981) Similarities and differences between the fibronectins of normal and transformed hamster cells. J Biol Chem 256:11708–11715.

Wang FF, Hirs CHW: (1977) Influence of the heterosaccharides in porcine pancreatic ribonuclease on the conformation and stability of the protein. J Biol Chem 252:8358–8364.

Warren L, Buck CA, Tuszkinsky GP: (1978) Glycopeptide changes and malignant transformation: A possible role for carbohydrate in malignant behavior. Biochim Biophys Acta 516:97–127.

Warren L, Fuhrer JP, Buck CA: (1973) Surface glycoproteins of cells before and after transformation by oncogenic viruses. Fed Proc 32:80–85.

Watanabe K, Hakomori S, Powell ME, Yokota M: (1980) The amphipathic membrane proteins associated with gangliosides: The Paul-Bunnell antigen is one of the gangliophilic proteins. Biochim Biophys Res Commun 92:638–648.

Watkins WM: (1980) Biochemistry and genetics of the ABO, Lewis, and P blood group systems. In Harris H, Hirschorn K, eds. Advances in Human Genetics, Volume 10. New York: Plenum. 1–136.

Weber K, Osborn M: (1969) The reliability of molecular weight determinations by dodecyl sulfate-polyacrylamide gel electrophoresis. J Biol Chem 244:4406–4412.

Weigel PH, Schnarr RL, Kuhlenschmidt MS, Schmell E, Lee RT, Lee Y-C, Roseman S: (1979) Adhesion of hepatocytes to immobilized sugars. A threshold phenomenon. J Biol Chem 254:10830–10838.

Wicha MS, Lowrie G, Kohn E, Bagavandoss P, Mohn T: (1982) Extracellular matrix promotes mammary epithelial growth and differentiation in vitro. Proc Natl Acad Sci USA 79:3213–3217.

Williams EC, Janmey PA, Ferry JD, Mosher DF: (1982) Conformational states of fibronectin. J Biol Chem 257:14,973–14,978.

Winzler RJ, Harris ED, Pekas DJ, Johnson CA, Weber P: (1967) Studies on glycopeptides released by trypsin from intact human erythrocytes. Biochemistry 6:.2195–2202.

Wolff I, Timpl R, Pecker I, Steffan C: (1967) A two-component system of human serum agglutinating gelatin-coated erythrocytes. Vox Sang 12:443–456.

Wood E, Hounsell EF, Feizi T: (1981) Preparative affinity chromatography of sheep gastric mucins having blood group Ii activity and release of antigenically active oligosaccharides by alkaline-borohydride degradation. Carbohyd Res 90:269–282.

Yamada KM: (1983) Cell surface interactions with extracellular materials. Ann Rev Biochem 52: 761–799.

Yamada KM, Kennedy DW, Kimata K, Pratt RM: (1980) Characterization of fibronectin interactions with glycosaminoglycans and identification of active proteolytic fragments. J Biol Chem 255:6055–6063.

Yamada KM, Olden K: (1978) Fibronectins: Adhesive glycoproteins of cell surface and blood. Nature 275:179–184.

Yamada KM, Weston JA: (1974) Isolation of a major cell surface glycoprotein from fibroblasts. Proc Natl Acad Sci USA 71:3492–3496.

Yamada KM, Yamada SS, Pastan I: (1976) Cell surface protein partially restores morphology, adhesiveness, and contact inhibition of movement to transformed fibroblasts. Proc Natl Acad Sci 73:1217–1221.

Yamaguchi H, Ikenaka T, Matsushima Y: (1971) The complete sequence of a glycopeptide obtained for Taka-amylase A. J Biochem (Tokyo) 70:587–594.

Yamashita K, Tachibana Y, Kobuta A: (1978) The structure of the galactose-containing sugar chains of ovalbumin. J Biol Chem 254:3862–3869.

Zardi L, Siri A, Carnemolla B, Santi L, Gardner WD, Hoch SO: (1979) Fibronectin: A chromatin-associated protein? Cell 18:649–657.

8

Structure and Metabolism of Proteoglycans

Dick Heinegård and Mats Paulsson

8.1. Introduction and Overview

Proteoglycans are macromolecules present in varying amounts in all connective tissues. They are primarily extracellular, although both intracellular as well as cell membrane proteoglycans have been described. Studies of their structure were initiated in Germany in the middle of the nineteenth century. Those early investigations were mainly concerned with the components unique to proteoglycans, the glycosaminoglycans, now known to be polysaccharides covalently attached as side chains to a protein core. They may represent from 50 to 95% of the molecular weight of the proteoglycan, the remainder being protein and oligosaccharides. The structures of the glycosaminoglycans were to a large extent established by the work of Karl Meyer and collaborators (*See* Rodén 1980).

Glycosaminoglycans are linear polymers of repeated disaccharides, in most cases containing an O-sulfated N-acetylhexosamine and a uronic acid, the specific type of which determines the type of glycosaminoglycan. The abbreviations used for carbohydrates can be found in the table at the beginning of the book. The number of repeat disaccharides varies, but typical values are on the order of 50. Consequently, each glycosaminoglycan chain contributes about 100 negatively charged groups to the proteoglycan. The glycosaminoglycans are attached to a protein core by specific carbohydrate sequences, containing three or four monosaccharides. Each proteoglycan molecule contains one or two different types of glycosaminoglycans and the total number of glycosaminoglycan chains may vary from one or two to more than 100, potentially giving as many as 10,000

Department of Physiological Chemistry, University of Lund, Lund, Sweden
This work has been financially supported by the Swedish Medical Research Council (grants 5739 and 05668) and the Faculty of Medicine, University of Lund, Sweden

negatively charged groups per proteoglycan molecule. In addition the protein core is substituted with a number of oligosaccharides similar to those found in glycoproteins. The molecular weights of proteoglycans may vary from about 50,000 to several million. A still higher level of organization is provided by the ability of some types of proteoglycans to form aggregates having molecular weights exceeding 100 million.

A striking feature of the proteoglycans is their variability. First, even in a single tissue there are often several different populations of proteoglycans. Second, each population is polydisperse, containing molecules of varying size and charge. The number of glycosaminoglycans may vary, possibly as a consequence of differences in the size of the protein core. Furthermore, the number of disaccharide units within a particular type of glycosaminoglycan varies, as well as its sulfate content. As a result, the charge density and molecular weight of a particular proteoglycan population can only be expressed as the average for the whole population and the figure obtained may differ greatly depending on the technique used for preparation and measurement. The large number of charged groups gives the proteoglycans a very extended structure occupying a large space in the tissue, a property important for their function.

One function of proteoglycans is to regulate diffusion and flow of macromolecules through connective tissues, thereby providing a barrier. To some tissues, like cartilage, proteoglycans also provide resilience. Because of their polyanionic character, they can bind other molecules, especially cationic amines, peptides, and proteins, which are known to be essential for regulation of cell function. Some glycosaminoglycan chains contain more than one type of repeat disaccharide and have a highly variable structure both chemically and sterically, perhaps required for specific interactions not yet recognized. Examples of such molecules are the glycosaminoglycans isolated from cell surfaces.

The aim of this review is to present current knowledge of proteoglycan structure and turnover. Many areas of proteoglycan research are still in their infancy and only limited information is available. To date most studies have focused on structure, synthesis, and catabolism of the glycosaminoglycans present in different tissues or produced by various cell types. The proteoglycans as such have been studied only in selected tissues, mainly cartilage. Knowledge of the biosynthesis and catabolism of the whole proteoglycan is very incomplete.

Recent reviews should be consulted for more detailed background information and references to the early literature (Módis 1978; Comper and Laurent 1978; Lindahl and Höök 1978; Lowther and Handley 1979; Rosenberg et al. 1979; Hascall and Heinegård 1979; Stockwell 1979; Rodén 1980; Maroudas 1980; Muir 1980; Mason 1981; Hascall 1981).

8.2. Structure of Glycosaminoglycan Side Chains

Glycosaminoglycans are an integral part of the proteoglycans and, with one possible exception, require a protein acceptor for their biosynthesis, as is discussed below. Historically, however, glycosaminoglycans were extracted from tissues by proteolytic digestion, which degraded the proteoglycans to the extent of leaving only a few amino acids from the protein core still attached to the glycosaminoglycan. Consequently, many structural features of the glycosaminoglycans were well known before it was established that they were covalently attached to protein. Much of the structure of the glycosaminoglycans has been elucidated through the work of Karl Meyer and coworkers. More recently enzymes for specific degradation of glycosaminoglycans (Table 8.1) have become available, simplifying structural studies.

Basically the glycosaminoglycans consist of a number of repeat disaccharide units in a linear arrangement. The constituent monosaccharide residues are usually in the chair C-1 conformation, although the iduronic acid residues in dermatan sulfate and heparan sulfate may possibly switch into the 1-C conformation. The C-1 conformation is favorable for the D-monosaccharides, because most of the substituents will then occupy equatorial positions and be at the largest possible distance from one another. The length of the disaccharide unit measured by x-ray crystallography varies somewhat from 0.93 to 0.97 nm for the different glycosaminoglycans and depends on the ionic environment. Isolated glycosaminoglycans form helices of various types stabilized by hydrogen bonds.

8.2.1. Hyaluronic Acid

This polysaccharide is the largest glycosaminoglycan, with a molecular weight from a few hundred thousand to several million. It is believed to be unbranched and contains from 500 to several thousand of the repeat disaccharides (Figure 8.1) and differs from the other glycosaminoglycans in that it does not contain any sulfate ester groups. A central question, which has not yet been settled, is whether or not hyaluronic acid is covalently bound to protein and has a specific oligosaccharide linkage region. Some preparations of hyaluronate have extremely low protein contents (less than 0.1% of dry weight), but one should realize that such a preparation, with a molecular weight of 5×10^6, still may contain a protein with a molecular weight of 5000 for each hyaluronate molecule.

The conformation of hyaluronate is highly dependent on whether the counter ion is monovalent or divalent. Based on data obtained with x-ray crystallography and nuclear magnetic resonance (NMR) it has been suggested that the hyaluronic acid molecule contains stiff segments and that various portions of the molecule may interact with one another (Morris et al. 1980). The result is an extended structure in solution, occupying a large domain and thereby rendering the solution viscous.

The very large molecule can easily be depolymerized if not handled with proper precautions. Reducing conditions or high temperature result in depolymerization. Even treatment with ascorbate will result in degradation of hyaluronic acid.

Hyaluronate is present in all connective tissues and in a few, e.g., rooster comb, umbilical cord, and vitreous body, it is the main glycosaminoglycan. Cartilage, on the other hand, has a very low relative hyaluronate content, about 1% of total glycosaminoglycan, while the corresponding figure for skin and aorta is 10–20%. Fibroblasts in culture usually produce more than 50% of total glycosaminoglycan as hyaluronate. Hyaluronate is also produced by bacteria such as group A streptococci. Hyaluronic acid is bound to the cell surface of several types of cells in culture (Mason 1981). Binding is, at least in some cases, mediated

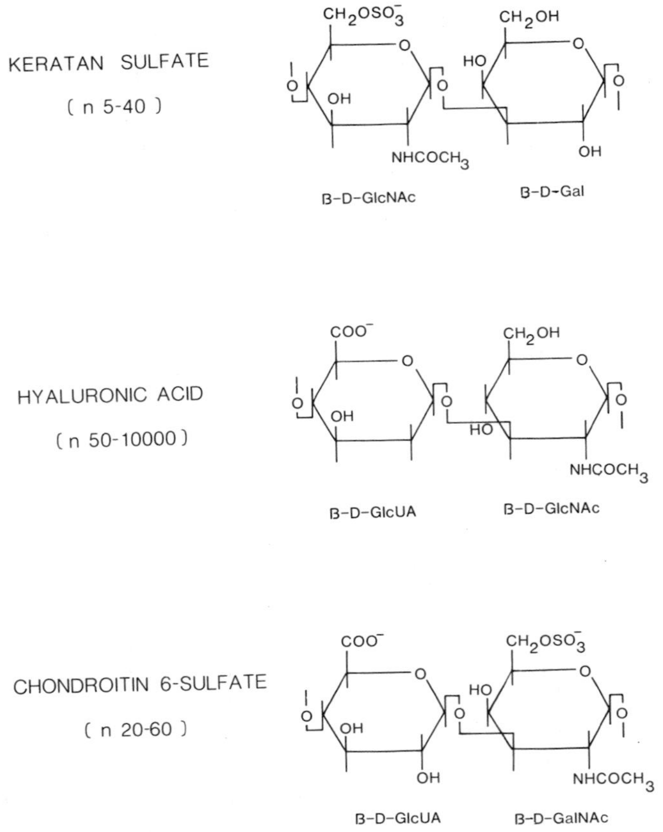

KERATAN SULFATE

(n 5-40)

$CH_2OSO_3^-$ CH_2OH O OH HO NHCOCH$_3$ OH

ß-D-GlcNAc ß-D-Gal

HYALURONIC ACID

(n 50-10000)

COO$^-$ CH_2OH O OH HO NHCOCH$_3$

ß-D-GlcUA ß-D-GlcNAc

CHONDROITIN 6-SULFATE

(n 20-60)

COO$^-$ $CH_2OSO_3^-$ O HO OH OH NHCOCH$_3$

ß-D-GlcUA ß-D-GalNAc

Figure 8.1. Structures of repeating disaccharide units of the glycosaminoglycans. *n* Denotes number of disaccharide units in the chains. *GlcNAc* = N-acetylglucosamine; *GalNAc* = N-acetylgalactosamine; *Gal* = galactose; *GlcUA* = glucuronic acid; *IdUA* = iduronic acid.

---

by specific receptors. Such receptors have been found on liver cells (Truppe et al. 1977) and on transformed fibroblasts and chondrocytes (Underhill and Toole 1981).

A primary role of hyaluronate, at least in some tissues, is to retain water and regulate water flow, important for the structure of the tissue. Furthermore, hyaluronate participates in proteoglycan aggregation (Section 8.3.2), exerts feedback regulation on glycosaminoglycan biosynthesis, and influences phagocytosis by white blood cells. Hyaluronate may represent the earliest evolutionary form of the glycosaminoglycans; it is abundant in embryonic matrices and is replaced by other proteoglycans during development (Toole et al. 1977).

8.2.2. Chondroitin Sulfate

A chondroitin sulfate is defined as a galactosaminoglycan containing only one type of uronic acid, i.e., glucuronic acid (Figure 8.1). Two types of chondroitin sulfate can be distinguished, differing in having the ester sulfate group on the 4 or 6 carbon of the N-acetylgalactosamine residue, respectively. The number

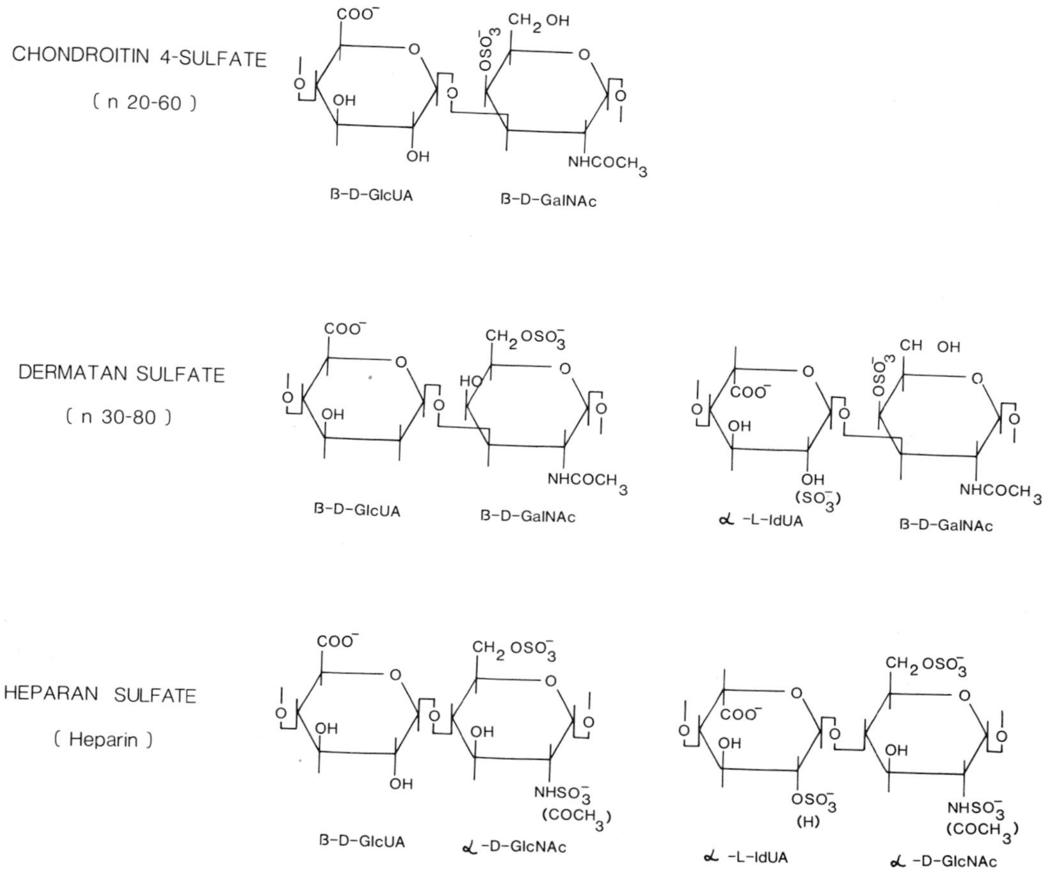

CHONDROITIN 4-SULFATE (n 20-60)

β-D-GlcUA β-D-GalNAc

DERMATAN SULFATE (n 30-80)

β-D-GlcUA β-D-GalNAc α-L-IdUA β-D-GalNAc

HEPARAN SULFATE (Heparin)

β-D-GlcUA α-D-GlcNAc α-L-IdUA α-D-GlcNAc

---

of repeat disaccharides (Figure 8.1) within a preparation of chains varies from 20 (or perhaps even 10) to 60 with an average of about 40, corresponding to a molecular weight of about 20,000. The number of ester sulfate groups also varies, with an average of about 0.8 sulfate groups per disaccharide. Available data suggest that one glycosaminoglycan chain may contain ester sulfate groups attached both to the 4 position and to the 6 position of the galactosamine, but on separate disaccharide units. It is likely that most, if not all, chondroitin sulfate chains are copolymers of segments of one to several 6-sulfated disaccharides interrupted by segments of one to several 4-sulfated disaccharides. Aging cartilage is known to contain a higher proportion of 6-sulfated residues than young cartilage. Because of the abundance of charged groups the glycosaminoglycan is rather extended and forms a random coil in solution.

As early as 1889, Mörner showed that chondroitin sulfate isolated after autolysis (proteolysis) of cartilage had a higher nitrogen content than that isolated after alkaline treatment. This and similar observations were, however, not explained until 1958, when Muir demonstrated that chondroitin sulfate was attached to protein via an alkali labile bond to serine. At that time it became clear

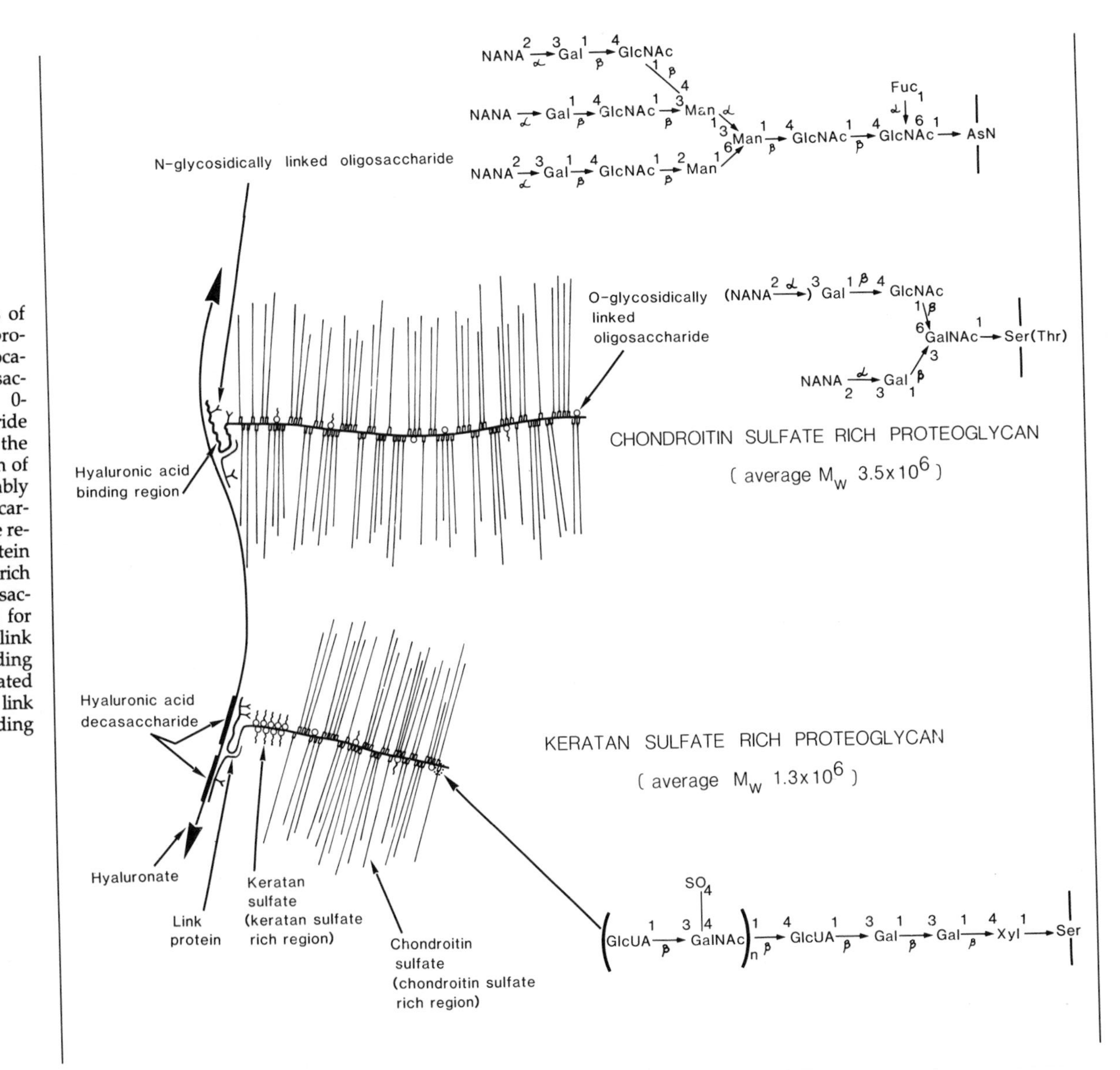

Figure 8.2. Tentative structures of four subpopulations of cartilage proteoglycans. The structure and locations of the two types of oligosaccharides are shown. The 0-glycosidically linked oligosaccharide minus the sialic acid *(NANA)* of the side chain linked to the 6-position of the N-acetylgalactosamine probably represents the linkage region of cartilage keratan sulfate. The linkage region of chondroitin sulfate to protein is indicated on the keratan sulfate rich proteoglycan. The specific decasaccharide of hyaluronate required for the specific interaction between link protein and hyaluronic acid-binding region with hyaluronate is indicated as well as the interaction between link protein and hyaluronic acid binding region.

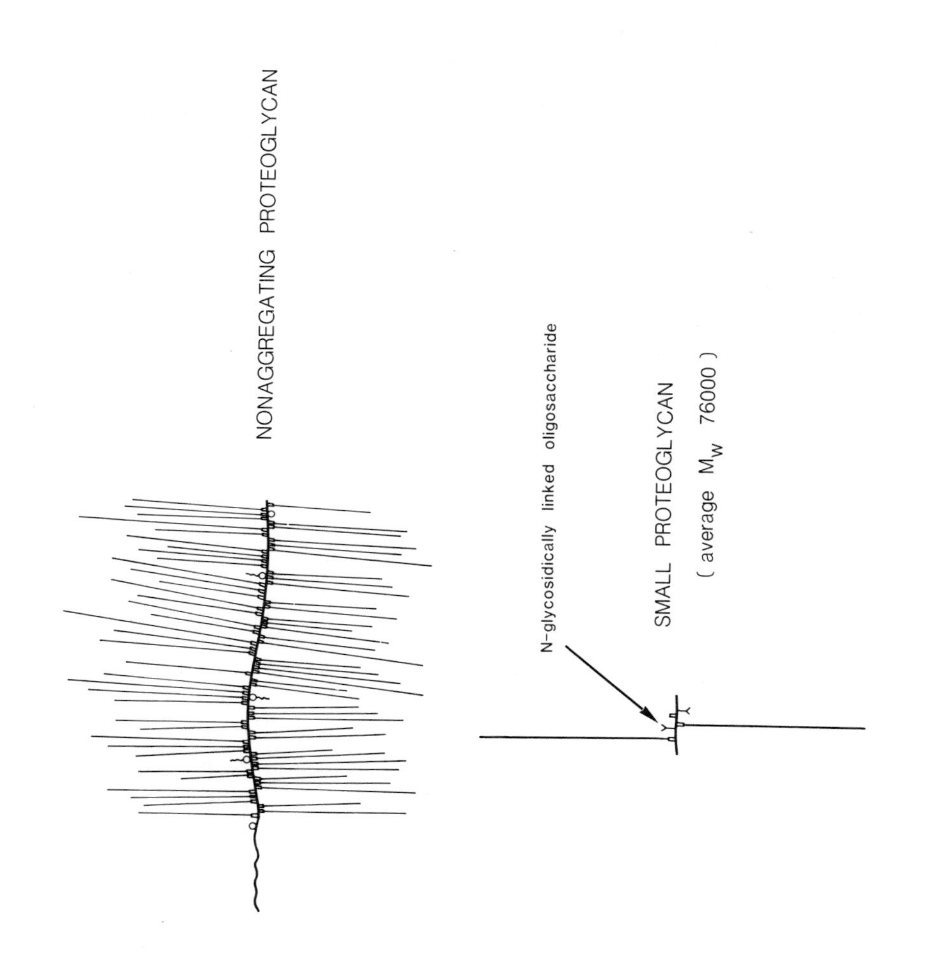
NONAGGREGATING PROTEOGLYCAN
N-glycosidically linked oligosaccharide
SMALL PROTEOGLYCAN
(average M_W 76000)

that chondroitin sulfate was actually part of a much larger complex—a proteoglycan. The work of Lindahl and Rodén during the 1960s (*See* Rodén 1980) demonstrated that the linkage between several glycosaminoglycans, including chondroitin sulfate, and protein is actually an 0-glycosidic bond between serine and a xylose residue, which is part of a specific linkage region sequence of xylose-galactose-galactose-glucuronic acid (Figure 8.2). This type of linkage is common to most of the glycosaminoglycans, with the exception of keratan sulfate (Section 8.2.5) and possibly hyaluronic acid.

The tissues having the highest content of chondroitin sulfate are cartilage and intervertebral disc, where as much as 10% of the wet weight may be chondroitin sulfate. Nasal cartilage, epiphyseal cartilage, and the Swarm rat chondrosarcoma have a high proportion of position 4 ester sulfate, while articular cartilage, particularly of older individuals, and especially the nucleus pulposus of the intervertebral disc, have high relative contents of position 6 ester sulfate. Fibrous connective tissues contain little, if any, chondroitin sulfate. Most of the galactosaminoglycans in these tissues probably consist of iduronic acid as well as glucuronic acid-containing disaccharides. The corneal galactosaminoglycans are unique in that they have lower sulfate contents (less than 0.5 ester sulfate per disaccharide), while a chondroitin sulfate isolated from shark cartilage contains more than one sulfate group per disaccharide.

8.2.3. Dermatan Sulfate

The nomenclature of this glycosaminoglycan is confusing, because it may be viewed as a modified chondroitin sulfate (Section 8.7.2). The designation dermatan sulfate will be used for all galactosaminoglycans containing an additional type of monosaccharide, L-iduronic acid, which is the C-5 epimer of D-glucuronic acid. One type of disaccharide unit (Figure 8.1) consists of L-iduronic acid and N-acetylgalactosamine. The hexosamine contains an ester sulfate group at position 4. Fransson and Rodén (*See* Rodén 1980) showed that dermatan sulfate also contains the type of disaccharide unit previously identified in chondroitin sulfate (Figure 8.1). The two disaccharide units containing L-iduronic acid and D-glucuronic acid, respectively, are distributed in a copolymeric fashion, with several alternating segments, each containing one to several disaccharide units of either type. The number of disaccharide units in the glycosaminoglycan chain is variable, but on the average is somewhat higher than that of chondroitin sulfate (about 50 to 60). The proportion of L-iduronic acid-containing disaccharides in a single chain varies greatly from a few percent to almost 100%, even within the same tissue. More recently Fransson and collaborators (*See* Rodén 1980) have shown that some of the iduronic acid residues contain an ester sulfate group, probably at the 2 position.

It is apparent that the structure of dermatan sulfate is extremely variable with regard to the following parameters: (a) total number of disaccharides; (b) proportion of L-iduronic acid; (c) distribution of L-iduronic and D-glucuronic acid containing disaccharides; (d) distribution of sulfate ester groups on the N-acetylgalactosamine; and (e) number of sulfated L-iduronic acid residues. It is attractive to speculate that a carbohydrate polymer of such complexity may be able to participate in various specific interactions and serve in a regulatory capacity. The average sulfation of dermatan sulfate is higher than that of chondroitin sulfate and close to one per disaccharide. It has been shown that dermatan sulfate has the same sequence of monosaccharides forming the linkage to protein, i.e., the O-glycosidic linkage between xylose and serine, (Figure 8.2) as was first described for heparin and chondroitin sulfate.

Dermatan sulfate has not been identified in cartilage, except in the meniscus. It is present in fibrous connective tissues like tendon, skin, aorta, sclera, and joint capsule, and also in cornea. Its relative content of L-iduronic acid is usually high (more than 50%) although exact figures are lacking. The exception is cornea, where probably less than 20% of the uronic acid is L-iduronic acid. The functions of dermatan sulfate are unclear, but it has been shown that the glycosaminoglycan can interact with several matrix components. Tissues containing dermatan sulfate have notably low overall contents of glycosaminoglycans (as parts of proteoglycans), usually 0.1–1%. Little is known, however, about tissue distribution. Local concentrations may be quite high.

8.2.4. Heparan Sulfate; Heparin

These two glycosaminoglycans cannot be clearly distinguished according to structure. It should be stressed, however, that they are synthesized on different protein acceptors and therefore represent distinct entities (Section 8.7.3). Their structure will be discussed together and differences in degree will be pointed out where relevant. In general, heparin is more highly charged than heparan sulfate.

They are copolymers of the two types of disaccharides shown in Figure 8.1 and are the most complex of the glycosaminoglycans. The number of such disaccharides in a chain usually varies from 10 to 60. Segments of one or several disaccharides, containing one type of uronic acid (L-iduronic or D-glucuronic acid) are interrupted by segments containing the other. The glucosamine residues contain ester sulfate at position 6 and in some instances also at position 3. A large proportion of the L-iduronic acid residues contain ester sulfate at position 2, while none of the D-glucuronic acid residues contain ester sulfate. Unique for heparin and heparan sulfate is that in many of the glucosamine residues the N-acetyl group is replaced by an N-sulfate group, as indicated in Figure 8.1. This N-sulfate group is a prerequisite for formation of the L-iduronic acid residue, which can only be found in a position linked to the reducing end of the N-sulfated glucosamine. Heparin contains a higher proportion of N-sulfate and therefore also of iduronic acid residues (both sulfated and not sulfated) than heparan sulfate. As many as 90% of the uronic acid residues in heparin may be iduronic acid. To summarize, heparan sulfate and heparin may vary with respect to: (a) chain size, i.e., number of disaccharide units; (b) proportion of L-iduronic acid containing disaccharide; (c) distribution of L-iduronic acid and D-glucuronic acid containing disaccharides; (d) number of ester sulfate groups at C-6 and C-3 of the hexosamine residues; (e) number of ester sulfate groups at the C-2 of the iduronic acid residues and (f) proportion of N-sulfate and N-acetyl groups on the hexosamine.

Heparan sulfate appears always to contain the same specific linkage oligosaccharide to protein as chondroitin sulfate (Figure 8.2), forming an O-glycosidic bond between xylose and serine. Heparin on the other hand is degraded after synthesis so that only some of the heparin molecules contain the linkage oligosaccharide. The linkage oligosaccharide was first identified and characterized in heparin by Lindahl and Rodén (*See* Rodén 1980).

Heparin is often isolated from lung or intestine. The molecules are derived from mast cells and have also been isolated from mast cell tumors. Heparan sulfate is present on the cell surface of many cells, but may in addition be a matrix component in certain fibrous connective tissues, like lung, aorta, and basement membrane. The physiologic function of these polymers is not known. Heparin particularly can form very specific interactions with proteins dependent on its complex copolymeric structure. Its activation of antithrombin is an example. It is, however, not known if such interactions occur in nature or only when heparin is administered intraveneously as a drug. Heparan sulfate is probably also involved in interactions with a number of matrix constituents, further discussed below.

8.2.5. Keratan Sulfate

Keratan sulfate has many features in common with the complex oligosaccahrides found in glycoproteins. Examples are the structure of the linkage region to protein and the sialic acid residues in nonreducing terminal positions. Furthermore, unlike the other glycosaminoglycans, keratan sulfate contains oligosaccharide branches. It is still considered a glycosaminoglycan as it contains a repeating disaccharide of a 6-sulfated N-acetylglucosamine linked to galactose (Figure 8.1). This repetitive portion is unbranched and extremely variable in size. Cartilage keratan sulfate contains 5 to 10 disaccharide units, while keratan sulfate from intervertebral disc may contain 20 to 30 disaccharides. Corneal keratan sulfate contains 30 to 50 disaccharide units of the same type as those in the skeletal tissues.

Keratan sulfate is linked to protein, skeletal keratan sulfate having a different linkage than corneal keratan sulfate. The skeletal keratan sulfate is linked via N-acetylgalactosamine by an O-glycosidic bond to serine or threonine residues in the protein core. The N-acetylgalactosamine is part of a specific linkage oligosaccharide, similar to the O-glycosidically linked oligosaccharide (Figure 8.2). The linkage is sensitive to alkaline treatment and undergoes a β-elimination reaction, as shown by Meyer and collaborators (*See* Rodén 1980). The galactosamine residue is substituted with a side chain of galactose-sialic acid at carbon 3 (Figure 8.2) while the glycosaminoglycan chain proper is attached to carbon 6 of the galactosamine (Hopwood and Robinson 1974). Some of the linkage oligosaccharides appear not to be substituted with a glycosaminoglycan, but rather with a disaccharide terminated by a sialic acid residue (Figure 8.2). This oligosaccharide is particularly prominent in young tissue and in proteoglycans from the transplantable Swarm rat chondrosarcoma (Lohmander et al. 1980).

Corneal keratan sulfate has an entirely different linkage to a different core protein. The linkage is N-glycosidic from asparagine to glucosamine (Section 8.3.6) and is similar to the biantennary, mannose-containing oligosaccharide of glycoproteins (Keller et al. 1981). More recently Nilsson et al. (1983) have conclusively shown that the linkage region of corneal keratan sulfate to protein has the structure depicted in Figure 8.6. It is apparent that skeletal and corneal keratan sulfate represent distinct entities, having different synthetic pathways.

Skeletal keratan sulfate is found in tissues related to cartilage. In hyaline cartilage 5 to 20% of the glycosaminoglycans are keratan sulfate, while nucleus pulposus contains between 30 and 40%, depending on the age. Fibrous connective tissues probably do not contain keratan sulfate. The other type of keratan sulfate is found exclusively in the cornea of the eye, where it forms more than 50% of the glycosaminoglycans. No physiologic role has been assigned to keratan sulfate. Interestingly, its concentration in the developing chick cornea increases with increasing transparency, and the nontransparent cornea of patients with macular dystrophy does not synthesize keratan sulfate (Section 8.3.6).

8.2.6. Isolation and Analysis of Glycosaminoglycans

The glycosaminoglycans are often isolated from papain digests of tissues. They can best be purified and separated by methods employing their polyanionic character. Examples are ion exchange chromatography and precipitation with long aliphatic chain quarternary amines, e.g., cetylpyridinium chloride. Alternatively, fractional precipitation with ethanol may be used. An excellent account of methods used for preparation and analysis of glycosaminoglycans has been presented by Rodén et al. (1972). Sensitive identification and fragmentation of glycosaminoglycans can be obtained by using the enzymes listed in Table 8.1. Specific fragments produced by enzymic digestion of the glycosaminoglycans can be identified by paper or thin layer chromatography, or, more recently, using high pressure liquid chromatography. (For further information on the use of these enzymes, see Suzuki 1972; Linker and Hovingh 1972; Nakazawa et al. 1975).

8.3. Structure of the Proteoglycans

Proteoglycans are complex molecules containing from a few to several hundred glycosaminoglycan chains covalently attached to a protein core via their reducing terminals. The polyanionic glycosaminoglycans radiate out from the central protein core and create, because of their high charge density, an expanded structure. Each proteoglycan contains one or two types of glycosaminoglycan chains and, in addition, N-glycosidically linked oligosaccharides of the types found in glycoproteins (Chapter 7) containing N-acetylglucosamine, mannose, galactose, and sialic acid. The composition of proteoglycans isolated from various tissues differs considerably. To date studies have focused on cartilage proteoglycans, mainly because of cartilage's high proteoglycan content and low content of other matrix components. Comparatively little is known about the structure of proteoglycans in most other tissues. Studies on cartilage proteoglycans will be described below in some detail and can be taken as models for the study of proteoglycans from other tissues. It should, however, be stressed that proteoglycans from fibrous tissues do differ from cartilage proteoglycans, for example in their glycosaminoglycan side chains.

Table 8.1. Degradation of Glycosaminoglycans with Specific Enzymes (Products of Limit Digestion)

Enzyme	HA	CS-4 and CS-6[a]	DS[b]	HS-Hep[c]	KS
Streptomyces hyaluronidase[d]	ΔUA-GlcNAc	—	—	—	—
Leech hyaluronidase	GlcNAc-UA-GlcNAc-UA	—	—	—	—
Testicular hyaluronidase[d]	UA-GlcNAc-UA-GlcNAc	S S UA-GalNAc-UA-GalNAc	—	—	—
Chondroitinase AC[d]	ΔUA-GlcNAc	S[e] ΔUA-GalNAc	—	—	—
Chondroitinase ABC[d]	ΔUA-GlcNAc	S[e] ΔUA-GalNAc	S[e] ΔUA-GalNAc	—	—
Heparinases	—	—	—	variably long oligosaccharides	—
Keratanase[d]	—	—	—	—	oligosaccharides

HA = Hyaluronic acid; KS = keratan sulfate; CS = chondroitin sulfate; DS = dermatan sulfate; HS = heparan sulfate.
[a]Includes glucuronic acid containing sequences of dermatan sulfate.
[b]Iduronic acid containing sequences.
[c]Cleavage with nitrous acid (Cifonelli and King 1972) is often used for structural studies.
[d]Commercially available from Miles Chemicals.
[e]The 4- or 6-sulfate ester group can be removed with chondrosulfatases[d] specific for the 4 and 6 position, respectively.

8.3.1. Cartilage Proteoglycan Monomers

The most recent data indicate that cartilage contains at least four populations of distinct proteoglycans. Two populations of high molecular weight, aggregating proteoglycans (Section 8.3.2) form the bulk of the proteoglycans in the tissue. Some 10% of the proteoglycans are large but not capable of forming aggregates with hyaluronate. The fourth population is a class of low molecular weight proteoglycans representing some 2 to 3% (by weight) of the total (Figure 8.2).

Most of the data on cartilage proteoglycan has been acquired using preparations containing mixtures of all the populations of large proteoglycans. Analyses of such materials have provided a great deal of information on proteoglycan structure and will therefore be discussed, followed by the more recent data on individual populations.

Early information was provided by Schubert, Partridge, Mathews, Muir and their collaborators during the 1950s and 1960s. The breakthrough, however, came with the introduction of a new elegant method for the extraction and purification of proteoglycans from cartilage (Figure 8.3) in two papers by Hascall and Sajdera (Sajdera and Hascall 1969; Hascall and Sajdera 1969). The extraction was much improved by the use of 4 *M* guanidine-HCl, which dissociates all noncovalent bonds between the components of proteoglycan aggregates and between proteoglycans and other matrix components. An additional advantage is that a chaotropic agent like guanidine-HCl inhibits tissue proteases from degrading the proteoglycans. More recent modifications of the extraction procedure involve inclusion, as extra insurance, of protease inhibitors such as phenylmethyl-sulphonyl-fluoride, N-ethylmaleimide, EDTA, 6-amino-hexanoic acid, pepstatin and benzamidine-HCl in the extraction solvent. The separation of proteoglycans from other components in the extract was achieved by density gradient centrifugation in self-forming CsCl-gradients. Because of their high charge density the proteoglycans have a much higher buoyant density than proteins and are therefore recovered in the bottom of the gradient (Figure 8.3). Such procedures were originally used by Franek and Dunstone for the purification of proteoglycans extracted from cartilage with water (*See* Dunstone 1969).

Hascall and Sajdera showed that when the extract was brought to associative conditions, i.e., 0.5 *M* guanidine-HCl, the proteoglycans formed aggregates. These aggregates have been shown to contain hyaluronate, specific link proteins, and proteoglycan monomers, as outlined in Figures 8.2 and 8.3. Aggregate fractions can be isolated under associative conditions by CsCl-density gradient centrifugation, as shown in Figure 8.3, (*left*). The components of the aggregates can subsequently be separated under dissociative conditions in 4 *M* guanidine-HCl (Figure 8.3, *left*). Alternatively, the polyanionic proteoglycan monomers can be isolated using direct CsCl-gradient centrifugation of the extract in 4 *M* guanidine-HCl (Figure 8.3, *right*). Since small, low buoyant density, aggregating proteoglycan monomers are recovered as high buoyant density aggregates, the starting density can be high in the associative scheme. When direct dissociative preparations are made, the starting density should be lower to prevent losses of low buoyant density proteoglycans to the top of the gradient.

In most studies of proteoglycan monomer structure the Al-Dl fraction (Figure 8.3, legend), prepared by sequential associative-dissociative centrifugation, has

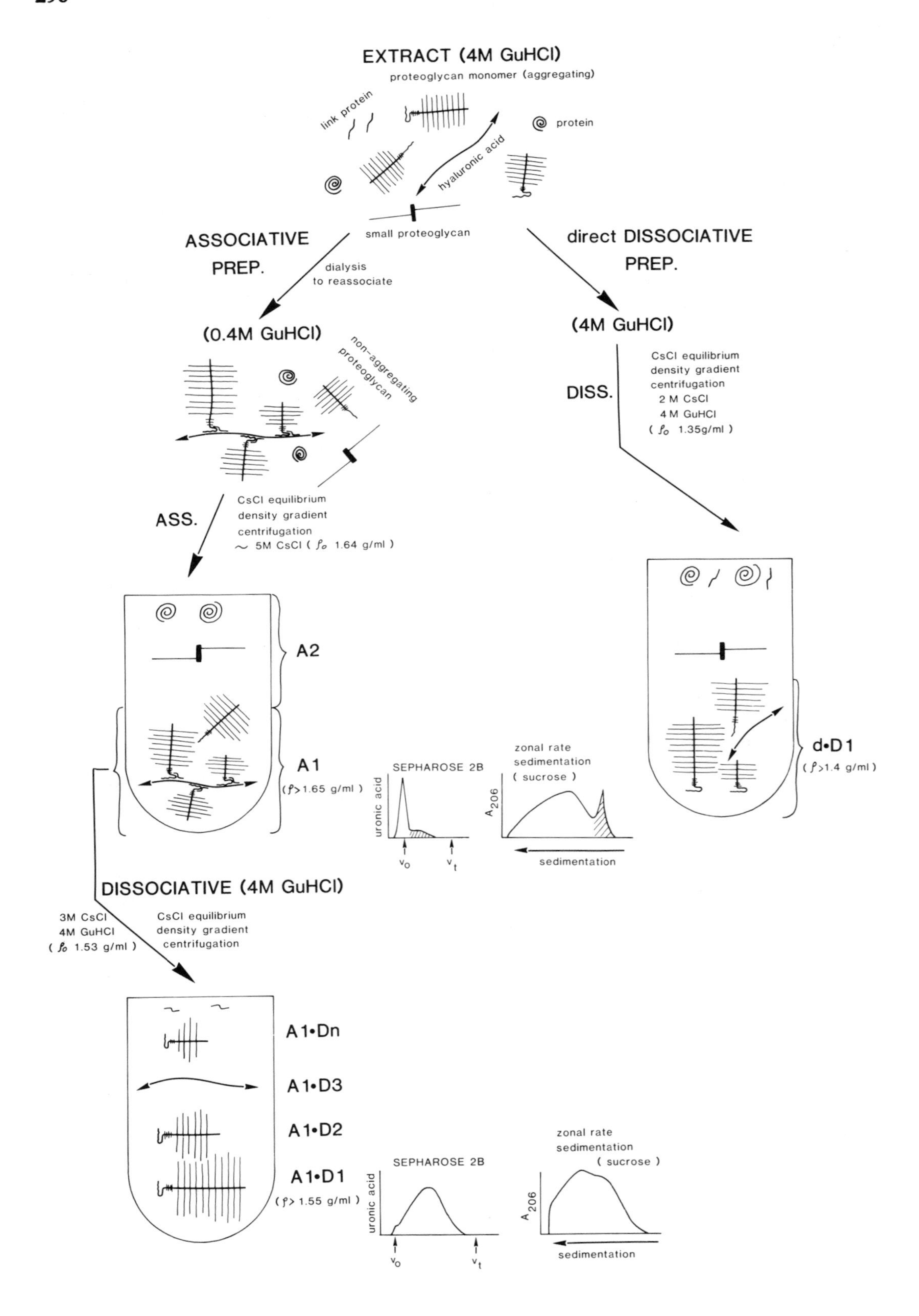
EXTRACT (4M GuHCl)
proteoglycan monomer (aggregating)
link protein
protein
hyaluronic acid
small proteoglycan
ASSOCIATIVE PREP.
dialysis to reassociate
direct DISSOCIATIVE PREP.
(0.4M GuHCl)
non-aggregating proteoglycan
(4M GuHCl)
DISS.
CsCl equilibrium density gradient centrifugation
2 M CsCl
4 M GuHCl
(ρo 1.35g/ml)
ASS.
CsCl equilibrium density gradient centrifugation
~ 5M CsCl (ρo 1.64 g/ml)
A2
A1
(ρ> 1.65 g/ml)
SEPHAROSE 2B
uronic acid
vo
vt
zonal rate sedimentation (sucrose)
A206
sedimentation
d•D1
(ρ>1.4 g/ml)
DISSOCIATIVE (4M GuHCl)
3M CsCl
4M GuHCl
(ρo 1.53 g/ml)
CsCl equilibrium density gradient centrifugation
A1•Dn
A1•D3
A1•D2
A1•D1
(ρ> 1.55 g/ml)
SEPHAROSE 2B
uronic acid
vo
vt
zonal rate sedimentation (sucrose)
A206
sedimentation

Figure 8.3. Schematic illustration of preparative schemes for the proteoglycans. Nonaggregating monomers present in the A1 fraction are indicated by the dashed area in the sedimentation and gel chromatographic profiles. The nomenclature used is operational and defines the purification steps and the resulting fractions. A refers to associative and D to dissociative solvents.

---

been analyzed. Typically, such a proteoglycan preparation contains from 7 to 15% protein, depending on the source. Keratan sulfate represents 5 to 30% of the dry weight, the remainder being chondroitin sulfate attached to the same protein core. Cartilage proteoglycans have high contents of serine; 60% of these residues being substituted with glycosaminoglycan chains. Also glycine and glutamic acid or glutamine are enriched in cartilage proteoglycans.

Hardingham and Muir (1972a) demonstrated that proteoglycan aggregation was due to an interaction with hyaluronate. Subsequently it was shown that the binding to hyaluronate was mediated by a specific domain of the proteoglycan core not containing any glycosaminoglycan (Heinegård and Hascall 1974a). This region is referred to as the hyaluronic acid-binding region and is located at one end of the protein core (Figure 8.2). It has a protein composition with relatively high contents of glutamic acid or glutamine and arginine and a molecular weight of about 60,000, which represents approximately one fourth of the protein.

Later, two additional domains of the protein core were identified (Heinegård and Axelsson 1977). In the opposite end from the hyaluronic acid-binding region, a chondroitin sulfate-rich region is located (Figure 8.2). It contains all the approximately 100 chondroitin sulfate chains and 20 to 30% of the proteoglycan's 50 to 80 keratan sulfate chains. This portion of the core contains more than one half of the total protein in the proteoglycan. It is, however, more extended than the folded hyaluronic acid binding region and therefore forms most of the length of the protein core. Serine, glycine, and glutamic acid or glutamine represent about 60% of the amino acid residues. The third region of the proteoglycan is the keratan sulfate-rich region, located between the other two regions (Figure 8.2). It contains the majority of the keratan sulfate chains, very closely spaced. The protein portion of this region has a molecular weight of about 20,000 when isolated by proteolytic digestion (Figure 8.4, II and III). Consistent with the close spacing of side chains, which does not allow even papain to liberate single glycosaminoglycan chains, the region is markedly enriched in a few amino acids; 50% of the residues are glutamic acid, glutamine, and proline, while serine constitutes another 10 to 15%.

These regions of the proteoglycan can be isolated after combined digestion with trypsin and chondroitinase, as outlined in Figure 8.4, II and III. The purest preparations are obtained by the procedure using fractionation by CsCl-gradient centrifugation of fragments produced by trypsin digestion, followed by chondroitinase digestion and gel chromatography, as outlined in Figure 8.4, III. The sequential chondroitinase–trypsin digestion outlined in Figure 8.4, II is best suited for analytic purposes.

More recently, identification of oligosaccharide side chains has added to the

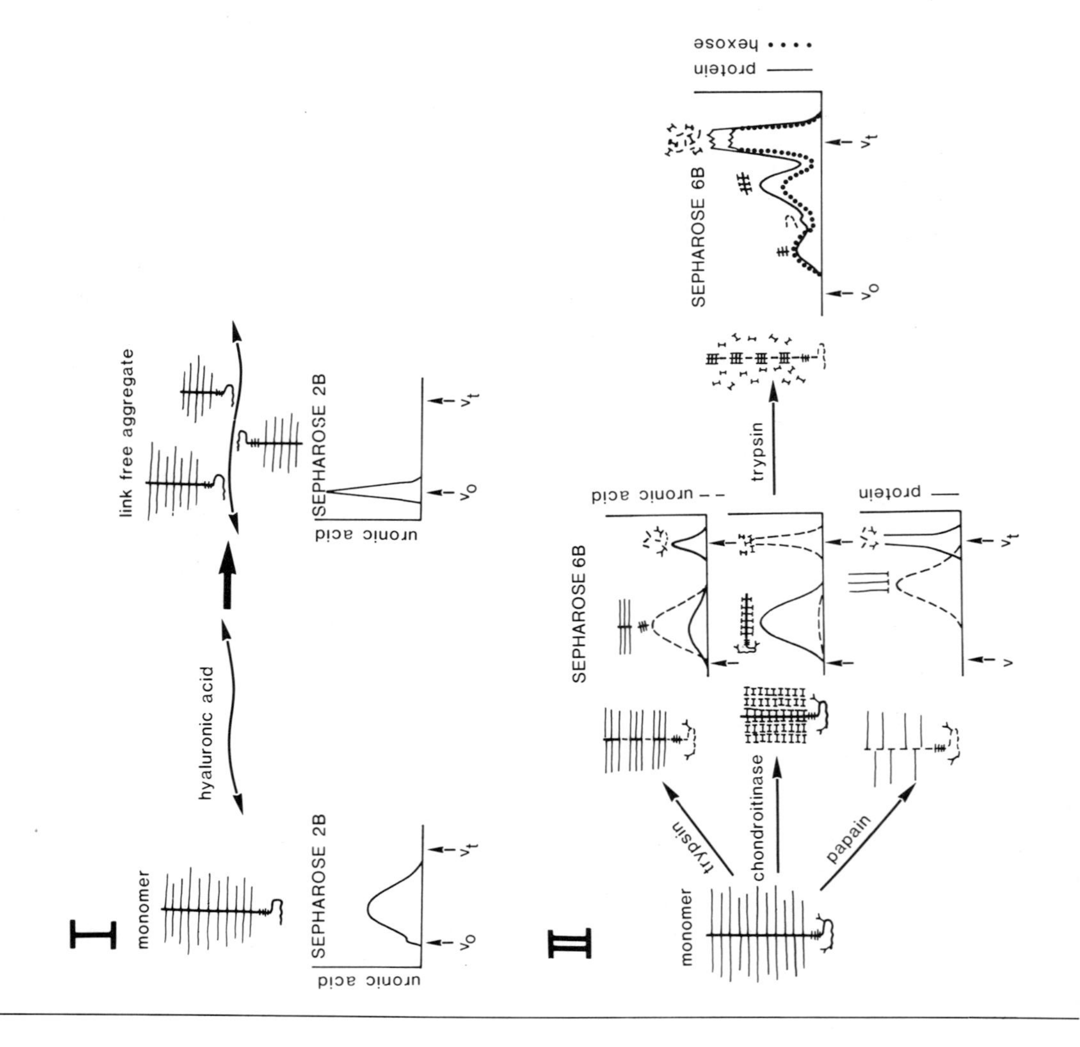

I
monomer
SEPHAROSE 2B
uronic acid
V_o
V_t
hyaluronic acid
link free aggregate
SEPHAROSE 2B
uronic acid
V_o
V_t
II
monomer
trypsin
chondroitinase
papain
SEPHAROSE 6B
uronic acid – –
protein —
V_t
trypsin
SEPHAROSE 6B
V_o
V_t
protein —
hexose • • • •

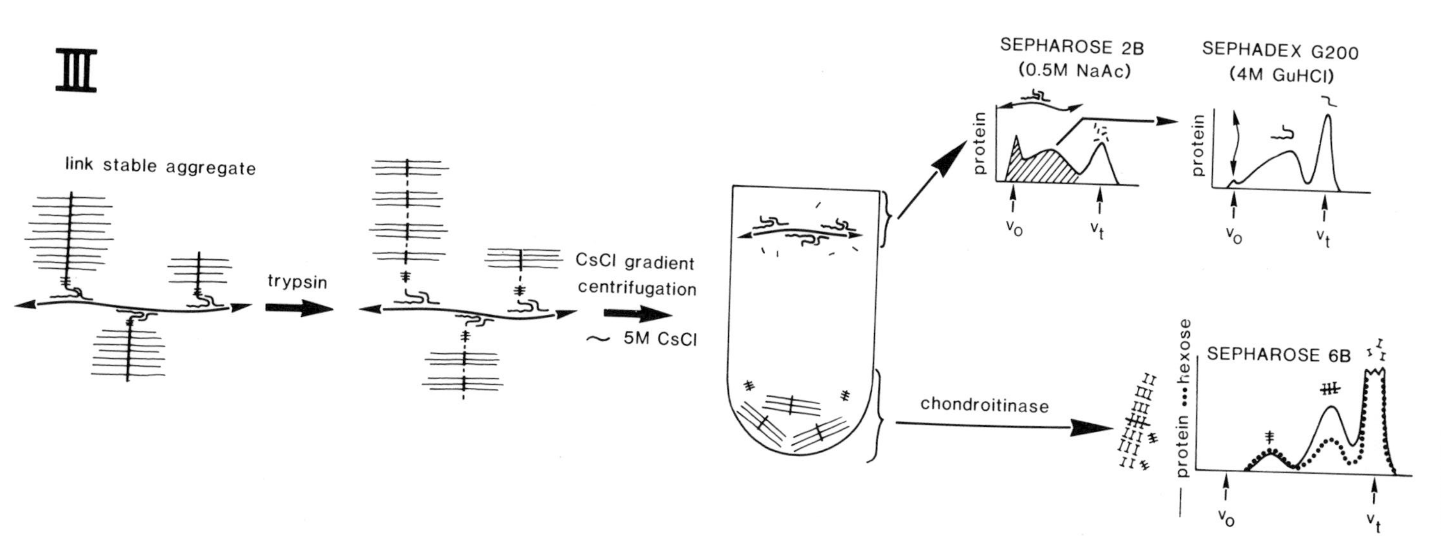

Figure 8.4. Schematic illustration of gel chromatographic patterns of monomeric and aggregated proteoglycans (I), fragmented proteoglycan monomers (II), and fragmented proteoglycan aggregates (III). The schematic protocol (III) can be used to prepare link protein and hyaluronic acid-binding region. When a particular peak is indicated, the subsequent chromatogram represents only material from that peak.

complexity of the cartilage proteoglycan structural model (Thonar and Sweet 1979; DeLuca et al. 1980). The hyaluronic acid-binding region has been shown to contain N-glycosidically linked oligosaccharide of the type shown in Figure 8.2 (Lohmander et al. 1980, 1981). This oligosaccharide is N-asparagine-linked, and has a structure similar to those in glycoproteins (Chapter 7). Both the keratan sulfate-rich region and the chondroitin sulfate-rich region contain, in addition, O-glycosidically linked oligosaccharides with a structure similar to that of the linkage region of keratan sulfate, but terminated with sialic acid residues (Figure 8.2) (Lohmander et al. 1980). These oligosaccharides are of the type found in mucins. An additional substituent of proteoglycans has been reported to be ester phosphate groups probably attached to the linkage region carbohydrates in 1 out of 20 chondroitin sulfate chains (Schwermann et al. 1981).

The chondroitin sulfate chains are attached to the protein core in clusters (Mathews 1971) containing up to 10 chains per cluster, with an average of four chains (Heinegård and Hascall 1974b). The chains within the cluster are separated by a few amino acids, while the clusters are separated by longer sequences containing several nonsubstituted amino acids. The clusters are liberated by sequential digestion with trypsin and chymotrypsin, while single chondroitin sulfate chains are liberated by papain digestion, as indicated in Figure 8.4, II. The size distribution of the various fragments can be monitored by gel chromatography (Figure 8.4, II). Data supporting the clustering of glycosaminoglycans have also been obtained by electron microscopy of proteoglycan molecules spread on cytochrome C films (Thyberg et al. 1975).

The hyaluronic acid-binding region and the keratan sulfate-rich region of the proteoglycans appear constant, while the length of the chondroitin sulfate-rich region is variable. Molecules with fewer chains have a proportionally shorter protein core. Since the hyaluronic acid-binding region constitutes a higher relative proportion of smaller proteoglycans, they have a higher protein content and therefore a lower buoyant density in CsCl gradients, as illustrated in Figure 8.3 (Al-D-gradient). It is possible that shorter chondroitin sulfate chains may also contribute to the lower buoyancy of some proteoglycan molecules. Supporting data for this model have been obtained by structural analysis of proteoglycans subfractionated to greater homogeneity. Such studies have included (a) quantitation of substructures of proteoglycans after specific cleavage essentially as outlined in Figure 8.4 (Heinegård 1977); (b) quantitation of substructures using immunological methods (Wieslander and Heinegård 1980); (c) variations in amino acid and glycosaminoglycan composition of subfractions (*See* Hascall 1981); (d) electron microscopy (*See* Hascall 1980). There is some controversy as to which portion of the proteoglycan represents the N-terminal. Rosenberg and collaborators (1979) found that valine was the N-terminal amino acid of isolated hyaluronic acid-binding region as well as of several proteoglycan subfractions containing proteoglycans of different sizes, which were assumed to have core proteins of different length. They concluded that the hyaluronic acid binding region was probably located in the N-terminal portion of the molecule. Biosynthesis data (Section 8.8), in contrast, indicate that the hyaluronic acid-binding region may represent the C-terminal portion.

Variability of physical properties of cartilage proteoglycans has hampered their study. To obtain a relevant average molecular weight, only a limited degree of polydispersity of the preparation can be accepted. Hascall and Sajdera (1970) obtained centrifugal data indicating average molecular weights of about 2.5×10^6. Pasternack et al. (1974), using light scattering in different solvents, obtained an average molecular weight of 2.3×10^6 and radius of gyration from 159 nm in 0.05 *M* guanidine hydrochloride to 57 nm in 4 *M* guanidine hydrochloride. Sheehan et al. (1978), using light scattering, also obtained molecular weights of $2.5\text{-}3 \times 10^6$ at sodium chloride concentrations of 0.1, 0.2, and 0.3 *M*. At 0.15 *M* sodium chloride, however, the observed molecular weight was about 5×10^6, taken to indicate dimerization. Reihanian et al. (1979) used quasielastic light scattering to show average molecular weight values of 3.97×10^6 and assumed a prolate ellipsoid shape of semiaxis a = 190 nm and b = 40 nm. Their data supported those of Sheehan et al. in indicating formation of proteoglycan dimers and oligomers. Electron microscopy (*See* Hascall 1980) has shown that proteoglycans have a variably large central filament with an average, extended length of about 290 nm. To this core filament, an average of 25 side chain filaments (representing chondroitin sulfate clusters) with an average length of about 46 nm are attached (numbers from Thyberg et al. 1975). The molecules, then, appear extended in solution, and more so when the ionic strength is lowered. The core protein may actually contain stiff regions due to effects of the charged glycosaminoglycan side chains (Kitchen and Cleland 1978). Torchia et al. (1981), using ^{13}C-NMR, demonstrated that the core protein of proteoglycans was segmentally flexible.

Electron microscopy of proteoglycans, a technique introduced by Rosenberg et al. (1970), allows measurements of individual molecules, in contrast to the data on the average molecule obtained by chemical and physical methods. Results indicate continuous variations of parameters like length of core filament and length of chondroitin sulfate side chains.

Cartilage proteoglycans should be characterized by physical and chemical criteria including gel chromatography and sedimentation velocity centrifugation, (Figure 8.3), identification of chondroitin sulfate and keratan sulfate, amino acid analysis, and demonstration of ability to interact specifically with hyaluronic acid (Figure 8.4, I). Degradation and subsequent identification of domains (hyaluronic acid-binding, keratan sulfate-rich, and chondroitin sulfate-rich) and clustering of chondroitin sulfate chains aid in the further characterization and classification of the proteoglycans (Figure 8.4, II, III).

Identifying subpopulations in polydisperse systems, like that of cartilage proteoglycans, presents several problems. Due to the continuous structural variations within a polydisperse population, it is difficult to distinguish the extremes of one population from another. Another complication is the degradation that occurs in vivo, postmortem, or due to the methods of isolation and fractionation. Subpopulations produced by such degradation should share structural features with intact molecules, but should be smaller.

Over the years, indications that there may be more than one population of proteoglycans in cartilage have accumulated. Agarose-polyacrylamide composite gel electrophoresis (introduced by McDevitt and Muir 1971) yields 2 to 4 surprisingly distinct bands with somewhat different mobilities (*See* Stanescu et al. 1977). Subfractionation of proteoglycans by repeated CsCl-density gradient cen-

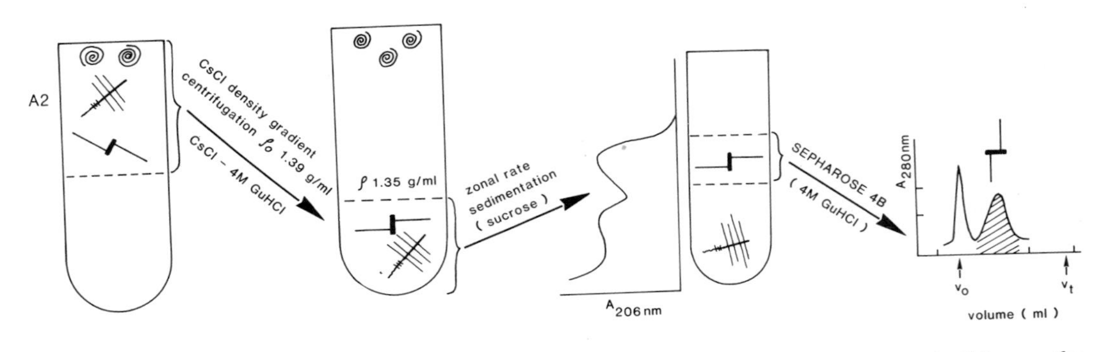

Figure 8.5. Schematic representation of the preparation of small proteoglycans, in this case the small cartilage proteoglycan. Sedimentation velocity centrifugation patterns as well as gel chromatographic profiles (Sepharose 4B) are shown.

Table 8.2. Amino Acid Composition of Proteoglycans Isolated from Several Tissues

	Cartilage proteoglycans[a,b]			Aorta proteoglycans[c]		Sclera proteoglycans[d]	
	KS-rich[a]	CS-rich[a]	Small[b]	CS-rich	DS-rich	CS-rich	DS-rich
Aspartic acid	67	72	129	102	102	96	123
Threonine	58	62	39	86	86	58	49
Serine	119	139	66	98	88	92	68
Glutamic acid	151	138	98	141	136	138	122
Proline	104	96	72	92	93	83	74
Glycine	123	136	71	68	68	110	84
Alanine	72	68	50	61	62	65	54
Cysteine	—	—	19	5	8	8	9
Valine	62	57	53	54	55	62	59
Methionine	—	—	1	11	12	—	7
Isoleucine	41	41	55	37	39	41	55
Leucine	74	77	143	85	87	88	115
Tyrosine	24	21	28	20	22	16	15
Phenylalanine	38	33	34	37	36	43	34
Histidine	12	13	36	25	24	19	25
Lysine	21	20	64	48	49	45	76
Arginine	33	28	44	31	34	37	32
Hexosamine/protein	3.14	3.86	0.62	1.29	1.04	0.44	0.21
Glucosamine/protein	0.31	0.21	0.04	0.20	0.08	—	—
Galactosamine/protein	2.83	3.65	0.58	1.09	0.96	—	—
Iduronic acid/ glucuronic acid	0	0	0	~0.1	—	0.25	1.08

KS = keratan sulfate; CS = chondroitin sulfate; DS = dermatan sulfate; HS = heparan sulfate.
[a]Data from Heinegård et al., unpublished observations.
[b]Heinegård et al. 1981a.
[c]Oegema et al. 1979b.
[d]Cöster and Fransson 1981.

trifugation (Swann et al. 1979) or zonal rate centrifugation in NaCl gradients (Hoffman 1979) has yielded preparations with differing compositions. Analytic ultracentrifugation has occasionally allowed demonstration of two somewhat differently sedimenting populations of proteoglycans (Pearson and Mason 1979; Pal et al. 1981). Kimata et al. (1974) have identified rapidly as well as very slowly sedimenting, newly synthesized proteoglycans from epiphyseal cartilage incubated in vitro.

More recently, subpopulations of cartilage proteoglycans have been isolated using—as criteria for the subpopulation—electrophoretic mobility, functionality (binding to hyaluronate), or differences in immunologic reactivity. One population of small, low buoyant density proteoglycans having a unique amino acid composition and containing no keratan sulfate has been isolated (Stanescu and Sweet 1981; Heinegård et al. 1981a) (Figure 8.5, Table 8.2). It probably corresponds to the slowly sedimenting proteoglycan previously observed. The amino

Table 8.2. (*continued*)

Bone proteoglycan[e]	Basement membrane[f] HS-proteoglycan	Liver cell membrane[g] HS-proteoglycan	Follicular fluid[h] DS-proteoglycan	Cornea proteoglycans[i]	
				DS-rich	KS-rich
128	81	125	84	101	139
45	68	78	98	37	40
76	84	142	131	58	77
105	154	161	168	106	110
77	81	77	34	84	67
86	106	100	98	178	48
49	70	57	66	87	47
—	—	0	2	1	8
52	63	33	55	58	53
—	13	15	14	2	3
58	26	34	38	39	51
128	81	68	72	84	154
29	24	19	19	19	38
32	32	16	35	30	41
31	32	12	23	53	65
69	29	41	40	18	23
36	54	21	22	48	36
0.46	0.066	~1.44	1.29	0.22	0.28
0.03	0.62	1.44	0.17	0.06	0.27
0.43	0.04	—	1.12	0.16	0.01
—	—	—	0.10	0.10	—

[e]Franzén and Heinegård, unpublished observations.
[f]Hassel et al. 1980b.
[g]Oldberg et al. 1979.
[h]Yanagishita et al. 1979.
[i]Axelsson and Heinegård 1975.

acid composition and the electrophoretic mobility of the monodisperse core protein (obtained after chondroitinase digestion, apparent molecular weight 45,000) is similar to that found for proteoglycans isolated from several other tissues, e.g., aorta (Gardell and Heinegård, unpublished observations), sclera, and cornea. It is possible that it represents a ubiquitous proteoglycan. Another population of proteoglycans is large but cannot interact with hyaluronate—the nonaggregating proteoglycans (Heinegård and Hascall 1979a). It has a distinct composition, with a comparatively high protein content. Also, the large aggregating proteoglycans can be separated into two immunologically different populations. These aggregating proteoglycans can be separated by zonal rate centrifugation (indicated by the bimodality of the Al-Dl fraction, Figure 8.3) and each corresponds to one of the two major bands observed on agarose-polyacrylamide composite gel electrophoresis. It appears that only one of these proteoglycans contains a prominent keratan sulfate-rich region, in addition to the chondroitin sulfate-rich region. This population is therefore called the keratan sulfate-rich proteoglycan. The other proteoglycan has only a chondroitin sulfate-rich region and by analogy is called the chondroitin sulfate-rich proteoglycan (Figure 8.2) (Heinegård et al. 1981b; Heinegård et al. unpublished observations). Future studies on cartilage proteoglycan composition, polydispersity, and biosynthesis will have to take into account these examples of proteoglycan heterogeneity.

8.3.2. Cartilage Proteoglycan Aggregates

Cartilage proteoglycans form aggregates, as originally described by Hascall and Sajdera (1969). Later Hardingham and Muir (1972a) identified the aggregating factor as hyaluronic acid. It is now known that the proteoglycans specifically interact with a minimal decasaccharide sequence of hyaluronic acid, and they do not bind to any other glycosaminoglycan. The portion of the core protein involved is the hyaluronic acid-binding region. The K_d of this noncovalent interaction is 10^{-7} to 10^{-8} and carboxyl groups of hyaluronate are required. Blocking of lysine residues of the hyaluronic acid binding region, however, does not prevent interaction (Heinegård and Hascall 1979b; Muir 1980). Several proteoglycan monomers of different sizes bind to one hyaluronic acid molecule, forming an aggregate with a length of as much as 4000 nm and a diameter of 500–600 nm, as revealed by electron microscopy. Each aggregate contains some 50 proteoglycan monomers. Even though the binding of proteoglycan to hyaluronate by itself is strong, a group of molecules—the link proteins—provides additional stability. Three link proteins with somewhat different apparent molecular weights of 40,000 to 50,000 have been identified (*See* Baker and Caterson 1979), although only one molecule is present at each proteoglycan–hyaluronate linkage. An interesting possibility is that each type of link protein may specifically bind to only one subpopulation of aggregating proteoglycans. The insolubility of isolated link proteins presents a problem in attempts to study the specificity of the interactions. Data on proteoglycan aggregate stability indicate that link proteins interact both with proteoglycan and with hyaluronate (Hardingham 1979). Results obtained using protein modifications indicate that link proteins bind to the hyaluronic acid binding region of the proteoglycan (Heinegård and Hascall 1979b).

Competition experiments show that the link proteins can bind specifically to the hyaluronic acid-binding region (Franzén et al. 1981a). Other studies have demonstrated that link proteins also bind to hyaluronic acid, with a specificity for a decasaccharide sequence (Tengblad 1981; Wieslander and Heinegård, unpublished observations). It appears that aggregates are stabilized because of the capability of the link proteins to lock the bond by interaction both with the hyaluronic acid and with the hyaluronic acid-binding region of the proteoglycan monomers. The bonds of such stable aggregates appear irreversible in nondenaturing conditions and at normal pH (about pH 5–7). The stabilization of aggregates by link proteins protects the hyaluronic acid binding region from degradation by proteases and makes it possible to isolate an intact hyaluronic acid binding region after trypsin digestion of aggregates, as is outlined in Figure 8.4, III.

Recent data (discussed below) indicate that aggregate formation is an extracellular process. It is possible that one of the major functions of the aggregate is to prevent proteoglycans from diffusing out of the cartilage by sterically restraining them in the network of collagen. Therefore, in order to obtain a good yield during extraction, the aggregates have to be dissociated. Proteoglycan aggregates can, however, be isolated from dissociative cartilage extracts after reassociation, as indicated in Figure 8.3. The proportion of aggregate can be quantitated by Sepharose 2B chromatography or by centrifugal techniques (Figure 8.3). Some of the proteoglycan molecules do, however, remain monomeric—the nonaggregating proteoglycans.

One of the major functions of the proteoglycans is to provide elastic stiffness to cartilage by furnishing a large number of fixed charged groups. As proteoglycans occupy a large volume, they also contribute a barrier function, preventing large molecules from entering their domain. Other functions may include regulation of collagen fibril deposition and interactions with several other matrix components.

The cartilage type of proteoglycan is present in all cartilages. Young cartilage appears to contain more of large chondroitin sulfate-rich proteoglycans and old cartilage contain more of somewhat smaller keratan sulfate-rich proteoglycans. The proportions of the proteoglycan subpopulations change with age (Inerot and Heinegård 1983). The character of the proteoglycan also changes with tissue localization. The proteoglycans near the surface of articular cartilage are quite different—smaller and containing more protein—than those of deeper layers (*See* Franzén et al. 1981b). Proteoglycans from the articulating area of minimum contact have a somewhat higher relative content of chondroitin sulfate than those from areas of maximum contact, although their sizes are similar (Sweet et al. 1978).

8.3.3. Aorta Proteoglycans

Over the years several attempts have been made to isolate and purify proteoglycans from aorta (*See* Hascall 1981). Acceptable extraction yields of 85 to 90% were obtained after the introduction of 4 *M* guanidinium chloride for extraction. Subsequent purification has presented problems, mainly because of the high relative contents of collagen and noncollagenous proteins in the extract. Furthermore, a large proportion of proteoglycans from noncartilage tissues are of the low molecular weight type, having low buoyant densities, due to high protein content, and therefore do not separate well from the nonproteoglycan proteins. In most studies only the intimal layer of the aorta has been used, mainly because of the important biological function and distinct character of this part of the tissue. One of the early techniques for preparing aorta proteoglycans involved extraction with 4 *M* guanidine-HCl followed by DEAE-cellulose ion exchange chromatography of extracts in 7–8 *M* urea. The highly charged proteoglycans were retained on the resin, while proteins eluted early. Purer preparations were obtained using CsCl-density gradient centrifugation. Preferably, such centrifugations should be performed in the 4 *M* guanidine-HCl used for extraction, with a low starting density of 1.30 to 1.35 g/ml. Thereby, denaturated proteins are kept in solution, and also low buoyant density proteoglycans are recovered at the bottom of the gradient. This type of proteoglycan preparation can be further fractionated by gel chromatography and/or zonal rate sedimentation techniques. The preparations are heterogenous and 2 or 3 proteoglycan subpopulations can be recognized. One population contains predominantly dermatan sulfate of low iduronic acid content (about 10%) and additional oligosaccharides. The protein content is 10 to 15%, with a composition as listed in Table 8.2. The molecular weight is about 1.5×10^6 by sedimentation equilibrium centrifugation (Oegema et al. 1979b). This proteoglycan probably interacts with hyaluronate to form aggregates and it has been shown that aorta proteoglycans react with antibodies directed against the hyaluronic acid-binding region of cartilage proteoglycans (Gardell et al. 1980). Furthermore, link proteins, similar to those from cartilage, have been isolated from aorta. Other aorta proteoglycans are considerably smaller (molecular weights in the order of 10^5) and have higher protein contents. They appear to contain dermatan sulfate copolymers enriched in iduronic acid. Core preparations of such proteoglycans migrate on sodium dodecyl sulfate (SDS)-polyacrylamide gel electrophoresis, with the same mobility as the core of the small proteoglycan isolated from cartilage (Gardell and Heinegård, unpublished observations). In addition, the tissue contains heparan sulfate proteoglycans, which have not been characterized. The distribution of proteoglycans within the tissue, i.e., whether or not the two or more populations are differently located, is not known.

Aorta proteoglycans appear to be capable of interaction with low density lipoproteins, which may be relevant in the pathogenesis of atherosclerosis. Furthermore, it has been reported that aorta proteoglycans retard blood coagulation. It is likely that they are an important component in the tissue barrier, but too little is known of the tissue distribution to qualify such statements.

8.3.4. Skin and Sclera Proteoglycans

Proteoglycans have been isolated from a number of loose connective tissues using extraction with 4 *M* guanidine-HCl, followed by CsCl-density gradient centrifugation to separate according to buoyancy. The two best studied fibrous tissues are skin (Damle et al. 1979) and sclera (Cöster and Fransson 1981; Cöster et al. 1981). Both of these tissues contain proteoglycans of similar composition. The description below focuses on data obtained with sclera proteoglycans, although data from skin proteoglycans, when available, are similar. One population of proteoglycans is of low molecular weight (less than 10^5) and contains dermatan sulfate side chains containing equal proportions of iduronic acid and glucuronic acid. It is rich in protein (about 60% of dry weight) and has an amino acid composition similar to that of the small cartilage proteoglycan (Table 8.2). The apparent molecular weight is about 90,000 by sedimentation equilibrium. Core preparations from the small sclera and cartilage proteoglycans have an identical mobility on SDS gel electrophoresis. A major difference, however, is that the cartilage proteoglycan does not contain any dermatan sulfate. Another population of sclera proteoglycans is of higher molecular weight and is substituted with dermatan sulfate with a very high proportion of glucuronic acid (more than 80% of the uronic acid). The protein content is 40 to 50% and the amino acid composition (Table 8.1) is different from the low molecular weight proteoglycans but similar to that of the large aorta proteoglycans. The apparent molecular weight is about 170,000 by sedimentation equilibrium. The core protein preparation of the high molecular weight proteoglycan has an apparent molecular weight by SDS gel electrophoresis in excess of 10^5. Both sclera proteoglycans contain mannose-rich oligosaccharides, possibly N-glycosidically linked.

Again, little is known of the tissue distribution of these proteoglycan populations and their concentrations are extremely low, on the order of 0.1%. Their function is unknown, but it appears that they are common to most fibrous connective tissues. It is therefore possible that they play a role in organizing the tissue matrix components, such as collagen fibrils. As is discussed below, there are several reports indicating that, in particular, dermatan sulfate-rich glycosaminoglycans and proteoglycans interact with type I collagen.

8.3.5. Bone Proteoglycans

One difficulty in the study of bone proteoglycans has been to prepare intact proteoglycans from a tissue which is rich in proteases. Bone proteoglycans can, however, be extracted with EDTA-containing 4 *M* guanidine-HCl, provided proteases and other molecules that are not integral components of the calcified matrix have first been removed by preextraction in 4 *M* guanidine-HCl at −18°C (Franzén and Heinegård, unpublished observations). Further purification is achieved by a sequence of CsCl-density gradient centrifugation and gel chromatography in 4 *M* guanidine-HCl. The purified proteoglycan has a size similar to that of the small cartilage proteoglycan; its core protein, prepared by chondroitinase digestion, has an identical mobility on SDS gel electrophoresis. Furthermore, the amino acid composition of the bone proteoglycan is similar (Table 8.2). Immunological data, however, indicate only partial identity (Franzén and Heinegård, unpublished). In a recent publication similar data have been presented by Fisher et al. (1983).

8.3.6. Cornea Proteoglycans

Cornea contains at least two types of proteoglycan (Axelsson and Heinegård 1975; Axelsson and Heinegård 1978; Hassel et al. 1979). One is a proteoglycan containing only dermatan sulfate (Table 8.2) and mannose-rich oligosaccharide

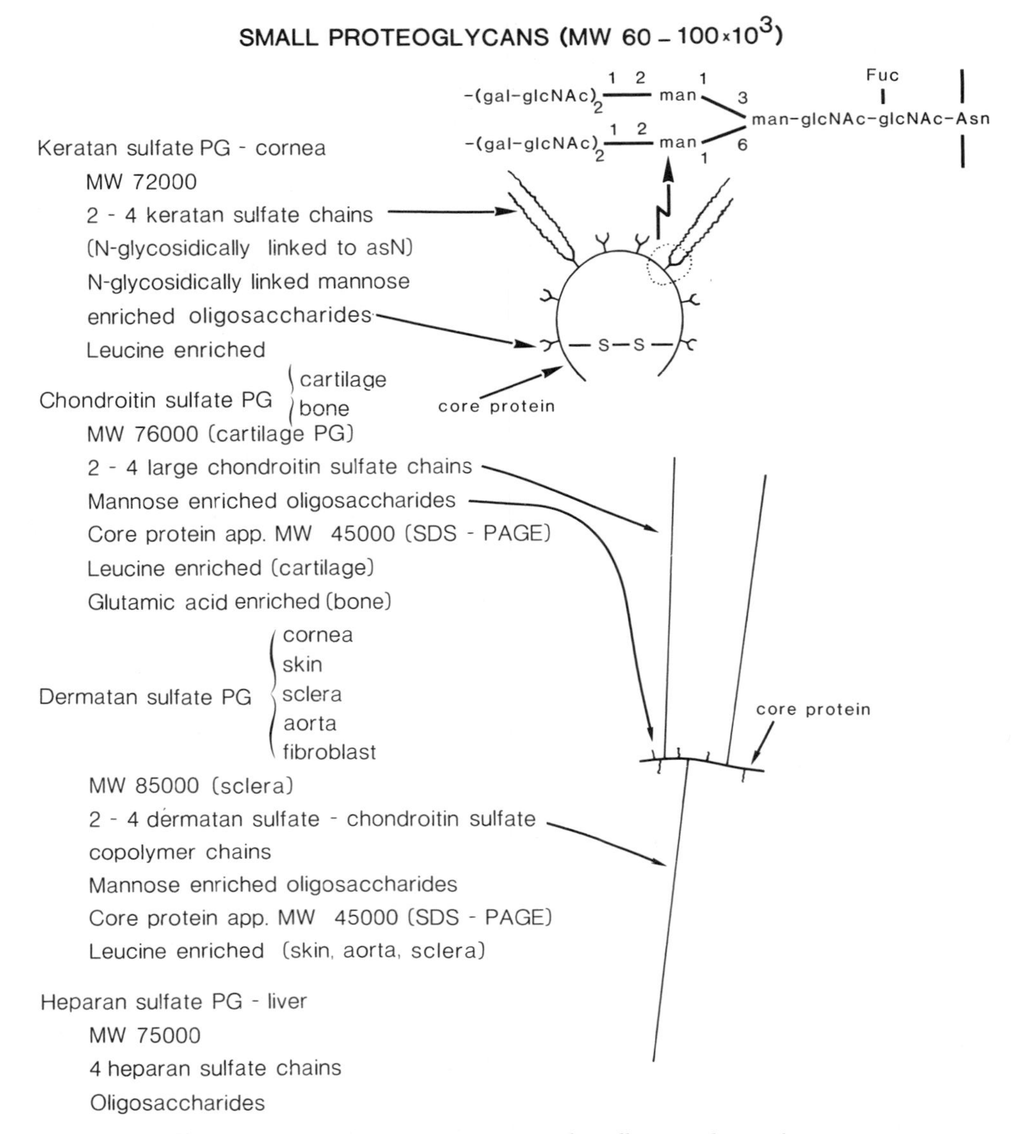

Figure 8.6. Characteristics and tentative structures of small proteoglycans from various sources.

side chains. The average molecular weight of this population is probably about 100,000 and its protein content is approximately 50%. Another major proteoglycan population contains only keratan sulfate chains plus mannose-rich oligosaccharides (Figure 8.6). On the average, this proteoglycan contains two keratan sulfate chains, making up about half of the apparent molecular weight of 72,000. Most of the remaining half is protein, with an amino acid composition very similar to that of the small cartilage proteoglycan (Table 8.2) with a notably high content of leucine.

In addition, some corneal proteoglycans appear to contain both dermatan sulfate and keratan sulfate side chains (Axelsson and Heinegård 1980). Molecular dimensions and oligosaccharide contents are probably similar to those of the other corneal proteoglycans.

Proteoglycans can be prepared from corneal stroma by extraction with 4 *M* guanidine-HCl (80% extracted) and purified either by ion exchange chromatography or by CsCl-density gradient centrifugation. Subpopulations can be separated by ethanol precipitation (Axelsson and Heinegård 1975) or by gel chromatography (Hassel et al. 1979), depending on the source of the cornea. They appear to be essential for maintaining the specific arrangement of the collagen fibrils. Corneal opacity can be related to changes in proteoglycan composition (Hassel et al. 1980a), but the primary cause is unknown.

8.3.7. Liver Cell Surface and Basement Membrane Proteoglycans

Proteoglycans isolated from liver parenchymal cells and from liver plasma membranes, as well as those isolated from a basement membrane-producing mouse sarcoma, contain heparan sulfate side chains and oligosaccharide (Figures 8.6 and 8.7). The one isolated from liver plasma membranes has a molecular weight of 75,000 and contains four heparan sulfate chains, each with a molecular weight of 14,000 (Oldberg et al. 1979). The protein content is less than 20%, with high relative contents of acidic amino acids (Table 8.2).

The heparan sulfate proteoglycan prepared from basement membranes has a higher apparent molecular weight, probably about 0.75×10^6. Its heparan sulfate side chains have apparent molecular weights of 70,000 and each proteoglycan probably contains 6 to 12 heparan sulfate chains (Hassel et al. 1980b). The protein content, about 50%, is higher than that of the cell surface proteoglycan and the amino acid composition is different (Table 8.2) (Hassel et al. 1980b). A high molecular weight heparan sulfate proteoglycan (molecular weight about 2×10^6) containing about 15% protein has been isolated from gas exchange tissue of the lungs (Radhakrishnamurthy et al. 1980) and is possibly also of basement membrane origin. It is likely that many basement membranes contain the same type of proteoglycan, as antibodies to the mouse sarcoma proteoglycan crossreact with a number of basement membranes. Furthermore, heparan sulfate has been isolated from glomerular basement membrane (Kanwar and Farquhar 1979; Parthasarathy and Spiro 1981).

It is well documented that most cells contain heparan sulfate bound at the surface and that the quantity may vary during the cell cycle. It has been shown that a fraction of the cell surface heparan sulfate proteoglycans can be displaced by competition with heparin. It appears that the specific bond is via the glycosaminoglycans to a receptor at the cell surface (Kjellén et al. 1980). Another fraction of the cell surface heparan sulfate proteoglycan appears to be an integral part of the cell membrane, i.e., a portion of the core protein is intercalated (Kjellén et al. 1981).

Heparan sulfate proteoglycans are isolated after extraction of tissue with 4 *M* guanidine-HCl or after extraction of cell membranes with detergent. Purification is obtained by CsCl-gradient centrifugation, gel chromatography, and ion exchange chromatography.

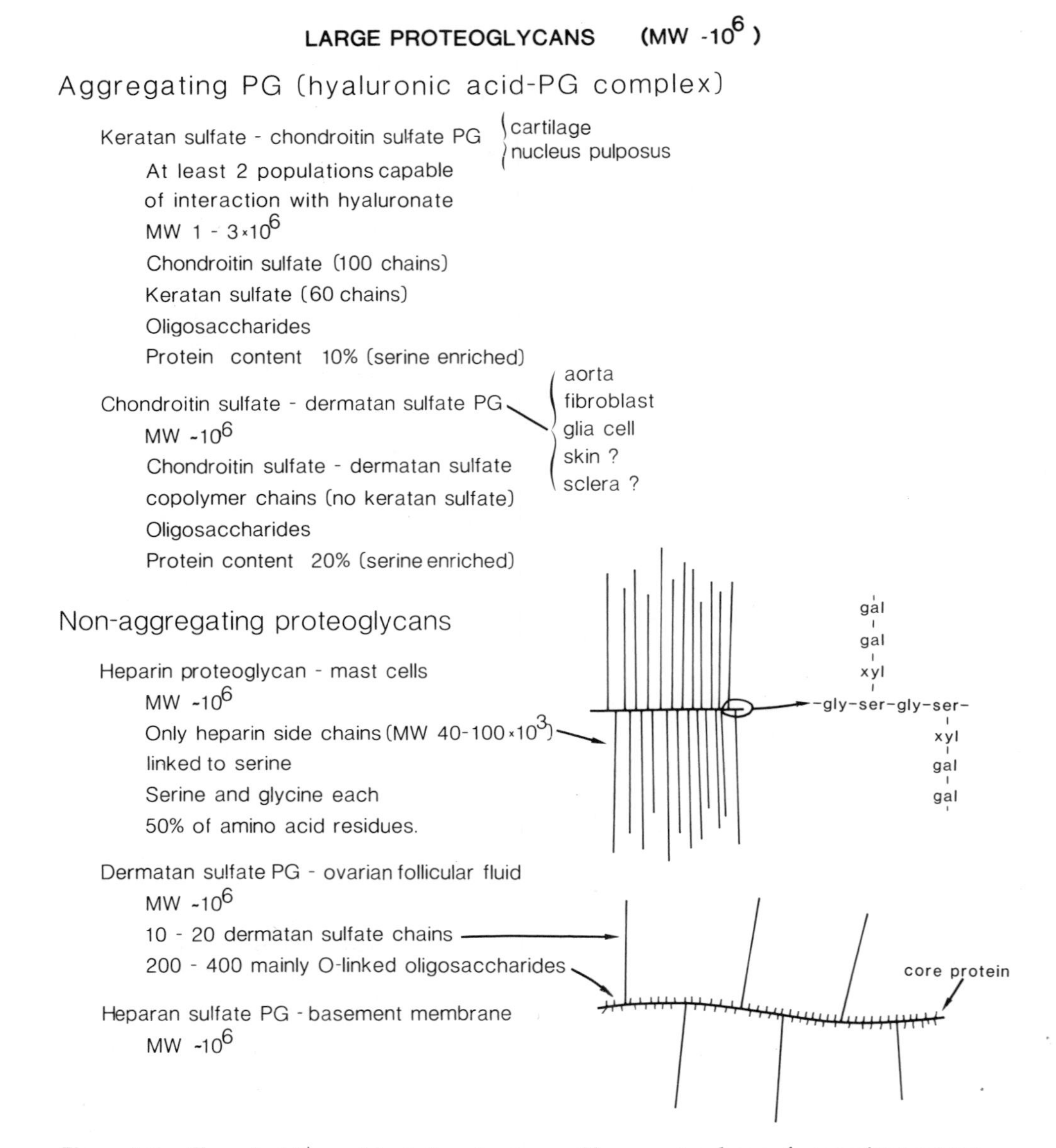

Figure 8.7. Characteristics and tentative structures of large proteoglycans from various sources.

The function of heparan sulfate proteoglycans is not clear. They may be involved in cell communications. Heparan sulfate chains interact with a number of proteins, like fibronectin and low density lipoproteins.

8.3.8. Heparin Proteoglycan

Heparin is produced by mast cells. A proteoglycan form of heparin was originally identified by Horner (1971) in extracts of skin. The proteoglycan has a molecular weight in excess of 900,000 (*See* Robinson et al. 1978), with heparin side chains of molecular weights 60,000 to 100,000 (Figure 8.7). The protein content is low, 1 to 2%, with a unique amino acid composition consisting of approximately equimolar amounts of probably alternating serine and glycine residues. Each core contains about 15 heparin chains and two out of three serine residues are substituted. The proteoglycan is degraded by endogenous glycosidases and fragments having molecular weights of 7000 to 25,000 are stored in the mast cell granules. Similar heparin proteoglycans have been characterized after isolation from rat peritoneal mast cells (Metcalfe et al. 1980).

8.3.9. Proteoglycans from Other Sources

Follicular fluid contains a large proteoglycan with a molecular weight of about 2×10^6. The central protein core has a molecular weight of about 400,000 with high contents of serine and glutamic acid or glutamine (Table 8.2). The proteoglycan contains about 20 dermatan sulfate copolymer side chains with average molecular weights M_r of about 56,000. Most notably, however, the proteoglycan contains a very large number—about 350—of at least three types of sialic acid-containing oligosaccharides. A schematic illustration of the proteoglycan is shown in Figure 8.7. The proteoglycan is sensitive to digestion with a number of proteases and is degraded by plasmin, unlike cartilage proteoglycans. The physiologic role of the proteoglycan is unclear, but it is likely that it plays an essential part in creating the proper environment for the ovum (Yanagishita et al. 1979).

Glial cells in culture produce a large chondroitin sulfate-rich proteoglycan. It can interact specifically with hyaluronate to form aggregates (Norling et al. 1978).

Fibroblasts in culture likewise appear to produce a high molecular weight proteoglycan containing glucuronic acid-rich dermatan sulfate chains. It has the capability to interact with hyaluronate (Carlstedt et al. 1981). In addition, cultured fibroblasts produce heparan sulfate proteoglycans and a small proteoglycan containing iduronic acid-rich dermatan sulfate chains. The latter proteoglycan appears similar to the small proteoglycans identified in aorta, bone, cartilage, cornea, and sclera.

Brain contains a cytoplasmic proteoglycan. It has a sedimentation coefficient of 6.55 and contains 56% protein, 20% oligosaccharide, and 24% glycosaminoglycan, essentially all chondroitin sulfate. Some of the oligosaccharides contain large amounts of mannose and a novel linkage to protein via mannose has been suggested (*See* Finne et al. 1979). Proteoglycans containing typical glycosaminoglycans have been isolated from sea urchins. They contain dermatan sulfate linked to protein in addition to unidentified glycosaminoglycans resistant to digestion with chondroitinase (Oguri and Yamagata 1978).

8.4. Immunology of Proteoglycans

Much of the early data published on the specificity and immunologic reactivity of proteoglycan antibodies are contradictory. More consistent results have been obtained with antibodies against proteoglycans prepared by CsCl gradient centrifugation. To date only proteoglycans from cartilage have been studied in any detail.

Antibodies raised against intact or fragmented cartilage proteoglycans react with antigenic sites on the protein core. Exceptions are the antibodies reacting with unsaturated disaccharides produced by digestion of galactosaminoglycans with chondroitinase (Christner et al. 1980) or the oligosaccharides produced by hyaluronidase digestion of chondroitin sulfate (Jenkins et al. 1982). These antibodies have been raised against proteoglycans digested with the respective enzyme. It is possible that the oligosaccharide stub that remains attached to the core protein acts as a hapten.

The different regions of the cartilage proteoglycan are antigenically different and do not crossreact immunologically. For example, antibodies raised against the hyaluronic acid-binding region neither crossreact with any other core regions nor with the link proteins. However, each isolated region (as outlined in Figure 8.4) contains several antigenic sites. The various populations of large cartilage proteoglycans share some of the antigenic determinants and differ with respect to others.

Removal of chondroitin sulfate side chains by chondroitinase digestion is a prerequisite for antibody reaction with the chondroitin sulfate-rich region of the cartilage proteoglycans. Immunization with such core preparations should be avoided, however, in view of the strong antibody response to unsaturated disaccharides left on the core. When injected into rabbits, proteoglycans give rise to antibodies to core protein in the chondroitin sulfate-rich region, even though that region is not available for in vitro binding of antibodies without prior removal of glycosaminoglycan.

Antibodies raised against bovine nasal cartilage proteoglycans show strong crossreaction with cartilage proteoglycans from other species. Most notably, the chondroitin sulfate-rich region shows a strong or enhanced reaction when proteoglycans from a number of other mammals, as well as chick, are tested. In contrast, antibodies to the hyaluronic acid-binding region show variable crossreactivity, and in some cases, no crossreaction (Wieslander and Heinegård 1981). Interestingly, proteoglycan aggregates isolated from bovine aorta crossreact with antibodies to cartilage proteoglycans and link proteins, indicating structural and functional similarities (Gardell et al., 1980). Recent work has indicated that there are antigenic similarities among aggregating proteoglycans present in several tissues like cartilage, aorta, and sclera (Figure 8.7). Similarly, there are antigenic similarities among the group of low molecular weight proteoglycans present in most tissues, for example the dermatan sulfate and chondroitin sulfate proteoglycans depicted in Figure 8.6. (Heinegård, Cöster and Malmström, unpublished).

Fluorescent antibodies have been used to study the localization of proteoglycan in bovine articular cartilage. The distribution is rather even throughout the whole matrix, although somewhat stronger staining was obtained in the pericellular area (Poole et al. 1980). The degree of staining, however, reflects several properties of the matrix and not just the concentration of proteoglycan. In other experiments, differences in the staining patterns between rat xiphosternal cartilage and bovine nasal cartilage were obtained using antibodies to intact proteoglycan. Identical results were obtained with antibodies to the chondroitin sulfate-rich region and to the hyaluronic acid-binding region. Bovine nasal cartilage showed strongest staining in the interterritorial matrix, whereas rat sternal cartilage showed strongest staining in the pericellular and territorial matrix (Minor and Heinegård, unpublished observations). The differences observed probably represent variations in matrix composition.

Antibodies raised against mouse sarcoma basement membrane proteoglycans react specifically with basement membranes from several tissues (Hassel et al. 1980b).

8.5. Interactions of Proteoglycans and Glycosaminoglycans

A number of reports have dealt with in vitro interactions of various matrix components with proteoglycans and glycosaminoglycans. A central problem is the specificity and physiologic relevance of such interactions. Many nonspecific interactions can be mediated because of the strong polyanionic nature of the glycosaminoglycans. The spatial structure of the molecules is likely to be altered under the experimental conditions of dilute solution as compared with the very high local concentrations in many tissues. The conformation and expansion of proteoglycans largely depends on their concentration. Many of the interactions reported have no proven specificity. Until the interactions have been shown to be specific for certain structures or demonstrated in the tissue, data should be viewed with care.

8.5.1. Glycosaminoglycan–Glycosaminoglycan

Many of these studies have used isolated glycosaminoglycans, or fragments thereof, that have been allowed to react with other glycosaminoglycans covalently bound to a matrix. Retardation of a molecule has been taken as proof for interaction and the profile obtained by elution with salt solutions has been used to indicate the strength of the interaction. Using this type of system, modified for agglutination studies, it was shown that chondroitin sulfate can interact with hyaluronic acid, although there is no self-interaction with other chains of like kind (Turley and Roth 1980). In a series of experiments, Fransson and collaborators (1979, 1980) used affinity methods to study binding between dermatan sulfate chains and between heparan sulfate chains, respectively. The studies were extended using light scattering and viscosity measurements and showed that in both glycosaminoglycans, segments containing an unusually high frequency of alternating iduronic acid and glucuronic acid residues are prone to self-interaction. Furthermore, both the small and the large proteoglycan isolated from sclera appear to interact to form multimers with average molecular weights of about 3×10^6, compared with molecular weights of about 70,000 and 200,000, respectively, of the monomer proteoglycans using light scattering measurements (Cöster et al. 1981). In these studies, all interactions were shear dependent, and consistently larger complexes were observed using light scattering than seen in analytical ultracentrifugation.

8.5.2. Protein–Glycosaminoglycan and Protein–Proteoglycan

One much studied interaction is that occurring in vitro between collagen and glycosaminoglycan. The conformations of dermatan sulfate and heparan sulfate, both containing iduronic acid, appear to be much more favorable for collagen interaction than those of other glycosaminoglycans. However, little data are available on interactions between collagen and purified, intact dermatan sulfate containing proteoglycan. The best evidence is indirect, obtained in electron microscopy studies showing that collagen and proteoglycan have a specific distribution pattern in the tissue, indicative of interactions (Scott et al. 1981; Scott and Orford 1981). In addition, it has been shown that glycosaminoglycans and proteoglycans affect collagen fibrillogenesis in vitro (*See* Lindahl and Höök 1978).

In a recent study (Vogel, Paulsson, and Heinegård, unpublished) it was shown that a small proteoglycan isolated from tendon inhibited in vitro fibrillogenesis of type I collagen isolated from tendon. None of the large proteoglycans from tendon or cartilage had this effect on the collagen. It appeared that the core protein was essential, since chondroitinase-digested proteoglycan was an equally efficient inhibitor. Relevance of these and other results to in vivo fibrillogenesis has yet to be shown.

Since the discovery that heparin is an efficient and clinically useful inhibitor of blood coagulation, much effort has been devoted to establish how this effect is mediated. It has been shown that heparin activates a protease inhibitor—antithrombin—that inhibits several of the proteases involved in the blood coagulation cascade (*See* Lindahl and Höök 1978). Heparin contains a specific antithrombin-binding oligosaccharide, having a minimal size of an octasaccharide. It carries a unique 3-O-sulfate ester group on the fourth residue, i. e., a glucosamine. The suggested structure of the oligosaccharide is shown in Figure 8.8 (Lindahl et al. 1980). The interaction between heparin chains and antithrombin is rather stable, with an average binding constant of 10^{-7} M^{-1}. Heparin chains can be fractionated on the basis of their variable affinity for antithrombin. High affinity heparin contains the specific oligosaccharide sequence and binds strongly to antithrombin, while low affinity heparin binds only weakly to antithrombin and probably does not contain the specific oligosaccharide sequence. Heparin has also been shown to bind lipoprotein lipase. In this case, heparin with low and high affinity for antithrombin is equally effective. It is possible that a specific oligosaccharide also is essential for this interaction, but if so, it is different from that mediating the binding of antithrombin. Oligosaccharide structures active in binding of lipoprotein lipase are also present in dermatan sulfate and heparan sulfate. The observed in vitro effects can be reproduced in vivo by intravenous injection of heparin, which slows or prevents blood coagulation and liberates lipoprotein lipase.

In addition to binding a number of extracellular proteins, heparin binds several proteins located intracellularly. An example is the insoluble complex formed between heparin and a basic protein in the granules of mast cells. This protein in turn binds and immobilizes histamine. When the granules are emptied, the

Figure 8.8. Tentative structure of an oligosaccharide of heparin required for antithrombin binding (Lindahl et al. 1980).

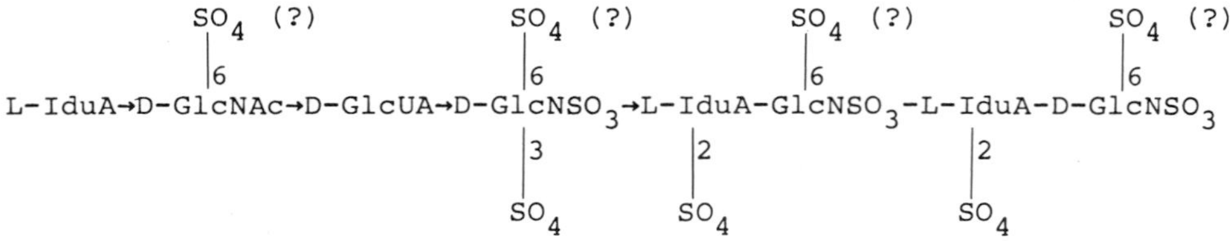

```
  SO4 (?)                   SO4 (?)                  SO4 (?)                  SO4 (?)
  |6                        |6                       |6                       |6
L-IduA→D-GlcNAc→D-GlcUA→D-GlcNSO3→L-IduA-GlcNSO3-L-IduA-D-GlcNSO3
                            |3              |2                       |2
                            SO4             SO4                      SO4
```

extracellular cations liberate histamine. Intracellular storage of basic proteins by binding to glycosaminoglycan probably also occurs in other cells, such as granulocytes. The specific interactions between cartilage proteoglycans, link proteins, and hyaluronic acid have been discussed above (Section 8.3.2).

Much interest has been focused on the interactions of the matrix protein fibronectin with other matrix components. As discussed in Chapter 7, fibronectin contains specific interaction sites for heparin and hyaluronic acid (Yamada et al. 1980; Ruoslahti et al. 1981) with the K_d of the interactions being 10^{-7} to 10^{-8} (Yamada et al. 1980). However, little is known about the carbohydrate structure involved in the interactions. In a more complex test system, also involving collagen, the high charge density of the polyanionic heparin was an essential factor (Johansson and Höök 1980; Ruoslathi and Engvall 1980).

It has been shown that isolated glycosaminoglycans bind lipoproteins. Iverius (see 1973) demonstrated binding of both low density lipoproteins (LDL) and very low density lipoproteins (VLDL), most strongly to the iduronic acid-containing glycosaminoglycans (heparin, heparan sulfate, and dermatan sulfate). The assays used often employ insolubilized glycosaminoglycan that is allowed to bind lipoprotein from solution, although studies of the reversed binding of glycosaminoglycan in solution to insolubilized LDL have provided similar results (Fransson and Johansson 1981). Others have used precipitation techniques, where mixtures of glycosaminoglycan and lipoprotein are allowed to form insoluble precipitates. Binding is mediated by interactions between cationic groups in the LDL and VLDL apoproteins and anionic groups of the glycosaminoglycan and may be stabilized by divalent ions that form bridges from phosphate groups in the lipoprotein (phospholipid) to sulfate groups in the glycosaminoglycan.

It has been suggested that LDL–glycosaminoglycan interactions are important in the genesis of arteriosclerotic lesions. The demonstration of binding between LDL and a dermatan sulfate-containing proteoglycan isolated from arterial wall (Vijayagopal et al. 1980) shows that intact proteoglycan can also interact.

Extracellular matrix contains proteins that appear to interact with proteoglycans under physiological conditions. For example, cartilage contains comparatively large amounts of a protein of 148,000 molecular weight that appears to be present in the tissue as a proteoglycan complex. The protein consists of three subunits, each with an apparent molecular weight of 50,000 (Paulsson and Heinegård 1979, 1981). Other important interactions are those between cell surface receptors and glycosaminoglycans like hyaluronate (Underhill and Toole 1981) and heparan sulfate (discussed above).

8.6. Functions of Glycosaminoglycans and Proteoglycans

Much information on the functional role of proteoglycans and glycosaminoglycans has been gained from studies of cartilage (*See* Comper and Laurent 1978; Maroudas 1980). The proteoglycans have an important role in retaining water and producing a swelling pressure in the tissue. An important consequence is that considerable force is required to shift water within the tissue. Because of this, compression by load is resisted, a property referred to as compressive stiffness. The compressive stiffness is low in the degenerated and osteoarthrotic articular cartilage, which is partially depleted of proteoglycans, and can be experimentally decreased in normal cartilage by removal of glycosaminoglycans using specific enzymes (*See* Kempson 1980).

More information on the important role of proteoglycan as a space-creating material has been obtained by studies of the skeletal manifestations of various genetic diseases. The nanomelic chick has severe defects of skeletal development and its epiphyseal cartilage has a markedly decreased width, as it lacks the aggregating proteoglycans (McKeown and Goetinck 1979). Similar defects have been observed in a mouse mutant (Kimata et al. 1981).

Several studies indicate that proteoglycans have an important role in limiting diffusion of macromolecules through connective tissue (*See* Comper and Laurent 1978; Maroudas 1980). Small molecules may, however, actually show accelerated diffusion, particularly if anionic. These findings should be taken into account when physiologic regulation of connective tissue cells is studied, especially with regard to factors derived from serum or other external sources.

8.7. Biosynthesis of Glycosaminoglycan Side Chains

Being the precursor for the proteoglycan, the protein core is synthesized first and serves as an acceptor for the monosaccharide transferases involved in glycosaminoglycan biosynthesis (*See* Muir 1980; Rodén 1980). The relatively few studies on the biosynthesis of the core protein will be discussed separately. The various steps involved in the biosynthesis of a single glycosaminoglycan are well characterized, while the sequence of events in the synthesis of a complete proteoglycan is less well known. Examples of relevant questions that remain to be answered are: In what order are the different carbohydrate substituents added? Are glycosaminoglycan chains on one core protein completed one at a time or are several chains completed simultaneously?

The glycosaminoglycans are formed by sequential addition of the respective monosaccharides to the precursor. Much information on the process of chain elongation has been gained through the work of Dorfman and collaborators (*See* Rodén 1980). The energy for forming the glycosidic bonds between the monosaccharides is provided by nucleotide sugar precursors. In view of the lack of effect of tunicamycin on glycosaminoglycan biosynthesis, it appears that dolichol intermediates are not involved in the formation of these molecules. An exception is corneal keratan sulfate (Section 8.7.5), which is N-glycosidically linked through a dolichol intermediate rich in mannose (Chapter 7).

8.7.1. Chondroitin Sulfate

The first step in the formation of a chondroitin sulfate chain is the addition of xylose to a serine residue on the protein core forming the protein-carbohydrate linkage. This and succeeding steps are shown in Figure 8.9. It is not known if this process is initiated before synthesis of the protein core is completed. It probably occurs in the rough endoplasmatic reticulum, and it is therefore possible that it does not require a completed core protein. The specificity of the enzyme involved, xylosyltransferase, is not known, although it interacts with the core protein and appears to recognize sequences containing glycine residues next to the serine acceptor. Xylosyltransferase has been purified and found to be a soluble enzyme. It appears that xylosyltransferase interacts with and binds to the next enzyme in the synthetic sequence, the first galactosyltransferase. This enzyme, probably located in the smooth endoplasmatic reticulum, is much more insoluble. It requires detergent for extraction and is therefore believed to be membrane bound. These first two steps in the synthesis were originally described by Robinson et al. (1966). The third step, the addition of the second galactose residue, was shown by Helting and Rodén (1969a) to be catalyzed by a second galactosyltransferase. This enzyme appears to recognize both the acceptor galactose and the penultimate xylose. The fourth step, addition of the first glucuronic acid, is catalyzed by a glucuronyltransferase different from that acting in chain elongation (Helting and Rodén, 1969b). It probably only recognizes the acceptor galactose of the linkage region. The subsequent steps in chain elongation are the alternating additions of N-acetylgalactosamine and glucuronic acid to the growing chain, catalyzed by two transferases. These transferases function in vitro with other acceptors like hyaluronic acid oligosaccharides. It is, however, likely that in vivo only the core protein carrying a linkage region oligosaccharide will be presented in an appropriate manner to the membrane-bound transferase. The final process in completing the glycosaminoglycan chain is the transfer of the sulfate group from the energy rich 3′-phosphoadenosine-5′-phosphosulfate (PAPS), catalyzed by two enzymes specific for transfer of position 4- and 6-sulfate, respectively. Available data indicate that sulfation is initiated well before chain elongation has stopped.

The factors determining the length of the chondroitin sulfate chains are not known. It has been suggested that sulfation, which starts prior to the completion of the glycosaminoglycan chain, proceeds faster than chain elongation. With time, sulfation catches up and the nonreducing terminal galactosamine is sulfated. It has been shown in vitro that a 4-sulfated nonreducing terminal galactosamine cannot function as an acceptor for glucuronic acid, and the result therefore is termination of chain elongation. Alternatively, simple sterical factors may determine the size of the chondroitin sulfate chains. It is possible that the strength of the attachment of the core precursor to the membrane-bound transferases is an important factor.

Several studies have demonstrated that the biosynthesis of the chondroitin sulfate chain does not necessarily require a primer protein core (*See* Rodén 1980). Xylosides and galactosides of various types can serve as efficient acceptors. Examples are p-nitrophenyl-β-D-xyloside or phenyl-β-D-galactoside. Even xylose at high concentrations will serve as a primer for chondroitin sulfate (Helting and Rodén 1969a; Robinson et al. 1975). The higher the concentration of β-xyloside, the more efficiently it compete with the core protein as the acceptor, and the shorter becomes the newly synthesized chondroitin sulfate chains on both the β-xyloside and the core protein. It is therefore possible that the growing chain is bound to the membrane mainly via the glycosyl transferases.

The biosynthesis of chondroitin sulfate chains appears to be regulated by the level of UDP-xylose. The enzyme UDP-D-glucose dehydrogenase, which catalyzes the formation of UDP-glucuronic acid, is inhibited by UDP-xylose. Consequently, high concentrations of UDP-xylose prevent formation of UDP-glucuronic acid, which in turn is a precursor for UDP-xylose. The result is decreasing concentrations of both UDP-glucuronic acid and UDP-xylose, and the rate of chondroitin sulfate synthesis diminishes accordingly.

8.7.2. Dermatan Sulfate

Dermatan sulfate is formed by a modification at the polymer level of the non-sulfated chondroitin precursor to chondroitin sulfate. The work of Malmström et al. (1975) showed that iduronic acid residues are formed by C-5-epimerization of internal glucuronic acid residues in chondroitin, (Figure 8.9) catalyzed by an epimerase different from that involved in heparin biosynthesis (Malmström et

Figure 8.9. Schematic illustration of sequence of events in glycosaminoglycan biosynthesis. The epimerization of D-glucuronic acid to L-iduronic acid is indicated within brackets. X = xylose; G = galactose; GA = glucuronic acid; IA = iduronic acid; GN = hexosamine.

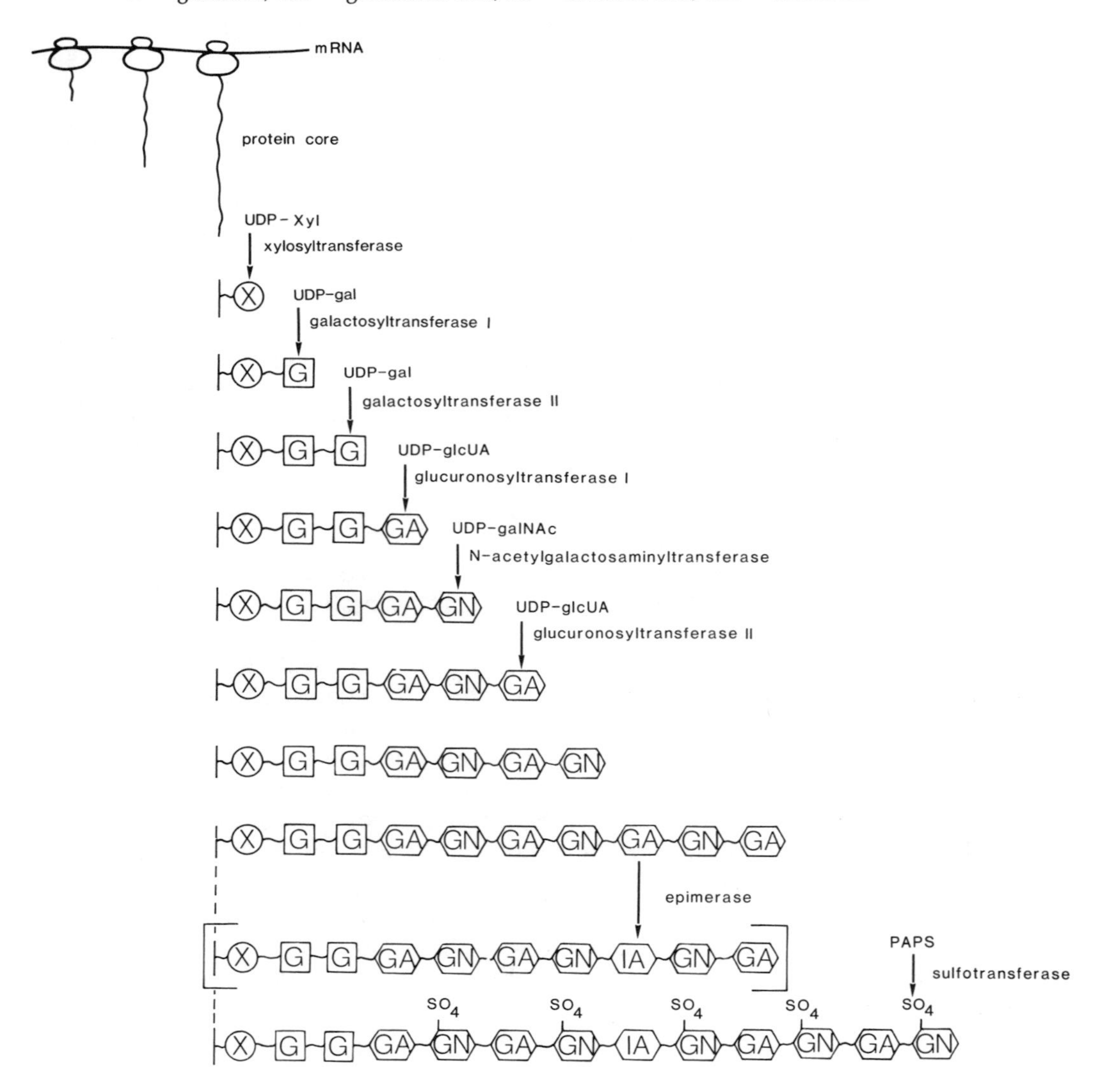

al. 1981). The epimerization is tightly coupled with a subsequent sulfation. When sulfation is not allowed to occur (PAPS is not included), the equilibrium between glucuronic acid and iduronic acid is much in favor of glucuronic acid. When sulfation is allowed to proceed concomitantly with the epimerization, the number of resulting iduronic acid residues is greatly increased. The factors governing sulfation of the iduronic acid residues have not been studied.

8.7.3. Heparin and Heparan Sulfate

Heparin and heparan sulfate differ in the higher content of sulfated residues and iduronic acid residues in heparin. Their biosynthesis, therefore, should involve the same steps, although their protein core acceptors are structurally different. Most data on their biosynthesis have been obtained through studies of the synthesis of heparin in a transplantable mouse mastocytoma. Much of the basic information on chain elongation and sulfation was provided by Silbert and coworkers (*See* Lindahl and Höök 1978; Rodén 1980). They showed that the process is similar to that of chondroitin sulfate biosynthesis in that alternating N-acetylglucosamine and glucuronic acid residues are transferred to the precursor and that sulfation occurs later, both at the amino and the hydroxy groups of glucosamine. The first four monosaccharides of the linkage region of heparin are transferred in a fashion analogous to that described for chondroitin sulfate (Helting 1971). Lindahl and coworkers (*See* Rodén 1980) have since provided much essential knowledge both on the structure of heparin and on its biosynthesis through elegant work on polymer modifications.

The primary product of heparin biosynthesis is a nonsulfated polymer of D-glucuronic acid and N-acetyl-D-glucosamine. As outlined in Figure 8.10, the first step in the series of polymer modifications is N-deacetylation of some of the N -acetylglucosamine residues. The free amino groups are rapidly sulfated, the reaction being catalyzed by a specific N-sulfotransferase. Glucuronic acid residues linked to the 1-position of N-sulfated glucosamine residues are subsequently converted by an epimerase to the C-5 epimer L-iduronic acid. The reaction appears to have an equilibrium in favor for glucuronic acid. However, in concert with the epimerization, the newly formed L-iduronic acid residues are sulfated at the 2-position by an O-sulfotransferase. Sulfated iduronic acid residues are probably not recognized by the epimerase and therefore do not participate in the equilibrium, which is shifted in favor of iduronic acid. The final step in the polymer modification is O-sulfation of the glucosamine residues. It is likely that all iduronic acid residues in heparin are formed via polymer modification, as discussed, because oligosaccharides with a nonreducing terminal L-iduronic acid do not function as an acceptor for subsequent N-acetylglucosamine residues. The epimerase is specific for heparin–heparan sulfate and does not catalyze epimerization of glucuronic acid residues in chondroitin (Malmström et al. 1981).

Heparan sulfate contains fewer N-sulfate groups than heparin. As a consequence, the number of L-iduronic acid residues and the number of O-sulfate groups (of iduronic acid) is lower. The factors governing the location of the N-sulfate and the iduronic acid residues in heparin and heparan sulfate are not known. It is possible that the positioning of enzymes on membranes is an important regulating factor.

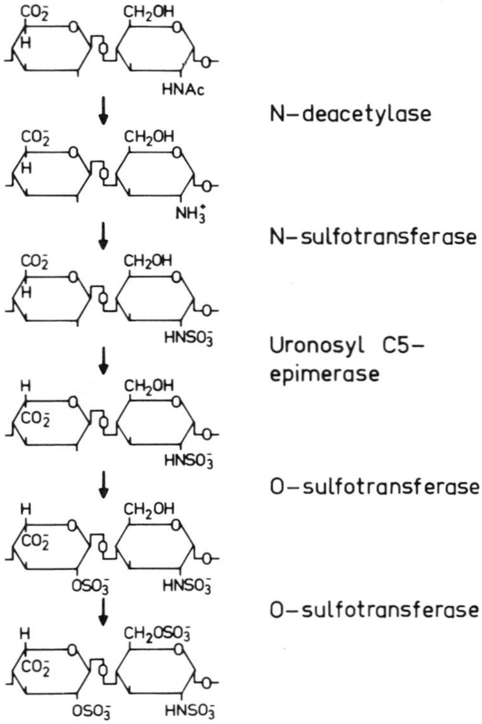

N-deacetylase

N-sulfotransferase

Uronosyl C5-epimerase

O-sulfotransferase

O-sulfotransferase

Figure 8.10. Steps involved in the biosynthesis of heparin. The enzymes involved are separately indicated. *(Source: Reproduced from Lindahl and Höök (1978), Annual Review of Biochemistry, Volume 47C 1978, Annual Reviews, Inc, with permission.)*

8.7.4. Hyaluronic Acid

Chain elongation of hyaluronic acid requires the concerted action of a glucuronosyltransferase and an N-acetylglucosaminyltransferase in the same fashion as described for the other glycosaminoglycans and outlined for chondroitin in Figure 8.9. Little is known about the primer of hyaluronic acid biosynthesis, partly because it constitutes such a minor portion of the completed molecule. It is not known whether the hyaluronate chain is formed on a protein precursor or requires a lipid intermediate precursor. A third alternative would be a carbohydrate precursor by analogy to glycogen. Tunicamycin, an inhibitor of the formation of dolichol–carbohydrate intermediates, fails to inhibit hyaluronic acid biosynthesis. It is therefore likely that lipid intermediates are not involved in the biosynthesis. Cycloheximide, a potent inhibitor of protein synthesis, also fails to inhibit hyaluronic acid biosynthesis. However, the major portion of hyaluronic acid biosynthesis is directed to chain elongation of the very large molecule. Therefore, any effects of inhibitors of acceptor synthesis would be much delayed. Hyaluronic acid synthesized by group A streptococci is bound to membranes and not released until completed. The nature of the structure mediating this attachment is not known.

Biosynthesis of hyaluronate by mammalian cells, e.g., chondrosarcoma cells, proceeds at a rate somewhat different from that of proteoglycan biosynthesis.

It is therefore possible that there is more than one pool of newly synthesized hyaluronate, one of which would be associated with aggregating proteoglycans (*See* Mason 1981). Hyaluronate synthesized by these cells appears not to contain protein (Mason et al. 1982).

Further studies of hyaluronate biosynthesis will be stimulated by identification of a specific linkage to a primer.

8.7.5. Keratan Sulfate

There are two types of keratan sulfate having different linkages to protein (Section 8.2.5). The corneal keratan sulfate has an N-glycosidic linkage to protein. Oligosaccharides having such protein linkages are synthesized via a lipid intermediate precursor, dolichol, and then transferred to protein (Chapter 7). Tunicamycin, a potent inhibitor of the formation of dolichol intermediates, inhibits the synthesis of corneal keratan sulfate, showing that corneal keratan sulfate is also synthesized via lipid intermediates. The complex oligosaccharides in glycoproteins are first synthesized as a large precursor, which is then trimmed down to a smaller acceptor for the various monosaccharides found in the final product. To date, however, there is no further information on the synthesis of the linkage regions of the two types of keratan sulfate. The reactions involved in the glycosyl transfer of galactose and N-acetylglucosamine have not been studied. It is likely that chain elongation occurs by alternate transfer to the primer of the respective monosaccharide. It is known, for example, that UDP-galactose is an essential precursor. UDP-galactose is formed from UDP-glucose and requires the enzyme UDP-glucose-4-epimerase. Consequently UDP-glucose is used for the formation of both UDP-glucuronic acid for chondroitin sulfate biosynthesis and UDP-galactose for keratan sulfate biosynthesis. It appears that UDP-glucose dehydrogenase, and thereby the formation of UDP-glucuronic acid, is regulated by feedback of UDP-xylose, whereas the UDP-glucose-4-epimerase is not regulated by a feedback mechanism. As a result, keratan sulfate biosynthesis appears less well regulated, although some coordination with chondroitin sulfate biosynthesis may occur, via inhibition by NADH of the two key enzymes.

8.8. Biosynthesis of Proteoglycans

There are several key questions regarding the biosynthesis and processing of the whole proteoglycan and its regulation. An important problem to be taken into account is the polydispersity and heterogeneity of proteoglycans. Does this arise from posttranslational modifications of the protein core or does it result from the existence of different gene products? What is the polarity of the core synthesis? What is the timing and spatial relationship between synthesis of the various side chains on the protein core, i.e., oligosaccharides, keratan sulfate, and chondroitin sulfate? At what stage do polymer modifications of the glycosaminoglycan chains occur? How are the proteoglycans transported through the cell and what are the mechanisms for export of these large macromolecules? Are proteoglycan aggregates formed extra- or intracellularly? What extracellular modifications take place?

Many of these questions have not yet been answered. Most studies have been limited to cartilage and the production of proteoglycan by chondrocytes, largely because the structure of other proteoglycans has been less well characterized. Therefore the following discussion will only concern cartilage proteoglycan biosynthesis.

8.8.1. Cartilage Proteoglycans

Both in vitro incubation of cartilage slices (Hardingham and Muir 1972b) and in vivo experiments (Lohmander 1977) have indicated that there are at least two metabolic pools of proteoglycan, one with turnover times of 60 to several hundred days and one with a turnover time of 3 days. The somewhat smaller, lower buoyant density proteoglycans appear to have a faster turnover. No product–precursor relationship has been established between proteoglycans of different size. Triphaus et al. (1980) showed that there are two metabolically distinct populations of proteoglycans in human xiphoid cartilage and that they differ with respect to their relative content of chondroitin sulfate and keratan sulfate. The proteoglycans containing more keratan sulfate are smaller and appear to represent a larger proportion of newly synthesized proteoglycan in older cartilage. It appears that the synthesis of these two proteoglycan pools may be differentially regulated. Inhibition experiments with xylosides give further support to the presence of at least two metabolically different populations. The synthesis of a small proteoglycan, probably corresponding to the small cartilage proteoglycan discussed above, is much less inhibited than that of a large proteoglycan (Kato et al. 1978). It should be stressed, however, that the synthesis of the core protein is only slightly affected by treatment of cells or tissues with xylosides. A normal core is produced and substituted with fewer chondroitin sulfate chains, but a normal number of keratan sulfate chains (Lohmander et al. 1979). All chains are somewhat shorter, depending on the concentration of xylosides. The resulting proteoglycan has a much smaller size. It seems to be transported and secreted at a slower rate than intact proteoglycan (Schwarz 1979).

The polarity of the core has been studied in chick limb bud cultures after labeling of proteoglycan with radioactive amino acids for different lengths of time and subsequent degradation with proteases. Most of the peptides liberated and separated from chondroitin sulfate were assumed to be derived from the hyaluronic acid binding region. The experiments showed that, at short times, a large proportion of the radioactivity was liberated with those peptides; after longer times an increasing proportion of the radioactivity remained with the chondroitin sulfate (DeLuca et al. 1978). The results were taken to indicate that the chondroitin sulfate-rich region represents the N-terminal portion of the protein core. Chemical determination of N-terminal amino acids, however, indicates that the hyaluronic acid-binding region may be the N-terminal part (*See* Rosenberg et al. 1979). No explanation for the conflicting data can as yet be offered.

Using cell-free translation of isolated messenger RNA (mRNA), followed by immunoprecipitation of the products with anticartilage proteoglycan antibodies, a component probably representing a core precursor and having an apparent molecular weight on SDS-polyacrylamide gel electrophoresis of 340,000 has been identified (Upholt et al. 1979; Treadwell et al. 1980). Subsequently, Kimura et al. (1981) used binding to hyaluronic acid, as well as immunoprecipitation with specific antibodies, to isolate from chondrosarcoma cells a protein core with an apparent molecular weight of 370,000. With prolonged incubation, the core precursor was converted to proteoglycan. These studies indicate the presence of a specific protein core precursor.

Other recent studies have indicated that chondrocytes contain a large pool of proteoglycan precursors at various stages of completion (Kimura et al. 1981; Mitchell and Hardingham 1981). The proteoglycans are completed and secreted at a rate indicating a half-life of about 90 minutes for the intracellular pool. Sulfation occurs immediately prior to secretion.

These data indicate that the proteoglycan polydispersity results from post-translational modifications and show that the average precursor spends considerable time in the cell. Further information on the origin of the polydispersity has been provided by Fellini et al. (1981). They showed that newly synthesized proteoglycans are polydisperse with respect to size and that this polydispersity may, at least in part, result from chondroitin sulfate side chains of variable length. From studies of cartilage proteoglycans isolated from tissue it appears that a variation in the size of the core protein is a major reason for variable size. It is therefore likely that the proteoglycans are continuously slowly degraded while in the extracellular matrix, starting with the chondroitin sulfate-rich region. The fragments formed are not attached to hyaluronate and therefore are free to diffuse out of the cartilage and escape detection. In support, McDevitt et al. (1981) and Oegema et al. (1979a) have shown that proteoglycans in the extracellular matrix become progressively smaller with time. None of these studies, however, have taken into account the added complication of different proteoglycan subpopulations.

Another interesting problem is the export of proteoglycan and assembly of proteoglycan aggregates. By an indirect approach, Kimura et al. (1979) showed that proteoglycan aggregation is an extracellular event in rat chondrosarcoma cell monolayer cultures. Endogenous proteoglycan aggregation was inhibited by the addition of hyaluronate oligosaccharides or proteoglycan monomers. Similarly, it was shown that newly secreted proteoglycans are monomeric and bind to hyaluronate to form link-stabilized aggregates only with time (Björnsson and Heinegård 1981). These experiments used zonal rate centrifugation of medium from suspension cultures of fetal bovine chondrocytes. Available data indicate that the components of the aggregate are secreted from the chondrocyte separately. The first step appears to be binding of link protein to the hyaluronic acid binding region of the proteoglycan monomer (Björnsson and Heinegård, unpublished observations). Subsequently, the proteoglycan monomer—link protein complex will bind to hyaluronic acid, resulting in formation of stable aggregates (Kimura et al. 1980; Björnsson and Heinegård 1981).

The structural information available on proteoglycans from noncartilage sources has not allowed study of biosynthetic mechanisms. A number of investigations have been directed to the differential accumulation of various glycosaminoglycans and proteoglycans intracellularly, pericellularly, and in the medium of cultured fibroblasts. The proteoglycans in the three compartments have different half-lives. Most prominently, heparan sulfate proteoglycans are retained with the cells, while dermatan sulfate-containing proteoglycans are released to the medium. Future studies should involve the kinetics of secretion and accumulation in different pools of the two types of proteoglycan in fibroblast cultures, as described by Carlstedt et al. (1981).

8.8.2. Regulation

Little is known about regulating factors important in vivo. Removal of cartilage matrix proteoglycans by incubation with hyaluronidase or protease results in a rapid increase of proteoglycan synthesis (Hardingham et al. 1972). Within a few days, the concentration of proteoglycans in the tissue returns to normal. Sandy et al. (1980) showed that cartilage slices, incubated in vitro, respond to matrix losses with increased synthesis of proteoglycan. The chondrocytes appear to sensitively monitor the matrix for loss of proteoglycan and to respond adequately to such loss. The synthetic response depends on the degree of proteoglycan depletion and can be partially prevented if high concentrations of exogenous proteoglycan are included in the medium. Finally, it appears that the correction does not overshoot. Synthesis is stopped when proteoglycan levels have returned to normal. The nature of the factors involved in such feedback regulation is not known, although hyaluronic acid is one candidate. There are several reports that addition of even very low concentrations of hyaluronic acid to chondrocyte suspension cultures will specifically inhibit sulfate incorporation into proteoglycan (Wiebkin and Muir 1973; Solursh et al. 1980). It appears that the inhibition occurs at the level of core protein synthesis, as incorporation of sulfate into glycosaminoglycans in xyloside treated cells is not altered. Various growth factors, including somatomedins, will increase proteoglycan synthesis in cell cultures, probably by increasing core protein synthesis. High concentrations of proteoglycan or a related polymer in the medium of cultured chondrocytes will result in inhibition of proteoglycan biosynthesis, probably as a nonspecific effect of the high concentration of polymer.

8.9. Catabolism

Matrix proteoglycans are probably first partially degraded in the matrix. There is little information on the catabolic process, although potential individual steps have been studied. Some data indicate that cartilage proteoglycans are gradually degraded by proteases (Sandy et al. 1978; Barret 1981). The fragments representing chondroitin sulfate attached to peptides are small and are therefore no longer trapped in the collagen network. The fragment containing the hyaluronic acid binding region remains in the matrix, bound to hyaluronate. Indeed, hyaluronic acid-binding region fragments not containing glycosaminoglycan have been identified in cartilage, possibly representing degradation products. Fragments not bound to hyaluronate either diffuse out of the cartilage or are turned over by chondrocytes. It appears that similar mechanisms, i.e., proteolytic fragmentation of the core protein, operate in degenerative diseases like osteoarthrosis. Although much less is known about the metabolism of proteoglycan in other tissues, it is likely that initial proteolytic fragmentation in the matrix, followed by transport of released fragments to the cells or out of the tissue, is a general mechanism.

All connective tissue cells appear to have the capacity for synthesizing a variety of proteolytic enzymes. Proteases with neutral pH-optimum have been isolated from cartilage matrix. Some of these are metalloproteases. Cathepsin G and a number of other proteases can be derived from granulocytes and macrophages and have been shown to degrade proteoglycan in vitro. More recently, a factor (catabolin) capable of stimulating chondrocytes to degrade matrix proteoglycans has been identified (*See* Dingle et al. 1979; Steinberg et al. 1979). The factor is derived from synovium and appears to be a protein. The effect on proteoglycan is probably not mediated by diffusible proteolytic enzymes. Possibly chondrocyte cell processes are involved.

Other factors involved in proteoglycan processing are the low molecular weight protease inhibitors, so far only identified in cartilage (*See* Kuettner et al. 1981; Walton et al. 1981). It is possible that the rate of proteoglycan catabolism is determined by the balance between proteases and inhibitors.

Proteoglycan fragments are probably further degraded in the pericellular region by proteases with a somewhat more acid pH-optimum. One example is Cathepsin B, with a pH-optimum of about 5. The glycosaminoglycan peptides are then pinocytosed by the cells. It is also possible that intact proteoglycan can be taken up by some connective tissue cells. Kresse and coworkers (Kresse et al. 1975; Truppe and Kresse 1978) showed that arterial and skin fibroblasts in culture are able to specifically bind and internalize proteoglycans. Dermatan sulfate proteoglycans appear to undergo pinocytosis more efficiently and via a different route than heparan sulfate proteoglycans. Optimal pinocytosis is observed with intact proteoglycans. Fragments produced by trypsin digestion, or chains produced either by alkaline treatment or biosynthetically on xyloside primers, are also taken up by cells. Other studies showed that liver cells have a specific receptor for and can internalize hyaluronic acid (Truppe et al. 1977).

The internalized glycosaminoglycan peptides are further degraded in the lysosomes. The peptides are degraded by cathepsins with acid pH-optima of about 3, especially Cathepsin D. The glycosaminoglycans can be depolymerized by endoglycosidases (Wasteson et al. 1975; Glaser and Conrad 1979) and further degraded by a number of exoenzymes (Figure 8.11). Many of these enzymes were discovered by Neufeld and collaborators in pioneering studies of the mucopolysaccharidoses. This term, derived from an older name for the glycosaminoglycans, refers to a group of hereditary diseases resulting from defective degradative enzymes. All cleavage occurs from the nonreducing terminal. Initially, sulfate groups have to be removed by a number of sulfatases, each specific for the particular sulfate group and location. N-desulfation is followed by N-

(A)

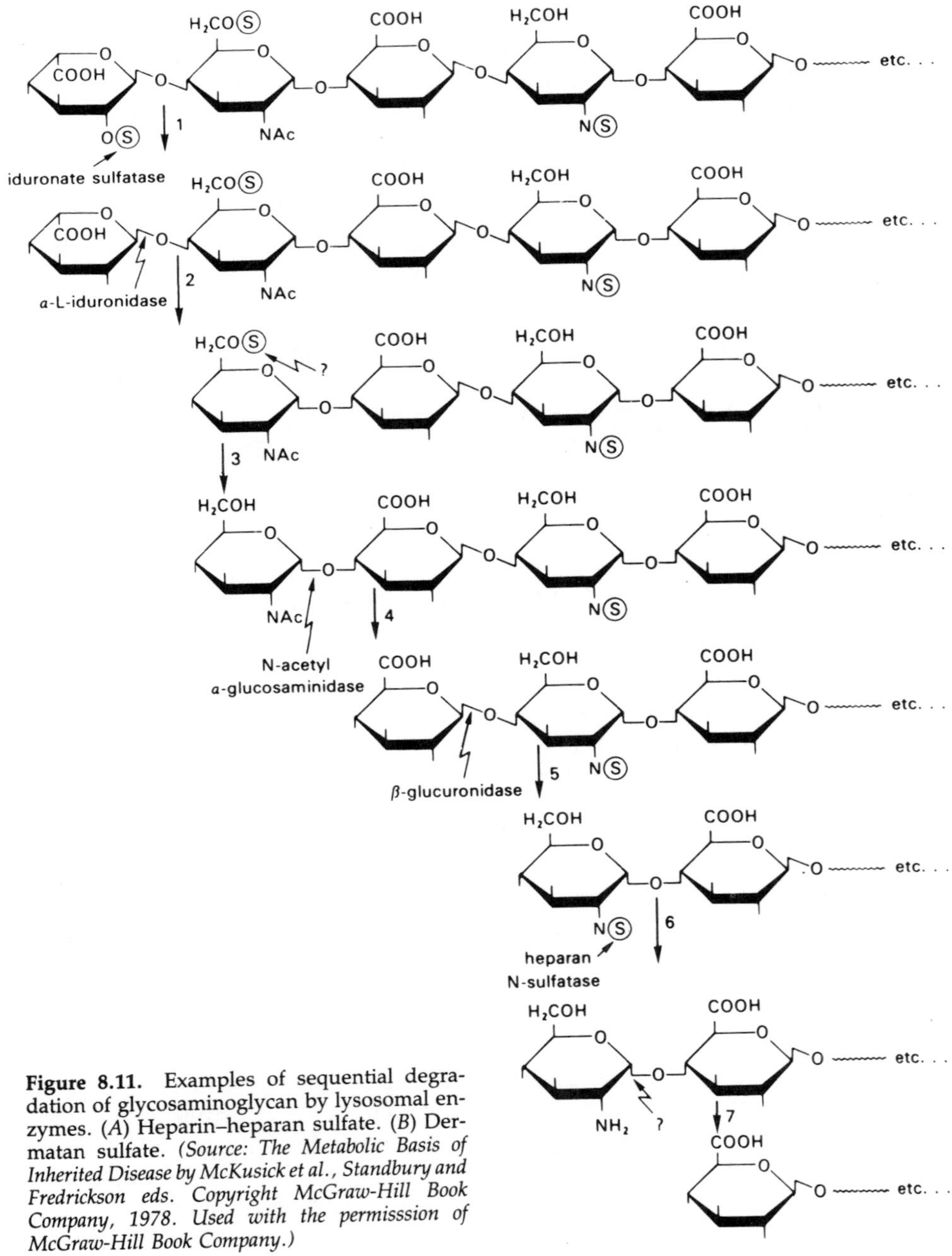

Figure 8.11. Examples of sequential degradation of glycosaminoglycan by lysosomal enzymes. (*A*) Heparin–heparan sulfate. (*B*) Dermatan sulfate. *(Source: The Metabolic Basis of Inherited Disease by McKusick et al., Standbury and Fredrickson eds. Copyright McGraw-Hill Book Company, 1978. Used with the permisssion of McGraw-Hill Book Company.)*

acetylation, prior to further depolymerization. Monosaccharide residues are then removed by the alternating action of specific iduronidases or glucuronidases, and specific N-acetylglucosaminidases or N-acetylgalactosaminidases. By the concerted action of these enzymes, all glycosaminoglycans can be depolymerized to free sulfate and the respective monosaccharide residues. Inherited enzyme

(B)

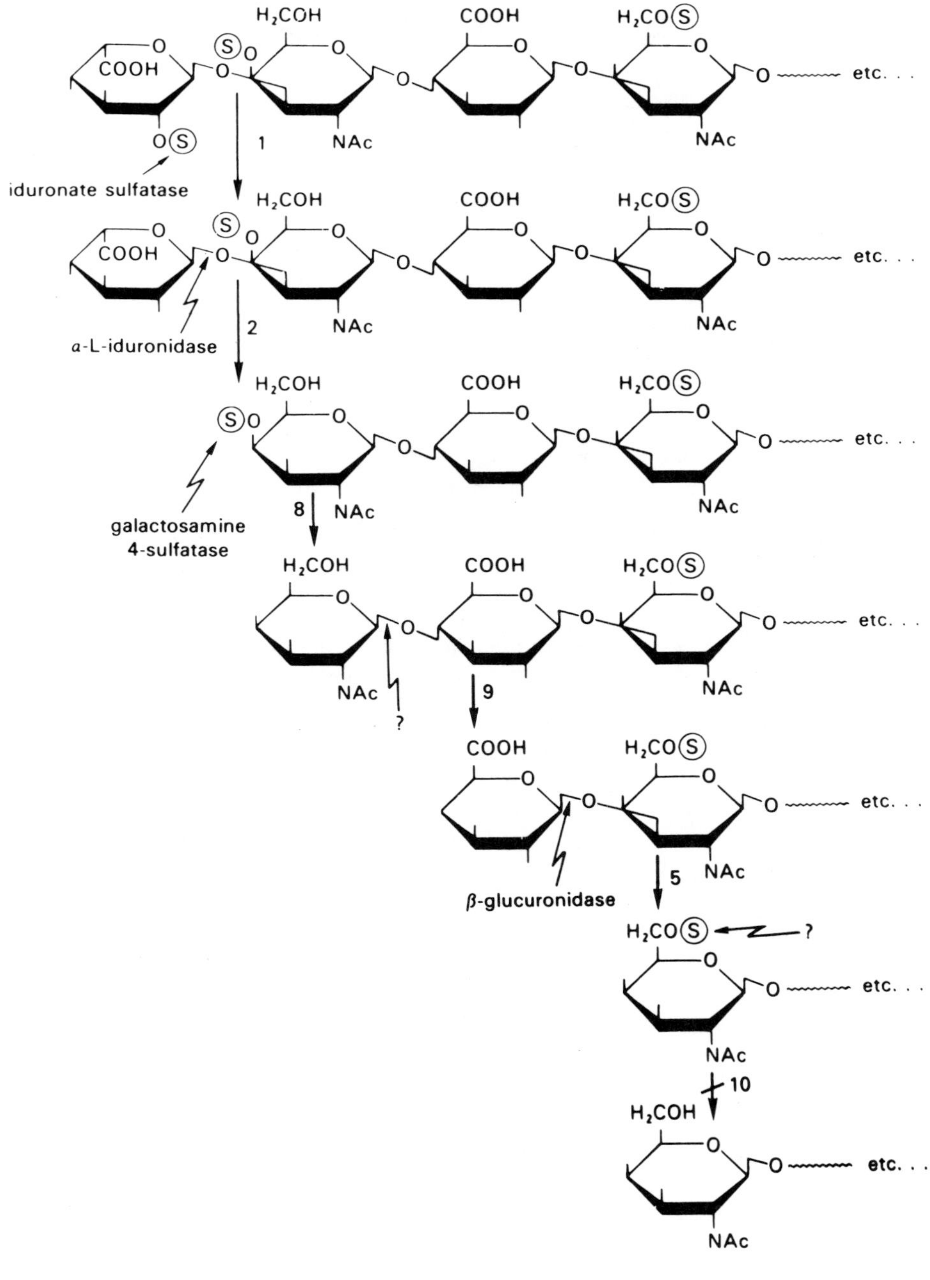
H₂COH
COOH
H₂CO(S)
COOH
(S)O
O(S)
NAc
etc. . .
1
iduronate sulfatase
2
a-L-iduronidase
galactosamine
4-sulfatase
8
?
9
β-glucuronidase
5
10
H₂COH

deficiencies result in the intracellular accumulation of glycosaminoglycan or glycosaminoglycan fragments that cannot be further degraded.

To date, the regulation of the catabolic process has not been studied. Most work has been aimed at describing the individual degradative steps, where much is still left to be done. The liver probably plays an important role in taking up and degrading proteoglycan fragments released to the circulation from various tissues.

The constructive criticism of early versions of this chapter by Drs. Sven Gardell, Stefan Lohmander, and Anders Malmström is greatly appreciated. Secretarial assistance by Marianne Larsson and Rut Lovén has been essential; Birgitta Jönsson's assistance in preparing the figures is appreciated.

References

Axelsson I, Heinegård D: (1975) Fractionation of proteoglycans from bovine corneal stroma. Biochem J 145:491–500.

Axelsson I, Heinegård D: (1978) Characterization of the keratan sulphate proteoglycans from bovine corneal stroma. Biochem J 169:517–530.

Axelsson I, Heinegård D: (1980) Characterization of chondroitin sulfate-rich proteoglycans from bovine corneal stroma. Exp Eye Res 31:57–66.

Baker JR, Caterson B: (1979) The isolation and characterization of the link proteins from proteoglycan aggregates of bovine nasal cartilage. J Biol Chem 254:2387–2393.

Barret AJ: (1981) Which proteinases degrade cartilage matrix? Semin Arthr Rheum 11:1(Suppl 1)52–56.

Björnsson S, Heinegård D: (1981) Assembly of proteoglycan aggregates in cultures of chondrocytes from bovine tracheal cartilage. Biochem J 199:17–29.

Carlstedt I, Cöster L, Malmström A: (1981) Isolation and characterization of dermatan sulphate and heparan sulphate proteoglycans from fibroblast culture. Biochem J 197:217–225.

Christner JE, Caterson B, Baker JR: (1980) Immunological determinants of proteoglycans. Antibodies against the unsaturated oligosaccharide products of chondroitinase ABC-digested cartilage proteoglycans. J Biol Chem 255:7102–7105.

Cifonelli JA, King J: (1972) The distribution of 2-acetamido-2-deoxy-D-glucose residues in mammalian heparins. Carbohydr Res 21:173–186.

Comper WD, Laurent TC: (1978) Physiological function of connective tissue polysaccharides. Physiol Rev 58:255–315.

Cöster L, Fransson L-Å: (1981) Isolation and characterization of dermatan sulphate proteoglycans from bovine sclera. Biochem J 193:143–153.

Cöster L, Fransson L-Å, Sheehan J, Nieduszynski IA, Phelps CF: (1981) Self-association of dermatan sulphate proteoglycans from bovine sclera. Biochem J 197:483–490.

Damle SP, Kieras FJ, Tzeng WK, Gregory JD: (1979) Isolation and characterization of proteochondroitin sulfate from pig skin. J Biol Chem 254:1614–1620.

DeLuca S, Caplan AI, Hascall VC: (1978) Biosynthesis of proteoglycans by chick limb bud chondrocytes. J Biol Chem 253:4713–4720.

DeLuca S, Lohmander S, Nilsson B, Hascall VC, Caplan A: (1980) Proteoglycans from chick limb bud chondrocyte cultures. Keratan sulfate and oligosaccharides which contain mannose and sialic acid. J Biol Chem 255:6077–6083.

Dingle JT, Saklatvala J, Hembry R, Tyler J, Fell HB, Jubb R: (1979) A cartilage catabolic factor from synovium. Biochem J 184:177–180.

Dunstone JR: (1969) Application of equilibrium density gradient sedimentation to the separation and purification of protein-polysaccharides. Separation Science 4:267–285.

Fellini SA, Kimura JH, Hascall VC: (1981) Polydispersity of proteoglycans synthesized by chondrocytes from the Swarm rat chondrosarcoma. J Biol Chem 256:7883–7889.

Finne J, Krusius T, Margolis RK, Margolis RU: (1979) Novel mannitol-containing oligosaccharides obtained by mild alkaline borohydride treatment of a chondroitin sulfate proteoglycan from brain. J Biol Chem 254:10295–10300.

Fisher LW, Termine JD, Dejter SW, Whitson SW, Yanagishita M, Kimura JH, Hascall VC, Kleinman HK, Hassel JR, Nilsson B: (1983) Proteoglycans of developing bone. J Biol Chem 258:6588–6594.

Fransson L-Å, Havsmark B, Nieduszynski IA, Huckerby TN: (1980) Interaction between heparan sulphate chains. II. Structural characterization of iduronate- and glucuronate-containing sequences in aggregating chains. Biochim Biophys Acta 633:95–104.

Fransson L-Å, Johansson BG: (1981) Glycosaminoglycan-lipoprotein interactions: 1. Effect of uronic acid composition and charge density of the glycan. Int J Biol Macromol 3:25–30.

Fransson L-Å, Nieduszynski IA, Phelps CF, Sheehan JK: (1979) Interactions between dermatan sulphate chains. III. Light-scattering and viscometry studies of self-association. Biochim Biophys Acta 586:179–188.

Franzén A, Björnsson S, Heinegård D: (1981a) Cartilage proteoglycan aggregate formation. Role of link protein. Biochem J 197:669–674.

Franzén A, Inerot S, Hejderup S-O, Heinegård D: (1981b) Variations in the composition of bovine hip articular cartilage with distance from the articular surface. Biochem J 195:535–543.

Gardell S, Baker J, Caterson B, Heinegård D, Rodén L: (1980) Link protein and a hyaluronic acid-binding region as components of aorta proteoglycan. Biochem Biophys Res Commun 95:1823–1831.

Glaser JH, Conrad HE: (1979) Chondroitin SO_4 catabolism in chick embryo chondrocytes. J Biol Chem 254:2316–2325.

Hardingham TE: (1979) The role of link-protein in the structure of cartilage proteoglycan aggregates. Biochem J 177:237–247.

Hardingham, TE, Fitton-Jackson S, Muir H: (1972) Replacement of proteoglycans in embryonic chick cartilage in organ culture after treatment with testicular hyaluronidase, Biochem J 129:101–112.

Hardingham TE, Muir H: (1972a) The specific interaction of hyaluronic acid with cartilage proteoglycans. Biochim Biophys Acta 279:401–405.

Hardingham TE, Muir H: (1972b) Biosynthesis of proteoglycans in cartilage slices. Fractionation by gel chromatography and equilibrium density-gradient centrifugation. Biochem J 126:791–803.

Hascall GK: (1980) Cartilage proteoglycans: Comparison of sectioned and spread whole molecules. J Ultrastruct Res 70:369–375.

Hascall VC: (1977) Interaction of cartilage proteoglycans with hyaluronic acid. J Supramol Str 7:101–120.

Hascall VC: (1981) Proteoglycans: Structure and function. In: Ginsburg V, Robbins P, eds. Biology of Carbohydrates, Vol 1. New York: Wiley. 1–49.

Hascall VC, Heinegård D: (1979) Structure of cartilage proteoglycans. In: Gregory J, Jeanloz R, eds. Glycoconjugate Research, Vol 1. New York: Academic. 341–374.

Hascall VC, Sajdera SW: (1969) Proteinpolysaccharide complex from bovine nasal cartilage. The function of glycoprotein in the formation of aggregates. J Biol Chem 244:2384–2396.

Hascall VC, Sajdera SW: (1970) Physical properties and polydispersity of proteoglycan from bovine nasal cartilage. J Biol Chem 245:4920–4930.

Hassel JR, Newsome DA, Hascall VC: (1979) Characterization and biosynthesis of proteoglycans of corneal stroma from Rhesus monkey. J Biol Chem 254:12,346–12,354.

Hassel JR, Newsome DA, Krachmer JH, Rodrigues MM: (1980a) Macular corneal dystrophy: Failure to synthesize a mature keratan sulfate proteoglycan. Proc Natl Acad Sci USA 77:3705–3709.

Hassel JR, Robey PG, Barrach H-J, Wilczek J, Rennard SI, Martin GR: (1980b) Isolation of a heparan sulfate-containing proteoglycan from basement membrane. Proc Natl Acad Sci USA 77:4494–4498.

Heinegård D: (1977) Polydispersity of cartilage proteoglycans. Structural variations with size and buoyant density of the molecules. J Biol Chem 252:1980–1989.

Heinegård D, Axelsson I: (1977) Distribution of keratan sulfate in cartilage proteoglycans. J Biol Chem 252:1971–1979.

Heinegård D, Hascall VC: (1974a) Aggregation of cartilage proteoglycans. III. Characteristics of the proteins isolated from trypsin digests of aggregates. J Biol Chem 249:4250–4256.

Heinegård D, Hascall VC: (1974b) Characterization of chondroitin sulfate isolated from trypsin-chymotrypsin digests of cartilage proteoglycans. Arch Biochem Biophys 165:427–441.

Heinegård D, Hascall VC: (1979a) Characteristics of the nonaggregating proteoglycans isolated from bovine nasal cartilage. J Biol Chem 254:927–934.

Heinegård D, Hascall VC: (1979b) The effects of dansylation and acetylation on the interaction between hyaluronic acid and the hyaluronic acid-binding region of cartilage proteoglycans. J Biol Chem 254:921–926.

Heinegård D, Paulsson M, Inerot S, Carlström C: (1981a) A novel low-molecular-weight chondroitin sulphate proteoglycan isolated from cartilage. Biochem J 197:355–366.

Heinegård D, Paulsson M, Wieslander J: (1981b) Heterogeneity and polydispersity of cartilage proteoglycans. Semin Arthr Rheum 11:1(Suppl 1)31–33.

Helting TH: (1971) Biosynthesis of heparin. Solubilization, partial separation, and purification of uridine diphosphate-galactose: Acceptor galactosyltransferases from mouse mastocytoma. J Biol Chem 246:815–822.

Helting TH, Rodén L: (1969a) Biosynthesis of chondroitin sulfate. I. Galactosyl transfer in the formation of the carbohydrate-protein linkage region. J Biol Chem 244:2790–2798.

Helting TH, Rodén L: (1969b) Biosynthesis of chondroitin sulfate. II. Glucuronosyl transfer in the formation of the carbohydrate-protein linkage region. J Biol Chem 244:2799–2805.

Hoffman P: (1979) Selective aggregation of proteoglycans with hyaluronic acid. J Biol Chem 254:11,854–11,860.

Hopwood JJ, Robinson HC: (1974) The alkali-labile linkage between keratan sulphate and protein. Biochem J 141:57–69.

Horner AA: (1971) Macromolecular heparin from rat skin isolation, charatcterization, and depolymerization with ascorbate. J Biol Chem 246:231–239.

Inerot S, and Heinegård D: (1983) Bovine tracheal cartilage proteoglycans. Variations in structure and composition with age. Collagen Rel Res 3:245–262.

Iverius P-H: (1973) Possible role of the glycosaminoglycans in the genesis of atherosclerosis. In: Porter R, Knight J., eds. Atherogenesis: Initiating factors. Ciba Foundation Symposium 12. Associated Scientific Publishers, Amsterdam. 185–196.

Jenkins RB, Hall T, Dorfman A: (1981) Chondroitin 6-sulfate oligosaccharides as immunological determinants of chick proteoglycans. J Biol Chem 256:8279–8282.

Johansson S, Höök M: (1980) Heparin enhances the rate of binding of fibronectin to collagen. Biochem J 187:521–524.

Kanwar YS, Farquhar MG: (1979) Isolation of glycosaminoglycans (heparan sulfate) from glomerular basement membranes. Proc Natl Acad Sci USA 76:4493–4497.

Kato Y, Kimata K, Ito K, Karasawa K, Suzuki S: (1978) Effect of β-D-xyloside and cycloheximide on the synthesis of two types of proteochondroitin sulfate in chick embryo cartilage. J Biol Chem 253:2784–2789.

Keller R, Stein T, Stuhlsatz HW, Greiling H, Obst E, Mueller E, Scharf HD: (1981) Studies on the characterization of the linkage-region between polysaccharide chain and core protein in bovine corneal proteokeratan sulfate. Hoppe Seylers Z Physiol Chem 362:327–336.

Kempson GE: (1980) Mechanical properties of articular cartilage. In: Freeman MAR, ed. Adult Articular Cartilage, 2nd ed. Tunbridge Wells, UK: Pitman Medical. 333–414.

Kimata K, Barrach HJ, Brown KS, Pennypacker JP: (1981) Absence of proteoglycan core protein in cartilage from the cmd/cmd (cartilage matrix deficiency) mouse. J Biol Chem 256:6961–6968.

Kimata K, Okayama M, Oohira A, Suzuki S: (1974) Heterogeneity of proteochondroitin sulfates produced by chondrocytes at different stages of cytodifferentiation. J Biol Chem 249:1646–1653.

Kimura JH, Caputo CB, Hascall VC: (1981a) The effect of cycloheximide on synthesis of proteoglycans by cultured chondrocytes from the Swarm rat chondrosarcoma. J Biol Chem 256:4368–4376.

Kimura JH, Hardingham TE, Hascall VC, Solursh M: (1979) Biosynthesis of proteoglycans and their assembly into aggregates in cultures of chondrocytes from the Swarm rat chondrosarcoma. J Biol Chem 254:2600–2609.

Kimura JH, Hardingham TE, Hascall VC: (1980) Assembly of newly synthesized proteoglycan and link protein into aggregates in cultures of chondrosarcoma chondrocytes. J Biol Chem 255:7134–7143.

Kimura JH, Thonar EJ -MA, Hascall VC, Reiner LA, Poole AR: (1981b) Identification of core protein, an intermediate in proteoglycan biosynthesis in cultured chondrocytes from the Swarm rat chondrosarcoma. J Biol Chem 256:7890–7897.

Kitchen RG, Cleland RL: (1978) Dilute solution properties of proteoglycan fractions from bovine nasal cartilage. Biopolymers 17:759–783.

Kjellén L, Oldberg Å, Höök M: (1980) Cell surface heparan sulfate. Mechanisms of proteoglycan–cell association. J Biol Chem 255:10,407–10,413.

Kjellén L, Pettersson I, Höök M: (1981) Cell-surface heparan sulfate: An intercalated membrane proteoglycan, Proc Natl Acad Sci USA 78:5371–5375.

Kresse H, Tekolf W, von Figura K, Buddecke E: (1975) Metabolism of sulfated glycosaminoglycans in cultured bovine arterial cells. II. Quantitative studies on the uptake of $^{35}SO_4$-labeled proteoglycans. Hoppe Seylers Z Physiol Chem 356:943–952.

Kuettner KE, Memoli VA, Croxen RL, Madsen L, Pauli BU: (1981) Antiinvasion factor mediates avascularity of hyaline cartilage. Semin Arthr Rheum 11:1(Suppl 1)67–69.

Lindahl U, Bäckström G, Thunberg L, Leder IG: (1980) Evidence for a 3-O-sulfated D-glucosamine residue in the antithrombin-binding sequence of heparin. Proc Natl Acad Sci USA 77:6551–6555.

Lindahl U, Höök M: (1978) Glycosaminoglycans and their binding to biological macromolecules. Ann Rev Biochem 47:385–417.

Linker A, Hovingh P: (1972) Heparinase and heparitinase from flavobacteria. Methods Enzymol 28:902–911.

Lohmander S: (1977) Turnover of proteoglycans in guinea pig costal cartilage. Arch Biochem Biophys 180:93–101.

Lohmander S, DeLuca S, Nilsson B, Hascall VC, Caputo CB, Kimura JH, Heinegård D: (1980) Oligosaccharides on proteoglycans from the Swarm rat chondrosarcoma. J Biol Chem 255:6084–6091.

Lohmander S, Nilsson B, DeLuca S, Hascall V: (1981) Structures of N- and O-linked oligosaccharides from chondrosarcoma proteoglycan. Semin Arthr Rheum 11:1(Suppl 1)12–13.

Lohmander SL, Hascall VC, Caplan AI: (1979) Effects of 4-methyl umbelliferyl-β-D-xylopyranoside on chondrogenesis and proteoglycan synthesis in chick limb bud mesenchymal cell cultures. J Biol Chem 254:10,551–10,561.

Lowther DA, Handley CJ: (1979) Regulatory effect of extracellular molecules on matrix biosynthesis by chondrocytes cultured in vitro. In: Yagi K, ed. Structure and Function of Biomembranes. Tokyo: Japan Scientific Societies Press. 51–61.

Malmström A, Åberg L: (1981) Biosynthesis of dermatan sulphate. Assay and properties of the uronosyl C-5 epimerase. Biochem J 201:489–493.

Malmström A, Fransson L-Å, Höök M, Lindahl U: (1975) Biosynthesis of dermatan sulfate. I. Formation of L-iduronic acid residues. J Biol Chem 250:3419–3425.

Maroudas A: (1980) Physicochemical properties of articular cartilage. In: Freeman MAR, ed. Adult Articular Cartilage, 2nd ed. Tunbridge Wells, UK: Pitman Medical. 215–290.

Mason MR: (1981) Recent advances in the biochemistry of hyaluronic acid in cartilage. In: Deyl Z, Adam M, ed. Connective Tissue Research: Chemistry, Biology, and Physiology. New York: Alan R Liss, 87–112.

Mason RM, d'Arville C, Kimura JH, Hascall VC: (1982) Absence of covalently linked core protein from newly synthesized hyaluronate. Biochem J 207:445–457.

Mathews MB: (1971) Comparative biochemistry of chondroitin sulfate-proteins of cartilage and notochord. Biochem J 125:37–46.

McDevitt CA, Billingham MEJ, Muir H: (1981) In vivo metabolism of proteoglycans in experimental osteoarthritic and normal canine articular cartilage and the invertebral disc. Semin Arthr Rheum 11:1(Suppl 1)17–18.

McDevitt CA, Muir H: (1971) Gel electrophoresis of proteoglycans and glycosaminoglycans on large pore composite polyacrylamide-agarose gels. Anal Biochem 44:612–622.

McKeown PJ, Goetinck PF: (1979) A comparison of the proteoglycans synthesized in Meckel's and sternal cartilage from normal and nanomelic chick embryos. Dev Biol 71:203–215.

McKusick VA, Neufeld EF, Kelly TE: (1978) The mucopolysaccharide storage diseases. In: Stanbury JB, Wyngaarden JB, Fredrickson DS, eds. The Metabolic Basis of Inherited Disease. New York: McGraw-Hill. 1282–1307.

Metcalfe DD, Smith JA, Austen KF, Silbert JE: (1980) Polydispersity of rat mast cell heparin. Implications for proteoglycan assembly. J Biol Chem 255:11,753–11,758.

Mitchell D, Hardingham T: (1981) The effects of cycloheximide on the biosynthesis and secretion of proteoglycans by chondrocytes in culture. Biochem J 196:521–529.

Módis L: (1978) The molecular structure of the interfibrillar matrix in connective tissue. Acta Biol Acad Sci Hung 29:197–226.

Mörner CT: (1889) Chemische studien über den Trachealknorpel. Skand Arch f Physiol 1:210–243.

Morris ER, Rees DA, Welsh EJ: (1980) Conformation and dynamic interactions in hyaluronate solutions. J Mol Biol 138:383–400.

Muir H: (1958) The nature of the link between protein and carbohydrate of a chondroitin sulphate complex from hyaline cartilage. Biochem J 69:195–204.

Muir IHM: (1980) The chemistry of the ground substance of joint cartilage. In: Sokoloff L, ed. The Joints and the Synovial Fluid, Vol 2. New York: Academic. 27–94.

Nakazawa K, Suzuki N, Suzuki S: (1975) Sequential degradation of keratan sulfate by bacterial enzymes and purification of a sulfatase in the enzymatic system. J Biol Chem 250:905–911.

Nilsson B, Nakazawa K, Hassell J, Newsome D, and Hascall VC: (1983) Structures of oligosaccharides and linkage region between keratan sulfate and the core protein on proteoglycans from monkey cornea. J Biol Chem 258:6056–6063.

Norling B, Glimelius B, Westermark B, Wasteson Å: (1978) A chondroitin sulphate proteoglycan from human cultured glial cells aggregates with hyaluronic acid. Biochem Biophys Res Commun 84:914–921.

Oegema TR, Bradford DS Jr, Cooper KM: (1979a) Aggregated proteoglycan synthesis in organ cultures of human nucleus pulposus. J Biol Chem 254:10,579–10,581.

Oegema TR, Hascall VC, Eisenstein R: (1979b) Characterization of bovine aorta proteoglycan extracted with guanidine hydrochloride in the presence of protease inhibitors. J Biol Chem 254:1312–1318.

Oguri K, Yamagata T: (1978) Appearance of a proteoglycan in developing sea urchin embryos. Biochim Biophys Acta 541:385–393.

Oldberg Å, Kjellén L, Höök M: (1979) Cell surface heparan sulfate. Isolation and characterization of a proteoglycan from rat liver membranes. J Biol Chem 254:8505–8510.

Pal S, Tang L-H, Choi H, Habermann E, Rosenberg L, Roughley P, Poole AR: (1981) Structural changes during development in bovine fetal epiphyseal cartilage. I. Isolation and characterization of proteoglycan monomers and aggregates. Collagen Rel Res 1:151–176.

Parthasarathy N, Spiro RG: (1981) Characterization of the glycosaminoglycan component of the renal glomerular basement membrane and its relationship to the peptide portion. J Biol Chem 256:507–513.

Pasternack SG, Veis A, Breen M: (1974) Solvent-dependent changes in proteoglycan subunit conformation in aqueous guanidine hydrochloride solutions. J Biol Chem 249:2206–2211.

Paulsson M, Heinegård D: (1979) Matrix proteins bound to associatively prepared proteoglycans from bovine cartilage. Biochem J 183:539–545.

Paulsson M, Heinegård D: (1981) Purification and structural characterization of a cartilage matrix protein. Biochem J 197:367–375.

Pearson JP, Mason RM: (1979) Proteoglycan aggregates in adult human costal cartilage. Biochim Biophys Acta 583:512–526.

Poole AR, Pidoux I, Reiner A, Tang L-H, Choi H, Rosenberg L: (1980) Localization of proteoglycan monomer and link protein in the matrix of bovine articular cartilage: An immunohistochemical study. J Histochem Cytochem 28:621–635.

Radhakrishnamurthy B, Smart F, Dalferes ER Jr, Berenson GS: (1980) Isolation and characterization of proteoglycans from bovine lung. J Biol Chem 255:7575–7582.

Reihanian H, Jamieson AM, Tang LH, Rosenberg L: (1979) Hydrodynamic properties of proteoglycan subunit from bovine nasal cartilage. Self-association behavior and interaction with hyaluronate studied by laser light scattering. Biopolymers 18:1727–1747.

Robinson HC, Brett MJ, Tralaggan PJ, Lowther DA, Okayama M: (1975) The effect of D-xylose, β-D-xylosides and β-D-galactosides on chondroitin sulphate biosynthesis in embryonic chicken cartilage. Biochem J 148:25–34.

Robinson HC, Horner AA, Höök M, Ögren S, Lindahl U: (1978) A proteoglycan form of heparin and its degradation to single-chain molecules. J Biol Chem 253:6687–6693.

Robinson HC, Telser A, Dorfman A: (1966) Studies on biosynthesis of the linkage region of chondroitin sulfate-protein complex. Proc Natl Acad Sci USA 56:1859–1866.

Rodén L: (1980) Structure and metabolism of connective tissue proteoglycans. In: Lennarz WJ, ed. The Biochemistry of Glycoproteins and Proteoglycans. New York: Plenum. 267–371.

Rodén L, Baker JR, Cifonelli JA, Mathews MB: (1972) Isolation and characterization of connective tissue polysaccharides. Methods Enzymol 28:73–140.

Rosenberg L, Choi H, Pal S, Tang L: (1979) Carbohydrate–protein interactions in proteoglycans. In: Goldstein IJ, ed. Carbohydrate–Protein Interaction. ACS Symposium, Series 88. Washington: American Chemical Society. 186–216.

Rosenberg L, Hellman W, Kleinschmidt AK: (1970) Macromolecular models of protein polysaccharides from bovine nasal cartilage based on electron microscopic studies. J Biol Chem 245:4123–4130.

Ruoslahti E, Engvall E: (1980) Complexing of fibronectin, glycosaminoglycans and collagen. Biochim Biophys Acta 631:350–358.

Ruoslahti E, Hayman EG, Engvall E, Cothran WC, Butler WT: (1981) Alignment of biologically active domains in the fibronectin molecule. J Biol Chem 256:7277–7281.

Sajdera SW, Hascall VC: (1969) Proteinpolysaccharide complex from bovine nasal cartilage. A comparison of low and high shear extraction procedures. J Biol Chem 244:77–87.

Sandy JD, Brown HLG, Lowther DA: (1978) Degradation of proteoglycan in articular cartilage. Biochim Biophys Acta 543:536–544.

Sandy JD, Brown HLG, Lowther DA: (1980) Control of proteoglycan synthesis. Studies on the activation of synthesis observed during culture of articular cartilages. Biochem J 188:119–130.

Schwartz NB: (1979) Synthesis and secretion of an altered chondroitin sulfate proteoglycan. J Biol Chem 254:2271–2277.

Schwermann CP, Schmidt A, Buddecke E: (1981) Phosphate ester groups in proteoglycans from bovine nasal cartilage. Biochim Biophys Acta 673:270–278.

Scott JE, Orford CR: (1981) Dermatan sulphate-rich proteoglycan associates with rat tail-tendon collagen at the d band in the gap region. Biochem J 197:213–216.

Scott JE, Orford CR, Hughes EW: (1981) Proteoglycan–collagen arrangements in developing rat tail tendon. An electron-microscopical and biochemical investigation. Biochem J 195:573–581.

Sheehan JK, Nieduszynski IA, Phelps CF, Muir H, Hardingham TE: (1978) Self-association of proteoglycan subunits from pig laryngeal cartilage. Biochem J 171:109–114.

Solursh M, Hardingham TE, Hascall VC, Kimura JH: (1980) Separate effects of exogenous hyaluronic acid on proteoglycan synthesis and deposition in pericellular matrix by cultured chick embryo limb chondrocytes. Dev Biol 75:121–129.

Stanescu V, Sweet MBE: (1981) Characterization of a proteoglycan of high electrophoretic mobility. Biochim Biophys Acta 673:101–113.

Stanescu V, Maroteaux P, Sobczak E: (1977) Proteoglycan populations of baboon (Papio papio) articular cartilage. Gel-electrophoretic analysis of fractions obtained by density-gradient centrifugation and by sequential extraction. Biochem J 163:103–109.

Steinberg J, Sledge CB, Noble J, Stirrat CR: (1979) A tissue-culture model of cartilage breakdown in rheumatoid arthritis. Quantitative aspects of proteoglycan release. Biochem J 180:403–412.

Stockwell RA: (1979) Biology of Cartilage Cells. Cambridge: Cambridge University Press.

Suzuki S: (1972) Chondroitinase from *Proteus vulgaris* and *Flavobacterium heparinum*. Methods Enzymol 28:911–921.

Swann DA, Powell S, Sotman S: (1979) The heterogeneity of cartilage proteoglycans. Isolation of different types of proteoglycans from bovine articular cartilage. J Biol Chem 254:945–954.

Sweet MBE, Thonar EJ-MA, Immelman AR: (1978) Anatomically determined polydispersity of proteoglycans of immature articular cartilage. Arch Biochem Biophys 189:28–36.

Tengblad A: (1981) A comparative study of the binding of cartilage link protein and the hyaluronate-binding region of the cartilage proteoglycan to hyaluronate-substituted Sepharose gel. Biochem J. 199:297–305.

Thonar EJ-MA, Sweet MBE: (1979) An oligosaccharide component in proteoglycans of articular cartilage. Biochim Biophys Acta 584:353–357.

Thyberg J, Lohmander S, Heinegård D: (1975) Proteoglycans of hyaline cartilage. Electron-microscopic studies on isolated molecules. Biochem J 151:157–166.

Toole BP, Okayama M, Orkin RW, Yoshimura M, Muto M, Kaji A: (1977) Developmental roles of hyaluronate and chondroitin sulfate proteoglycans. In: Lash JW, Burger MM, eds. Cell and Tissue Interactions. New York: Raven. 139–154.

Torchia DA, Hasson MA, Hascall VC: (1981) ^{13}C nuclear magnetic resonance suggests a flexible proteoglycan core protein. J Biol Chem 256:7129–7138.

Treadwell BV, Mankin DP, Ho PK, Mankin HJ: (1980) Biochemistry 19:2269–2275.

Triphaus GF, Schmidt A, Buddecke E: (1980) Age-related changes in the incorporation of [^{35}S]sulfate into two proteoglycan populations from human cartilage. Hoppe Seylers Z Physiol Chem 361:1773–1779.

Truppe W, Basner R, Von Figura K, Kresse H: (1977) Uptake of hyaluronate by cultured cells. Biochem Biophys Res Comm 78:713–719.

Truppe W, Kresse H: (1978) Uptake of proteoglycans and sulfated glycosaminoglycans by cultured skin fibroblasts. Eur J Biochem 85:351–356.

Turley EA, Roth S: (1980) Interactions between the carbohydrate chains of hyaluronate and chondroitin sulphate. Nature 283:268–271.

Underhill CB, Toole BP: (1981) Receptors for hyaluronate on the surface of parent and virus-transformed cell lines. Exp Cell Res 131:419–423.

Upholt WB, Vertel BM, Dorfman A: (1979) Translation and characterization of messenger RNAs in differentiating chicken cartilage. Proc Natl Acad Sci USA 76:4847–4851.

Yamada KM, Kennedy DW, Kimata K, Pratt RM: (1980) Characterization of fibronectin interactions with glycosaminoglycans and identification of active proteolytic fragments. J Biol Chem 255:6055–6063.

Yanagishita M, Rodbard D, Hascall VC: (1979) Isolation and characterization of proteoglycans from porcine ovarian follicular fluid. J Biol Chem 254:911–920.

Vijayagopal P, Radhakrishnamurthy B, Srinivasan SR, Berenson GS: (1980) Studies of biological properties of proteoglycans from bovine aorta. Lab Invest 42:190–196.

Walton EA, Upfold LI, Stephens RW, Gosh P, Taylor TKF: (1981) The role of serine protease inhibitors in normal and osteoarthritic human articular cartilage. Semin Arthr Rheum 11:1(Suppl 1)73–74.

Wasteson Å, Amado R, Ingmar B, Heldin C -H: (1975) Degradation of chondroitin sulphate by lysosomal enzymes from embryonic chick cartilage. Protides Biol Fluids Proc Colloq Bruges Vol. 22. 431–435.

Wiebkin OW, Muir H: (1973) The inhibition of sulphate incorporation in isolated adult chondrocytes by hyaluronic acid. FEBS Letters 37:42–46.

Wieslander J, Heinegård D: (1980) Immunochemical analysis of cartilage proteoglycans. Radioimmunoassay of the molecules and the substructures. Biochem J 187:687–694.

Wieslander J, Heinegård D: (1981) Immunochemical analysis of cartilage proteoglycans. Cross - reactivity of molecules isolated from different species. Biochem J 199:81–87.

9

Bones and Teeth

Arthur Veis

9.1. Introduction

Calcified tissues are widely distributed in nature and appear in interestingly diverse and beautiful forms. Their functions fall into two categories, structural and physiologic. As structural elements, calcified tissues may act as protective armor or as part of the musculoskeletal system, creating the form and shape of the organism and providing the rigid stress-bearing elements of the skeleton utilized in locomotion. Physiologically, the calcified tissues act as a depot for the storage of a variety of cations and anions, principally calcium and phosphate. By both active and passive processes, they participate in the crucial homeostatic regulation of the levels of these ions in body fluids.

Although a great deal of important information has been gained from the study of mineralized tissues in invertebrates, and bearing in mind that every species is part of an evolutionary continuum, discussion in this chapter is restricted to vertebrates and focuses on the three major hard tissues: bone, dentin, and enamel. These three have distinctly different compositions and ultrastructures. Their physical properties and metabolic activities are also markedly different. Nevertheless, they share several features in common.

Three phases or compartments can be distinguished in all hard tissues. First, there is a hydrated extracellular organic matrix constructed from several macromolecular, principally protein, components. Next, the extracellular crystalline phase—principally calcium hydroxyapatite—is embedded in or lies in juxtaposition with the organic matrix. The third phase or compartment is that of the cellular components, including the nerves and vascular elements that permeate

---

From the Department of Oral Biology, Northwestern University, Medical and Dental Schools, Chicago, Illinois

the tissue and provide entry for the nutrients supporting metabolic activity via the blood vessels and capillaries. The metabolic reactivity and responsiveness of the hard tissues cannot be overemphasized. In general such activity is greatest in bone, but in developing teeth both dentinogenesis and amelogenesis are very active processes. Mature dentin-forming odontoblasts remain capable of matrix synthesis at all times, and are highly reactive in the continuously erupting incisors of some species. In bone, no cell is more than 300 μm from a blood vessel, and in teeth the odontoblasts sit on the surface of well vascularized dental pulp.

It is tempting to treat the fluids surrounding bone cells as a fourth distinct phase, as many reports have stressed that ion concentrations within the extracellular fluids in the pericellular regions are different from their concentrations in serum. There is, however, no convincing evidence that concentrations of Ca^{2+} and phosphate in bone fluid are very different from that in serum, and there is strong evidence that only minor barriers to free diffusion exist. For our purposes, we assume that there are no fixed membranes, and any barriers to diffusion of low molecular weight molecules and ions between extracellular matrix and serum are considered the result of the macromolecular nature of the organic matrix components. Thus, both water and diffusable substances are constituents of the matrix and not a separate phase. Nevertheless, it is clear that all parts of the matrix may not be uniformly accessible to all ions or small molecules.

Regions comprising matrix, mineral, and cellular compartments can be easily identified at microscopic and ultrastructural levels, but the intimate intermixing of these structural elements makes the biochemical dissection of the hard tissues exceedingly difficult. Each hard tissue must be treated from the perspective of its particular organization. Very little generally applicable data exist in the literature concerning the composition of bone and dentin, even in terms of the relative amounts of cellular and noncellular space. Very early studies showed that the relative volume occupied by tubules in dentin varied in different parts of teeth, and in the same region of the same types of teeth in different age groups. In the region of the dentino–enamel junction, about 10% of the dentin is occupied by the tubules, whereas 40% of the space may be taken up by cell processes within predentin at the predentin–odontoblast border. Bone is even more complex. Depending on its origin, a specimen of fresh, hydrated bone may vary in density from 1.7 to 2.05 g/cm^3. Table 9.1 summarizes data painstakingly collected by R. A. Robinson (1960). The distribution of water is particularly interesting. Of the 15% of the volume of compact bone occupied by cells, their processes and the cannaliculi in which these lie, some 91% is water. That is, cellular organic matter contributes approximately only 1% of the mass of the bone. The organic matter of the extracellular matrix contributes 22% of the mass; thus the organic component of the cells can, at most, account for approximately 4% of the total organic content of the bone. As cells are a complex mixture of nucleic acids, proteins, lipids, and polysaccharides it seems very unlikely that any single cellular component could account for as much as 1% of the total organic matter. This is important in deciding if a component that has been isolated by extraction from intact tissue is, or is not, intra- or extracellular in origin.

Compact cortical bone is the highest density material. As one proceeds to less dense bone specimens (*See* Table 9.1), relatively more of the tissue is cellular.

Table 9.1. Distribution of Matrix, Mineral, and Cellular Components in Dog Tibias

	Compact bone[a]	Less dense bone[b]
Density range (g/cm^3)	2.05[c]	1.95[c]
	2.32[d]	2.35[d]
Cellular components		
Weight (%)	7.6	13.2
Volume (%)	15.0	25.0
Matrix		
Weight (%)	25.7	22.8
Volume (%)	39.4	33.4
Mineral		
Weight (%)	66.7	63.9
Volume (%)	45.6	41.6

[a]Average of 228 specimens.
[b]Average of 92 specimens.
[c]Hydrated.
[d]Dry.

However, the ratio of matrix to mineral is nearly the same, indicating that matrix that mineralizes does so to the same extent in each case. Hydrated human dentin from permanent teeth has a density ranging from 1.71 to 2.23 and a dry density on the order of 2.33 g/cm^3. Thus, although organized differently, bone and dentin matrices are comparable with respect to mineral content. Mature enamel is quite different, with a hydrated density of 2.76 g/cm^3, reflecting its acellular, highly mineralized nature.

Bone formation and resorption in remodeling are highly localized activities, and individual cells may act independently. To understand the biochemistry one needs a clear picture of the local cellular environment and ultrastructure. We shall proceed first to a brief discussion of ultrastructure before we consider the biochemistry of the matrix components. Additional discussion of endochondral bone differentiation will be found in Chapter 10.

9.2. Ultrastructure

9.2.1. Bone

Bone tissue grows only by the apposition of matrix on existing surfaces. Endochondral ossification is the process whereby bone matrix is added onto the hypertrophic cartilage matrix of the epiphysial growth plate; intramembranous ossification takes place by appositional growth on the periosteal bone surface or in the osteons of the haversian system. Insofar as the production of extracellular matrix is concerned, all osteoblasts, differentiated bone-forming cells, relate to the matrix in the same way. Figure 9.1 depicts the typical situation of an active, matrix-producing osteoblast. Such cells have a very rich endoplasmic reticulum, a large Golgi complex lying near the nucleus, and a high content of mitochondria—all indicating active metabolism and extensive production of secretory protein. Between the mineralized matrix and the main cell body there is a region of unmineralized matrix called osteoid. The main constituent of the osteoid is collagen in fibrous network form. Cell processes extend through this

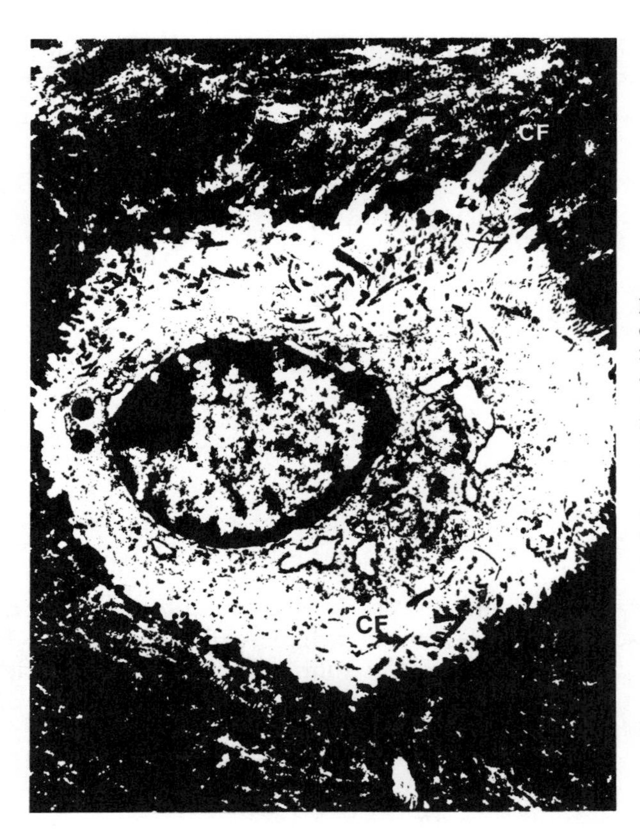

Figure 9.1. An osteoblast, almost completely surrounded by mineralized matrix. The dark, outer mineralized region contains collagen fibrils (CF) organized in the same way as in the osteoid-containing region surrounding the cell. Engorged endoplasmic reticulum is prominent at this stage. Although difficult to discern in this picture, processes extend from the osteoblast into the osteoid. (Courtesy of Melvin J. Glimcher.)

network deep into the osteoid. Membrane-enclosed vesicles may be seen in the osteoid, but mineral crystals are not evident. In fact, there does not appear to be any gradient in calcium concentration in the osteoid between the cell membrane and the mineralized matrix. The sharply defined boundary between bone and osteoid, the mineralization front, is the most striking aspect of bone tissue.

The dynamics of advancement of the mineralization front are of special interest. Consider a fixed reference point P in the calcified zone. As the adjacent osteoblast secretes osteoid matrix, the cell moves away from P on the base of osteoid it has constructed. The mineralization front advances into the osteoid, converting it to bone at the same rate as new osteoid is deposited at the cell surface. The width of the osteoid seam between mineralization front and cell membrane is constant. The processes that coordinate advance of the front and deposition of matrix several microns removed are not understood. The clarification of the molecular mechanisms involved in these pericellular, mineralization-related events is one of the central areas of current investigation in bone biochemistry. In certain osteomalacias, the width of the osteoid seam becomes greater, indicating that mineralization and matrix deposition may be uncoupled.

In haversian systems the osteoblasts ultimately surround themselves with mineral phase. As self-entrapment proceeds, the osteoblasts become less active. Their endoplasmic reticulum is less pronounced, the Golgi apparatus is reduced, and there are fewer mitochondria. When these changes become prominent, that is, when matrix production is no longer the primary role of the cell, it is termed

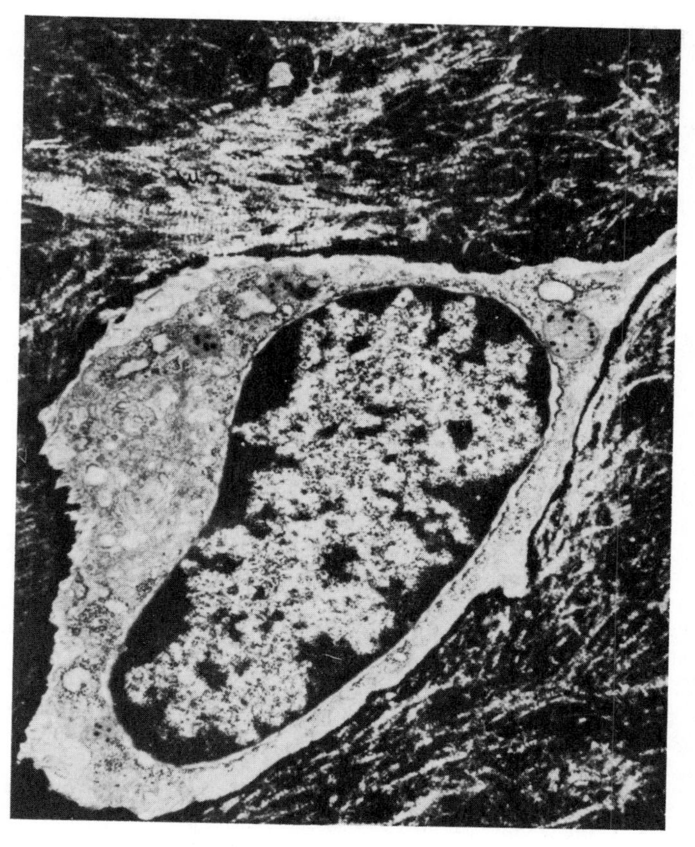

Figure 9.2. An osteocyte, completely surrounded by mineralized matrix. There is very little osteoid. As compared with the osteoblast of Figure 9.1, there is much less endoplasmic reticulum and fewer organelles related to secretion. Note the prominent osteocytic process on the right. (Courtesy of Melvin J. Glimcher).

an osteocyte. As shown in Figure 9.2, the osteocyte maintains an osteoid-containing space or lacunae between itself and the encroaching bone mineral. A special characteristic of osteocytes is their long processes which course through the haversian canaliculi. These processes form junctions with other osteocytic and osteoblastic processes. These contacts and the canals they traverse provide a communication pathway linking all parts of a mineralized bone. Material fluxes to maintain localized cell and tissue viability move via these contact networks. The precise roles of the osteocyte population have not been determined. Young osteocytes, osteoblasts in the early stages of entrapment, share intermediate osteoblastic and osteocytic function and structure.

Osteoclasts, the final bone cell type, probably originate from monocytes unrelated to the progenitors of osteoblasts. This point has been debated extensively in the literature. However, it is clear that osteoclasts are specialized resorptive cells and that osteocytes do not take on resorptive properties and turn into osteoclasts. As depicted in Figure 9.3, resorptive surfaces of osteoclasts have a ruffled brush border of microvilli and a very large number of mitochondria. The brush border region is highly invaginated and the membranes are frequently

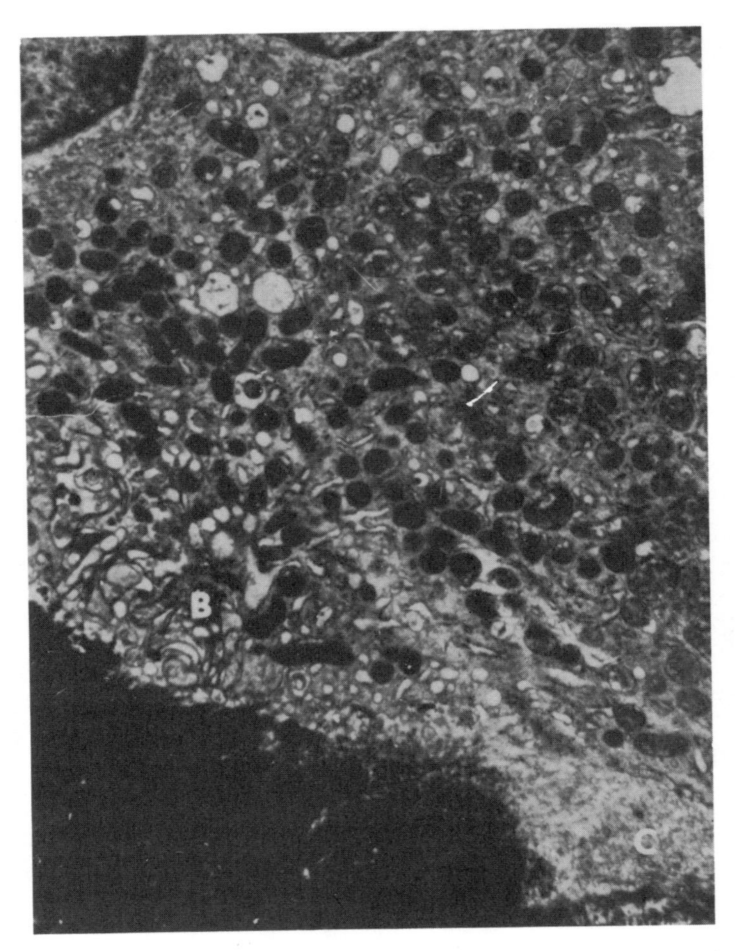

Figure 9.3. The resorptive region of an osteoclast (**B**) adjacent to a clear region (**C**), in which no resorption takes place. There is an extensive amount of microvilli and membrane invaginations at *B* and many filled resorptive vesicles. Osteoclasts are usually multinucleated cells *(not shown)*. *(Source: Matthews. In: Urist M (ed): Fundamental and Clinical Bone Physiology, with permission.)*

pinched off into vacuoles. The contact surface between mineral and osteoclast is not entirely brush border. Regions of active resorption are enclosed by tight osteoclast–bone surface contacts, isolating the resorptive site. The mineral solubilization and matrix removal reactions involved in resorption take place in these isolated, localized regions. Localization is such that osteoblastic appositional deposition of matrix can take place along the same surface and adjacent to a site of osteoclastic resorption.

9.2.2. Dentin

Dentin formation is initiated in the tooth primordium following an apparently epigenetic interaction between undifferentiated tooth epithelial and mesenchymal cells. These cell layers are initially in apposition, but separated by a basement membrane. The epigenetic interaction, as yet not understood, leads to degradation of the basement membrane and differentiation of the epithelial cells into polarized ameloblasts, enamel-producing cells; and the differentiation of the mesenchymal cells into polarized odontoblasts, dentin-producing cells. The arrangement is such (Figure 9.4) that the nuclei of these cell types are maximally separated at the distal boundaries of the contact region. The matrix-producing endoplasmic reticulum, Golgi apparatus, and secretory granule systems are directed towards the boundary, known as the dentino–enamel junction. The mature matrix-secreting odontoblast remains connected to the dentino–enamel junction (Figure 9.4). Like the osteoblast, the odontoblast secretes its collagenous matrix (dentin) and recedes from the site of deposition using the base it has constructed. A well defined matrix region adjacent to the main cell body remains as unmineralized predentin (the counterpart of osteoid). Mineralization is marked by a distinct mineralization front. The mineralized matrix, dentin, fills the region between dentino–enamel junction and the mineralization front. It is important to note, as indicated earlier, that the odontoblastic processes anchoring the odontoblast to the dentino–enamel junction are continuous and take up a considerable volume prior to mineralization. The processes create the system of dentinal tubules. There is considerable argument as to contents of the dentinal tubule in the oldest mineralized portions of mature dentin, but it appears that viable cellular material penetrates deeply into the mineralized dentin.

The principal function of the odontoblast is to secrete the extracellular matrix and control its mineralization. Consequently, odontoblasts have a very rich

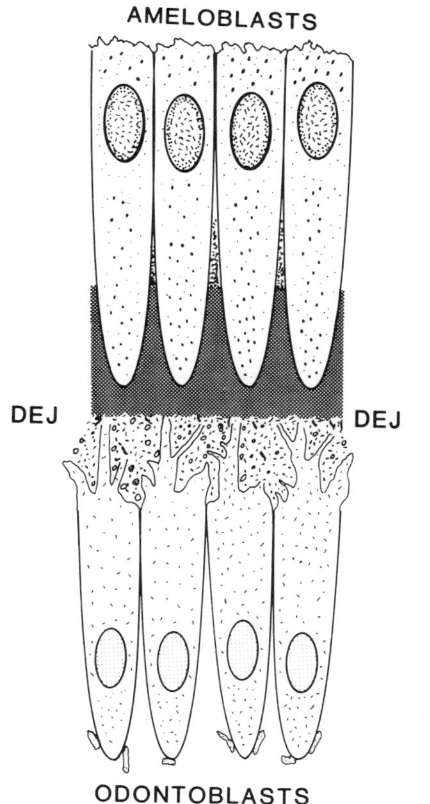

Figure 9.4. A schematic depiction of the dentino–enamel junction *(DEJ)* shortly after the epigenetic events leading to odontogenesis and amelogenesis have begun. The newly formed odontoblasts become polarized and secrete dentin. Their processes anchor at the DEJ and the cell bodies recede. The ameloblasts secrete their amelogenins and enamelins (mineral forms large crystals), and the cells retreat, but do not maintain contact with the DEJ.

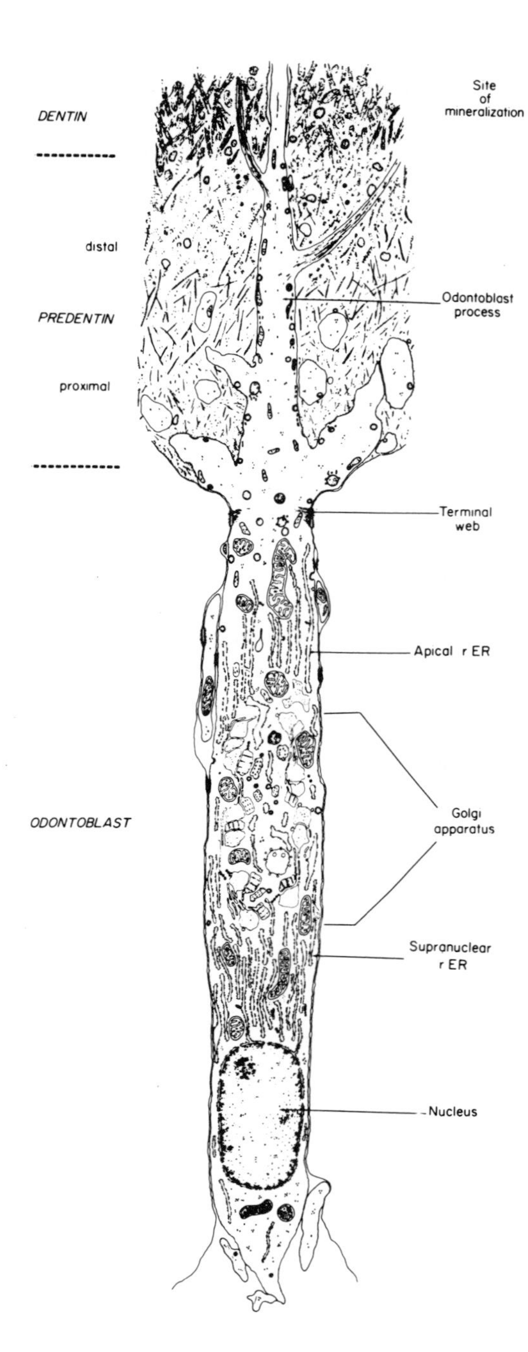

Figure 9.5. A schematic view of a rat incisor odontoblast showing the intracellular organization and the multibranched nature of the odontoblastic process. rER = rough endoplasmic reticulum. (*Source: Weinstock M, Leblond, (1973), with permission.*)

endoplasmic reticulum and, as illustrated in Figure 9.5, many Golgi vesicles and other secretory granules. Most research has focused on the structure of odontoblasts in active matrix deposition. Although the evidence is limited, mature, nonmatrix-secreting odontoblasts appear to retain their secretory apparatus. That is, these cells do not acquire the appearance of osteocytes. Dentin does not remodel and the tooth has no counterpart to the resorptive osteoclast. Thus, the tooth is a particularly appropriate organ in which to study collagen matrix

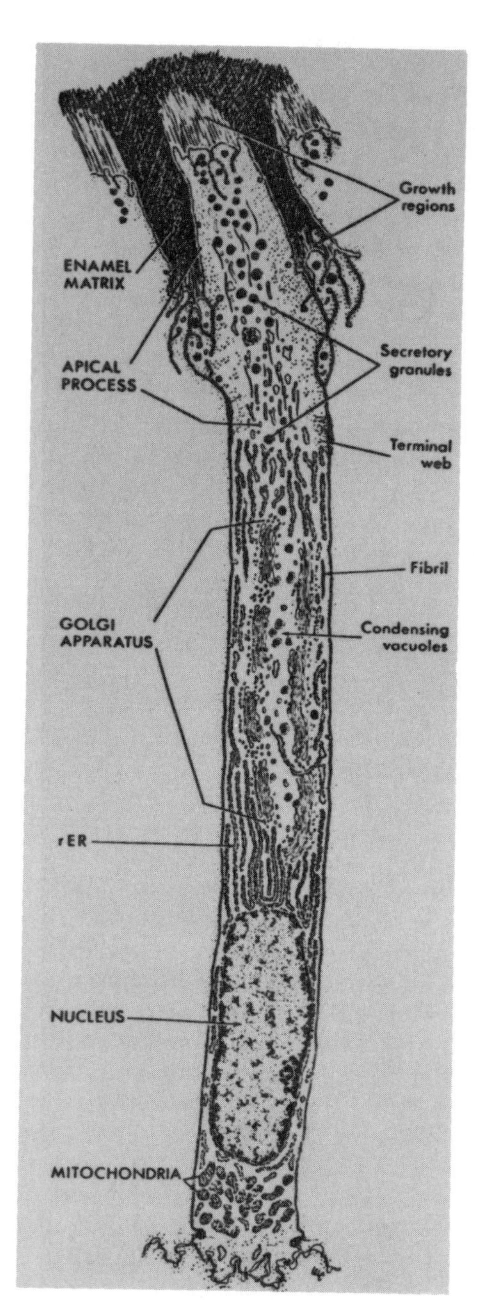

Figure 9.6. Schematic view of rat incisor ameloblast. (*Source: Weinstock A and Leblond (1971), with permission.*)

formation and mineralization processes without excessive complications of turnover and resorption.

9.2.3. Enamel

Once the signals initiating differentiation have been passed, ameloblasts become elongated polarized cells, as depicted in Figures 9.4 and 9.6. During their active life these cells have a very rich endoplasmic reticulum and secretory apparatus.

Although mature enamel contains very little matrix, being more than 95% mineral, it is obvious that the major role of the ameloblast is to secrete a protein matrix in which mineral crystal growth can take place in controlled fashion. The enamel organ wears off after tooth eruption and has no role in any process other than initial matrix formation and mineralization. All posteruption enamel changes are either degradative or arise from passive chemical interactions with exogenous materials in the oral cavity. The early matrix is comprised of two classes of proteins, enamelins and amelogenins, very different in all respects from collagen.

9.3. The Organic Matrix

The extracellular matrix components of mineralized bone, dentin, and enamel cannot be studied readily as almost all components are inaccessible until after removal of the mineral phase. Even after demineralization, three factors complicate all studies: (a) the intensely interactive nature of the matrix components; (b) the intrinsic lability of several matrix proteins; and (c) the presence of degradative enzymes that might act during prolonged demineralization. These problems have been perceived by most investigators, but differing emphasis has been placed on them depending on the investigators' objectives or judgements. Consequently, results from different laboratories cannot always be reconciled easily. The most crucial part of a report in this field may well be the initial approach to demineralization, and thus, discussion of matrix proteins begins with an analysis of demineralization schemes.

9.3.1. Demineralization Protocols

Though collagenous and noncollagenous proteins have been found in fossilized mammoth bone, the first stage of tissue stabilization is important. As illustrated in Figure 9.7, the tissue must be chilled to 4°C and surface washed with a mixture of proteinase inhibitors active at the wash pH prior to mechanical removal of adhering tissue. Reduction to small chips, shards, or powders before demineralization must be accomplished with minimal heating. As bone and dentin are porous, small chips, 1–2 mm on an edge, are sufficient and subsequent handling is much easier than if particle size was reduced further. The dissociative scheme recommended by Termine et al. (1980) is especially useful for enamel and dentin separation because most of the enamel protein and very little dentin protein appear in fractions DS2 (Figure 9.7). Fraction DS3 contains a mixture of dentin and enamel noncollagenous proteins. Very little extracellular matrix protein appears in bone fraction DS2. In either case, fraction DS3 contains all the matrix-associated but nonmatrix-bound proteins. The inclusion of both denaturant and protease inhibitors yields a mixture of individual, minimally degraded components. Removal of the guanidine HCl prior to fractionation permits associative interactions. Removal of the protease inhibitors prior to fractionation permits some degradation.

Under associative conditions, direct demineralization with 0.5 M EDTA yields fractions AS2-3 and AI2-3, which are probably very similar to DS3 and DI3, but not identical. In tooth preparations AS2-3 would include some of DS2, but because of associative interactions some DS2 components might be included in

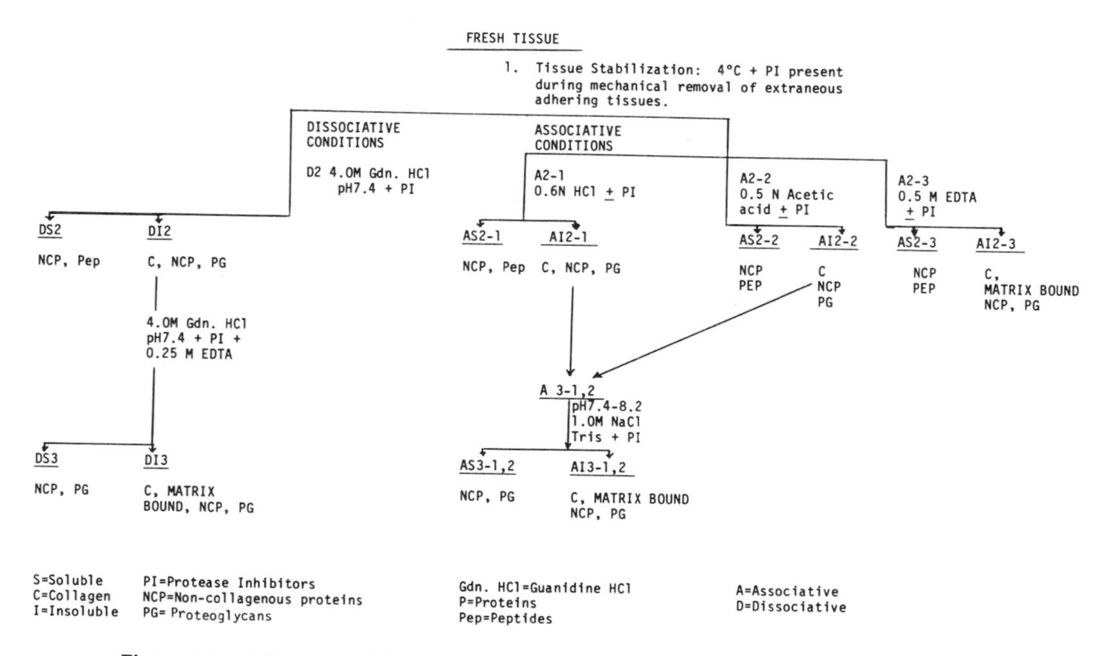

Figure 9.7. Schema used for bone and tooth demineralization. Fractions DI2, AI3-1,2, and AI2-3 are essentially equivalent. A dissociative step added to AI3-1,2 or AI2-3 can also be used to obtain the matrix bound components.

AS2-3. Subsequent fractionation would be required. The dissociative scheme is the most direct route to isolation of pure components, but the associative extraction utilizing EDTA at pH 7.4 (A-2-2), as practiced by Herring and his colleagues (Herring and Kent 1963) has an advantage: One can retain and isolate fractions potentially complexed in vivo in their native conformation. Several enzyme activities can be detected in the EDTA extract. Studies focused on tissue proteases should obviously delete the inhibitors after the initial surface wash.

Both EDTA extraction procedures are slow and require several days to approach completion. Large volumes of EDTA solution are required. To avoid sample dilution, extraction may be carried out with the bone or tooth shards in dialysis bags.

Butler et al. (1972) initially used procedure A2-2 for the isolation of dentin proteins without protease inhibitors. This permitted substantial degradation. Even in the presence of inhibitors, prolonged exposure to dilute acetic acid may lead to degradation of some proteins. Procedure A2-1 is also useful. It is rapid, apparently minimally degradative, and particularly amenable to the analysis of small specimens, such as an individual tooth or a bone biopsy specimen. Surprisingly, fraction AS2-1 is a clean source for the major dentin noncollagenous protein components.

9.3.2. The Collagen Matrix

Collagen comprises more than 90% of the organic matrix of both bone and dentin. Within the limits of detection, normal tissues contain entirely type I collagen. As discussed in Chapter 11, bone from some individuals with osteogenesis imperfecta has been shown to contain some type III collagen in nonmineralized regions (Mueller et al. 1977). The most striking properties of bone and dentin collagen are their near complete insolubility and dimensional stability (Veis and Schlueter 1964; Glimcher and Katz 1965; Glimcher et al. 1965). Dentin collagen does not swell in acid in the range from pH 1 to 5 (Veis and Schlueter 1964), unlike skin and tendon which may also be partially solubilized (Chapter 2).

At the electron microscopic level, demineralized bone and dentin collagen fibrils cannot be distinguished from soft tissue fibrils. The basic molecular packing schemes must be the same. Density measurements, combined with x-ray diffraction data, led Katz and Li (1972, 1973) to postulate that the intermolecular spaces, principally involved with the "hole zones" in the overlapping molecular packing scheme (Chapter 1), were larger in the mineralized collagens. The average distance between molecules is approximately 0.6 nm in hard tissues, compared to approximately 0.3 nm in soft tissues. Because they are less densely packed, it seems that the stability of bone and dentin collagens must arise either from some subtle difference in chemistry or perhaps be related to covalent crosslinking.

Posttranslational Modifications. Although microheterogeneity is evident, there is a very narrow range of degree of hydroxylation of proline residues in type I collagen (Chapter 2). It is unlikely that the hard tissue collagens can be distinguished on this basis. On the other hand, there is substantial evidence that the hard tissue collagens have higher contents of hydroxylysine, both in the nonhelical terminal regions and along the triple helix (Table 9.2). The α2(I) chain is more heavily hydroxylated than the α1(I) chain, but is essentially the same in hard and soft tissues. The α1(I) chains of bone and dentin are somewhat increased in lysine hydroxylation, but data are variable for different bone specimens and decrease with age. A more distinctive difference is in the subsequent glycosylation reactions. As discussed in Chapter 2, all the type I collagens are glycosylated at the same sites, principally in regions represented by cyanogen bromide peptides α1-CB5 and α2-CB4 at Hyl 87 and to a lesser extent in the equivalent sequence regions in α1-CB6 and α2-CB5 at Hyl 930. However, skin collagen contains glucosylgalactosyl hydroxylysine as the predominant modi-

Table 9.2. Lysine Hydroxylation and Glycosylation in Type I Collagen Chains[a]

Bovine skin[b]	Lysine	Hydroxylysine	Gal	Glc-Gal
α1(I)	32	4	—	1
α2(I)	19	10	—	2
Chick bone[c]				
α1(I)	20	15		
α2(I)	21	10		
α1(I) + α2(I)[d]	29	7		
Bovine dentin[b]				
α1(I)	29	8	1	1
α2(I)	10	11	1	1

[a]Residues per chain (1000 amino acid residues).
[b]Volpin and Veis (1973), from sum of cynanogen bromide peptides.
[c]Fujii and Tanzer (1974).
[d]Bovine bone, unfractionated mixture of chains (Shuttleworth and Veis 1972).

fication, whereas dentin and bone have a higher content of the monosaccharide derivative, galactosylhydroxylysine.

Volpin and Veis (1973) determined that dentin collagen contains covalently bound phosphate, one residue on the α1 chain in the α1-CB6 region, and two residues on the α2 chain in the α2-CB4 and α2-CB5 regions. Chicken bone contains four moles of organic phosphate per mole collagen, principally on the α2 chains. These phosphates are reported to be glutamine linked, γ-glutamyl phosphates (Cohen-Solal et al. 1979).

Portions of bone and dentin collagens are linked to noncollagenous protein moieties which remain in the insoluble fraction at demineralization stages DI3, AI3-1,2, and AI2-3. In dentin these moieties are highly phosphorylated. We shall return to this topic after consideration of the noncollagenous proteins.

The consequences of any of these differences in posttranslational modification are unclear at this time.

Crosslinking. If structure stabilization is due to covalent crosslinks, then one must inquire if the effect is the result of more crosslinks, intrinsically more stable crosslinks, or some peculiar arrangement of crosslinkage sites. As discussed in Chapter 2, aldehyde-mediated crosslinking is the only known option for type I collagen. The number of reducible crosslinks does not appear to be increased, but it has been determined that in reduced bone and dentin collagen the most abundant crosslink is dihydroxylysinonorleucine (DHLNL), although hydroxylysinonorleucine (HLNL) is also present. Intermolecular crosslinks form between molecules overlapped by 4D, and the aldehyde donors must be in the terminal regions (*See* Figure 1.18, Chapter 1). Kuboki et al. (1981) have examined α1(I)-CB6 × α1(I)-CB5 from bone collagen and found residues 87 and 17^{C} to both be hydroxylysine, the former partly glycosylated. The major crosslink in dentin is in the corresponding region but involves α1(I) and α2(I) chain regions with the bond α1(I)-CB6 × α2(I)-CB4 also DHLNL (Scott et al. 1976). In dentin α1(I)-CB6 × α1(I)-CB0,1 is also present but in smaller amounts than α1(I)-CB6 × α2(I)-CB4. Skin and tendon collagen contain α1(I)-CB1 with HLNL as the major intermolecular crosslink.

Other nonreducible crosslinks, possibly the pyridinoline linkage recently discovered by Fujimoto et al. (1974), may be present but no evidence is available. The crosslinking data in hand, as well as the limited, but different, pepsin susceptibility of bone and dentin collagens, certainly indicate that the tissues do differ from each other, and from the soft tissue collagens in the mode of aldehyde-mediated crosslinking. In vivo the DHLNL structure formed as the initial crosslink can rearrange to the more stable hydroxylysino-5-oxonorleucine form (Chapter 2, Figure 2.6), providing part of the enhanced stability of the hard tissue collagens.

9.3.3. Noncollagenous Proteins

Although the presence of small quantities of macromolecules other than collagen has been recognized for many years, the major advance in this field began when Geoffrey M. Herring and his collaborators developed and applied systematic EDTA extraction and fractionation procedures to bone while Veis and colleagues initiated similar studies on dentin. Herring's two early reviews are well worth reading (Herring 1964, 1968).

Bone. The complexity of the noncollagenous protein is evident when the dissociative procedures of Figure 9.7 are applied to bone. Termine et al. (1981) extracted fetal bovine bone and quantified the composition of the extract. Fraction DS2 yielded approximately 4% of the total bone protein, of which about one quarter was soluble collagen. The remaining components of this nonmineral-bound fraction were proteoglycan, fibronectin, α-2HS-glycoprotein, serum albumin, and nonprotein containing γ-carboxyl glutamic acid (Gla protein). These account for about one half of DS2. Fetal bone is highly cellular and less mineralized than mature bone, 23% mineral by weight compared with more than 60% (Table 9.1), and it is likely that many of the remaining, unidentified components of DS2 arise from cellular contents rather than matrix. The mineral-bound, EDTA-released fraction DS3, approximately 6% of the total matrix protein, contains little collagen. Figure 9.8 shows the result of dissociative gel filtration. Protein and peptide components range in molecular weight from 2×10^6 to 2.5×10^3. Bone Gla protein accounts for 10% of this mixture, α-2HS-glycoprotein for 25%, and serum albumin for less than 6%. These components, with molecular weights of 62,000, 32,000, and 24,000, appear to be bone-specific and readily adsorbed onto hydroxyapatite surfaces.

One immediate observation arises from these data—bone contains two pools of noncollagenous protein, one arising from serum (e.g., α-2HS-glycoprotein) and the other from the bone itself. Any detailed investigation of bone protein must therefore question the origin of each constituent. Bone is obviously a sink or trap for certain proteins produced elsewhere. The mere presence of a constituent in substantial quantity may not imply any special function for that constituent in the tissue.

Figure 9.8. Gel filtration chromatography of the components of fraction DS3 (*See* Figure 9.7) from fetal bovine bone. (*Source: Termine et al.(1981), with permission.*)

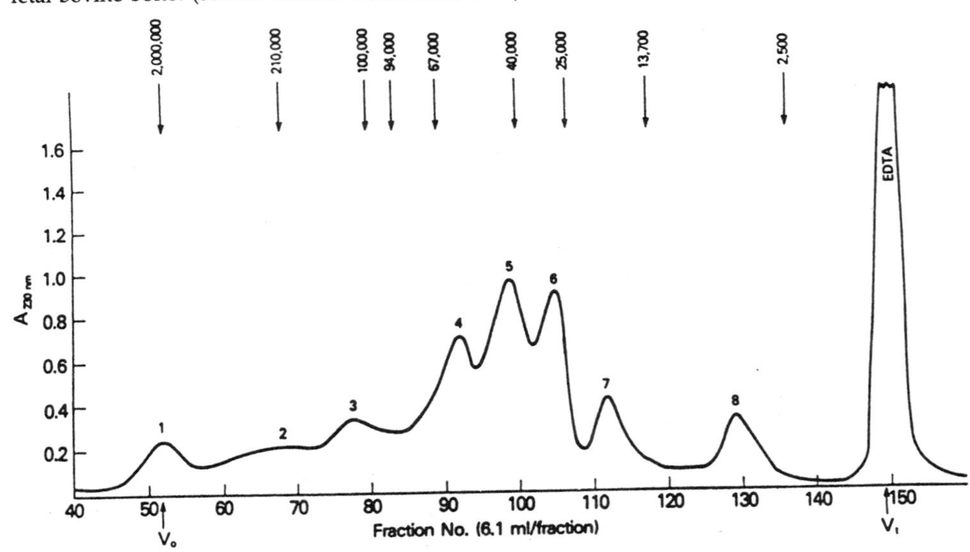

Fraction DI3, the residue from exhaustive dissociative, but nondegradative, extraction is not entirely collagen. Some noncollagenous protein components remain bound to the collagen. Bacterial collagenase digestion (Leaver 1979), with appropriate inhibitors of nonspecific proteases, releases these bound components. Digestion of the demineralized and thoroughly extracted matrix with periodate (Shuttleworth and Veis 1972) and cyanogen bromide (Veis et al. 1977) also yield differing distributions of these components, and can potentially show the nature of the collagen-binding sites.

The compositions determined for several bovine bone-specific noncollagenous proteins are listed in Table 9.3. Some comparable data exist for rabbit and chicken bone. In every case, aspartic acid and glutamic acid residues are the major constituents, as free acids rather than as amides. These comprise at least 25% of the residues of each molecule. Not shown in the table is the fact that each of these proteins contains covalently bound phosphate. In all early studies reported in this table (*See* Table 9.3, footnotes *b, c, f, h,* and *i*), degradation during extraction was a distinct possibility, so the compositions must be treated with some caution. Nevertheless, the EDTA nondissociatively extracted mature bovine cortical bone fractions each contain 20 to 30 residues organic phosphate per 1000 total amino acid residues as serine phosphate. The organic phosphate contents of the dissociatively extracted, protease-inhibited preparations of fetal bone with molecular weights of 62,000, 32,000, and 24,000 were 13, 16, and 17 residues per 1000 residues, respectively, presumably all as serine phosphate. Bone proteins are thus phosphoproteins. A chick bone noncollagenous protein of molecular weight 12,000 has been found (Lee and Glimcher 1981) to have approximately five residues of O-phosphoserine and approximately two residues of O-phosphothreonine per molecule.

Another striking observation is that at least one bovine noncollagenous protein (*See* Table 9.3) contains hydroxyproline. The list of noncollagenous connective tissue matrix proteins containing a few hydroxyprolyl residues continues to grow.

In addition to bone proteoglycan, all bone noncollagenous proteins are probably glycosylated. The sialoprotein, the principal objective of Herring's early studies, is relatively well characterized. It has a molecular weight of about 25,000, of which a single, highly branched carbohydrate moiety accounts for nearly 40%. Herring (1972) depicts the carbohydrate moiety as in Figure 9.9, with each branch terminating in one or more sialic acid residues. This portion of the bone sialoprotein thus establishes an extensive single region of high negative charge density. As no N-terminal residue could be detected, this polysaccharide region might be near the protein N-terminal end. The carbohydrate-binding region isolated after pronase digestion is not rich in acidic residues; hence the bulk of the aspartic and glutamic acid-containing sequence is not directly contiguous with this region and occupies another molecular domain. Even this sialoprotein is phosphorylated with two to four moles phosphate per mole of protein.

All bone noncollagenous proteins show small amounts of hexosamine, with more glucosamine than galactosamine (Veis et al. 1977). Eventually, the exact nature and distribution of these carbohydrate moieties must be explored. The 62,000 and 32,000 molecular weight proteins described in Table 9.3 contain two and three moles of sialic acid per mole. They are thus distinctly different from

Table 9.3. Compositions of Non-Collagenous Proteins from Bovine Bone[a]

	Proteoglycan[b]	Sialoprotein[c]	Gla protein[d,e]	Cortical bone[f]	
				0.4I	0.8I
Hydroxyproline	—	—	20 (1)	—	8
Aspartic acid	162	174	120 (6)	118	116
Threonine	87	111	— (0)	56	58
Serine	97	80	— (0)	81	91
Glutamic acid	194	229	60 (3)	117	119
Proline	63	49	120 (6)	73	72
Glycine	140	122	60 (3)	94	87
Alanine	38	40	80 (4)	67	75
1/2 Cystine	—	13	40 (2)	4	13
Valine	31	31	40 (2)	53	67
Methionine	T	—	— (0)	19	4
Isoleucine	19	25	20 (1)	35	32
Leucine	34	29	100 (5)	84	77
Tyrosine	19	22	80 (4)	24	21
Phenylalanine	16	11	40 (2)	33	35
Histadine	20	12	40 (2)	29	24
Hydroxylysine	—	—	— (0)	—	—
Lysine	34	29	20 (1)	58	61
Arginine	42	12	60 (3)	49	41
Tryptophane	ND	ND	20 (1)	ND	ND
γ-Carboxyglutamic acid	ND	ND	60 (3)	ND	ND

I = ionic strength; T = trace; — = absent; ND = not determined
[a]Residues per 1000 amino acid residues.
[b]Herring 1968a.
[c]Herring 1968b.
[d]Price et al. 1976.
[e]Residues per molecule in parentheses.
[f]Veis et al. 1977.
[g]Termine et al. 1981; molecular weight in thousands.
[h]Leaver et al. 1975b.
[i]Shuttleworth and Veis 1972.

the bone sialoprotein. In fact, Termine et al. (1981) found no comparable sialoprotein in the fetal bone samples examined.

The Gla protein, also known as osteocalcin, is the major endogenous protein of bone. It is synthesized by bone cells, utilizing a process requiring vitamin K, bicarbonate, and a carboxylase to add the γ-carboxy groups to specific glutamic acid residues. The function of the Gla protein is not known, and bone appears to be entirely normal in spite of its nearly complete absence in the presence of vitamin K antagonists. Nevertheless, a comparison of Gla protein from bones of divergent species shows a very high degree of sequence homology, particularly in the positions of the γ-carboxy-glutamic acid and cysteine residues (Figure 9.10), suggesting an important function. Gla protein is also found in plasma and its content there is elevated in individuals with metabolic bone disease.

The name osteocalcin implies that the protein has a specific role in matrix mineralization. This does not seem to be the case, because in fetal bone the appearance of the Gla protein occurs 1 or 2 weeks after initiation of mineral deposition. Further, as pointed out above, vitamin K-depleted animals, fetal rabbits included, mineralize their bone normally in all respects except for the

Table 9.3 (*continued*)

Fetal bone[g]			Collagenase released[h]	Cynanogen bromide released[f]	Periodate released[i]
62K	32K	24K			
—	—	27	—	30	—
92	130	110	130	99	105
49	59	48	64	50	47
94	42	92	73	69	137
122	148	147	166	113	148
75	72	96	75	83	52
105	65	190	223	110	200
93	58	44	68	76	61
2	6	8	—	11	—
82	69	60	T	62	45
6	11	3	T	4	—
33	37	26	39	34	22
75	90	41	62	73	43
26	24	14	T	20	8
29	44	13	T	34	26
29	44	19	36	24	14
—	—	—	20	T	8
50	71	29	29	51	54
38	28	33	15	54	28
ND	ND	ND	ND	ND	ND
ND	ND	ND	ND	ND	ND

content of Gla protein. In vivo studies also show that Gla protein synthesis is vitamin D regulated. Because vitamin D is so intimately involved in calcium homeostasis, it has been argued that the Gla protein might be involved in this process.

As a typical extracellular, secreted protein the Gla protein is probably synthesized in a precursor form. As yet fragmentary evidence suggests that the precursor is very large (molecular weight >50,000) compared with the Gla protein (molecular weight 5200–5900, depending upon species). At neutral pH in the absence of divalent ions, the Gla protein has a predominantly random chain conformation. Upon addition of Ca^{2+} ions, the circular dichroic spectrum shows changes indicative of partial α-helix formation. The Gla residues are crucial; decarboxylated Gla protein does not show this conformational transition. Furthermore, the two cysteine residues that form an internal disulfide bond also are important. The disulfide bond must be intact for Ca^{2+} ion binding to induce the conformational transition.

Bone proteoglycans appear to a small extent in extracts such as DS3, AS3-1,2 or AS2-3. The remainder can only be removed from the insoluble DI3, AI3, or AI2-3 fractions by degradation of the collagenous matrix, usually with bacterial

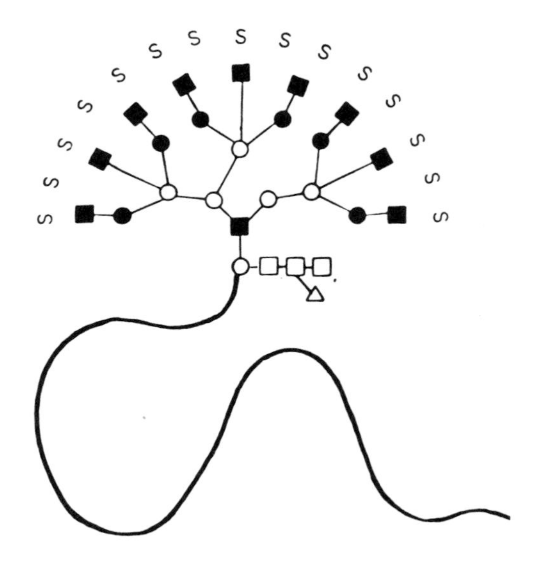

Figure 9.9. The structure of the polysaccharide region of the bovine sialoprotein. S is the terminal sialic acid moiety. ■ = galactose; ● = galactosamine; ○ = glucosamine; □ = mannose; and △ = fucose. *(Source: Herring (1972), with permission.)*

collagenase. Acetyl pyridinium chloride precipitate of an EDTA extract of mature bovine cortical bone contains about 11% proteoglycan core protein. Chromatography of the precipitate fraction on an anion exchange column, usually diethylaminoethyl (DEAE) cellulose, shows marked heterogeneity of the proteoglycan. Three distinct subfractions can be separated. These have similar contents of protein but different amounts of glycosaminoglycan, in each case identified as chondroitin-4-sulfate (Herring 1968). The most acidic fraction is almost protein free, essentially pure chondroitin-4-sulfate. The core protein of the other two fractions is virtually identical to the bone sialoprotein described above, including the presence of the unique sialic acid-containing polysaccharide. Compared to cartilage proteoglycan, the bone proteoglycan is smaller and has only a few glycosaminoglycan chains. These chains are lower in average molecular weight

Figure 9.10. The sequences of swordfish, chicken, and calf γ-carboxyglutamic acid-containing proteins (Gla protein, osteocalcin). Note the conservation of the positions of the Gla residues over this wide species range.

1				5					10					
				Ala	Thr	Arg	Ala	Gly	Asp	Leu	Thr	Pro	Leu	Swordfish
His	Tyr	Ala	Gln	Asp	Ser	Gly	Val	Ala	Gly	Ala	Pro	Pro	Asn	Chicken
Tyr	Leu	Asp	His	Trp	Leu	Gly	Ala	Hyp	Ala	Pro	Tyr	Pro	Asp	Calf

15		17			20	21			24	25					30					35
Gln	Leu	Gla	Ser	Leu	Arg	Gla	Val	Cys	Gla	Leu	Asn	Val	Ser	Cys	Asp	Glu	Met	Ala	Asp	Thr
Pro	Ile	Gla	Ala	Gln	Arg	Gla	Val	Cys	Gla	Leu	Ser	Pro	Asp	Cys	Asn	Glu	Leu	Ala	Asp	Glu
Pro	Leu	Gla	Pro	Lys	Arg	Gla	Val	Cys	Gla	Leu	Asn	Pro	Asp	Cys	Asp	Glu	Leu	Ala	Asp	His

36				40					45					50		
Ala	Gly	Ile	Val	Ala	Ala	Tyr		Ile	Ala	Tyr	Tyr	Gly	Pro	Ile	Gln	Phe
Leu	Gly	Phe	Gln	Glu	Ala	Tyr	Gln	Arg	Arg	Phe	Tyr	Gly	Pro	Val		
Ile	Gly	Phe	Gln	Glu	Ala	Tyr		Arg	Arg	Phe	Tyr	Gly	Pro	Val		

than the comparable ones from cartilage. Collagenase-released proteoglycan also contains primarily chondroitin-4-sulfate.

A difficulty in the study of bone proteins and proteoglycans is the problem of endogenous degradation in addition to that taking place artifactually during extraction, as described earlier. Bone particles of different density have different contents of proteoglycan, with the more mineralized regions containing lesser amounts and lower molecular weight proteoglycan. The average size of the glycosaminoglycan chains is small. Thus, a mechanism for the degradation and removal of protein and glycosaminoglycan components from the extracellular matrix must be a part of the normal mineralization process.

Leaver and his colleagues (1975a) found that bovine cortical bone contains a substantial quantity of small peptides, about 8% of the noncollagenous proteins. These peptides could be extracted with either 0.6 N HCl or 0.5 *M* EDTA. Ranging in molecular weight from 750 to 5000, the peptides have the same acidic compositions as the larger bone proteins. The serine residues are frequently phosphorylated. No serious study of the origins of the bone peptides has been undertaken. If one takes the rather easily supported position that the peptides might be markers for specific extracellular protein turnover during active mineralization, then such studies are very evidently in order.

Bone extracellular matrix contains a small amount of lipid. Quantitation has been difficult because bone marrow, the vascular system, and the cell processes that pervade the tissues all are lipid rich. Nevertheless it is generally considered that somewhat less than 0.1% of bone and perhaps 3 to 4% of demineralized organic matrix is extracellular mineral-phase-related lipid. Lipids comparable to those of all tissues (triglycerides, cholesterol, sphingomyelin, lecithin, phosphatidyl ethanol amine) can be extracted from bone powders with chloroform–methanol. Acidic phospholipids, principally phosphatidyl serine and phosphatidylinositol, remain and are extractable only after demineralization. Boskey (1981) and Boskey and Posner (1976) have isolated these acidic phospholipids in the form of calcium–phospholipid–phosphate complexes. The content of this complex in rabbit diaphyseal bone ranges from 10 μg/mg demineralized matrix in 1-day-old bone to 3 μg/mg in adult bone. All of the complexed phospholipid is membrane-bound, but the precise location, on cell processes or extracellular membrane-bound vesicles is not known. Boskey (1981) has argued that the calcium–phospholipid–phosphate complexes are required for all biological calcification processes involving hydroxyapatite, but no specific mechanism has been offered.

Complexes of nonpolar protein and lipid (proteolipids) have also been isolated from mineralizing matrices. The protein moiety is very rich in leucine and isoleucine and is complexed to phosphatidyl serine and a variety of phosphoinositides, particularly mono- and diphosphophorylated phosphoinositides. As proteolipids are probably integral membrane proteins, bone cell membranes may be the source of the phospholipid in bone.

Dentin. Dentin noncollagenous proteins are comparable in most respects to those of bone: phosphoproteins, glycoproteins, Gla protein, proteoglycans, and peptides are all present. The proteins are all markedly acidic. However, the principal phosphoproteins are strikingly different in composition.

Application of the dissociative extraction and demineralization scheme of Figure 9.7 to fetal bovine teeth removes all of the enamel proteins into fractions DS2 and DS3. As discussed later, the amelogenins are removed entirely in DS2, while the enamelins are mineral bound and are extracted with the dentin proteins in DS3. If the enamel is removed by dissection, dentin-rich regions of the fetal tooth are contaminated with the enamelins to only a small extent. Termine et al. (1980b) applied this procedure to dissected chips of fetal bovine dentin and found that a number of proteins were present, ranging in molecular weight from more than 100,000 to less than 14,000. This finding is essentially identical to earlier (Veis et al. 1977) extractions of isolated fetal bovine dentin using associative EDTA extraction to obtain fraction AS2-3. Comparison shows that either procedure extracts all of the noncollagen-bound matrix proteins. The soluble components from rat incisors, washed in a step comparable to D2 and then demineralized by step A2-3 (Figure 9.7) to yield noncollagenous protein comparable to fraction AS2-3, have been fractioned further, as depicted in Figure 9.11 (Butler 1981a). Table 9.4 lists the nature and relative amounts of these fractions, and Figure 9.12 illustrates their gel electrophoretic patterns.

Fraction I contains the phosphophoryns, which comprise more than 50% of both rat incisor and bovine molar noncollagenous protein, and are unique to

Figure 9.11. Schema for the fractionation of the noncollageneous protein (NCP) of dentin. Data adapted from fractionation schemes of Butler et al. (1981a); Dimuzio and Veis (1978a); Stetler-Stevenson and Veis (1983).

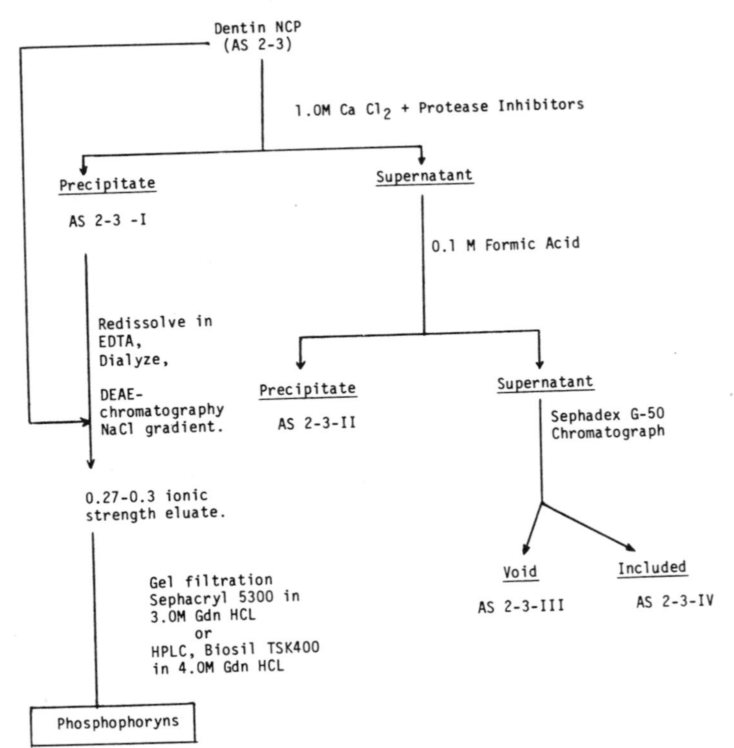

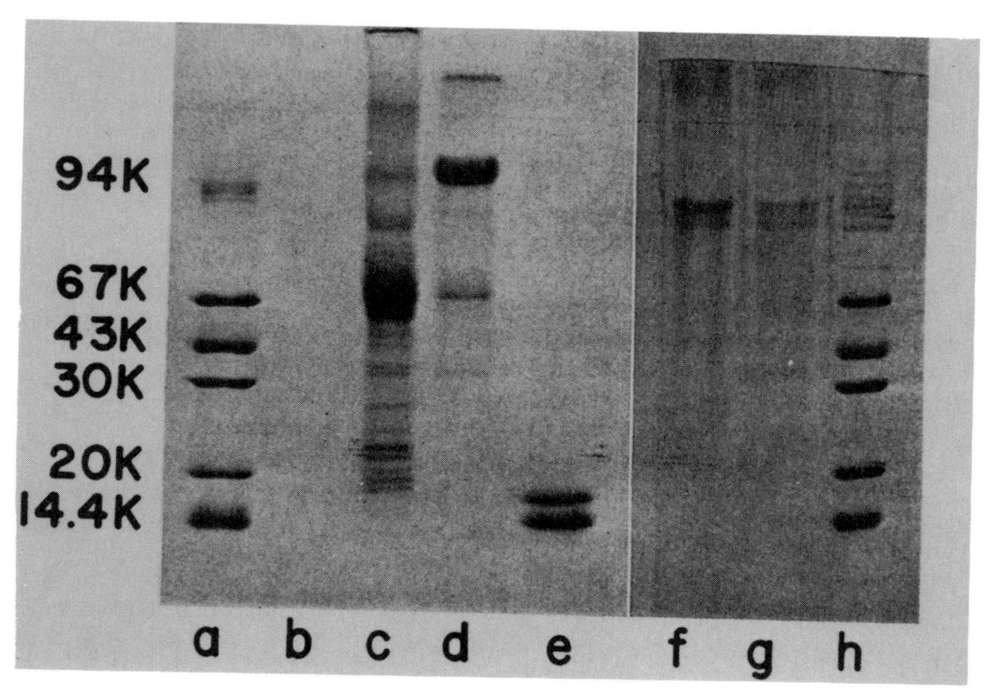

Figure 9.12. SDS-gel electrophoretic analysis of the proteins extracted from rat incisor dentin. a–e = Coomassie brilliant blue stain; a,h = standard proteins; b, f, g = 1.0 *M* $CaCl_2$ supernatant, 0.1 *M* formic acid precipitate; d, e = Sephadex G50 fractions of 0.1 *M* formic acid supernate of DS3. *(Source: Butler et al. 1981a, with permission.)*

dentin. Typical compositions are shown in Table 9.5. Only one phosphophoryn component appears to exist in human and bovine molars but the rat incisor contains two phosphophoryns, called α and β (Dimuzio and Veis 1978a,b).

The unique character of the phosphophoryns is clear. They are exceedingly rich in aspartic acid and serine residues, with their sum comprising a minimum of 75% of the residues. Although exact quantitation is difficult, 85–90% of the serine residues are phosphorylated. Recent results from several laboratories are

Table 9.4. Nature and Relative Yields of the Components of Rat Incisor Noncollagenous Protein Obtained from Associative EDTA Demineralization Fraction AS2-3[a]

Fraction	Character of fraction	Relative yield (%)
I	Phosphophoryns	58.5 ± 4.5
II	Mixture of serum proteins, proteoglycans, and phosphoproteins with low phosphate content	24.5 ± 2.9
III	95,000 and 75,000 molecular weight glycoproteins, albumin, and phosphoproteins with low phosphate content	12.0 ± 2.1
IV	Gla protein(s)	5.0 ± 1.6

[a]Butler et al 1981a

Table 9.5. The Compositions of Phosphophoryns from Teeth of Various Species[a]

	Fetal Bovine Molars[b]	Unerupted Bovine Molars[c]	Erupted Human Molars[d]	Rat Incisor			
				α-PP[e]	β-PP[e]	α-PP[f]	β-PP[f]
Hydroxyproline	—	—	—	—	—	—	—
Aspartic acid	384	335	360	363	360	375	347
Threonine	6	13	19.8	15.0	19.1	12	17
Serine + Phosphoserine	430	439	387	489	447	491	437
Glutamic acid	29	27	49	36	65	31	66
Proline	8	18	49	8.5	19.5	6	8
Glycine	39	35	46	39	71	31	37
Alanine	14	14	6.8	14	29	13	18
Valine	7	9	8.5	2.4	6.7	4	5
1/2 Cystine	T	1.4	15.4	0.6	—	—	—
Methionine	T	4.2	—	0.9	2.5	—	—
Isoleucine							
Leucine	8	12.6	8.2	2.4	5.7	4	4
Tyrosine	2	5.4	4.4	3.5	4.7	3	3
Phenylalanine	2	5.3	4.9	1.4	3.9	2	4
Hydroxylysine	—	—	—	—	—	—	—
Lysine	54	53	13.4	10.1	7.9	13	15
Histadine	7	7.4	14.7	9.5	10.2	8	9
Arginine	5	8.8	8.4	1.0	3.9	3	12
Glucosamine	2-3	4.4	1.2	—	—	—	—
Galactosamine	6	T	—	—	—	—	—
Sialic acid	2-3	ND	ND	ND	ND	ND	ND

PP = phosphophoryn; ND = not determined; — = absent; T = trace.
[a]Residues per 1000 amino acid residues.
[b]Termine et al. 1980b.
[c]Stetler-Stevenson and Veis 1983.
[d]Takagi and Veis 1983.
[e]Dimuzio and Veis 1978a.
[f]Butler et al. 1981a.

in agreement on phosphophoryn composition (Table 9.5), but this has not always been the case. Early differences (Richardson et al. 1978) were the result of the peculiar acid lability of the phosphophoryns. Peptide bonds containing aspartic acid residues are more labile to hydrolysis than other peptides because of the potential attack of the β-carboxyl on neighboring peptide bonds. In dilute acetic acid, aspartic acid residues and any other residue in sequence between two aspartic acid residues will be released preferentially. An analysis of the amino acids and peptides released under mild acid hydrolysis conditions has shown (Lechner et al. 1981) that there are several distinctly different domains in phosphophoryn. As expected from the high content of aspartic acid and phosphoserine, there are regions enriched in one or the other or both. However, amino acids other than glycine or alanine are released in larger peptides. Cysteine, valine, methonine, leucine, and arginine are not released as free amino acids. Thus, without regard for the order of the sequences, a phosphophoryn can be depicted as comprised of four types of regions or domains: $(Asp)_w$, $(PSer)_x$, $(Asp\text{-}PSer)_y$, and $(\text{other residues})_z$, where w, x, y, and z are unknown and variable.

Intact phosphophoryn does not appear to have a free terminal residue and hence has resisted direct sequencing attempts. The phosphophoryn of Linde et al. (1980) was reported to have a free N-terminal and to be capable of undergoing a few cycles of the Edman degradation. However, chromatography on sulfonated polystyrene at pH 1.75 was probably sufficient to cause release of one or more hydrophobic domains. Repeated ion exchange chromatography, even at neutral pH, leads to the breakdown of this very labile molecule. Surface catalysis must be involved, as a phosphophoryn solution standing for a comparable period under identical conditions does not degrade so rapidly. In any event, highly charged supporting media should be avoided in the chromatography of phosphophoryns.

The fetal bovine phosphophoryn and that isolated from both unerupted and erupted third molars from 1 to 2-year-old cattle are virtually identical in amino acid composition; hence there would not seem to be any changes that parallel development. No sufficiently complete studies of the hexoses and hexosamines have been carried out to warrant any conclusions concerning these components. However, the hexosamine contents of rat incisor α- and β-phosphophoryns are distinctly different. The presence of the hexosamines is real and not an indication of contamination.

Because of their high negative charge, the phosphophoryns do not migrate normally on SDS-gels. It appears that they do not bind SDS to the same extent as most proteins, and hence, in spite of their high intrinsic charge, migrate more slowly than expected from their denatured chain length. This effect is substantial: Mature bovine phosphophoryn has an apparent molecular weight of approximately 150,000 on a SDS-gel, but only approximately 100,000 on a calibrated gel filtration column. Rat incisor phosphophoryn is smaller, with apparent molecular weights of 100,000 and 70,000 by the same procedures. The α- and β-phosphophoryns are clearly separable on SDS-gel (Figure 9.12). Note in Figure 9.12 that the phosphophoryns do not stain at all with Coomassie brilliant blue. Stains-all is particularly useful, staining the phosphoproteins blue and other proteins a pinkish color.

Determination of the amount of phosphoserine is difficult. During acid hydrolysis a portion of the phosphoserine is hydrolyzed to serine but some is destroyed. Some serine is also destroyed, but at a different rate. Richardson et al. (1978) determined empirically that the true initial phosphoserine content could be equated to the number of moles or organic phosphate (P) and the initial content of serine determined from:

$$\mathrm{Ser} = (\mathrm{Ser}_{\mathrm{obs}} - 0.763\ \mathrm{P})/0.937,$$

where $\mathrm{Ser}_{\mathrm{obs}}$ is the amount directly determined by amino acid analysis. All of the phosphate is present as the orthomonophosphate ester, as determined by ^{31}P nuclear magnetic resonance studies (Lee et al. 1977).

The salient characteristic of the phosphophoryns is their ability to bind very large numbers of calcium ions with high affinity. As indicated above, 1.0 *M* $CaCl_2$ precipitates phosphophoryns selectively in comparison to the other dentin noncollagenous proteins. The phosphophoryns are also selectively bound by hydroxyapatite crystals and require concentrations of more than 1 *M* phosphate for elution from hydroxyapatite columns in the range of pH 7 to 8. Binding curves are biphasic, showing two classes of calcium binding sites. The high affinity sites have an association constant $K_a = 3.5 \times 10^4\ M^{-1}$ in 0.8 mM KCl at 21°C. The number of high affinity binding sites is equivalent to the total number of phosphate and carboxy groups on the molecule. That is, there is a one-to-one rather than the one-to-two relationship that might be expected based on charge. Additional binding sites of lower affinity, $K_a = 5 \times 10^2\ M^{-1}$, are present but are indefinite in number, indicating multilayer binding. The binding is non-ideal, depending upon the protein concentration. As the phosphophoryn concentration is increased, precipitates are formed that do have electrostatic equivalence between calcium ions and anionic groups of the protein.

In dilute solution in the presence of monovalent ions, phosphophoryn has a random chain conformation. Upon addition of calcium ion, the conformation changes to approximate an extended chain and, with precipitation, a network of β-structure. These phenomena are depicted in Figure 9.13. Manning's theory of counterion binding (Manning 1979) predicts that a linear chain polyanion of

Figure 9.13. Models for the interaction of dentin phosphophoryn with calcium ions in dilute and concentrated solution. Pse = phosphoserine.

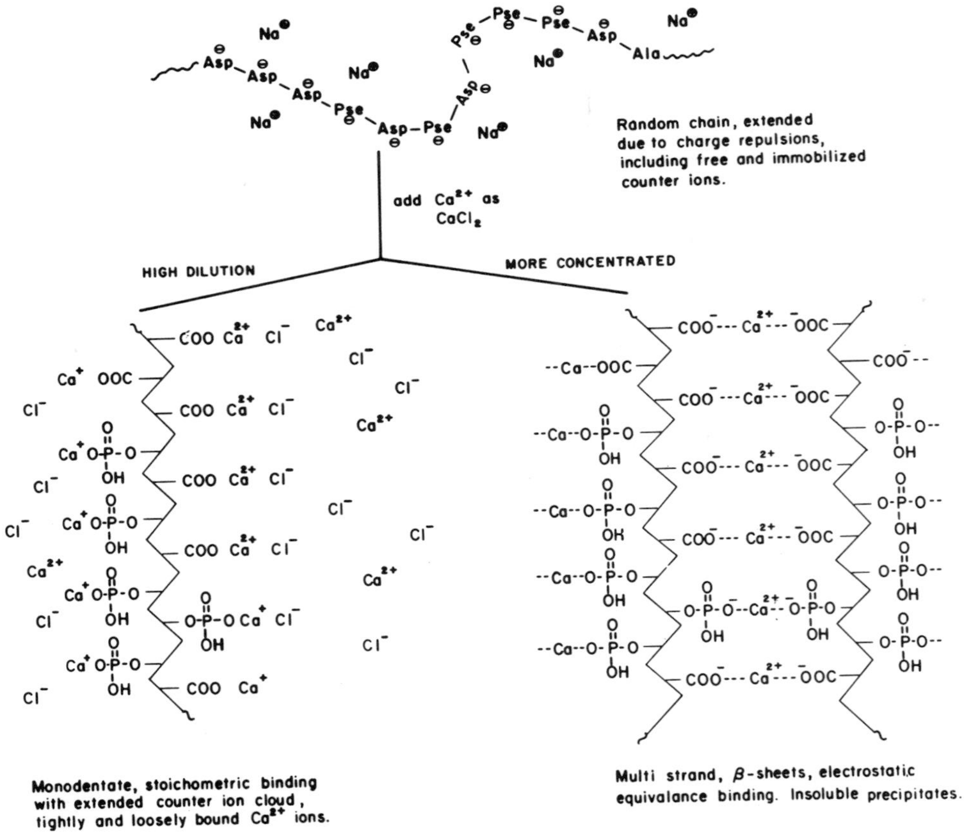

Table 9.6. Amino Acid Composition of Dentin Proteoglycan and Glycoprotein[a]

	Human dentin proteoglycan[b]	Rat incisor		Human molars		Rat incisor AS2-3,IV[e] Gla protein
		AS2-2,II[c]	AS2-3,III[d] (mol. wt. 95,000)	AS2-3,III[b] (mol. wt. 20,000)	AS2-3,III[b] (mol. wt. 14,000)	
Hyp	—	—	—	—	—	10
Asp	206	253	165	257	254	161
Thr	83	37	65	42	32	47
Ser	154	284[f]	111	180	190	73
Glu	154	140	180	138	180	149[g]
Pro	45	37	43	49	51	79
Gly	167	65	156	102	126	101
Ala	61	36	66	47	34	70
Val	37	20	34	37	32	38
Cys (1/2)	—	—	—	—	—	23
Met	T	—	T	T	T	5
Ile	14	12	26	18	23	29
Leu	19	23	36	35	39	67
Tyr	2	16	6	12	36	32
Phe	10	12	8	12	13	19
Hyl	—	—	—	—	—	—
Lys	27	25	43	31	12	30
His	23	14	33	9	10	34
Arg	17	27	28	27	18	35

— = absent; T = trace.
[a]Results expressed as residues per 1000 amino acid residues.
[b]Jones and Leaver.
[c]Butler et al. 1981a; fractions AS2-3,II, rechromatographed on DEAE cellulose.
[d]Butler et al. 1981b; contains hexosamine, sialic acid, and neutral sugar.
[e]Linde et al. 1980; Gla-containing protein fraction, Butler et al. 1981a.
[f]Ser + Pser
[g]Of these, 60 are γ-carboxyglutamic acid (Gla).

high charge, such as a phosphophoryn, with a distance of about 0.35 nm between pendant charged groups, must have at least 0.3 equivalents of counterion held within the charged domain region. Thus, even monovalent counterions would have high binding affinities for phosphophoryn. However, the calcium ion affinity is about 100-fold larger. The implications of differing domains in phosphophoryn, and their potential role in mineralization will be considered later.

Upon amino acid analysis of 6 N HCl hydrolysates of bovine phosphophoryn, a ninhydrin positive peak appears in the chromatographic effluent between proline and glycine identified by Hiraoka et al. (1980) as α-amino adipic acid. The significance of this unusual constituent is not known.

Dentin contains a small quantity of proteoglycan, representing about 2.5% of the organic matrix. All the usual glycosaminoglycan components have been detected in the proteoglycan—chondroitin-4-sulfate, chondroitin-6-sulfate, dermatan sulfate, keratan sulfate, heparan sulfate, and hyaluronic acid. Chondroitin-4-sulfate is the principal constituent, about 70% of the total glycosaminoglycan. Human dentin chondroitin-4-sulfate proteoglycan is different from bovine bone proteoglycan in two respects: the core protein comprises only 6% of the intact proteoglycan, and the core protein is distinctly different in composition (Table 9.6). The dentin core protein has much higher contents of aspartyl and seryl residues. No studies of bovine dentin proteoglycans comparable to those of bone have been carried out.

Predentin and dentin are not identical in proteoglycan content or glycosaminoglycan distribution. Dentin contains primarily the chondroitin-4-sulfate variety, whereas predentin is relatively enriched in the chondroitin-6-sulfate-containing proteoglycan. The total proteoglycan content is higher in predentin than in dentin. There is obviously a mechanism for degradation of the predentin proteoglycans as the mineral front advances and the tissue converts to dentin. On this basis many investigators have postulated that the proteoglycans have some mineralization regulatory function in predentin, and in a similar fashion, in osteoid.

There are no clear descriptions of the dentin glycoproteins. The situation can best be explained by examining lanes c, d, and e of Figure 9.12. These lanes depict rat incisor dentin factions AS2-3,II, III, and IV (Figure 9.11). There are at least 20 well-defined bands in these gel lanes. Some, including α2-HS-glycoprotein, transferrin, immunoglobulin, and albumin derive from blood serum. As in bone, these components appear to be trapped within the mineralized tissue as it forms. The entrapment or binding is selective. For example, in dentin the ratio of albumin to immunoglobulin is higher than in the serum.

The remaining glycoproteins range in molecular weight from more than 95,000 to less than 14,000, as in the bone system. Representative amino acid analyses of human and rat dentin proteins are listed in Table 9.6. Examination of the compositions of the comparable components in Table 9.3 shows a consistent difference between the bone and dentin systems. The dentin proteins are substantially richer in aspartic acid and serine residues, but all are quite acidic. Rat incisor fraction AS2-3,II contains about 73 phosophoserine residues per 1000 amino acid residues and is obviously related to the more completely phosphorylated phosphophoryns. It is of special interest that this major protein is not calcium precipitable. Its molecular weight is uncertain but its behavior on a SDS-gel is similar to that of the phosphophoryns, including poor staining with Coomassie brilliant blue.

The dentin glycoprotein of molecular weight 95,000 is nearly 33% carbohydrate by weight and may have two types of polysaccharide chains; one set linked by N-acetyl glucosamine to asparagine residues, and the second set linked O-glycosidically through N-acetyl galactosamine to serine or threonine residues. It appears to be unique among the dentin glycoproteins. The lower molecular weight glycoproteins found by Jones and Leaver (1974) are quite similar to the higher molecular weight AS2-3,II fraction in composition. They may be related proteins or in vivo degradation products of larger proteins.

The dentin Gla protein has not been isolated in purified form. The AS2-3,IV fraction containing the Gla protein yields three closely related components, with molecular weights in the 10,000 to 15,000 range. The first 16 terminal residues of an impure mixture of these components have been determined and have many homologies with the more complete sequences shown in Figure 9.10. It seems reasonable that the multiple Gla proteins, like the obvious phosphophoryn and glycoprotein degradation peptides, are products of degradation in vivo. These data point again to an active proteolytic degradation system operating within dentin presumably during growth.

Proteases have been detected in matrix extracts of dentin. Crude extract, comparable to A2-3 in Figure 9.7 but obtained in the absence of inhibitors, contains at least two proteases, one with an optimum at pH 8.6, and the other at pH 3.2. Upon dialysis of AS2-3 against distilled water to remove EDTA, a precipitate forms. This precipitate contains the major portion of the neutral protease activity. The supernatant contains all the cathepsin-like, pH 3.2 enzyme. The supernatant also contains trypsin-inhibitor activity. Cathepsin D can also be found extracellularly in the predentin matrix. Lysozyme, another degradative enzyme, is present in the crude EDTA extracts of dentin. Lysozyme hydrolyzes β-1,4-linkages between N-acetylmuramic acid and N-acetylglucosamine in bacterial cell wall polysaccharide, but other β-1,4 linkages in connective tissue polysaccharides may be cleaved as well. In cartilage, lysozyme is an extracellular matrix component as well as a lysosomal constituent. As a very cationic protein, the lysozyme in the dentin EDTA-solubilized complex is bound to several of the acidic dentin proteins and lysozyme activity has been found in almost every DEAE-cellulose fraction of crude EDTA dentin extracts. The presence of lysozyme and enzymes capable of degrading the core protein of proteoglycans in dentin and predentin further emphasizes, although it clearly does not prove, the protein and polysaccharide degradative activity of the matrix of the dentin, and presumably of all mineralized systems.

Calcium phospholipid phosphate complexes, similar to those in bone, have been found in fetal bovine dentin at about the same level (5 μg/mg dry weight) as in adult rabbit diaphyseal bone. It is probable that rabbit and rat incisors, which are continuously erupting, may have a comparable or somewhat higher content of proteolipids. As fetal bone has a higher phospholipid content than mature compact bone, the level of calcium phospholipid phosphate may be a reasonable measure of the relative cellularity of the tissue. However, these complexes are not distributed uniformly in dentin, and the region of the mineralization front can be stained selectively with Sudan black B, a dye specific for lipids. The significance of this observation, implicating extracellular phospholipids in mineralization, has not been determined.

Enamel. The extracellular matrix of enamel is distinctly different from the matrix of bone or of dentin. There is no collagen. The presence of proteoglycans, glycoproteins, serum proteins, and phospholipids has been verified, but the situation is confused because in the course of normal enamel maturation, most of the matrix is degraded. Two proteins are the primary focus of interest. In the fetal tooth, about 90% of the total protein can be solubilized by dissociative extraction (Figure 9.7) to yield fraction DS2, without demineralization of the matrix. The main components of the enamel surface proteins, the amelogenins, have characteristic proline-rich amino acid compositions (Table 9.7). Molecular weights range from about 28,000 to less than 5000 (Termine et al. 1980a). Early studies of the amelogenins and other enamel proteins were complicated by strong associative interactions between the several components, most evident

Table 9.7. The Composition of the Major Enamel Proteins from Fetal Bovine Teeth[a,b]

	Amelogenins—DS2 (mol. wt. range, × 10^{-3})			Enamelins—DS3 (mol. wt. range, × 10^{-3})	
	(21–38)	(12–20)	(5–10)	(42–72)	(8–30)
4-Hyp	—	—	—	—	—
Asp	35	38	64	115	105
Thr	33	31	36	67	48
Ser + Pser	34	43	71	77	62
Glu	235	220	140	150	136
Pro	304	287	203	166	144
Gly	23	40	114	112	176
Ala	41	37	58	56	94
Val	41	32	28	33	32
1/2 Cys	—	—	—	—	—
Met	—	—	—	—	—
Ile	34	33	31	21	16
Leu	110	117	117	23	53
Tyr	6	14	33	32	15
Phe	18	26	32	55	53
Hyl	—	—	—	—	—
Lys	7	10	15	25	19
His	76	60	34	32	16
Arg	4	11	19	36	23
Organic P, %	T	0.2	0.64	0.22	0.24
Sialic acid	0.9	0.82	0.61	4.71	2.82
GalN	0.17	0.26	0.40	1.08	1.07
GlcN	0.14	0.14	0.12	3.24	2.58
Total anthrone carbohydrate, %	8.2	8.6	19.2	15.2	14.8

— = absent; T = trace.
[a]Results in residues per 1000 amino acid residues.
[b]Termine et al. 1980a.

in gel electrophoretic systems. Chromatography on gel filtration media with 4.0 *M* guanidine HCl seems to be the best technique for fractionation and isolation.

A further intrinsic property of enamel is a developmental change in molecular weight of the amelogenin fraction. Figure 9.14 shows that the largest component, presumably a precursor amelogenin, disappears with time, while lower weight components increase in relative concentration. Moreover, as described many years ago, the molecular weight distribution depends on the region of the tooth surface examined; all regions do not mature at the same time or at the same rate. The same kinds of changes are seen longitudinally in the continuously erupting incisor. Enamel forming at the apical end of the tooth is principally amelogenin in nature, decreasing in average molecular size through a transitional zone and dropping to a very low content at the mature, erupted enamel surface.

Fraction DS3 contains hydroxyapatite bound enamelins. These are more acidic with lower proline content. Molecular weights are higher than that of the amelogenins, the main components being 72,000, 56,000, and 42,000 in molecular weight. Similar enamelins are present in mature teeth, where they are essentially the main protein constituents of the enamel, albeit present in very small amounts.

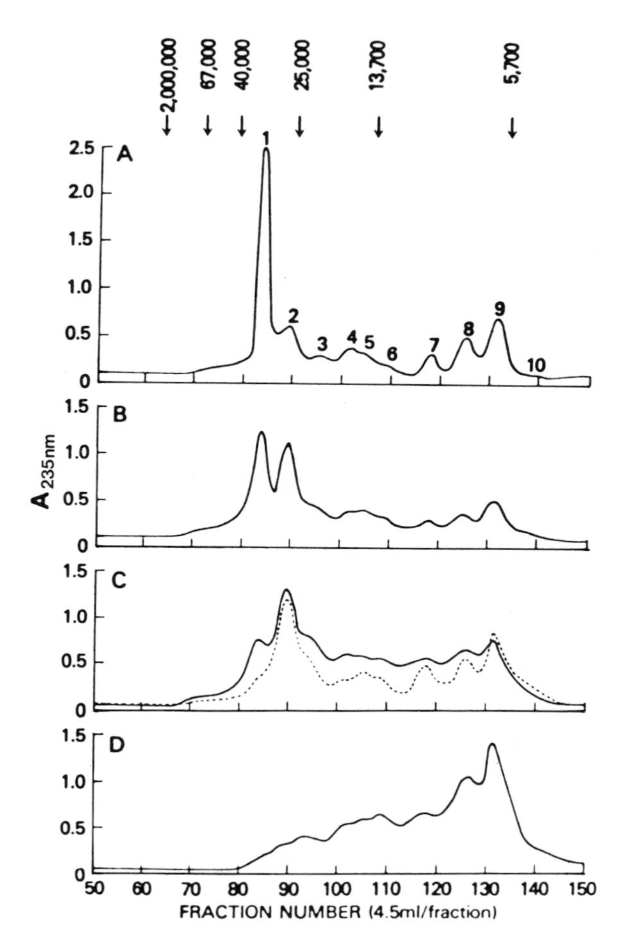

Figure 9.14. Gel filtration chromatography of fraction DS2 of fetal bovine enamel as a function fetal age: (**A**) 2–4 months; (**B**) 5–6 months; (**C**) *solid line* = 7–8 months, *dashed line* = heavily mineralized 5–6 month molar enamel scraped laterial segments; (**D**) cusp tips. *(Source: Termine et al. (1980a), with permission.)*

9.3.4. Collagen–Noncollagen Component Conjugates

Referring once again to Figure 9.7, fractions corresponding to DI3, AI3-1,2 and AI2-3 (the residual insoluble matrix that cannot be dissolved by nonhydrolytic means) have usually been treated as essentially pure collagen, with minor amounts of contaminants. Most of the data on collagen described in Section 9.3.2 has been derived from such fractions. In fact, such residues are not pure and contain a small portion of noncollagenous protein similar in composition to the soluble noncollagenous protein components.

Three procedures have been used to demonstrate the presence of such constituents, as illustrated in Figure 9.15. Each of these procedures has its problems and their application requires very careful attention to detail. Some authors have argued that the collagen–noncollagenous protein fractions illustrated in Figure 9.15 are artifacts. Although their significance is not fully understood, enough evidence has been obtained independently in several laboratories to conclude that these conjugates of collagen and noncollagenous proteins are real and ubiquitous constituents of the mineralized matrix.

The presumption is that the insoluble matrix is an essentially infinitely cross-

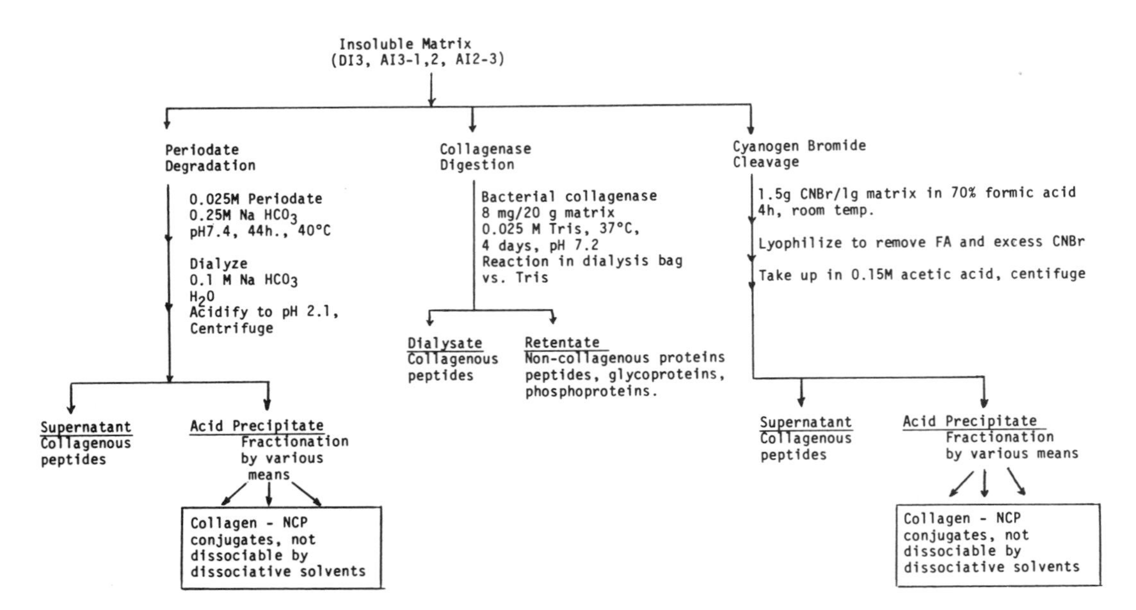

Figure 9.15. Isolation procedures used to demonstrate the presence of collagen–noncollagenous protein conjugates in bone and dentin. *NCP* = noncollagenous protein.

linked network of collagen molecules (Chapters 1 and 2), with covalently bound noncollagenous protein molecules placed within the matrix, as shown in Figure 9.16. One type (Figure 9.16, *A*) may be placed in or on the fibril surfaces at the "hole regions," crosslinked to nonhelical or helical region side chains. Possibly this type might bridge two adjacently packed chains. Conceivably, a molecule such as shown in Figure 9.16, *B* might lie along the molecular or fibril axis covering, and binding to, a portion of a collagen molecular surface. An extended bridging molecule (Figure 9.16, *C*) might interact at a number of equivalent packing points such as staggered telopeptides. Obviously, only a very long flexible chain molecule could serve in such a role.

Periodate oxidation under the conditions used leads to the selective destruction of peptide residues. Tyrosine and nonglycosylated hydroxylysine residues are completely destroyed. Some peptide bonds involving arginine, histidine, methionine, phenylalanine, and leucine are also cleaved, as these residues may be oxidized. Thus, as depicted in Figure 9.16, periodate oxidation produces an ill-defined mixture of collagen chain fragments, plus modified A, B, and C proteins, possibly linked to collagen peptides by periodate resistant bonds. Table 9.3 shows the composition of one such component isolated from bovine bone. It is remarkably similar to the 62,000 molecular weight extractable protein shown in the same table, but has a higher glycine content. Most interesting, it has a significant content of both hydroxylysine and tyrosine, both of which are destroyed by periodate treatment in model systems. Thus, these residues must be involved in, or protected by, the linkage region. Recall that tyrosine residues appear only in the nonhelical terminal regions of collagen (Chapter 2). Comparable components from bovine dentin are described in Table 9.8. Two fractions, obviously differing in the amount of collagen associated with a phosphophoryn-like peptide, are listed. The principal objection to concluding that the

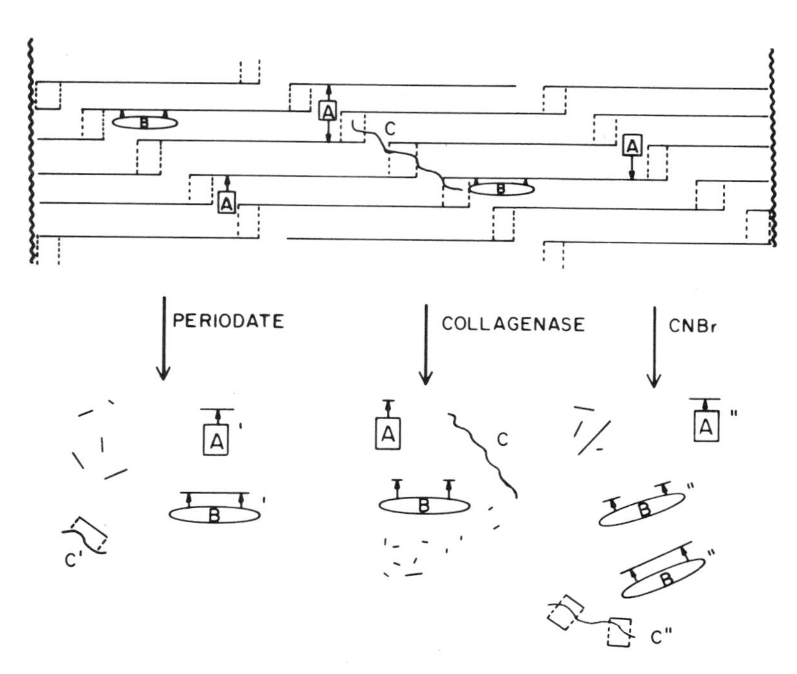

Figure 9.16. Hypothetical schemes of placement of noncollagenous protein *(A,B,* and *C)* within a collagen crosslinked network and possible network degradation products, using the methods of Figure 9.15.

isolated conjugate exists in vivo is that the oxidation of the hydroxylysine by periodate generates formaldehyde, which might have artifactually linked associated, but noncovalently bound, components.

The collagenase-released proteins (Tables 9.3 and 9.8) are less useful for study for two reasons. The collagenous portion is very small; the amount remaining bound to the noncollagenous protein may be only a few residues in length, making identification difficult. Furthermore, the very long digestion at 37°C leads to random hydrolysis, and small peptides may be lost. The isolated noncollagenous protein fragments are of low molecular weight and have compositions not truly representative of intact material. Nevertheless, the presence of hydroxylysine in this fraction is supportive of the concept that noncollagenous protein binds to collagen, presumably the nonhelical terminal regions containing hydroxylysine.

Cyanogen bromide cleavage avoids the problems of artifactual crosslinking and also produces specific collagen peptides split only at methionine residues. The conjugate peptides isolated by this technique are similar in all species examined (Table 9.8). Clearly, the noncollagenous protein portion is closely related to the soluble phosphophoryns in each case. In the bovine system the conjugate contains approximately 4% phosphorous by weight, consistent with a nearly completely phosphorylated phosphophoryn. It is equally clear that more than one conjugate is present in most preparations. At this time it is not known whether these are separate entities or if the heterogeneity is the result of multiple forms of the same noncollagenous protein.

Biosynthetic studies indicate that the conjugates are not formed in the odon-

Table 9.8. Amino Acid Composition of Collagen–Noncollagenous Protein Conjugates Released from Teeth[a]

	Periodate digest[b]		Human molars		Bovine molars		Rat incisors
	Fraction III	Fraction IV	Collagenase[c]	CNBr[d]	CNBr[e]	CNBr[f]	CNBr[g]
4-Hyp	12	31	—	84	38	63	21
Asp	339	239	138	86	228	137	189
Thr	15	20	29	21	16	20	27
Ser + Pser	408	274	77	64	282	155	289
Glu	29	47	133	77	51	65	99
Pro	17	62	90	96	53	86	55
Gly	68	135	207	286	151	229	113
Ala	23	49	67	110	54	76	55
Val	4	19	28	25	13	24	26
1/2 Cys	ND	ND	ND	ND	ND	ND	ND
Met	3	1	—	—	1	—	T
Ile	4	10	31	11	7	13	9
Leu	7	23	57	29	16	26	19
Tyr	—	1	—	6	4	7	—
Phe	3	11	24	15	8	13	10
Hyl	4	2	17	7	4	6	1
Lys	53	48	26	24	34	33	24
His	4	5	31	7	6	8	22
Arg	8	23	17	47	25	38	22

CNBr = cyanogen bromide; — = absent; ND = not determined; T = trace.
[a]Residues per 1000 amino acid residues.
[b]Carmichael et al. 1971.
[c]Leaver et al. 1975b.
[d]Takagi and Veis 1981.
[e]Curley-Joseph and Veis 1979.
[f]Fujisawa et al. 1981.
[g]Maier et al. 1981.

toblast, but that conjugation of collagen and phosphophoryn is an event related to mineralization of the predentin (Fujisawa et al. 1981). The role proposed for the conjugate will be discussed below.

9.4. Mineral–Matrix Relationships

Detailed discussion of the mineral phases of bone, dentin, and enamel is not possible in this chapter. However, some knowledge of the mineral phase and its dynamics is essential to understanding the mechanism of mineral crystal placement within the matrix. This problem—the regulation of the ordered intercalation of crystalline material into the collagen matrix to produce bone and dentin from osteoid and predentin—is the most intriguing aspect of mineralized tissue biology.

9.4.1. Nature of the Mineral Phase

Calcium phosphates can exist in aqueous systems in a variety of crystalline forms. These range from $CaHPO_4 \cdot 2H_2O$, dicalcium phosphate dihydrate (DCPD), with a Ca : P molar ratio of 1.0 to $Ca_5(PO_4)_3OH$, calcium hydroxyapatite (HAP) with a Ca : P ratio 1.66. Dicalcium phosphate dihydrate is quite soluble at physiologic pH, ionic strength, and temperature, and serum and extracellular fluids are undersaturated with respect to it. All other forms of calcium phosphate—amorphous calcium phosphate (ACP); octacalcium phosphate, $Ca_4 H(PO_4)_3$ (OCP); and tricalcium phosphate, $Ca_3(PO_4)_2$ (TCP)—are unstable with respect to serum levels of calcium and phosphate. Normal human serum has a total calcium concentration of 2.5 m*M*. Of this, about 40% is electrostatically complexed as part of the shielding counterions of serum macromolecules and an additional small quantity can be thought of as complexed with citrate, bicarbonate, inorganic phosphate, and pyrophosphate ions. The free calcium ion concentration, important in establishing the equilibrium binding to all forms, is very close to 1.25 m*M* Ca^{2+}. In fact, all animal species have free calcium concentrations close to this value. The phosphate levels are more variable, as less is involved with protein binding. However, at physiologic pH all three forms of phosphate are present in these percentages: (H_2PO_4), 20%; $(HPO_4)^{2-}$, 80%; and $(PO_4)^{3-}$, less than 0.01%. As the pH and temperature are constant, these forms are generally combined under the term phosphate, which totals 1 m*M*. Thus, the serum ion product (free Ca^{2+}) (phosphate) = 1.3 mM^2 under physiologic conditions. This value is sensitive to pH, ionic strength, and temperature but to pH in particular. In in vitro systems, OCP, TCP, and HAP all have ion products smaller than this value. Furthermore, as HAP has the lowest ion product, HAP is the thermodynamically stable form and all recrystallization is in this direction. The observed states in which crystals of OCP or TCP are present are quasiequilibrium systems controlled by various kinetic parameters.

The solid states of the calcium phosphates are not unrelated in the sense that a crystal face of one form may be equivalent to one face of a second form. For example, on one face OCP has phosphate ion and calcium ion spacings and dispositions equivalent to a HAP face. Thus, a crystal of OCP could seed the growth of HAP under appropriate conditions.

Although it is the thermodynamically most stable form, HAP cannot grow spontaneously from free calcium and phosphate because of the complexity of the unit cell. The minimum unit cell contains 18 atoms $[Ca_{10}(PO_4)_6 (OH)_2]$. Simpler, albeit less stable forms of calcium phosphate, such as ACP or OCP, require the initial organization of fewer ions and have a higher probability of forming nuclei for further growth. The kinetics of growth and transformation from one crystalline form to another are dependent upon the solution conditions and these provide the potential for the required regulation of HAP formation noted above.

Another key property of the calcium phosphates is their complexity. The phosphate ions, with diameters on the order of 0.4 nm, occupy large spaces over which the charge can be diffused. Thus carbonate and bicarbonate can be substituted. Similarly, Mg^{2+} and other cations can replace Ca^{2+} in the lattice, as F^- and other halogens can replace OH^-. The incorporation of such ions at the surface of a growing crystal may alter growth kinetics.

Enamel crystals are unique in their large dimensions. In bone and dentin, HAP is microcrystalline, the crystallites being 2 to 4 nm in width and about 30 to 40 nm in length. The surface area of HAP in bone and dentin is thus enormous. Surface binding of proteins and uptake of other ions must be the usual state.

Inhibitors of recrystallization can readily act if they gain access to the crystal surface. Because of the mixed crystalline nature and inclusion of other ions, the Ca : P ratio rarely matches 1.66 in natural apatites. Even in the large crystals of human enamel, the Ca : P ratio does not exceed 1.62.

9.4.2. Mineral Placement

The mineral phase of bone, although the major component on a weight basis, occupies less than 45% of the bone space on a volume basis (Table 9.1). Because the crystals are small and discrete units, it is obvious that it is the organic phase that is continuous and provides form and structure to the tissue. In the earliest stages, mineral deposition takes place within the collagen matrix. Katz and Li (1973) analyzed this situation in terms of the spaces between collagen molecules in D-staggered arrays (Chapter 1). Figure 9.17 defines their model. The majority of the intermolecular space within a collagen fibril is occupied by the hole regions rather than by the pores. However, the pores are crucial. The average pore diameter in a bone collagen fibril is approximately 0.6 nm, whereas in rat tail tendon fibrils, the critical pore dimension is approximately 0.3 nm. Because phosphate ions have diameters on the order of 0.4 nm, such ions would be prevented from diffusing within soft-tissue fibrils and thus, tendon could not mineralize. The measurements of density were made on demineralized bone matrix, however. Thus, the Katz–Li model may depict a situation resulting from prior mineralization rather than one regulating mineral placement. Analysis of structural rearrangements in predentin or osteoid upon mineralization is needed to verify this approach.

The turkey tendon system offers the most promise for resolution of the matrix–mineral relationship. The tibia tendon of the young turkey is a normal tendon, predominately type I collagen. As the turkey grows, a critical point is reached at about 17 weeks, when the tendon begins to mineralize. Mineralization begins at the extremities of the tendon and a mineralization front advances with

Figure 9.17. The Katz–Li model for mineralized tissue, defining the hole space and pore space volumes based on a hexagonal packing model.

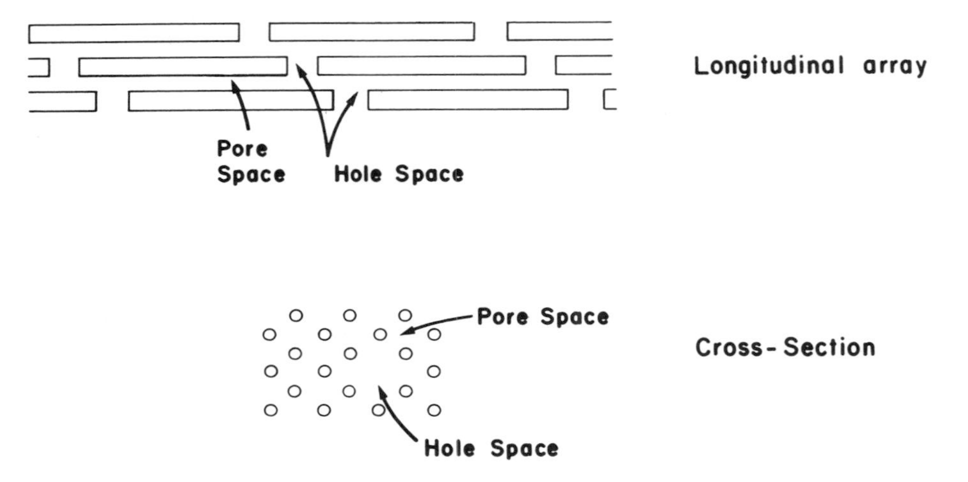

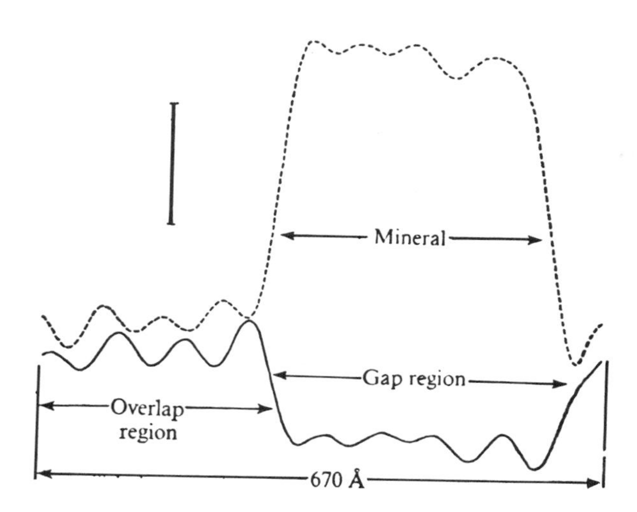

Figure 9.18. The electron distribution along a collagen 700 Å D-period during advancement of the mineralization front in a mineralizing turkey tendon. The mineral deposits in the hole region *(gap region),* as indicated by the increase in density through this region. Solid line = uncalcified tendon. Dashed line = calcified tendon. The vertical bar represents a height of 50 electrons per residue translation. *(Source: White et al. (1977), with permission.)*

a sharply defined boundary to the center of the tendon. X-ray diffraction and electron microscopic studies indicate that the mineralization is typical of bone in that discrete microcrystals are produced and, in the early stages, the crystals are entirely within the collagen fibrils. Initial electron micrographs led several investigators, particularly Glimcher (1976), to postulate that the crystals are deposited in the hole regions between D-staggered molecules. This postulate has been the subject of debate, principally on two grounds. First, crystals larger than a single hole region have been observed frequently. Second, many observers claim to see crystals outside the fibers at a very early stage in mineralization.

The most convincing work on this subject is that of White et al. (1977). Neutron scattering studies were made of tendons in regions bordering the advancing mineralization front. Water molecules, calcium phosphate, and protein all have different neutron scattering cross-sections. By contrast matching techniques one can determine the density of collagen peptide material, water, and mineral along the fibril axis. As shown in Figure 9.18, mineral accumulates in the gap or hole regions first. Moreover, it appears as if the crystals grow from an initiation site at one end of the hole. Growth is such, from other data, that the crystal c-axis is directly along the collagen fibril axis. The needlelike microcrystals of HAP thus exhibit the same 67 nm repeat periodicity as the collagen fibers in which they are deposited. That is, there is an apparent epitaxial relationship between the organic and mineral phases of bone and dentin. These two very different materials are organized so that one crystal axis of the mineral phase is directed to lie along the fiber axis of the collagen phase. Hydroxyapatite crystals cannot be induced to form an ordered array of this periodicity in the absence of collagen. As there is about 50% more mineral than collagen by volume in mature bone (Table 9.1), crystals must grow into the interfibrillar spaces at later times.

As in a composite like fiberglass, where crystallized regions are embedded discretely within a continuous polymer network, the presence of crystals and matrix creates the unique mechanical properties of bone. Most notably, bone has a high tensile as well as compressive strength and resistance to shearing forces. Bone is resilient. In sharp contrast, enamel with its very large crystals and very small matrix is brittle and has only high compressive strength. These uniquely large crystals are somehow produced by active ameloblasts and it may well be that the amelogenins or enamelins have, before their degradation, a role comparable to that of collagen. However, that problem has hardly been touched conceptually and is experimentally inaccessible at this time.

9.5. Matrix Elaboration

The features of collagen and proteoglycan biosynthesis have been discussed in detail in Chapters 3 and 8. Aside from specific aspects such as the higher level of hydroxylation of lysine and the prevalence of monosaccharide rather than disaccharide, discussed earlier, there are no biosynthetic processes unique to bone matrix collagen that need to be considered. Interest here is focused on the next steps, the secretory events that define the rates of exocytosis of the individual components and the extracellular loci for delivery and assembly of the matrix. Three techniques have been used to examine these events: ultrastructural, biochemical and, most recently, organ culture.

9.5.1. Ultrastructural Analysis

C.P. Leblond and his colleagues pioneered in the direct radiographic visualization of the incorporation of specific radioactive components into bone, dentin, and enamel. In essence, all of the experiments have been carried out in the same way. Rats or mice engaged in active tissue mineralization were given an intravenous or intracardiac radioisotope injection. The animals were then killed at selected intervals by intracardiac perfusion with a glutaraldehyde or formaldehyde solution that fixed the tissue in situ. Subsequently, the organs of interest were dissected and fixed in vitro, and then demineralized in neutral EDTA. The demineralized specimens were sectioned and stained for light or electron microscopy, and radioautography. Ultimately the radioautographs developed (often 270 days or more were required) and emulsion silver grains were matched with the positively stained tissue morphology. Following the early work of Carneiro and Leblond (1959), complete and now classic studies of dentin (Weinstock M and Leblond 1973, 1974) and enamel (Weinstock A and Leblond 1971) were carried out. Similar studies of bone, equally beautiful but not yet so complete (Weinstock M 1975, 1979) are in progress.

Consider dentin first. Within 2 minutes after introduction into the circulation, ^{3}H-proline label can be seen over the rough endoplasmic reticulum of the odontoblast. Label moves into the Golgi within 10 minutes and is seen diffusely on the secretory portion of the cells within 20 minutes. By 30 minutes, label appears in the predentin near the cell terminal web. The predentin is heavily labeled throughout at 4 hours, but very little radioactivity is seen within the mineralized phase. At 30 hours most of the label has been overtaken by the mineral front and the newly formed collagen lies well within the dentin. These events are depicted in Figure 9.19. In marked contrast ^{33}P-labeled teeth showed matrix-bound organic phosphate moving rapidly via the Golgi, into predentin within 30 minutes and accumulating at the mineralization front within 90 minutes. Labeling with ^{3}H-serine, which should act as a marker for both collagen and phosphophoryns, showed a displacement of labeling from cells to dentin at the

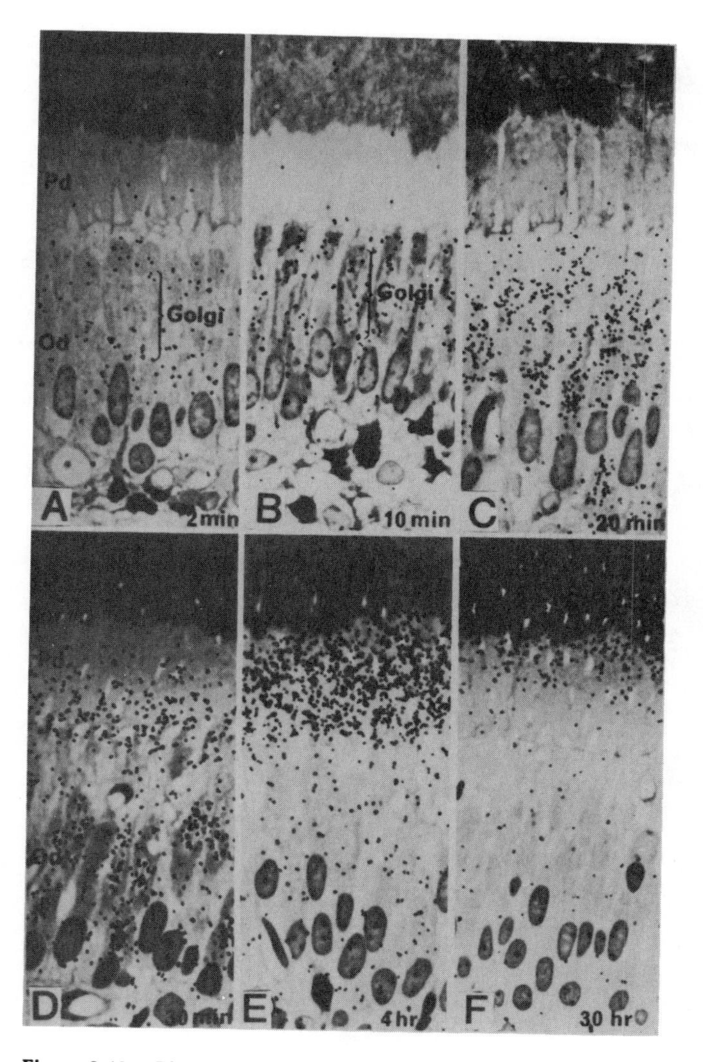

Figure 9.19. The movement of ^{3}H-proline through rat incisor odontoblasts and into the extracellular matrix as a function of time after administration of a single ^{3}H-proline injection. The dark grains represent regions of radiolabeling in the substrate, in this case, presumably collagen. *(Source: Weinstock M, Leblond (1974), with permission.)*

same rate as collagen, plus an additional independent wave of label clearly accumulating in the region of the mineralization front (Figure 9.20). These data make it abundantly clear that collagen and the phosphophoryns are not secreted at the same rate and do not follow the same secretory pathway. That is, in the rats studied, phosphate containing phosphophoryn appeared at the region of the mineralization front several hours earlier than collagen molecules synthesized at the same time. Moreover, the lack of a wave of ^{33}P-label within the predentin at any time indicates that phosphophoryns are not deposited in the predentin at the region of the terminal web. Rather they appear to be inserted directly at the mineralization front.

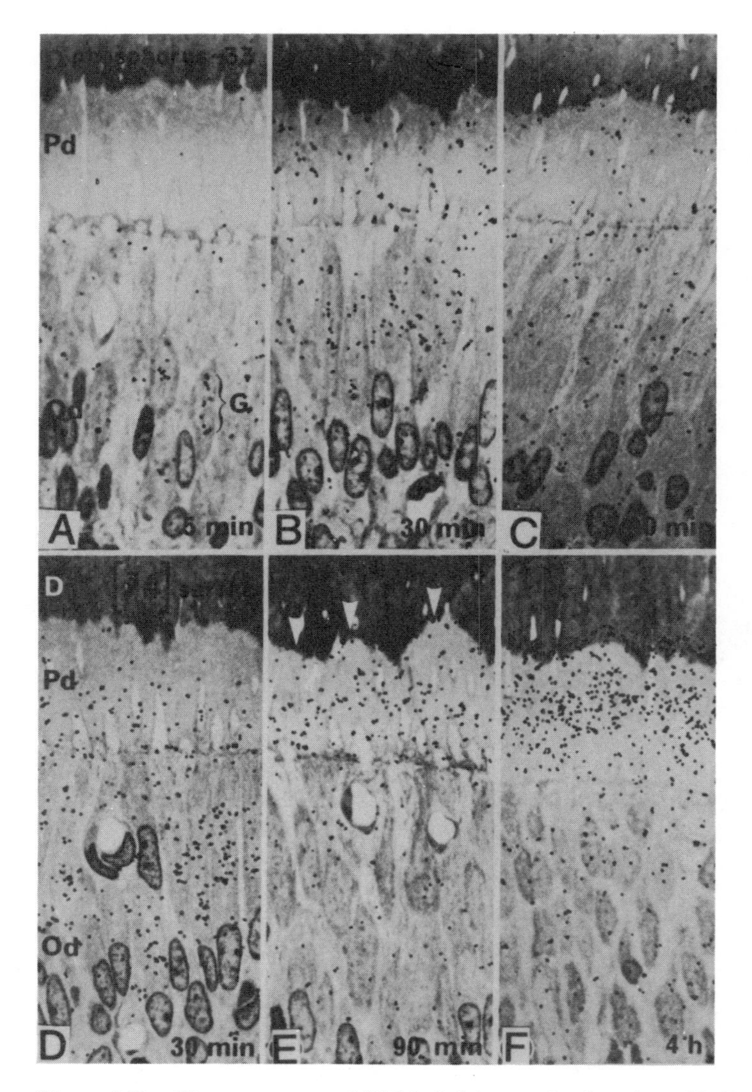

Figure 9.20. The movement of ^{33}P-labeled inorganic phosphate *(A, B, C)* and ^{3}H-serine labeled proteins through rat incisor odontoblasts as a function of time following a single injection of each substance. *(Source: Weinstock M, Leblond (1973), with permission.)*

The secretory processing of collagen in osteoblasts is equivalent to that in odontoblasts. On the same time scale, ^{3}H-proline label proceeds from the rough endoplasmic reticulum to the Golgi and finally, within 90 minutes can be observed in osteoid or prebone. After 4 hours, the prebone is heavily labeled, whereas the mineralized bone is not labeled. The situation is reversed within 24 hours, the bone being labeled and the prebone label-free (Figure 9.21) (Weinstock 1979).

Bone phosphoproteins do not have either uniquely high serine or phosphate levels and attempts to use these components as markers of noncollagenous protein secretion have failed. However, acting on the assumption that some

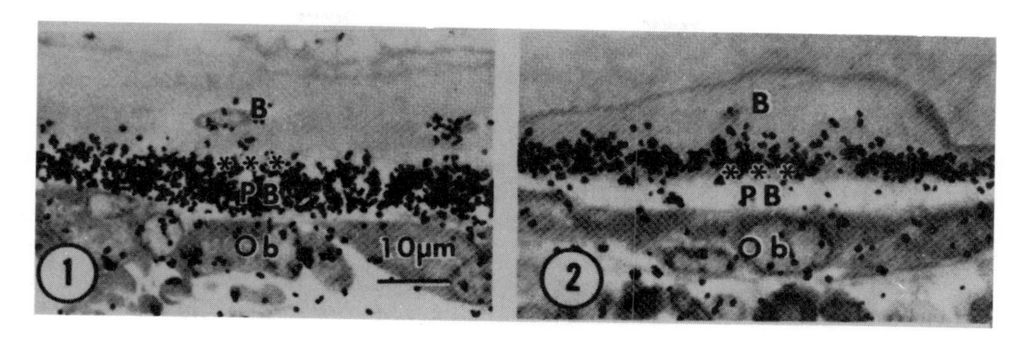

Figure 9.21. Radioautographic localization of ^{3}H-proline over prebone at 4 hours (*1*) and over bone at 24 hours (*2*) following a single injection of ^{3}H-proline into mice. *Ob* = osteoblast; *PB* = prebone; *B* = bone. *(Source: Cho, Garant (1981), with permission.)*

Figure 9.22. Radioautographic localization of ^{3}H-colchicine in mouse bone 10 minutes after injection (*3*) and in mouse incisor dentin at 10 minutes (*7*) and 50 minutes (*8*). *Ob* = osteoblast; *PB* = prebone; *B* = bone. *D* = dentin; *PD* = predentin; *Od* = odontoblast. *(Source: Cho, Garant. (1981), with permission.)*

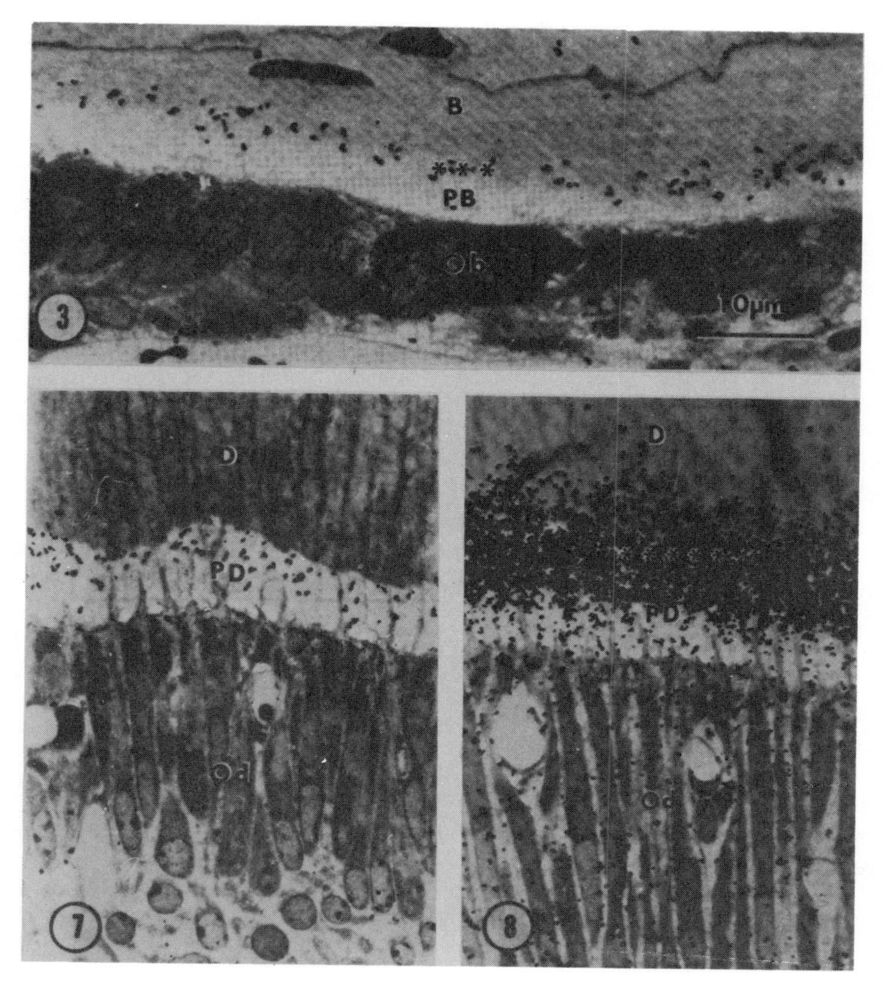

bone glycoproteins might contain fucose as one of the terminal residues of oligosaccharide chains, Weinstock (1979) used ^{3}H-fucose as a marker for noncollagenous protein secretion. The ^{3}H-fucose labeled substances behaved in a fashion similar to collagen. They accumulated primarily in the prebone region within 4 hours, but a distinct band of silver grains appeared in the radioautographs directly over the bone–prebone junction, again appearing there more rapidly than the collagen. Although bone and dentin are different in noncollagenous protein composition, on the basis of these observations it appears that the molecular events surrounding advancement of the mineralization front utilize collagen and noncollagenous protein in the same way.

Cho and Garant (1981) very elegantly demonstrated another crucial facet of the dynamics of mineralization. The extracellular matrix in dentin can be reached directly from the circulatory system. Cells do not have to mediate transport. In an experiment similar in design to those described above,^{3}H-colchicine was injected into mice, which were then killed only 10 minutes later—insufficient time for any of the observed cellular secretory pathways to have taken in and cleared the colchicine. Yet, as shown in Figure 9.22, within 10 minutes ^{3}H-colchicine accumulated specifically at the mineralization front. Possibly α-2-HS-glycoprotein could be incorporated in bone via the same noncell mediated route. The mechanism of this means of entry is not understood.

Ameloblasts secrete their matrix in the same fashion. For example, ^{3}H-fucose can be seen in the cell and as labeled protein in the growth regions of the enamel matrix adjacent to the apical process. Secretion is rapid, with extracellular components appearing within 20 to 30 minutes after introduction of the label, after following an intracellular route.

9.5.2. Biochemical Analysis

The revealing observations described above have one drawback. One sees localized silver grains developed by an underlying radionuclide, yet there is no direct way of knowing to what molecule that nuclide belongs. The alternative is to take the direct approach of labeling rats, isolating the incisor collagen and phosphophoryn (Dimuzio and Veis 1978) and phosphophoryn–collagen conjugate (Maier et al. 1981), and determining the specific activities of ^{3}H-serine and ^{3}H-proline. The resultant data show unequivocally that the collagen and soluble phosphophoryn components are handled differently by the odontoblasts. Secretion times for rat incisor α-phosphophoryn are shorter than for collagen. Hence they must be included in different sets of secretory granules. The rough endoplasmic reticulum that buds off to form the Golgi vesicles carries out a variety of posttranslational reactions, but it does not, in this case, mix different proteins.

Biochemical data from several sources agree strongly that predentin is relatively low in phosphophoryn content compared with dentin. Thus, as the ultrastructural studies also imply, the phosphophoryns must be secreted in the region of the mineralization front.

All of the phosphophoryn produced is not incorporated into the dentin. Hence, there must be, as suggested earlier, an active protein degradation system associated with the mineralization front. Whether this degradation is enzymic or hydroxyapatite surface-catalyzed is not known. Odontoblasts do not appear to be engaged in tissue removal. They have none of the characteristics of specialized resorptive cells such as osteoclasts. The pathway for removal of degraded extracellular matrix components has been obscure. However, as there is an apparent direct interchange between matrix and serum, degraded components may be removed via that same route. By analogy, it seems reasonable to assume that in both bone and enamel, a major portion of the degraded proteins produced during active matrix deposition are rapidly removed from the tissue by direct exchange with serum components.

9.6. Bone Remodeling

Remodeling is a cell-mediated process unique, among the mineralized tissues in vertebrates, to bone. Enamel and dentin may be eroded and degraded but specialized cells are not involved and the degraded tissue is not replaced. The factors regulating osteoclastic resorption or stimulating osteoblasts are beyond the scope of this chapter, but this subject does embrace many interesting cellular phenomena. Bone physiology is a complex but well-studied field of great practical value. For the purpose of this discussion, consider the problem faced by the osteoclast in removing bone. It has to dissolve the mineralized matrix, pumping calcium and phosphate into a medium already saturated with these ions; then it has to degrade and remove the collagen matrix.

The localized ruffled border regions of osteoclasts are probably regions of low pH. The osteoclasts contain relatively low levels of isocitric dehydrogenase activity and citrate accumulates within these cells. In addition, osteoclasts are rich in carbonic anhydrase activity, converting carbon dioxide to carbonic acid. How these acids reach the extracellular matrix—if they do—is unclear, but both citrate and carbonic acid are concentrated in the ruffled border region. Carbonic acid formation probably generates the localized low pH, enhancing hydroxyapatite dissolution. Citrate may complex with calcium and carry the complexed Ca^{2+} to the plasma. Citrate increases in plasma during periods of enhanced resorptive activity.

The action of collagenases in bone is probably similar to that in all other tissues, as considered in Chapter 4. It is unnecessary to comment further, except to say that collagenase activity has been demonstrated in bone. Other neutral proteinase activities are also prominent. Lysosomal enzyme activity may also be stimulated by the heightened carbonic acid production. Acid phosphatase activity is particularly high in bone and is stimulated at periods of high bone resorption rates. At least two acid phosphatase activities, probably isozymes, have been detected: acid β-glycerolphosphatase and acid phenylphosphatase. Attempts to assess the pathways of movement of degraded matrix by ultrastructural, autoradiographic methods have not been nearly as successful as the studies cited earlier relating to matrix formation. However, osteoclasts can be seen with membrane-enclosed mineral crystal fragments and collagen fiber fragment inclusions, indicating that while degradation begins in the extracellular space, removal of the matrix is distinctly cell mediated.

9.7. Summary and General Concepts

This discussion has focused primarily on the components of the extracellular matrix of bone, dentin, and enamel and the manner in which these components interact to form the matrix. What emerges is the concept that, although different in detail, each of these tissues utilizes the same strategy in mineralization.

Taking the dentin system as our model, Figure 9.23 depicts in schematic form the components and their flow through the system. The important points may be summarized as follows:

1. Actively mineralizing teeth are in direct communication with the components of plasma in two ways. The odontoblast can take up amino acids, Ca^{2+}, phosphate and many other constituents (natural or otherwise) within a few minutes after such substances appear in the plasma. Selected blood constituents (natural or otherwise) can move rapidly directly into the extracellular matrix. There are, therefore, two pathways of communication between blood and dentin, one cell mediated, the other bypassing the odontoblast.
2. The extracellular matrix components synthesized in, or traversing, the cell-mediated pathway are not necessarily treated by the cell in the same way and are not necessarily deposited at the same sites within the matrix. Collagen, proteoglycans, and glycoproteins appear to be deposited in the region of the odontoblast terminal web at the cell–predentin boundary. The phosphophoryns and perhaps alkaline phosphatases, activated proteinases, gly-

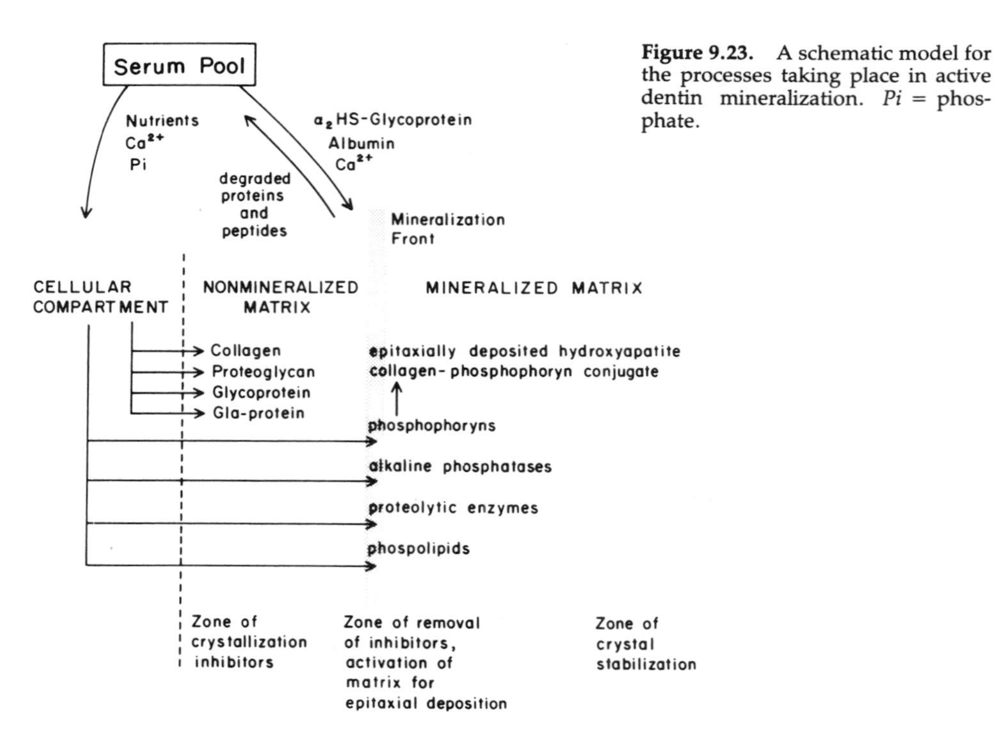

Figure 9.23. A schematic model for the processes taking place in active dentin mineralization. *Pi* = phosphate.

cosidases, and free phospholipids all move directly to, or appear in active form at, the region of the mineralization front.

3. The predentin can be considered as a zone inhibitory to precipitation of calcium phosphates. Collagen maturation, that is, stabilization by crosslinking, is completed rapidly in this region.
4. The mineralization front itself is a zone of both degradative and synthetic activity. Matrix proteinases and glycosidases degrade the inhibitory components, and turnover of some of the phosphophoryns by both dephosphorylation and chain degradation may take place, creating locally higher concentrations of phosphate. The products of these degradative steps may either be removed via the direct, noncell-mediated pathway or incorporated into the newly formed mineralized dentin. At the same time, phosphophoryns and other calcium-binding proteins associate with specific regions of the collagen and, in a complex fashion, bind Ca^{2+} as well. Binding data suggest a further conformational change in the phosphophoryn leading to a structure capable of nucleating growth of apatite in the crystal c-axis dimension, seeding epitaxial apatite deposition. A portion of the phosphophoryn may be covalently linked to the collagen fibers at this time.
5. Crystal growth inhibitors limit the size of hydroxyapatite crystals, possibly by surface binding. Surface-bound growth inhibitors also stabilize the mineralized dentin against direct exchange and solubilization by transitory variations in plasma composition.

This model, though tentative and incomplete in some aspects, provides for the orderly formation of a stabilized collagen fiber network, the priming of that network to accept apatite crystals, the directed initial growth of the crystals within or upon the fiber matrix, the inhibition of crystal growth, and ultimately the stabilization of the apatite surface. A similar set of processes occur in bone, utilizing a similar but not identical set of crystallization inhibitors, epitaxial activator proteins, and crystal growth and crystal stabilizers. Teleologically one may argue that less highly phosphorylated acidic proteins rather than acid-insoluble phosphophoryns are utilized in bone because bone must be responsive to remodeling.

It is more difficult to describe and understand enamel formation. From this modeling approach one might say that the amelogenins have the role of forming the inhibitory preenamel zone, whereas the calcium-binding enamelins form the mineral crystal directing phase. After degradation of the amelogenins there are no crystal growth inhibitors to limit crystal size or regulate solubility. In the erupted tooth the exogenous salivary phosphoprotein, statherin, may have this function.

Bone remodeling systems (Figure 9.24) have as their principal feature the highly localized osteoclast-mediated removal of matrix. The pumping of protons and citrate into the bone is an active, energy requiring process and the osteoclast is sensitively poised to carry out this activity under hormonal control.

This discussion has gone on at some length and in detail, yet the questions of regulation of matrix activities by parathyroid hormone, calcitonin, vitamin D metabolites, intestinal absorption of calcium, renal retention of calcium and phosphate, and so on have hardly been mentioned. These topics are indeed a

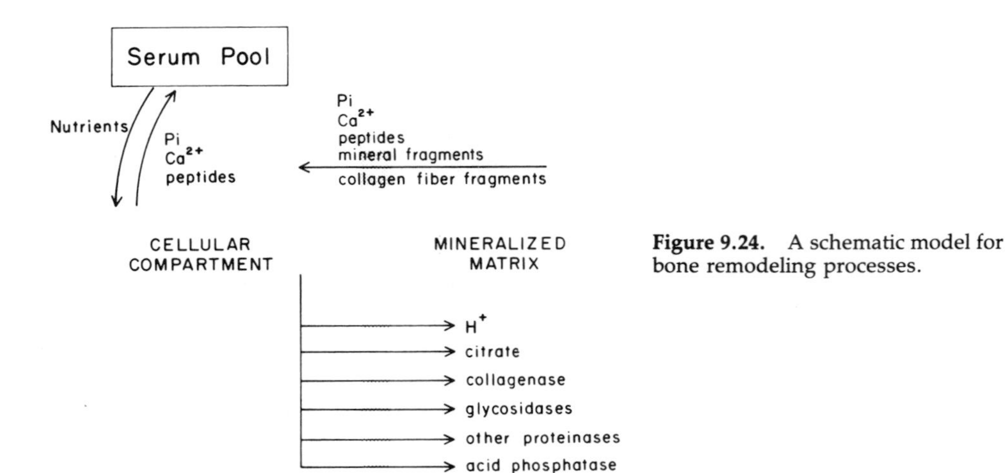

Figure 9.24. A schematic model for bone remodeling processes.

part of matrix biochemistry. The literature is vast, however, and many excellent reviews are available. Here, the extracellular components have been described at the molecular level and the system of interactions leading to the formation of the tissue has been discussed. It should be obvious that very little is known other than the uniquely acidic nature of the several constituents. The mode of tissue formation outlined in Figure 9.23 is hypothetical. The task of the next years is the more thorough delineation of this system.

References

Boskey AL: (1981) Current concepts of the physiology and biochemistry of calcification. Clin Orthop 157:227–257.

Boskey AL, Posner AS: (1976) Extraction of a calcium-phospholipid-phosphate complex from bone. Calcif Tissue Res 19:273–283.

Butler WT, Bhown M, Dimuzio MT, Linde A: (1981a) Noncollagenous proteins of dentin. Isolation and partial characterization of rat dentin proteins and proteoglycans using a three-step preparative method. Coll Res 1:187–199.

Butler WT, Bhown M, Thomana M, Fretwell B, Schrohenloher RE: (1981b) Characterization of a unique dentin glycoprotein. In: Veis A, ed. The Chemistry and Biology of Mineralized Connective Tissues. New York: Elsevier/North-Holland, 399–402.

Butler WT, Finch JE, DeSteno CV: (1972) Chemical character of proteins of rat incisors. Biochim Biophys Acta 257:161–171.

Carmichael DJ, Veis A, Wang ET: (1971) Dentin matrix collagen: Evidence for a covalently linked phosphoprotein attachment. Calcif Tissue Res 7:331–344.

Carneiro J, Leblond CP: (1959) Role of osteoblasts and odontoblasts in secreting the collagen of bone and dentin as shown by radioautography in mice given tritium labeled glycine. Exp Cell Res 18:291–300.

Cho MI, Garant PR: (1981) Autoradiographic demonstration of colchicine binding at the mineralization front of bone and dentin. In: Veis A, ed. Chemistry and Biology of Mineralized Tissues. New York: Elsevier/North-Holland, 303–307.

Cohen-Solal L, Cohen-Solal M, Glimcher MJ: (1979) Identification of δ-glutamyl phosphate in the α2 chains of chicken bone collagen. Proc Natl Acad Sci USA 76:4327–4330.

Curley-Joseph J, Veis A: (1979) The nature of covalent complexes of phosphoproteins with collagen in the bovine dentin matrix. J Dent Res 58:1625–1633.

Dimuzio MT, Veis A: (1978a) Phosphophoryns—Major noncollagenous proteins of rat incisor dentin. Calcif Tissue Res 25:169–178.

Dimuzio MT, Veis A: (1978b) The biosynthesis of phosphophoryns and dentin collagen in the continuously erupting rat incisor. J Biol Chem 253:6845–6852.

Fujii K, Tanzer ML: (1974) Age-related changes in the reducible crosslinks of human tendon collagen. FEBS Letters 43:300–302.

Fujimoto D, Akiba K, Nakamura N: (1974) Isolation and characterization of a fluorescent material in bovine achilles tendon collagen. Biochem Biophys Res Commun 76:1124–1129.

Fujisawa R, Takagi T, Kuboki Y, Sasaki S: (1981) The characterization of collagen-phosphophoryn complex in bovine dentin. In: Veis A, ed. Chemistry and Biology of Mineralized Connective Tissues. New York: Elsevier/North-Holland, 483–487.

Glimcher MJ: (1976) Composition, structure, and organization of bone and other mineralized tissues and the mechanism of calcification. In: Handbook of Physiology—Endocrinology VII. Baltimore, Williams & Wilkins, 25–116.

Glimcher MJ, Katz EP: (1965) The organization of collagen in bone: The role of the noncovalent bonds in the relative insolubility of bone collagen. J Ultrastruct Res 12:705–729.

Glimcher MJ, Katz EP, Travis DF: (1965) The solubilization and reconstitution of bone collagen. J Ultrastruct Res 13:163–171.

Herring GM: (1964) Chemistry of the bone matrix. Clin Orthop 36:169–183.

Herring GM: (1968a) Studies on the protein-bound chondroitin sulfate of bovine cortical bone. Biochem J 107:41–49.

Herring GM: (1968b) The chemical structure of tendon, cartilage, dentin and bone matrix. Clin Orthop 60:261–299.

Herring GM: (1972) The organic matrix of bone. In: Bourne GH, ed. The Biochemistry and Physiology of Bone, Vol 1. New York: Academic Press, 127–189.

Herring GM, Kent PW: (1963) Some studies on the mucosubstances of bovine cortical bone. Biochem J 89:405–414.

Hiraoka BY, Fukasawa K, Fukasawa KM, Harado M: (1980) Identification and quantitation of α-amino adipic acid in bovine dentine phosphoprotein. J Biochem 88:373–377.

Jones IL, Leaver AG: (1974) Studies on the minor matrix components of the organic matrix of human dentin. Arch Oral Biol 19:371–380.

Katz EP, Li ST: (1972) The molecular packing of collagen in mineralized and non-mineralized tissues. Biochem Biophys Res Commun 46:1368–1373.

Katz EP, Li ST: (1973) The intermolecular space of reconstituted collagen fibrils. J Mol Biol 80:1–15.

Kuboki Y, Takagi T, Shimokawa H, Oguchi H, Sasaki S, Mechanic GL: (1981) Location of an intermolecular cross-link in bovine bone collagen. Connect Tissue Res 9:107–114.

Leaver AG: (1979) Noncollagenous proteins. In: Simmons DJ, Kunin AS, eds. Skeletal Research. New York: Academic Press, 193–226.

Leaver AG, Holbrook IB, Jones IL, Thomas M, Sheil L: (1975b) Components of the organic matrices of bone and dentin isolated only after digestion with collagenase. Arch Oral Biol 20:211–216.

Leaver AG, Triffitt JT, Holbrook IB: (1975a) Newer knowledge of non-collagenous protein in dentin and cortical bone matrix. Clin Orthop 110:269–292.

Lechner JH, Veis A, Sabsay B: (1981) Domain sequences in dentin phosphophoryn. In: Veis A, ed. The Chemistry and Biology of Mineralized Connective Tissues. New York: Elsevier/North-Holland, 395–398.

Lee SL, Glimcher MJ: (1981) Purification, composition, and ^{31}P-NMR spectroscopic properties of a noncollagenous phosphoprotein isolated from chicken bone matrix. Calcif Tissue Int 33:385–394.

Lee SL, Veis A, Glonek T: (1977) Dentin phosphoprotein: An extracellular calcium-binding protein. Biochemistry 16:2971–2979.

Linde A, Bhown M, Butler WT: (1980) Noncollagenous proteins of dentin. A re-examination of proteins from rat incisor dentin utilizing techniques to avoid artifacts J Biol Chem 255:5931–5942.

Maier GD, Lechner JH, Veis A: (1981) Evidence for the conjugation of collagen to the phosphophoryn of rat incisor dentin in relation to the mineralization front. In: Veis A, ed. Chemistry and Biology of Mineralized Connective Tissues. New York: Elsevier/North-Holland. 477–481.

Manning GS: (1979) Counterion binding in polyelectrolyte theory. Accts Chem Res 12:443–449.

Mueller PK, Raisch K, Matzen K, Gay S: (1977) Presence of Type III collagen in bone from a patient with osteogenesis imperfecta. Eur J Pediat 125:29–37.

Price PA, Poser JW, Raman N: (1976) Primary structure of the γ-carboxyglutamic acid-containing protein from bovine bone. Proc Natl Acad Sci USA 73:3374–3375.

Richardson WS, Munksgaard ECh, Butler WT: (1978) Rat incisor phosphoprotein. The nature of the phosphate and quantitation of the phosphoserine. J Biol Chem 253:8042–8046.

Robinson RA: (1960) Chemical analysis and electron microscopy of bone. In: Rodahl K, Nicholson JT, Brown EM, eds. Bone as a Tissue. New York: McGraw-Hill, 186–250.

Scott PG, Veis A, Mechanic G: (1976) The identity of a cyanogen bromide fragment of bovine dentin collagen containing the site of an intermolecular cross-link. Biochemistry 15:3191–3198.

Shuttleworth A, Veis A: (1972) The isolation of anionic phosphoproteins from bovine cortical bone via the periodate solubilization of bone collagen. Biochem Biophys Acta 257:414–420.

Stetler-Stevenson WG, Veis A: (1983) Bovine Dentin Phosphophoryn: Composition and Molecular Weight. Biochemistry 22:4326–4335.

Takagi Y, Veis A: (1981) Matrix protein difference between human normal and dentinogenesis imperfecta dentin. In: Veis A, ed. Chemistry and Biology of Mineralized Connective Tissues. New York: Elsevier/North-Holland, 233–243.

Takagi Y, Veis A, Sauk JJ: (1983) Relation of Mineralization Defects in Collagen Matrices to Non-collagenous Protein Components. Identification of a Molecular Defect in Dentinogenis Imperfecta. Clinical Orthopaedics 176:282–289.

Termine JD, Belcourt AB, Conn KM, Kleinman HK: (1981) Mineral and collagen-binding proteins of fetal calf bone. J Biol Chem 256:10,403–10,408.

Termine JD, Belcourt AB, Christner PJ, Conn KM, Nylen MU: (1980a) Properties of dissociatively extracted fetal tooth matrix proteins. 1. Principal molecular species in developing enamel. J Biol Chem 255:9760–9768.

Termine JD, Belcourt AB, Miyamoto MS, Conn KM: (1980b) Properties of dissociatively extracted fetal tooth matrix proteins. II. Separation and purification of fetal bovine dentin phosphoprotein. J Biol Chem 255:9769–9772.

Veis A, Schlueter RJ: (1964) Macromolecular organization of dentine matrix collagen. I. Characterization of dentine collagen. Biochemistry 3:1650–1656.

Veis A, Sharkey M, Dickson I: (1977) Non-collagenous proteins of bone and dentin extracellular matrix and their role in organized mineral deposition. In: Wasserman et al. eds. Calcium Binding Proteins and Calcium Function. New York: Elsevier/North-Holland, 409–418.

Volpin D, Veis A: (1973) Cyanogen bromide peptides from insoluble skin and dentin bovine collagens. Biochemistry 12:1452–1464.

Weinstock A, Leblond CP: (1971) Elaboration of the matrix glycoprotein of enamel by the secretory ameloblasts of the rat incisor as revealed by radioautography after galactose-^{3}H injection. J Cell Biol 51:26–51.

Weinstock M: (1975) Elaboration of precursor collagen by osteoblasts as visualized by radioautography after ^{3}H-proline administration. In: Slavkin HC, Gruelich RC, eds. Extracellular Matrix Influences on Gene Expression. New York: Academic Press, 119–128.

Weinstock M: (1979) Radioautographic visualization of ^{3}H-fucose incorporation into glycoprotein by osteoblasts and its deposition into bone matrix. Calcif Tissue Int 27:177–185.

Weinstock M, Leblond CP: (1973) Radioautographic visualization of the deposition of a phosphoprotein at the mineralization front in the dentin of the rat incisor. J Cell Biol 56:838–845.

Weinstock M, Leblond CP: (1974) Synthesis, migration and release of precursor collagen by odontoblasts as visualized by radioautography after ^{3}H-proline administration. J Cell Biol 60:92–127.

White SW, Hulmes DJS, Miller A, Timmins PA: (1977) Collagen-mineral axial relationship in calcified turkey leg tendon by x-ray and neutron diffraction. Nature 266:421–425.

10

Extracellular Matrix and Development

A.H. Reddi

10.1. Introduction

The molecular basis of development is one of the central problems of modern biology. The major phases of embryonic development include cell differentiation, formation of tissues from differentiated cells (organogenesis), and finally, the controlled growth of tissues to generate the characteristic form of the developing embryo. How does the spherical fertilized egg differentiate and develop into the beautiful human being with perfect bilateral symmetry? The progressive differentiation of specialized cells and their oganization into tissues is accompanied by the prominent appearance of extracellular connective tissue matrix. In fact, the extracellular matrix functions as the "connective tissue" that binds together the different cell types in a complex tissue or organ.

An analysis of the mechanisms of differential gene expression during development will have important implications for cancer and developmental anomalies. While considerable attention has been focused on intracellular molecules involved in the transcription and translation of the genetic information, not much is known about the role of extracellular matrix macromolecules. Growing knowledge about the chemistry and properties of extracellular matrix has set the stage for investigations of the developmental role of these extracellular constituents. The aim of this chapter is to provide a selective survey of this area and to pose questions that will stimulate the spirit of inquiry, rather than to provide comprehensive coverage.

From the National Institute of Dental Research, National Institutes of Health, Bethesda, Maryland

10.2. Evolutionary Origin and Developmental Appearance

When did the extracellular matrix first appear during the course of evolution? In all likelihood the extracellular matrix is the hallmark of multicellularity (Figure 10.1). An examination of the phylogenetic relationships among animals reveals the first definite evidence of extracellular matrix in certain sponges and in coelenterates such as *Hydra* (Borradaile et al. 1958; Reddi 1976; Garrone 1978). The mesoglea in *Hydra* represents an extracellular matrix. Among the mammals, basement membranes in glandular structures such as prostate, mammary glands, and salivary glands are prominent and consist of collagenous proteins, laminin, and certain proteoglycans. In certain tissues, such as bones and teeth, the extracellular matrix is very highly developed and accounts for the bulk of the tissue mass. As described in detail in section 10.4.1, the extracellular bone matrix is a useful choice for studies on the role of extracellular matrix in differentiation and morphogenesis of tissue in vivo. The extracellular matrix is therefore a very early and necessary feature of multicellularity among animals.

When does the extracellular connective tissue matrix first appear in the developing embryo? Among the echinoderms, extracellular matrix synthesis begins

Figure 10.1. The origin and evolution of extracellular matrix. Note the discrete appearance of extracellular matrix in multicellular metazoa. Bones and teeth abound in extracellular matrix. Basement membranes are prominent in glandular structures such as salivary or mammary glands.

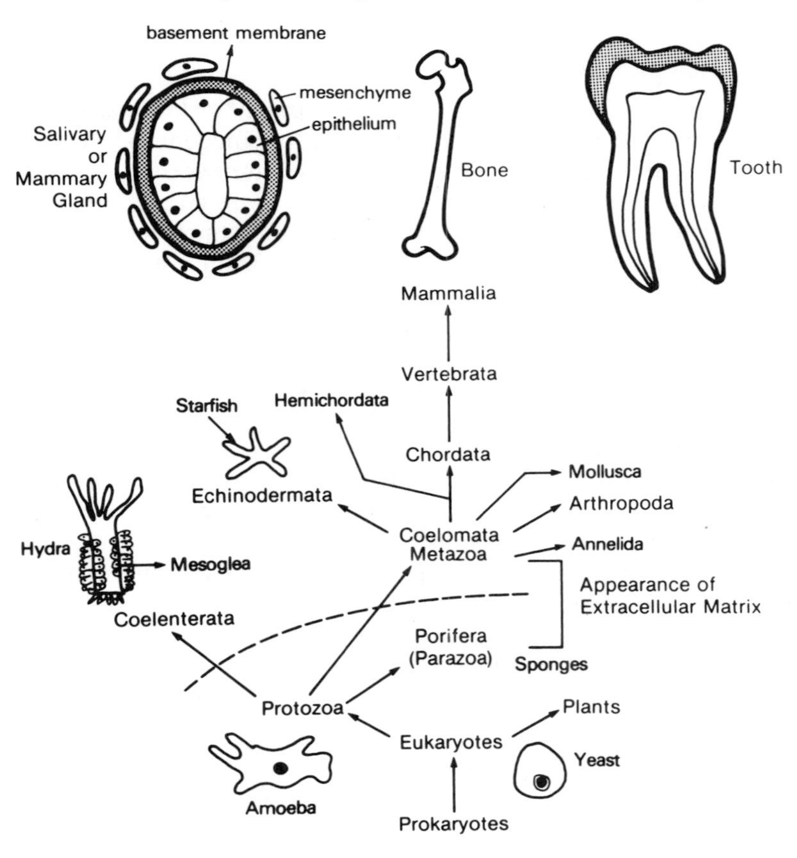

during gastrulation when endoderm and ectoderm first appear. Collagen synthesis has been demonstrated in the sea urchin *Paracentrotus lividus* during cleavage and increases several-fold during gastrulation and spicule formation (Pucci-Minafra et al. 1972; Golob et al. 1974). In two other genera of sea urchins, proteoglycan synthesis has been demonstrated during gastrulation (Kinoshita and Saiga 1979). The hyaline layer of sea urchin embryos is a collagenous matrix and it plays an important role in development (Spiegel and Spiegel 1979). Fibronectin is localized on all surfaces in the blastula and gastrula of developing sea urchin embryo (Spiegel et al. 1980). Collagen synthesis occurs in very early frog embryos (Edds 1958; Klose and Flickinger 1971). There is a pronounced increase in collagen synthesis during gastrulation in *Xenopus* (Green et al. 1968). Among chick embryos, collagen is present in the incomplete basal laminae of the epiblast as evidenced by ultrastructure (Trelstad et al. 1967) and hyaluronate synthesis occurs during early organogenesis by the ectoderm (Solursh et al. 1979). In the developing mouse embryo, collagen heterogeneity has been described by Adamson and Ayers (1979). Fibronectin synthesis in the developing chick embryo has also been reported (Critchley et al. 1979). Thus it can be concluded that extracellular matrix components play an early role in embryogenesis. There is a need for further studies on heterogeneity of collagens, proteoglycans, and glycoproteins in developing mammalian embryos, with special reference to mechanisms and function especially during gastrulation, neural crest migration, and primordial germ cell migration.

10.3. Tissue Interactions and Morphogenesis

It is well known that specific tissue interactions control morphogenesis. Much of our understanding has been derived from the study of development of salivary glands, lungs, skin, and kidney. The interacting components are epithelial and mesenchymal cells. This section will examine these interactions and describe organogenesis with special reference to extracellular matrix in several tissues.

10.3.1. Epithelial–Mesenchymal Interactions

Epithelial and mesenchymal components interact and influence each other during early phases of organogenesis (Grobstein 1967; Fleishmajer and Billingham 1968; Kratochwil 1972). The penetrating studies of Grobstein (1967) have delineated the underlying principles. His technique consisted of separating the epithelium and mesenchyme of embryonic tissue rudiments by mild trypsinization and then simply recombining the tissues or culturing the tissues after interposing a Millipore filter between the epithelium and the mesenchyme. There was no further epithelial morphogenesis when the separated epithelium and mesenchyme were cultured separately. However, in the recombinants, including those separated by a Millipore filter (20 μm thick; 1.0 μm pore size), morphogenesis was observed. Based on these transfilter epithelio–mesenchymal interactions and the histologic appearance of the interfacial material, Grobstein (1954) recognized the importance of extracellular matrix in tissue morphogenesis. The factors stimulating epithelial morphogenesis are not known, but the extracellular matrix of the mesenchyme is thought to contain the active components. For example, collagen is localized at the epithelial–mesenchymal interface by ultrastructural and autoradiographic techniques (Kallman and Grobstein 1964; 1965). Mesenchyme is the source of the collagen (Bernfield 1970). The branching morphogenesis of salivary epithelium is abolished by collagenase treatment, implying the importance of collagen in epithelial–mesenchymal interactions (Grobstein and Cohen 1965). However, the precise molecular mechanisms underlying epithelial–mesenchymal interactions are not known despite our growing knowledge about extracellular matrix biology. The difficulty is in part due to scanty amounts of matrix materials available for mechanistic investigations.

In view of the fact that epithelial–mesenchymal interactions continued in the presence of a Millipore filter, the putative messenger molecules were considered to be diffusible in early studies. However, the use of Nucleopore filters with uniform straight pores (produced by bombardment of polycarbonate by charged particle beams) have demonstrated the importance of cell contact during kidney tubulogenesis induced by spinal cord (Saxen 1971; Wartiovaara et al. 1974; Ekblom et al. 1980). In retrospect, it is likely that early studies may have overlooked the presence of cell processes in the Millipore filters as they have tortuous pores. In view of these findings, it is likely that cell–cell contact and cell–matrix interactions are crucial in epithelial–mesenchymal interactions.

10.3.2. Epithelial Collagens

Although mesenchyme is generally considered to be the source of extracellular matrix collagen (Bernfield and Wessels 1970), there is a growing realization of the importance of epithelial production of collagen. The elegant electron microscopic and biochemical studies of Hay and her collaborators (Cohen and Hay 1971; Trelstad et al. 1973; Hay, 1980) have demonstrated the secretion of collagen by neuroepithelium and corneal epithelium. It is safe to assume that both mesenchymal and epithelial cells have the potential for collagen biosynthesis and in many cases combine their capacities to produce gradients of different collagen types.

10.3.3. Organogenesis

Skeletal System. The vertebral cartilage is developmentally formed from the somite sclerotome. Chondrogenesis in the sclerotomal cells is stimulated by notochord and neural tube (Figure 10.2) and culminates in the cartilage surrounding the spinal cord in the form of a vertebra (Lash 1968; Lash and Vasan 1978). In vitro culture of scleratomal mesenchymal cells yields chondrocytes (Gordon and Lash 1974). Chondrogenesis can be quantitated by $^{35}SO_4$ incorporation into proteoglycans and is augmented by notochord, neural tube, proteoglycans, and collagens. Prior removal of the extracellular matrix from the notochord results in a decrease in the chondrogenic induction (Kosher and Lash 1975). In other experiments Kosher and Church (1975) have demonstrated the stimulatory effects of type II collagen on chondrogenesis as monitored by $^{35}SO_4$ incorporation into glycosaminoglycans. The biosynthesis of proteoglycans and their molecular size is increased by the presence of type II collagen (Lash and Vasan 1978). The induction of chondrogenesis by spinal cord in chick embryo

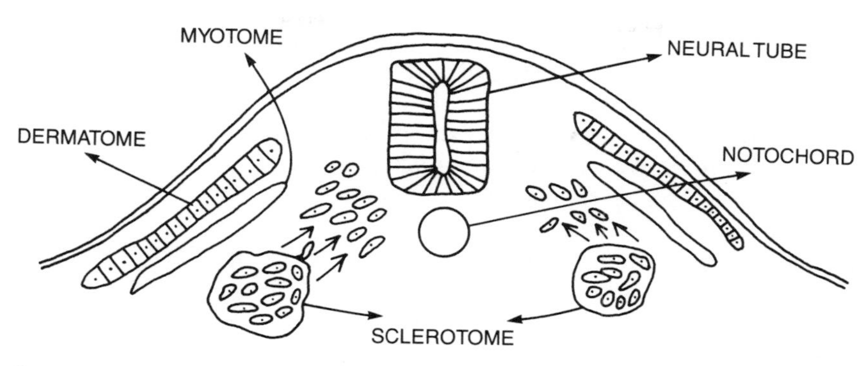

Figure 10.2. Schematic drawing to illustrate a cross-section through the axial region of the developing chick embryo. Note the direction of migration *(arrows)* of sclerotomal cells in relation to the notochord.

somites in chorioallantois cultures is abolished by collagenase and hyaluronidase together but not by either alone (O'Hare 1972). These observations imply a role for both collagens and glycosaminoglycans.

The developmental changes in the embryonic limb during endochondral bone differentiation consist of sequential transitions from mesenchymal cells to cartilage and finally bone formation. Linsenmeyer et al. (1973) have examined the changes in collagen types during this developmental sequence. Type I collagen was detected in the prechondrogenic mesenchyme. The appearance of a cartilaginous core at stage 25–26 in the chick embryo correlated with the appearance of type II collagen. With the onset of bone formation, type I collagen was detected. These biochemical observations were confirmed by indirect immunofluorescent techniques (von der Mark et al. 1976). Similar transitions have been detected during matrix-induced bone development (*See* section 10.4.1, below).

Muscle. It is well known that differentiation of chick myoblasts to myotubes in vitro is promoted by the addition of conditioned medium derived from culture of fibroblasts. The studies of Hauschka and Konigsberg (1966) demonstrated that rat tail tendon collagen can substitute for the conditioned medium. It is likely that the source of collagen in the developing muscle in vivo is the fibroblastic cells. The molecular specificity of the collagen-dependent myotube differentiation was explored by Hauschka and White (1972). Both α1(I) and α2(I) chains of rat skin collagen and cyanogen bromide fragments α1-CB7 and α1-CB8 support muscle differentiation. However, α1(I)-CB3 and α1(I)-CB6 are ineffective in this regard. Although the major collagen in muscle is type I, other types (II, III, and IV) are equally effective in promoting myoblast fusion (Ketley et al. 1976). Myotube formation from myoblasts in vitro is inhibited by analogs of proline such as *cis*-4-hydroxy-L-proline (de la Haba and Bricker 1981) and can be reversed by addition of collagen or gelatin. In view of the known affinity of fibronectin for gelatin (*See* Chapter 8), the role of this glycoprotein in muscle cell attachment and myogenesis has been investigated. Chiquet et al. (1979) have demonstrated a fibronectin-dependent attachment of chicken myoblasts to gelatin-coated substratum. Serum depleted of fibronectin by gelatin-affinity chromatography has low attachment-promoting activity. Using the myogenic cell line L-8, changes in the pattern of immunofluorescence of fibronectin were examined and Chen (1977) found a fibrillar staining pattern of fibronectin on the cell surface upon myoblast fusion. There was a change in the pattern to clusters and a quantitative reduction in fibronectin was found.

With the growing knowledge about collagen heterogeneity, the role of various collagen types has been examined in developing chick muscle. The presence of collagen types I, III, and V was established (Bailey et al. 1979). Immunofluorescence localization studies have demonstrated the presence of type I mainly in epimysium and perimysium, type III in perimysium, and type V in the endomysium. The experiments of Lipton (1977) suggest that cell–cell interaction between fibroblasts and myoblasts yields basement membrane around the myofibers. Future studies may provide the molecular and cellular basis of pattern formation in muscle with special reference to extracellular matrix (*See* Section 10.7) and may have implications for pathogenesis of muscular dystrophy (Duance et al. 1980).

Tooth (See also Chapter 6*).* The development of a tooth begins with an epithelial cell growth from the basal layer of the oral epithelium. The dental papilla, a condensation of mesenchymal cells, is in close proximity to the epithelium. The intricate epithelial–mesenchymal interactions lead to tooth morphogenesis (Slavkin 1972; Kollar 1981; Thesleff and Hurmerinta 1981). The mesenchymal odontoblasts and epithelial ameloblasts secrete and form dentin and enamel matrices respectively. The extracellular matrix and certain RNA-containing matrix vesicles have been implicated in tissue interactions (Slavkin 1972; Slavkin and Greulich 1975). Kollar and Baird (1969) have demonstrated in an elegant series of experiments the cardinal role of the mesenchymal tissue, dental papilla, in determining the final shape of the incisors or molars. The tooth germs from 13 to 16-day-old embryonic mice were separated into their component epithelium and mesenchyme by trypsin treatment and cultured as recombinants or in isolation. The dental papilla had a profound influence of the form of the developing tooth germ. When incisor papilla was cultured with molar enamel epithelium, the resulting tooth germ was in the shape of an incisor. In the opposite situation the tooth germ assumed the shape of a molar.

The biosynthesis and immunofluorescent localization of matrix molecules such as the collagens, fibronectin, and laminin have been examined (Thesleff et al. 1979; Lesot et al. 1981b; *See* Chapters 1–9). Types I, III, and IV collagen, fibronectin, and laminin have been localized in the epithelial–mesenchymal junction of the tooth. Mesenchyme is the source of fibronectin in the basal lamina and is thought to be important in the formation of the basement membrane (Brownell et al. 1981).

Skin. Tissue interactions also play an important role in skin morphogenesis. The epidermal epithelium and mesenchymal dermis interact to give rise to the skin and the resultant patterns of the cutaneous appendages such as scales in reptiles, feathers in birds, and hair in mammals (Sengel 1975; Dhouailly, 1977; Goetinck 1980). However, as frozen-killed dermis is sufficient for epidermal differentiation (Dodson 1963), living dermal cells are not required. Collagen gels or frozen-thawed dermis are effective substrates for epidermal cell polarity and proliferation (Wessells 1964). Feather morphogenesis in embryonic chick skin is dependent on the birefringent collagenous lattice in the dermis (Stuart et al. 1972) and is abolished by pretreatment of the dermis with collagenase. The collagenous lattice determines the pattern of distribution of the dermal papillae. Scaleless mutant chick skin lacks feathers and the fibrous lattice in the dermis is apparently absent (Goetinck and Sekellick 1972). Since the rate of collagen synthesis is unaltered, either the architecture or some other component is at fault. Further experiments have revealed that in the scaleless mutant the epidermis is unresponsive to the dermis (Brotman 1977). Sawyer (1979) has demonstrated the possible role of underlying dermis to induce histogenesis in epidermal placodes. Perhaps the best example of the role of dermis in determining epidermal structure is provided by recombination experiments with duck and chick (Figure 10.3). The epidermal (ectoderm) webbing on duck's foot is due to dermis (mesoderm). The chick epidermis, which is not normally webbed, undergoes webbing on contact with duck dermis. These experiments reveal clearly the role of dermis in epidermal morphogenesis. Human fetal skin consists of types III and I collagen (Epstein 1974), with the ratio of type III to type I decreasing with growth. In early embryonic chick skin, type III collagen has also been demonstrated (Vinson and Seyer 1974). Epidermal cells from adult guinea pig skin interact and differentiate on type IV collagen (Murray et al. 1979). Laminin mediates the attachment of epithelial cells to type IV collagen (Terranova et al. 1981). Although present in dermis (Weiss and Reddi 1981a), fibronectin is

Figure 10.3. The morphogenesis of epidermal structures is dependent on dermis. Mesoderm regulates webbing on a duck's foot. The reciprocal combination of epidermis of the chick and dermis in the duck results in the webbed morphology of chick ectoderm. However, when wing ectoderm is used, feather formation takes place on the developing scales. *(Source: Sengel (1975) CIBA Found Symp 29:51–70, with permission.)*

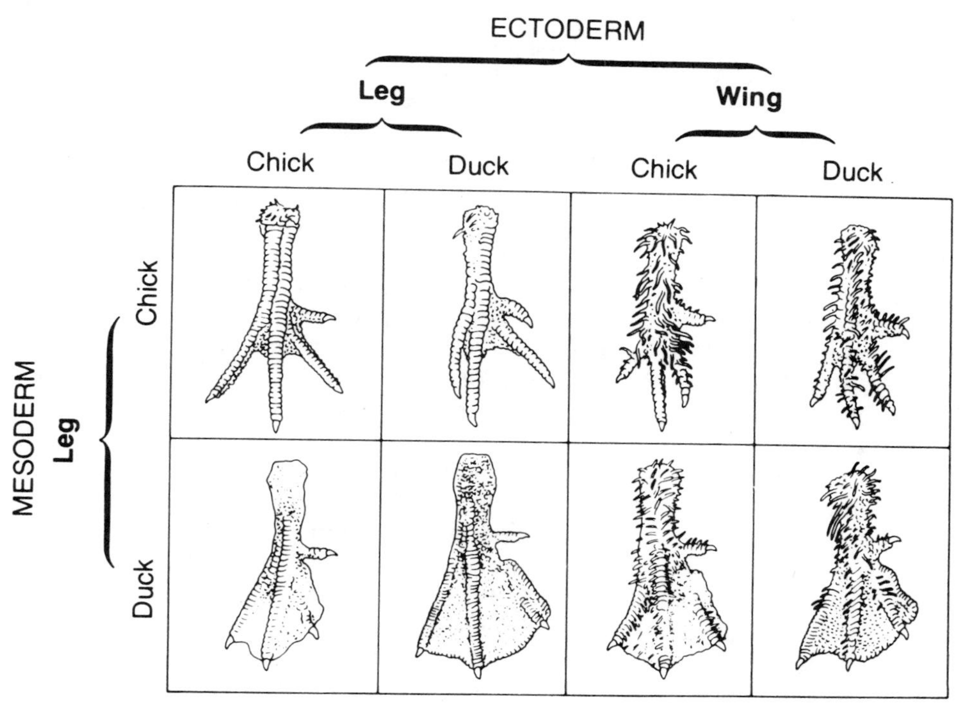

apparently not involved in attachment of epidermal cells to collagen. The role of proteoglycans in skin morphogenesis is not well understood (Goetinck 1980).

Nervous System. The developing nervous system consists of interacting cells including nerve cells and various supporting cells. Myelination of peripheral nerve tissue involves contact with extracellular matrix (Bunge and Bunge 1978). For normal Schwann cell development, contact with extracellular matrix and axons is necessary. There appears to be an interaction between nerve cells and Schwann cells in the production of basal lamina and extracellular matrix macromolecules (Bunge et al. 1980). The extracellular matrix fibrils are collagenase-sensitive and consist of types I, III, and V collagen. Basement membranes ensheathe the vertebrate muscle fiber and are present in the neuromuscular junction in the synaptic cleft. Fibronectin, laminin, and types IV and V collagen have been localized by immunohistochemical techniques (Schachner et al. 1978; Sanes 1982). In addition, fibronectin increases the growth rate of Schwann cells (Baron-van Evercooren et al. 1982). Laminin promotes Schwann cell adhesion and neurite outgrowth.

The specialized synaptic region of the basement membrane has a profound role in neuromuscular physiology and regeneration. The specificity and order among the synapses are well known to students of neurobiology. The reinnervation of the muscle after damage is precise. The damaged muscle fiber is degraded and removed by phagocytosis. The basement membrane persists and acts as a scaffold for muscle regeneration (Marshall et al. 1977; Burden et al. 1979). In addition, regional specialization of the basement membrane marks the synaptic sites and regenerating axons form synapses with regenerating myofibers. Apparently the synaptic basement membrane has the information to guide precise topographic reinnervation (Marshall et al. 1977). The accumulation of acetylcholine receptors in regenerating muscles has been localized to the synaptic sites by α-bungarotoxin binding (Burden et al. 1979). Acetylcholine receptors are clustered to regions of cell–substratum contact (Bloch and Geiger 1980), and collagen biosynthesis is involved in receptor aggregation (Kalcheim et al. 1983). The enzyme acetylcholinesterase is concentrated in the neuromuscular junction and it hydrolyzes acetylcholine released from the nerve terminal. It is associated with the basement membrane of the synaptic cleft (McMahan et al. 1978).

The molecular structure of acetylcholinesterase reveals a collagen-like tail (Chapter 1), possibly anchored in the basement membrane. The collagenous tail is susceptible to bacterial collagenase treatment, and yields a globular form of the enzyme (Lwebuga-Mukasa et al. 1976; Anglister and Silman 1978; Bon et al. 1979; Mays and Rosenberry 1981). These findings reveal the structural design of membrane-associated enzymes and possible role of collagen-like tails for anchorage in the cell membrane.

During vertebrate embryonic development, neural crest cells originate at the dorsal surface of the neural tube, migrate, and differentiate into numerous cell types (LeDouarin 1980). The pathways and mechanism of cell migration are not well understood. Hyaluronate is present in the extracellular space around the neural crest cells (Pratt et al. 1975; Toole 1976). The morphogenetic movement of neural crest cells in the embryonic salamander involves a network of fibrillar extracellular matrix (Lofberg et al. 1980). Cultured neural crest cells attach to various collagen types via fibronectin (Greenburg et al. 1981). Furthermore, fibronectin is chemotactic for neural crest cells. The glycosaminoglycans in the extracellular regions of neural crest cells also play a major role in cell migration and motility (Derby 1978; Pintar 1978; Toole 1976). The morphogenetic movement of neural crest cells in the embryonic axolotl also involves the fibrillar extracellular matrix (Lofberg et al. 1980).

Eye. The development of the eye has been reviewed by Coulombre (1965) and Hay and Revel (1969). Tissue interactions during morphogenesis of the eye lead to its precise geometrical structure (Figure 10.4). The intraocular pressure in the developing eye determines the geometry of the cornea. The outer sclera protects

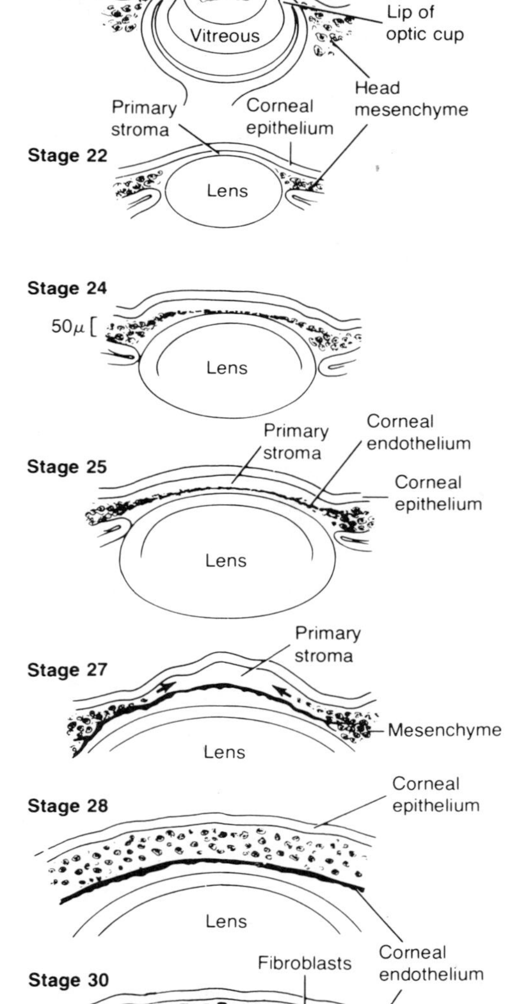

Figure 10.4. Camera lucida drawings of the developing avian eye. When the lens placode begins to invaginate, the presumptive corneal epithelium lies over the lip of the optic cup. At 3 days of incubation the lens vesicle is pinched off and the overlying ectoderm becomes corneal epithelium (stage 19). By stage 22 (4 days), the corneal epithelium produces the corneal stroma and the presumptive corneal endothelial cells invade the area. The migration is complete by stage 25 (4.5–5 days). By stage 27 (5–5.5 days), primary stroma cells and corneal fibroblasts begin to invade. By stage 30 (6.5–7 days), fibroblasts occupy all layers of the stroma. Data from Hay and Revel 1969. *(Source: Wessells (1977), with permission.)*

and maintains the shape of the eye. Collagenous matrix is present in cornea, lens, vitreous body, and sclera. The mature collagenous cornea is transparent and consists of type I collagen fibrils oriented at approximately right angles to each other in orthogonal layers (Trelstad and Coulombre 1971). The assembly of collagen in a morphogenetic system is best illustrated by avian corneal morphogenesis (Trelstad 1974; Hay, 1980). Trelstad and Kang (1974) isolated five different collagen types from the eye of 3-week-old chicks localized to various structures. The chick vitreous body collagen is synthesized by two different cell types at different stages of development. The neural retina is involved in early stages and cells in the vitreous body during later stages (Newsome et al. 1976). Embryonic chick corneal epithelium synthesizes both types I and II collagens (Linsenmeyer et al. 1977). The synthesis of type II collagen by chick neural retina is now well established (Smith et al. 1976; Lisenmeyer and Little 1978). The immunofluorescent localization of types I, II, and III collagen has been studied (von der Mark et al. 1977). Type I collagen can be detected in stage 19 chick embryo in the corneal stroma, vitreous body, and lens. Type II collagen is observed at stage 20 in the primary corneal stroma, neural retina, and vitreous body. At later stages, type II collagen is localized in the sclera, whereas the secondary corneal stroma contains largely type I collagen. Exogenous collagenous substrata—such as lens capsule basement membrane and rat tail tendon collagen—stimulated collagen synthesis by corneal epithelium (Dodson and Hay 1974; Meier and Hay 1974), suggesting that collagenous matrices may promote cell differentiation in addition to being a structural component.

Heart. Collagen and glycosaminoglycan synthesis has been detected in the developing heart of chick embryos (Johnson et al. 1974; Manasek et al. 1973). Type I collagen is the predominant type in early chick embryo heart. During morphogenesis of developing heart the formation of endocardial cushions in the atrioventricular canal is a critical event. Cell migration is a key step in the morphogenesis of cushion pad. Markwald et al. (1979) have demonstrated the role of extracellular matrix constituents in cell migration.

Lung. Epithelial–mesenchymal interactions play a pivotal role in lung morphogenesis. Bronchial mesenchyme directs the branching pattern of bronchial epithelial buds (Alescio and Cassini 1962; Spooner and Wessells 1970). Collagenase pretreatment abolishes the stabilizing influence of collagen on mouse lung morphogenesis. In addition, lung morphogenesis in vitro is inhibited by the presence of a proline analog L-azetidine-2-carboxylic acid that inhibits collagen synthesis (Alescio 1973; Spooner and Faubion 1980). Collagen synthesis in developing rabbit lungs is high in late gestation and declines to a lower level postnatally (Bradley et al. 1974).

Mammary Glands. As in the salivary glands (Grobstein 1967), epithelial–mesenchymal interactions are essential for mammary gland morphogenesis and development. The mammary mesenchyme determines the branching pattern of developing mammary epithelial buds (Kratochwil 1972). There is a sex hormone-dependent mesenchyme-directed morphogenesis of the epithelium (*See* Figure 10.5). The adult mammary epithelium retains its competence to respond to embryonic mesenchyme. Transplantation of 14-day-old embryonic mammary

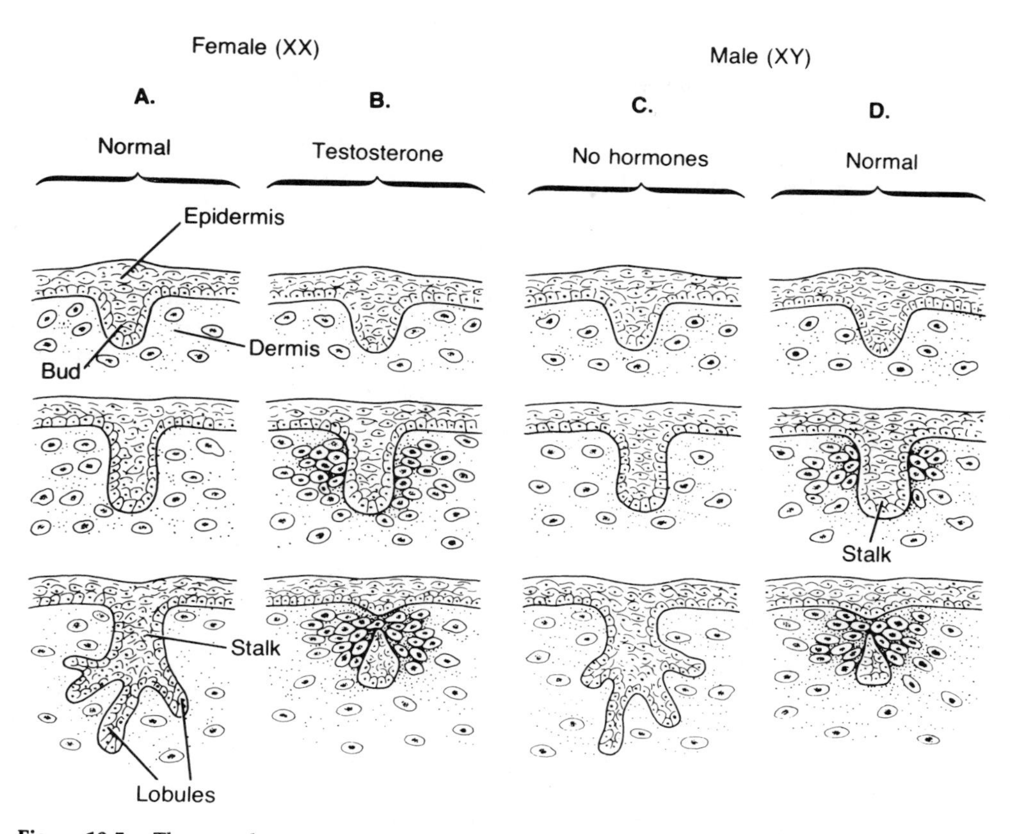

Figure 10.5. The complexity of interactions between hormones and genetic information is seen in development of early mammary glands. (*A*) In a female mammal, an epithelial bud grows downward into the dermis and branches into lobules. (*D*) In a normal male, the bud forms and elongates, but then mesenchyme cells accumulate on each side of the stalk. The stalk breaks free from the epidermis, and the lower portion of the mammary epithelial tissue regresses and may disappear. (*C*) If a male gland is cultured in medium lacking hormones, the stalk remains connected to the epidermis and lobules form. This shows that the genetic information required to construct an early mammary gland is present in an individual having the male XY sex chromosome complement. It also shows that female hormones are not required for the gland to develop in the female pattern. If a developing female gland is cultured either in the presence of testes tissue or with the male hormone testosterone in the medium (*B*), then regression identical to that in a male occurs. This means that a Y chromosome is not needed for the regression phenomenon to occur. It also implies that male hormones are normally responsible for regression of mammary glands during normal development of male individuals. Perhaps most important is that the genetic sex of the gland tissue itself has nothing to do with its developmental capabilities. Data from Kratochwil K: (1971) J Embryol Exp Morphol 25:141. *(Source: Wessells (1977), with permission.)*

mesenchyme to adult mammary glands resulted in a branching pattern similar to developing mammary gland (Sakakura et al. 1979).

Type IV collagen (basement membrane type) is synthesized by mammary ducts and alveoli (Wicha et al. 1980) in culture and is blocked by the proline analog *cis*-hydroxyproline. Treatment with this proline analog in vivo causes involution of glandular structures, implying a role for basement membrane collagen in the mammary epithelial cell function. The synthesis of type IV collagen by mammary epithelium is under the hormonal control of insulin, prolactin, hydrocortisone, progesterone, and estradiol (Liotta et al. 1979b). The mammary stroma produces both types I and III collagen. Thus, collagens are intimately involved in the development of mammary tissue.

Floating collagen substrates promote mammary epithelial cell growth and differentiation in culture (Emerman and Pitelka 1977). Also, embedding mammary tumor epithelial cells in collagen gels results in better growth and differentiation (Yang et al. 1979). There appears to be a preferential attachment of mammary epithelial cells to type IV collagen in comparison with types I, II, and III (Wicha et al. 1979). Type IV collagen is synthesized by rat mammary epithelial cells in a serum-free medium supplemented with epidermal growth factor, dexamethasone, insulin, transferrin, and fetuin. In the absence of epidermal growth factor, mammary epithelial cells thrive on a type IV collagen substratum but not on type I collagen or plastic (Salomon et al. 1981). Typical mammary morphogenesis occurs when cloned mammary epithelial cells are cultured on collagen gels (Bennett 1980). This finding implies a role for mesenchymal collagen in branching morphogenesis. Proteoglycans may also be involved in mammary development. David and Bernfield (1981) have demonstrated that the degradation of basement membrane proteoglycan is diminished by collagen and have proposed a role for collagen in stabilizing the proteoglycan during mammary gland morphogenesis.

Urogenital System. The development of kidney and the reproductive organs also involves precise epithelial–mesenchymal interactions and is regulated by connective tissue matrices. The metanephric kidney is the product of interactions between the epithelium from the Wolffian duct and the nephrogenic mesenchyme (Saxen 1971). Wartiovaara et al. (1974) have demonstrated the role of cell–cell apposition in kidney tubulogenesis induced by the spinal cord. Recent experiments have revealed a close correlation between the appearance of laminin and kidney tubule formation (Ekblom et al. 1980). Cunha (1976) has extensively studied epithelial–mesenchymal interactions in the development of prostate and other male accessory sexual glands. For example, glycosaminoglycans appear to play a role in seminal vesicle morphogenesis as ascertained by pretreatment with purified hyaluronidase. The androgen-induced morphogenesis of seminal vesicles is blocked by the proline analog L-azetidine-2-carboxylic acid, indicating the possible role of collagenous matrix.

Salivary Glands. The morphogenesis of salivary glands (*See* Section 10.3.3) represents a prototype for the analysis of factors involved in branching patterns of various tissues (Borghese 1950; Grobstein 1953). As discussed in the preceding sections on organogenesis, the recurrent theme underlying morphogenetic events is the role of mesenchyme in epithelial architecture. The mouse embryonic salivary gland is a favorite model for detailed investigation. Bernfield and coworkers (Bernfield and Banerjee 1972; Bernfield et al. 1972; Bernfield 1981) have done systematic studies on this system. Glycosaminoglycans and proteoglycans are present in the epithelial–mesenchymal interface and the rate of accumulation of labeled glycosaminoglycan is maximal at the distal ends of branching lobules. The enzymatic removal of hyaluronic acid by testicular hyaluronidase results in the loss of clefts in the epithelium. The proteoglycans in the basement membrane are crucial for the shape of the epithelium. The turnover of proteoglycan is modulated by mesenchymal enzymes and collagen. Collagen stabilizes and pre-

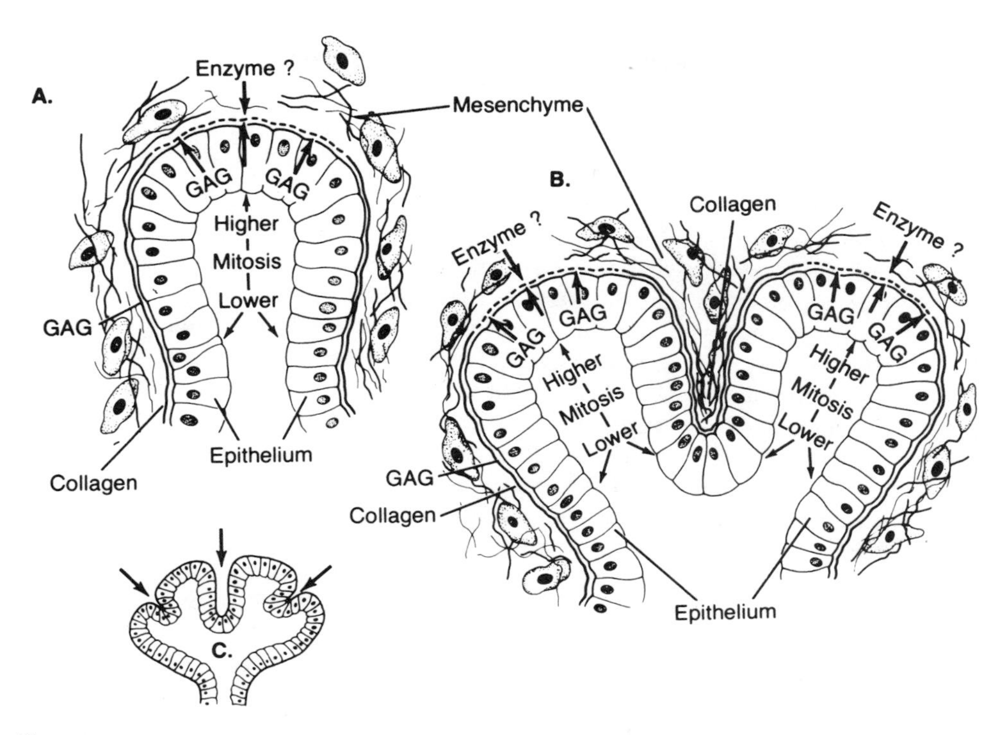

Figure 10.6. Hypothetical scheme for cleft formation in an epithelium. (*A*) The tip of the epithelium is a major site of new glycosaminoglycan (*GAG*) secretion and accumulation in the basal lamina region. Newly synthesized GAG tends to disappear from that tip region, perhaps because of enzyme-induced turnover. Along the stalk region of the same epithelium, the GAG and collagen are more stable, and accumulate in quantity. Mitotic activity is particularly intense near the tip, which is also the site of cleft formation, perhaps because of microfilament activity. (*B*) The cleft has formed and is stabilized with accumulated GAG and collagen. To each side of the cleft, the tips retain the ability to synthesize, secrete, and apparently turn over GAG. Mitosis and cleft formation occur in those regions, forming the shape depicted in *C*. Many other sites of intimate contact between epithelial and mesenchymal cells occur, and the surface of the epithelium itself is highly dynamic. (*Source: Wessells (1977), with permission.*)

vents proteoglycan degradation in the salivary epithelial basement membrane (Bernfield 1981). As salivary gland morphogenesis in vitro is blocked by L-azetidine-2-carboxylic acid, an inhibitor of collagen synthesis (Spooner and Faubion 1980), it is likely collagen plays a role in stabilizing the proteoglycan during development. Furthermore, the whole morphogenetic process is a dynamic cell–cell, cell–matrix interactive system undoubtedly connected to intracellular machinery such as microtubules and microfilaments, resulting in intricate shape changes and attendant lobule formation (Wessells 1977). The working model is depicted in Figure 10.6.

Hematopoietic System. Hematopoietic bone marrow is in close physical and metabolic proximity to the extracellular matrix of bone. It is likely that certain components of the matrix play a role in the maintenance of hematopoietic microenvironment (Weiss 1976). Types I and III collagen have been detected by indirect immunofluorescence staining in bone marrow stromal cells (Reddi et al. 1977; Bentley et al. 1981). Fibronectin is also present in bone marrow (Weiss and Reddi 1981c). Recently, collagen gels have been utlized for cloning granulocyte/macrophage progenitor cells and for maintenance of stem cell proliferation (Lanotte et al. 1981).

10.4. Wound Healing and Regeneration

The preceding sections discussed the emerging role of extracellular matrix during the development of a variety of tissues. In addition to the growth and development of the organism, extracellular matrix components play a continuing role in the repair and healing of both soft and skeletal tissues. In certain lower vertebrates, such as tailed amphibians, there is complete regeneration of limbs following amputation. In addition to the contributions of inflammatory cells and fibroblasts (Ross 1968), extracellular matrix figures prominently in this repair and regeneration of tissue. Plasma fibronectin functions as an opsonic glycoprotein and probably functions in wound debridement as shown by the observation that plasma fibronectin is depleted in patients with burn injuries. Levels return to normal values within 24 hours. The deficiency of plasma fibronectin and the resulting opsonic dysfunction can be ameliorated by intravenous infusion of a cryoprecipitate containing plasma fibronectin (Saba et al. 1978; Lanser et al. 1980). One of the earliest phases of wound repair is chemotaxis of fibroblasts; recent evidence indicates that proteolytic fragments of fibronectin are intensely chemotactic. It is likely that proteases in the vicinity of the wound generate peptides from a variety of proteins that are chemotactic for cells (Seppa et al. 1982). Collagen–derived peptides are chemotactic for human fibroblasts. Membrane receptors have been identified on fibroblasts for α1(I) chains (Postelthwaite et al. 1978; Chiang et al. 1978).

Although the predominant collagen in skin is type I, during the early phase of healing in skin wounds higher proportions of type III collagen may be present (Gay et al. 1978). Topical application of collagen to incision wounds in guinea pig skin has been reported to accelerate the healing process (Shoshan and Finkelstein 1970). In view of these findings, it is possible that collagen promotes chemotaxis and proliferation of progenitor cells prior to healing.

Among the vertebrates, limb regeneration potential is highest in newts and salamanders. On amputation, there is dedifferentiation of muscle and skeletal cells to the relatively undifferentiated blastema cells (Hay 1974). Hyaluronate is the predominant glycosaminoglycan in the blastema. With the onset of cartilage differentiation, there is an upswing in the hyaluronidase activity and concomitant fall in hyaluronic acid. Chondroitin sulfate is the predominant glycosaminoglycan in developing cartilage (Toole and Gross 1971). There is a decline in cartilage (type II) collagen during the formation of the blastema (Linsenmeyer and Smith 1976). Fibronectin may play an important role in blastemal cell aggregation and differentiation during limb regeneration (Gulati et al. 1983).

10.5. Extracellular Matrix-Induced Bone Development: A Prototype

Among vertebrate tissues, bone has the highest potential for regenerative healing. The remarkable phenomenon of fracture healing is known to most students. Furthermore, there is a considerable turnover of bone by remodeling throughout the life of the organism. It is likely that the extracellular matrix of the bone is a repository of factors initiating and modulating remodeling and regeneration of skeletal tissues.

Subcutaneous implantation of demineralized, diaphyseal, extracellular bone matrix results in the sequential development of cartilage and bone (Urist 1965; Reddi and Huggins 1972, 1975; Reddi and Anderson 1976; Reddi 1976, 1981). The sequential cellular and molecular changes in the developing cartilage and bone serve as a prototype of regenerative, as opposed to fibrotic (*See* Chapter 11), healing.

Subcutaneous or intramuscular implantation of autologous extracellular bone matrix results in instantaneous formation of a blood clot and migration of polymorphonuclear leukocytes into the site. On day 1, the implant is a discrete button-like, plano-convex plaque that is a conglomerate of implanted extracellular matrix, fibrin, and leukocytes. On day 3, fusiform fibroblast-like mesenchymal cells appear in the close vicinity of the matrix (Figure 10.7). Type III collagen synthesis is detected at this time (Steinmann and Reddi 1980). Mesenchymal cells proliferate as monitored by ^{3}H-thymidine incorporation and radioautography (Rath and Reddi 1979). Cell processes appear to interact with the implanted extracellular matrix (Figure 10.8). This initial matrix–membrane interaction, whether by contact or through soluble factors or both, sets in motion a series of sequential cellular changes, resulting in formation of cartilage (Figure 10.9), bone, and hematopoietic marrow (Figure 10.10). The earliest chondroblasts are seen on day 5; at days 7–8, the implant is a conglomerate of the matrix particles and the newly induced hyaline cartilage. Figure 10.11 summarizes the biochemical sequences during the cellular changes. The appearance of cartilage is marked by localization of type II collagen in the cartilage and an increase in the incorporation of $^{35}SO_4$ into proteoglycans (Reddi et al. 1977; Reddi et al. 1978). Calcification of the hypertrophic cartilage matrix, as occurs in endochondral bone formation, is evident on day 9. Vascular invasion is a prerequisite for bone formation and is accompanied by the appearance of type IV collagen, laminin, and blood coagulation factor VIII around invading endothelial cells (Foidart and Reddi 1980). Bone-forming osteoblasts are seen on days 10–11 and new bone formation is seen on the surface of cartilage spicules and implanted bone matrix. Bone formation is indicated by ^{45}Ca incorporation, alkaline phos-

Figure 10.7. Extracellular bone matrix-induced cell differentiation, day 3. Fibroblast-like cells in the vicinity of the implant. Magnification, × 330.

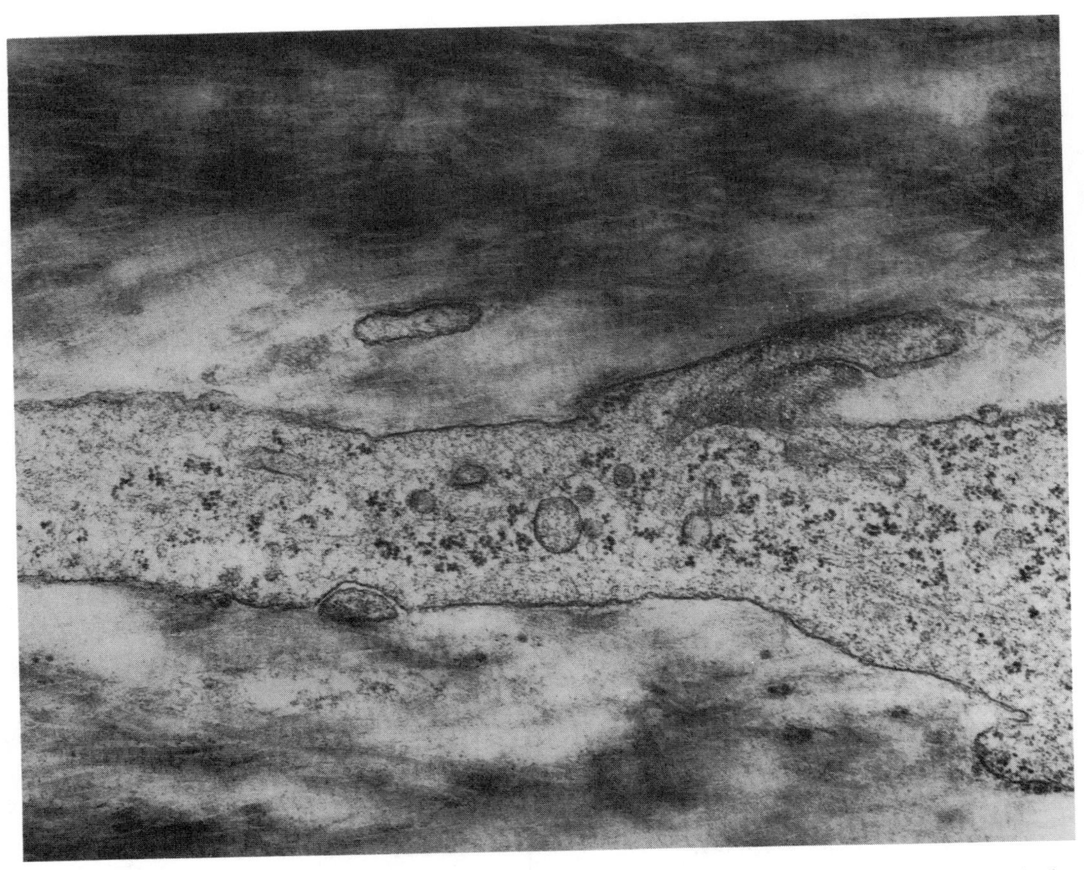

Figure 10.8. An electronmicrograph depicting a fibroblastic cell process in contiguity with the implanted collagenous diaphyseal bone matrix on day 5. Magnification, × 20,000.

phatase, and type I collagen localization (*See* Figure 10.11). There is extensive bone remodeling on days 12–18, resulting in the formation of an ossicle that is replete with hematopoietic marrow on day 21. Incorporation of ^{59}Fe into heme is a useful marker for erythropoiesis (Reddi and Anderson 1976; Reddi 1981).

The sequence can be partially duplicated in vitro. Organ culture of neonatal muscle on demineralized hemicylinders of bone matrix permits cartilage differentiation (Nogami and Urist 1974; Nathanson et al. 1978; Nathanson and Hay 1980).

The precise molecular mechanisms underlying the action of extracellular matrix on cells to initiate the developmental cascade of cartilage, bone, and marrow are not known. Some of the salient properties of the action of extracellular matrix are listed in Table 10.1. The response to the extracellular matrix is stringently specific and is elicted by only bone and tooth matrix, but not by tendon, skin, cartilage, or aorta (Reddi 1976). The surface charge on the matrix is critical. Perturbation of the charge by N-acetylation, carboxymethylation, or modification of the guanidino groups of arginine abolishes bone induction (Reddi 1982). Pretreatment of the matrix by polyanions such as heparin, dextran sulfate, and the anionic dye Evans blue also blocks bone development (Table 10.2) (Reddi

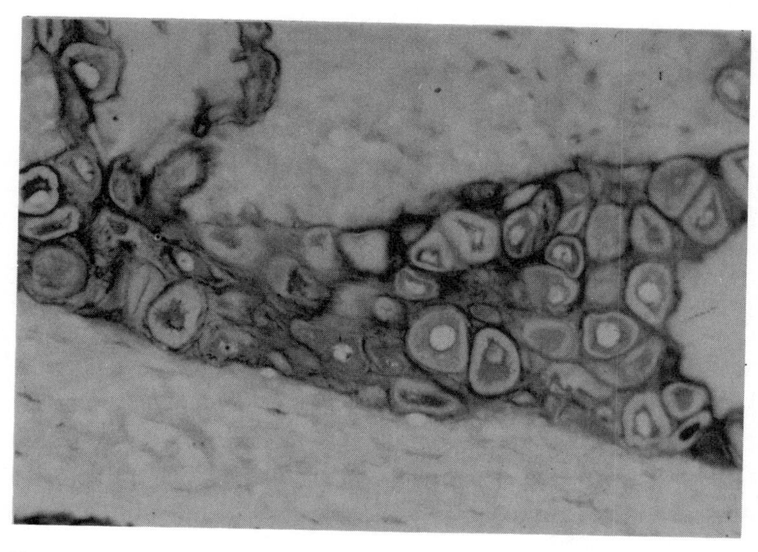

Figure 10.9. Extracellular matrix-induced cartilage development, day 7. Note the chondrocytes in apposition to the implanted matrix. Magnification, × 330.

1975). There is an excellent correlation between the degree of sulfation of various heparan sulfates and heparin with the extent of inhibition (Table 10.3) (Reddi 1975). The inductive property is acid stable but alkali labile. The biological property is heat stable at 65°C for 8 hours, trypsin-sensitive, and is likely to be a glycoprotein as suggested by its periodate sensitivity. Conformation stabilized

Figure 10.10. Extracellular matrix-induced bone and marrow development, day 21. Note bone and marrow in the ossicle. Magnification, × 220.

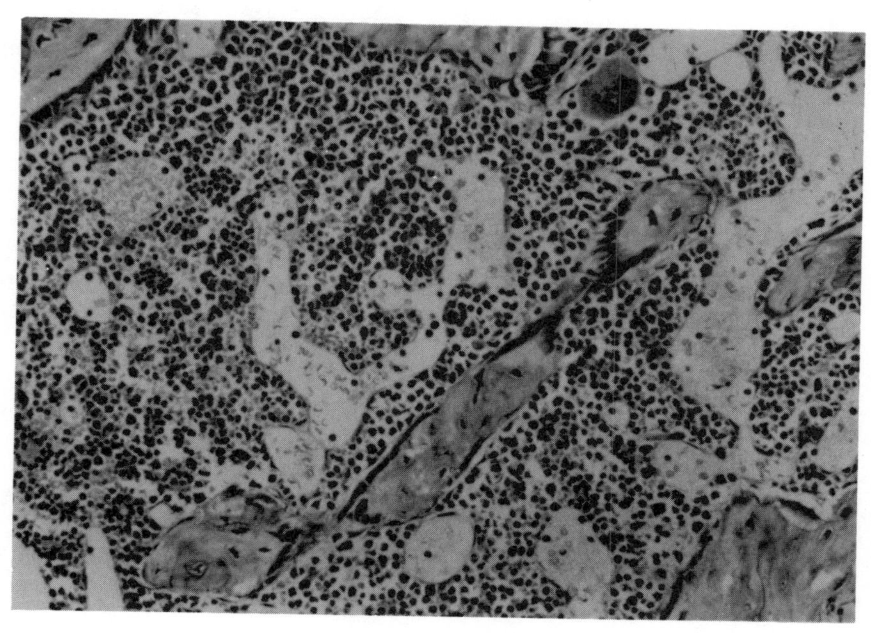

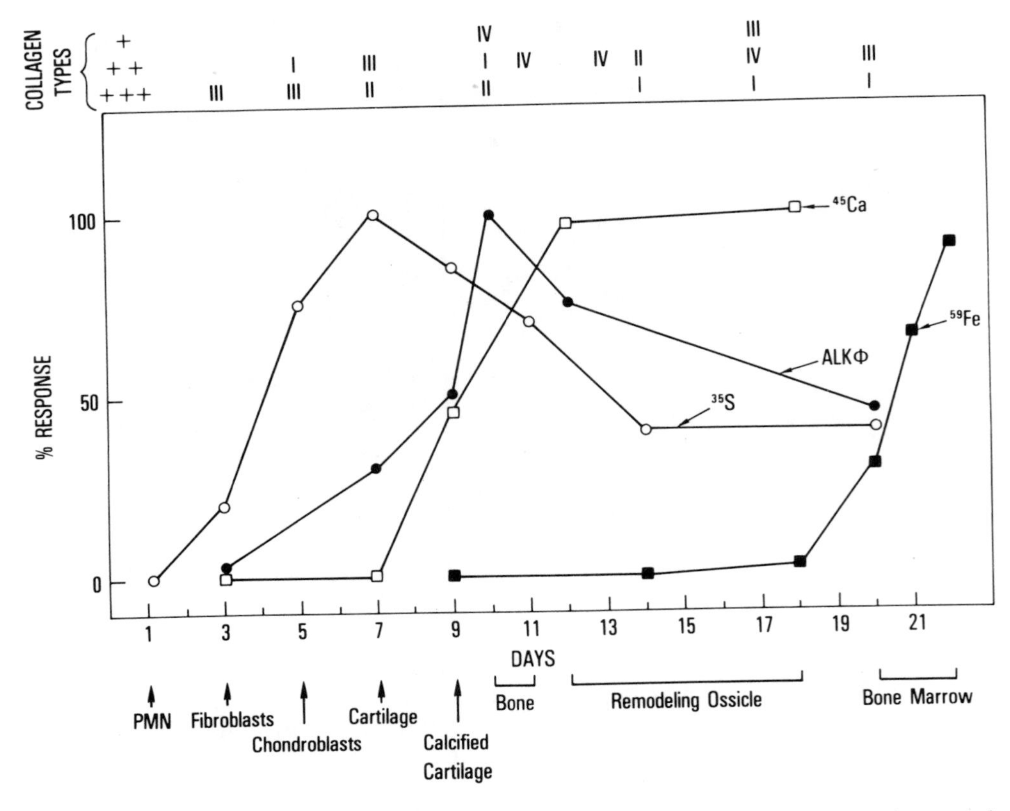

Figure 10.11. Developmental sequence of the extracellular matrix-induced cartilage, bone, and marrow formation. Changes in $^{35}SO_4$ incorporation into proteoglycans and ^{45}Ca incorporation into mineral phase indicate peaks of cartilage and bone formation, respectively (Reddi 1976). The ^{59}Fe incorporation into heme is an index of erythropoiesis, as plotted from the data of Reddi and Anderson (1976). The values for alkaline phosphatase indicate early stages of bone formation (Reddi and Huggins 1972). The transitions in collagen types I to IV are summarized on top of the figure and are based on immunofluorescent localization (Reddi et al. 1977; Foidart and Reddi 1980). *PMN* = polymorphonuclear leukocytes. *(Source: Reddi (1981), Collagen Res 1:209–226, with permission.)*

Table 10.1. Properties of Extracellular Matrix-Induced Bone Cell Differentiation

1. Specificity of the matrix
2. Surface charge characteristics
3. Geometry of the matrix
4. Acid stability
5. Alkali lability
6. Heat stability (65°C for 8 h)
7. Trypsin sensitivity
8. Loss of activity by periodate oxidation
9. Loss of activity by disulfide bond reduction
10. Dissociative extraction and reconstitution

Table 10.2. Inhibitory Effects of Polyanions on Bone Matrix-Induced Differentiation of Fibroblasts[a]

	Alkaline phosphatase (units/g)	^{45}Ca incorporation (cpm/mg)
Control	49.1 ± 6.6[b]	5243 ± 1147
Heparin 0.1%	6.1 ± 1.0	69 ± 10
Dextran sulfate 1.0%	5.8 ± 0.8	82 ± 17
Polyvinylsulfonate 1.0%	5.0 ± 0.9	83 ± 15

[a]Data from Reddi (1975).
[b]± standard error of mean of 8 observations.

by disulfide bonds is necessary for biological activity, as reducing agents abolish the inductive response (Reddi 1976).

The profound influence of matrix geometry on the rate and extent of bone formation has been observed. The size of the matrix determines the yield of bone. The optimal size is in the range of 74–850 μm (Figure 10.12). Fine powders (44–74 μm) are weak in bone induction. It is possible that the matrix dimensions must be larger than the responding cells, and there may be multiple sites of attachment over the entire surface of the cells. This experiment with varying sizes of collagenous extracellular bone matrix is an example of anchorage-dependent proliferation and differentiation of cells in vivo (Section 10.5.1).

As the extracellular matrix derived from bone is predominantly collagenous and fibronectin mediates the attachment of cells to collagen, the possible involvement of fibronectin in this biological sequence has been investigated. On implantation the matrix binds fibronectin as an important initial event for cell attachment to the matrix (Weiss and Reddi 1980; 1981b). The current working model is depicted in Figure 10.13. Orientation of the cell surface membrane along the surface of the matrix is by electrostatic interaction. This is a necessary but not sufficient condition for subsequent changes. The collagen–fibronectin

Table 10.3. Correlation Between the Sulfate Content and Inhibitory Effects of Heparan Sulfates on Bone Matrix-Induced Differentiation of Fibroblasts[a]

	Sulfate/hexosamine molar ratio	Alkaline phosphatase (units/g)	^{45}Ca incorporation (cpm/mg)
Control	—	47.5 ± 6.5[b]	5027 ± 1413
Heparan sulfate (low sulfate)	0.54	56.6 ± 6.5	4126 ± 941
Heparan sulfate (medium sulfate)	1.00	30.9 ± 8.5	2725 ± 1398
Heparan sulfate (high sulfate)	1.25	18.0 ± 3.5	2128 ± 688
Heparin	2.33	4.2 ± 1.2[c]	80 ± 27[c]

[a]Data from Reddi (1975).
[b]± standard error of mean of 9 observations.
[c]Significant difference, $P < 0.01$.

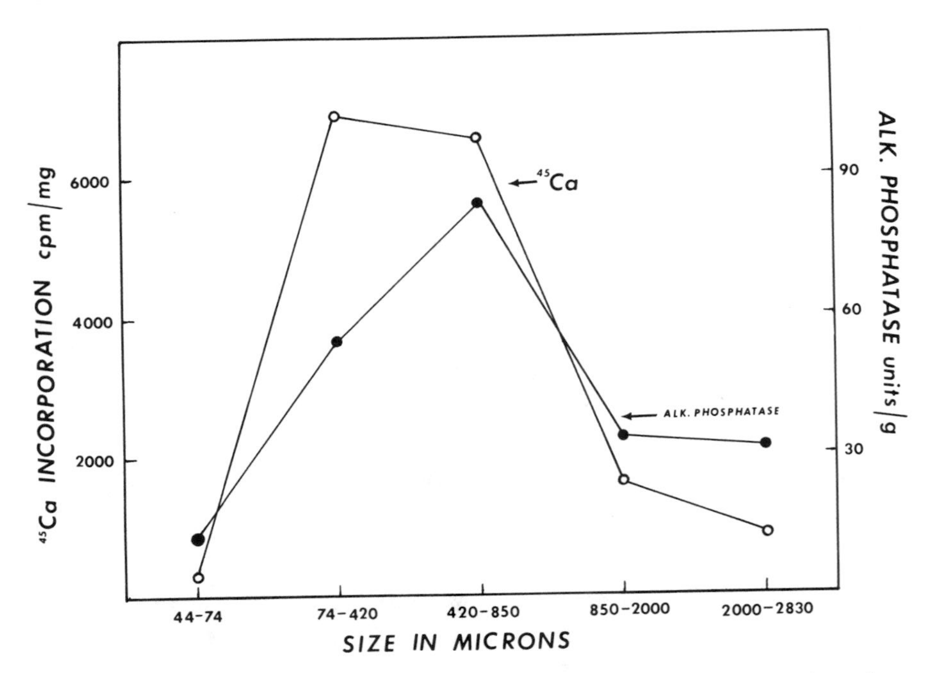

Figure 10.12. Influence of the geometry of extracellular matrix on anchorage-dependent proliferation and differentiation. The optimal size is in the range of 74 to 850 μm for bone differentiation. *(Source: Reddi (1976) In: Biochemistry of Collagen, with permission.)*

interaction may aid in the close contact of responding cell surface receptors to putative inducer molecules. This matrix–cell surface interaction sets into motion the developmental cascade of bone formation.

One of the first steps is chemotaxis of cells to the implanted extracellular matrix. This is followed by mitosis and differentiation into cartilage and bone. Dissociative extracts of the matrix have been examined for their potential to stimulate chemotaxis, mitosis, and differentiation. Bone inductive activity is lost by pretreatment of the matrix with 4 *M* guanidine hydrochloride, or 8 *M* urea containing 1 *M* NaCl or 1% (weight per volume) sodium dodecyl sulfate (Sampath and Reddi 1981). The lyophilized extract, and the residue when tested

Figure 10.13. A model of extracellular bone matrix–cell interaction. See text for details.

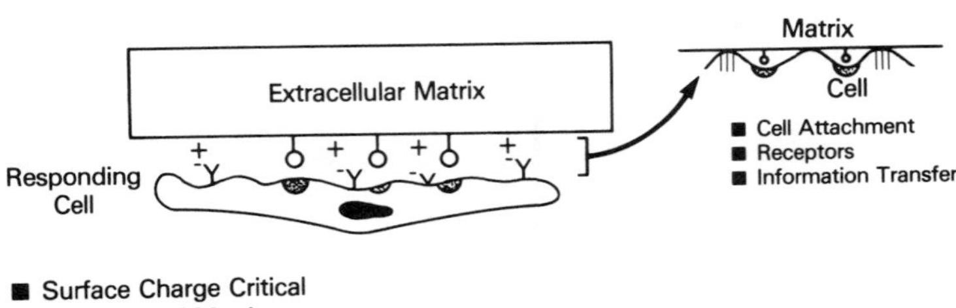

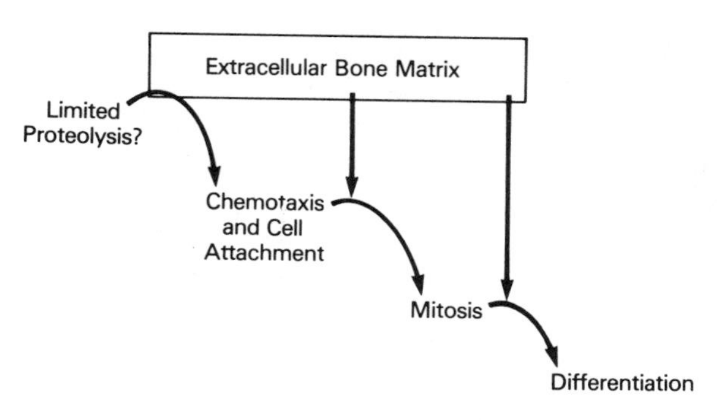

Figure 10.14. A cascade type mechanism regulates the local differentiation of bone by extracellular matrix. The major phases are chemotaxis, mitosis, and differentiation.

alone, are devoid of bone inductive potential. However, on reconstitution of the extract with the residue, the inductive property is regained. The active fraction consists of proteins less than 50,000 molecular weight. The 4 *M* guanidine extracts are intensely chemotactic for clonal osteoblasts (Somerman et al. 1982). Growth promoting activity is present in fractions of less than 50,000 molecular weight and is tightly associated with the extracellular matrix (Sampath et al. 1982). These observations imply that the bone matrix is a repository for chemotactic, mitogenic, and differentiation-inducing factors and a sequential cascade mechanism is involved in endochondral bone formation (Figure 10.14).

10.6. Anchorage Dependence and Cancer

> Snatched from a life of obscurity and installed in contemporary glass and plastic palaces, cells are in danger of becoming Pygmalion's proteges. Housed in more traditional residences constructed of water and collagen instead of plastic or glass, do cells lead primitive, less cultured lives?
>
> Elsdale and Bard (1972)

Animal cells recalcitrant to growth in culture in vitro frequently require collagen as a substratum to thrive (Ehrmann and Gey 1956). Most cells must attach to a solid substratum of plastic, glass, or collagen in order to spread and proliferate. This property has been termed anchorage dependence by Stoker et al. (1968). Inoculation of clonal cells in semisolid media results in no further growth until a solid substratum of optimal geometry is provided such as glass fibrils or beads. Glass filaments ranging in length from 250 to 750 μm or glass beads with diameters from 50 to 90 μm support proliferation as evidenced by attached colonies of cells. However, no colonies are seen in the presence of 5 μm fragments. The optimal size for growth of baby hamster kidney (BHK 21) C 13 cells is 50 μm (Maroudas 1972). Anchorage dependence is lost in transformed cells (Stoker et al. 1968; O'Neill et al 1979), a fact that may have important implications for malignant and metastatic potential.

10.6.1. Permissive Substratum

There is a growing body of knowledge implicating collagenous matrices in the adhesion and growth of cells (Kleinman et al. 1981). The effect appears to be permissive rather than instructive. The now classic experiments of Hauschka and Konigsberg (1966) (Section 10.3.3) on muscle differentiation revealed that collagen promotes myotube formation from myoblasts. Certain components in horse serum were observed to aid cell attachment to collagen (Hauschka and White 1972), and a collagen-dependent cell attachment factor was found to be present in calf serum (Klebe 1974) (*See* Chapter 7 for a detailed account of cell-attachment factors). Attachment proceeds in a stepwise manner: (1) binding of serum component to collagen; and (2) attachment of cells to the collagen-serum factor complex. The serum component is structurally similar to a membrane-associated glycoprotein that mediates fibroblast adhesion to collagen (Pearlstein 1976). This cell adhesion factor was subsequently designated fibronectin. Cells attached to collagen via fibronectin are flat and extended as opposed to the rounded cells found in the absence of fibronectin. The attachment and spreading of fibroblasts is enhanced by serum or plasma fibronectin even under conditions that are inhibitory to cells for new protein synthesis by agents such as cyclohexamide. Fibronectin binds to a variety of collagens, including types I through IV, but the specificity to aggregate forms in vivo is yet to be elucidated. The binding site of fibronectin to type I collagen has been determined and is in the vicinity of the site of mammalian collagenase cleavage (Kleinman et al. 1981). A peptide of 35 amino acid residues has been isolated from the α1(I) chain of type I collagen, which has affinity for fibronectin (Figure 10.15). Further experiments have revealed homologous sequences with binding activity in types II and III collagen α chains (Figure 10.16).

Not all adhesive phenomena are mediated by fibronectin (Linsenmeyer et al. 1978). Variant Chinese hamster ovary (CHO) cells have been isolated that appear to have lost their ability to bind fibronectin (Harper and Juliano 1981). Also, it appears that there may be two different mechanisms of fibroblast adhesion depending on whether the collagen substratum is native or denatured (Schor and Court 1979). Cell attachment to native collagen can occur via a serum-independent mechanism (Goldberg 1979). On the other hand, the serum-dependent mechanism is operative in the case of films of denatured collagen in culture. Other cell types, such as epithelial cells, may attach via laminin a basement membrane glycoprotein (*See* Chapter 8). Chondrocytes utilize chondronectin (Kleinman et al. 1981), a cartilage-specific attachment protein.

The growing information about collagenous matrices and cell attachment (Ehrmann and Gey 1956; Elsdale and Bard 1972; Rucker et al. 1972) has permitted

Figure 10.15. Amino acid sequence of the collagen α1(I) chain region, known to bind fibronectin. The secondary structure of the active region residues (774–785) is depicted in a more open conformation than the remainder of the molecule, whose tight helical coil is depicted with a regular repeat. The fibronectin binding site (residues 766–786) is shown in boldface. *(Source: Kleinman et al. (1981) J Cell Biol 88:473–485, with permission.)*

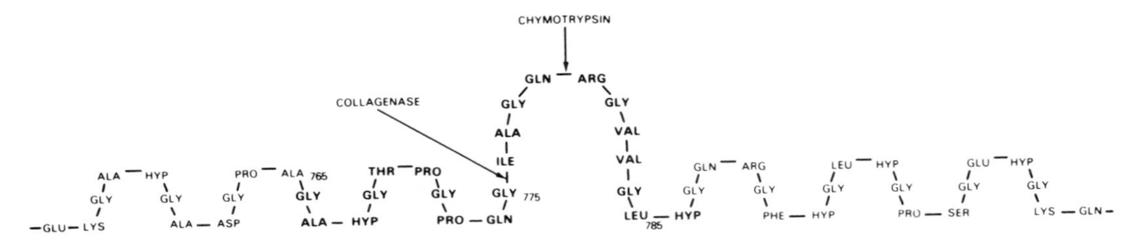

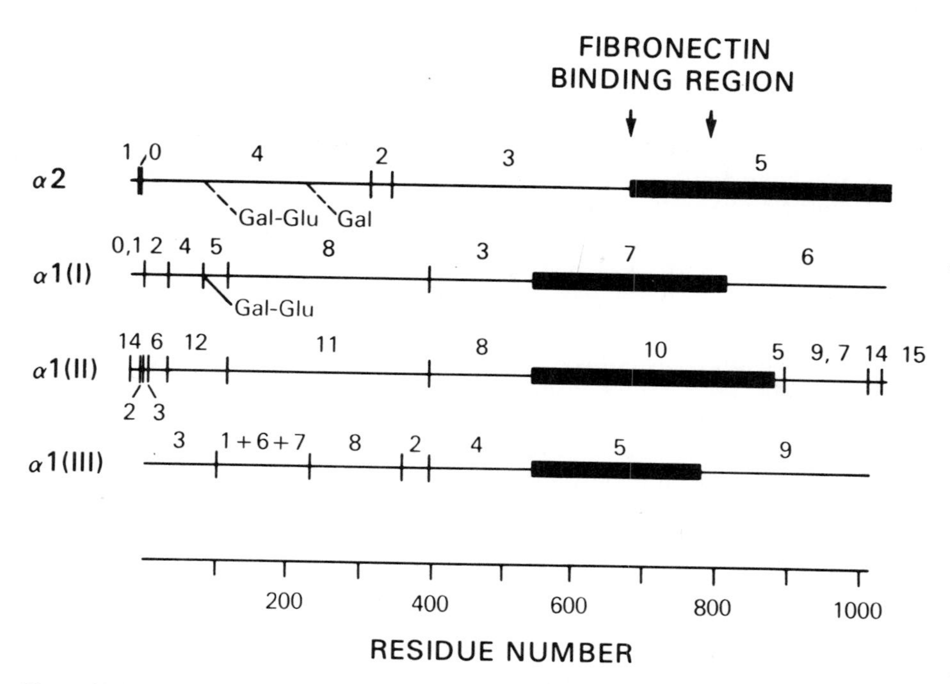

Figure 10.16. Activity of various cyanogen bromide-derived peptides of collagen α chains in binding fibronectin. The numbers between the vertical lines indicate the cyanogen bromide peptide number. The peptides active in binding are shown in solid bars. *(Source: Kleinman et al. (1981) J Cell Biol 88:473–485, with permission.)*

the study of more fastidious cell types such as hepatocytes in culture. Floating collagen membranes support the growth of parenchymal cells from adult liver (Michalopoulos and Pitot (1975). These hepatocytes respond to induction of tyrosine aminotransferase by steroids for up to 3 weeks in culture. Isolated connective tissue matrix (biomatrix) derived from liver permits hepatocyte survival up to 5 months (Rojkind et al. 1980). Hepatocytes do not discriminate among the various collagens and attach equally well to types I, II, III, IV, and V. The rate of cell attachment to native collagen is higher compared with that of denatured collagen (Rubin et al. 1981). Collagen-like synthetic polypeptides such as (Gly-Ala-Pro)$_n$, (Gly-Pro-Pro)$_n$, and (Gly-Pro-Hyp)$_n$ also permit low-affinity binding. However, no attachment is seen to polyproline, which cannot form a triple helix. Hepatocytes from regenerating mouse liver exhibit increased adhesion to laminin and fibronectin (Carlsson et al. 1981).

The use of hydroxyethylmethacrylate hydrogels for cell growth in conjunction with collagen (Civerchia-Perez et al. 1980) has been usefully applied in the study of nerve fiber growth. The extension of nerve fibers by neurons in vitro is dependent on the substratum. Incorporation of fibronectin into hydrogels stimulates neurite outgrowth from chick embryo dorsal root ganglia (Carbonetto et al. 1982).

What is the mechanism of action of the extracellular matrix on growth control? Folkman and Moscona (1978) have considered cell shape critical for DNA synthesis. To test this hypothesis the adhesivity of the tissue culture substratum was varied by incorporation of poly-2-hydroxyethylmethacrylate. Increasing concentrations decreased adhesion, and bovine aortic endothelial cells appeared increasingly rounded with small cell diameters (10 ± 2 μm) and a low rate of cell proliferation. On the other hand, cells grown on unmodified tissue culture plastic were flattened and elongated, with a cell diameter of 51 ± 7 μm and exhibited a brisk rate of DNA synthesis and growth. Gospodarowicz et al. (1978) have shown the importance of cell shape in cell growth and its dependence on extracellular matrix. Corneal epithelial cells are flattened on plastic but are tall and columnar on a collagen substratum. These examples illustrate the profound relationship between cell shape and cell physiology.

What is the biochemical explanation for this relationship? Employing 3T6 fibroblasts, an anchorage-dependent cell line, the influence of suspension in methocel has been investigated (Farmer et al. 1978; Ben-Zeev et al. 1980). The change from attached cells to suspension culture results in a decrease in messenger RNA production and protein synthesis. These decreases are reversible upon resumption of anchorage-dependent growth on solid substratum. Emerging information about the molecular basis of cell anchorage has important implications for development (Section 10.3.3), wound healing and bone regeneration (Section 10.4), and in pathogenesis of cancer (Section 10.2).

10.6.2. Role in Cancer and Metastasis

Collagen and other extracellular matrix components are increasingly implicated in development and differentiation. In view of this, it is not surprising that extracellular matrix is intimately involved in cancer. Cancer is a disease of differentiation in which normal growth control and cell structure are perturbed (Markert 1968). In an advanced state of tumor progression, the almost inevitable consequence is metastasis. In metastatic cells, the potential for crossing barriers between tissues is increased, possibly due to an increase in collagenases (Harper 1980) and other lytic enzymes and due to a decrease or loss of anchorage dependence. For example, it is conceivable that carcinoma cells (derived from epithelium) lose their dependence on basement membrane and can grow into other sites. It is noteworthy that metastatic breast and prostate carcinoma cells invariably migrate to an intrabony site in the skeleton. What mechanisms determine this seemingly specific osteotropism? Other tumors metastasize to lung, liver, and brain. Is this migration random or is it based on some navigational clues in the extracellular matrix? Recent work indicates that pathogenesis of metastasis is the result of selective growth of subpopulations of metastatic cells (Poste and Fidler 1980).

Basement membranes act as a scaffold for epithelial cells (Vracko 1974) and function as a selective filter for macromolecules. Liotta et al. (1979) have identified a collagenolytic enzyme in a metastatic tumor that is specific for type IV collagen of basement membrane. Furthermore, the metastatic potential of tumor cells correlates well with collagenolytic degradation of basement membrane (Liotta et al. 1980). Collagen and its component peptides are chemotactic for tumor cells and may explain in part the high incidence of metastases to organs rich in collagenous stroma such as bone, liver, and lungs (Mundy et al. 1981).

Anchorage-dependent phenomena involved in carcinogenesis are related to particle size. In man, exposure to asbestos is associated with an increased incidence of tumors in lung and in pleural and peritoneal mesothelioma. In experiments in the rat, asbestos fibers 40–320 μm long induced anchorage-dependent mesotheliomas. Particles smaller that 20 μm were phagocytosed by the monocyte-macrophage system and no tumors were elicited. This experiment reveals the critical role of the physical characteristics of asbestos in cancer, which is a reflection of cell–substratum interactions (Maroudas et al. 1973).

Collagen increases the colony forming capacity of normal cells, whereas the growth of transformed cells is not affected (Sanders and Smith 1970). The judicious use of analogs of proline, such as *cis*-hydroxyproline, that block collagen synthesis allows the design of experiments to examine the role of collagen in proliferation of cells as related to the substratum. The attachment, spreading, and proliferation of normal connective tissue cells are inhibited by *cis*-hydroxyproline. The inhibitory effects are overcome by transferring cells to a collagen-coated tissue culture dish (Liotta et al. 1978). However, the tumorigenic cells appear to be resistant to the inhibitory influence of *cis*-hydroxyproline, despite the fact that collagen synthesis is decreased by the proline analog. These experiments reveal the importance of collagen for proliferation of normal cells, but not transformed cells.

Various studies have demonstrated a role for hyaluronic acid in the aggregation of lymphoma cells (Pessac and Defendi 1972; Wasteson et al. 1973). The interaction of hyaluronic acid with lymphoma cells is blocked by tetra- and hexasaccharides derived from hyaluronic acid, suggesting specificity in the interactions (Wasteson et al. 1973). Furthermore, hyaluronate content of invasive tumors is elevated in comparison to that of noninvasive tumors (Toole et al. 1979). Similarly, it is likely that hyaluronate aids cell migration in development of a variety of embryonic tissues (Toole 1976). In this connection it is important to consider the case of certain fibrosarcoma cells apparently resistant to lymphocyte-mediated cytolysis. McBride and Bard (1979) have reported that this resistance was due to cell surface hyaluronic acid that appeared to sterically exclude lymphocytes. Hyaluronidase treatment released the blockage. These observations indicate a possible therapeutic importance of extracellular matrix components in regulatory immune surveillance of tumors. Thus, it is evident that our increased awareness of the role of extracellular matrix components in normal development can help us understand and possibly prevent the ravaging actions of malignant cells.

10.7. Cell Surface–Extracellular Matrix Interactions

10.7.1. A Continuum Model

Growing realization of the importance of extracellular matrix components in development and disease, together with rapid advances in membrane molecular biology, have led to the construction of hypothetical models of cell membrane–matrix interaction. The intricate interplay between the cell exterior and

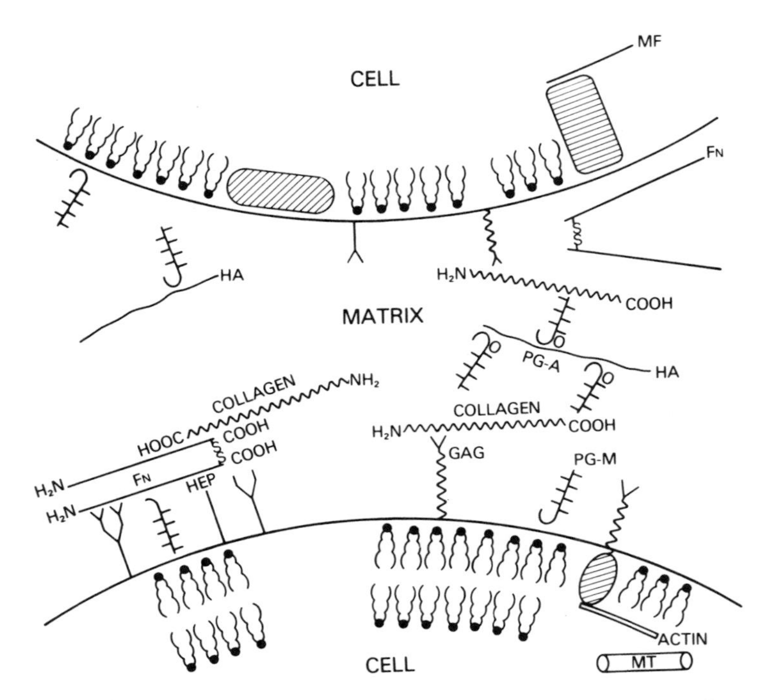

Figure 10.17. The emerging model for extracellular matrix–cell surface continuum. The cell surface/membrane glycosaminoglycans (*GAG*), proteogylcan monomers (*PG-M*), heparin and heparan sulfates (*HEP*) interact with fibronectin (*Fn*) and the collagen. Specific domains are involved in these interactions. In addition, electrostatic forces may play a role. The intracellular microfilaments (*MF*), microtubules (*MT*), and actin cables modulate and transduce external signals at the cell surface–matrix interface to the interior. This results in dynamic changes in cell shape, motility, and eventually, phenotype, as seen during embryonic development.

the interior is largely orchestrated by interaction between integral membrane proteins and components on both sides of the cell membrane. Furthermore, critical events in the realm of sociology of multicellular organisms such as cell–cell interactions, cell attachment, cell shape changes, chemotaxis, and motility require interaction with extracellular matrix components. It is difficult, perhaps impossible, to demarcate where a cell surface ends and the extracellular matrix begins. One working model is presented for a generalized eukaryotic cell in Figure 10.17; it emphasizes that the cell surface and extracellular matrix are a continuum. The ideas about the structure of the cell membrane are based on the fluid mosaic model of Singer and Nicolson (1972). The integral membrane proteins have their nonpolar groups buried in the hydrophobic interior and the polar and ionic groups are exposed to the aqueous phase on both sides of the membrane. The phospholipids are arranged as a discontinuous bilayer with their polar heads in contact with the aqueous phase. Certain integral proteins may span the width of the membrane with domains exposed to both the exterior and the interior or cytoplasmic side. There is a whole spectrum of variations in the conformation of proteins in the lipid bilayer. Lipids constitute the fluid phase in which integral membrane proteins have the potential for lateral mobility by diffusion in the plane of the membrane. The continuum model, which provides the framework for the present discussion, has been constructed by adapting the Singer and Nicolson model to include the extracellular matrix and interactions with intracellular organelles.

Extracellular matrix components in the immediate vicinity of the cell surface include collagens, proteoglycans, and glycoproteins such as fibronectin. The extracellular matrix is predominantly in the solid state, as the component macromolecules are in the form of large sparingly soluble aggregates formed by self-assembly. These include fibrils, as in the case of collagen; supramolecular aggregates formed by noncovalent interactions, as observed in proteoglycans (Chapter 8); or proteins such as fibronectin with a potential for multiple interactions or for self-crosslinking by disulfide bonds (Chapter 7). Fibronectin is known to interact with other macromolecules such as collagens, proteoglycans, heparin, and other glycosaminoglycans (Yamada et al. 1980; Ruoslahti and Engvall 1980; Rich et al. 1981; Hedman et al. 1982). An additional potential for covalent cross-linking is afforded by transglutaminase catalyzed reactions (Folk 1980; Mosher et al. 1979), a scheme still to be developed in detail. Thus, the extracellular matrix is a heterogenous supramolecular assembly with a physical form ranging from a sparingly soluble gel to a stable solid state. In view of this, the matrix may provide a heterogeneous structural and functional framework for the instruction of cells in various tissues during development, regeneration, and repair. It is conceivable that the interaction between peripheral and integral proteins of the cell membrane and extracellular matrix may alter the characteristics of the fluid mosaic. This could lead to altered or restricted mobility of cell surface receptors. Such changes lead to regional inhomogeneities in the cell membrane and may be crucial in transduction of cell surface events as discussed below (Section 10.7.3). Also, extracellular matrix may locally alter cell surface/membrane function and regulate cell motility, cell adhesion, cell–cell interactions, and finally, interpretation of positional information during embryonic development and morphogenesis (Section 10.7). Surface modulation has been implicated in cell recognition and cell growth (Edelman 1976). Surface events may be relayed by vinculin, an intracellular phosphoprotein located in the terminal foci of a group of microfilament bundles (Geiger et al. 1980; *See* Chapter 7), connected to intracellular machinery for cell motility and shape changes. At this point in the model a sequential response is set into motion that leads to changes in intracellular control mechanisms regulating function and ultimately phenotype. The relationship between cell shape, anchorage, and growth was discussed above (Section 10.5).

10.7.2. Receptors and Transduction of Cell Surface Events

The fluid mosaic model of the cell membrane stimulated the design and interpretation of studies on receptors for polypeptide hormones and growth factors. As the chemistry of these ligands is known, the availability of high specific activity probes led to attempts to isolate the hormone receptors and in turn resulted in new insights into mechanisms of the hormone action (Cuatrecasas et al. 1975; Kahn 1976). One can use similar approaches to the study of cell membrane receptors for extracellular matrix components. However, there is a major limitation in that, although the individual components of the matrix are well characterized, information about supramolecular aggregates is only now beginning to be understood.

The binding characteristics of soluble type I collagen to the plasma membrane of fibroblasts, studied by Goldberg (1979), is potentially important. Iodinated type I collagen, used as a probe, binds to fibroblasts but not to epithelial cells. The binding is saturable and reversible and is apparently highly specific. There is a single class of high affinity binding sites ($K_d = 1.2 \times 10^{-11}\ M$). About 500,000 molecules of type I collagen bind to a single fibroblast. This binding appears to be unrelated to fibronectin-mediated binding. It is not clear what role the putative receptors in the plasma membrane normally play.

Similarly, cell surfaces may possess receptors for glycosaminoglycans and proteoglycans. The interaction of hyaluronic acid with certain lymphoma cells has been considered (Pessac and Defendi 1972; Wasteson et al. 1973). It is possible that cell surface glycosaminoglycans such as chondroitin sulfate and heparan sulfate (Kjellen et al. 1981) may exclude proteins by a steric mechanism (Laurent 1977). Thus, matrix and cell surface proteoglycans may locally regulate the proteins in the immediate vicinity of cells.

The emerging information about receptors for matrix molecules may help to explain the transduction of cell surface events to the genome. For example, what is the mechanism of action of extracellular matrix proteins on local bone differentiation (Section 10.4.1)? The local restriction of cell surface mobility by extracellular matrix components may influence the function of receptors for hormones, growth factors, and a variety of developmental cues and signals as yet chemically undefined. Also, electrostatic and other interactions between developmental regulatory proteins and extracellular matrix may result in altered transport (Laurent 1977). Finally, the physical nature of extracellular matrix aggregates, which establishes the nature of the environment, must be critical in regulating cell function. For example, the piezoelectric properties of collagen are well known (Athenstaedt 1970).

10.7.3. Cell Polarity and Function

Numerous cell types in an organism have a constant polarity. Perhaps the best example is epithelial cells resting on a basement membrane. This confers a striking functional polarity. Epithelial cells in certain tissues such as kidney tubules have an apical surface termed brush border. The basal surface is involved in cell–substratum interactions and the lateral surface in cell–cell interactions by means of desmosomes and gap junctions. Epithelial cell monolayers in tissue culture have been employed to understand the polarity (Rodriguez-Boulan and Pendergast 1980). Analysis of the mechanism of epithelial cell polarity is possible by the use of enveloped viruses as an experimental probe. For example, influenza virus and two paramyxo viruses (Sendai and simian virus S) bud from the apical surface of epithelial cells whereas vesicular stomatitis virus (VSV) is released from basolateral regions. It has been shown by immunofluorescence and ferritin immunoelectronmicroscopy that polarized budding is preceded by segregation of viral envelope proteins. When the monolayers are dissociated into suspension cultures the influenza virus buds from the entire cell surface; however VSV budding is severely restricted. On replating the infected cells on a collagen layer under sparse conditions such that cell–cell contact is minimized, the correct budding polarity is restored. Thus the cell–substratum interaction appears to regulate polarized functions of epithelial cells. Culture of tongue epithelium on type I collagen rafts organizes the cells into stratified squamous epithelium at the tissue–liquid interface, which is maintained for up to 30 days (Lillie et al. 1982). Collagen in the basement membrane may determine the polarity of epithelial cells in general. However, mesenchymal cells involved in chick vertebral cartilage development appear to have an inherent polarity (Trelstad 1977). More studies are needed in this area to establish the role of extracellular matrix in cell polarity and function and to learn more about the mechanisms involved.

10.8. Pattern Formation and Positional Information

During embryonic development and in regeneration of limbs in amphibians, cellular activity ultimately determines the spatial pattern (Bryant et al. 1981). The molecular basis of pattern formation is not known. Patterns are specified when cellular differentiation is coordinated in time and space. A possible mechanism for pattern formation is that based on positional information (Wolpert 1969). The current model of specification of positional information is based on gradients of diffusible molecules. It is entirely possible that extracellular matrix in the form of a supramolecular assembly and region-specific molecules, whose rates of diffusion are restricted, determine specification. As discussed in Section 10.6, extracellular matrix potentially can regulate the availability and rate of diffusion of certain sparingly soluble morphogenetically important molecules. For example, Newman and Frisch (1979) have proposed that fibronectin may determine the spatial pattern of developing chick limb. In the current state of ignorance, it is premature to consider specific models for the molecular basis of positional information except heuristically. However, in view of the fact that several developmental phenomena are local events, the possibility that extracellular matrix may function as an affinity matrix to orient morphogenetically crucial molecules is very appealing; future studies will shed more light on this question. Studies on extracellular bone matrix-induced local differentiation of bone (Section 10.4.1) indicate that the above possibility may be close to reality.

10.9. Perspectives

The growing realization of the importance of extracellular matrix in development, disease, and pathogenesis of cancer indicates that the phenomenologic details are currently at hand. However, the mechanisms are not known. The importance of extracellular matrix in development has been considered for a long time (Baitsell 1925; Weiss 1933; Grobstein 1954; Harkness, 1961; Gross 1974). The furious pace of advances in molecular biology of extracellular matrix and its individual components—collagens, proteoglycans, and glycoproteins such as fibronectin and laminin—provides the needed chemical basis. With this background and armed with useful developmental models, the possibility for the eventual understanding of the molecular basis of extracellular matrix organization and of development is excellent. Rapid advances in this area of biology have important implications for the study of both normal and abnormal development.

References

Adamson ED, Ayers SE: (1979) The localization and synthesis of some collagen types in developing mouse embryo. Cell 16:953–965.

Alescio T: (1973) Effect of proline analogue, azetidine-2-carboxylic acid, on the morphogenesis in vitro of mouse embryonic lung. J Embryol Exp Morphol 29:439–451.

Alescio T, Cassini A: (1962) Induction in vitro of tracheal buds by pulmonary mesenchyme grafted on tracheal epithelium. J Exp Zool 150:83–94.

Anglister L, Silman I: (1978) Molecular structure of elongated forms of electric eel acetylcholinesterase. J Mol Biol 125:293–311.

Athenstaedt H: (1970) Permanent longitudinal electric polarization and pyroelectric behavior of collagenous structure and nervous tissue in man and other vertebrates. Nature 228:830–834.

Bailey AJ, Shellswell GB, Duance VC: (1979) Identification and change of collagen types in differentiating myoblasts and developing chick muscle. Nature 278:67–69.

Baitsell GA: (1925) On the origin of the connective-tissue ground substance in the chick embryo. Q J Micro Sci 69:571–587.

Baron-van Evercooren A, Kleinman HK, Seppa HEJ, Rentier B, Dubois-Dalcq M: (1982) Fibronectin promotes rat Schwann cell growth and motility. J Cell Biol 93:211–216.

Bennett DC: (1980) Morphogenesis of branching tubules in cultures of cloned mammary epithelial cells. Nature 285:657–659.

Bentley SA, Alabaster O, Foidart JM: (1981) Collagen heterogeneity in normal human bone marrow. Br J Haematol 48:287–291.

Ben-Zeev A, Farmer SR, Penman S: (1980) Protein synthesis requires cell surface contact while nuclear events respond to cell shape in anchorage-dependent fibroblasts. Cell 21:365–372.

Bernfield MR: (1970) Collagen synthesis during epitheliomesenchymal interactions. Dev Biol 22:213–231.

Bernfield MR: (1981) Organization and remodeling of the extracellular matrix in morphogenesis. In: Connelly TG, ed. Morphogenesis and Pattern Formation. New York: Raven Press. 139–162.

Bernfield MR, Banerjee SD: (1972) Acid mucopolysaccharide (glycosaminoglycan) at the epithelial mesenchymal interface of mouse embryo salivary glands. J Cell Biol 52:664–673.

Bernfield MR, Banerjee SD, Cohn RH: (1972) Dependence of salivary epithelial morphology and branching morphogenesis upon acid mucopolysaccharide-protein (proteogylcan) at the epithelial surface. J Cell Biol 52:674–689.

Bernfield MR, Wessells NK: (1970) Intra- and extracellular control of epithelial morphogenesis. In: Runner MN, ed. Changing Synthesis in Development. New York: Academic Press. 195–249.

Bloch RJ, Geiger B: (1980) The localization of acetylcholine receptor clusters in areas of cell substrate contact in cultures of rat myotubes. Cell 21:25–35.

Bon S, Vigny M, Massoulie J: (1979) Asymmetric and globular forms of acetylcholinesterase in mammals and birds. Proc Natl Acad Sci USA 76:2546–2550.

Borghese E: (1950) Explanation experiments on the influence of the connective tissue capsule on the development of the epithelial part of the submandibular gland of Mus musculus. J Anat 84:303–318.

Borradaile LA, Potts FA, Eastham LES, Saunders JT: (1958) The Invertebrata. 3rd Edition. Revised by Kerkut GA. Cambridge: Cambridge Univ Press.

Bradley K, McConnel-Breul S, Crystal R: (1974) Lung collagen heterogeneity. Proc Natl Acad Sci USA 71:2828–2831.

Brotman HF: (1977) Dermal tissue interactions between mutant and normal embryonic back skin: Site of mutant gene activity determining abnormal feathering in the epidermis. J Exp Zool 200:243–253.

Brownell AG, Besstien CC, Slavkin HC: (1981) Possible functions of mesenchyme cell-derived fibronectin during formation of basac lamina. Proc Natl Acad Sci USA 78:3711–3718.

Bryant SV, French V, Bryant PJ: (1981) Distal regeneration and symmetry. Science 212:993–1002.

Bunge MB, Williams K, Wood PM, Uitto J, Jeffery JJ: (1980) Comparison of nerve cells and nerve cells plus Schwann cells, with particular emphasis on basal lamina and collagen formation. J Cell Biol 84:184–202.

Bunge RP, Bunge MB: (1978) Evidence that contact with connective tissue matrix is required for normal interaction between Schwann cell and nerve fibers. J Cell Biol 78:945–950.

Burden SJ, Sargent PB, McMahan UJ: (1979) Acetylcholine receptors in regenerating muscle accumulate at original synaptic sites in the absence of the nerve. J Cell Biol 82:412–425.

Carbonetto ST, Gruver MM, Turner DC: (1982) Nerve fiber growth on defined hydrogel substrates. Science 216:897–899.

Carlsson R, Engvall E, Freeman A, Ruoslahti E: (1981) Laminin and fibronectin in cell adhesion: Enhanced adhesion of cells from regenerating liver to laminin. Proc Natl Acad Sci USA 78:2403–2406.

Chen LB: (1977) Alteration in cell surface LETS protein during myogenesis. Cell 10:393–400.

Chiang TM, Postlethwaite AE, Beachey EH, Seyer JM, Kang AH: (1978) Binding of chemotactic collagen-derived peptides to fibroblasts. J Clin Invest 62:916–922.

Chiquet M, Puri EC, Turner DC: (1979) Fibronectin mediates attachment of chicken myoblasts to a gelatin-coated substratum. J Biol Chem 254:5475–5482.

Civerchia-Perez L, Faris B, LaPointe G, Beldekas J, Leibowitz H, Franzblau C: (1980) Use of collagen -hydroxyethyl-methacrylate hydrogels for cell growth. Proc Natl Acad Sci USA 77:2064–2068.

Cohen AM, Hay ED: (1971) Secretion of collagen by embryonic neuroepithelium at the time of spinal cord-somite interaction. Dev Biol 26:578–605.

Coulombre A: (1965) The eye. In: DeHaan RL, Ursprung H, eds. Organogenesis. New York: Holt, Rinehart, Winston. 219–251.

Critchley DR, England MA, Wakely J, Hynes RO: (1979) Distribution of fibronectin in the ectoderm of gastrulating chick embryos. Nature 280:498–500.

Cuatrecasas P, Hollenberg MD, Chang KJ, Bennett V: (1975) Hormone receptor complexes and their modulation of membrane function. Rec Prog Horm Res 31:37–110.

Cunha GR: (1976) Epithelial-stromal interactions in development of the urogenital tract. Int Rev Cytol 47:137–194.

David G, Bernfield M: (1981) Type I collagen reduces the degradation of basal lamina proteoglycan by mammary epithelial cells. J Cell Biol 91:281–286.

de la Haba G, Bricker JV: (1981) Formation of striated muscles from myoblasts in vitro: Inhibition of myotube formation by cis-4-hydroxy-L-proline and its reversal by native or denatured collagen (gelatin). Mol Cell Biochem 40:61–63.

Derby M: (1978) Analysis of glycosaminoglycans within the extracellular environments encountered by migrating neural crest cells. Dev Biol 66:321–336.

Dhouailly D: (1977) Dermo-epidermal interaction during morphogenesis of cutaneous appendages in amniotes. Front Matrix Biol 4:86–121.

Dodson JW: (1963) On the nature of tissue interactions in embryonic skin. Exp Cell Res 31:233–240.

Dodson JW, Hay ED: (1974) Secretion of collagen by corneal epithelium. Effect of the underlying substratum on secretion and polymerization of epithelial products. J Exp Zool 189:51–72.

Duance VC, Stephens HR, Dunn M, Bailey AJ, Dubowitz V: (1980) A role for collagen in the pathogenesis of muscular dystrophy? Nature 284:470–472.

Edds MV: (1958) Development of collagen in the frog embryo. Proc Natl Acad Sci USA 44:296.

Edelman GM: (1976) Surface modulation in cell recognition and cell growth. Science 192:218–226.

Ehrmann RL, Gey GO: (1956) The growth of cells on a transparent gel of reconstituted rat-tail collagen. J Natl Cancer Inst 16:1375–1403.

Ekblom P, Alitalo K, Vaheri A, Timpl R, Saxen L: (1980) Induction of a basement membrane glycoprotein in embryonic kidney: Possible role of laminin in morphogenesis. Proc Natl Acad Sci USA 77:485–489.

Elsdale T, Bard J: (1972) Collagen substrates for studies on cell behavior. J Cell Biol 54:626–637.

Emerman JT, Pitelka DR: (1977) Maintenance and induction of morphological differentiation in dissociated mammary epithelium on floating collagen membranes. In Vitro 13:316–328.

Epstein EH: (1974) $(\alpha 1(III))_3$ Human skin collagen. Release by pepsin digestion and preponderance in fetal life. J Biol Chem 249:3225–3228.

Farmer SR, Ben-Zeev A, Benecke BJ, Penman J: (1978) Altered translatability of messenger RNA from suspended anchorage-dependent fibroblasts: Reversal upon cell attachment to a cell surface. Cell 15:627–637.

Fleischmajer R, Billingham RE: (1968) Epithelial–Mesenchymal Interactions. Baltimore: Williams and Wilkins.

Foidart JM, Reddi AH: (1980) Immunofluorescent localization of type IV collagen and laminin during endochondral bone differentiation and regulation by pituitary growth hormone. Dev Biol 75:130–136.

Folk JE: (1980) Transglutaminases. Ann Rev Biochem 49:517–591.

Folkman J, Moscona A: (1978) Role of cell shape in growth control. Nature 273:345–349.

Garrone R: (1978) Sponges. Berlin: Karger.

Gauss-Miller V, Kleinman HK, Martin GR, Schiffman E: (1980) Role of attachment factors and attractants in fibroblast chemotaxis. J Lab Clin Med 96:1071–1080.

Gay S, Viljamo J, Rakeallio J, Pentinen R: (1978) Collagen types in early phases of wound healing in children. Acta Chir Scand 144:205–211.

Geiger B, Tokuyasu KT, Dutton AH, Singer SJ: (1980) Vinculin, an intracellular protein localized at specialized sites where microfilament bundles terminate at cell membranes. Proc Natl Acad Sci USA 77:4127–4131.

Goetinck PF: (1980) Genetical aspects of skin differentiation. In: Spearman RIC, Riley PA, eds. The Skin of Vertebrates. 169–184.

Goetinck PF, Sekellick MJ: (1972) Observations on collagen synthesis, lattice formation and morphology of scaleless and normal embryonic skin. Dev Biol 28:636–648.

Goldberg B: (1979) Binding of soluble type I collagen molecules to the fibroblast plasma membrane. Cell 16:265–275.

Golob R, Chetsanga CJ, Doty P: (1974) The onset of collagen synthesis in sea urchin embryos. Biochim Biophys Acta 349:135–141.

Gordon JS, Lash JW: (1974) In vitro chondrogenesis and cell viability. Dev Biol 36:88–104.

Gospodarowicz D, Greenburg G, Birdwell CR: (1978) Determination of cellular shape by the extracellular matrix and its correlation with the control of cellular growth. Cancer Res 38:4155–4171.

Green H, Goldberg B, Schwartz M, Brown DD: (1968) The synthesis of collagen during the development of *Xenopus laevis*. Dev Biol 18:391–402.

Greenburg JH, Seppa S, Seppa H, Hewitt AT: (1981) Role of collagen and fibronectin in neural crest cell adhesion and migration. Dev Biol 87:259–266.

Grobstein C: (1953) Analysis in vitro of the early organization of the rudiment of the mouse submandibular gland. J Morphol 93:19–43.

Grobstein C: (1954) Tissue interaction in the morphogenesis of mouse embryonic rudiments in vitro. In: Rudnick D, ed. Aspects of Synthesis and Order in Growth. Princeton: Princeton Univ Press. 233–256.

Grobstein C: (1967) Mechanisms of organogenetic tissue interaction. Natl Cancer Inst Monogr 26:279–299.

Grobstein C, Cohen J: (1965) Collagenase: Effect on the morphogenesis of embryonic salivary epithelium in vitro. Science 150:626–628.

Gross J: (1974) Collagen biology: Structure, degradation and disease. Harvey Lect 68:351–432.

Gulati AK, Zalewski AA, Reddi AH: (1983) An immunofluorescent study of the distribution of fibronectin and laminin during limb regeneration in the adult newt. Dev Biol 96:355–365.

Harkness RD: (1961) Biological functions of collagen. Biol Rev 36:399–403.

Harper E: (1980) Collagenases. Ann Rev Biochem 49:1063–1078.

Harper PA, Juliano RL: (1981) Two distinct mechanisms of fibroblast adhesion. Nature 290:136–138.

Hauschka SD, Konigsberg IR: (1966) The influence of collagen on the development of muscle clones. Proc Natl Acad Sci USA 55:119–126.

Hauschka SD, White NK: (1972) Studies of myogenesis in vitro. In: Banker BQ, Przbylski RJ, Van der Meulen JP, Victor M, eds. Research in Muscle Development and Muscle Spindle. Amsterdam: Excerpta Medica. 53–71.

Hay ED: (1973) Origin and role of collagen in the embryo. Am Zool 13:1085–1107.

Hay ED: (1974) Cellular basis of regeneration. In: Lash JW, Whittaker JR, eds. Concepts of Development. Stamford, CT: Sinaner Assoc. 404–428.

Hay ED: (1980) Development of the vertebrate cornea. Int Rev Cytol 63:263–322.

Hay ED, Revel JP: (1969) Fine Structure of the Developing Arian Cornea. Berlin: Karger.

Hedman K, Johansson S, Vartio T, Kjellen L, Vaheri A, Hook M: (1982) Structure of the pericellular matrix: Association of heparan and chondroitin sulfates with fibronectin–procollagen fibers. Cell 28:663–671.

Johnson RC, Manasek FJ, Vinson WC, Seyer JM: (1974) The biochemical and ultrastructural demonstration of collagen during early heart development. Dev Biol 36:252–267.

Kahn CR: (1976) Membrane receptors for hormones and neurotransmitters. J Cell Biol 70:261–286.

Kalcheim C, Vogel Z, Duksin D: (1982) Embryonic brain extract induces collagen biosynthesis in cultured muscle cells: Involvement in acetylcholine receptor aggregation. Proc Natl Acad Sci USA 79:3077–3081.

Kallman F, Grobstein C: (1964) Fine structure of differentiating mouse pancreatic exocrine cells in transfilter culture. J Cell Biol 20:399–411.

Kallman F, Grobstein C: (1965) Source of collagen at epitheliomesenchymal interfaces during inductive interactions. Dev Biol 11:109–176.

Ketley JN, Orkin RW, Martin GR: (1976) Collagen in developing chick muscle in vivo and in vitro. Exp Cell Res 99:261–268.

Kinoshita S, Saiga H: (1979) The role of proteoglycans in the development of sea urchins. I. Abnormal development of sea urchin embryos caused by the disturbance of proteoglycan synthesis. Exp Cell Res 123:229–236.

Kjellén L, Petterson I, Höök M, (1981) Cell-surface heparan sulfate: An intercalated membrane proteoglycan. Proc Natl Acad Sci USA 78:5371–5375.

Klebe RJ (1974) Isolation of a collagen-dependent cell attachment factor. Nature 250:248–251.

Kleinman HK, Klebe RJ, Martin GR: (1981) Role of collagenous matrices in the adhesion and growth of cells. J Cell Biol 88:473–485.

Klose J, Flickinger RA: (1971) Collagen synthesis in frog endoderm cells. Biochim Biophys Acta 232:207–210.

Kollar EJ: (1981) Tooth development and dental patterning. In: Connelly TG, ed. Morphogenesis and Pattern Formation. New York: Raven. 87–101.

Kollar EJ, Baird GR: (1969) The influence of the dental papilla on the development of tooth shape in embryonic mouse tooth germs. J Embryol Exp Morphol 21:131–148.

Kosher RA, Church RL: (1975) Stimulation of in vitro somite chondrogenesis by procollagen and collagen. Nature 258:327–330.

Kosher RA, Lash JW: (1975) Novochord stimulation of in vitro somite chondrogenesis before and after enzymatic removal of perinotochordal materials. Dev Biol 42:362–368.

Kratochwil K: (1972) Tissue interactions during embryonic development: Grand properties. In: Tarin D, ed. Tissue Interactions in Carcinogenesis. New York: Academic Press. 1–47.

Lanotte M, Schor S, Dexter TM: (1981) Collagen gels as a matrix for hemopoiesis. J Cell Physiol 106:269–277.

Lanser ME, Saba RM, Scoville WA: (1980) Opsonic glycoprotein (plasma fibronectin) levels after burn injury: Relationship to burn and development of sepsis. Ann Surg 192:776–782.

Lash JW: (1968) Chondrogenesis: Genotypic and phenotypic expression. J Cell Physiol 72:35–45.

Lash JW, Vasan NS: (1978) Somite chondrogenesis in vitro. Stimulation by exogenous matrix components. Dev Biol 66:151–171.

Laurent TC: (1977) Interaction between proteins and glycosaminoglycans. Fed Proc 36:24–27.

Le Douarin NM: (1980) The ontogeny of the neural crest in avian embryo chimaeras. Nature 286:663–669.

Lesot H, Karcher-Djoricic V, Ruch JV: (1981a) Synthesis of collagen type I, type I trimer and type III by embryonic mouse dental epithelial and mesenchymal cells in vitro. Biochim Biophys Acta 656:206–212.

Lesot H, Osman M, Ruch JV: (1981b) Immunofluorescent localization of collagens, fibronectin, and laminin during terminal differentiation of odontoblasts. Dev Biol 82:371–381.

Lillie JH, MacCallum DK, Jepson A: (1982) The role of defined matrices on growth and differentiation of mammalian stratified squermocis epithelium. In: Sawyer R, Fallon J, eds. Epithelial–Mesenchymal Interactions. New York: Praeger Press, (in press).

Linsenmeyer TF, Gibney E, Toole BP, Gross J: (1978) Cellular adhesion to collagen. Exp Cell Res 116:470–474.

Linsenmeyer TF, Little CD: (1978) Embryonic neural retinal collagen: In vitro synthesis of high molecular weight forms of type II plus a new genetic type. Proc Natl Acad Sci USA 75:3235–3239.

Linsenmeyer TF, Smith GN Jr: (1976) The biosynthesis of cartilage type collagen during limb regeneration in the larval salamander. Dev Biol 52:19–30.

Linsenmeyer TF, Smith GN Jr, Hay ED: (1977) Synthesis of two collagen types by embryonic chick corneal epithelium in vitro. Proc Natl Acad Sci USA 74:39–43.

Linsenmeyer TF, Toole BP, Trelstad RL: (1973) Temporal and spatial transitions in collagen types during embryonic chick development. Dev Biol 35:232–239.

Liotta LA, Abe S, Robey PG, Martin GR: (1979a) Preferential digestion of basement membrane collagen by an enzyme derived from a metastatic murine tumor. Proc Natl Acad Sci USA 76:2268–2272.

Liotta LA, Trygvasson K, Garbisa S, Hart I, Foltz CM, Shafie SA: (1980) Metastatic potential correlates with enzymatic degradation of basement membrane collagen. Nature 284:67–68.

Liotta LA, Vembu D, Kleinman HK, Martin GR, Boone C: (1978) Collagen required for proliferation of cultured connective tissue cells but not their transparent counterparts. Nature 272:602–624.

Liotta LA, Wicha MS, Foidart JM, Rennard SI, Garbisa J, Kidwell WR: (1979b) Hormonal requirements for basement membrane collagen deposition by cultured rat mammary epithelium. Lab Invest 41:511–518.

Lipton BH: (1977) Collagen synthesis by normal and bromodeoxyuridine-modulated cells in myogenic culture. Dev Biol 61:153–165.

Lofberg J, Ahlfors K, Fallstrom C: (1980) Neural crest cell migration in relation to extracellular matrix organization in the embryonic axolotl trunk. Dev Biol 75:148–167.

Lwebuga-Mukasa JS, Lappi J, Taylor P: (1976) Molecular forms of acetylcholinesterase from *Torpedo California:* Their relationship to synaptic membranes. Biochemistry 15:1425–1434.

Manasek FJ, Reis M, Vinson W, Seyer J, Johnson R: (1973) Glyosaminoglycan synthesis by the early embryonic chick heart. Dev Biol 35:332–348.

Markert CL: (1968) Neoplasia: A disease of cell differentiation. Cancer Res 28:1908–1914.

Markwald RR, Fitzharris TP, Bolender DL, Bernanke DH: (1979) Structure analysis of cell: Matrix associated during the morphogenesis of atrioventricular cushion tissue. Dev Biol 69:634–654.

Maroudas NG: (1972) Anchorage dependence: Correlation between amount of growth and diameter of bead, for single cells grown on individual glass beads. Exp Cell Res 74:337–342.

Maroudas NG, O'Neill CH, Stanton MF: (1973) Fibroblast anchorage in carcinogenesis by fibres. Lancet 1:807–809.

Marshall LM, Sanes JR, McMahan UJ: (1977) Reinnervation of original synaptic sites on muscle fiber basement membrane after disruption of the muscle cells. Proc Natl Acad Sci USA 74:3073–3077.

Mays C, Rosenberry TL: (1981) Characterization of pepsin-resistant collagen-like tail subunit fragments of 18S and 14S acetylcholinesterase from *Electrophorus electricus*. Biochemistry 20:2810–2817.

McBride WH, Bard JBL: (1979) Hyaluronidase-sensitive halos around adherent cells. J Exp Med 149:507–515.

McMahan UJ, Sanes JR, Marshall LM: (1978) Cholinesterase is associated with the basal lamina at the neuromuscular junction. Nature 271:172–174.

Meier S, Hay ED: (1974) Control of corneal differentiation by extracellular materials. Collagen as a promoter and stabilizer of epithelial stroma production. Dev Biol 38:249–270.

Michalopoulos G, Pitot HC: (1975) Primary culture of parenchymal liver cells on collagen membranes. Exp Cell Res 94:70–78.

Mosher DF, Schad PE, Kleinman HK: (1979) Cross-linking of fibronectin to collagen by blood coagulation factor XIII_a. J Clin Invest 64:781–787.

Mundy GR, DeMartino S, Rowe DW: (1981) Collagen and collagen-derived fragments are chemotactic for tumor cells. J Clin Invest 68:1102–1105.

Murray JC, Stingl G, Kleinman HK, Martin GR, Kate SI: (1979) Epidermal cells adhere preferentially to type IV (basement membrane) collagen. J Cell Biol 80:97–202.

Nathanson MA, Hay ED: (1980) Analysis of cartilage differentiation from skeletal muscle grown on bone matrix. I. Ultrastructural aspects. Dev Biol 78:301–331.

Nathanson MA, Hilfer SR, Searls RL: (1978) Formation of cartilage by non-chondrogenic cell types. Dev Biol 64:99–117.

Newman S, Frisch HL: (1979) Dynamics of skeletal pattern formation in developing chick limb. Science 205:662–668.

Newsome DA, Linsenmayer TF, Trelstad RL: (1976) Vitreous body collagen: Evidence for dual origin from the neural retina and hyacocytes. J Cell Biol 71:59–67.

Nogami H, Urist MR: (1974) Substrate prepared from bone matrix for chondrogenesis in tissue culture. J Cell Biol 62:510–519.

O'Hare MJ: (1972) Aspects of spinal cord induction of chondrogenesis in chick embryo somites. J Embryol Exp Morphol 27:235–243.

O'Neill CH, Riddle PN, Jordan PW: (1979) The relation between surface area and anchorage dependence of growth in hamster and mouse. Cell 16:909–918.

Pearlstein E: (1976) Plasma membrane glycoprotein which mediates adhesion of fibroblasts to collagen. Nature 202:497–499.

Pessac B, Defendi V: (1972) Cell aggregation: Role of acid mucopolysaccharides. Science 175:898–900.

Pintar JE: (1978) Distribution and synthesis of glycosaminoglycans during quail neural crest morphogenesis. Dev Biol 67:444–464.

Poste G, Fidler IJ: (1980) The pathogenesis of cancer metastasis. Nature 283:139–145.

Postelthwaite AE, Seyer JM, Kang AH: (1978) Chemotactic attraction of human fibroblasts to type I, II and III collagen and collagen-derived peptides. Proc Natl Acad Sci USA 75:871–875.

Pratt RM, Larsen MA, Johnston MC: (1975) Migration of cranial neural crest cells in a cell-free hyaluronate-rich matrix. Dev Biol 44:298–305.

Pucci-Minafra I, Casano C, La Rosa C: (1972) Collagen synthesis and spicule formation in sea urchin embryos. Cell Different 1:157–165.

Rath NC, Reddi AH: (1979) Collagenous bone matrix is a local mitogen. Nature 278:855–857.

Reddi AH: (1974) Bone matrix in the solid state: Geometric influence on differentiation of fibroblasts. In: Lawrence J, Gofman JW, eds. Advances in Biological and Medical Physics, Vol 15. New York: Academic Press. 1–18.

Reddi AH: (1975) Collagenous bone matrix and gene expression in fibroblasts. In: Slavkin H, Greulich RC, eds. Extracellular Matrix Influences on Gene Expression. New York: Academic Press. 619–625.

Reddi AH: (1976) Collagen and cell differentiation. In: Ramachandran GN, Reddi AH, eds. Biochemistry of Collagen. New York: Plenum. 449–478.

Reddi AH: (1981) Cell Biology and biochemistry and endochondral bone development. Collagen Res 1:209–226.

Reddi AH (1982) Regulation of local differentiation of cartilage and bone by extracellular matrix: A cascade type mechanism. In: Kelley RO, Goetinck P, McCabe JA, eds. Limb Development and Regeneration. New York: Liss. 261–268.

Reddi AH, Anderson WA: (1976) Collagenous bone matrix-induced endochondral ossification and hemopoiesis. J Cell Biol 69:557–572.

Reddi AH, Gay R, Gay S, Miller EJ: (1977) Transitions in collagen types during matrix-induced cartilage, bone and bone marrow formation. Proc Natl Acad Sci USA 74:5589–5592.

Reddi AH, Hascall VC, Hascall GK: (1978) Changes in proteoglycan types during matrix-induced cartilage and bone development. J Biol Chem 253:2429–2436.

Reddi AH, Huggins CB: (1972) Biochemical sequences in the transformation of normal fibroblasts in adolescent rats. Proc Natl Acad Sci USA 69:1601–1605.

Reddi AH, Huggins CB: (1975) Formation of bone marrow in fibroblast-transformation ossicles. Proc Natl Acad Sci USA 72:2212–2216.

Rich AM, Pearlstein E, Weissmann G, Hoffstein ST: (1981) Cartilage proteoglycans inhibit fibronectin-mediated adhesion. Nature 293:225–226.

Rodriguez-Boulan E, Pendergast M: (1980) Polarized distribution of viral envelope proteins in the plasma membrane of infected epithelial cells. Cell 20:45–54.

Rojkind M, Gatmaitan Z, Mackensen S, Giambrone MA, Ponce P, Reid LM: (1980) Connective tissue biomatrix: Its isolation and utilization for long term cultures of rat hepatocytes. J Cell Biol 87:255–263.

Ross R: (1968) The fibroblast and wound repair. Biol Rev 43:51–96.

Rubin K, Höök M, Öbrink B, Timpl R: (1981) Substrate adhesion of rat hepatocytes: Mechanism of attachment to collagen substrates. Cell 24:463.

Rucker I, Kettrey R, Zeleznick LD: (1972) Microcrystalline collagen (Avitene) film as a substrate for outgrowth or primary rabbit kidney fibroblasts. Proc Soc Exp Biol Med 139:749–752.

Ruoslahti E, Engvall E: (1980) Complexing of fibronectin glycosaminoglycans and collagen. Biochim Biophys Acta 631:350–358.

Saba TM, Blumenstock FA, Scovill WA, Bernard H: (1978) Cryoprecipitate reversal of opsonic α 2-surface binding glycoprotein deficiency in septic surgical and trauma patients. Science 201:622–624.

Sakakura T, Sakagami Y, Nishizuka Y: (1979) Persistence of responsiveness of adult mouse mammary gland to induction by embryonic mesenchyme. Dev Biol 72:201–210.

Salomon DS, Liotta LA, Kidwell WR: (1981) Differential response to growth factor by rat mammary epithelium plated on different collagen substrata in serum-free medium. Proc Natl Acad Sci USA 78:382–386.

Sampath TK, DeSimone DP, Reddi AH: (1982) Extracellular bone matrix-derived growth factor. Exp Cell Res 142:460–464.

Sampath TK, Reddi AH: (1981) Dissociative extraction and reconstitution of extracellular matrix components involved in local bone differentiation. Proc Natl Acad Sci USA 78:7599–7603.

Sanders FK, Smith JD: (1970) Effect of collagen and acid polysaccharides on the growth of BHK/21 cells in semi-solid media. Nature 227:513–515.

Sanes JR: (1982) Laminin, fibronectin and collagen in synaptic and extrasynaptic portions of the muscle fiber basement membrane. J Cell Biol 93:442–451.

Sawyer RH: (1979) Avian scale development: Effect of the scaleless gene on morphogenesis and histogenesis. Dev Biol 68:1–15.

Saxen L: (1971) Inductive interactions in kidney development. In: Control Mechanisms of Growth and Differentiation. Symposia of the Society for Experimental Biology. Cambridge: Cambridge Univ Press. 207–221.

Schachner M, Schoonmaker G, Hynes RO: (1978) Cellular and subcellular localization of Lets protein in the nervous system. Brain Res 158:149–158.

Schor SL, Court J: (1979) Different mechanisms in the attachment of cells to native and denatured cells. J Cell Sci 38:267–281.

Sengel P: (1975) Feather pattern development. In: Cell Patterning. CIBA Foundation Symposium 29:51–70.

Seppa H, Seppa S, Grotendorst GR, Gauss-Muller V, Kleinman HK, Schiffman E, Martin GR: (1982) Matrix glycoproteins as mediators of cell adhesion and tissue repair. In: Germach O, Pott G, Rautherberg J, Yoss B, eds. Connective Tissue of the Normal and Fibrotic Human Liver. Shattover, Stuttgart.

Shoshan S, Finkelstein S: (1970) Acceleration of wound healing induced by enriched collagen solutions. J Surg Res 10:485–491.

Singer SJ, Nicolson G: (1972) The fluid mosaic model of the structure of cell membranes. Science 175:720–731.

Slavkin HC: (1972) Intercellular communication during odontogenesis. In: Slavkin HC, Bavetta LA, eds. Developmental Aspects of Oral Biology. New York: Academic Press. 165–199.

Slavkin HC, Greulich RC (eds.): (1975) Extracellular Matrix Influence on Gene Expression. New York: Academic Press.

Smith GN Jr, Linsenmeyer TF, Newsome DA: (1976) Synthesis of type II collagen in vitro by embryonic neural retina tissue. Proc Natl Acad Sci USA 73:4420–4423.

Smith GN Jr, Toole BP, Gross J: (1975) Hyaluronidase activity and glycosaminoglycan synthesis in the amputated Newt limb: Comparison of denervaties nonregenerating limbs with regeneraties. Dev Biol 43:221–232.

Solursh M, Fisher M, Singley CT: (1979) The synthesis of lyaluronic acid by ectoderm during early organogenesis in the chick embryo. Differentiation 14:77–85.

Somerman M, Hewitt AT, Reddi AH, Seppa H, Varner H, Termine JD, Schiffman E: (1983) The role of chamotaxis in bone induction. In: Silbermann M, ed. Proceedings International Workshop on Calcified Tissues. Amsterdam: Excerpta Medica.

Spiegel E, Burger M, Spiegel M: (1980) Fibronectin in the developing sea urchin embryo. J Cell Biol 87:309–313.

Spiegel E, Speigel M: (1979) The hyaline layer is a collagen-containing extracellular matrix in sea urchin embryos and reaggregating cells. Exp Cell Res 123:434–441.

Spooner BS, Faubion JM: (1980) Collagen involvement in branching morphogenesis of embryonic lung and salivary gland. Dev Biol 77:84–102.

Spooner BS, Wessells NK: (1970) Mammalian lung development: Interactions in primordium formation and bronchial morphogenesis. J Exp Zool 175:445–461.

Steinmann BU, Reddi AH: (1980) Changes in synthesis of types-I and -III collagen during matrix-induced endochondral bone differentiation in rat. Biochem J 186:919–924.

Stoker M, O'Neill C, Berryman S, Waxman V: (1968) Anchorage and growth regulation in normal and virus-transformed cells. Int J Cancer 3:683–693.

Stuart ES, Garber B, Moscona AA: (1972) An analysis of feather germ formation in the embryo and in vitro, in normal development and in skin treated with hydrocortisone. J Exp Zool 179:97–118.

Terranova VP, Rohrbach DH, Martin GR: (1981) Role of laminin in the attachment of PAM 212 (epithelial) cells to basement membrane collagen. Cell 22:719–726.

Thesleff I, Hurmerinta K: (1981) Tissue interactions in tooth development. Differentiation 18:75–88.

Thesleff I, Stenman S, Vaheri A, Timpl R: (1979) Changes in the matrix proteins, fibronectin and collagen during differentiation of mouse tooth germ. Dev Biol 70:116–126.

Toole BP: (1976) Morphogenetic role of glycosaminoglycans (acid mucopolysaccharides) in brain and other tissues. In: Barondes SH, ed. Neural Recognition. New York: Plenum. 275–329.

Toole BP, Biswas C, Gross J: (1979) Hyaluronate and invasiveness of the rabbit V2 carcinoma. Proc Natl Acad Sci USA 76:6299–6303.

Toole BP, Gross J: (1971) The extracellular matrix of the regenerating newt limb: Synthesis and removal of hyaluronate prior to differentiation. Dev Biol 25:57–77.

Trelstad RL: (1977) Mesenchymal cell polarity and morphogenesis of chick cartilage. Dev Biol 59:153–163.

Trelstad RL, Coulombre AJ: (1971) Morphogenesis of the collagenous stroma in the chick cornea. J Cell Biol 50:840–858.

Trelstad RL, Hay ED, Revel JP: (1967) Cell contact during early embryogenesis in the chick embryo. Dev Biol 16:78–91.

Trelstad RL, Hayashi K, Toole BP: (1974) Epithelial collagens and glycosaminoglycans in the embryonic cornea. Macromolecular order and morphogenesis in the basement membrane. J Cell Biol 62:815–830.

Trelstad RL, Kang AH: (1974) Collagen heterogeneity in the avian eye: Lens, vitreous body, cornea and sclera. Exp Eye Res 18:395–406.

Trelstad RL, Kang AH, Cohen AM, Hay ED: (1973) Collagen synthesis in vitro by embryonic spinal cord epithelium. Science 179:295–297.

Urist MR: (1965) Bone: Formation by autoinduction. Science 150:893–899.

Vinson WC, Seyer JM: (1974) Synthesis of type III collagen by embryonic chick skin. Biochem Biophys Res Commun 58:58–66.

von der Mark H, von der Mark K, Gay S: (1976) Study of differential collagen synthesis during development of the chick embryo by immunofluorescence. Dev Biol 48:237–248.

von der Mark K, von der Mark H, Timpl R, Trelstad RL: (1977) Immunofluorescent localization of collagen types I, II and III in the embryonic chick eye. Dev Biol 59:75–85.

Vracko R: (1974) Basal lamina scaffold—Anatomy and significance for maintenance of orderly tissue structure. Am J Pathol 77:314–346.

Wartiovaara J, Nordling S, Lehtonen E, Saxen L: (1974) Transfilter induction of kidney tubules: Correlation with tyocplasmic penetration into Nucleopore filters. J Embryol Exp Morphol 31:667–682.

Wasteson A, Westermark R, Lindahl U, Ponten J: (1973) Aggregation of feline lymphoma cells by hyaluronic acid. Int J Cancer 12:169–178.

Weiss L: (1976) The hematopoietic microenvironment of the bone marrow: An ultrastructural study of the stroma in rats. Anat Rec 186:161–184.

Weiss P: (1933) Functional adaptation and the role of ground substances in development. American Nationalist 67:322–340.

Weiss RE, Reddi AH: (1980) Synthesis and localization of fibronectin during collagenous matrix-mesenchymal cell interaction and differentiation of cartilage and bone in vivo. Proc Natl Acad Sci USA 77:2074–2078.

Weiss RE, Reddi AH: (1981a) Isolation and characterization of rat plasma fibronectin. Biochem J 197:529–534.

Weiss RE, Reddi AH: (1981b) Role of fibronectin in collagenous matrix-induced mesenchymal cell proliferation and differentiation in vivo. Exp Cell Res 133:247–254.

Weiss RE, Reddi AH: (1981c) Appearance of fibronectin during the differentiation of cartilage, bone and bone marrow. J Cell Biol 88:630–636.

Wessells NK: (1964) Substrate and nutrient effects upon epidermal basal cell differentiation and proliferation. Proc Natl Acad Sci USA 52:252–259.

Wessells NK: (1977) Tissue Interactions and Development. Menlo Park, Calif.: W. A. Benjamin Inc. 276.

Wicha MS, Liotta LA, Garbisa S, Kidwell WR: (1979) Basement membrane collagen requirements for attachment and growth of mammary epithelium. Exp Cell Res 124:181–190.

Wicha MS, Liotta LA, Vonderhaar BK, Kidwell WR: (1980) Effects of inhibition of basement membrane collagen deposition on rat mammary gland development. Dev Biol 80:253–266.

Wolpert L: (1969) Positional information and the spatial pattern of cellular differentiation. J Theor Biol 25:1–40.

Yamada K, Kennedy DW, Kimata K, Pratt RM: (1980) Characterization of fibronectin interactions with glycosaminoglycans and identification of active proteolytic fragments. J Biol Chem 255:6055–6063.

Yang J, Richards J, Bowman P, Guzman R, Enami J, McCormick K, Hamamoto J, Pitelka D, Nandi S: (1979) Sustained growth and three-dimensional organization of primary mammary tumor epithelial cells embedded in collagen gels. Proc Natl Acad Sci USA 76:3401–3405.

11

Genetic and Acquired Disorders of Collagen Deposition

Stephen M. Krane

11.1. Introduction

In disease, aberrations in collagen deposition may be manifested by (a) insufficient collagen content in regions where collagen is normally deposited; (b) presence of chemically and/or morphologically abnormal collagen, in normal or decreased concentrations; (c) excessive collagen content, generally or in specific locations, of normal or abnormal type; (d) insufficient collagen resorption: and (e) excessive collagen resorption. These aberrations could result from genetic or acquired causes. This chapter will consider selected genetic and acquired disorders in humans in which some information is available concerning the mechanisms of abnormalities in collagen synthesis.

11.2. Genetic Diseases of Collagen

McKusick (1972) defined the heritable disorders of connective tissue as generalized defects that clinically involve primarily one element of the connective tissues (e.g., collagen or elastin) and are transmissible in mendelian patterns. Syndromes usually considered to belong to this group include the Ehlers-Danlos syndrome, Marfan syndrome, forms of cutis laxa, and osteogenesis imperfecta. Certain diseases whose manifestations do not appear until adulthood, such as idiopathic osteoporosis, floppy mitral valves, and idiopathic dissection of the aorta, may eventually be shown to have their basis in genetically determined

---

From the Department of Medicine, Harvard Medical School, and Medical Services (Arthritis Unit), Massachusetts General Hospital, Boston, Massachusetts

Original work referred to here was supported by NIH grants AM-03564, AM-07258 and grants from the Massachusetts Chapter and National Arthritis Foundation. This is publication number 913 of the Robert W. Lovett Memorial Group for the study of Diseases Causing Deformities.

abnormalities in the biosynthesis of a major connective tissue macromolecular component such as a collagen.

Several of the diseases generally accepted as heritable disorders of connective tissue are distinct entities, although the clinical phenotype may vary considerably. Although this variability had been traditionally explained by differences in penetrance and expressivity of single genes, it is likely that genetic heterogeneity is responsible. Each clinically distinguishable subtype might thus have a unique biochemical basis. Although the literature on the heritable disorders of connective tissue is relatively enormous, it has been possible in only a few instances to obtain any information concerning specific biochemical defects. Even in these disorders, a direct connection cannot always be established between the biochemical defect and the clinical abnormality.

Understanding genetic disorders of collagen is particularly difficult, however, in view of the large size of collagen molecules compared to other proteins (Krane 1980). Furthermore, each component polypeptide chain of the different collagen types is the product of a different gene. As a further complexity, it is presumed but not proven that a collagen type from different tissues is identical and, therefore, that the genes (for example, skin and bone type I procollagen) are identical. There are tissue-specific differences in posttranslational modifications of these proteins that serve as markers, although the functional significance of such differences has not yet been demonstrated. The collagen genes, whose structure has been partially elucidated, are complex and contain many intervening sequences. Genetic abnormalities in collagen could therefore result from several different types of mutations that lead to amino acid substitutions or deletions or to deficient synthesis of a portion or the entire product coded for by the gene. These abnormalities could result in insufficient messenger RNA (mRNA), or a defective mRNA that might not be processed or translated normally. Because the mature, completed collagens also undergo a large number of critical posttranslational modifications, diseases could also result from defects in enzymes essential for these posttranslational events. Despite such complexities, it is worthwhile considering in some detail certain human diseases where there is information available on the genetic control of collagen synthesis.

11.2.1 Ehlers-Danlos Syndrome

Ehlers-Danlos syndrome (EDS) is heterogeneous. It is comprised of a group of clinical manifestations, the most characteristic of which are joint hypermobility (Figure 11.1) and skin hyperextensibility (Figure 11.2). The skin may show a tendency to split with minor trauma, leaving thin gaping scars or violaceous nodules called molluscoid pseudotumors (Figures 11.3 and 11.4). Calcified spherules (spheroids) may be found in deep subcutaneous tissues. A tendency to bruise easily is common and rarely major arteries rupture, apparently spontaneously. Tissue friability may be so extensive that major viscera rupture resulting in sudden death. On the basis of a survey of 100 patients with features of EDS, Beighton et al. (1969) observed that whereas the manifestations of the condition were similar in affected members of the same kindred, there was wide variation among different families. It was therefore proposed to classify the syndrome

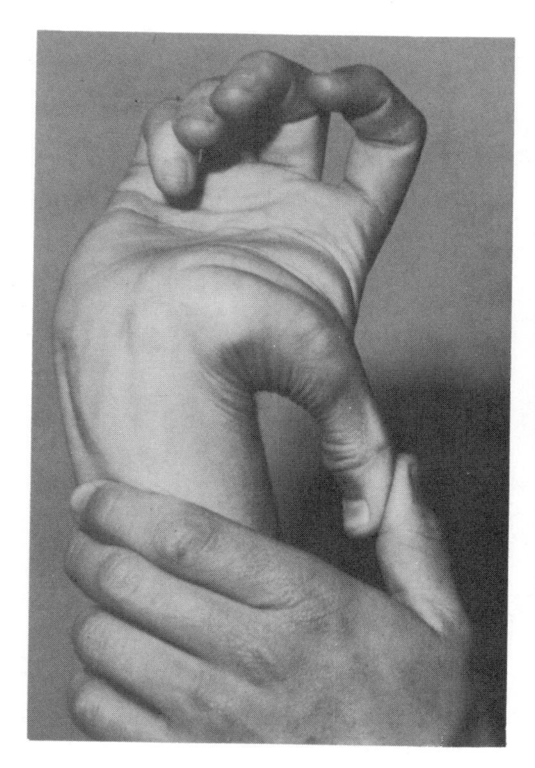

Figure 11.1. Typical hypermobility of the thumb joint in an individual with Ehlers-Danlos syndrome, probably type I. With minimal pressure and no pain, the thumb can be touched to the forearm.

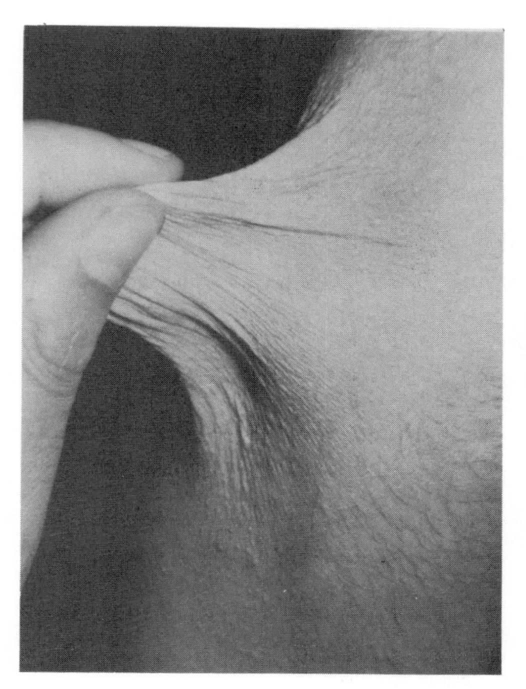

Figure 11.2. Hyperextensibility of the skin in an individual with Ehlers-Danlos syndrome (probably type I).

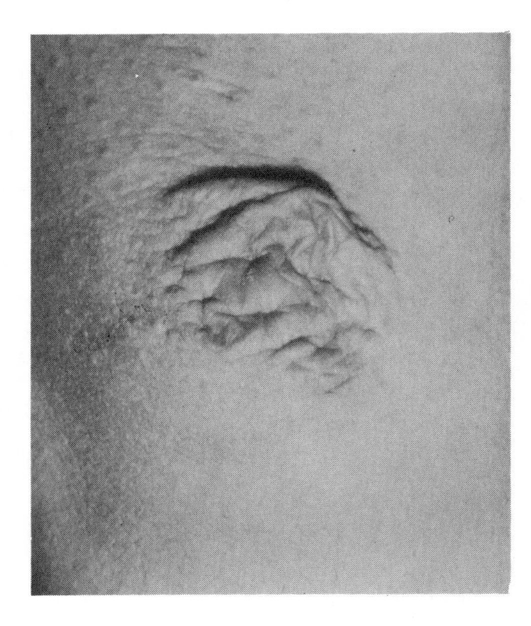

Figure 11.3. An example of thinning of the dermis leading to redundancy of the skin in an individual with Ehlers-Danlos syndrome (probably type I).

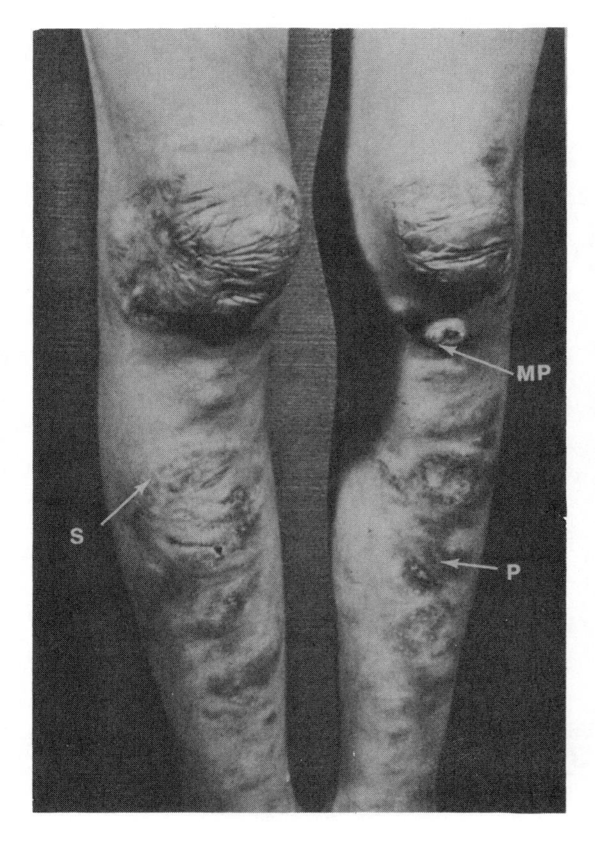

Figure 11.4. Extreme changes in the skin and subcutaneous tissues in a girl with type I Ehlers-Danlos syndrome. Note thin scar (*S*) and pigmentation (*P*) as a result of old hemorrhage, and markedly thin skin over the knees and nodules consistent with molluscoid pseudotumors (*MP*).

Table 11.1. Classification of Ehlers-Danlos Syndrome (1969)[a]

Type	Name	Features
I	Gravis	Gross joint hypermobility, skin hyperextensibility, skin splitting, easy bruising
II	Mitis	All of above, moderate
III	Benign hypermobile	Gross joint hypermobility, variable skin hyperextensibility, minimal skin splitting and bruising
IV	Ecchymotic	Minimal joint hypermobility (digits) and skin hyperextensibility, moderate skin splitting, marked bruising
V	X-linked	Moderate skin hyperextensibility minimal joint hypermobility (digits), minimal skin splitting and bruising

[a]Beighton et al. 1969.

into five types using clinical and genetic criteria (Table 11.1). The most common types in Beighton's survey were types I and II (gravis and mitis). In the type V EDS, in which skin hyperextensibility was striking, the inheritance appeared to be X-linked, whereas the other forms of EDS described were thought to be inherited as an autosomal dominant trait.

Since Beighton proposed this classification of the EDS, biochemical abnormalities have been defined in a small number of patients and the classification has been revised to include these forms of the condition. However, there are still proposals, such as that of Hollister (1978), to include other forms of the syndrome still defined solely on a clinical basis. A revised classification of EDS, including the form associated with marked periodontal disease (Stewart et al. 1977) and that thought to be due to defective fibronectin (Arneson et al. 1980)—both tentatively placed into a separate category rather than included in types

Table 11.2. Classification of Ehlers-Danlos Syndrome (1982)

Type	Name	Biochemical defect
I	Gravis	Unknown
II	Mitis	Unknown
III	Benign hypermobile	Unknown
IV	Ecchymotic	Decreased type III collagen synthesis and/or secretion
V	X-linked	Peptidyl lysine oxidase defect (?)
VI	Hydroxylysine deficient (ocular)	Peptidyl lysine hydroxylase defect
VII	Arthrochalasis multiplex congenita	Decreased procollagen N protease or resistant cleavage site
VIII	Periodontal disease	Unknown
IX	Skeletal and urinary tract dysplasia	Peptidyl lysine oxidase defect (?)
X	Fibronectin deficient	Abnormal fibronectin (?)

I–III—is given in Table 11.2. It has become apparent from examination of different affected kindreds that more than one type of mutation may be responsible for an apparently identical biochemical defect. Furthermore, the origin of the defects could range from mutations in structural portions of genes, possible regulatory regions, or in genes coding for posttranslational modifying enzymes.

Ehlers-Danlos Syndrome I–III. Although the first three forms of EDS are the most common, a biochemical mechanism accounting for the phenotype has not been elucidated. The possibility that features of the disorder could be explained on the basis of abnormal intermolecular crosslinking of collagen seemed reasonable at one time. In 1972, Mechanic described a markedly abnormal chromatographic pattern of reducible crosslinks following reduction with sodium borotritide and hydrolysis. Unfortunately, no clinical details or amino acid analysis of the dermal sample were reported. Since then Black et al. (1980) have performed detailed studies on skin from eight patients with EDS I and three with EDS II, all of whom fulfilled the clinical and genetic criteria for classification. The amino acid analysis and the pattern of reducible crosslinks of the skin collagen were indistinguishable from controls. The collagen fibers also had normal ultrastructure as assessed by transmission electron microscopy (usual axial periodicity and organization). In those samples from EDS skin, where the fibers were homogenized briefly, and the specimens examined using negative staining, broken-ended fibers were found that were not usually observed in control samples. The mean collagen fiber diameters were slightly smaller than control (gravis, 81 ± 10nm; mitis, 78 ± 10 nm; normal controls, 101 ± 10 nm). These findings differed from those of Vogel et al. (1979), who found that the fibrils were larger than normal (110–140 nm compared to controls of 90–100 nm). As the controls were similar in both studies, the explanation for the differences might be due to some artifact of preparation. It is also possible (and likely) that what is called EDS I is also heterogeneous and that different patients have different sized collagen fibers and fibrils. The most consistent abnormalities observed by Black et al. (1980) were seen by scanning electron microscopy. The collagen fibers from EDS I were "grossly disorganized and appeared to have dried down to a flat felt work" and had a coating of "ground substance" on the flattened bundles. Black et al. concluded that these differences could account for the skin hyperextensibility and joint hypermobility, assuming that under tension considerable extension would be required before the majority of collagen fiber bundles would be aligned in the direction of the tension. At that point they would be tightly packed and resist further extension. It is possible that the problem of EDS I–III will be better approached when more is known about the detailed mechanisms of collagen fiber formation.

Ehlers-Danlos Syndrome IV. In the type IV syndrome, affected individuals have only slight joint hypermobility, usually limited to the digits, and skin hyperextensibility is minimal or absent (Beighton et al. 1969). It was pointed out by Barabas in 1967 that the tendency towards bruising, which in some individuals is extensive, and the extreme arterial fragility distinguishes this variant from other forms of EDS. This syndrome is, however, clinically heterogeneous. Pope et al. (1980b) included patients with so-called acrogeria. The characteristic facies of acrogeria consists of a pinched, delicate nose, prominent eyes, and unusually thin skin with easily visible superficial veins. Minor trauma may produce extensive ecchymoses, particularly over bony prominences, which eventually result in thin, pigmented scars. The musculoskeletal features of other forms of EDS are usually not encountered. Sudden death in EDS is usually due to rupture of an artery or perforation of a major viscus, most often the colon, or rupture of the uterus in pregnant women (Byers et al. 1981a; Beighton 1968; Krane and Trelstad 1979). These catastrophies occur almost exclusively in EDS I and IV. The clinical heterogeneity in EDS IV is accompanied by genetic heterogeneity; evidence has been presented for autosomal dominant and autosomal recessive inheritance (Byers et al. 1981a; Pope et al. 1977). In view of the clinical and genetic heterogeneity of the syndrome, it is not surprising that a single biochemical abnormality could not account for all instances of EDS IV.

The possibility that the manifestations of EDS IV could be explained by an absence of type III collagen was first proposed by Pope et al. in 1975. Specimens of several tissues were obtained within 12 hours after death in one individual and skin by biopsy from four others. In the tissues from the patient who died, amino acid analysis showed a marked decrease in collagen concentration, indicated mainly by decreases in the levels of 4-hydroxyproline and hydroxylysine (Table 11.3). The contents of valine and alanine were higher than controls, suggesting that the proportion of elastin was higher in this EDS IV aorta. Cyanogen bromide cleavage products of skin and other tissues examined by (sodium

Table 11.3. Amino Acid Analysis of Aorta from a Patient with Ehlers-Danlos Syndrome IV Compared with Controls[a]

Amino acid	EDS IV	Controls
	residues/1000	
4-Hydroxyproline	10	30
Aspartic acid	47	56
Threonine	26	29
Serine	22	34
Glutamic acid	56	76
Proline	93	91
Glycine	226	236
Alanine	168	131
Cystine (half)	6	13
Valine	117	89
Methionine	2.2	3.6
Isoleucine	21	25
Leucine	76	53
Tyrosine	22	18
Phenylalanine	37	24
Hydroxylysine	0.5	3.3
Histidine	17	14
Lysine	36	32
Arginine	23	35

[a]Data from Pope et al. (1975); data expressed as residues per 1000.

dodecyl sulfate) SDS polyacrylamide gel electrophoresis (Chapter 2) differed significantly from controls in that peptides from type III collagen were not detected. Furthermore, no bands consistent with type III collagen were detected after limited pepsin digestion and SDS polyacrylamide gel electrophoresis. Fibroblasts grown from the skin of these patients incubated with ^{14}C-amino acids did not incorporate label into medium proteins that eluted on *O*-(diethylaminoethyl) (DEAE)-cellulose column chromatography in regions expected for type III procollagen. Furthermore, cells in culture from two of the five patients studied by Pope et al. (1975) failed to stain by immunofluorescence with antibodies specific for type III procollagen (the defect was therefore not a posttranslational abnormality of procollagen secretion) (Gay et al. 1976b). Moreover, as some normal fetal fibroblasts reacted with both antibodies to type III procollagen and type I collagen, it was considered unlikely that in these subjects with EDS IV there was a deficiency of a specific type III procollagen-synthesizing cell.

Subsequent studies on members of the families described initially were reported (Pope et al. 1977). It was suggested for the kindred of the individual who had died of spontaneous rupture of the aorta (Pope et al. 1975) that the putative heterozygote had decreased type III collagen in skin compared to controls. In addition, cultured fibroblasts released levels of labeled type III procollagen lower than in controls, whereas fibroblasts from the affected member of that kindred released no detectable labeled type III procollagen. These results suggested that the inheritance of the defect (in this kindred) was autosomal recessive.

On the basis of these observations it was concluded that no type III collagen (procollagen) was synthesized and that the defect in EDS IV could be analogous to the disturbances in hemoglobin synthesis in forms of thalassemia. It is not clear from these reports, however, that the techniques used initially by Pope et al. (1975) to detect type III collagen in tissues and type III procollagen synthesis by fibroblasts were sufficiently sensitive to permit the conclusion that *no* type III collagen is synthesized. For example, recovery from DEAE-cellulose columns is not complete. In addition, it was not demonstrated definitively that the defect was the same in all affected subjects. It soon became apparent that EDS IV is clinically heterogeneous. For example, a 27-year-old man with a lifelong history of easy bruising; fragile varicose veins, which were operated on twice at ages 18 and 21 years; and bilateral spontaneous pneumothoraces died suddenly of rupture of a hepatic artery. He was observed to have thin, wrinkled scars and a prominent superficial venous pattern over the chest, abdomen, and upper extremities (Krane and Trelstad 1979). Although this individual had many clinical features of EDS IV, at autopsy he had demonstrable type III collagen in skin and other tissues.

Subsequently, Pope et al. (1980b) described additional individuals who could be considered to have EDS IV. One of these patients had acrogeria, a history of spontaneous rupture of the spleen, recurrent pneumothoraces, and rupture of a renal artery aneurysm. Another patient was a 7-year-old girl with persistent unexplained bruising. Her father had abnormal hands and feet consistent with acrogeria but a normal facies. He also had recurrent spontaneous pneumothoraces. It was stated that, in all three subjects, no type III collagen was synthesized by their cultured fibroblasts, although the data were not shown and the detection sensitivity in controls was not given. The first patient had detectable but low levels of cyanogen bromide peptides from type III collagen. In contrast, the two individuals from the second kindred had only traces of these peptides. The type of inheritance in this kindred would be consistent with autosomal dominant. It was thus proposed to classify EDS IV as shown in Table 11.4.

The problem is more complex than this, however, in view of the recent reports of Byers et al. (1981b) and Holbrook and Byers (1981). They studied in detail skin and dermal fibroblasts from a 26-year-old woman with clinical features of EDS IV associated with decreased content of type III collagen in skin. The levels of labeled procollagen III in medium proteins following incubation with [^{3}H]proline were markedly reduced (10–15% of control) as analyzed directly by DEAE-cellulose chromatography or after limited pepsin digestion and CM-cellulose chromatography. The total labeled collagenase-digestible protein was *increased* in the cells, however, and cell proteins analyzed by SDS polyacrylamide gel electrophoresis following pepsin digestion showed the presence of a disulfide-bonded collagenous trimer consistent with type III procollagen. However, the intracellular type III procollagen band was of higher molecular weight than the normal type III procollagen band. Type I procollagen was secreted normally by these cells. The cellular retention of type III procollagen was also demonstrated using immunofluorescence with antibodies directed against type III procollagen. The finding that the retained intracellular band, consistent with type III procollagen, was of higher molecular weight than expected suggested that the procollagen might have extensive posttranslational modifications possibly secondary to some alteration in the amino acid sequence of the component polypeptide chains. It was proposed further that the relatively low amounts of apparently normal type III procollagen would be approximately those predicted if only one allele at a single locus were abnormal. As type III procollagen is a homopolymer containing three identical chains, only one eighth of the assembled molecules would be normal were the molecules assembled randomly from a pool containing equal numbers of normal and abnormal chains. The normal–abnormal heteropolymers or the abnormal homopolymers might then be retained or degraded intracell-

Table 11.4. Proposed Classification of Ehlers-Danlos Syndrome IV[a]

Type	Life expectancy	Phenotype	Inheritance	Biochemical lesion
A	Short	"Typical;" Pinched face, prominent eyes, thin skin, short stature, arterial rupture	Autosomal recessive	No type III collagen in tissues or cells
B	Average	Slightly atypical face, less severe arterial disease	Autosomal recessive	Some type III in tissues, none in culture
C	Average	Normal face, thin skin, no arterial problems	Autosomal dominant	Little or no type III in culture

[a]*Source: Pope, Nicholls, et al. (1980) J R Soc Med 73:180–186, modified, with permission.*

ularly. It could be predicted that such an abnormality would be inherited as an autosomal dominant trait.

The ultrastructural findings of the skin from this patient (Byers et al. 1981b; Holbrook and Byers 1981) were consistent with the results of the in vitro studies. Dermal fibroblasts contained a markedly dilated rough endoplastic reticulum that might have resulted from storage of the nonsecreted type III procollagen. The failure to find this dilated endoplastic reticulum in endothelial or smooth muscle cells from the same individual suggested to Holbrook and Byers (1981) that there might be multiple genes coding for type III procollagen or different mechanisms for processing the same type III procollagen in different tissue cells. Another individual with EDS IV with decreased tissue levels (lung) of type III collagen and decreased synthesis of type III procollagen by lung fibroblasts had prominent dilation of the endoplasmic reticulum in the cells from the affected tissue (Clark et al. 1980a). A biopsy of pulmonary tissue from this patient was done at the time of surgery for recurrent pneumothoraces. Increased collagen was solubilized by acetic acid or pepsin, although the proportion of type III collagen in the solubilized fraction was markedly reduced.

Thus, EDS IV should be considered to encompass a group of patients with common clinical features. In almost all individuals who fit the clinical criteria for EDS IV there is a defect in the synthesis of type III collagen. In view of the association with vascular lesions, type III collagen must be critical in some way for the mechanical properties of blood vessel walls. An artery lacking type III collagen (or possibly containing an abnormal type III collagen) is not able to withstand the repeated force of pulsatile pressure and eventually ruptures, although the nature of this postulated mechanical defect is unknown. Some studies have indicated that type III collagen is a major fibrillar component beneath endothelial cells (Gay et al. 1975). Perhaps a deficiency in this collagen, which interacts well with platelets to cause their aggregation (Balleisen et al. 1975), contributes to the bleeding diathesis.

The defect in type III collagen synthesis is obviously not the same in all affected subjects. It is possible that some individuals, such as those initially reported by Pope et al. (1975) make *no* type III collagen, but most make some, although at levels perhaps 10–30% normal (Byers et al. 1981a; Aumailley et al. 1980; Byers et al. 1979). In other individuals some alteration in type III collagen structure (primary and/or posttranslational) leads to altered secretion (Byers et al. 1981a, 1981b; Holbrook and Byers 1981). The genetic mechanisms for deficient synthesis could involve deletion of the gene or portions of it, coding for type III collagen, insertion of stop codons, or mutations in regulatory regions. A classification of EDS IV taking all of these observations into consideration is given in Table 11.4. The possibility that defective deposition of type III collagen results in disorders much more widespread than EDS IV has been proposed more recently by Pope et al. (1981). They found, using limited pepsin digestion, that skin and temporal arteries from some patients with ruptured cerebral aneurysms were deficient in type III collagen. Furthermore, cultured dermal fibroblasts from these individuals also synthesized and secreted decreased amounts of type III collagen and procollagen. Thus, defects in type III collagen deposition, while not extensive enough to produce the EDS IV syndrome, might enhance aneurysm formation at points of known arterial weakness, such as branches of the circle of Willis.

Ehlers-Danlos Syndrome V. Beighton et al. (1969) distinguished this form of the Ehlers-Danlos syndrome from the others on the basis of an apparent X-linked inheritance pattern. Clinically, skin hyperextensibility is prominent, whereas joint hypermobility is mild, and cutaneous fragility and poor scarring moderate. No definite biochemical abnormality was demonstrated in affected subjects until DiFerrante et al. (1975) reported deficiency (decrease to approximately 35%) in peptidyl lysine oxidase activity in fibroblast culture media. Subsequently, tissues and cells from four subjects with a phenotypic and inheritance pattern consistent with EDS V were studied by Siegel et al. (1979). In the latter individuals, reducible and thermally stable collagen crosslinks were normal. Furthermore, peptidyl lysine oxidase activity was not reduced (and was actually twice that of controls) and peptidyl lysine oxidase antigen, quantitated by immunodiffusion, was identical to that in control skin extracts. The discrepancy of the results in the two studies may be explained by poor solubility of the enzyme protein assayed after lyophilization, the conditions utilized by DiFerrante et al. (1975). Alternatively, EDS V may also be biochemically heterogeneous. In other individuals with EDS V, where morphologic and biochemical studies were performed, the collagen fibril bundles were found by scanning electron microscopy to be thinner than normal and many were disorganized (Black et al. 1980). These results suggest that the disorder is not due to defective crosslinking (in agreement with the conclusions of Siegel et al. [1979]), but due to defective aggregation of collagen fibrils into bundles.

It is probable that there are abnormalities in lysine- and hydroxylysine-derived crosslinks in heritable disorders of collagen metabolism. Such defects have been postulated for homocystinuria (Kang and Trelstad 1973), Menke's syndrome (Danks et al. 1972), and X-linked forms of cutis laxa (Byers et al. 1976, 1980). In homocystinuria, it has been proposed that homocysteine, which accumulates behind the enzymatic block, forms a stable ring structure with collagen aldehydes and prevents these aldehydes from functioning in Schiff base type crosslinks (Kang and Trelstad 1973). In Menke's syndrome, abnormal copper metabolism is assumed to be responsible for a hypothetical decrease in peptidyl lysine oxidase activity, a situation analogous to that described in the so-called mottled mice (Rowe et al. 1974, 1977). It has been suggested, based on these and other studies, that the gene for peptidyl lysine oxidase is located on the X chromosome. In the X-linked cutis laxa syndrome, affected males have soft droopy hyperextensible skin and a characteristic droopy facies in addition to skeletal and urinary tract abnormalities, but lack typical features of EDS (Byers et al. 1976, 1980). The activity of peptidyl lysine oxidase in extracts of skin from two such subjects was 13 and 26% of control, respectively, and that in culture medium from fibroblasts, 15 and 20% of control, respectively (Byers et al. 1980). Moreover, there was no detectable immunoreactive enzyme by immunodiffusion using antibodies directed against enzyme protein. Although no studies of crosslinks in the tissues were performed, it was assumed that the defect in peptidyl lysine oxidase activity was the biochemical basis of this heritable form of the cutis laxa syndrome. The possibility that this disorder is linked to a defect in copper metabolism was also raised, as levels of circulating total and ceruloplasmin copper were reduced in the affected subjects. McFarlane et al. (1980) have described an X-linked clinical syndrome in males with features of the EDS as well as abnormalities in the genitourinary tract. The biochemical basis for the syndrome may be similar to that postulated by Byers et al. (1980) for X-linked cutis laxa.

Ehlers-Danlos Syndrome VI. EDS VI was first identified in two sisters with similar clinical abnormalities, including severe kyphoscoliosis (Figures 11.5 and 11.6) apparent since early childhood, striking hypermobility of almost all joints with recurrent dislocations (Figures 11.7 and 11.8), microcorneas, soft hyperextensible skin with a characteristic velvety feel, and poor wound healing (Pinnell et al. 1972). The mother had ruptured her membranes while 3 months pregnant with the older child, yet still managed to carry the pregnancy to term. The younger affected girl required enucleation following relatively mild trauma to one eye. Biopsy specimens of skin from both children contained collagen that was normally soluble in nondenaturing solvents such as cold dilute acetic acid, but was abnormally soluble in denaturing solvents such as 4 *M* $CaCl_2$ or 9 *M* KSCN (Table 11.5). In view of the increased solubility in denaturing solvents, it was assumed that there was some defect in intermolecular crosslinking of the dermal collagen. The striking chemical abnormality revealed by amino acid analysis of the dermis was the decrease in hydroxylysine content to approximately 5–7% of normal, i.e., 0.2 to 0.3 residues per 1000 amino acids (normal 4.1 $\pm$ 0.4) (Pinnell et al. 1972). As the 4-hydroxyproline content of the dermis was similar to controls, the molar ratio of hydroxylysine : 4-hydroxyproline in biopsy samples in the affected children was also low (0.002 and 0.003 respectively, compared

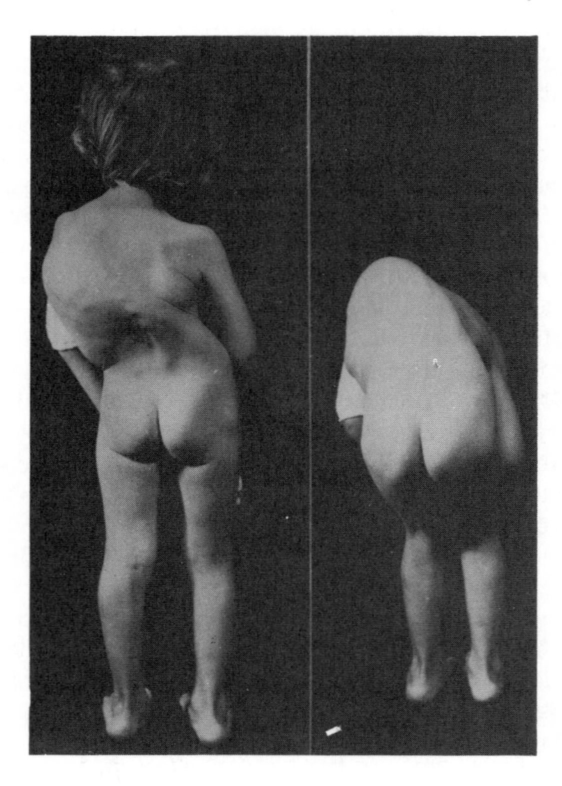

Figure 11.5. Severe kyphoscoliosis in a 4-year-old girl, with type VI Ehlers-Danlos syndrome. The deformity of the chest wall is brought out by forward bending.

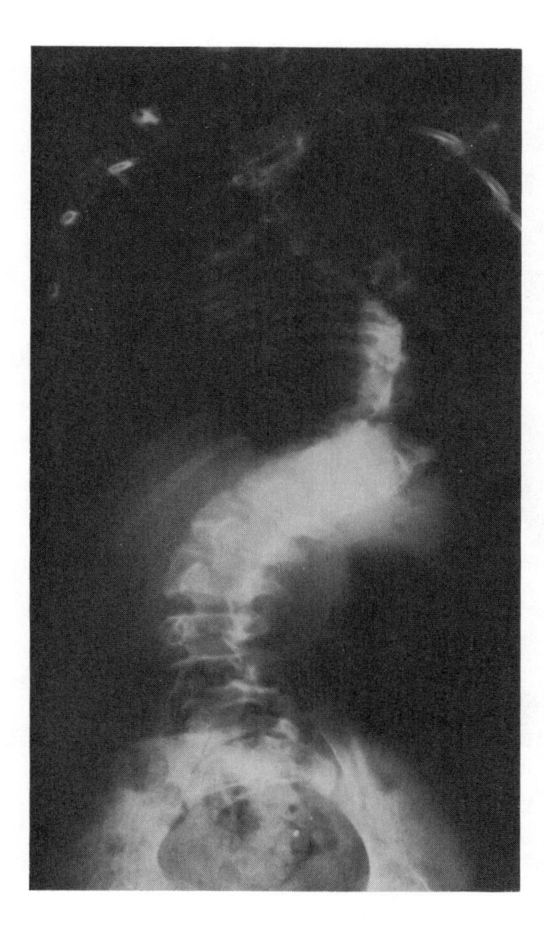

Figure 11.6. An anteroposterior roentgenogram of the spine in the same individual as in Figure 11.4, obtained at age 9 years. The curvature of the spine had progressed in these five years.

with controls of 0.048 ± 0.005). As this ratio was within the normal range in the clinically normal parents (0.045 and 0.039, respectively) and the clinically normal older sister (0.048), it was assumed that the inheritance of this disorder was autosomal recessive. Because these children both had features of the EDS but a distinctive biochemical defect, it was suggested by McKusick (1972) that they be classified as a new subtype, EDS VI. Ocular abnormalities were present in these children and others subsequently reported; EDS IV was therefore also termed the ocular type of EDS (McKusick 1972; Sussman et al. 1974).

Since the initial communication, descriptions of five other kindreds in which collagen hydroxylysine deficiency has been demonstrated have been published. (Sussman et al. 1974; Steinmann et al. 1975; Hanson et al. 1977; Elsas et al. 1978; Krieg et al. 1979). Although each of the reported cases has not been examined in the same detail with identical methodology, it seems reasonable on the basis of recorded data that the defect could be due to a different mutation in each case. The abnormal phenotype of these individuals is remarkably similar, however (Krane 1982).

The reduced hydroxylysine content in the dermal collagen could not be accounted for by a mutation resulting in an amino acid substitution for a precursor

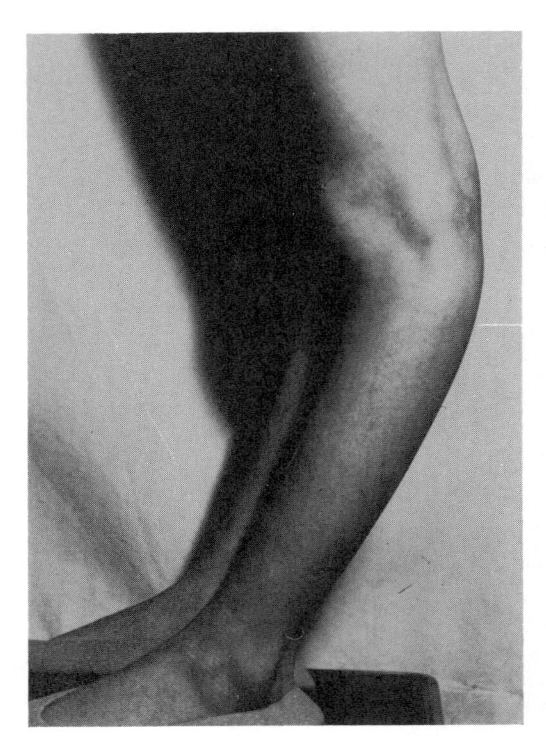

Figure 11.7. Recurvature of the knee due to hypermobility of the joint and ligamentous laxity with type VI Ehlers-Danlos syndrome in the same individual as in Figure 11.6.

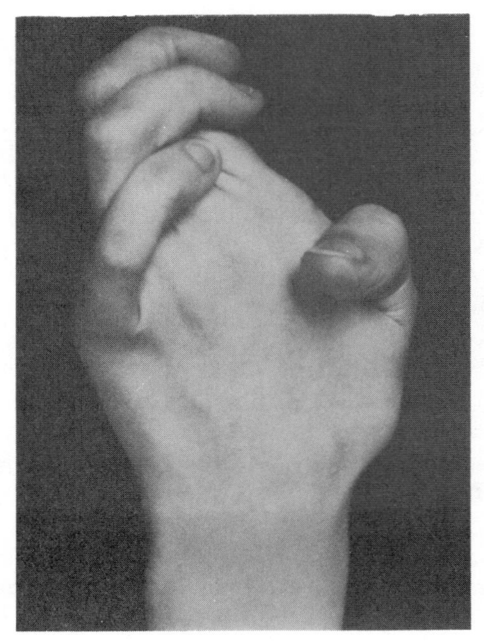

Figure 11.8. An example of extreme joint hypermobility in another individual with type VI Ehlers-Danlos syndrome.

Table 11.5. Solubility of Skin Collagen in Patients with Ehlers-Danlos Syndrome V Compared with Controls[a]

Experiment	Solvent	EDS VI	Controls
1	1 *M* NaCl, then 0.5 *M* acetic acid	11.3	2.1;2.8
	9 *M* KSCN	74.0	18.1;18.7
2	1 *M* NaCl, then 0.5 *M* acetic acid	5.3	4.4–6.6
	4 *M* $CaCl_2$	76.0	15.3–19.2

[a]Data from Pinnell et al. (1972). Experiment 1 shows results for the older affected sibling in a family; Experiment 2 shows results for the younger siblings. Collagen solubility was measured at 2–4° on fresh skin biopsies, expressed as a percent of total.

lysine residue in the polypeptide chain, because the reduction in hydroxylysine content approached 4 residues per 1000 amino acid residues. Furthermore, it was subsequently reported by Krieg et al. (1979) that type III as well as type I collagen was deficient in hydroxylysine. Decreased hydroxylysine content in type III collagen-derived cyanogen bromide peptides from skin samples of EDS IV has also been found (Byrne and Krane, unpublished observations). It was observed that the lysine content of the dermis was not reduced; rather, it was slightly increased (31.3 and 33.0 residues per 1000 compared to controls of 28.1 ± 1.6) (Pinnell et al. 1972). Because the possibility of multiple point mutations substituting other amino acids for lysine residues was eliminated, a decreased enzymatic hydroxylation of several lysine residues was postulated as the basic defect. It was also to be expected that an abnormality in a posttranslational modification would be inherited as an autosomal recessive trait.

Attempts were therefore made to assay peptidyl lysine hydroxylase in homogenates of the small amounts of skin available, but enzyme activity could not be adequately measured. Activity was measurable in sonicates of normal dermal fibroblasts, however. Skin fibroblasts were therefore cultured from the two cases initially described, as well as from controls and other family members (Krane et al. 1972). The nonhydroxylated substrate was derived from embryonic chick tibias labeled with [^{3}H] or [^{14}C] lysine in the presence of α, α′-dipyridyl, to bind Fe^{2+}, essential for the hydroxylase reaction. Fibroblast sonicates from control subjects contained easily measurable peptidyl lysine hydroxylase activity by quantitation of labeled formaldehyde following periodate treatment of hydroxylysine, quantitation of labeled hydroxylysine using the automatic amino acid analyzer following acid hydrolysis, or ^3H_2O arising from the [^{3}H] substrate (Krane et al. 1972.) Peptidyl lysine hydroxylase activity in fibroblast sonicates from the two affected girls averaged 10–14% of controls based on cell number, cell protein, or other cellular enzyme activities (peptidyl proline hydroxylase or lactic acid dehydrogenase). These differences from controls were demonstrable at all levels of cell density following passage of the cells by trypsinization. Furthermore, the decrease in activity of the peptidyl lysine hydroxylase was roughly proportional to the observed decrease in hydroxylysine content of the dermis in the affected subjects. This has been noted in all reported instances of the disorder, with the exception of the kindred of Steinmann et al. (1975). The

Table 11.6. Hydroxylysine Content of Skin Collagen (Expressed as Hydroxylysine: 4-Hydroxyproline Ratio) and Peptidyl Lysine Hydroxylase Activity in Skin Fibroblasts from Reported Instances of Ehlers-Danlos Syndrome VI

		Hydroxylysine : 4-hydroxyproline in skin		Peptidyl lysine hydroxylase activity in cultured skin fibroblasts[a]
Reference	Case no.	Mole/mole	% control	% control
Pinnell et al. 1972; Krane et al. 1972; Quinn and Krane 1976	1	0.002	4	2–14
	2	0.003	6	3–10
Sussman et al. 1974	1(IL)	0.003	6	12
Steinmann et al. 1975	1(NS)	0.027	66	3–12
	2(AS)	0.039	95	14
Hanson et al. 1977	1	0.006	12	9–11
Elsas et al. 1978	1	0.011	22	17
Krieg et al. 1979; Risteli et al. 1980	1	0	0	10

[a]Substrates used were predominantly type I collagen.

hydroxylysine content in collagen from the skin of the affected individuals in this kindred was hardly altered although peptidyl lysine hydroxylase activity was markedly reduced in fibroblast lysates. A comparison of the hydroxylysine content of skin (expressed as hydroxylysine : 4-hydroxyproline ratio) and peptidyl lysine hydroxylase activity in fibroblasts in the reported cases of EDS VI is shown in Table 11.6.

Human peptidyl lysine hydroxylase has not been extensively purified, and kinetic studies must be interpreted with caution. Peptidyl lysine hydroxylase from chick embryos has been purified to homogeneity and has been found to have substrate requirements similar to peptidyl proline hydroxylase except, of course, for the amino acid hydroxylated (Puistola et al. 1980a, 1980b; Turpeenniemi et al. 1977). The purified hydroxylases also have similar affinities for α-ketoglutarate (2-oxoglutarate), Fe^{2+}, ascorbate, and molecular oxygen (Chapter 3). Some properties of the hydroxylases in crude extracts of normal fibroblasts as well as EDS VI fibroblasts have been examined (Quinn and Krane 1976), and are summarized in Table 11.7. The major conclusion from these studies is that the K_m for ascorbate for the EDS VI peptidyl lysine hydroxylase is significantly higher than controls. This raises the possibility that low tissue levels of ascorbate could have relatively greater effects on lysine hydroxylation in these EDS VI tissues than in normal tissues. These differences in kinetic properties between normal and mutant cell extracts suggest that an abnormal protein could be responsible for the low level of enzyme activity in the mutant cells. Other observations supporting this concept are that the optimal temperature for the EDS VI lysine hydroxylase is lower than controls (32°C compared to 37°C) and, in contrast to controls, the mutant enzyme fails to form high molecular weight aggregates in low ionic strength buffers. Furthermore, when fibroblast sonicates

Table 11.7. Apparent Substrate Affinities[a] of Peptidyl Proline and Lysine Hydroxylases in Fibroblast Sonicates from Patients with Ehlers-Danlos Syndrome VI and from Controls[b]

Enzyme	Substrate	EDS VI	Controls
Lysine hydroxylase	Ascorbate	20	4
	α-Ketoglutarate	20	20
Proline hydroxylase	Ascorbate	100	100
	α-Ketoglutarate	4	4

[a]K_m (μ*M*)
[b]Data from Quinn and Krane (1976).

are dialyzed against buffers containing dithiothreitol (10 μ*M*), a variable increase in peptidyl lysine hydroxylase activity is observed with the mutant enzyme preparations but not with the control preparations. Mutations leading to abnormal protein structure are common in recessively inherited disorders.

Although it is likely that EDS VI is inherited as an autosomal recessive trait, this has not yet been proven. Preliminary data suggest the presence of intermediate levels of enzyme in the mother and clinically unaffected sister from the initially reported kindred (Pinnell et al. 1972, Krane et al. 1972). Similar findings were reported by Elsas et al. (1978). However, decreases in peptidyl lysine hydroxylase levels in cultured cells from putative heterozygotes have not regularly been found. Furthermore, it has not yet been shown definitively that clinically affected individuals inherit the identical mutant allele from both parents. In what is presumably a rare mutation, it is equally plausible that the abnormal phenotype could result from the product of two different mutant alleles.

The biochemical heterogeneity in EDS VI reflects the genetic heterogeneity. The idea that the defect in each separate kindred results from a different mutation is supported by contrasting the findings of Steinmann et al. (1975) and Krieg et al. (1979) (Table 11.6). In the cases of Steinmann et al. (1975), the hydroxylysine content of the skin collagen was normal in one instance and borderline low in the other. Yet lysine hydroxylase levels in cultured fibroblasts ranged from 3 to 14% of controls. At the other extreme are the patients initially reported by Krieg et al. (1979) and studied further by Risteli et al. (1980). These patients had undetectable hydroxylysine in types I and III collagen in the skin but only minor alterations in other tissues. Using partially hydroxylated interstitial collagens, the lysine hydroxylase activity was about 10% that of controls, a value similar to that obtained by Steinmann et al. (1975). The enzyme assays are usually performed with a substrate derived from the chick that could be sufficiently different from the natural human substrate to affect the apparent rate of enzymatic hydroxylation. It could well be that different mutations in the gene coding for the enzyme protein could lead to variably altered kinetic parameters for amino acid sequences surrounding lysines and catalyze the hydroxylation of some of these lysines at rates different from others. This might lead to heterogeneity in the extent of lysine hydroxylation rather than to uniformly low levels of hydroxylation. Indeed, the hydroxylysine content in skin from EDS VI is not uniformly depressed in isolated cyanogen bromide peptides, but absent from some peptides and measurable, although reduced, in others (Krane and

Table 11.8. Hydroxylysine Content of Several Tissues and of $C1_q$ from Patients with Ehlers-Danlos Syndrome VI Compared to that of Controls[a]

Tissue	Hydroxylysine (% control)
Dermis	~5
Fascia	~20
Bone	43–100
Cartilage	~90
$C1_q$	77

[a]Data from Pinnell et al. 1972; Pinnell 1975; Eyre and Glimcher 1972.

Byrne, unpublished observations). One might speculate that the abnormality in lysine hydroxylase in the cases of Steinmann et al. (1975) could result in underhydroxylation of a critical lysine residue, yet not be detected as a decrease in total hydroxylysine content. Support for this concept is also provided by the observations of Eyre and Glimcher (1972) that the pattern of reducible crosslinks in EDS VI is markedly abnormal even in a tissue such as cartilage, in which the total hydroxylysine content is not significantly different from normal.

In EDS VI, not only is there heterogeneity of lysine hydroxylation within the interstitial collagens of the dermis, but there is, in addition, a striking difference in the extent of hydroxylation in presumably the same collagen type in different tissues. Some of these are listed in Table 11.8. There is only a slight decrease, compared to normal, in hydroxylysine content of Clq, a subcomponent of the first component of complement (Pinnell 1975). Hanauske-Abel and Röhm (1980) were also unable to find any compositional or functional alteration in serum Clq from the propositus and family members of the kindred reported by Krieg et al. (1979).

Several possible explanations have been proposed to account for the variation in hydroxylysine content of different tissue collagens (Krane 1982; Quinn and Krane 1976):

1. A tissue difference affects the rate of polypeptide chain elongation. Slow growth of the polypeptides with correspondingly slower helix formation would favor lysine hydroxylation (Chapter 3). Indeed, the rate of hydroxylation of collagen lysines in cultured EDS VI cells is not reduced to the same extent as lysine hydroxylase activity (Quinn and Krane 1979).
2. A tissue difference affects the rate of proline hydroxylation. As stable helix formation depends upon 4-hydroxylation of proline residues (Chapter 1), decreased hydroxyproline content would permit increased time for lysine hydroxylation.
3. A tissue difference affects a critical cofactor concentration. The higher K_m for proline than lysine hydroxylase (Table 11.7) suggests that the ascorbate concentration could differentially influence rates of these reactions. Furthermore, in human fibroblasts, high levels of α-ketoglutarate (e.g., 1 mM) produce substrate inhibition of proline but not lysine hydroxylase (Quinn and Krane 1976).

4. The same mutant enzyme might catalyze a relatively greater rate of hydroxylation of one type of collagen over another. This possibility was examined by Risteli et al. (1980) who used underhydroxylated type I (chick) and type IV (mouse sarcoma) collagens as substrates for peptidyl lysine hydroxylase of control and EDS VI skin and corneal fibroblasts from the kindred of Krieg et al. (1979). They found that the levels of lysine hydroxylase in both types of EDS VI fibroblasts were 10 to 11% that of controls with the type I collagen substrate, but 27 and 32% that of controls with the type IV collagen substrate.
5. There might be multiple forms of lysine hydroxylase, each with a characteristic tissue distribution. The results of Risteli et al. (1980) could also be interpreted on this basis.

Although indirect, additional evidence supporting explanation number 5 is provided by further studies of one of the children originally reported by Pinnell et al. (1972). Additional tissue samples were obtained 5 years after the initial analysis, including dermis as well as bone stripped of periosteum (Krane et al. 1980). The skin sample still had a hydroxylysine content of less than 5% of normal. Although the absolute levels of lysine hydroxylase were lower in normal bone than skin cells, the ratios of lysine/proline hydroxylase were still approximately 8 to 10% that of controls in dermal fibroblasts but were approximately 46 to 49% that of controls in bone cells. Another distinction between these cultured bone and skin cells included the cyclic AMP (cAMP) response to calcitonin, suggesting that phenotypically distinct cells can be cultured from different connective tissues. These tissue differences were also seen in the hydroxylysine content of the bone collagen, which was within the range of controls. The possibility must also be considered that such differences in lysine hydroxylase activity in cells from different tissues could be explained by variations in the rate of degradation or synthesis of abnormal enzyme protein. One must also account for the severe kyphoscoliosis, a characteristic feature of EDS VI, in view of the observations that bone collagen is nearly normal in hydroxylysine content. Presumably the kyphoscoliosis is caused by extraosseous factors such as abnormal ligaments, muscles, and other supporting structures of the vertebral column during periods of growth.

A critical role for hydroxylysine in collagen is evident from the severe problems that result from its deficiency. It is presumed that aberrations in intermolecular crosslinking are the consequences of this deficiency, based on the increased solubility of the collagen in denaturing solvents (Pinnell et al. 1972; Sussman et al. 1974) and the pattern of [^{3}H]-labeled reducible crosslinks (Eyre and Glimcher 1972). Because of the Amadori rearrangement and formation of ketoamines (Chapter 2), crosslinks derived from hydroxylysine aldehydes are chemically more stable than those derived from lysine aldehydes. There is also evidence that the stable 3-hydroxypyridinium crosslink described by Fujimoto et al. (1978, 1979) and Eyre and Oguchi (1980) is derived from hydroxylysine residues. If the ketoamine compounds are critical intermediates in the generation of the 3-hydroxypyridinium crosslink, then the generation of hydroxylysine is rate-limiting (Eyre and Oguchi 1980; Eyre 1980). Indeed, Eyre (1982) has found decreased levels of the 3-hydroxypyridinium crosslink in collagen from the annulus fibrosus of the propositus from the initially reported kindred of EDS VI (Pinnell et al.

1972) and the presence of a newly observed, more basic hydroxypyridinium compound that could be a lysine analog of the usual compound. It is also obvious that the content of hydroxylysine glycosides in collagen would be markedly reduced, proportional to the reduction in content of hydroxylysine. However, the function of the glucose and galactose residues normally linked to hydroxylysine is not known.

Attempts have been made to increase lysine hydroxylation in subjects with EDS VI by the administration of ascorbic acid (Miller et al. 1979), as ascorbic acid is required for collagen hydroxylation and the mutant enzyme has a higher than normal K_m for this substrate (Table 11.7). Elsas et al. (1978) treated one patient with up to 4 g ascorbic acid daily for 2 years and observed gross and histologic improvement in scars and increased muscle strength, but no change in joint laxity or skin friability. Although the urinary ratio of hydroxylysine : 4-hydroxyproline increased slightly with therapy, there was no increase in the hydroxylysine content of the dermis.

Studies of EDS VI have thus revealed several previously unappreciated aspects of collagen metabolism. For example, it had earlier been postulated that hydroxylation of collagen lysines and glycosylation of the hydroxylysines would be required for secretion of collagen molecules from the cell. However, it became obvious that collagen containing less than 0.6 lysine residues per molecule is secreted and deposited as fibrils. The idea that collagens of the same type in different tissues undergo separately controlled posttranslational modifications also emerged from studies of EDS VI. Suggestions that urinary markers for collagen degradation are derived from sources other than skin collagen were also made in individuals affected with EDS VI (Pinnell et al. 1972; Krane et al. 1977). Finally, the role of hydroxylysine in crosslinking is emphasized by the observed abnormalities in conditions of hydroxylysine deficiency.

Ehlers-Danlos Syndrome VII. Individuals have been described with a disorder called arthrochalasis multiplex congenita, characterized by multiple joint subluxations, ligamentous tears, a peculiar scooped-out facies with epicanthal folds and hypertelorism, and thin velvety skin occasionally accompanied by poor healing of wounds with thin, atrophic scars (Lichtenstein et al. 1973). This disease is now considered a form (type VII) of EDS. In 1973, it was found that patients with EDS VII accumulate incompletely processed collagen (probably pN collagen; *See* Chapters 2 and 3) in their skin, and it was therefore thought that this disorder resembled dermatosparaxis (Lichtenstein et al. 1973a, 1973b). The syndrome of dermatosparaxis described in cattle (Lenaers et al. 1971; Stark et al. 1971; Lapière et al. 1971; Kohn et al. 1974) had been shown to be due to low activity of procollagen N-protease, which resulted in the deposition of type I pN collagen in the extracellular matrix. An apparently similar disorder in sheep has not yet been shown to be due to the enzyme deficiency, although this is likely (Becker et al. 1976). It is postulated that the extremely fragile skin in the animal disorder (*dermatosparaxis* means torn skin) is accounted for by distorted packing of collagen molecules in the fibril due to the persistence of uncleaved peptides on the N-terminal ends. This, in turn, would not permit normal contacts of critical side chains involved in intermolecular crosslinks.

When tissues from the human disease were first analyzed, it was not appreciated that type I procollagen contains extensions at both the N- and C-terminal ends and that cleavage of these extensions is the function of different proteases. Nevertheless, using SDS polyacrylamide gel electrophoresis, it was found that the denatured collagen, initially extracted at 4°C in 0.5 *M* acetic acid, from the skin of three patients with arthrochalasis multiplex congenita contained chains with an apparent molecular size larger than normal (Lichtenstein et al. 1973b). As it was postulated that the accumulation of procollagen was due to a defect in conversion of procollagen to collagen, assays of the converting peptidase were performed in medium from cultured fibroblasts derived from these patients. In the three cases of EDS VII, levels of enzyme activity were 17, 12, and 22% respectively, of the average of two controls.

Subsequent additional studies have been performed in one of the cases originally reported, and the results lead to a different conclusion with respect to the mechanism of the defect (Steinmann et al. 1980). When the collagen from a subsequent biopsy was extracted either into cold neutral salt solutions containing protease inhibitors or into 0.5 *M* acetic acid, a component not present in extracts of normal skin was revealed on SDS polyacrylamide gel electrophoresis. This component was identified as the pN $\alpha2(I)$ chain. No pN $\alpha1(I)$, pC $\alpha1(I)$, or pC $\alpha2(I)$ chains were detected, but normal $\alpha2(I)$ chains were present. Furthermore, the pN $\alpha2(I)$ chains were resistent to digestion with pepsin, α-chymotrypsin, or purified procollagen N protease. Nevertheless, incubation of extracts of the patient's cultured fibroblasts with normal procollagen revealed normal N protease activity. These findings were at variance with those in the original report (Lichtenstein et al. 1973b), where cells from all three patients had low levels of what was considered to be procollagen peptidase and *both* pro $\alpha1(I)$ and pro $\alpha2(I)$ chains in skin extracts. Whatever the explanation for the discrepancies, the detailed observations in the subsequent report (Steinmann et al. 1980) are consistent with a structural mutation in that portion of the pro $\alpha2(I)$ chain that is normally cleaved by the procollagen N protease. Although the pro $\alpha1(I)$ chains are cleaved at the normal site, the N-terminal extensions do not dissociate from the pro $\alpha2(I)$ extensions and pN collagen molecules are therefore deposited extracellularly. This possibilty can be proven, however, only by demonstration of an altered amino acid sequence in the abnormal $\alpha2(I)$ chain or altered sequences of bases in the DNA that code for this portion of the protein. As normal $\alpha2$ chains were found in addition to the pN $\alpha2$ chains and both parents were phenotypically (biochemically) normal, a further conclusion from these studies is that the affected individual is likely to be a sporadic heterozygote arising from a new mutation, with one normal and one abnormal allele coding for each pro $\alpha2$ chain.

Thus, EDS VII is also biochemically heterogeneous and, as further detailed analyses of the two cases other than that restudied by Steinmann et al. (1980) have yet to be reported, it has not been proven to be biochemically identical to dermatosparaxis in animals. It should not be surprising that the human and animal diseases are basically different in view of the clinical (phenotypic) differences described earlier. The skin changes are relatively mild in humans but severe in animals.

Observations of increased collagen biosynthesis in fibroblasts from patients with EDS VII (Lichtenstein et al. 1973b) led to studies of a possible role of the procollagen extension peptides in the regulation of collagen biosynthesis. The hypothesis had been proposed (Lichtenstein 1973b) that the extension peptides might act as feedback inhibitors of collagen synthesis. Evidence for this is discussed in Chapter 3. Thus, the study of EDS VII provides another example of findings in a disease leading to further general insight into biological control mechanisms.

Other Possible Forms of Ehlers-Danlos Syndrome. McKusick (1972) pointed out the coexistence of absorptive periodontosis with early loss of teeth in a woman with lesions on the legs resembling those of EDS. The periodontal disease and skin changes (pretibial scarring) were also present in several members of her paternal family. Stewart et al. (1977) subsequently described an autosomal dominant disorder characterized by fragile skin, abnormal scarring, and generalized periodontitis with loss of teeth by the early twenties. This syndrome was felt to be a unique variant of EDS and it was proposed that it be termed EDS VIII (Hollister 1978; Stewart et al. 1977). Another case with autosomal dominant inheritance has also been reported (Nelson and King 1981). Unfortunately, no information is available with regard to any biochemical abnormality in these patients, and in view of the clinical and biochemical heterogeneity of the other forms of EDS, it may be premature to place these patients in a unique category.

It has also been suggested that a form of EDS is associated with a structurally defective fibronectin (Arneson et al. 1980). A kindred was described with mild clinical features of EDS, including easy bruising and mitral valve prolapse. The patients were thought to clinically fit best into the EDS II or III categories. A platelet aggregation defect was noted that was partially corrected in vitro by the addition of normal plasma or cryoprecipitate. It was therefore postulated that there was a defect in fibronectin responsible for the abnormal platelet aggregation and possibly other features of the syndrome, although this was not demonstrated directly. No confirmation of these findings has yet been reported. Another individual, a 30-year-old woman with marked joint hypermobility, died following ventricular fibrillation and was found to have cardiac injury and other visceral abnormalities (Cupo et al. 1981). Dermal collagen fibrils were found to be heterogeneous in size and irregular and frayed in appearance. No biochemical data are available to place this patient's disorder among the different types of EDS.

11.2.2. Osteogenesis Imperfecta

The major feature of the heterogeneous group of heritable disorders considered under the classification of osteogenesis imperfecta (OI) is the tendency of bones to fracture (osseous fragility). The frequent association of fragile bones with abnormalities in other tissues (skin, teeth, sclerae) has suggested that some defect in collagen could underlie these disorders (McKusick 1972; Smith et al. 1975). It is likely that the variability in the clinical manifestations of OI is accounted for by genetic heterogeneity. Thus, each of the OI syndromes must be examined individually for possible abnormalities because no one biochemical feature could explain all the different phenotypes. Any suggested classifications of OI should therefore be considered as tentative until further detailed clinical, biochemical, and genetic data provide the basis for a more definitive nosological

Table 11.9. Suggested Classification of Osteogenesis Imperfecta Syndrome[a]

Type	Phenotype	Mode of inheritance
I	Disease moderate; early onset of fractures; not usually progressively deforming; sclerae blue; hearing impairment as young adults; no dentinogenesis imperfecta	Autosomal dominant
II	Onset in utero; often stillborn or death soon after birth; marked long bone bowing and severe skull involvement	Autosomal recessive
III	Progressively deforming with multiple fractures and kyphoscoliosis; normal sclerae; hearing loss infrequent; dentinogenesis imperfecta prominent	Autosomal recessive
IV	Variable age of onset and variable deformity; sclerae pale blue or white; dentinogenesis imperfecta variable	Autosomal dominant

[a]Classification according to Sillence et al. 1979.

approach. Table 11.9 shows the clinical and genetic classification proposed by Sillence et al. (1979), but others are also useful (Rowe and Shapiro 1982). There are undoubtedly individuals with some features of OI that cannot yet be included in this classification (Hollister et al. 1982). Although there have been suggestions of abnormalities in extracellular matrix components other than collagen in OI (Dickson et al. 1975), these will not be considered in this discussion.

The most common form of OI is type I (Table 11.9) (Figure 11.9). This is usually considered the classical phenotype of the disease, although here too there undoubtedly is biochemical and genetic heterogeneity. Although other collagen types may also be found in bone around cells and in relationship to blood vessels, the essentially exclusive collagenous component of the mineralized matrix is type I collagen. It is assumed that the primary structures of bone and skin type I collagens are identical (not yet proven for human bone) but there are differences

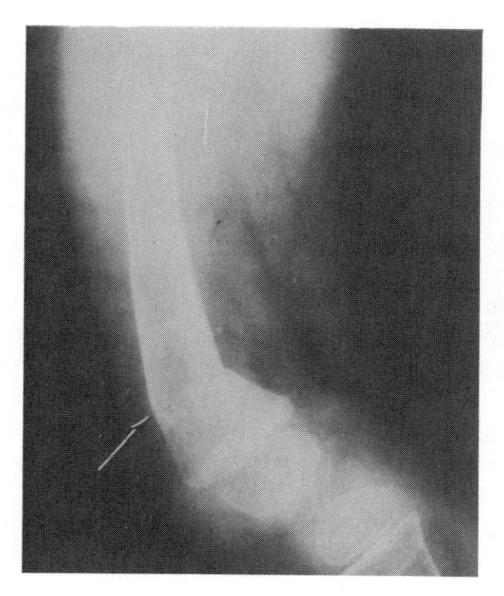

Figure 11.9. Roentgenogram of the right femur of a 3-year-old girl with type I osteogenesis imperfecta. A fracture of the distal femur is shown *(arrow)*. This was the second fracture to occur with only minor trauma. Deep blue sclerae were present. By the age of 10, only three minor fractures had been sustained.

in posttranslational modifications (Pinnell et al. 1971; Krane and Simon 1981). So far, there is no evidence for more than one set of structural genes for α1(I) and α2(I) chains for type I collagen, and the tissue distribution of the abnormalities in type I OI (and other types) could be explained by defects in these genes. A reduction in total and dispersible collagen is the major abnormality detected in analyses of skin (Smith et al. 1975; Stevenson et al. 1970; Cetta et al. 1977). In two cases of type I OI, bone collagen was found to have a normal amino acid composition, although 4-hydroxyproline and hydroxylysine contents were increased slightly (Fujii and Tanzer 1977). Increased levels of the major reducible crosslink in bone, dihydroxylysinonorleucine, were also found (Fujii and Tanzer 1977). Sykes et al. (1977) observed that seven of nine subjects with OI possibly classified as Sillence type I had increased ratios of type III : type I collagen in skin extracted with pepsin for 16 hours at 4°C, precipitated with 2 *M* NaCl, and analyzed using a method of interrupted polyacrylamide gel electrophoresis. The amount of collagen solubilized was recorded as approximately 70–80% of the total. Sykes et al. found that in normal individuals after age 20, the type III : type I collagen ratio in the skin was 0.14 ± 0.06. In seven patients with OI, the ratio was greater than 0.26. The major implication of these studies is that, as bone contains only type I collagen, decreased amounts of this component in other tissues such as dermis and sclera could cause a relative increase in type III collagen. It was also proposed that as type I OI is inherited as an autosomal dominant trait, affected individuals could be heterozygous for a mutant gene concerned with the expression of type I collagen; the situation could therefore superficially resemble that of some forms of thalassemia.

When cultured fibroblasts from patients with type I OI were studied after 4–8 passages, it was found that total collagen synthesis was normal as measured by incorporation of labeled proline into medium collagens in the presence of β-aminopropionitrile (Steinmann et al. 1979). Furthermore, intracellular collagen degradation, measured by the methods of Bienkowski et al. (1978a, 1978b) was also normal. Similar results were obtained by Rowe and Shapiro (1982). However, the relative proportion of collagen types made by OI cells was different from that made by control cells; the ratio of type III : type I collagen was increased in the OI cells and, therefore, the amount of type I collagen synthesized was decreased. Krieg et al. (1981) also found that patients with a phenotype consistent with type I OI released (synthesized) decreased amounts of type I procollagen. Thus, the situation in these individuals with OI may be analogous to that in some individuals with type IV EDS, in that there is a genetically determined decrease in the synthesis of a specific collagen type. However, these observations in OI could also reflect a basic, as yet unexplained, abnormality in the function of osteoblasts (and fibroblasts that also make type I collagen) of which collagen synthesis is merely one expression. The histologic changes in bones in some instances of OI may be profound and not explained simply as the consequence of a mutation in a single gene governing the formation of a single product, i.e., collagen. The morphology of bone and skin cells may be very abnormal in OI (Bullough et al. 1981) and in some individuals there are even alterations in the growth plate, although this is encountered more often in severe forms, rather than type I OI. It would be difficult at this time to explain how the function of chondrocytes (which synthesize the type II collagen in the growth plate) would be disturbed as a result of a defect in type I collagen synthesis.

Type II OI is inherited as an autosomal recessive trait and is characterized by extreme bone fragility leading to intrauterine or early infant death as a consequence of fractures (Sillence et al. 1979). Several studies have now been performed on tissues from individuals with this syndrome, the results of which indicate that even type II OI is biochemically heterogeneous (Pope and Nicholls 1980). Penttinen et al. (1975) measured synthesis of procollagens by dermal fibroblasts from an infant with severe OI (type II) and found that the total amount of type I procollagen was decreased while the ratio of type III : type I procollagen was increased. The decrease in total collagen synthesis by these cells was subsequently confirmed by Steinmann et al. (1979). A marked relative increase in type III collagen synthesized by dermal fibroblasts from a patient with a form of OI, not described in detail, was also found using chemical as well as immunologic techniques by Müller et al. (1975).

Observations have also been made on the distribution in bone of collagens other than type I (Pope et al. 1980a). Normal fetal bones contain small amounts of type V collagen as determined by SDS polyacrylamide gel electrophoresis of pepsin-solubilized demineralized bone. Bones from lethal OI (type II) contain more type V than normal and some type III collagen as well. Müller et al. (1977) noted type III collagen within holes in compact bone from a patient with a milder form of OI. Type III collagen may be associated with marrow and vascular elements and not part of mineralized bone collagen proper. The relative increase in marrow–vascular tissue relative to mineralized bone collagen in lethal OI might therefore account for the presence of type III collagen in bone in this condition.

Further studies by Barsh and Byers (1981) on fibroblasts cultured from a patient with type II OI previously reported by Penttinen et al. (1975) have revealed additional abnormalities that might account for the decrease in type I collagen synthesis. Barsh and Byers (1981) found, in short-term labeling experiments, that these cells produced two distinct α1(I) chains at the same rate. These chains differed in their primary structure as evidenced by analysis of the cyanogen bromide peptides. In longer-term labeling studies, type I collagen was secreted more slowly and accumulated intracellularly. As there was no increased intracellular degradation (Steinmann et al. 1979), it was proposed that abnormal pro α1(I) chains result from a structural gene mutation that is responsible for an inability of the chains to be secreted normally from the cell. Such an abnormality might reside in amino acid sequences that determine interactions with the cellular machinery regulating the export of secreted proteins.

Trelstad et al. (1977) subsequently analyzed several tissues from another infant with the type II OI phenotype. The dermis contained slightly increased amounts of type III relative to type I collagen, but only 25% of the total was solubilized, and the relative amounts of types III and I collagens synthesized by cultured dermal fibroblasts from this individual were normal. The most significant finding was an increase of 1.5- to 2.5-fold in hydroxylation of lysines in the α1(I) and α2(I) chains of the bone collagen. Smaller increases in hydroxylation of lysine of cartilage (type II) collagen were also found, whereas hydroxylation of skin collagen was in the normal range. The mechanisms for this increase in hydroxylysine content of skeletal collagens in this patient are not known. There is the possibility that overhydroxylation of lysines might occur if the rate of chain elongation, molecular assembly, or export from the cell were very low (Eyre

1981). Furthermore, it is known that bone collagen lysine hydroxylation is decreased when serum calcium concentrations are low, e.g., as a result of hypoparathyroidism or vitamin D deficiency (Barnes and Lawson 1978). Changes in the content of the 3-hydroxypyridinium crosslinks, which are derived from hydroxylysine, have also been found in vitamin D deficiency (Fujimoto et al. 1979). It is possible, therefore, that hypocalcemia in utero could account for the overhydroxylation of lysines, and explain the findings of Trelstad et al. (1977).

In the severe, progressive forms of OI (types III and IV) (Figure 11.10), the amounts of type III collagen relative to type I in pepsin digests of skin may be normal or increased (Sykes et al. 1977) and the synthesis of type I collagen by cultured dermal fibroblasts may be normal or decreased (Rowe and Shapiro 1982). Detailed studies are essential to clarify possible abnormalities in collagen structure and biosynthesis in each affected individual with OI, in view of the probable phenotypic and genetic heterogeneity. In fibroblasts from one such individual, initially reported by Penttinen et al. (1975) and found to have an increased type III : I ratio in secreted procollagens, additional abnormalities were found in the structure of the type I procollagen (Peltonen et al. 1980a). The type I procollagen has been noted to be unusually sensitive to pepsin, but was indistinguishable from normal based on electrophoretic mobility and helicity (circular dichroism). However, it was found that this collagen tended to aggregate more readily than normal, suggesting that it might be structurally abnormal (Peltonen et al. 1980a). Further studies of these fibroblasts showed that the procollagens incorporated more mannose into the C-terminal precursor region than controls (Peltonen et al. 1980b). These results were interpreted as indicating a structural change in the propeptides that might impair secretion. Alternatively, the increased glycosylation could result from a decreased rate of procollagen secretion, allowing more time for glycosylation to take place.

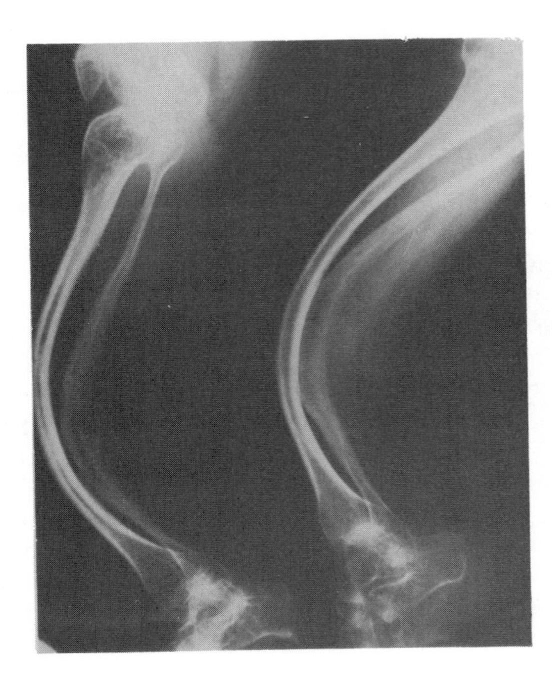

Figure 11.10. Severe deformity of the bones of the lower extremities in a man with osteogenesis imperfecta, probably type III (Table 11.9). The skeletal disorder is more severe in this type as compared to type I, yet sclerae are usually white and hearing loss infrequent.

Another striking observation has been made by Nicholls et al. (1979) in fibroblasts cultured from a 21-month-old male with "moderately severe OI born to consanguineous parents." The collagens were labeled biosynthetically by the fibroblasts cultured from this child and the secreted proteins digested with pepsin and analyzed by SDS polyacrylamide gel electrophoresis. Whereas control cells showed a normal α1(I) : α2(I) ratio of approximately 2.0, the collagen from the patient contained no detectable α2(I) chains. Detailed studies revealing the nature of the defect in this individual with OI have yet to be reported, i.e., whether the abnormality resides in the structural gene or some posttranscriptional event. Observations similar to those of Nicholls et al. (1979) had been made earlier by Meigel et al. (1974), who measured collagen biosynthesis in fibroblasts from a 10-year-old male with moderately severe OI, without blue sclerae. The boy, who was also the child of consanguineous parents, also had a marfanoid appearance, but no ectopia lentis. No α2(I) chains were detected in the medium or cell layer of cultured fibroblasts. Unfortunately, analyses of tissues, particularly bone, were not mentioned in either report. Müller et al. (1975) found subsequently that new cultures of fibroblasts from the skin of the case reported earlier by Meigel et al. (1974) produced α2(I) chains. The discrepancy has not been fully explained. It was also stated (Pope and Nicholls 1980), without details, that the parents of the patient reported by Nicholls et al. (1979) produced about half as much α2(I) as normal and could, therefore, be heterozygous for the putative mutant gene.

Thus, osteogenesis imperfecta syndromes include individuals with different defects in collagen biosynthesis, although available information is incomplete. Some of these defects might account for the abnormal phenotype. It is probable that the study of this group of diseases will provide important information with regard to structure of the human collagen genes and control of their expression (Prockop 1980).

11.2.3. Marfan Syndrome

Major features of Marfan syndrome include abnormalities of the eye (particularly eclopia lentis with upward displacement of the lens); blood vessels (aneurysm of the ascending aorta and aortic regurgitation); and skeleton (long limbs relative to the height of the trunk, arachnodactyly, kyphoscoliosis, deformities of the chest wall, and mild joint laxity (Figure 11.11) (McKusick 1972; Pyeritz and McKusick 1979). When these features are all present, clinical impressions may be sufficient criteria for diagnosis. The prevalence of the Marfan syndrome is relatively high and has been estimated to be 4–6 per 100,000 individuals. Because all of the features of the syndrome may not be striking, it is difficult to distinguish phenotypic variations due to differences in expression of the putative Marfan gene from genetic heterogeneity, such as that seen in EDS and OI. It is probable that different mutations lead to the Marfan phenotype (Pyeritz and McKusick 1981); homocystinuria (Section 11.2.4) is just one example.

It is reasonable to propose that the clinical features of Marfan syndrome could be accounted for by abnormalities in collagen biosynthesis. The discovery of the molecular defect in animal lathyrism (Gross 1974) and the similarities between the animal disease (kyphoscoliosis and dissecting aneurysms) and human Marfan syndrome provide further support for this concept. Indeed, it is also rea-

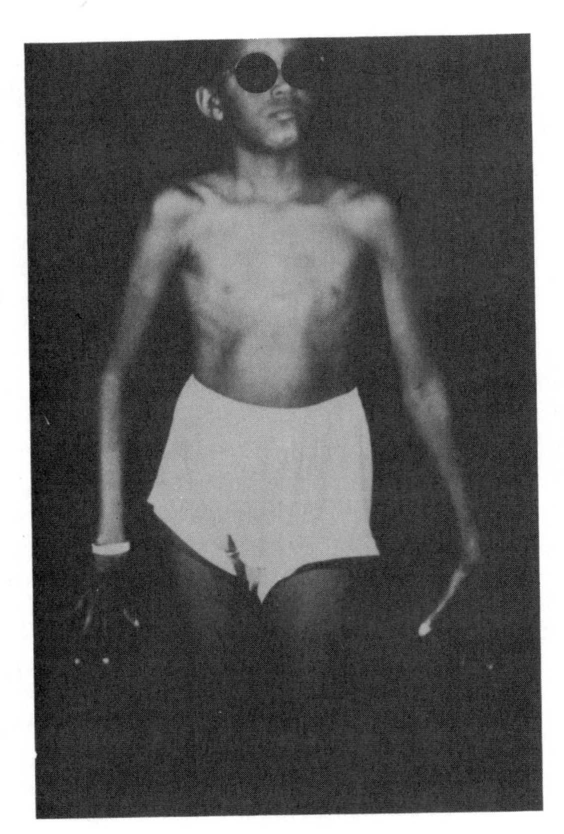

Figure 11.11. Typical habitus of a boy with Marfan syndrome. The extremities are long and the trunk is relatively short. The chest is deformed with a prominent anterior depression (pectus excavatum). The fingers are long and thin (arachnodactyly).

sonable to conclude that the defect involves crosslinking of collagen. Observations that skin collagen from some patients with Marfan syndrome is more soluble than controls are consistent with a crosslinking defect (Laitinen et al. 1968; Priest et al. 1973), although increased solubility is not uniformly observed (Harris and Sjoerdsma 1966a). However, detailed chemical information has become available only recently. Boucek et al. (1981) measured crosslinks following reduction with sodium borotritide and found a significant decrease in the levels of dihydroxylysinonorleucine relative to hydroxylysinonorleucine in the skin from three patients with Marfan syndrome compared with those of controls. The levels of reduced reactive aldehydes were actually higher in the samples from the patients compared with the controls. In the aorta of a patient who died from an aneurysm of a sinus of Valsalva, there was no significant change in these reducible crosslinks, in contrast with the skin. However, there was a decrease of approximately 50% in the content of the 3-hydroxypyridinium compound in the several samples of aortic tissue from the patients compared with those from two matched controls. These data were interpreted to indicate adequate lysine and hydroxylysine aldehyde formation (and therefore peptidyl lysine oxidase activity), but an altered generation of crosslinks. It was proposed that the abnormal crosslinks could be accounted for by misalignment of collagen molecules in fibrils and therefore wrong contacts between side chains. It was proposed, based on the prior reports of Siegel and Chang (1978) and Scheck et al. (1979) that the defect would most likely reside in the α2(I) chain.

Siegel and Chang (1978) found that the collagen from the skin of two patients with typical Marfan syndrome had increased solubility, particularly in nondenaturing solvents. In pepsin digests of the skin from one of these patients, there were two separate bands in the region of migration of α2(I) on SDS polyacrylamide gel electrophoresis. One of these patients had aortic tissue and a mitral valve removed at surgery, which provided the basis for the further study by Scheck et al. (1979). Pepsin digests of skin again showed the decreased total α2(I) accompanied by two α2(I) bands. Similar results were obtained when the aortic collagen was extracted with 4 *M* $CaCl_2$. Morphologically the aortic collagen was sometimes abnormal (unraveled) in longitudinal section. Changes interpreted as disintegration of elastic fibers were thought to be secondary to decreased tensile strength resulting from the deposition of collagen fibrils comprised of an abnormal type I collagen. An abnormality in the synthesis of the α2(I) chain was proposed as the primary defect. Subsequently, the synthesis of type I collagen by fibroblasts from this woman with Marfan syndrome was analyzed by Byers et al. (1981c). They interpreted the duplicate α2(I) chain bands as consisting of one normal α2(I) chain and a larger abnormal α2(I) chain. Corresponding changes in precursor forms secreted into the medium were observed. When RNA was extracted from these cells and the mRNA translated in a reticulocyte lysate system, a labeled band was noted, which was considered to be a more slowly migrating prepro α2(I) chain. When the medium proteins were digested with mammalian collagenase following pepsin treatment to remove the precursor regions, only the TC^A fragment of the α2(I) chain (*See* Chapter 4) was of higher than normal molecular weight. Analysis of cyanogen bromide peptides narrowed the region and it was concluded that there was an insert of about 20–25 amino acids in the α2(I) chain, which might have as one effect the shifting of the normal crosslinking sites out of register. Such an explanation could also account for the observations of Boucek et al. (1981) of the reduced content of the 3-hydroxypyridinium crosslink in the aorta from their patient. However, such an insert would be expected to produce complex changes. It should be emphasized that Marfan syndrome is a relatively common heritable disorder of connective tissue and the findings described above need to be confirmed and extended to a larger number of affected subjects to determine the frequency of this defect.

It may well be shown that the problem in some affected subjects with Marfan syndrome does not primarily involve collagen at all. For example, evidence has been presented indicating that the regulation of hyaluronate biosynthesis is abnormal. Fibroblasts cultured from patients with Marfan syndrome demonstrate metachromasia accounted for by the accumulation of excessive amounts of hyaluronic acid (Matalon and Dorfman 1968). The increased cellular content of hyaluronic acid is due to an increased rate of hyaluronate synthesis (Lamberg and Dorfman 1973), which may be contrasted with the decreased degradation of glycosaminoglycans in cells from patients with the so-called mucopolysaccharidoses, such as Hurler's syndrome. It has now been demonstrated that cell-free extracts of fibroblasts from individuals with the Marfan syndrome have three- to ten-fold more hyaluronate synthetase activity than comparable fibroblasts from normal individuals (Appel et al. 1979). Whether abnormal hyaluronate synthesis would also be present in individuals with a defined defect in collagen structure, such as that described by Scheck et al. (1979) and Byers et al. (1981c), remains to be shown.

11.2.4. Other Genetic Disorders

Homocystinuria. This term refers to a group of metabolic disorders, usually inherited as an autosomal recessive trait, which are characterized by a deficiency of cystathionine β-synthase (Mudd et al. 1970). This form of homocystinuria is manifested clinically by ectopia lentis (with progressive downward displacement of the lens), osteoporosis, occasionally arachnodactyly, and dilation and thrombosis in medium-sized arteries and veins (McKusick 1972). As previously discussed, some of these clinical features overlap those of Marfan syndrome. Because of the enzymatic block in the formation of cystathionine from homocysteine and serine, which normally is catalyzed by cystathione β-synthase, homocysteine accumulates and its oxidation product, homocystine, is excreted in the urine. It has been proposed that high tissue levels of homocystine, homocysteine, or homocysteic acid could be responsible for some of the manifestations of the disorder, including the connective tissue abnormalities. There is evidence that D-penicillamine, a compound related to homocysteine, interferes with collagen crosslinking by binding to aldehydes and forming a stable thiazolidine ring structure (Nimni 1968; Deshmukh and Nimni 1969). As increased solubility of dermal collagen had been found in some individuals with homocystinuria (Harris and Sjoerdsma 1966a), it was reasonable to postulate that a similar structure might be formed with homocysteine (Kang and Trelstad 1973). In a model system in vitro, Kang and Trelstad (1973) found that collagen in solutions containing 10–100m*M* homocysteine, but not methionine or homocystine, failed to form insoluble fibrils and normal amounts of reducible crosslinks. However, concentrations of homocysteine used in these experiments were extremely high with reference to the human disorder, as plasma levels of the oxidation product of homocysteine, homocystine, do not exceed 0.1 m*M* (Mudd et al. 1970). Another mechanism was proposed by Lindberg et al. (1976) based on observations that homocysteine, albeit at similarly high concentrations (1–10 m*M*), inhibits peptidyl lysine oxidase through some mechanism that does not involve chelation of the enzyme copper or substrate depletion. However, the correct mechanism is probably much more complex. Siegel (1977) showed that D-penicillamine at a very low concentration (0.1 m*M*) blocks the formation of polyfunctional crosslinks in which allysine (the crosslink precursor aldehyde derived from lysine; *See* Chapter 2) but not hydroxyallysine participates. These results suggest that homocysteine may function in the same way, explaining the connective tissue defects in homocystinuria.

Alcaptonuria. This rare hereditary disease is characterized by the presence of homogentistic acid in the urine and pigmentation of connective tissues, particularly cartilage, with the subsequent development of degenerative joint disease. The metabolic defect is a deficiency in the enzyme, homogentisic acid oxidase, which catalyzes the conversion of homogentisic acid to maleylacetoacetic acid (LaDu 1978). The homogentisic acid forms a polymer, possibly containing other constituents as well, which is deposited in the connective tissues, although the mechanism whereby the pigment deposition produces degenerative joint disease is unknown. Murray et al. (1977) proposed another mechanism by which individuals with alcaptonuria may develop connective tissue disorders. They found that homogentisic acid inhibits the activity of chick embryo peptidyl lysine hydroxylase with a K_i of 120–180 μM. In organ cultures, homogentisic acid at 0.5–5 mM inhibits, in a dose-dependent manner, the formation of hydroxylysine-derived intermolecular crosslinks (Murray et al. 1977). As blood levels of homogentisic acid of 0.1 mM or greater have been observed in patients with alcaptonuria, it was thought that the in vitro observation had implications in the pathophysiology of the disease. However, no data on levels of hydroxylysine or its crosslinks in tissues from patients with alcaptonuria were reported.

Epidermolysis Bullosa. Epidermolysis bullosa is a term used to describe a group of disorders in which blisters arise from apparently trivial trauma to the skin. Within the group, different types can be defined on the basis of clinical, biological, and genetic features and the site in the skin where the blisters occur. For example, there is a so-called dystrophic type, inherited as a recessive trait, in which the blisters are subepidermal. There is evidence that patients with this type synthesize increased amounts of a structurally altered collagenase (Bauer 1977; Bauer and Eisen 1978; Valle and Bauer 1980), the production of which can be decreased by therapy with phenytoin (Bauer et al. 1980).

Another group of individuals with so-called epidermolysis bullosa simplex, inherited as a dominant trait, have intraepidermal bullae. Even this disorder is heterogeneous. In screening for changes in the levels of enzymes involved in posttranslational modifications of collagens in serum of patients with dermatologic disorders, low activity of galactosylhydroxylysine glucosyltransferase (GGT; *See* chapter 3) was found in several members of a large kindred (Savolainen et al. 1981). Dominant epidermolysis bullosa simplex was also present in this kindred. It was found that when levels of GGT were low in serum, they were also low in skin tissue and cultured fibroblasts. Distinctly low levels of GGT were found in 6 of 12 affected family members with the skin disease, and five others had either suggestive abnormalities in GGT and/or abnormal urinary excretion patterns of hydroxylysine glycosides. In contrast, 13 of 15 family members without epidermolysis bullosa had no indication of abnormal GGT. It was proposed that the abnormality in GGT might somehow be involved in the pathogenesis of the blistering. One possibility could involve the attachment of epidermal cells to basement membrane collagens that are normally highly glycosylated. Alternatively the findings could reflect some linkage between the gene coding for the enzyme and that responsible for the disease.

11.3. Acquired Disorders of Collagen

The obvious heritable disorders that affect collagen synthesis, despite their biological interest, are probably relatively rare. It may well be, however, that disorders generally considered to be more common and not apparent until relatively late in life, e.g., the idiopathic osteoporosis that appears in postmenopausal women may also have a genetic basis. Nevertheless, there are acquired diseases where problems result from the deposition of an insufficient or altered collagenous matrix, and others in which collagen deposition is excessive (fibrosis). Three of the conditions associated with fibrosis as a major event will be considered in more detail. They serve as examples for understanding abnormal fibrosis as a general problem.

11.3.1. Control of Synthesis

Before discussing possible mechanisms underlying abnormal collagen deposition, some consideration of what is known about control of collagen synthesis is pertinent. Much of our knowledge in this area is derived from studies of collagen synthesis in vitro. It would be highly desirable to have quantitative information about rates and patterns of collagen synthesis in vivo, but although some progress has been made using such markers as urinary excretion of hydroxyproline and hydroxylsine and its glycosides (Krane et al. 1977; Prockop et al. 1979; Kivirikko 1970; Askenasi 1974), and measurement of procollagen fragments by radioimmunoassay (Taubman et al. 1974, 1976; Rohde et al. 1979), information is incomplete. Measurements of incorporation of labeled amino acids administered in vivo into tissue collagens has yielded useful data in animals (Prockop et al. 1979), but such techniques are limited in investigations of human disease. Collagen synthesis has been quantitated in cultures of tissues such as tendons, small bones (Jeffrey and Martin 1966), and even fragments of human lung (Bradley et al. 1975). Freshly isolated cells from tissues such as tendons and cartilage synthesize collagens in a fashion similar to that in organ culture and presumably reflect patterns of synthesis in vivo (Dehm and Prockop 1971). However, many of the studies providing a basis for understanding collagen synthesis in the whole organism have involved observations on cells following passage in monolayer culture. Whereas such studies can yield reproducible data, the isolation of the component cells alters the biological system, and thus the data cannot be applied with certainty to the interpretation of physiologic processes or pathologic states. Some of the factors that influence rates of collagen synthesis by cells in culture have been considered in the review by Müller et al. (1981). These include selection of populations of cells from the tissue of origin, effects of serum concentration, presence of inhibitory and stimulatory molecules, the requirement for ascorbate, and effects of cell density or aging. Control of collagen synthesis has been demonstrated at the level of transcription and at the level of translation. For example, the suppression of collagen synthesis in cells transformed with Rous sarcoma virus (Adams et al. 1977; Howard et al. 1978; Sobel et al. 1981) is associated with decreases in the amounts of translatable and hybridizable mRNA in cell-free translation systems. Similarly, the decrease in collagen synthesis in calvaria incubated with parathyroid hormone is associated with a decrease in levels of mRNA coding for type I procollagen (Kream et al. 1980). In contrast, exposure of the calvaria to insulin, which increases collagen synthesis, is accompanied by an increase in translatable mRNA.

Evidence has also been presented for the control of collagen synthesis at the level of translation. Levels of specific amino acids in the intracellular pool, e.g., proline, influence the levels of collagen synthesis in experimental liver fibrosis (Rojkind and Diaz de Léon 1970; Dunn et al. 1977; Ehrinpreis et al. 1980). The possible role of the N-terminal procollagen peptides in regulating collagen synthesis has been discussed earlier with respect to type VII EDS. The importance of these peptide fragments in inhibiting collagen synthesis under physiologic conditions or in other pathologic states has yet to be elucidated, however. There may also be posttranslational controls. It has also been proposed that the amount of collagen released by cells is regulated through the degradation of a portion of the newly synthesized collagen within the cells, prior to secretion (Steinmann et al. 1979; Bienkowski et al. 1978a, 1978b; Baum et al. 1978, 1980; Berg et al. 1980). This has been demonstrated by measurements of low molecular weight hydroxyproline peptides labeled within minutes after exposure of cells or tissues to labeled proline. The rate of intracellular degradation is in turn controlled by factors such as levels of cAMP. Thus, in cells with receptors for ligands such as β-adrenergic agonists or prostaglandins, both of which increase intracellular content of cAMP, exposure to these ligands also increases intracellular degradation. The rate of intracellular degradation, which is higher in early log-phase culture than in confluent cultures, can be suppressed with inhibitors of lysosomal proteases, suggesting that the lysosome is the site of this degradation. This process may be particularly important in regulating levels of abnormal collagen peptides, but it has also been implicated in alterations of collagen synthesis that might result, for example, from the use of β-adrenergic antagonists.

Collagen synthesis must therefore be regulated at several levels. The amount and type of collagen produced is directly related to the number of active collagen synthesizing cells and is subject to controls exerted at the levels of transcription, translation, and posttranslation. The type and character of the collagen produced is also affected by cell form, density, and many genetically determined tissue-specific influences. The extracellular matrix itself influences the amount and type of collagen synthesized by cells (Gospodarowicz et al. 1981).

At the clinical level it has been possible to identify many conditions characterized by abnormal scarring. Local trauma, often accompanied by bleeding into tissues, ischemia, and tissue death, may all be associated with fibrosis. The products resulting from hemostasis (e.g., polypeptide factors and serotonin released from platelets) and blood coagulation (e.g., fibrin, fibronectin) may all be important in attracting cells such as those of the polymorphonuclear leukocyte and monocyte-lymphocyte series and fibroblasts to the site of the lesion. Once fibroblasts accumulate, their proliferation and function are further modulated by the blood- and platelet-derived products as well as those provided by cells infiltrating the lesion. Such scarring may be considered as a physiologic response to injury. The fibrosis may then be only temporary and subsequently regress. The mechanisms involved in persistence and hypertrophy of local scars are not completely understood. Scarring that is more generalized and involves parenchymal organs not obviously exposed to mechanical trauma is even more poorly understood. Some of the conditions associated with fibrosis will be considered to illustrate how these abnormalities are currently interpreted.

11.3.2. Systemic Sclerosis (Scleroderma)

Scleroderma is a disorder of unknown etiology that most frequently affects the skin and subcutaneous tissue and is associated clinically with Raynaud's phenomenon and ectopic calcification (LeRoy 1981). Internal organs may be involved in some instances (systemic sclerosis). In the characteristic skin lesion, the epidermis is thin and atrophic and the dermis is bound to deeper connective tissue structures because of excessive deposition of collagen. A major feature of the disorder is the presence of dilated, distorted capillaries, especially in distal portions of the extremities, accompanied by obliteration of small arteries. Mononuclear cell infiltrates in the skin and other organs and the frequent presence of antinuclear antibodies in the serum are evidence of altered immune responses in affected individuals. Abnormalities in the lung, gastrointestinal tract, kidneys, and heart are also present in individuals with systemic involvement.

Considerable information has been accumulated to explain certain aspects of the excessive fibrosis in scleroderma, most of which has been obtained from studies of skin and dermal fibroblasts. Although the ratio of collagen to dry weight in scleroderma skin may be normal, the total collagen of the skin is increased. The relative amount of collagen that can be solubilized into nondenaturing solvents is distinctly less than normal, however (Harris and Sjoerdsma 1966b). The fibrosis is accounted for by increases in types I and III collagens and possibly basement membrane components as well (Fleischmajer et al. 1978, 1980; Lovell et al. 1979; Cooper et al. 1979). The levels of sodium borohydride-reducible labile crosslinks in scleroderma skin are also increased, consistent with increased rates of collagen synthesis (Herbert et al. 1974). Furthermore, direct evidence for high rates of collagen synthesis has been provided by the demonstration of increased levels of peptidyl proline hydroxylase activity in scleroderma skin extracts compared with that of controls (Uitto et al. 1969; Keiser et al. 1971) and increased incorporation of labeled proline into collagen hydroxyproline (Keiser and Sjoerdsma 1969; Uitto et al. 1970).

It appears, based on studies of cultured scleroderma fibroblasts, that each affected cell synthesizes more collagen than normal and that this abnormality persists through several passages in culture (LeRoy 1972; LeRoy et al. 1974; Buckingham et al. 1978; Perlish et al. 1976). There is, however, contradictory evidence (Uitto et al. 1979). Several factors have been suggested as explanations for the discrepant results, such as differences in culture conditions, variations in the site of origin of the biopsy samples, stage of the disease at the time of biopsy, and the number of cell passages when collagen synthesis is measured. Attempts have been made to culture cells from different levels of normal and scleroderma skin and measure protein synthesis in these primary cultures at a time when confluent density is achieved (Fleischmajer et al. 1981). Increases compared to controls in total collagen (incorporation of labeled proline into hydroxyproline) type III procollagen (radioimmunoassay), and fibronectin (radioimmunoassay) were found in cultured fibroblasts from five patients studied. The increases were more marked, however, in fibroblasts from the deeper layers of the dermis (reticular and reticular-fat) compared to the superficial layers (papillary). In some instances, the rates of synthesis were found to be more than 40 times normal. Although type III procollagen synthesis is also increased in cultured scleroderma fibroblasts, the ratio of type I : type III procollagens produced is not different from controls, nor is the helical stability of the synthesized collagen altered (Uitto et al. 1979). Furthermore, the net increase in collagen synthesis cannot be accounted for by decreased levels of collagenase (Uitto et al. 1979), as claimed in an earlier report (Brady 1975).

There is also a possibility that serum factors are important in determining the rate at which collagen is synthesized by scleroderma fibroblasts. For example, it was found that when cloned preputial skin fibroblasts are exposed to serum from scleroderma patients, there is selective growth of high collagen-producing cells (Botstein et al. 1982). Furthermore, other factors such as platelet-derived growth factor may alter the proliferation of fibroblasts (LeRoy et al. 1982). Vascular injury, which is prominent in scleroderma, may result in adherence and aggregation of platelets with subsequent release of the platelet-derived growth factor among other substances. These factors could then indirectly stimulate proliferation of fibroblasts among which high collagen-synthesizing cells might be abundant.

Thus, the continued increased synthesis of collagen by scleroderma fibroblasts after several passages implies either the selection of a population of cells of a high collagen-producing phenotype or the persistence in culture of a disease-induced alteration. Several observations suggest that immunologic mechanisms may be involved (LeRoy 1981; Fleischmajer and Perlish 1982). Infiltration of the scleroderma lesion with mononuclear cells has been commented upon. These cellular infiltrates have been observed in the deeper layers of the dermis where collagen deposition is particularly prominent; and this region is the source of cells that are high collagen producers. Mononuclear cell products such as interleukin 1 or related polypeptides not only alter rates of collagen synthesis in several systems but are also chemotactic for fibroblasts and can stimulate their proliferation (Fleischmajer and Perlish 1982; Postelthwaite et al. 1976; Wahl et al. 1978; Schmidt et al. 1982). Mononuclear cell supernates from patients with scleroderma, when activated in a one-way mixed lymphocyte reaction, enhance collagen synthesis by lung fibroblasts without altering their rate of proliferation (Cathcart and Krakauer 1981). Mononuclear cell supernates also stimulate prostaglandin synthesis and secretion in several culture systems. Prostaglandins inhibit fibroblast proliferation and collagen synthesis and counteract direct effects of the mononuclear cell products (Krane et al. 1982).

The possible mechanisms discussed could also account for excessive collagen deposition in several other disorders such as forms of pulmonary fibrosis, hepatic fibrosis, or the scarring of the joint that occurs late in rheumatoid arthritis. As each of these disease processes is distinctive there must be tissue- and disease-specific factors that determine the pathologic pattern. In scleroderma, the prominent early involvement of vascular elements may somehow be related to subsequent events. Observations that some patients with scleroderma have a circulating serum factor specifically cytotoxic for endothelial cells (Kahaleh et al. 1979) provide a possible link between vascular injury and subsequent scarring. However, this factor is not detectable in all patients with the disease and it is still necessary to understand its origin and pathogenetic role. It is likely that scleroderma is a heterogeneous syndrome with different etiologies and pathogenesis, but with similar eventual expression. Support for this concept is provided by observations of a scleroderma-like syndrome in workers exposed to polyvinyl chloride (Maricq et al. 1976).

Clinical features of scleroderma have also been associated with disorders of tryptophan metabolism, such as those found in the carcinoid syndrome (Fries et al. 1973). Indeed, abnormalities in the 5-hydroxytryptamine and tryptamine pathways of tryptophan metabolism have been documented in scleroderma

(Stachow et al. 1977). Elevations in urinary excretion of kynurenine have been observed, which are accounted for by decreased activity of enzymes involved in oxidation of kynurenine. In part, these abnormalities may be explained by pyridoxine deficiency. The use of drugs such as isoniazide has also been associated with sclerodermatous changes and increased urinary excretion of kynurenine (Price et al. 1957). More recently, the development of a scleroderma-like illness, associated with increased plasma levels of kynurenine, has been described in a patient treated for intention myoclonus with a combination of the drugs L-5-hydroxytryptophan and the aromatic L-amino acid decarboxylase inhibitor, carbidopa (Sternberg et al. 1980). The plasma levels of kynurenine remained high when the drugs were withdrawn but rose further when the drugs were reintroduced, suggesting that an abnormality in this patient in one of the enzymes that catabolizes kynurenine was unmasked by therapy. Plasma kynurenine levels were also high in 7 of 15 additional patients with scleroderma. High levels of serotonin and further increases in kynurenine in plasma were also produced by the drug combination, suggesting that concentrations of both of these metabolites might be important in the pathogenesis of the sclerodermatous changes, perhaps directly stimulating fibroblasts.

Drugs such as methysergide, used for treatment of headache, have also been associated with fibrosis, not usually of the sclerodermatous type, but manifested mainly by retroperitoneal fibrosis (Lepor and Walsh 1979) and occasionally cardiac and pulmonary fibrosis (Graham 1967). Methysergide has structural homologies with serotonin and it is possible that it produces its effects by altering tryptophan metabolism.

It is also possible to distinguish a possible variant of scleroderma characterized by fibrosis of deep subcutaneous tissue and fascia associated with peripheral eosinophilia (Shulman 1975). Mononuclear cell infiltrates of the lesions are common as in scleroderma, but several prominent clinical features of scleroderma such as Raynaud's phenomenon are lacking in this syndrome.

11.3.3. Interstitial Lung Disease (Pulmonary Fibroses)

Interstitial lung diseases are a group of heterogeneous disorders characterized by diffuse involvement of the lung parenchyma (Crystal et al. 1981). A common histologic feature is fibrosis of the alveolar interstitium, i.e., that portion of the alveolar structure bound by the epithelial and endothelial basement membranes. The term interstitial lung disease is synonymous with fibrotic-lung disease. Most of the diseases included in this category are progressive and often fatal, with loss of alveolar-capillary units, right-sided cardiac hypertrophy, and eventually cor pulmonale, a syndrome caused by failure of the right side of the heart.

In approximately one third of patients with interstitial lung disease, a specific etiology can be found. The most common include chronic inhalation of dusts containing silica and silicates, such as occurs in miners and asbestos workers. A number of different organic dusts may also be responsible, as well as drugs such as methotrexate and bleomycin and poisons such as paraquat. In most instances of interstitial lung disease, the etiology is unknown, although there is occasionally an association with conditions such as rheumatoid arthritis or systemic lupus erythematosus. Most cannot be classified and are called idiopathic pulmonary fibrosis.

In interstitial lung disease, the initial event presumably is an alveolitis characterized by an infiltration of cells not normally present in the lung (Crystal et al. 1981, 1976; Reynolds et al. 1977). Whereas the alveolar macrophage is the predominant inflammatory or immune effector cell of the lung parenchyma, in fibrotic disorders there is a striking infiltration with other cells such as lymphocytes and polymorphonuclear leukocytes, depending upon the type of disease. At some later stage, interstitial fibrosis occurs accompanied by a shift in cell populations. In disorders such as sarcoidosis, granulomas are formed and the alveolar structures are distorted. Eventually the aleveolar structures may be totally replaced by cystic spaces separated by wide bands of collagenous connective tissue interspersed with chronic inflammatory cells.

The appearance of the collagen is variable in interstitial lung disease (Cassan et al. 1974; Fulmer et al. 1980). In some areas there is deposition of thick, typically cross-striated collagen fibers. These fibers may be abnormally whorled or frayed and are comprised mostly of type I collagen. Normally, type III collagen represents about a third of the total lung parenchymal collagen. Lesser amounts of types IV and V collagen are present predominantly in the basement membranes. In interstitial lung disease such as idiopathic pulmonary fibrosis, type I collagen predominates and the amount of type III is proportionately reduced (Seyer et al. 1976).

It has been proposed that some form of alveolitis is the initial event in interstitial lung disease (Crystal et al. 1981, 1976; Reynolds et al. 1977). This may be triggered by defined stimuli such as silica or asbestos, but in the case of idiopathic pulmonary fibrosis, the stimulus is not known. Whereas persistence of the alveolitis is critical for progression of the lesion, the characteristics of the lesion are thought to be determined by the type of cell that predominates in the cellular infiltrate. In idiopathic pulmonary fibrosis, polymorphonuclear leukocytes (neutrophils) are abundant. A sequence of events proposed to account for the pathogenesis of interstitial fibrosis is as follows: The alveolar macrophages are activated, perhaps by immune complexes through the Fc receptor on the cells, and release a chemotactic factor that preferentially leads to influx of neutrophils (Schoenberger et al. 1980; Gadek et al. 1980; Hunninghake et al. 1980). The specific antigen contained in the putative immune complexes has not been defined, however. Evidence to support this hypothesis has been provided by examination of alveolar macrophages present in bronchial-alveolar washes from patients with idiopathic lung fibrosis and compared to that of controls. Macrophages from patients actively release the factor chemotactic for neutrophils (Hunninghake et al. 1981). These neutrophils release a collagenase that is capable of degrading type I collagen (Gadek et al. 1979). Another product of the alveolar macrophage is fibronectin (Villiger et al. 1981; Rennard et al. 1981), a glycoprotein that is not only chemotactic for fibroblasts but also plays a role in attachment of the cells to the interstitial collagens of the matrix (Chapter 7). Aleveolar macrophages from patients with interstitial lung disease may produce up to 100 times more fibronectin per cell than controls, which results in several-fold increases in fibronectin present in lavage fluid (Rennard and Crystal 1982). Furthermore, the fibronectin produced by these alveolar macrophages is orders of magnitude more potent as a chemotactic agent than the fibronectin present in plasma (Rennard et al. 1982).

Alveolar macrophages also secrete factors that stimulate proliferation of fibroblasts (Bitterman et al. 1981). One of these factors could be related to interleukin 1, which also promotes fibroblast proliferation (Schmidt et al. 1982). It has been suggested that the increased number of fibroblasts present in the lesion is sufficient to account for the increased collagen deposition, without invoking increased collagen production per cell. This latter remains a possibility, however. Also, cell number influences the amount and type of collagens produced. For example, high cell density dermal fibroblasts produce proportionally more type III collagen relative to type I (Abe et al. 1979). This has also been found in cultures of lung fibroblasts (Kelley et al. 1981), a result somewhat contrary to what might be anticipated based on the findings in interstitial lung disease. Alveolar macrophages when cocultured with lung fibroblasts also increase the relative proportion of type III collagen synthesized (Kelley et al. 1981). Such alterations might reflect a phase in the sequence of events in lung fibrosis, with fibroblasts predominantly synthesizing type I collagen late in the disease to account for the pattern of collagen fiber deposition.

Alveolar macrophages also release reactive oxygen species (Hoidal et al. 1979) such as superoxide anion and hydroxyl free radical, which could be involved in many aspects of tissue injury (Nathan et al. 1979). These substances may play such a role by inactivating α_1-antitrypsin, a major protease inhibitor in the lung (Carp and Janoff 1979). The α_1-antitrypsin would have its predominant effect in inhibiting enzymes such as elastases, not the collagenases. Superoxide species may also indirectly alter collagen synthesis. Paraquat, an agent known to produce pulmonary fibrosis, increases collagen accumulation in lung organ culture (Hussain and Bhatnagar 1979), which is reversed by the addition of superoxide dismutase, an enzyme that catalyzes the degradation of superoxide radical (Fridovich and Hassan 1979). Thus, superoxide could mediate increased collagen synthesis in specific instances such as paraquat poisoning and oxygen toxicity. Of further interest is the finding that superoxide may also be the active form of O_2 in the peptidyl proline and lysine hydroxylase reactions (Myllylä et al. 1979).

Mechanisms of pathogenesis of fibrosis in disorders such as sarcoidosis and chronic hypersensitivity pneumonitis are thought to be somewhat different from those that pertain to idiopathic pulmonary fibrosis or silicosis (Crystal et al. 1981; Bitterman et al. 1981). The central role in the former disorders is assigned to T lymphocytes. These cells, by processes yet to be defined, localize within the alveolar structures and respond specifically to inhaled complex antigens. The T lymphocytes then release soluble products (lymphokines), which include a factor chemotactic for monocytes. Monocytes then localized in the lesion become part of the granuloma and contribute their mediators (e.g., factors inducing proliferation of fibroblasts), which may also alter collagen production.

Although it is assumed that these mechanisms account for most of the fibrosis of the various types of interstitial lung disease, other possibilities exist. For example, pulmonary fibrosis is a complication of chemotherapy with the drug bleomycin. Alveolitis is a feature of bleomycin therapy, and as an infiltration with polymorphonuclear and mononuclear cells precedes other abnormalities, it has been suggested that this infiltrate is responsible for the fibrosis as discussed (Crystal et al. 1976). Models of bleomycin fibrosis have been described in experimental animals, where it has been shown that the increase in lung collagen is correlated with an increase in collagen synthesis in explant cultures (Clark et al. 1980b). However, collagen synthesis by cultured lung fibroblasts is stimulated directly by adding bleomycin and the increase is primarily in type I collagen (Clark et al. 1980c). This direct effect of the drug, as observed in vitro, could be one of the mechanisms responsible for the accumulation of lung collagen, if a similar mechanism is operative in vivo.

11.3.4. Hepatic Fibrosis

Increase in connective tissue in the liver (hepatic fibrosis) is a relatively common occurrence (Popper and Becker 1975; Gay and Miller 1978; Popper 1977; Rojkind and Dunn 1979). Depending upon the extent and distribution of this connective tissue and the underlying disease process responsible, hepatic fibrosis may either be of little clinical consequence or a devastating, fatal disorder—cirrhosis. Although there are numerous causes of hepatic fibrosis, there is evidence that liver cell injury is a frequent event. Infiltration with chronic inflammatory and immunocompetent cells is also seen in many of these disorders, and, in a fashion perhaps similar to what has been discussed with reference to scleroderma and interstitial lung disease, these cells must contribute to the fibrotic process. The major etiologies of hepatic fibrosis in the United States include alcoholic-nutritional diseases and viral hepatitis. Other underlying diseases are iron overload (hemochromatosis), Wilson's disease (hepatolenticular degeneration secondary to a heritable disorder of copper metabolism), extrahepatic biliary obstruction, primary biliary cirrhosis, and chronic congestive heart failure. Another important cause of hepatic fibrosis afflicting millions of individuals in many areas of the world is hepatic schistosomiasis.

The patterns of collagen deposition may be distinctive for each of these diseases. For example, in alcoholic liver injury, the cirrhosis is characterized by small nodules up to 3 mm in diameter (micronodular) with regular bands of septal connective tissue. In contrast, the so-called postnecrotic cirrhosis, a consequence of viral hepatitis, is characterized by nodules of different sizes, occasionally several centimeters in diameter with irregular septa. It is also envisioned that the formation of the excessive connective tissue in hepatic fibrosis may be an "active" or "passive" process. Active septa result from the formation of new connective tissue, whereas passive septa result from hepatocellular necrosis and collapse of the hepatic parencyma (Popper and Becker 1975; Gay and Miller 1978). With death of the hepatocytes and loss of the area occupied by these cells, there is passive formation of thin septa containing mostly the thin fibers of type III collagen, presumably derived from the preexisting intercellular connective tissue. During later stages of tissue injury, however, there is thickening of these septa associated with an increase in number of lipocytes or Ito cells, which resemble sinusoidal cells, and appear in the spaces between the hepatocytes and the sinusoidal endothelium (Popper and Becker 1975). Using antibodies to type III collagen and type III procollagen it has been shown with immunofluorescence that fine fibrillar material, which reacts with these antibodies, accumulates around the lipocytes (Gay and Miller 1978; Kent et al. 1976). It has therefore been suggested that these cells initiate fibrogenesis following liver injury. Subsequently, however, these septa increase in thickness consequent to the deposition of type I collagen.

Analysis of the type of collagen deposited in the cirrhotic liver has been based either on immunologic methodology utilizing specific antibodies to helical regions of the collagen molecules, or on chemical methodology utilizing cyanogen bromide cleavage or selective precipitation following pepsin solubilization. By immunofluorescence, type I collagen is deposited late in the disease and accounts for the thickened septa (Gay and Miller 1978). The amount of collagen solubilized by pepsin in cirrhotic livers is almost twice that in normal livers (Seyer et al. 1977). However, whereas in normal livers, type III collagen represents approximately 47% of the total collagen, type III represents only 18 to 34% of the collagen in cirrhotic livers (Seyer et al. 1977).

It has been demonstrated that fibrosis in a tissue is generally accompanied by increased levels of enzymes, such as peptidyl proline hydroxylase or peptidyl lysine oxidase, which are involved in posttranslational modifications of collagen (Chapter 3). In experimental liver disease and in several instances of human liver disease, it has been found that the activities of these enzymes as well as immunospecific enzyme proteins are increased even before the collagen content of the tissue is increased (Popper and Becker 1975; Mezey et al. 1976; Kuutti-Savolainen et al. 1979; McPhie 1980). Moreover, the increased tissue levels of these enzymes may be accompanied by increased levels in plasma or serum, as shown either by measurements of activity (Mezey et al. 1976; Kuutti-Savolainen et al. 1979) or radioimmunoassay (Tuderman et al. 1977). These changes reflect collagen-synthetic activity, although none of these enzymes has been proven to be rate-limiting for collagen synthesis in tissues. Recent studies of acute and chronic liver injury produced in rats by carbon tetrachloride have shown that only chronic liver injury is accompanied by fibrosis as well as increased serum levels of enzymes (Carter et al. 1982). Patients with liver injury also have increased serum levels of the N-terminal precursor fragment of type III procollagen as well the C-terminal precursor fragment of type I procollagen, as measured by radioimmunoassay (Taubman et al. 1974; Rohde et al. 1979). These increases presumably reflect increases in the rate of collagen synthesis; cleavage of the procollagens takes place prior to fibril formation, and the precursor peptides are released into the circulation.

It has been suggested that lipocytes (Ito cells) are the precursors of parenchymal fibroblasts and the source of the collagen early in hepatic fibrosis. Whereas in this view the liver cell injury initiates the fibrosis, it is also assumed that under some circumstances excessive collagen deposition compresses the hepatic parenchymal cell plate and, perhaps by interfering with cellular nutrition, leads to liver cell injury. The entire process may be aggravated by the deposition of new basement membranes forming layers around the fibrous bridges in the sinusoidal spaces. Interference with the microcirculation may then result in the establishment of venous bypasses. The synthesis of these basement membrane collagens results from the activity of myofibroblastic cells, probably related to smooth muscle cells (Gay et al. 1976a). The disabling consequences of the deposition of this disorganized connective tissue in cirrhosis are portal hypertension, decreased biliary secretion, and impairment of hepatic cell function.

Four different cell types normally populate the hepatic sinusoids: hepatocytes, macrophages (Kupfer cells), endothelial cells, and fat-storing (Ito) cells. Although, as discussed above, a role for the fat-storing cells in collagen synthesis has been proposed, attention has also been directed towards possible synthetic functions of the epithelial hepatocytes. For example, it has been found that several cell lines, not of connective tissue origin, possess peptidyl proline hydroxylase activity and incorporate labeled proline into protein hydroxyproline (Langness and Udenfriend 1974). Among these lines is the Chang liver cell. When rat liver cells are separated into a hepatocyte-enriched population and an endothelial cell-enriched population, it is found that the peptidyl proline hydroxylase activity in the hepatocytes is 100-fold that in the endothelial cells (Ohuchi and Tsurufiji 1972). This enzyme is also localized to the hepatocyte in other studies using immunofluorescent (Ooshima 1977) and biochemical techniques (Guzelian and Diegelmann 1979). Cultures of an established rat liver cell line, ARL6, also synthesize primarily type I collagen (75%) and smaller amounts of type III collagen (25%) (Berman and Foidart 1980). Findings that hepatocytes from rat liver produce substantially more collagen than the sinusoidal cells (Tseng et al. 1982) and other observations described above are consistent with a possible role of this cell in interstitial-type collagen synthesis in liver fibrosis. The hepatocyte has not traditionally been assigned this role, and there are as yet no compelling data that hepatocytes synthesize collagen in vivo. Nevertheless, it appears that the capacity for collagen synthesis by the hepatocyte is increased in association with liver regeneration in vivo (Guzelian and Diegelmann 1979; Guzelian et al. 1981). It has been proposed that postreplication synthesis of collagen could be important for anchoring regenerating parenchymal cells. This function might also result in collagen deposition in the sinusoidal spaces and lead to the distorted architecture that is a feature of hepatic cirrhosis.

The pathogenetic mechanisms that might be responsible for hepatic fibrosis (Rojkind and Dunn 1979; Rojkind and Kershenobich 1981) are generally similar to several of those discussed with regard to scleroderma or lung fibrosis. Chemotactic factors must be important for attracting cells to the sites of liver cell injury. Infiltrating lymphocytes and monocytes or resident cells having such functions (e.g., Kupfer cells) when specifically stimulated could then release products that induce or increase proliferation of collagen-synthesizing cells such as fibroblasts, fat-storing cells, or possibly even hepatocytes. Furthermore, collagen production by cells already committed to synthesis of this protein could be stimulated by these or other cell-derived factors (McGee et al. 1973). Epithelial cells have been found to produce substances that affect connective tissue cells (Johnson-Wint 1980), some of which have analogies with factors such as interleukin 1, traditionally thought to be derived from monocytes (Krane et al. 1982; Oppenheim et al. 1982). In addition, cells that normally may not be actively synthesizing collagen (e.g., fat-storing cells) may do so when exposed to soluble products of the type discussed or when their cell membranes are somehow altered as a result of injury. Additional influences on cellular synthesis of collagen involve responses to environmental factors such as epidermal growth factor, which affects cells in tissues other than the liver. Prostaglandins, which are released by many types of cells, may also affect collagen production not only by altering cellular cAMP and modulating intracellular degradation, but through a suppression of collagen synthesis. These concepts have been reviewed with respect to inflammatory joint disease (Krane et al. 1982), but the general principles apply to liver fibrosis as well.

The finding of increased levels of collagenase in the liver in experimental fibrosis has also raised the possibility that this enzyme somehow reflects collagen accumulation (Carter et al. 1982). Collagenase and peptidyl lysine oxidase activities serve as markers for the fibrotic process. Perhaps degradation of collagen by collagenase releases cleavage fragments that affect the ingress (chemotaxis), proliferation, or synthetic function of collagen-producing cells.

Some aspects of these potential pathogenetic mechanisms may also apply to specific forms of hepatic fibrosis such as schistosomiasis (Dunn et al. 1977). Extracts of cells and cell culture supernatants from granulomas obtained from livers of mice injected with *Schistosoma mansoni* stimulate the proliferation of cultured fibroblasts and alter collagen synthesis (Wyler et al. 1978). The inflammatory cells, particularly the monocytes, are the likely source of the stimulating activity.

11.4. Perspectives

It is apparent from this discussion that diseases can be identified in which the quality or quantity of collagen deposition is a prominent feature. Genetic disorders, usually the result of a single mutation, can be understood in terms of alterations in collagen structure or collagen processing enzymes. Some of the genetic diseases that result in decreased deposition of one type of collagen, such as forms of osteogenesis imperfecta, are beginning to be understood with the growth of our knowledge concerning structure and control of gene expression. Understanding more complex disorders, e.g., idiopathic osteoporosis, requires information about control of general cellular function that is not yet available. On the other hand, several of the fibrotic disorders can best be interpreted in terms of cellular interactions that result in perturbations in cellular functions, one manifestation of which is excessive collagen deposition. Ultimately these controls will also be understood at the level of control of expression, not only of the collagen structural genes but of the genes that determine posttranslational modifications (Prockop 1981).

The author is grateful to Drs. E.A. Bauer, P.H. Byers, R.G. Crystal, A.Z. Eisen, D.R. Eyre, R. Fleichmajer, S. Gay, S.I. Rennard, and M. Rojkind for copies of reviews and articles prior to publication and to D. Malcuit for preparation of the manuscript.

References

Abe S, Steinmann BU, Wahl LM, Martin GR: (1979) High cell density alters the ratio of type III to I collagen synthesis by fibroblasts. Nature 279:442–444.

Adams SL, Sobel ME, Howard BH, Olden K, Yamada KM, deCrombrugghe B, Pastan I: (1977) Levels of translatable mRNAs for cell surface protein, collagen precursors and two membrane proteins are altered in Rous sarcoma virus-transformed chick embryo fibroblasts. Proc Natl Acad Sci USA 74:3399–3405.

Appel A, Horwitz AL, Dorfman A: (1979) Cell-free synthesis of hyaluronic acid in Marfan syndrome. J Biol Chem 254:12199–12203.

Arneson MA, Hammerschmidt DT, Furcht LT, King RA: (1980) A new form of Ehlers-Danlos syndrome. Fibronectin corrects defective platelet function. JAMA 244:144–147.

Askenasi R: (1974) Urinary hydroxylysine and hydroxylysine glycoside excretions in normal and pathological states. J Lab Clin Med 83:673–679.

Aumailley M, Krieg T, Dessau W, Müller PK, Timpl R, Bricaud H: (1980) Biochemical and immunological studies of fibroblasts from a patient with Ehlers-Danlos syndrome, type IV demonstrate reduced type III collagen synthesis. Arch Dermatol Res 269:169–177.

Balleisen L, Gay S, Marx R, Kühn K: (1975) Comparative investigation of the influence of human bovine collagen types I, II and III on the aggregation of human platelets. Klin Wochenschr 53:903–905.

Barabas AP: (1967) Heterogeneity of the Ehlers-Danlos syndrome: Description of three clinical types and a hypothesis to explain the basic defect. Br Med J 2:612–614.

Barnes MJ, Lawson DEM: (1978) Biochemistry of bone in relation to the function of vitamin D. In: Lawson DEM, ed. The Vitamins. Vitamin D. New York: Academic.

Barsh GS, Byers PH: (1981) Reduced secretion of structurally abnormal type I procollagen in a form of osteogenesis imperfecta. Proc Natl Acad Sci USA 78:5142–5146.

Bauer EA: (1977) Recessive dystrophic epidermolysis bullosa: Evidence for an altered collagenase in fibroblast cultures. Proc Natl Acad Sci USA 74:4646–4650.

Bauer EA, Cooper TW, Tucker DR, Esterly NB: (1980) Phenytoin therapy of recessive epidermolysis bullosa. Clinical trial and proposed mechanism of action on collagenase. N Engl J Med 303:776–781.

Bauer EA, Eisen AZ: (1978) Recessive dystrophic epidermolysis bullosa: Evidence for increased collagenase as a genetic characteristic in cell culture. J Exp Med 148:1378–1387.

Baum BJ, Moss J, Breuel SD, Berg RA, Crystal RG: (1980) Effect of cyclic AMP on the intracellular degradation of newly synthesized collagen. J Biol Chem 255:2843–2847.

Baum BJ, Moss J, Breul SD, Crystal RG: (1978) Association in normal human fibroblasts of elevated levels of adenosine 3′:5′-monophosphate with a selective decrease in collagen production. J Biol Chem 253:3391–3394.

Becker U, Timpl R, Helle O, Prockop DJ: (1976) NH_2-terminal extensions on skin collagen from sheep with a genetic defect in conversion of procollagen into collagen. Biochemistry 15:2853–2862.

Beighton P: (1968) Lethal complications of the Ehlers-Danlos syndrome. Br Med J 3:656–659.

Beighton P, Price A, Lord J, Dickson E: (1969) Variants of the Ehlers-Danlos syndrome. Clinical, biochemical, haematological, and chromosomal features of 100 patients. Ann Rheum Dis 28:228–245.

Berg RA, Schwartz ML, Crystal RG: (1980) Regulation of the production of secretory proteins: Intracellular degradation of newly synthesized "defective" collagen. Proc Natl Acad Sci USA 77:4746–4750.

Berman JJ, Foidart JM: (1980) Synthesis of collagen by rat liver epithelial cultures. Ann NY Acad Sci 349:153–164.

Bienkowski RS, Baum BJ, Crystal RG: (1978a) Fibroblasts degrade newly synthesized collagen within the cell before secretion. Nature 276:413–416.

Bienkowski RS, Cowan MJ, McDonald JA, Crystal RG: (1978b) Degradation of newly synthesized collagen. J Biol Cham 253:4356–4363.

Bitterman PB, Rennard SI, Crystal RG: (1981) Environmental lung disease and the interstitium. Chest in Med 2:393–412.

Black CM, Gathercole LJ, Bailey AJ, Beighton P: (1980) The Ehlers-Danlos syndrome: An analysis of the structure of the collagen fibres of the skin. Br J Dermatol 102:85–96.

Botstein BR, Sherer GK, LeRoy EC: (1982) Fibroblast selection in scleroderma. An alternative model of fibrosis. Arthritis Rheum 25:189–195.

Boucek RJ, Noble NL, Gunja-Smith Z, Butler WT: (1981) The Marfan syndrome: A deficiency in chemically stable collagen cross-links. N Engl J Med 305:988–991.

Bradley K, McConnell-Breul S, Crystal RG: (1975) Collagen in the human lung. Quantitation of rates of synthesis and partial characterization of composition. J Clin Invest 55:543–550.

Brady AH: (1975) Collagenase in scleroderma. J Clin Invest 56:1175–1180.

Buckingham R, Prince RK, Rodnan GP, Taylor F: (1978) Increased collagen accumulation in dermal fibroblast cultures from patients with progressive systemic sclerosis (scleroderma). J Lab Clin Med 92:5–21.

Bullough PG, Davidson DD, Lorenzo JC: (1981) The morbid anatomy of the skeleton in osteogenesis imperfecta. Clin Ortho Rel Res 159:42–57.

Byers PH, Barsh GS, Holbrook KA: (1981a) Molecular mechanisms of connective tissue abnormalities in the Ehlers-Danlos syndrome. Collagen Rel Res 1:475–489.

Byers PH, Holbrook KA, Barsh GS, Smith LT, Bornstein P: (1981b) Altered secretion of type III procollagen in a form of type IV Ehlers-Danlos syndrome. Biochemical studies in cultured fibroblasts. Lab Invest 44:336–341.

Byers PH, Holbrook KA, McGillivray B, MacLeod PM, Lowry RB: (1979) Clinical and ultrastructural heterogeneity of type IV Ehlers-Danlos syndrome. Hum Genet 47:141–150.

Byers PH, Narayanan AS, Bornstein P, Hall JG: (1976) An X-linked form of cutis laxa due to deficiency of lysyl oxidase. Birth Defects 12:293–298.

Byers PH, Siegel RC, Peterson KE, Rowe DW, Holbrook KA, Smith LT, Chang Y-H, Fu JCC: (1981c) Marfan syndrome: Abnormal α2 chain in type I collagen. Proc Natl Acad Sci USA 78:7745–7749.

Byers PH, Siegel RC, Holbrook KA, Narayanan AS, Bornstein P, Hall JG: (1980) X-linked cutis laxa. Defective cross-link formation due to decreased lysyl oxidase activity. N Engl J Med 303:61–65.

Carp H, Janoff A: (1979) In vitro suppression of serum elastase-inhibitory capacity by reactive oxygen species generated by phagocytosing polymorphonuclear leukocytes. J Clin Invest 63:793–797.

Carter EA, McCarron MJ, Alpert E, Isselbacher KJ: (1982) Lysyl oxidase and collagenase in experimental acute and chronic liver injury. Gastroenterology 82:526–534.

Cassan SM, Divertie MD, Brown AL: (1974) Fine structural morphometry on biopsy specimens of human lung, 2, diffuse idiopathic pulmonary fibrosis. Chest 65:275–278.

Cathcart MK, Krakauer RS: (1981) Immunological enhancement of collagen accumulation in progressive systemic sclerosis. Clin Immunol Immunopathol 21:128–133.

Cetta G, Lenzi L, Rizzotti M, Ruggeri A, Vall M, Boni M: (1977) Osteogenesis imperfecta: Morphological histochemical and biochemical aspects. Modifications induced by (+)-catechin. Conn Tissue Res 5:51–58.

Clark JG, Kuhn C III, Uitto J: (1980a) Lung collagen in type IV Ehlers-Danlos syndrome: Ultrastructural and biochemical studies. Am Rev Resp Dis 122:971–978.

Clark JG, Overton JE, Marino BA, Uitto J, Starcher BC: (1980b) Collagen biosynthesis in bleomycin-induced pulmonary fibrosis in hamsters. J Lab Clin Med 96:943–953.

Clark JG, Starcher BC, Uitto J: (1980c) Bleomycin-induced synthesis of type I procollagen by human lung and skin fibroblasts in culture. Biochim Biophys Acta 631:359–370.

Cooper SM, Keyser AJ, Beaulieu AD, Ruoslahti E, Nimni M, Quismoro FP Jr: (1979) Increase in fibronectin in the deep dermis of involved skin in progressive systemic sclerosis. Arthritis Rheum 22:983–987.

Crystal RG, Gadek JE, Ferrans VJ, Fulmer JD, Line BR, Hunninghake GW: (1981) Interstitial lung disease: Current concepts of pathogenesis, staging and therapy. Am J Med 70:542–568.

Crystal RG, Fulmer JD, Roberts WC, Moss ML, Line BR, Reynolds HY: (1976) Idiopathic pulmonary fibrosis: Clinical, histologic, radiographic, scintographic, cytologic, and biochemical aspects. Ann Int Med 85:769–788.

Cupo LN, Pyeritz RE, Olson JL, McPhee SJ, Hutchins GM, McKusick VA: (1981) Ehlers-Danlos syndrome with abnormal collagen fibrils, sinus of Valsalva aneurysms, myocardial infarction, panacinar emphysema and cerebral heterotopias. Am J Med 71:1051–1058.

Danks DM, Stevens BJ, Campbell PE, Gillespie JM, Walker-Smith J, Bloomfield J, Turner B: (1972) Menkes' kinky-hair syndrome. Lancet 1:1100–1103.

Dehm P, Prockop DJ: (1971) Synthesis and extrusion of collagen by freshly isolated cells from chick embryo tendon. Biochim Biophys Acta 240:358–369.

Deshmukh K, Nimni ME: (1969) A defect in the intramolecular and intermolecular cross-linking of collagen caused by penicillamine. II. Functional groups involved in the interaction process. J Biol Chem 244:1787–1795.

Dickson IR, Millar EA, Veis A: (1975) Evidence for abnormality of bone-matrix proteins in osteogenesis imperfecta. Lancet 2:586–587.

DiFerrante N, Leachman RD, Angelini P, Donnelly PV, Francis G, Almazan A: (1975) Lysyl oxidase deficiency in Ehlers-Danlos syndrome type V. Conn Tissue Res 3:49–53.

Dunn MA, Rojkind M, Warren KS, Hait PK, Rifas L, Seifter S: (1977) Liver collagen synthesis in murine schistosomiasis. J Clin Invest 59:666–674.

Ehrinpreis MN, Giambrone MA, Rojkind M: (1980) Liver proline oxidase activity and collagen synthesis in rat with cirrhosis induced by carbon tetrachloride. Biochim Biophys Acta 629: 184–193.

Elsas LJ II, Miller RL, Pinnell SR: (1978) Inherited human collagen lysyl hydroxylase deficiency: Ascorbic acid response. J Pediatr 92:378–384.

Eyre DR: (1980) Collagen: Molecular diversity in the body's protein scaffold. Science 207:1315–1322.

Eyre DR: (1981) Concepts in collagen biochemistry: Evidence that collagenopathies underlie osteogenesis imperfecta. Clin Ortho Rel Res 159:97–107.

Eyre DR: (1982) Collagen crosslinking. In: Akeson WH, Bornstein P, Glimcher MJ, eds. AAOS: Symposium on Heritable Disorders of Connective Tissue, St Louis: CV Mosby.

Eyre DR, Glimcher MJ: (1972) Reducible crosslinks in hydroxylysine-deficient collagens of a heritable disorder of connective tissue. Proc Natl Acad Sci USA 69:2594–2598.

Eyre DR, Oguchi H: (1980) The hydroxypyridinium crosslinks of skeletal collagens: Their measurement, properties and proposed pathway of formation. Biochem Biophys Res Commun 92:403–410.

Fleischmajer R, Dessau W, Timpl R, Krieg T, Lunderschmidt C, Wiestner M: (1980) Immunofluorescence analysis of collagen, fibronectin, and basement membrane protein in scleroderma skin. J Invest Dermatol 75:270–275.

Fleischmajer R, Gay S, Meigel WN, Perlish JS: (1978) Collagen in the cellular and fibrotic stages of scleroderma. Arthritis Rheum 21:418–428.

Fleischmajer R, Perlish JS: (1983) The pathophysiology of the fibrosis in scleroderma skin. In: Berk PD, Wasserman LR eds. Myelofibrosis and the Biology of Connective Tissue. New York: A.R. Liss.

Fleischmajer R, Perlish JS, Krieg T, Timpl R: (1981) Variability in collagen and fibronectin synthesis by scleroderma fibroblasts in primary culture. J Invest Dermatol 76:400–403.

Fridovich I, Hassan HM: (1979) Paraquat and the exacerbation of oxygen toxicity. Trends Biochem Sci 4:113–115.

Fries JF, Lindgren JA, Bull JM: (1973) Scleroderma-like lesions and the carcinoid syndrome. Arch Intern Med 131:550–553.

Fujii K, Tanzer ML: (1977) Osteogenesis imperfecta: Biochemical studies of bone collagen. Clin Orthop Rel Res 124:271–277.

Fujimoto D, Fujie M, Abe E, Suda T: (1979) Effect of vitamin D on the content of the stable crosslink, pyridinoline, in chick bone collagen. Biochem Biophys Res Commun 91:24–28.

Fujimoto D, Moriguchi T, Ishida T, Hayashi H: (1978) The structure of pyridinoline, a collagen crosslink. Biochem Biophys Res Commun 84:52–57.

Fulmer JD, Bienkowski RS, Cowan MJ, Breuel SD, Bradley KM, Ferrans VJ, Roberts WC, Crystal RG: (1980) Collagen concentration and rates of synthesis in idiopathic pulmonary fibrosis. Amer Rev Respir Dis 122:289–301.

Gadek JE, Hunninghake GW, Zimmerman R, Crystal RG: (1980) Regulation of release of the alveolar macrophage-derived chemotactic factor. Am Rev Resp Dis 121:723–733.

Gadek JE, Kelman JA, Fells G, Weinberger SE, Horwitz AL, Reynolds HY, Fulmer JD, Crystal RG: (1979) Collagenase in the lower respiratory tract of patients with idiopathic pulmonary fibrosis. N Engl J Med 301:737–742.

Gay S, Balleisen L, Remberger K, Fietzek P, Adelmann BC, Kühn K: (1975) Immunohistochemical evidence for the presence of collagen type III in human arterial walls, arterial thrombi, and in leukocytes incubated with collagen in vitro. Klin Wochenschr 53:899–902.

Gay S, Inouye T, Minick OT, Kent G, Popper H: (1976a) Basement membrane formation in experimental hepatic injury. Gastroenterology 71:907.

Gay S, Martin GR, Müller PK, Timpl R, Kühn K: (1976b) Simultaneous synthesis of types I and III collagen by fibroblasts in culture. Proc Natl Acad Sci USA 73:4037–4040.

Gay S, Miller EJ: (1978) Collagen in the Physiology and Pathology of Connective Tissue. Stuttgart: Gustav Fischer Verlag.

Gospodarowicz D, Cheng J, Hirabayashi K, Tauber J-P: (1981) The extracellular matrix and the control of vascular endothelial and smooth muscle cell proliferation. In: Dingle JT, Gordon JL, eds. Cellular Interactions. Amsterdam: Elsevier/North Holland 135–165.

Graham JR: (1967) Cardiac and pulmonary fibrosis during methysergide therapy for headache. Am J Med Sci 254:1–12.

Gross J: (1974) Collagen biology: Structure, degradation, and disease. Harvey Lect 68:351–432.

Guzelian PS, Diegelmann RF: (1979) Localization of collagen prolyl hydroxylase to the hepatocyte. Studies in primary monolayer cultures of parenchymal cells from adult rat liver. Exp Cell Res 123:269–279.

Guzelian PS, Qureshi GD, Diegelmann RF: (1981) Collagen synthesis by the hepatocyte: Studies in primary cultures of parenchymal cells from adult rat liver. Collagen Rel Res 1:83–93.

Hanauske-Abel HM, Röhm K-H: (1980) The collagenous part of Clq is unaffected in the hydroxylysine-deficient collagen disease. FEBS Letters 110:73–76.

Hanson PA, Quinn RS, Krane SM: (1977) Hydroxylysine deficient collagen in a floppy baby. Pediatr Res 11:562 (abstr).

Harris ED Jr, Sjoerdsma A: (1966a) Collagen profile in various clinical conditions. Lancet 2:707–711.

Harris ED Jr, Sjoerdsma A: (1966b) Effect of penicillamine on human collagen and its possible application to treatment of scleroderma. Lancet 2:996–999.

Herbert CM, Lindberg KA, Hayson MIV, Bailey AJ: (1974) Biosynthesis and maturation of skin collagen in scleroderma and effect of D-penicillamine. Lancet 1:187–192.

Hoidal JR, Beall GD, Repine JE: (1979) Production of hydroxyl free radical by human alveolar macrophages. Infect Immunol 26:1088–1094.

Holbrook KA, Byers PH: (1981) Ultrastructural characteristics of the skin in a form of the Ehlers-Danlos syndrome type IV. Storage in the rough endoplastic reticulum. Lab Invest 44:342–349.

Hollister DW: (1978) Heritable disorders of connective tissue: Ehlers-Danlos syndrome. Pediatr Clin North Am 25:575–591.

Hollister DW, Byers PH, Holbrook KA: (1982) Genetic disorders of collagen metabolism. Adv Hum Genet. 12:1–87.

Howard BH, Adams SL, Sobel ME, Pastan I, deCrombrugghe B: (1978) Decreased levels of collagen mRNA in Rous-sarcoma virus-transformed chick embryo fibroblasts. J Biol Chem 253:5869–5874.

Hunninghake GW, Gadek JE, Crystal RG: (1980) Human alveolar macrophage chemotactic factor for neutrophils: Stimuli and partial characterization. J Clin Invest 66:473–483.

Hunninghake GW, Gadek JE, Lawlay TJ, Crystal RG: (1981) Mechanisms of neutrophil accumulation in the lungs of patients with idiopathic pulmonary fibrosis. J Clin Invest 68:259–269.

Hussain MZ, Bhatnagar RS: (1979) Involvement of superoxide in the paraquat-induced enhancement of lung collagen synthesis in organ culture. Biochem Biophys Res Commun 89:71–76.

Jeffrey JJ, Martin GR: (1966) The role of ascorbic acid in the biosynthesis of collagen. I. Ascorbic acid requirement by embryonic chick tibia in tissue culture. Biochim Biophys Acta 121:269–280.

Johnson-Wint B: (1980) Regulation of stromal cell collagenase production in adult rabbit cornea: In vitro stimulation and inhibition by epithelial cell products. Proc Natl Acad Sci USA 77: 5331–5335.

Kahaleh MB, Sherer GK, LeRoy EC: (1979) Endothelial injury in scleroderma. J Exp Med 149:1326–1335.

Kang AH, Trelstad R: (1973) A collagen defect in homocystinuria. J Clin Invest 52:2571–2578.

Keiser HR, Sjoerdsma A: (1969) Direct measurement of the rate of collagen synthesis in skin. Clin Chim Acta 23:341–346.

Keiser HR, Stein HD, Sjoerdsma A: (1971) Increased protocollagen proline hydroxylase activity in sclerodermatous skin. Arch Dermatol 104:57–60.

Kelley J, Trombley L, Kovacs EJ, Davis GS, Absher M: (1981) Pulmonary macrophages alter the collagen phenotype of lung fibroblasts. J Cell Physiol 109:353–361.

Kent G, Gay S, Inouye T, Bahu R, Minick OT, Popper H: (1976) Vitamin A-containing lipocytes and formation of type III collagen in liver injury. Proc Natl Acad Sci USA 73:3719–3722.

Kivirikko KI: (1970) Urinary excretion of hydroxyproline in health and disease. Int Rev Connect Tissue Res 5:93–163.

Kohn LD, Isersky C, Zupnik J, Lenaers A, Lee G, Lapière CM: (1974) Calf tendon procollagen peptidase: Its purification and endopeptidase mode of action. Proc Natl Acad Sci USA 71:40–44.

Krane SM: (1980) Genetic diseases of collagen. In: Prockop DJ, Champe PC, eds. Gene Families of Collagen and Other Proteins. New York: Elsevier/North Holland.

Krane SM: (1982) Hydroxylysine-deficient collagen disease: A form of Ehlers-Danlos syndrome (type VI). In: Akeson WH, Bornstein P, Glimcher MJ eds. AAOS: Symposium on Heritable Disorders of Connective Tissue. St Louis: CV Mosby.

Krane SM, Goldring SR, Dayer J-M: (1982) Interactions among lymphocytes, monocytes, and other synovial cells in the rheumatoid synovium. Lymphokines 7:75–136.

Krane SM, Goldring SR, Dayer J-M, Byrne MH, Quinn RS: (1980) Cells cultured from human bone and skin express in vivo phenotype. Calcif Tissue Int 31:57 (abstr).

Krane SM, Kantrowitz FG, Byrne M, Pinnell SR, Singer FR: (1977) Urinary excretion of hydroxylysine and its glycosides as an index of collagen degradation. J Clin Invest 59:819–827.

Krane SM, Pinnell SR, Erbe RW: (1972) Lysyl-protocollagen hydroxylase deficiency in fibroblasts from siblings with hydroxylysine-deficient collagen. Proc Natl Acad Sci USA 69:2899–2903.

Krane SM, Simon LS: (1981) Organic matrix defects in metabolic and related bone diseases. In: Veis A, ed. The Chemistry and Biology and Mineralized Connective Tissues. New York: Elsevier/North Holland.

Krane SM, Trelstad RL: (1979) Case records of the Massachusetts General Hospital. Case 3-1979. N Engl J Med 300:129–135.

Kream BE, Rowe DW, Gworek SC, Raisz LG: (1980) Parathyroid hormone alters collagen synthesis and procollagen mRNA levels in fetal rat calvaria. Proc Natl Acad Sci USA 77:5654–5658.

Krieg T, Feldmann U, Kessler W, Müller PK: (1979) Biochemical characteristics of Ehlers-Danlos syndrome type VI in a family with one affected infant. Hum Genet 46:41–49.

Krieg T, Kirsch E, Matzen K, Müller PK: (1981) Osteogenesis imperfecta: Biochemical and clinical evaluation of 13 cases. Klin Wochenschr 59:91–93.

Kuutti-Savolainen ER, Risteli J, Miettinen TA, Kivirikko KI: (1979) Collagen biosynthesis enzymes in serum and hepatic tissue in liver disease. I. Prolyl hydroxylase. Eur J Clin Invest 9:89–95.

LaDu BN: (1978) Alcaptonuria. In: Stanbury JB, Wyngaarden JB, Fredrickson DS, eds. The Metabolic Basis of Inherited Disease. New York: McGraw-Hill.

Laitinen O, Uitto J, Iivanainen M, Hannuksela M, Kivirikko KI: (1968) Collagen metabolism of the skin in Marfan's syndrome. Clin Chim Acta 21:321–326.

Lamberg SI, Dorfman A: (1973) Synthesis and degradation of hyaluronic acid in the cultured fibroblasts of Marfan's disease. J Clin Invest 52:2428–2433.

Langness U, Udenfriend S: (1974) Collagen biosynthesis in nonfibroblastic cell lines. Proc Natl Acad Sci USA 71:50–51.

Lapière CM, Lenaers A, Kohn LD: (1971) Procollagen peptidase: An enzyme excising the coordination peptides of procollagen. Proc Natl Acad Sci USA 68:3054–3058.

Lenaers A, Ansay M, Nusgens BV, Lapière CM: (1971) Collagen made of extended α-chains, procollagen, in genetically-defective dermatosparaxic calves. Eur J Biochem 23:533–543.

Lepor H, Walsh PC: (1979) Idiopathic retroperitoneal fibrosis. J Urol 122:1–6.

LeRoy EC: (1972) Connective tissue synthesis by scleroderma skin fibroblasts in culture. J Exp Med 135:1351–1362.

LeRoy EC: (1981) Scleroderma (systemic sclerosis). In: Kelley WN, Harris ED Jr, Ruddy S, Sledge CB, eds. Textbook of Rheumatology. Philadelphia: WB Saunders. 1211–1230.

LeRoy EC, McGuire M, Chen N: (1974) Increased collagen synthesis by scleroderma skin fibroblasts in vitro. A possible defect in the regulation or activation of the scleroderma fibroblasts. J Clin Invest 54:880–889.

LeRoy EC, Mercurio S, Sherer GK: (1982) Replication and phenotypic expression of control and scleroderma human fibroblasts: Responses to growth factors. Proc Natl Acad Sci USA 79: 1286–1290.

Lichtenstein JR, Kohn LD, Martin GR, Byers P, McKusick VA: (1973) Procollagen peptidase deficiency in a form of the Ehlers-Danlos syndrome. Trans Assoc Am Physicians 86:333–339.

Lichtenstein JR, Martin GR, Kohn LD, Byers PH, McKusick VA: (1973b) Defect in conversion of procollagen in a form of the Ehlers-Danlos syndrome. Science 182:298–300.

Lovell CR, Nicholls AC, Duance VC, Bailey AJ: (1979) Characterization of dermal collagen in systemic sclerosis. Br J Dermatol 100:359–369.

MacFarlane JD, Hollister DW, Weaver DD, Brandt KD, Luzzatti LL, Biegel AA: (1980) A new Ehlers-Danlos syndrome with skeletal dysplasia. Am J Hum Genet 32:118A (abstr).

Maricq HR, Johnson MN, Whetstone CL, LeRoy E: (1976) Capillary abnormalities in polyvinyl chloride production workers. Examination by in vivo microscopy. JAMA 236:1368–1371.

Matalon R, Dorfman A: (1968) The accumulation of hyaluronic acid in cultured fibroblasts of the Marfan syndrome. Biochem Biophys Res Commun 32:150–154.

McGee JO'D, O'Hare RP, Patrick RS: (1973) Stimulation of the collagen biosynthetic pathway by factors isolated from experimentally injured livers. Nature (New Biol) 243:121–123.

McKusick VA: (1972) Heritable Disorders of Connective Tissue. St Louis: CV Mosby.

McPhie JL: (1980) Hepatic prolyl hydroxylase activity in human liver disease. Hepatogastroenterology 27:277–282.

Mechanic G: (1972) Cross-linking of collagen in a heritable disorder of connective tissue: Ehlers-Danlos syndrome. Biochem Biophys Res Commun 47:267–272.

Meigel WN, Müller PK, Pontz BF, Sorensen N, Spranger J: (1974) A constitutional disorder of connective tissue suggesting a defect in collagen biosynthesis. Klin Wochenschr 52:906–912.

Mezey E, Potter JJ, Maddrey WC: (1976) Hepatic collagen proline hydroxylase activity in alcoholic liver disease. Clin Chim Acta 68:313–320.

Miller RL, Elsas LJ, Priest RE: (1979) Ascorbate action on normal and mutant lysyl hydroxylases from cultured dermal fibroblasts. J Invest Dermatol 72:241–247.

Mudd SH, Edwards WA, Loeb PM, Brown MS, Laster L: (1970) Homocystinuria due to cystathionine synthase deficiency: The effect of pyridoxine. J Clin Invest 49:1762–1773.

Müller PK, Kirsch E, Gauss-Müller V, Krieg T: (1981) Some aspects of the modulation and regulation of collagen synthesis in vitro. Molec Cell Biochem 34:73–85.

Müller PK, Lemmen C, Gay S, Meigel WN: (1975) Disturbance in the regulation of the type of collagen synthesized in a form of osteogenesis imperfecta. Eur J Biochem 59:97–104.

Müller PK, Raisch K, Matzen K, Gay S: (1977) Presence of type III collagen in bone from a patient with osteogenesis imperfecta. Eur J Pediatr 125:29–37.

Murray JC, Lindberg KA, Pinnell SR: (1977) In vitro inhibition of chick embryo lysyl hydroxylase by homogentisic acid. A proposed connective tissue defect in alkaptonuria. J Clin Invest 59:1071–1079.

Myllylä R, Schubotz LM, Weser U, Kivirikko KI: (1979) Involvement of superoxide in the prolyl and lysyl hydroxylase reactions. Biochem Biophys Res Commun 89:98–102.

Nathan CF, Brukner LH, Silverstein SC, Cohn ZA: (1979) Extracellular cytolysis by activated macrophages and granulocytes. I. Pharmacologic triggering of effector cells and the release of hydrogen peroxide. J Exp Med 149:84–99.

Nelson D, King RA: (1981) Ehlers-Danlos syndrome type VIII. J Am Acad Dermatol 5:297–303.

Nicholls AC, Pope FM, Schloon H: (1979) Biochemical heterogeneity of osteogenesis imperfecta: New variant. Lancet 1:1193.

Nimni ME: (1968) A defect in the intramolecular and intermolecular cross-linking of collagen caused by penicillamine. I. Metabolic and function abnormalities in soft tissues. J Biol Chem 243:1457–1466.

Ohuchi K, Tsurufiji S: (1972) Protocollagen proline hydroxylase in isolated rat liver cells. Biochim Biophys Acta 258:731–740.

Ooshima A: (1977) Immunohistochemical localization of prolyl hydroxylase in rat tissues. J Histochem Cytochem 25:1297–1302.

Oppenheim JJ, Stadler BM, Siranganian RP, Mage M, Mathieson B: (1982) Lymphokines: Their role in lymphocyte responses. Properties of interleukin 1. Fed Proc 41:257–262.

Peltonen L, Palotie A, Hayashi T, Prockop DJ: (1980a) Thermal stability of type I and type III procollagens from normal human fibroblasts and from a patient with osteogenesis imperfecta. Proc Natl Acad Sci USA 77:162–166.

Peltonen L, Palotie A, Prockop DJ: (1980b) A defect in the structure of type I procollagen in a patient who has osteogenesis imperfecta: Excess mannose in the COOH-terminal propeptide. Proc Natl Acad Sci USA 77:6179–6183.

Penttinen RP, Lichtenstein JR, Martin GR, McKusick VA: (1975) Abnormal collagen metabolism in cultured cells in osteogenesis imperfecta. Proc Natl Acad Sci USA 72:586–589.

Perlish JS, Bashey RI, Stephens RE, Fleischmajer R: (1976) Connective tissue synthesis by cultured scleroderma fibroblasts. I. *In vitro* collagen synthesis by normal and scleroderma fibroblasts. Arthritis Rheum 19:891–901.

Pinnell SR: (1975) Abnormal collagens in connective tissue diseases. Birth Defects 11:23–30.

Pinnell SR, Fox R, Krane SM: (1971) Human collagens: Differences in glycosylated hydroxylysines in skin and bone. Biochim Biophys Acta 229:119–122.

Pinnell SR, Krane SM, Kenzora JE, Glimcher MJ: (1972) A heritable disorder of connective tissue. Hydroxylysine-deficient collagen disease. N Engl J Med 286:1013–1020.

Pope FM, Martin GR, Lichtenstein JR, Penttinen R, Gerson B, Rowe DW, McKusick VA: (1975) Patients with Ehlers-Danlos syndrome type IV lack type III collagen. Proc Natl Acad Sci USA 72:1314–1316.

Pope FM, Martin GR, McKusick VA: (1977) Inheritance of the Ehlers-Danlos syndrome. J Med Genet 14:200–204.

Pope FM, Nicholls AC: (1980) Heterogeneity of osteogenesis imperfecta congenita. Lancet 1:820–821.

Pope FM, Nicholls AC, Eggleton C, Narcissi P, Hey E, Parkin JM: (1980a) Osteogenesis imperfecta (lethal) bones contain types III and V collagens. J Clin Pathol 33:534–538.

Pope FM, Nicholls AC, Jones PM, Wells RS, Lawrence D: (1980b) EDS IV (acrogeria): New autosomal dominant and recessive types. J R Soc Med 73:180–186.

Pope FM, Nicholls AC, Narcissi P, Bartlett J, Neil-Dwyer G, Doshi B: (1981) Some patients with cerebral aneurysms are deficient in type III collagen. Lancet 1:973–975.

Popper H: (1977) Pathologic aspects of cirrhosis. Am J Pathol 87:228–258.

Popper H, Becker K, eds: (1975) Collagen Metabolism in the Liver. New York: Stratton Intercontinental Medical Book Corp.

Postelthwaite AE, Snyderman R, Kang AH: (1976) The chemotactic attraction of human fibroblasts to a lymphocyte-derived factor. J Exp Med 144:1188–1203.

Price JM, Brown RR, Larson FC: (1957) Quantitative studies on human urinary metabolites of tryptophan as affected by isoniazid and deoxypyridoxine. J Clin Invest 36:1600–1607.

Priest RE, Moinuddin JF, Priest JH: (1973) Collagen of Marfan syndrome is abnormally soluble. Nature 245:264–266.

Prockop DJ: (1980) Consequences of recombinant DNA and other new technologies for future research on human genetic diseases. In: Prockop DJ, Champe PC, eds. Gene Families of Collagen and Other Proteins. New York: Elsevier/North Holland.

Prockop DJ: (1981) Recombinant DNA and collagen research. Is amino acid sequencing obsolete? Can we study diseases involving collagen by analysis of the genes? Collagen Rel Res 1:129–135.

Prockop DJ, Kivirikko KI, Tuderman L, Guzman NA: (1979) The biosynthesis of collagen and its disorders. N Engl J Med 301:13–34, 77–85.

Puistola U, Turpeenniemi-Hujanen TM, Myllylä R, Kivirikko KI: (1980a) Studies on the lysyl hydroxylase reaction. I. Initial velocity kinetics and related aspects. Biochim Biophys Acta 611:40–50.

Puistola U, Turpeennieimi-Hujanen TM, Myllylä R, Kivirikko KI: (1980b) Studies on the lysyl hydroxylase reaction. II. Inhibition kinetics and the reaction mechanism. Biochim Biophys Acta 611:51–60.

Pyeritz RE, McKusick VA: (1981) Basic defects in the Marfan syndrome. N Engl J Med 305:1011–1012.

Pyeritz RE, McKusick VA: (1979) The Marfan syndrome: Diagnosis and management. N Engl J Med 300:772–777.

Quinn RS, Krane SM: (1976) Abnormal properties of collagen lysyl hydroxylase from skin fibroblasts of siblings with hydroxylysine-deficient collagen. J Clin Invest 57:83–93.

Quinn RS, Krane SM: (1979) Collagen synthesis by cultured skin fibroblasts from siblings with hydroxylysin deficient collagen. Biochim Biophys Acta 585:589–598.

Rennard S, Hunninghake G, Davis W, Mortiz E, Crystal R: (1982) Macrophage fibronectin is 1000-fold more potent as a fibroblast chemoattractant than plasma fibronectin. Clin Res 30:356A.

Rennard SI, Crystal RG: (1982) Fibronectin in human bronchopulmonary lavage fluid. Elevation in patients with interstitial lung disease. J Clin Invest 69:113–122.

Rennard SI, Hunninghake GW, Bitterman PB, Crystal RG: (1981) Production of fibronectin by the human alveolar macrophage: A mechanism for the recruitment of fibroblasts to sites of tissue injury in the interstitial lung diseases. Proc Natl Acad Sci USA 78:7141–7151.

Reynolds HY, Fulmer JD, Kazmierowski JA, Roberts WC, Frank MM, Crystal RG: (1977) Analysis of bronchoalveolar lavage fluid from patients with idiopathic pulmonary fibrosis and chronic hypersensitivity pneumonitis. J Clin Invest 159:165–175.

Risteli L, Risteli J, Ihme A, Krieg T, Müller PK: (1980) Preferential hydroxylation of type IV collagen by lysyl hydroxylase from Ehlers-Danlos syndrome type VI fibroblasts. Biochem Biophys Res Commun 96:1778–1784.

Rohde H, Vargas L, Hahn E, Kalbfleisch H, Brugera M, Timpl R: (1979) Radioimmunoassay for type III procollagen peptide and its application to human liver disease. Eur J Clin Invest 9:451–459.

Rojkind M, Diaz de Léon L: (1970) Collagen biosynthesis in cirrhotic rat liver slices: A regulatory mechanism. Biochim Biophys Acta 217:512–522.

Rojkind M, Dunn MA: (1979) Hepatic fibrosis. Gastroenterology 76:849–863.

Rojkind M, Kershenobich D: (1981) Hepatic fibrosis. Clin Gastroenterol 10:737–754.

Rowe DW, McGoodwin EB, Martin GR, Grahn D: (1977) Decreased lysyl oxidase activity in the aneurysm-prone, mottled mouse. J Biol Chem 252:939–942.

Rowe DW, McGoodwin EB, Martin GR, Sussman MD, Grahn D, Faris B, Franzblau C: (1974) A sex-linked defect in the cross-linking of collagen and elastin associated with the mottled locus in mice. J Exp Med 139:180–192.

Rowe DW, Shapiro J: (1982) Biochemical features of cultured skin fibroblasts from patients with osteogenesis imperfecta. In: Akeson WH, Bornstein P, Glimcher MJ eds. AAOS: Symposium of Heritable Disorders of Connective Tissue. St Louis: CV Mosby.

Savolainen E-R, Kero M, Pihlajaniemi T, Kivirikko KI: (1981) Deficiency of galactosylhydroxylysyl glucosyltransferase, an enzyme of collagen synthesis, in a family with dominant epidermolysis bullosa simplex. N Engl J Med 304:197–204.

Scheck M, Siegel RC, Parker J, Chang Y-H, Fu JCC: (1979) Aortic aneurysm in Marfan's syndrome: Changes in the ultrastructure and composition of collagen. J Anat 129:645–657.

Schmidt JA, Oliver CA, Green I, Gery I: (1982) Silica stimulated macrophages (Mϕ) release a fibroblast proliferation factor identical to interleukin-1 (IL-1). Fed Proc 41:438.

Schoenberger C, Hunninghake G, Gadek J, Crystal RG: (1980) Role of alveolar macrophages in asbestosis: Modulation of neutrophil migration to the lung following asbestos exposure. Am Rev Resp Dis 121(suppl):257.

Seyer JM, Hutcheson ET, Kang AH: (1976) Collagen polymorphism in idiopathic chronic pulmonary fibrosis. J Clin Invest 57:1498–1507.

Seyer JM, Hutcheson ET, Kang AH: (1977) Collagen polymorphism in normal and cirrhotic human liver. J Clin Invest 59:241–248.

Shulman LE: (1975) Diffuse fasciitis with eosinophilia: A new syndrome? Trans Assoc Am Physicians 88:70–86.

Siegel RC: (1977) Collagen cross-linking: Effect of D-penicillamine on cross-linking in vitro. J Biol Chem 252:254–259.

Siegel RC, Chang YH: (1978) Defective α2 chain synthesis in patients with sporadic Marfan syndrome. Clin Res 26:501A (abstr).

Siegel RC, Black CM, Bailey AJ: (1979) Cross-linking of collagen in the X-linked Ehlers-Danlos type V. Biochem Biophys Res Commun 88:281–287.

Sillence DO, Rimoin DL, Danks DM: (1979) Clinical variability in osteogenesis imperfecta—variable expressivity or genetic heterogeneity. Birth Defects 15:113–129.

Smith R, Francis MJO, Bauze RJ: (1975) Osteogenesis imperfecta. A clinical and biochemical study of a generalized tissue disorder. Q J Med 44:555–573.

Sobel ME, Yamamoto T, deCrombrugghe B, Pastan I: (1981) Regulation of procollagen messenger ribonucleic acid levels in Rous sarcoma virus transformed chick embryo fibroblasts. Biochemistry 20:2678–2684.

Stachow A, Jablonska S, Skiendzielewska A: (1977) 5-hydroxytryptamine and tryptamine pathways in scleroderma. Br J Dermatol 97:147–154.

Stark M, Lenaers A, Lapiere CM, Kühn K: (1971) Electronoptical studies of procollagen from the skin of dermatosparaxic calves. FEBS Letters 18:225–227.

Steinmann B, Tuderman L, Peltonen L, Martin GR, McKusick VA, Prockop DJ: (1980) Evidence for a structural mutation of procollagen type I in a patient with the Ehlers-Danlos syndrome type VII. J Biol Chem 255:8887–8893.

Steinmann BU, Martin GR, Baum B, Crystal RG: (1979) Synthesis and degradation of collagen by skin fibroblasts from controls and from patients with osteogenesis imperfecta. FEBS Letters 101:269–272.

Steinmann B, Gitzelmann R, Vogel A, Grant ME, Harwood R, Sear CHJ: (1975) Ehlers-Danlos syndrome in two siblings with deficient lysyl hydroxylase activity in cultured skin fibroblasts but only mild hydroxylysine deficiency in skin. Helv Paediatr Acta 30:255–274.

Sternberg EM, Van Woert MH, Young SN, Magnussen I, Baker H, Gauthier S, Osterland CK: (1980) Development of a scleroderma-like illness during therapy with L-5-hydroxytryptophan and carbidopa. N Engl J Med 303:782–787.

Stevenson CJ, Bottoms E, Shuster S: (1970) Skin collagen in osteogenesis imperfecta. Lancet 1:860–861.

Stewart RE, Hollister DW, Rimoin DL: (1977) A new variant of Ehlers-Danlos syndrome: An autosomal dominant disorder of fragile skin, abnormal scarring and generalized periodontitis. Birth Defects 13:85–93.

Sussman M, Lichtenstein JR, Nigra TP, Martin GR, McKusick VA: (1974) Hydroxylysine-deficient skin collagen in a patient with a form of the Ehlers-Danlos syndrome. J Bone Joint Surg 56A:1228–1234.

Sykes B, Francis MJO, Smith R: (1977) Altered relation of two collagen types in osteogenesis imperfecta. N Engl J Med 296:1200–1203.

Taubman MB, Goldberg B, Sherr CJ: (1974) Radioimmunoassay for human procollagen. Science 186:1115–1117.

Taubman MB, Kammerman S, Goldberg B: (1976) Radioimmunoassay of procollagen in serum of patients with Paget's disease of bone (39380). Proc Soc Exp Biol Med 152:284–287.

Trelstad RL, Rubin D, Gross J: (1977) Osteogenesis imperfecta congentia. Evidence for a generalized molecular disorder of collagen. Lab Invest 36:501–508.

Tseng SC, Lee PC, Ells PF, Bissell DM, Smuckler EA, Stern R: (1982) Collagen production by rat hepatocytes and sinusoidal cells in primary monolayer culture. Hepatology 2:13–18.

Tuderman L, Risteli J, Miettinen TA, Kivirikko KI: (1977) Serum immunoreactive prolyl hydroxylase in liver disease. Eur J Clin Invest 7:537–541.

Turpeenniemi TM, Puistola U, Anttinen H, Kivirikko KI: (1977) Affinity chromatography of lysyl hydroxylase on concanavalin A-agarose. Biochim Biophys Acta 483:215–219.

Uitto J, Bauer EA, Eisen AZ: (1979) Scleroderma. Increased biosynthesis of triple-helical type I and type III procollagens associated with unaltered expression of collagenease by skin fibroblasts in culture. J Clin Invest 64:921–930.

Uitto J, Helin P, Rasmussen O, Lorenzen I: (1970) Skin collagen in patients with scleroderma: Biosynthesis and maturation in vitro and the effect of D-penicillamine. Ann Clin Res 2:228–234.

Uitto J, Halem J, Hannuksela M, Peltokallio P, Kivirikko KI: (1969) Protocollagen proline hydroxylase activity in the skin of normal subjects and of patients with scleroderma. Scand J Clin Lab Invest 23:241–247.

Valle K-J, Bauer EA: (1980) Enhanced biosynthesis of human skin collagenase in fibroblast cultures from recessive epidermolysis bullosa. J Clin Invest 66:176–187.

Villiger B, Kelley DG, Engelman W, Kuhn C III, McDonald JA: (1981) Human alveolar macrophage fibronectin: Synthesis, secretion, and ultrastructural localization during gelatin-coated latex particle binding. J Cell Biol 90:711–760.

Vogel A, Holbrook KA, Steinmann B, Gitzelmann R, Byers PH: (1979) Abnormal collagen fiber structure in the gravis form (type I) of the Ehlers-Danlos syndrome. Lab Invest 40:201–206.

Wahl SM, Wahl LM, McCarthy JB: (1978) Lymphocyte-mediated activation of fibroblast proliferation and collagen production. J Immunol 121:942–946.

Wyler DJ, Wahl SM, Wahl LM: (1978) Hepatic fibrosis in schistosomiasis: Egg granulomas secrete fibroblast stimulating factor in vitro. Science 202:438–440.

Index